Operator Theory: Advances and Applications
Vol. 123

Operator Theory, System Theory and Related Topics

The Moshe Livšic Anniversary Volume

D. Alpay
V. Vinnikov
Editors

Birkhäuser Verlag
Basel · Boston · Berlin

Editors:

Daniel Alpay
Victor Vinnikov
Department of Mathematics
Ben-Gurion University of the Negev
P.O. Box 653
Beer Sheva 84105
Israel

2000 Mathematics Subject Classification 47-06; 30-06, 93-06

A CIP catalogue record for this book is available from the
Library of Congress, Washington D.C., USA

Deutsche Bibliothek Cataloging-in-Publication Data
Operator theory, system theory and related topics : the Moshe Livšic anniversary volume /
D. Alpay ; V. Vinnikov ed.. - Basel ; Boston ; Berlin : Birkhäuser, 2001
(Operator theory ; Vol. 123)
ISBN 3-7643-6523-4

ISBN 3-7643-6523-4 Birkhäuser Verlag, Basel – Boston – Berlin

Member of the BertelsmannSpringer Publishing Group
Printed on acid-free paper produced from chlorine-free pulp. TCF ∞
Cover design: Heinz Hiltbrunner, Basel
Printed in Germany
ISBN 3-7643-6523-4

9 8 7 6 5 4 3 2 1

Contents

Editorial Introduction

This volume presents the refereed proceedings of the Conference in Operator Theory in Honour of Moshe Livšic 80th Birthday, held June 29 to July 4, 1997, at the Ben-Gurion University of the Negev (Beer-Sheva, Israel) and at the Weizmann Institute of Science (Rehovot, Israel). The volume contains papers in operator theory and its applications (understood in a very wide sense), many of them reflecting, directly or indirectly, a profound impact of the work of Moshe Livšic.[1]

Moshe (Mikhail Samuilovich) Livšic was born on July 4, 1917, in the small town of Pokotilova near Uman, in the province of Kiev in the Ukraine; his family moved to Odessa when he was four years old. In 1933 he enrolled in the Department of Physics and Mathematics at the Odessa State University, where he became a student of M. G. Krein and an active participant in Krein's seminar — one of the centres where the ideas and methods of functional analysis and operator theory were being developed. Besides M. G. Krein, M. S. Livšic was strongly influenced by B. Ya. Levin, an outstanding specialist in the theory of analytic functions. A deep understanding of operator theory as well as function theory and a penetrating search of connections between the two, were to become one of the landmarks of M. S. Livšic's work.

M. S. Livšic defended his Ph. D. thesis in 1942 at the University of Odessa which was evacuated to Maikop in the beginning of the war. In the thesis he had introduced a generalized problem of moments and solved it using the theory of Hermitian operators in a Hilbert space. Both the problem itself and the method of solution have started a long line of investigations, including some well-known works of M. G. Krein. The generalized problem of moments and Livšic's theorem are discussed in the article of M. Briskin, J.-P. Françoise, and Y. Yomdin in this volume.

In 1943–1944 M. S. Livšic discovered the notion of the characteristic function. Following the completion by J. von Neumann of the spectral theory of selfadjoint

[1]Two other volumes in the series *Operator Theory: Advances and Applications* dedicated to Moshe Livšic are *Topics in Operator Theory and Interpolation* (Essays dedicated to M. S. Livšic on the occasion of his 70th birthday), **OT**29 (1988), and *Nonselfadjoint Operators and Related Topics* (Workshop on Operator Theory and Its Applications, Beersheva, February 24–28, 1992), **OT**73 (1994). The first one of these contains a biography of Moshe Livšic by H. Dym, I. Gohberg, and N. Kravitsky that supplements a sketch of the scientific path of Moshe Livšic below. It also contains a list of publications of Moshe Livšic till 1987; unfortunately this list does not include references to English translations of papers published in Russian or to *Mathematical Reviews*, so we have decided to include a complete list of publications of Moshe Livšic up to the present (with all the relevant information) in this volume as well. For additional biographical details see also the reminiscences of Moshe Livšic about V. P. Potapov in *Topics in Interpolation Theory*, **OT**95 (1997).

operators in the 1930's, the issue of constructing a spectral theory for nonselfadjoint operators was enthusiastically discussed. Alternatively the question was how to extend Weierstrass' theory of elementary divisors from matrices to operators; this was one of the initial motivations of I. M. Gelfand and his collaborators for the development of the theory of normed rings (Banach algebras). A closely related problem is to study nonselfadjoint extensions of Hermitian operators. For instance, one can consider a selfadjoint differential expression with nonselfadjoint boundary conditions; such nonselfadjoint boundary conditions (containing an arbitrary complex number as a parameter) appear in the well-known book of the physicist G. Gamow on the structure of atomic nucleus that was an important source of motivation for M. S. Livšic.

Starting with von Neumann's theory of selfadjoint extensions of Hermitian operators, M. S. Livšic has introduced the notion of the characteristic function of a nonselfadjoint extension of a Hermitian operator (with finite equal defficiency indices) and used it to describe all such nonselfadjoint extensions.[2] The groundbreaking paper of 1946 and the Doklady note of 1947 contain already the main ideas of the theory of characteristic functions: the fact that the characteristic function determines the corresponding operator essentially uniquely up to unitary equivalence and the relation between factorizations of the characteristic function and invariant subspaces of the operator. The general notion of the characteristic function of a nonselfadjoint operator and its applications to the study of invariant subspaces and the construction of canonical triangular models were developed by M. S. Livšic and his collaborators (most notably M. S. Brodskii) during the following decade. An important role was also played by the investigations of V. P. Potapov on the multiplicative structure of J-contractive matrix valued functions which were motivated by the work of M. S. Livšic. Following a suggestion of M. S. Livšic in the early 1960's, M. S. Brodskii introduced the notion of an operator colligation (or node) which provides a natural and convenient framework for the study of nonselfadjoint operators and the theory of characteristic functions.

The theory of characteristic functions is the cornerstone of the spectral theory of nonselfadjoint and nonunitary operators; it allows to apply the entire theory of bounded analytic functions to operator theoretic problems. Of equal and perhaps even greater importance, operator theory leads to new problems in function theory and provides new methods for a solution of old ones. The theory of characteristic functions was recast in the language of functional models in the work of B. Sz.-Nagy-C. Foiaş and L. de Branges-J. Rovnyak in the 1960's.[3]

[2]These results form the second part of the second (doctoral) thesis of M. S. Livšic that he defended at the Steklov Institute in Moscow in 1945; his opponents were S. Banach, I. M. Gelfand, M. A. Naimark, and A. I. Plessner. The first part of this thesis is dedicated to the two dimensional problem of moments; these results were never published. They are closely related to recent investigations of M. Bakonyi, G. Naevdal and others.

[3]The original perspectives of these authors were quite different from those of M. S. Livšic and his collaborators. The work of B. Sz.-Nagy and C. Foiaş grew out of von Neumann's inequality for contractions, Beurling's theorem on invariant subspaces of the shift operator, and Sz.-Nagy's

A physical interpretation of a mathematical theory has always been for M. S. Livšic an integral part of the theory itself. One would refer to him as a "natural philosopher" in the English sense of the word encompassing the natural sciences, mathematics and philosophy insofar as it relates to science and nature. M. S. Livšic came to devote much of his scientific energy to the investigation of a proper physical counterpart to the theory of nonselfadjoint operators and their characteristic functions. Already in the middle of 1950's he showed that the Heisenberg scattering matrix is a characteristic function of a certain nonselfadjoint operator; these ideas were later formalized by P. D. Lax and R. S. Phillips in their well-known approach to scattering theory, with numerous important applications in mathematical physics. M. S. Livšic proceeded during the 1960's to introduce and study open systems — in modern terminology, linear time-invariant input-output dynamical systems[4]; the characteristic function of a nonselfadjoint operator (more precisely, of an operator colligation) is the transfer function of the corresponding system. It took some time before the fundamental significance of this work by M. S. Livšic was to be understood. Somewhat later in mathematical systems theory, many similar results were obtained from a more algebraic perspective by R. Kalman and his collaborators. As for operator theory and function theory, a realization of a function as the transfer function of a linear system became an essential tool during the last two decades. The article of V. Katsnelson in this volume contains a historical survey of the theory of characteristic functions from the point of view of realization theory.

In the early 1970's M. S. Livšic started considering the problem of linking system theory and the theory of nonselfadjoint operators with Riemannian geometry. This led naturally to developing a spectral theory for tuples of commuting nonselfadjoint operators. The first major breakthrough here was achieved in a note published in 1978 in the Proceedings of the Georgian Academy of Sciences: M. S. Livšic proved that a pair of commuting nonselfadjoint operators with finite nonhermitian ranks satisfy an algebraic equation with constant coefficients. It seems then natural to expect that the spectral theory of such a pair of operators leads to function theory on an algebraic curve, the so called discriminant curve of the pair of operators (rather than say to function theory of two independent complex variables as one would assume a priori). In the early 1980's M. S. Livšic discovered that the proper analogue of the notion of the characteristic function of a single nonselfadjoint operator is the so called joint characteristic function of a pair of operators, which is a mapping of certain vector bundles (or more generally, of certain sheaves) on the discriminant curve. The theory of commuting nonselfadjoint operators, their discriminant curves and joint characteristic functions as well as the theory of corresponding multidimensional systems were further developed by M. S. Livšic and

dilation theorem. The work of L. de Branges and J. Rovnyak evolved within the framework of de Branges' theory of Hilbert spaces of entire functions.

[4]In the last chapter of his book on open systems published in 1966 (English translation in 1973), M. S. Livšic considers also linear time-varying systems and suggests to study non-linear systems as well; all of these became important topics of recent research.

his collaborators during the 1980's and 1990's. These theories combine the ideas of operator theory, function theory and system theory with fundamental notions of algebraic geometry such as line bundles and vector bundles on an algebraic curve, Jacobian varieties and theta functions.

There can be little doubt that the notions of the discriminant curve and the joint characteristic function will play an essential role in the decades to come as the usual notion of the characteristic function did during the last fifty years. The article of M. S. Livšic in this volume is dedicated to yet a further generalization of these ideas, towards time-varying multidimensional systems and noncommuting operators.

Beyond the obvious contributions of M. S. Livšic many in the mathematical community have benefited from his generosity of thoughts and ideas. As an example, V. A. Marchenko recently shared with us that it was M. S. Livšic who suggested to him in the early 1950's to consider the problem of recovering the potential of a Sturm-Liouville operator from the scattering data.

Many of us who are well acquainted with the mathematical achievements of M. S. Livšic are, however, unaware of the adverse circumstances in which he pursued his endeavors. After the hardships of the war and the evacuation he returned to Odessa in 1945. And yet, still, in the postwar years M. G. Krein and the Odessa school of functional analysis and operator theory became one of the most prominent victims of the new ethnic policies of the communist party and the zeal of the local antisemites. M. S. Livšic moved to Kharkov in 1957. He remained in Kharkov till 1975 when planning his immigration to Israel he moved to Tbilisi as it was impossible to carry out from Kharkov. After immigrating to Israel in 1978, M. S. Livšic joined the Department of Mathematics and Computer Science at Ben-Gurion University of the Negev in Beer-Sheva, where he is now Professor Emeritus.

As we wish Moshe Livšic many more years of sharing his wonderful insights with us, it seems appropriate to conclude this brief sketch of his mathematical path up to the present and to introduce the articles that follow with the words of Tyutchev:

Нам не дано предугадать,
Как слово наше отзовется, –
И нам сочувствие дается,
Как нам дается благодать, ...

Daniel Alpay, Victor Vinnikov

Acknowledgements

It is a pleasure to thank all those who have provided the funding for the conference in June–July 1997 and were instrumental in helping with the organization:

- Department of Mathematics and Computer Science, Ben-Gurion University,
- Faculty of Mathematical Sciences, Weizmann Institute,
- The Dozor Foundation at Ben-Gurion University,
- The Maurice and Gabriela Goldschleger Conference Foundation at the Weizmann Institute of Science,
- Israel Mathematical Union,
- last but not least our colleagues Professors A. Feintuch, P. Fuhrmann, and M. Lin (Ben-Gurion University), and Professors H. Dym and V. Katsnelson (Weizmann Institute).

We would also like to thank Ms. Miriam Abraham of Faculty of Mathematical Sciences, Weizmann Institute, for putting her excellent typing skills (and a great patience) at our disposal for several papers in this volume.

List of Publications of M. S. Livšic

Papers

1. *On some questions concerning the determinate case of Hamburger's moment problem*, Rec. Math. [Mat. Sbornik] N. S. **6:48** (1939), 293–306, **MR** 1:229c. In Russian.
2. (with B. Ya. Levin) *Quasi-analytic functions represented by Fourier series*, Rec. Math. [Mat. Sbornik] N. S. **9:51** (1941), 693–711, **MR** 3:106c. In Russian.
3. *On some special classes of quasi-analytic functions*, Trans. Odessa Univ. **3** (1941), 23–24. In Russian.
4. *On an application of the theory of Hermitian operators to the generalized problem of moments*, C. R. (Doklady) Acad. Sci. URSS (N.S.) **44** (1944), 3–7, **MR** 6:131b.
5. *On a class of linear operators in Hilbert space*, Rec. Math. [Mat. Sbornik] N.S. **19:61** (1946), 239–262, **MR** 8:588d. In Russian; English translation in Amer. Math. Soc. Transl. (2) **13** (1960), 61–83.
6. *On the theory of isometric operators with equal deficiency indices*, Doklady Akad. Nauk SSSR (N.S.) **58** (1947), 13–15, **MR** 9:446c. In Russian.
7. *On the theory of the elementary divisors of non-Hermitian operators*, Doklady Akad. Nauk SSSR (N.S.) **60** (1948), 17–20, **MR** 9:446d. In Russian.
8. (with M. S. Brodskiĭ) *On linear operator functions invariant with respect to the translation group*, Doklady Akad. Nauk SSSR (N.S.) **68** (1949), 213–216, **MR** 11:371b. In Russian.
9. *Isometric operators with equal deficiency indices, quasi-unitary operators*, Mat. Sbornik N.S. **26:68** (1950), 247–264, **MR** 11:669e. In Russian; English translation in Amer. Math. Soc. Transl. (2) **13** (1960), 85–103.
10. (with V. P. Potapov) *A theorem on the multiplication of characteristic matrix functions*, Doklady Akad. Nauk SSSR (N.S.) **72** (1950), 625–628, **MR** 11:669f. In Russian.
11. *On the theory of self-adjoint systems of differential equations*, Doklady Akad. Nauk SSSR (N.S.) **72** (1950), 1013–1016, **MR** 13:747d. In Russian.
12. *On the reduction of linear non-self-adjoint operators to triangular form*, Uspehi Mat. Nauk (N. S.) **7:3** (1952), 110–112. In Russian.
13. *On the reduction of linear non-Hermitian operators to "triangular" form*, Doklady Akad. Nauk SSSR (N.S.) **84** (1952), 873–876, **MR** 14:184d. In Russian.

14. *On the resolvent of a linear nonsymmetric operator*, Doklady Akad. Nauk SSSR (N.S.) **84** (1952), 1131–1134, **MR** 14:185a. In Russian.

15. *On spectral decomposition of linear non-self-adjoint operators*, Mat. Sbornik N.S. **34:76** (1954), 145–199, **MR** 16:48f. In Russian; English translation in Amer. Math. Soc. Transl. (2) **5** (1957), 67–114.

16. *On an inverse problem of the theory of operators. I*, Doklady Akad. Nauk SSSR (N.S.) **97** (1954), 399–402, **MR** 16:836c. In Russian.

17. *On an inverse problem of the theory of operators. II*, Doklady Akad. Nauk SSSR (N.S.) **97** (1954), 589–592, **MR** 16:836d. In Russian.

18. *The theory of non-selfadjoint operators and its applications*, Proceedings of the Third All-Union Mathematical Conference, Moscow, June–July, 1956, Izdat. Akad. Nauk SSSR, Moscow, 1956, Volume 2 (Summary of survey and sectional reports), pp. 40–41, and Volume 3 (Survey reports), pp. 269–276. In Russian.

19. *The application of non-self-adjoint operators to scattering theory*, Zh. Eksper. Teoret. Fiz. **31** 1956, 121–131, **MR** 19:221d. In Russian; English translation in Soviet Phys. JETP **4** (1957), 91–98.

20. *On the intermediate system formed in the scattering of elementary particles*, Doklady Akad. Nauk SSSR (N.S.) **111** (1956), 799–802, **MR** 18:856c. In Russian.

21. *On the scattering matrix of an intermediate system*, Doklady Akad. Nauk SSSR (N.S.) **111** (1956), 67–70, **MR** 19:216a. In Russian; English translation in Soviet Phys. Dokl. **1** (1957), 620–624.

22. *The method of non-selfadjoint operators in scattering theory*, Uspehi Mat. Nauk (N.S.) **12** (1957), no. 1(73), 212–218, **MR** 19:216b. In Russian; English translation in Amer. Math. Soc. Transl. (2) **16** (1960), 427–434.

23. (with M. S. Brodskiĭ) *Spectral analysis of non-self-adjoint operators and intermediate systems*, Uspehi Mat. Nauk (N.S.) **13** (1958), no. 1(79), 3–85, **MR** 20:7221. In Russian; English translation in Amer. Math. Soc. Transl. (2) **13** (1960), 265–346.

24. *Intermediate systems in quantum electrodynamics*, Doklady Akad. Nauk SSSR (N.S.) **126** (1959), 550–553, **MR** 22:1348. In Russian; English translation in Soviet Phys. Dokl. **4** (1959), 609–613,

25. *A mathematical problem connected with the theory of longitudinally polarized particles*, Doklady Akad. Nauk SSSR (N.S.) **131** (1960), 797–800, **MR** 22:4339. In Russian; English translation in Soviet Phys. Dokl. **5** (1960), 341–344.

26. (with M. Š. Flekser) *Expansion of a reactive four-terminal network into a chain of simplest four-terminal networks*, Doklady Akad. Nauk SSSR (N.S) **135** (1960), 542–544, **MR** 22:11818. In Russian; English translation in Soviet Phys. Dokl. **5** (1960), 1150–1152.

27. *Chebyshev matrix and Hamiltonian of a free particle*, Functional analysis and its applications (Proceedings of the Fifth All-Union Conference on Functional Analysis and its Applications), Izdat. Akad. Nauk Azerbaĭdžan. SSR, Baku, 1961, p. 157. In Russian.

28. (with M. Š. Flekser) *The synthesis of a transmission line from given frequency characteristics*, Zapiski Meh.-Mat. Fak. i Har'kov. Mat. Obšč. (4) **28** (1961), 149–162, **MR** 40:2431.

29. *Method of nonselfadjoint operators in the theory of waveguides*, Radiotehn. i Èlektron. **7** (1962), 291–297. In Russian; English translation in Radio Eng. Electron. Phys. **7** (1962), 260–276.

30. *On linear physical systems connected with the external world by coupling channels*, Izv. Akad. Nauk SSSR Ser. Mat. **27** (1963), 993–1030. In Russian.

31. *Open systems as linear automata*, Izv. Akad. Nauk SSSR Ser. Mat. **27** (1963), 1215–1228, **MR** 31:6739. In Russian.

32. *Some problems in the theory of linear open systems*, Third All-Union Symposium on Wave Diffraction (Tbilisi, September 24–30, 1964), Izdat. "Nauka", Moscow, 1964, pp. 41–42. In Russian.

33. *Nonunitary representations of groups*, Funkcional. Anal. i Priložen. **3** (1969), no. 1, 62–70, **MR** 39:7464. In Russian.

34. (with L. L. Vaksman) *Open geometry, and operator colligations*, Ukrain. Geometr. Sb. (1974), no. 15, 16–35, i, **MR** 51:11323. In Russian.

35. *Linear discrete systems and their connection with M. M. Džrbašjan's theory of the factorization of meromorphic functions*, Doklady Akad. Nauk SSSR **219** (1974), 793–796, **MR** 52:1385. In Russian; English translation in Soviet Math. Dokl. **15** (1974), no. 6, 1657–1662.

36. *Commuting nonselfadjoint operators and solutions of systems of partial differential equations generated by them*, Soobshch. Akad. Nauk Gruzin. SSR **91** (1978), no. 2, 281–284, **MR** 80m:47008. In Russian.

37. *Operator waves in Hilbert space and related partial differential equations*, Integral Equations Operator Theory **2** (1979), no. 1, 25–47, **MR** 80j:47022.

38. *The inverse problem for the characteristic functions of several commuting operators*, Integral Equations Operator Theory **2** (1979), no. 2, 264–286, **MR** 80i:47015.

39. *Operator waves in Hilbert space and related partial differential equations*, International Symposium on Mathematical Theory of Networks and Systems, Volume 3 (Delft, 1979), edited by P. Dewilde, Western Periodicals Co., North Hollywood, CA, 1979, pp. 32–33.

40. *A method for constructing triangular canonical models of commuting operators based on connections with algebraic curves*, Integral Equations Operator Theory **3** (1980), no. 4, 489–507, **MR** 82h:47013.

41. *An introductory lecture to the theory of open systems*, Signal processing: theories and applications (Lausanne, 1980), edited by M. Kunt and F. de Coulon, North-Holland, Amsterdam, 1980, pp. 105–109.

42. *On scattering and transfer matrices for fields of systems*, Circuit theory and design (The Hague, 1981), edited by R. Boite and P. Dewilde, North-Holland, Amsterdam, 1981, **MR** 84h:47009, pp. 1022–1025.

43. *Commuting operators and fields of systems, distributed in Euclidean space*, Toeplitz centennial (Tel Aviv, 1981), edited by I. Gohberg, Operator Theory Adv. Appl. **4**, Birkhäuser, Basel, 1982, **MR** 84c:47012, pp. 377–413.

44. *Cayley-Hamilton theorem, vector bundles and divisors of commuting operators*, Integral Equations Operator Theory **6** (1983), no. 2, 250–373, **MR** 84j:47005.

45. *System theory and wave dispersion*, Mathematical theory of networks and systems (Beer Sheva, 1983), edited by P. A. Fuhrmann, Lecture Notes in Control Info. Sci. **58**, Springer, Berlin, 1984, **MR** 87a:47025, pp. 663–675.

46. *Collective motions of spatio-temporal systems*, J. Math. Anal. Appl. **116** (1986), no. 1, 22–41, **MR** 88a:47009.

47. *Commuting nonselfadjoint operators and mappings of vector bundles on algebraic curves*, Operator theory and systems (Amsterdam, 1985), edited by H. Bart, I. Gohberg and M. A. Kaashoek, Operator Theory Adv. Appl. **19**, Birkhäuser, Basel, 1986, **MR** 88h:47013, pp. 255–277.

48. *On commuting nonselfadjoint operators*, Integral Equations Operator Theory **9** (1986), no. 1, 121–133, **MR** 87h:47023.

49. *Commuting nonselfadjoint operators and collective motions of systems*, Commuting nonselfadjoint operators in Hilbert space: two independent studies, Lecture Notes in Math. **1272**, Springer, Berlin, 1987, **MR** 89a:47014, pp. 1–38.

50. (with Y. Avishai) *A study of solitonic combinations based on the theory of commuting nonselfadjoint operators*, Linear Algebra Appl. **122/123/124** (1989), 357–414, **MR** 90m:58089.

51. *What is a particle from the standpoint of systems theory?*, Integral Equations Operator Theory **14** (1991), no. 4, 552–563, **MR** 92k:47141.

52. (with A. S. Markus) *Joint spectrum and discriminant varieties of commuting nonselfadjoint operators*, Nonselfadjoint operators and related topics (Beer Sheva, 1992), edited by A. Feintuch and I. Gohberg, Operator Theory Adv. Appl. **73**, Birkhäuser, Basel, 1994, **MR** 95m:47004, pp. 1–29.

53. *The Blaschke-Potapov factorization theorem and the theory of nonselfadjoint operators*, Topics in interpolation theory (Leipzig, 1994), edited by H. Dym et al., Operator Theory Adv. Appl. **95**, Birkhäuser, Basel, 1997, **MR** 98i:47009, pp. 391–396.

54. *Commuting nonselfadjoint operators and a unified theory of waves and corpuscles*, New results in operator theory and its applications: the Israel M. Glazman memorial volume, edited by I. Gohberg and Yu. Lyubich, Operator Theory Adv. Appl. **98**, Birkhäuser, Basel, 1997, **MR** 98j:47018, pp. 163–185.
55. *Vortices of 2D systems*, this volume.

Memorial

1. (with I. Gohberg and I. Piatetski-Shapiro) *David Milman (1912–1982)*, Integral Equations Operator Theory **9** (1986), no. 1, i, 1–10, **MR** 87f:01050.

2. (with D. Z. Arov, L. A. Sakhnovich, A. A. Nudelman, V. K. Dubovoj, and V. E. Katsnelson) *Vladimir Petrovich Potapov as remembered by colleagues, friends and former students*, Topics in interpolation theory (Leipzig, 1994), edited by H. Dym et al., Operator Theory Adv. Appl. **95**, Birkhäuser, Basel, 1997.

Books

1. *Операторы, колебания, волны. Открытые системы*, Izdat. "Nauka", Moscow, 1966, **MR** 38:1922. Translated as *Operators, oscillations, waves (open systems)*, Transl. Math. Monographs **34**, American Mathematical Society, Providence, R.I., 1973, ISBN 0-8218-1584-1.

2. (with A. A. Yantsevich) *Теория операторных узлов в гильбертовых пространствах*, Izdat. Har'kov. Univ., Kharkov, 1971, **MR** 58:30342a. Translated as *Operator colligations in Hilbert spaces*, V. H. Winston & Sons, Washington, D.C., 1979, ISBN 0-470-26541-8.

3. (with N. Kravitsky, A. S. Markus, and V. Vinnikov) *Theory of commuting nonselfadjoint operators*, Mathematics and its Applications **332**, Kluwer Academic Publishers Group, Dordrecht, 1995, **MR** 96m:47003, ISBN 0-7923-3588-0.[1]

[1] A list of errata maintained at http://www.math.bgu.ac.il/~vinnikov/lkmverrata.html

Operator Theory:
Advances and Applications, Vol. 123

Vortices of 2D Systems

M. S. Livšic

Abstract. We consider "vortical" systems, generated by a linear 2D system in the case of infinitely many revolutions along an infinitesimal circle. The equations of such vortical system are obtained and the role played by operator vessels in the theory of 2D systems is revealed. It turns out that open 2D systems with zero curvature generate certain remarkable relations between partial differential equations at the input and at the output.

1. Curvature operators and related vortical systems

1.1. Infinitely many revolutions along an infinitesimal circle

Let H and E be two Banach spaces and let $A_j(x) : H \to H$ and $\varphi_j(x) : E \to H$ $(x \in \mathbb{R}^2;\ j = 1,2)$ be two pairs of operator valued functions mapping H into H and E into H respectively. We assume that $A_j(x)$ and $\varphi_j(x)$ are differentiable functions of $x = (x_1, x_2)$ in some domain of $\mathbb{R}^2$.

To a given (piecewise smooth) arc Γ: $x = x(t)$ $(t_0 \le t \le t_1)$, there corresponds the following linear system:

$$\begin{cases} i\dfrac{df}{dt} + \left(\dfrac{dx}{dt}, A\right) f = \left(\dfrac{dx}{dt}, \varphi\right) u, \\ f(t_0) = f_0, \quad v = \varphi' u + \varphi'' f, \end{cases} \tag{1}$$

where $u = u(t) \in E$ $(t_0 \le t \le t_1)$ is the input of the system, $f = f(t) \in H$ is the corresponding inner state, and $v = v(t)$ is the output; the scalar products are

$$\left(\frac{dx}{dt}, A\right) = \frac{dx_1}{dt} A_1 + \frac{dx_2}{dt} A_2, \quad \left(\frac{dx}{dt}, \varphi\right) = \frac{dx_1}{dt}\varphi_1 + \frac{dx_2}{dt}\varphi_2.$$

The operator valued functions $\varphi' = \varphi'(x) : E \to E$ and $\varphi'' = \varphi''(x) : H \to E$ mapping E into E and H into E respectively are given (and are differentiable functions of x).

If Γ is a circle with center $M_0(0,0)$ and radius r: $x_1 = r\cos\theta$, $x_2 = r\sin\theta$, then the equations (1) can be written in the form:

$$\frac{df}{d\theta} = ir[C(r,\theta) f - \hat{C}(r,\theta) u], \tag{2}$$

where

$$C(r,\theta) = -A_1(r\cos\theta, r\sin\theta)\sin\theta + A_2(r\cos\theta, r\sin\theta)\cos\theta,$$
$$\hat{C}(r,\theta) = -\varphi_1(r\cos\theta, r\sin\theta)\sin\theta + \varphi_2(r\cos\theta, r\sin\theta)\cos\theta,$$

and $f = f(r,\theta)$, $u = u(r,\theta)$.

Let us assume that the point $M(x_1, x_2)$ revolves around the center M_0 and that the number n of revolutions tends to infinity, while the radius r tends to zero. More precisely, our goal is to find the limit

$$\lim_{r\to 0} f\left(r, 2\pi\left[\frac{q}{\pi r^2}\right]\right) = \tilde{f}(q),$$

where $[\xi]$ is the entire part of ξ, and q $(0 \le q < \infty)$ is a given number. The vector valued function $f(r,\theta)$ is defined uniquely in some region $(0 \le r \le \bar{r},\ 0 \le \theta < \infty)$ if the input $u(r,\theta)$ and the initial condition $f(r,0)$ are given. The corresponding integral equation is:

$$f(r,\theta) = f(r,0) + ir\int_0^\theta [C(r,\theta')f(r,\theta') - \hat{C}(r,\theta')u(r,\theta')]\,d\theta'. \tag{3}$$

THE CASE $u(t) = 0$ (OR, USING A GEOMETRICAL LANGUAGE, THE *"parallel" translation* CASE). In this case the corresponding integral equation is:

$$f(r,\theta) = f(r,0) + ir\int_0^\theta C(r,\theta')f(r,\theta')\,d\theta'. \tag{4}$$

The solution of this equation has the form $f(r,\theta) = T(r,\theta)f(r,0)$, where $T(r,\theta)$ is a linear operator. From $C(r,\theta+2\pi) = C(r,\theta)$ it follows that

$$T(r, 2\pi n + \alpha) = T(r,\alpha)[T(r,2\pi)]^n \quad (0 \le \alpha \le 2\pi).$$

From (4) it follows that $|f(r,\theta) - f(r,0)| \le ar2\pi \ \max|f(r,\theta)|$, where $a = \max|C(r,\theta)|$ $(0 \le \theta \le 2\pi,\ 0 \le r \le \bar{r})$. Therefore, if $\bar{r} < (2\pi a)^{-1}$ then $|T(r,\theta)| \le (1 - 2\pi a\bar{r})^{-1}$, $|T(r,\theta) - I| \le 2\pi ar(1 - 2\pi a\bar{r})^{-1}$. Considering the series for the Volterra equation (4):

$$T(r,\theta) = I + ir\int_0^\theta C(r,\theta')\,d\theta' + (ir)^2\int_0^\theta C(r,\theta')\,d\theta'\int_0^{\theta'} C(r,\theta'')\,d\theta'' + \cdots,$$

and using the representation

$$C(r,\theta) = C(0,\theta) + r\frac{\partial C(0,\theta)}{\partial r} + r\varepsilon(r,\theta),$$

where $\lim_{r\to 0}\varepsilon(r,\theta) = 0$ uniformly $(0 \le \theta \le 2\pi)$, we obtain (we use the same ε to denote different infinitesimals)

$$ir\int_0^{2\pi} C(r,\theta')\,d\theta' = ir\int_0^{2\pi}\left[C(0,\theta') + r\frac{\partial C(0,\theta')}{\partial r} + r\varepsilon(r,\theta')\right]d\theta',$$

where $C(0,\theta) = -A_1(M_0)\sin\theta + A_2(M_0)\cos\theta$ and

$$\frac{\partial C(0,\theta)}{\partial r} = \left[-\frac{\partial A_1(M_0)}{\partial x_1}\sin\theta + \frac{\partial A_2(M_0)}{\partial x_1}\cos\theta\right]\cos\theta + \left[-\frac{\partial A_1(M_0)}{\partial x_2}\sin\theta + \frac{\partial A_2(M_0)}{\partial x_2}\cos\theta\right]\sin\theta.$$

The equalities $\int_0^{2\pi}\cos\theta\,d\theta = \int_0^{2\pi}\sin\theta\,d\theta = 0$ imply that the first order term with respect to r vanishes, and the equalities $\int_0^{2\pi}\cos^2\theta\,d\theta = \int_0^{2\pi}\sin^2\theta\,d\theta = \pi$, $\int_0^{2\pi}\cos\theta\,\theta\,d\theta = 0$ imply

$$ir\int_0^{2\pi} C(r,\theta')\,d\theta' = i\pi r^2\left[\frac{\partial A_2(M_0)}{\partial x_1} - \frac{\partial A_1(M_0)}{\partial x_2}\right] + r^2\varepsilon(r).$$

The next term in the series is

$$(ir)^2\int_0^{2\pi} C(r,\theta')\,d\theta'\int_0^{\theta'} C(r,\theta'')\,d\theta''.$$

Taking the second order term only (with respect to r) we obtain:

$$(ir)^2\int_0^{2\pi} C(0,\theta')\,d\theta'\int_0^{\theta'} C(0,\theta'')\,d\theta'' = \pi(ir)^2(A_2A_1 - A_1A_2)(M_0).$$

Therefore, for the operator $T(r,2\pi)$ we have obtained the representation

$$T(r,2\pi) = I + i\pi r^2(R_{12} + \varepsilon(r)), \tag{5}$$

where

$$R_{12} = \frac{\partial A_2(M_0)}{\partial x_1} - \frac{\partial A_1(M_0)}{\partial x_2} + \frac{1}{i}[A_1, A_2](M_0) \tag{6}$$

is the *curvature operator* at the point M_0.

We are going to prove the following relation

$$(T(r,2\pi))^{[q/\pi r^2]} = e^{iqR_{12}} + \varepsilon(r,q), \tag{7}$$

where $\lim_{r\to 0}\varepsilon(r,q) = 0$ uniformly ($0 \le q \le \bar{q}$). Denote $h = \pi r^2$, $Q(h) = i(R_{12} + \varepsilon(\sqrt{\frac{h}{\pi}}))$. Then

$$(T(r,2\pi))^{q/h} = e^{q/h(hQ - \frac{h^2Q^2}{2} + \cdots)} = e^{qQ}e^{qh\omega},$$

where $\omega = -\frac{Q^2}{2} + \frac{Q^3h}{3} - \cdots$ ($|h|$ is sufficiently small). Hence, $|\omega(h)| \le \text{const}$ and

$$|(I+hQ)^{q/h} - e^{qQ}| = |e^{qQ}e^{qh\omega} - e^{qQ}| = |e^{qQ}|\,|e^{qh\omega} - I| \le a(\bar{q})h,$$

and

$$|[I + i\pi r^2(R_{12} + \varepsilon(r))]^{q/\pi r^2} - e^{iq(R_{12}+\varepsilon(r))}| \le a(\bar{q})\pi r^2.$$

Therefore,

$$|(T(r,2\pi))^{q/\pi r^2} - e^{iqR_{12}}| \leq \text{const} \cdot \varepsilon(r)$$

and the relation (7) follows from $|[q/\pi r^2] - q/\pi r^2| < 1$.

Using the relation (7) we obtain:

Theorem 1. *Let $f(r,\theta) = T(r,\theta)f(r,0)$ be the parallel translation along a circle with the center M_0 and let $\lim_{r\to 0} f(r,0) = f_0$. Then there exists*

$$\lim_{r\to 0} f(r, 2\pi[q/\pi r^2]) = \tilde{f}(q), \tag{8}$$

and the function $\tilde{f}(q)$ satisfies the equation

$$i\frac{d\tilde{f}}{dq} + R_{12}(M_0)\tilde{f} = 0, \quad \tilde{f}(0) = f_0. \tag{9}$$

Now we want to evaluate the parallel translation

$$f(r, 2\pi[q/\pi r^2] + \alpha) = T(r,\alpha)T(r, 2\pi[q/\pi r^2])f(r,0),$$

where $0 \leq \alpha \leq 2\pi$. The series for the Volterra equation imply:

$$T(r,\alpha) = I + ir\int_0^\alpha C(r,\theta')\,d\theta' + r\varepsilon(r,\alpha)$$

$$\left(\lim_{r\to 0}\varepsilon(r,\alpha) = \lim_{r\to 0}\frac{\partial\varepsilon}{\partial\alpha}(r,\alpha) = 0, \quad 0 \leq \alpha \leq 2\pi\right).$$

Using $C(r,\theta) = C(0,\theta) + \varepsilon(r,\theta)$ we obtain

$$\begin{aligned} T(r,\alpha) &= I + ir\int_0^\alpha [-A_1(M_0)\sin\theta + A_2(M_0)\cos\theta]\,d\theta + r\varepsilon(r,\alpha) \\ &= I + ir[A_1(M_0)(\cos\alpha - 1) + A_2(M_0)\sin\alpha] + r\varepsilon(r,\alpha). \end{aligned}$$

Then from

$$T(r, 2\pi\ [q/\pi r^2] + \alpha) = T(r,\alpha)[T(r,2\pi)]^{[q/\pi r^2]} = T(r,\alpha)(e^{iqR_{12}} + \varepsilon(r,q))$$

it follows that

Theorem 2. *If $f(r,\theta) = T(r,\theta)f(r,0)$ is the parallel translation along a circle and if $\lim_{r\to 0} f(r,0) = f_0$ then the following representation holds:*

$$\begin{aligned} &f(r, 2\pi[q/\pi r^2] + \alpha) = \\ &e^{iqR_{12}}f_0 + ir[A_1(M_0)\cos\alpha + A_2(M_0)\sin\alpha]e^{iqR_{12}}f_0 + \varepsilon(r,q) + r\eta(r,q,\alpha), \end{aligned} \tag{10}$$

where

$$\lim_{r\to 0}\varepsilon(r,q) = \lim_{r\to 0}\eta(r,q,\alpha) = \lim_{r\to 0}\frac{\partial\eta(r,q,\alpha)}{\partial\alpha} = 0$$

uniformly $(0 \leq q \leq \bar{q},\ 0 \leq \alpha \leq 2\pi)$.

1.2. Cyclodromic functions

Generalizing the relation (10) for parallel translations we introduce the notion of a "cyclodromic" function:

Definition. Let $\chi(r,\theta)$ $(0 \le \theta < \infty, 0 \le r \le \bar{r})$ be a differentiable function. Denote

$$\bar{\chi}(r,q,\alpha) = \chi(r, 2\pi[q/\pi r^2] + \alpha).$$

The function $\chi(r,\theta)$ is said to be *cyclodromic* if $\bar{\chi}$ admits a representation

$$\bar{\chi}(r,q,\alpha) = \tilde{\chi}(q) + r[\tilde{\chi}_1(q)\cos\alpha + \tilde{\chi}_2(q)\ \alpha] + \varepsilon(r,q) + r\eta(r,q,\alpha), \tag{11}$$

where

$$\lim_{r\to 0} \varepsilon(r,q) = \lim_{r\to 0} \eta(r,q,\alpha) = \lim_{r\to 0} \frac{\partial \eta(r,q,\alpha)}{\partial \alpha} = 0$$

uniformly $(0 \le q \le \bar{q},\ 0 \le \alpha \le 2\pi)$.

The relation (10) implies that the inner state $f(r,\theta)$ in the case $u(r,\theta) = 0$ is a cyclodromic function. In this case the vortical "intensity" $\tilde{f}(q) = \exp(iqR_{12})f_0$.

If χ is a cyclodromic function then

$$\frac{\partial \bar{\chi}}{r\partial \alpha} = -\chi_1(q)\sin\alpha + \chi_2(q)\cos\alpha + \frac{\partial \eta}{\partial \alpha}.$$

Definition. The limit

$$\lim_{r\to 0} \frac{\partial \bar{\chi}}{r\partial \alpha} = -\tilde{\chi}_1(q)\sin\alpha + \tilde{\chi}_2(q)\cos\alpha \tag{12}$$

is said to be the *vortical derivative* in the dirction $\tau(\alpha)$:

$$\lim_{r\to 0} \frac{\partial \bar{\chi}}{r\partial \alpha} = \frac{\widetilde{\partial \chi}}{\partial \tau(\alpha)},$$

where $\tau(\alpha)$ is the corresponding tangent vector to the circle (Figure 1).

If $\alpha = 0$ then $\dfrac{\widetilde{\partial \chi}}{\partial \tau(0)} = \dfrac{\widetilde{\partial \chi}}{\partial x_2}(M_0)$ and if $\alpha = \frac{3\pi}{2}$ then $\dfrac{\widetilde{\partial \chi}}{\partial \tau(\frac{3\pi}{2})} = \dfrac{\widetilde{\partial \chi}}{\partial x_1}(M_0)$, where $\dfrac{\widetilde{\partial \chi}}{\partial x_j}(M_0)$ is the vortical derivative in the direction x_j $(j = 1,2)$. The equality (12) implies

$$\tilde{\chi}_1(q) = \frac{\widetilde{\partial \chi}}{\partial x_1}, \quad \tilde{\chi}_2(q) = \frac{\widetilde{\partial \chi}}{\partial x_2}.$$

The equality (11) has, therefore, the form:

$$\bar{\chi} = \tilde{\chi}(q) + r\left[\frac{\widetilde{\partial \chi(q)}}{\partial x_1}\cos\alpha + \frac{\widetilde{\partial \chi(q)}}{\partial x_2}\sin\alpha\right](M_0) + \varepsilon(r,q) + r\eta(r,q,\alpha), \tag{13}$$

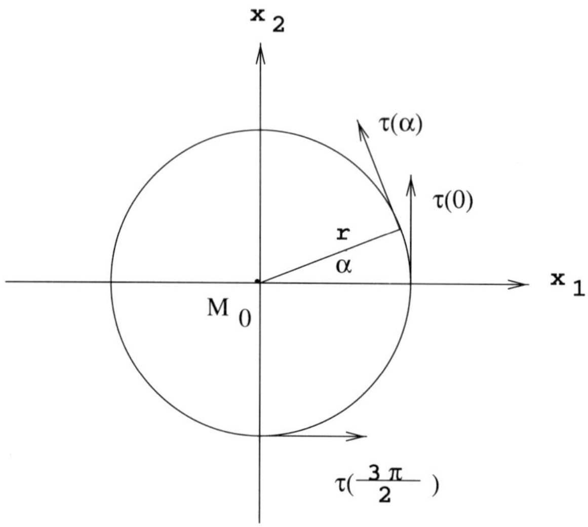

FIGURE 1

where

$$\tilde{\chi}(q) = \lim_{r\to 0} \bar{\chi}, \quad \frac{\widetilde{\partial \chi}}{\partial x_1} = \lim_{r\to 0} \left.\frac{\partial \bar{\chi}}{r \partial \alpha}\right|_{\alpha = 3\pi/2}, \quad \text{and} \quad \frac{\widetilde{\partial \chi}}{\partial x_2} = \lim_{r\to 0} \left.\frac{\widetilde{\partial \chi}}{r \partial \alpha}\right|_{\alpha = 0}.$$

For the inner state $f(r,\theta)$ in the case $u(r,\theta) = 0$ the vortical derivatives are:

$$\frac{\widetilde{\partial f(q)}}{\partial x_j} = i A_j(M_0) \exp(iqR_{12}(M_0)) f_0.$$

Example 1. Let $\chi(r,\theta) = F(\frac{1}{2}r^2\theta)$, where $F(z)$ is a differentiable function of z $(0 \leq z \leq z_1)$. Then $\bar{\chi} = F\left(\frac{1}{2}r^2(2\pi[q/\pi r^2] + \alpha)\right)$ and $[q/\pi r^2] = q/\pi r^2 - \delta$ $(0 \leq \delta < 1)$. Hence

$$\bar{\chi} = F\left(q + (-\delta\pi + \frac{\alpha}{2})r^2\right) = F(q) + r\eta \quad (\lim_{r\to 0} \eta = \lim_{r\to 0} \frac{\partial \eta}{\partial \alpha} = 0).$$

Therefore $F(\frac{1}{2}r^2\theta)$ is a cyclodromic function and

$$\tilde{\chi}(q) = F(q), \quad \frac{\widetilde{\partial \chi}}{\partial x_j} = 0 \quad (j = 1, 2).$$

Example 2. Let $F(x_1, x_2)$ be a differentiable function of two real variables (x_1, x_2). Consider

$$\chi(r,\theta) = F(r\cos\theta, r\sin\theta).$$

In this case

$$\bar{\chi} = F(r\cos\alpha, r\sin\alpha) = F(M_0) + r\left(\frac{\partial F}{\partial x_1}\cos\alpha + \frac{\partial F}{\partial x_2}\sin\alpha\right)(M_0) + r\eta.$$

Therefore

$$\tilde{\chi}(q) = F(M_0), \quad \widetilde{\frac{\partial \chi}{\partial x_j}} = \frac{\partial F}{\partial x_j}(M_0).$$

Example 3. Let $\chi(r,\theta) = F_0(\frac{1}{2}r^2\theta) + [x_1 F_1(\frac{1}{2}r^2\theta) + x_2 F_2(\frac{1}{2}r^2\theta]$, where $x_1 = r\cos\theta$, $x_2 = r\sin\theta$. Using Example 1 we obtain

$$\bar{\chi} = F_0(q) + r[F_1(q)\cos\alpha + F_2(q)\sin\alpha] + r\eta,$$

and therefore

$$\tilde{\chi} = F_0(q), \quad \widetilde{\frac{\partial \chi}{\partial x_j}} = F_j(q).$$

1.3. The Case $u(r,\theta) \neq 0$

Assume that the input $u(r,\theta)$ is a cyclodromic function. The solution of the equation

$$\frac{df}{d\theta} = irC(r,\theta)f - ir\hat{C}(r,\theta)u,$$

satisfying the initial condition $f(r,\theta) = 0$, is given by

$$\hat{f}(r,\theta) = -iT(r,\theta)\int_0^\theta T^{-1}(r,\theta')r\hat{C}(r,\theta')u(r,\theta')\,d\theta'.$$

Hence

$$\hat{f}(r, 2\pi n) = -i[T(r,2\pi)]^n \sum_{k=0}^{n-1} \int_{2k\pi}^{2(k+1)\pi} T^{-1}(r,\theta')\hat{C}(r,\theta')u(r,\theta')r\,d\theta'. \tag{14}$$

Putting $\theta' = 2k\pi + \alpha$ we obtain

$$\begin{aligned}(15)\quad & \hat{f}(r,2\pi n) \\ & = -i[T(r,2\pi)]^n \sum_{k=0}^{n-1}\int_0^{2\pi} [T(r,2\pi)]^{-k} T^{-1}(r,\alpha)\hat{C}(r,\alpha)u(r,2k\pi+\alpha)r\,d\alpha.\end{aligned}$$

Using the relation (7) with $q_k = k\pi r^2$ $(0 \leq q_k \leq q)$ we conclude that

$$[T(r,2\pi)]^k = e^{iq_k R_{12}} + \varepsilon(r, q_k),$$

where $\lim_{r\to 0}\varepsilon(r,q_k) = 0$ uniformly if $0 \leq q_k \leq \bar{q}$, and

$$[T(r,2\pi)]^{-k} = e^{-iq_k R_{12}} + \varepsilon(r,q_k),$$

$q_k - q_{k-1} = \pi r^2$. To evaluate the integral

$$I_k(r) = \int_0^{2\pi} [T(r,2\pi)]^{-k} T^{-1}(r,\alpha)\hat{C}(r,\alpha)u(r,2\pi k+\alpha)r\,d\alpha,$$

we will use the relations:

$$T^{-1}(r,\alpha) = I - ir[A_1(\cos\alpha - 1) + A_2\sin\alpha](M_0) + r\varepsilon(r,\alpha),$$

$$\hat{C}(r,\alpha) = \hat{C}(0,\alpha) + r\frac{\partial\hat{C}(0,\alpha)}{\partial r} + r\varepsilon(r,\alpha),$$

(16)

$$u(r,2k\pi+\alpha) = \tilde{u}(q_k) + r\left[\frac{\widetilde{\partial u(q_k)}}{\partial x_1}\cos\alpha + \frac{\widetilde{\partial u(q_k)}}{\partial x_2}\sin\alpha\right] + \varepsilon(r,q_k) + r\eta(r,q_k,\alpha).$$

After calculations we obtain:

$$f(r,2\pi[q/\pi r^2]) = -\, iT(r,2\pi\cdot[q/\pi r^2])\sum_{k=0}^{n-1} e^{-iq_k R_{12}}(L_{12}(q_k) + \varepsilon(r,q_k))\Delta q_k,$$

where $\Delta q_k = q_k - q_{k-n} = \pi r^2$ and

(17)

$$L_{12}(q) = \left[(\frac{\partial}{\partial x_1} - iA_1)\varphi_2 - (\frac{\partial}{\partial x_2} - iA_2)\varphi_1\right]_{M_0}\tilde{u}(q) + \left(\varphi_2\frac{\widetilde{\partial u(q)}}{\partial x_1} - \varphi_1\frac{\widetilde{\partial u(q)}}{\partial x_2}\right)_{M_0}.$$

Hence

$$\tilde{f}(q) = \lim_{r\to 0} f(r,2\pi[q/\pi r]) = -ie^{iqR_{12}}\int_0^q e^{-iq'R_{12}}L_{12}(q')\,dq',$$

and therefore $\tilde{f}(q)$ satisfies the equation

$$i\frac{d\tilde{f}}{dq} + R_{12}(M_0)\tilde{f} = L_{12}(q). \tag{18}$$

To conclude this section we will show:

Theorem 3. *If $u(r,\theta)$ is cyclodromic then $f(r,\theta)$ is also cyclodromic.*

For this purpose consider

$$\begin{aligned} &f(r,2\pi[q/\pi r^2]+\alpha) \\ &\quad= -irT(r,2\pi[q/\pi r^2]+\alpha)\int_0^{2\pi[q/\pi r^2]+\alpha} T^{-1}(r,\theta)\hat{C}(r,\theta)u(r,\theta)\,d\theta \\ &\quad= -iT(r,\alpha)T(r,2\pi[q/\pi r^2])r\left[\int_0^{2\pi[q/\pi r^2]} T^{-1}(r,\theta)\hat{C}(r,\theta)u(r,\theta)\,d\theta\right. \\ &\qquad\left. + \int_0^{\alpha} T^{-1}(r,\theta'+2\pi[q/\pi r^2])\hat{C}(r,\theta')u(r,2\pi[q/\pi r^2]+\theta')\,d\theta'\right]. \end{aligned}$$

After calculations we obtain

(19)

$$f(r,2\pi[q/\pi r^2]+\alpha) = \tilde{f}(q) + r\left[\frac{\widetilde{\partial f(q)}}{\partial x_1}\cos\alpha + \frac{\widetilde{\partial f(q)}}{\partial x_2}\sin\alpha\right] + \varepsilon(r,q) + r\eta(r,q,\alpha),$$

where *the vortical derivatives* $\dfrac{\widetilde{\partial f}}{\partial x_j}$ *can be found from the relations*:

$$i\frac{\widetilde{\partial f(M_0,q)}}{\partial x_j} + A_j(M_0)\tilde{f}(M_0,q) = \varphi_j(M_0)\tilde{u}(M_0,q). \tag{20}$$

2. Vortices of open systems

2.1. Assimilative vortical systems

Open systems form a remarkable subclass among the general linear systems. Let $A_j(x)$ ($j = 1,2$; $x \in \mathbb{R}^2$) be a pair of linear operators in a *Hilbert* space H and let E be a finite dimensional space equipped with a scalar product. Let $\varphi(x)$ be a linear mapping of H into E. Then we can find the adjoint mapping $\varphi^*(x) : E \to H$ using the relations

$$(\varphi(x)f, u) = (f, \varphi^*(x)u) \quad (u \in E, f \in H).$$

Definition. A collection

$$X(x) = (A_j(x);\ H, \varphi(x), E;\ \sigma_j(x)) \quad (j = 1,2)$$

is said to be a *colligation* if $\sigma_j(x)$ are selfadjoint operators in E and the relations

$$\frac{1}{i}(A_j(x) - A_j^*(x)) = \varphi^*(x)\sigma_j(x)\varphi(x) \quad (j = 1,2)$$

hold.

A colligation $X(x)$ is said to be *strict* at a point x if the following conditions are fulfilled:

1. $\varphi(x)H = E$,
2. $\bigcap_j \ker \sigma_j(x) = 0$.

To a given colligation there corresponds a family of systems, defined by the equation (1) with the mappings $\varphi_j(x) = \varphi^*(x)\sigma_j(x)$, $\varphi'(x) = I$, $\varphi''(x) = -i\varphi(x)$:

$$i\frac{df}{dt} + \left(\frac{dx}{dt}, A\right) f = \varphi^* \left(\frac{dx}{dt}, \sigma\right) u, \tag{21}$$
$$v = u - i\varphi f.$$

It turns out [3] that for such systems — we call them *open systems* — the law of "energy conservation" holds:

$$|f_\Gamma(M_1)|^2 - |f_\Gamma(M_0)|^2 = \int_\Gamma [(\sigma_1 u, u) - (\sigma_1 v, v)]\, dx_1 + [(\sigma_2 u, u) - (\sigma_2, v, v)]\, dx_2.$$

Therefore in this case the change of the inner scalar product (f, f) along the path is equal to the difference between the flows of "energy" through the input and the output. The mapping $\varphi(x) : H \to E$ is said to be the *window* of the colligation

$X(x)$. In the case $\varphi(x) = 0$ the operators $A_k(x) = A_k^*(x)$ are selfadjoint and the corresponding system is closed:

$$(f_\Gamma(M_0), f_\Gamma(M_0)) = (f_\Gamma(M_1), f_\Gamma(M_1)).$$

In the case $u = 0$ the equations (21) define the *parallel translation* of the state $f(M)$ along the line Γ. These translations are accompanied by emission and absorption of energy through the output. In the case $u \neq 0$ there is also an interaction through the input. We will use the term "open geometry" for the family of all possible open systems (21), where Γ is an arbitrary (piecewise smooth) arc. It is easy to see that *in the case of open geometry the equations* (18) *of the "vortical" system can be written in the form*

$$i\frac{d\tilde{f}}{dq} + R_{12}\tilde{f} = \varphi^*\left(\sigma_2\widetilde{\frac{\partial u}{\partial x_1}} - \sigma_1\widetilde{\frac{\partial u}{\partial x_2}} + \left(\frac{\partial \sigma_2}{\partial x_1} - \frac{\partial \sigma_1}{\partial x_2}\right)\tilde{u}\right) + i(\psi^{in})^*\tilde{u}, \tag{22}$$

where

$$\psi^{in}(x) = (\sigma_1(x)\nabla_2^\times - \sigma_2(x)\nabla_1^\times)\varphi(x), \tag{23}$$

$$\nabla_j^\times \varphi = \frac{1}{i}\frac{\partial \varphi}{\partial x_j} + \varphi A_j^*, \tag{24}$$

$$R_{12} = \frac{\partial A_2}{\partial x_1} - \frac{\partial A_1}{\partial x_2} + \frac{1}{i}[A_1, A_2].$$

In the equations (22) there appears, besides the window $\varphi(x)$, the mapping $\psi^{in}(x) : H \to E$ which is said to be the *vortical window*. We will say that the vortical window is *adaptable to the window* $\varphi(x)$ if there exists an operator $\gamma^{in}(x) : E \to E$ such that

$$\psi^{in}(x) = \gamma^{in}(x)\varphi(x). \tag{25}$$

Our final goal is to bring the equations (22) to a "colligational" form of an open system with the window $\varphi(x)$:

$$i\frac{d\tilde{f}}{dq} + R_{12}\tilde{f} = \varphi^*\sigma_{12}\tilde{u}, \tag{26}$$

where $\sigma_{12}(x) : E \to E$ is a selfadjoint operator such that:

$$\frac{1}{i}(R_{12}(x) - R_{12}^*(x)) = \varphi^*(x)\sigma_{12}(x)\varphi(x).$$

Then the "conservation" law (continuity condition) in the form

$$\frac{d}{dq}(\tilde{f}, \tilde{f}) = (\sigma_{12}\tilde{u}, \tilde{u}) - (\sigma_{12}\tilde{v}, \tilde{v})$$

will be applicable to the vortical system related to the original open geometry. The adaptability condition of $\psi^{in}(x)$ to $\varphi(x)$ can be written in the form

$$(\sigma_1(x)\nabla_2^\times - \sigma_2(x)\nabla_1^\times)\varphi(x) = \gamma^{in}(x)\varphi(x).$$

When the equations (26) hold, we will say that the vortical system is *assimilated* with the original open geometry.

Lemma. *If $X(x)$ is a colligation then the following relations hold:*

(27)
$$\operatorname{Im} R_{12} = \operatorname{Im}(\varphi^* \psi^{in}) + \frac{1}{2}\varphi^* \left(\frac{\partial \sigma_2}{\partial x_1} - \frac{\partial \sigma_1}{\partial x_2}\right)\varphi,$$
$$\operatorname{Im} R_{12} = \operatorname{Im}(\varphi^* \psi^{out}) + \frac{1}{2}\varphi^* \left(\frac{\partial \sigma_2}{\partial x_1} - \frac{\partial \sigma_1}{\partial x_2}\right)\varphi,$$

where

(28)
$$\psi^{in} = (\sigma_1 \nabla_2^{\times} - \sigma_2 \nabla_1^{\times})\varphi,$$
$$\psi^{out} = (\sigma_1 \nabla_2 - \sigma_2 \nabla_1)\varphi,$$

and

$$\nabla_j \varphi = \frac{1}{i}\frac{\partial \varphi}{\partial x_j} + \varphi A_j, \quad \nabla_j^{\times} \varphi = \frac{1}{i}\frac{\partial \varphi}{\partial x_j} + \varphi A_j^*,$$
$$R_{12} = \frac{\partial A_2}{\partial x_1} - \frac{\partial A_1}{\partial x_2} + \frac{1}{i}(A_1 A_2 - A_2 A_1).$$

Proof. From the colligation conditions it follows that

$$\frac{\partial}{\partial x_j}(A_k - A_k^*) = i\frac{\partial}{\partial x_j}(\varphi^* \sigma_k \varphi) = i\varphi^* \sigma_k \frac{\partial \varphi}{\partial x_j} - (i\varphi^* \sigma_k \frac{\partial \varphi}{\partial x_j})^* + i\varphi^* \frac{\partial \sigma_k}{\partial x_j}\varphi.$$

Hence

$$\operatorname{Im} R_{12} = \operatorname{Im}\left(\frac{\partial A_1}{\partial x_2} - \frac{\partial A_2}{\partial x_1} + \frac{1}{i}[A_1, A_2]\right)$$
$$= \operatorname{Im}\left(i\varphi^* \sigma_2 \frac{\partial \varphi}{\partial x_1} - i\varphi^* \sigma_1 \frac{\partial \varphi}{\partial x_2} + \frac{1}{i}[A_1, A_2]\right) + \frac{1}{2}\varphi^* \left(\frac{\partial \sigma_2}{\partial x_1} - \frac{\partial \sigma_1}{\partial x_2}\right)\varphi.$$

It is evident that

$$\operatorname{Im}(\frac{1}{i}[A_1, A_2]) = \operatorname{Im}\left(\frac{1}{i}(A_1 A_2 - A_2 A_1) - \frac{1}{i}(A_1^* A_2 - A_2^* A_1)\right)$$
$$= \operatorname{Im}(\varphi^* \sigma_1 \varphi A_2 - \varphi^* \sigma_2 \varphi A_1).$$

Hence

$$\operatorname{Im} R_{12} = \operatorname{Im}\left(\varphi^* (\sigma_1 \nabla_2 - \sigma_2 \nabla_1)\varphi\right) + \frac{1}{2}\varphi^* \left(\frac{\partial \sigma_2}{\partial x_1} - \frac{\partial \sigma_1}{\partial x_2}\right)\varphi,$$

which is the second relation (27).

To obtain the first relation (27) consider the adjoint colligation

$$X^*(x) = (A_1^*(x), A_2^*(x);\ H, -\varphi(x), E;\ -\sigma_1(x), -\sigma_2(x)).$$

It is easy to check that $\psi^{in}(X^*) = \psi^{out}(X)$ and $\psi^{out}(X^*) = \psi^{in}(x)$. Hence the first relation follows from the second relation by considering X^* instead of X. □

It is easy to see that

(29)
$$\psi^{out} - \psi^{in} = i(\sigma_1 \varphi \varphi^* \sigma_2 - \sigma_2 \varphi \varphi^* \sigma_1)\varphi.$$

Using the adaptability condition $\psi^{in} = \gamma^{in}\varphi$ we conclude that $\psi^{out} = \gamma^{out}\varphi$, where $\gamma^{out} = \gamma^{in} + i(\sigma_1\varphi\varphi^*\sigma_2 - \sigma_2\varphi\varphi^*\sigma_1)$. Hence if the vortical window $\psi^{in}(X)$ is adaptable then the vortical window $\psi^{in}(X^*) = \psi^{out}(X)$ is also adaptable to the same $\varphi(x)$.

2.2. Fields of vessels

Definition. A collection

$$V(x) = (A_1(x), A_2(x);\ H, \varphi(x), E;\ \sigma_1(x), \sigma_2(x), \gamma^{in}(x), \gamma^{out}(x))$$

is said to be a vessel if the following conditions are fulfilled:

$$\frac{1}{i}(A_j - A_j^*) = \varphi^*\sigma_j\varphi \quad (j = 1, 2), \tag{30}$$

$$(\sigma_1\nabla_2^\times - \sigma_2\nabla_1^\times)\varphi = \gamma^{in}\varphi, \tag{31}$$

$$\gamma^{out} = \gamma^{in} + i(\sigma_1\varphi\varphi^*\sigma_2 - \sigma_2\varphi\varphi^*\sigma_1), \tag{32}$$

$$(\sigma_1\nabla_2 - \sigma_2\nabla_1)\varphi = \gamma^{out}\varphi. \tag{33}$$

The relation (29) implies that (30), (31) and (32) imply (33), and that (30), (32) and (33) imply (31). The conditions (31) and (33) are the *adaptability* conditions of the vortical windows, the condition (30) is the *colligation* condition, and the condition (32) is the *linkage* condition [3].

Let $V(x)$ be a vessel. The relations (27), (28) and (31) imply

$$\begin{aligned}\operatorname{Im} R_{12} &= \operatorname{Im}(\varphi^*\psi^{in}) + \frac{1}{2}\varphi^*\left(\frac{\partial\sigma_2}{\partial x_1} - \frac{\partial\sigma_1}{\partial x_2}\right)\varphi \\ &= \varphi^*\left(\operatorname{Im}\gamma^{in} + \frac{1}{2}\left(\frac{\partial\sigma_2}{\partial x_1} - \frac{\partial\sigma_1}{\partial x_2}\right)\right)\varphi.\end{aligned}$$

Therefore if $V(x)$ is a vessel then

$$\frac{1}{i}(R_{12} - R_{12}^*) = \varphi^*\sigma_{12}\varphi, \tag{34}$$

where

$$\sigma_{12} = 2\operatorname{Im}\gamma^{in} + \frac{\partial\sigma_2}{\partial x_1} - \frac{\partial\sigma_1}{\partial x_2} = \sigma_{12}^*. \tag{35}$$

In the general case $\gamma^{out} \neq \gamma^{in}$, but the linkage condition implies that always $\operatorname{Im}\gamma^{in} = \operatorname{Im}\gamma^{out}$.

Definition. The collection

$$Y(x) = (A_1(x), A_2(x), R_{12}(x);\ H, \varphi(x), E;\ \sigma_1(x), \sigma_2(x), \sigma_{12}(x))$$

is said to be a colligation of the second order if the following conditions hold:

$$\frac{1}{i}(A_j - A_j^*) = \varphi^*\sigma_j\varphi, \quad \frac{1}{i}(R_{12} - R_{12}^*) = \varphi^*\sigma_{12}\varphi,$$

where $\sigma_j = \sigma_j^*, \sigma_{12} = \sigma_{12}^*$ (σ_{12} is not necessarily defined by (35)).

If $V(x)$ is a given vessel and σ_{12} is defined by (35) then the colligation $Y(x)$ is said to be the *derivative* of V. In this case we will write $Y = \delta V$.

Remark 1. If $X(x)$ is a colligation then the subspace

$$G_X(x) = \sum_j (A_j(x) - A_j^*(x))H$$

is said to be the non-Hermitian subspace of $X(x)$ (or of the pair $A_j(x)$, $j = 1, 2$) at the point $x \in \mathbb{R}^2$ (see [3, Chapter 2]). If $X(x)$ is strict then the window $\varphi(x)$ has the following properties [3]:

1. $\varphi^*(x)E = G_X(x)$;
2. $\varphi(x)G_X(x) = E$;
3. $\varphi(x)(H \ominus G_X(x)) = 0$.

Remark 2. Let $A_j(x)$ be continuous operator valued functions in some domain $\mathcal{D}$ in $\mathbb{R}^2$. Consider the non-Hermitian subspace $G(x)$ of the pair $A_j(x)$ ($j = 1, 2$; $x \in \mathcal{D}$). Assume that $\delta(x) = \dim G(x)$ ($x \in \mathcal{D}$) is bounded. Let us consider

$$\max_{x \in \mathcal{D}} \delta(x) = \delta_0.$$

It is evident that the set $\mathcal{D}'$ of points x' such that $\dim G(x') = \delta_0$ is not empty and that $\mathcal{D}' \subseteq \mathcal{D}$ is a subdomain of $\mathcal{D}$. Let E be a δ_0-dimensional space. Then there exists an operator valued function $\tau(x')$ such that $\tau(x')G(x') = E$. Evidently $\frac{1}{i}(A_j(x') - A_j^*(x')) = j_{G(x')} \, \frac{1}{i}(A_j(x') - A_j^*(x'))\Big|_{G(x')} P_{G(x')}$, where $P_{G(x')}$ is the orthogonal projection of H onto $G(x')$ and $j_{G(x')}$ is the embedding operator of $G(x')$ into H. Therefore

$$\frac{1}{i}(A_j(x) - A_j^*(x)) = \varphi^*(x)\sigma_j(x)\varphi(x) \quad (x \in \mathcal{D}'),$$

where $\varphi(x) = \tau(x)P_{G(x)}$, $\varphi^*(x) = j_{G(x)}\tau^*(x)$, and

$$\sigma_j(x) = \tau^{*-1}(x) \, \frac{1}{i}(A_j(x) - A^*(x))\Big|_{G(x)} \tau^{-1}(x)$$

is a self-adjoint operator in E. Hence *the given pair of operators $A_j(x)$ can always be embedded in a strict colligation $X(x)$, where $x = (x_1, x_2)$ belongs to some subdomain $\mathcal{D}'$.*

Remark 3. Let $A_j(x)$ ($x \in \mathcal{D}$) be a pair of differentiable operator valued functions such that $\dim G(x)$ is bounded in the domain $\mathcal{D}$. Then the relations (27) imply that $\dim(R_{12}(x) - R_{12}^*(x))$ is bounded in the same domain. Therefore *the operator valued functions $A_j(x)$ ($j = 1, 2$) can be embedded in a strict colligation of the second order* if x belongs to some subdomain $\mathcal{D}'$ of $\mathcal{D}$. (A colligation of the second order $Y(x)$ is strict at a point x if $\varphi(x)H = E$ and $\ker \sigma_1(x) \cap \ker \sigma_2(x) \cap \ker \sigma_{12}(x) = 0$.)

Theorem 4. *If*

$$Y(x) = (A_1(x), A_2(x), R_{12}(x);\ H, \varphi(x), E;\ \sigma_1(x), \sigma_2(x), \sigma_{12}(x))$$

is a strict colligation of the second order then there exists a unique vessel $V(x)$ such that $Y = \delta V$.

Proof. The first relation (27) implies that $\operatorname{Im} R_{12} = \operatorname{Im} L$, when

$$L = \varphi^* \psi^{in} + \frac{i}{2}\varphi^* \left(\frac{\partial \sigma_2}{\partial x_1} - \frac{\partial \sigma_1}{\partial x_2}\right)\varphi.$$

Using the condition

$$2 \operatorname{Im} R_{12} = \varphi^* \sigma_{12} \varphi.$$

we conclude that $\operatorname{Im}(L - \frac{1}{i}\varphi^*\sigma_{12}\varphi) = 0$. Hence, the operator $L_1 = L - \frac{1}{i}\varphi^*\sigma_{12}\varphi = L_1^*$ is a selfadjoint operator mapping H into G_Y,

$$L_1(H) \subseteq G_Y, \quad L_1(H \ominus G_Y) = 0$$

(here $G_Y(x) = (A_1(x) - A_1^*(x))H + (A_2(x) - A_2^*(x))H + (R_{12}(x) - R_{12}^*(x))H$ is the non-Hermitian subspace of the colligation of the second order $Y(x)$). From the strictness of Y it follows that $\varphi G_Y = E$, $\varphi(H \ominus G_Y) = 0$, and $\varphi^* E = G_Y$. Therefore the operator L_1 can be represented in the form $L_1 = \varphi^* \alpha^{in} \varphi$, where $\alpha^{in} = (\varphi_0^*)^{-1} L_1 \varphi_0^{-1}$ ($\varphi_0 = \varphi|_{G_Y}$) is a self-adjoint operator in E. Hence

$$\varphi^* \psi^{in} + \frac{i}{2}\varphi^* \left(\frac{\partial \sigma_2}{\partial x_1} - \frac{\partial \sigma_1}{\partial x_2} - \sigma_{12}\right)\varphi = \varphi^* \alpha^{in} \varphi,$$

and therefore $\psi^{in} = \gamma^{in}(x)\varphi(x)$, where

$$\gamma^{in}(x) = \alpha^{in} + \frac{i}{2}\left(\sigma_{12} + \frac{\partial \sigma_1}{\partial x_2} - \frac{\partial \sigma_2}{\partial x_1}\right).$$

Analogously, using the second relation (27) we obtain $\psi^{out} = \gamma^{out}(x)\varphi(x)$, where

$$\gamma^{out}(x) = \alpha^{out} + \frac{i}{2}\left(\sigma_{12} + \frac{\partial \sigma_1}{\partial x_2} - \frac{\partial \sigma_2}{\partial x_1}\right)$$

and $\alpha^{out} = (\alpha^{out})^*$. □

2.3. Vascular representation of vortical systems

Let $V(x)$ be a given two-operator vessel and let

$$Y = \delta V = (A_1, A_2, R_{12};\ H, \varphi, E;\ \sigma_1, \sigma_2, \sigma_{12})$$

be the derivative colligation of the second order:

$$\frac{1}{i}(A_j - A_j^*) = \varphi^* \sigma_j \varphi, \quad \frac{1}{i}(R_{12} - R_{12}^*) = \varphi^* \sigma_{12} \varphi,$$

where

$$\sigma_{12} = 2 \operatorname{Im} \gamma^{in} + \frac{\partial \sigma_2}{\partial x_1} - \frac{\partial \sigma_1}{\partial x_2}.$$

Consider the equations (22) of the vortical system:

$$i\frac{d\tilde{f}}{dq} + R_{12}\tilde{f} = \varphi^*\left(\sigma_2\widetilde{\frac{\partial u}{\partial x_1}} - \sigma_1\widetilde{\frac{\partial u}{\partial x_2}} + \left(\frac{\partial\sigma_2}{\partial x_1} - \frac{\partial\sigma_1}{\partial x_2}\right)\tilde{u}\right) + i(\psi^{in})^*\tilde{u}.$$

Using the relation $\psi^{in} = \gamma^{in}\varphi$ and

$$\frac{\partial\sigma_2}{\partial x_1} - \frac{\partial\sigma_1}{\partial x_2} = \sigma_{12} - \frac{\gamma^{in} - (\gamma^{in})^*}{i} = \sigma_{12} + i\gamma^{in} - i(\gamma^{in})^*,$$

we may bring the equations of the vortical system to the "vascular" form:

$$i\frac{d\tilde{f}}{dq} + R_{12}\tilde{f} = \varphi^*\sigma_{12}\tilde{u} + \varphi^*\tilde{m}^{in}(u),$$

$$\tilde{v} = \tilde{u} - i\varphi\tilde{f},$$

where

$$\tilde{m}^{in}(u) = \sigma_2\widetilde{\frac{\partial u}{\partial x_1}} - \sigma_1\widetilde{\frac{\partial u}{\partial x_2}} + i\gamma^{in}\tilde{u}$$

is the vortical "moment" at the input. If the condition $\varphi^*\tilde{m}^{in}(u) = 0$ is fulfilled then the equations of the vortical system have the canonical form

$$i\frac{d\tilde{f}}{dq} + R_{12}\tilde{f} = \varphi^*\sigma_{12}\tilde{u}$$

of the equations of an open system with the window $\varphi(x)$. In this case the vortical system is *assimilative* to the original open geometry. The condition $\tilde{m}^{in}(u) = 0$ is *sufficient* for the *assimilation* of the vascular system to the original geometry. If the window $\varphi(x)$ maps the inner space H onto the "coupling" space E then the condition $\tilde{m}^{in}(u) = 0$ is also *necessary* for the *assimilation* of the vascular system. The condition $\tilde{m}^{in}(u) = 0$ is said to be the *strong* assimilation condition.

Theorem 5. *The input and output vortical moments are equal:*

$$\tilde{m}^{in}(u) = \tilde{m}^{out}(v),$$

where $\tilde{m}^{out}(v) = \sigma_2\widetilde{\frac{\partial v}{\partial x_1}} - \sigma_1\widetilde{\frac{\partial v}{\partial x_2}} + i\gamma^{out}\tilde{v}.$

Proof. We use the relations for the vortical derivatives

$$i\widetilde{\frac{\partial f}{\partial x_1}} + A_1\tilde{f} = \varphi^*\sigma_1\tilde{u},$$

$$i\widetilde{\frac{\partial f}{\partial x_2}} + A_2\tilde{f} = \varphi^*\sigma_2\tilde{u}.$$

Multiplying by $\sigma_2\varphi$ and $\sigma_1\varphi$ (from the left) the first and the second equality respectively and subtracting we obtain:

$$i\left(\sigma_2\varphi\widetilde{\frac{\partial f}{\partial x_1}} - \sigma_1\varphi\widetilde{\frac{\partial f}{\partial x_2}}\right) + (\sigma_2\varphi A_1 - \sigma_1\varphi A_2)\tilde{f} = (\sigma_2\varphi\varphi^*\sigma_1 - \sigma_1\varphi\varphi^*\sigma_2)\tilde{u}.$$

Using the linkage condition we obtain

$$-i\left(\sigma_2\varphi\widetilde{\frac{\partial f}{\partial x_1}}-\sigma_1\varphi\widetilde{\frac{\partial f}{\partial x_2}}\right)=(\sigma_2\varphi A_1-\sigma_1\varphi A_2)\tilde{f}+i(\gamma^{in}-\gamma^{out})\tilde{u}.$$

From $\tilde{v}=\tilde{u}-i\varphi\tilde{f}$ it follows that

$$\begin{aligned}\left(\sigma_2\widetilde{\frac{\partial v}{\partial x_1}}-\sigma_1\widetilde{\frac{\partial v}{\partial x_2}}\right)-\left(\sigma_2\widetilde{\frac{\partial u}{\partial x_1}}-\sigma_1\widetilde{\frac{\partial u}{\partial x_2}}\right)&=-i\left(\sigma_2\widetilde{\frac{\partial(\varphi f)}{\partial x_1}}-\sigma_1\widetilde{\frac{\partial(\varphi f)}{\partial x_2}}\right)\\&=-i\left(\sigma_2\frac{\partial\varphi}{\partial x_1}-\sigma_1\frac{\partial\varphi}{\partial x_2}\right)\tilde{f}-i\left(\sigma_2\varphi\widetilde{\frac{\partial f}{\partial x_1}}-\sigma_1\varphi\widetilde{\frac{\partial f}{\partial x_2}}\right)\\&=((\sigma_2\nabla_1-\sigma_1\nabla_2)\,\varphi)\,\tilde{f}+i(\gamma^{in}-\gamma^{out})\tilde{u}\\&=-\gamma^{out}\varphi\tilde{f}+i\gamma^{in}\tilde{u}-i\gamma^{out}\tilde{u}\\&=-i\gamma^{out}(\tilde{u}-i\varphi\tilde{f})+i\gamma^{in}-\tilde{u}=-\gamma^{out}\tilde{v}+i\gamma^{in}\tilde{u}.\end{aligned}$$

Therefore

$$\sigma_2\widetilde{\frac{\partial v}{\partial x_1}}-\sigma_1\widetilde{\frac{\partial v}{\partial x_2}}+i\gamma^{out}\tilde{v}=\sigma_2\widetilde{\frac{\partial u}{\partial x_1}}-\sigma_1\widetilde{\frac{\partial u}{\partial x_2}}+i\gamma^{in}\tilde{u}.$$

□

2.4. Couplings of vascular systems

Let $X^{(k)}=(A_1^{(k)}(x),A_2^{(k)}(x);\ H^{(k)},\varphi^{(k)}(x),E;\ \sigma_1(x),\sigma_2(x))$ be two given colligations with the same coupling space E and with the same interaction rates $\sigma_1(x),\sigma_2(x)$. The colligation $X(x)$ is said to be the coupling [3] of $X^{(1)}$ and $X^{(2)}$: $X=X^{(1)}\vee X^{(2)}$ if $H=H^{(1)}\oplus H^{(2)}$, $\varphi=\varphi^{(1)}P^{(1)}+\varphi^{(2)}P^{(2)}$, and

$$A_j=A_j^{(1)}P^{(1)}+A_j^{(2)}P^{(2)}+i\varphi^{(2)}\sigma_j\varphi^{(1)}P^{(1)}.$$

It is not difficult to show that

$$\begin{aligned}\psi^{out}(X_1\vee X_2)=\psi^{out}(X^{(1)})P^{(1)}&+\psi^{out}(X^{(2)})P^{(2)}\\&+i(\sigma_1\varphi^{(2)}\varphi^{(2)*}\sigma_2-\sigma_2\varphi^{(2)}\varphi^{(2)*}\sigma_1)\varphi^{(1)}P^{(1)}\end{aligned}$$

and

$$\begin{aligned}\psi^{in}(X_1\vee X_2)=\psi^{in}(X^{(1)})P^{(1)}&+\psi^{out}(X^{(2)})P^{(2)}\\&+i(\sigma_2\varphi^{(1)}\varphi^{(1)*}\sigma_1-\sigma_1\varphi^{(1)}\varphi^{(1)*}\sigma_2)\varphi^{(2)}P^{(2)}.\end{aligned}$$

Decomposition Theorem. *Let $V=(X;\ \gamma^{in},\gamma^{out})$ be a vessel and let $X=X^{(1)}\vee X^{(2)}$. Then*

$$V^{(k)}=(X^{(k)};\ \gamma^{(k)in},\gamma^{(k)out})\quad(k=1,2),$$

where

$$\gamma^{(1)in} = \gamma^{in},\ \gamma^{(1)out} = \gamma^{in} + i(\sigma_1\varphi^{(1)}\varphi^{(1)*}\sigma_2 - \sigma_2\varphi^{(1)}\varphi^{(1)*}\sigma_1),$$
$$\gamma^{(2)in} = \gamma^{out} - i(\sigma_1\varphi^{(2)}\varphi^{(2)*}\sigma_2 - \sigma_2\varphi^{(2)}\varphi^{(2)*}\sigma_1),\ \gamma^{(2)out} = \gamma^{out},$$

are vessels. The output gyration of the first vessel coincides with the input gyration of the second vessel:

$$\gamma^{(1)out} = \gamma^{(2)in} \qquad \text{(the matching condition).}$$

Proof. Using $\varphi = \varphi^{(1)}P^{(1)} + \varphi^{(2)}P^{(2)}$ and the notations of the form

$$\sigma\varphi \wedge \varphi^*\sigma = \sigma_1\varphi\varphi^*\sigma_2 - \sigma_2\varphi\varphi^*\sigma_1,$$

we obtain

$$\begin{aligned}\gamma^{out} - \gamma^{in} &= i(\sigma\varphi \wedge \varphi^*\sigma) = i(\sigma\varphi^{(1)} \wedge \varphi^{(1)*}\sigma) + i(\sigma\varphi^{(2)} \wedge \varphi^{(2)*}\sigma)\\ &= \gamma^{(1)out} - \gamma^{in} + \gamma^{out} - \gamma^{(2)in}.\end{aligned}$$

Therefore $\gamma^{(1)out} = \gamma^{(2)in}$. To prove that $V^{(k)}$ $(k = 1, 2)$ are vessels consider the relation $\psi^{out}(X) = \gamma^{out}\varphi$:

$$\begin{aligned}\psi^{out}(X) &= \psi^{out}(X^{(1)})P^{(1)} + \psi^{out}(X^{(2)})P^{(2)} + i(\sigma\varphi^{(2)} \wedge \varphi^{(2)*}\sigma)\varphi^{(1)}P^{(1)}\\ &= \gamma^{out}(\varphi^{(1)}P^{(1)} + \varphi^{(2)}P^{(2)}).\end{aligned}$$

Hence,

$$\begin{aligned}\psi^{out}(X^{(2)}) &= \gamma^{out}\varphi^{(2)},\\ \psi^{out}(X^{(1)}) &= (\gamma^{out} - i(\sigma\varphi^{(2)} \wedge \varphi^{(2)*}\sigma))\varphi^{(1)} =\\ &= \gamma^{(2)in}\varphi^{(1)} = \gamma^{(1)out}\varphi^{(1)}.\end{aligned}$$

□

The vessels $V^{(k)}$ are said to be the *projections* of the vessel V onto the subspaces $H^{(k)}$ (notice that $H^{(2)}$ is an invariant subspace of A_j $(j = 1, 2)$).

COUPLINGS OF VESSELS. Let $V^{(k)} = (X^{(k)};\ \gamma^{(k)in}, \gamma^{(k)out})$ $(k = 1, 2)$ be two vessels. The collection $V = (X^{(1)} \vee X^{(2)};\ \gamma^{in}, \gamma^{out})$ is said to be the coupling $V = V^{(1)} \vee V^{(2)}$ of $V^{(1)}$ and $V^{(2)}$ if $\gamma^{in} = \gamma^{(1)in}$ and $\gamma^{out} = \gamma^{(2)out}$.

Theorem 6. *The coupling of two vessels is a vessel if and only if* $\gamma^{(1)out} = \gamma^{(2)in}$.

Proof. Let $\gamma^{(1)out} = \gamma^{(2)in}$. Then

$$\begin{aligned}\psi^{out}(X^{(1)} \vee X^{(2)}) &= \gamma^{(1)out}\varphi^{(1)}P^{(1)} + \gamma^{(2)out}\varphi^{(2)}P^{(2)} + (\gamma^{(2)out} - \gamma^{(2)in})\varphi^{(1)}P^{(1)}\\ &= \gamma^{(2)out}\varphi + (\gamma^{(1)out} - \gamma^{(2)in})\varphi^{(1)}P.\end{aligned}$$

From $\gamma^{(1)out} = \gamma^{(2)in}$ it follows that $\psi^{out}(X) = \gamma^{out}\varphi$. Consider next

$$\begin{aligned}\gamma^{(2)out} - \gamma^{(2)in} + \gamma^{(1)out} - \gamma^{(1)in} &= i(\sigma\varphi^{(2)} \wedge \varphi^{(2)*}\sigma + \sigma\varphi^{(1)} \wedge \varphi^{(1)*}\sigma)\\ &= i(\sigma\varphi \wedge \varphi^*\sigma).\end{aligned}$$

From the condition $\gamma^{(2)in} = \gamma^{(1)out}$ the linkage condition $\gamma^{out} - \gamma^{in} = i(\sigma\varphi \wedge \varphi^*\sigma)$ follows.

If $V = V^{(1)} \vee V^{(2)}$ is a vessel then

$$\begin{aligned}\gamma^{out} - \gamma^{in} &= i(\sigma\varphi \wedge \varphi^*\sigma) = i(\sigma\varphi^{(1)} \wedge \varphi^{(1)*}\sigma + \sigma\varphi^{(2)} \wedge \varphi^{(2)*}\sigma)\\ &= \gamma^{(1)out} - \gamma^{(1)in} + \gamma^{(2)out} - \gamma^{(2)in}.\end{aligned}$$

Hence, $\gamma^{(1)out} = \gamma^{(2)in}$. □

Let $V^{(k)} = (X^{(k)};\ \gamma^{(k)in}, \gamma^{(k)out})$ $(k = 1, 2)$ be two vessels and let $V = V^{(1)} \vee V^{(2)}$ be the coupling of these vessels. Assume that the matching condition $\gamma^{(1)out} = \gamma^{(2)in}$ holds. In this case V is a vessel. Consider the two corresponding vortical systems in the vascular form:

$$\tilde{\mathcal{F}}(V^{(1)}) : \begin{cases} i\dfrac{d\tilde{f}^{(1)}}{dq} + R(X^{(1)})\tilde{f}^{(1)} = \varphi^{(1)*}\sigma_{12}\tilde{u}^{(1)} + \varphi^{(1)*}\tilde{m}^{(1)in}(u^{(1)}),\\ \tilde{v}^{(1)} = \tilde{u}^{(1)} - i\varphi^{(1)}\tilde{f}^{(1)} \end{cases}$$

and

$$\tilde{\mathcal{F}}(V^{(2)}) : \begin{cases} i\dfrac{d\tilde{f}^{(2)}}{dq} + R(X^{(2)})\tilde{f}^{(2)} = \varphi^{(2)*}\sigma_{12}\tilde{u}^{(2)} + \varphi^{(2)*}\tilde{m}^{(2)in}(u^{(2)}),\\ \tilde{v}^{(2)} = \tilde{u}^{(2)} - i\varphi^{(2)}\tilde{f}^{(2)}. \end{cases}$$

To obtain the coupling $\tilde{\mathcal{F}}(V^{(1)}) \vee \tilde{\mathcal{F}}(V^{(2)})$ of these vortical systems we have to identify the output of the first system with the input of the second system:

$$\left(\tilde{u}^{(2)}, \widetilde{\frac{\partial u^{(2)}}{\partial x_k}}\right) = \left(\tilde{v}^{(1)}, \widetilde{\frac{\partial v^{(1)}}{\partial x_2}}\right).$$

The input and the output of the coupling are equal to the input of the first system and to the output of the second system respectively. Using the equality of the vortical moments at the input and at the output and the matching condition we obtain:

$$\tilde{m}^{(1)in}(u) = \tilde{m}^{(1)out}(v^{(1)}) = \tilde{m}^{(2)in}(u^{(2)}) = \tilde{m}^{(2)out}(v).$$

From the equality $\tilde{u}^{(2)} = \tilde{v}^{(1)} = \tilde{u}^{(1)} - i\varphi^{(1)}\tilde{f}^{(1)}$ it follows that

$$i\frac{d\tilde{f}^{(2)}}{dq} + R(X^{(2)})\tilde{f} + i\varphi^{(2)*}\sigma_{12}\varphi^{(1)}\tilde{f}^{(1)} = \varphi^{(2)*}(\sigma_{12}\tilde{u} + \tilde{m}^{(1)}(\tilde{u})).$$

Hence taking the sum of the equations for $\tilde{\mathcal{F}}(V^{(j)})$ we obtain:

$$\tilde{\mathcal{F}}(V) : \begin{cases} i\dfrac{d\tilde{f}}{dq} + [R(X^{(1)}) \vee R(X^{(2)})]\tilde{f} = \varphi^*\sigma_{12}\tilde{u} + \varphi^*\tilde{m}(u),\\ \tilde{v} = \tilde{u} - i\varphi\tilde{f}, \end{cases}$$

where

$$R(X^{(1)}) \vee R(X^{(2)}) = R(X^{(1)})P^{(1)} + R(X^{(2)})P^{(2)} + i\varphi^{(2)*}\sigma_{12}\varphi^{(1)}P^{(1)}$$

is the coupling of the curvatures, and $\tilde{m}(u) = \tilde{m}^{(1)}(u)$. Therefore *if* $\gamma^{(1)out} = \gamma^{(2)in}$ *then the curvature of the coupling equals to the coupling of the curvatures:*

$$R(X^{(1)} \vee X^{(2)}) = R(X^{(1)}) \vee R(X^{(2)}).$$

In the matrix form

$$R(X^{(1)} \vee X^{(2)})\tilde{f} = \begin{pmatrix} R(X^{(1)}) & 0 \\ i\varphi^{(2)*}\sigma_{12}\varphi^{(1)} & R(X^{(2)}) \end{pmatrix} \begin{pmatrix} \tilde{f}^{(1)} \\ \tilde{f}^{(2)} \end{pmatrix}.$$

THE CASE $\gamma^{(1)out} \neq \gamma^{(2)in}$. In this case the coupling of vessels is not a vessel. We will say in this case that the coupling is *imperfect*. In the following we assume that $\operatorname{Im}\gamma^{(1)out} = \operatorname{Im}\gamma^{(2)in}$, so that $\sigma_{12}^{(1)} = \sigma_{12}^{(2)} = \sigma_{12}$.

Consider the equations of the vascular systems in the case $\left(\tilde{u}^{(1)}, \widetilde{\frac{\partial u^{(1)}}{\partial x_k}}\right) = 0$:

$$\tilde{\mathcal{F}}(V^{(1)}): \begin{cases} i\dfrac{d\tilde{f}^{(1)}}{dq} + R(X^{(1)})\tilde{f}^{(1)} = 0, \\ \tilde{v}^{(1)} = -i\varphi^{(1)}\tilde{f}^{(1)}, \end{cases}$$

$$\tilde{\mathcal{F}}(V^{(2)}): \begin{cases} i\dfrac{d\tilde{f}^{(2)}}{dq} + R(X^{(2)})\tilde{f}^{(2)} = \varphi^{(2)*}\sigma_{12}\tilde{u}^{(2)} + \varphi^{(2)*}\tilde{m}^{(2)in}(u^{(2)}), \\ \tilde{v}^{(2)} = \tilde{u}^{(2)} - i\varphi^{(2)}\tilde{f}^{(2)}. \end{cases}$$

To obtain the coupling we assume that

$$\left(\tilde{u}^{(2)}, \widetilde{\frac{\partial u^{(2)}}{\partial x_k}}\right) = \left(\tilde{v}^{(1)}, \widetilde{\frac{\partial v^{(1)}}{\partial x_k}}\right), \quad \left(\tilde{v}, \widetilde{\frac{\partial v}{\partial x_k}}\right) = \left(\tilde{v}^{(2)}, \widetilde{\frac{\partial v^{(2)}}{\partial x_k}}\right).$$

Using the conservation law of the vortical moment we obtain

$$\tilde{m}^{(1)out}(v^{(1)}) = \tilde{m}^{(1)in}(u^{(1)}) = 0.$$

Then

$$\tilde{m}^{(2)in}(u^{(2)}) = \tilde{m}^{(2)in}(v^{(1)}) - \tilde{m}^{(1)out}(v^{(1)}) = i(\gamma^{(2)in} - \gamma^{(1)out})(-i\varphi^{(1)}\tilde{f}^{(1)}).$$

Hence the equations of $\tilde{\mathcal{F}}(V^{(2)})$ can be written in the form

$$i\frac{d\tilde{f}^{(2)}}{dq} + R(X^{(2)})\tilde{f}^{(2)} + i\varphi^{(2)*}\sigma_{12}\varphi^{(1)}\tilde{f}^{(1)} + \varphi^{(2)*}(\gamma^{(1)out} - \gamma^{(2)in})\varphi^{(1)}\tilde{f}^{(1)} = 0.$$

Taking the sum of the equations for $\tilde{\mathcal{F}}(V^{(1)})$ and $\tilde{\mathcal{F}}(V^{(2)})$ we obtain the equation of the coupling:

$$i\frac{d\tilde{f}}{dq} + [R(X^{(1)}) \vee R(X^{(2)})]\tilde{f} + \bar{R}\tilde{f} = 0,$$

where

$$\bar{R} = \begin{pmatrix} \varphi^{(1)*} & 0 \\ 0 & \varphi^{(2)*} \end{pmatrix} \begin{pmatrix} 0 & 0 \\ \beta^{12} & 0 \end{pmatrix} \begin{pmatrix} \varphi^{(1)} & 0 \\ 0 & \varphi^{(2)} \end{pmatrix} \qquad (\beta^{12} = \gamma^{(1)out} - \gamma^{(2)in}).$$

Therefore we came to the following result:

Accommodation Phenomenon. *If the coupling of vessels is imperfect then the vortical system acquires the additional curvature* $\bar{R}$*:*

$$R(X^{(1)} \vee X^{(2)}) = R(X^{(1)}) \vee R(X^{(2)}) + \bar{R}.$$

Imperfect chains of vascular systems. Broken vessels. Let us consider a chain of vascular systems $\tilde{\mathcal{F}}(V^{(1)} \vee \tilde{\mathcal{F}}(V^{(2)}) \vee \cdots \vee \tilde{\mathcal{F}}(V^{(n)})$. If $\gamma^{(k)out} = \gamma^{(k+1)in}$ $(k = 1, 2, \ldots, n-1)$ then this chain is perfect and the coupling is a vascular system with the vessel $V = V^{(1)} \vee V^{(2)} \vee \cdots \vee V^{(n)}$. Let us consider a more general case. Assume that $\gamma^{(k)out} \neq \gamma^{(k+1)in}$ but $\operatorname{Im} \gamma^{(k)out} = \operatorname{Im} \gamma^{(k)in}$

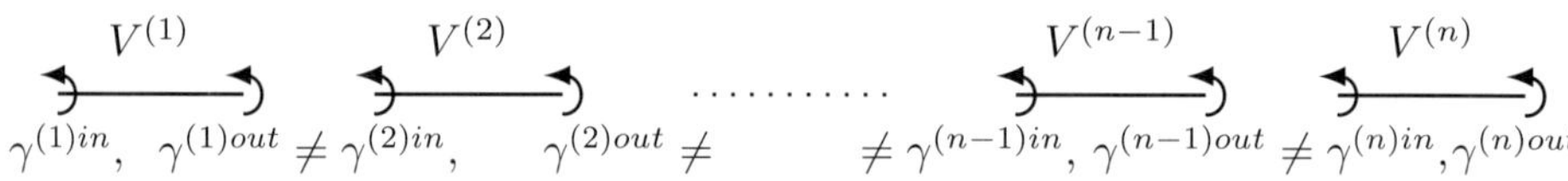

Figure 2. Broken Vessel

Let us write the equations of the corresponding chain of vascular systems in the case $\left(\tilde{u}, \widetilde{\frac{\partial u}{\partial x_k}}\right) = \left(\tilde{u}^{(1)}, \widetilde{\frac{\partial u^{(1)}}{\partial x_k}}\right) = 0$:

$$\begin{aligned}
& i\frac{d\tilde{f}^{(1)}}{dq} + R^{(1)}\tilde{f}^{(1)} = 0, \\
& \tilde{u}^{(2)} = \tilde{v}^{(1)} = -i\varphi^{(1)}\tilde{f}^{(1)}, \qquad \tilde{m}^{(1)out}(u^{(2)}) = \tilde{m}^{(1)in}(u^{(1)}) = 0, \\
& i\frac{d\tilde{f}^{(2)}}{dq} + R^{(2)}\tilde{f}^{(2)} = \varphi^{(2)*}\sigma_{12}\tilde{u}^{(2)} + \varphi^{(2)*}\tilde{m}^{(2)in}(u^{(2)}), \\
& \tilde{u}^{(3)} = \tilde{u}^{(2)} - i\varphi^{(2)}\tilde{f}^{(2)}, \qquad \tilde{m}^{(2)out}(u^{(3)}) = \tilde{m}^{(2)in}(u^{(2)}), \\
& \vdots \\
& i\frac{d\tilde{f}^{(n)}}{dq} + R^{(n)}\tilde{f}^{(n)} = \varphi^{(n)*}\sigma_{12}\tilde{u}^{(n)} + \varphi^{(n)*}\tilde{m}^{(n)in}(u^{(n)}).
\end{aligned}$$

These equations imply

$$\tilde{u}^{(k)} = -i\sum_{j=1}^{k-1} \varphi^{(j)}\tilde{f}^{(j)} \quad (k = 2, \cdots, n),$$

and

$$\tilde{m}^{(k+1)in}(u^{(k+1)}) - \tilde{m}^{(k)in}(u^{(k)}) = \tilde{m}^{(k+1)in}(u^{(k+1)}) - \tilde{m}^{(k)out}(u^{(k+1})$$
$$= (\gamma^{(k+1)in} - \gamma^{(k)out})\tilde{u}^{(k+1)} = (\gamma^{(k+1)in} - \gamma^{(k)out})\sum_{j=1}^{k}\varphi^{(j)}\tilde{f}^{(j)}.$$

Therefore

$$\tilde{m}^{(l)in}(u^{(l)}) = \sum_{k=1}^{l-1}(\gamma^{(k+1)in} - \gamma^{(k)out})\sum_{j=1}^{k}\varphi^{(j)}\tilde{f}^{(j)}$$
$$= \sum_{j=1}^{l-1}\left(\sum_{k=1}^{l-1}(\gamma^{(k+1)in} - \gamma^{(k)out})\right)\varphi^{(j)}\tilde{f}^{(j)} \quad (l = 2, \cdots, n).$$

Hence

$$i\frac{d\tilde{f}^{(l)}}{dq} + R^{(l)}\tilde{f}^{(l)} + i\sum_{j=1}^{l-1}\varphi^{(l)*}\sigma_{12}\varphi^{(j)}\tilde{f}^{(j)} + \varphi^{(l)*}\sum_{j=1}^{l-1}\alpha^{l,j}\varphi^{(j)}\tilde{f}^{(j)} = 0,$$

where

$$\alpha^{l,j} = \sum_{j=1}^{l-1}\beta^{k,k+1}, \qquad \beta^{k,k+1} = \gamma^{(k)out} - \gamma^{(k)in}.$$

Therefore the equation of the corresponding vortical system is:

$$i\frac{d\tilde{f}}{dq} + (R^{(1)} \vee R^{(2)} \vee \cdots \vee R^{(n)})\tilde{f} + \bar{R}\tilde{f} = 0,$$

where

$$\bar{R} = \begin{pmatrix} \varphi_1^* & & & \\ & \varphi^{(2)*} & & \\ & & \ddots & \\ & & & \varphi^{(n)*} \end{pmatrix} \begin{pmatrix} 0 & 0 & 0 & \dots & 0 \\ \alpha^{21} & 0 & 0 & \dots & 0 \\ \alpha^{31} & \alpha^{32} & 0 & \dots & 0 \\ \vdots & \vdots & \vdots & \dots & \vdots \\ \alpha^{n1} & \alpha^{n2} & \alpha^{n3} & \dots & 0 \end{pmatrix}$$
$$\times \begin{pmatrix} \varphi^{(1)} & & & \\ & \varphi^{(2)} & & \\ & & \ddots & \\ & & & \varphi^{(n)} \end{pmatrix}.$$

It is interesting to notice that $\bar{R}^n = 0$.

The accommodation formula is:

$$R(X^{(1)} \vee X^{(2)}) \vee \cdots \vee X^{(n)}) = R(X^{(1)}) \vee R(X^{(2)}) \vee \cdots \vee R(X^{(n)}) + \bar{R}.$$

If all the vessels have zero curvature then the curvature of the coupling is

$$R(X^{(1)} \vee X^{(2)} \vee \cdots \vee X^{(n)}) = \bar{R}.$$

2.5. Curvature of real parts

Let $V = (X;\ \gamma^{in}, \gamma^{out})$ be a vessel and let

$$\mathcal{F}_\Gamma(V): \begin{cases} i\dfrac{df}{dt} + \left(\dfrac{dx}{dt}, A\right) f = \varphi^* \left(\dfrac{dx}{dt}, \sigma\right) u, \\ v = u - i\varphi f, \end{cases}$$

where

$$\left(\frac{dx}{dt}, A\right) = \frac{dx_1}{dt} A_1 + \frac{dx_2}{dt} A_2, \quad \left(\frac{dx}{dt}, \sigma\right) = \frac{dx_1}{dt}\sigma_1 + \frac{dx_2}{dt}\sigma_2,$$

be the open system corresponding to a given arc $\Gamma : x_k = x_k(t)$. It is known [3] that the equations $\mathcal{F}_\Gamma(V)$ are equivalent to the equations of the adjoint system:

$$\mathcal{F}_\Gamma(V^*): \begin{cases} i\dfrac{df}{dt} + \left(\dfrac{dx}{dt}, A^*\right) f = \varphi^* \left(\dfrac{dx}{dt}, \sigma\right) v, \\ u = v + i\varphi f, \end{cases}$$

where $V^* = (X^*;\ -\gamma^{out}, -\gamma^{in})$ and $X^* = (A_k^*;\ H, -\varphi, E;\ -\sigma_k)$. From the equations for the systems $\mathcal{F}_\Gamma(V), \mathcal{F}_\Gamma(V^*)$ it follows that

$$i\frac{df}{dt} + \left(\frac{dx}{dt}, A'\right) f = \varphi^* \left(\frac{dx}{dt}, \sigma\right) \xi,$$

where $A'_k = \frac{1}{2}(A_k + A_K^*)$ and $\xi = \frac{1}{2}(u+v)$.

The vascular systems corresponding to the families of open systems $\mathcal{F}_\Gamma(V)$, $\mathcal{F}_\Gamma(V^*)$ are

$$\begin{aligned} \tilde{\mathcal{F}}(V) &: i\frac{d\tilde{f}}{dq} + R(V)\tilde{f} = \varphi^*\sigma_{12}\tilde{u} + \varphi^*\tilde{m}^{in}(u), \\ \tilde{\mathcal{F}}(V^*) &: i\frac{d\tilde{f}}{dq} + R(V^*)\tilde{f} = \varphi^*\sigma_{12}\tilde{v} - \varphi^*\tilde{m}_*^{in}(v), \\ &\quad \tilde{u} - \tilde{v} = i\varphi\tilde{f}, \end{aligned}$$

where

$$\tilde{m}^{in}(u) = \sigma_2\widetilde{\frac{\partial u}{\partial x_1}} - \sigma_1\widetilde{\frac{\partial u}{\partial x_2}} + i\gamma^{in}\tilde{u}, \quad \tilde{m}_*^{in}(v) = \sigma_2\widetilde{\frac{\partial v}{\partial x_1}} + \sigma_1\widetilde{\frac{\partial v}{\partial x_2}} - i\gamma^{out}\tilde{v}.$$

From the equations for $\tilde{\mathcal{F}}(V), \tilde{\mathcal{F}}(V^*)$ it follows that

$$i\frac{d\tilde{f}}{dq} + [R(V)]'\tilde{f} = \varphi^*\sigma_{12}\xi + \varphi^*\left(\sigma_2\widetilde{\frac{\partial\xi}{\partial x_1}} - \sigma_1\widetilde{\frac{\partial\xi}{\partial x_2}} + \frac{i}{2}\gamma^{in}\tilde{u} + \frac{i}{2}\gamma^{out}\tilde{v}\right),$$

where $[R(V)]' = \frac{1}{2}[R(V) + R^*(V)]$ is the real part of the curvature operator

$$R(v) = \frac{\partial A_2}{\partial x_1} - \frac{\partial A_1}{\partial x_2} + \frac{1}{i}(A_1A_2 - A_2A_1).$$

Using the new input $\tilde{\xi} = \frac{1}{2}(\tilde{u} + \tilde{v})$ and the new output $\tilde{\eta} = \frac{1}{2i}(\tilde{u} - \tilde{v}) = \varphi\tilde{f}$ we obtain:

$$i\frac{d\tilde{f}}{dq} + \left([R(v)]' + \frac{1}{4}\varphi^*(\gamma^{in} - \gamma^{out})\varphi\right)\tilde{f}$$
$$= \varphi^*\sigma_{12}\tilde{\xi} + \varphi^*\left(\sigma_2\widetilde{\frac{\partial\xi}{\partial x_1}} - \sigma_1\widetilde{\frac{\partial\xi}{\partial x_2}} + i\frac{\gamma^{in} + \gamma^{out}}{2}\tilde{\xi}\right).$$

Hence for the curvature operator

$$R(A_1', A_2') = \frac{\partial A_2'}{\partial x_1} - \frac{\partial A_1'}{\partial x_2} + \frac{1}{i}(A_1'A_2' - A_2'A_1')$$

we obtain the following *accommodation formula:*

$$R(A_1', A_2') = [R(A_1, A_2)]' + \frac{1}{4}\varphi^*(\gamma^{in} - \gamma^{out})\varphi.$$

If the gyration γ^{in} at the input is equal to the gyration γ^{out} at the output then the curvature $R(A_1', A_2')$ of the real parts of the operators A_1, A_2 is equal to the real part of the curvature $R(A_1, A_2)$ of the operators A_1, A_2. If the colligation X is strict then the condition $\gamma^{in} = \gamma^{out}$ is also necesary for equality

$$R(A_1', A_2') = [R(A_1, A_2)]'.$$

If $R(V) = 0$ and $\gamma^{in} \neq \gamma^{out}$ then the curvature of the real parts is

$$R(A_1', A_2') = \frac{1}{4}\varphi^*(\gamma^{in} - \gamma^{out})\varphi.$$

3. Vessels with zero curvature

3.1. Compatibility equations

Consider a vessel $V(x) = (A_j(x);\ H, \varphi(x), E;\ \sigma_j(x), \gamma^{in}(x), \gamma^{out}(x))$ $(j = 1, 2;$ $x = (x_1, x_2))$ with zero curvature $R(x) = 0$, $x \in \mathcal{D}$. In this case the parallel translation does not depend on the path Γ, and therefore the equations

$$\begin{cases} i\dfrac{\partial f}{\partial x_1} + A_1(x_1, x_2)f = 0, \\ i\dfrac{\partial f}{\partial x_2} + A_2(x_1, x_2)f = 0, \\ f(M_0) = f_0, \end{cases}$$

where $f(x_1, x_2)$ is a vector field, are compatible. The input $u(x_1, x_2)$ vanishes and the output is given by the equality $v(x_1, x_2) = -i\varphi(x_1, x_2)f(x_1, x_2)$. The conservation law for vortical moments (Theorem 5) implies that $\tilde{m}^{out}(v) = \tilde{m}^{in}(u) = 0$. But in the case of vector-fields the vortical intensity $\tilde{v} = v(x_1, x_2)$ and the vortical derivatives $\widetilde{\dfrac{\partial v}{\partial x_j}}$ coincide with the usual derivatives $\dfrac{\partial v}{\partial x_j}$ of the field. Therefore we came to the following result:

Let $V(x)$ be a field of vessels with zero curvature. Then the output $v(x) = -i\varphi(x)f(x)$, that corresponds to the field $f(x)$ of parallel translations, satisfies the following system of PDE:

$$\sigma_2(x)\frac{\partial v}{\partial x_1} - \sigma_1(x)\frac{\partial v}{\partial x_2} + i\gamma^{out}(x)v = 0.$$

If the input $u(x_1, x_2) \neq 0$, then the following question arises: when are the equations of the corresponding open field

$$\begin{cases} i\dfrac{\partial f}{\partial x_1} + A_1(x)f = \varphi^*(x)\sigma_1(x)u(x), \\ i\dfrac{\partial f}{\partial x_2} + A_2(x)f = \varphi^*(x)\sigma_2(x)u(x), \end{cases} \tag{36}$$

where $f = f(x)$ is a vector field, compatible? The equation of the vascular system, in the case $R(x) \equiv 0$ and $\sigma_{12}(x) \equiv 0$, is

$$i\frac{d\tilde{f}}{dq} = \varphi^* m^{in}(u).$$

Since $f(x)$ is a *field*, the vortical intensity $\tilde{f} = f(x)$ does not depend on q. Hence $\dfrac{d\tilde{f}}{dq} = 0$ and therefore the input field $u(x)$ satisfies the PDE $\varphi^* m^{in}(u) = 0$. From the conservation law for vortical moments we conclude that also $\varphi^* m^{out}(v) = 0$.

Theorem 7. *The equations* (36) *are compatible if and only if the input field $u(x)$ satisfies the PDE*

$$\varphi^*(x)\left[\sigma_2(x)\frac{\partial u}{\partial x_1} - \sigma_1(x)\frac{\partial u}{\partial x_2} + i\gamma^{in}(x)u\right] = 0. \tag{37}$$

If this equation is satisfied then the output $v(x) = u(x) - i\varphi(x)f(x)$ satisfies the equation

$$\varphi^*(x)\left[\sigma_2(x)\frac{\partial v}{\partial x_1} - \sigma_1(x)\frac{\partial v}{\partial x_2} + i\gamma^{out}(x)v\right] = 0. \tag{38}$$

Proof. The necessity of the condition (37) already has been proved. To prove that (37) is also sufficient for the compatibility of the equations (36) let us consider the operator field $K(x_1, x_2)$ which satisfies the equations

$$i\frac{\partial K}{\partial x_j} + A_j(x)K = 0.$$

Putting $f = K\bar{f}$ in (36) we obtain

$$\frac{\partial \bar{f}}{\partial x_j} = -iK^{-1}\varphi^*\sigma_j u.$$

These equations are compatible if and only if

$$\frac{\partial}{\partial x_1}\left(K^{-1}\varphi^*\sigma_2 u\right) = \frac{\partial}{\partial x_2}\left(K^{-1}\varphi^*\sigma_1 u\right).$$

Using $\dfrac{\partial K^{-1}}{\partial x_j} = -K^{-1}\dfrac{\partial K}{\partial x_j}K^{-1} = -iK^{-1}A_j$, we obtain the following condition:

$$\left[\left(i\frac{\partial \varphi^*}{\partial x_2} + A_2\varphi^*\right)\sigma_1 - \left(i\frac{\partial \varphi^*}{\partial x_1} + A_1\varphi^*\right)\sigma_2\right]u$$

$$+ i\varphi^*\left(\frac{\partial \sigma_1}{\partial x_2} - \frac{\partial \sigma_2}{\partial x_1}\right)u + i\varphi^*\left(\sigma_1\frac{\partial u}{\partial x_2} - \sigma_2\frac{\partial u}{\partial x_1}\right) = 0.$$

From the adjoint vessel condition

$$[(\sigma_1\nabla_2^\times - \sigma_2\nabla_1^\times)\varphi]^* = \varphi^*(\gamma^{in})^*$$

and the relation

$$(\gamma^{in})^* + i\left(\frac{\partial \sigma_1}{\partial x_1} - \frac{\partial \sigma_2}{\partial x_2}\right) = \gamma^{in}$$

it follows that the compatibility condition can be written in the form

$$\varphi^*(x)\left[\sigma_2(x)\frac{\partial u}{\partial x_1} - \sigma_1(x)\frac{\partial u}{\partial x_2} + i\gamma^{in}(x)u\right] = 0.$$

□

Remark 1. The strong compatibility condition

$$\sigma_2(x)\frac{\partial u}{\partial x_1} - \sigma_1(x)\frac{\partial u}{\partial x_2} + i\gamma^{in}(x)u = 0 \tag{39}$$

is sufficient for the compatibility of the equations (36) and in this case the output $v(x)$ satisfies the equation

$$\sigma_2(x)\frac{\partial v}{\partial x_1} - \sigma_1(x)\frac{\partial v}{\partial x_2} + i\gamma^{out}(x)v = 0. \tag{40}$$

Remark 2. If the vessel $V(x)$ is strict then the strong compatibility condition is also necessary for the compatibility of the equations (36).

Remark 3. If the compatibility conditions are fulfilled then the corresponding open geometry is said to be an *open field.*

3.2. Unitary equivalence

Let $X(x) = (A_j(x);\ H, \varphi(x), E;\ \sigma_j(x))$ be a given field of colligations:

$$\frac{1}{i}(A_j(x) - A_j^*(x)) = \varphi^*(x)\sigma_j(x)\varphi(x),$$

and let $U(x) : \bar{H} \longrightarrow H$ be a field of isometric transformations of $\bar{H}$ onto H. Consider the differential operators

$$D_j f = i\frac{\partial f}{\partial x_j} + A_j(x)f,$$

where $f = f(x)$ is a (differentiable) function with values in H. Let $f(x) = U(x)\bar{f}(x)$ $(U^*(x) = U^{-1}(x))$. Then

$$g_j = D_j f = i\frac{\partial}{\partial x_j}(U(x)\bar{f}) + A_j(x)U(x)\bar{f} = U(x)\bar{g}_j,$$

and therefore $\bar{g}_j = \bar{D}_j\bar{f}$, where

$$\bar{D}_j\bar{f} = i\frac{\partial \bar{f}}{\partial x_j} + \bar{A}_j(x)\bar{f}, \qquad \bar{A}_j(x) = U^{-1}(x)A_j(x)U(x) + iU^{-1}(x)\frac{\partial U}{\partial x_j}.$$

It is worth noticing that the additional term $iU^{-1}(x)\dfrac{\partial U(x)}{\partial x_j}$ in the transformation law of D_j is a selfadjoint operator. Hence the collection

$$\bar{X}(x) = (\bar{A}_j(x);\ \bar{H}, \bar{\varphi}(x), E;\ \sigma_j(x))$$

is a colligation:

$$\frac{1}{i}(\bar{A}_j(x) - \bar{A}_j^*(x)) = \frac{1}{i}U^*(x)(A_j(x) - A_j^*(x))U(x) = \bar{\varphi}^*(x)\sigma_j(x)\bar{\varphi}(x),$$

where $\bar{\varphi}(x) = \varphi(x)U(x)$. Colligations $X(x)$ and $\bar{X}(x)$ are *unitary equivalent.*

The vessel conditions are also invariant under the transformations $U(x)$. Indeed,

$$\begin{aligned}\bar{\nabla}_j\bar{\varphi} = \frac{1}{i}\frac{\partial\bar{\varphi}}{\partial x_j} + \bar{\varphi}\bar{A}_j &= \frac{1}{i}\frac{\partial}{\partial x_j}(\varphi U) + \varphi U\bar{A}_j \\ &= \frac{1}{i}\frac{\partial\varphi}{\partial x_j}U + \frac{1}{i}\varphi\frac{\partial U}{\partial x_j} + \varphi U\Big(U^{-1}A_jU + iU^{-1}\frac{\partial U}{\partial x_j}\Big) \\ &= \Big(\frac{1}{i}\frac{\partial\varphi}{\partial x_j} + \varphi A_j\Big)U = (\nabla_j\varphi)U.\end{aligned}$$

Hence $\bar{\nabla}_j\bar{\varphi} = (\nabla_j\varphi)U$, and

$$(\sigma_1\bar{\nabla}_2 - \sigma_2\bar{\nabla}_1)\bar{\varphi} = ((\sigma_1\nabla_2 - \sigma_2\nabla_1)\varphi)U = \gamma^{out}\bar{\varphi}.$$

3.3. Finite dimensional case

Let $V(x)$ be a field of vessels with zero curvature. Assume that $\dim\ H = N < \infty$. Let $H'(x_0)$ ($\dim\ H'(x_0) = N'$) be a subspace of H and let $h_1(x_0), \dots, h_{N'}(x_0)$ be a basis of $H'(x_0)$. Let $h_1(x), \dots, h_{N'}(x)$ be the vector fields obtained by parallel translations:

$$\begin{cases} i\dfrac{\partial h_k}{\partial x_j} + A_j(x)h_k = 0, \quad k = 1, 2, \dots, N',\ j = 1, 2, \\ h_k(x)\Big|_{x=x_0} = h_k(x_0). \end{cases}$$

The field of subspaces $H'(x)$ with the basis $h_k(x)$ $(k = 1, 2, \cdots, N')$ is said to be an *invariant subbundle* of the vessel. In the case $N' = N$ the space $H'(x) = H$.

Performing the successive orthonormalization:

$$f_k(x) = \sum_{l=1}^{N} C_{lk}(x)h_l(x), \qquad h_l(x) = \sum_{m=1}^{N} C_{ml}^{(-1)}(x)f_m(x),$$

where $C_{lk}(x) = 0$ if $l > k$, we obtain a new basis $f_k(x)$ such that $(f_k(x), f_\ell(x)) = \delta_{k\ell}$. Let us calculate

$$\begin{aligned} D_j f_k &= i\frac{\partial f_k}{\partial x_j} + A_j f_k = i\sum_l \frac{\partial C_{lk}(x)}{\partial x_j} h_l(x) \\ &= i\sum_l \frac{\partial C_{lk}}{\partial x_j} \sum_m C_{ml}^{(-1)} f_m = i\sum_m \left(\sum_l \frac{\partial C_{lk}}{\partial x_j} C_{ml}^{(-1)}\right) f_m. \end{aligned}$$

Hence $D_j f_k = \sum_{m=1}^{k} L_{j;mk} f_m$, $(j = 1, 2)$, where the matrix $L_j = iC^{-1}(x)\dfrac{\partial C(x)}{\partial x_j}$, $L_j = (L_{j;mk})$, $C = (C_{mk})$, and $L_{j;mk} = 0$ $(m > k)$. We can represent an arbitrary vector function $f(x)$ with values belonging to H using the orthonormal basis $f_k(x)$:

$$f(x) = \sum_{k=1}^{N} \xi_k(x) f_k(x).$$

The transformation $U(x)\vec{\xi}(x) = f(x)$, where $\vec{\xi}(x) = (\xi_k(x))_{k=1}^{N} \in \mathbb{C}^N$, is an isometric transformation of $\bar{H} = \mathbb{C}^N$ onto H. The representation $\bar{D}_j$ of D_j on $\mathbb{C}^N$ is: $\vec{\eta}_j(x) = \bar{D}_j \vec{\xi}(x)$, where

$$\vec{\eta}_j(x) = i\frac{\partial \vec{\xi}(x)}{\partial x_j} + L_j(x)\vec{\xi}(x),$$

and $L_j(x)$ is a triangular matrix. The subspaces

$$\bar{H}^0 = \mathbb{C}^N, \ \bar{H}^1 = (\xi_1, \xi_2, \dots, \xi_{N-1}, 0), \ \dots, \ \bar{H}^{N-1} = (\xi_1, 0, \dots, 0), \ \bar{H}^N = 0,$$

form a decreasing chain of fixed invariant subspaces:

$$\bar{H} = \bar{H}^0 \supset \bar{H}^1 \supset \cdots \supset \bar{H}^{N-1} \supset \bar{H}^N = 0,$$

and therefore the vessel field $\bar{V}(x)$ can be represented as a coupling $\bar{V} = \bar{V}^{(1)} \vee \bar{V}^{(2)} \vee \cdots \vee \bar{V}^{(N)}$, where $\bar{V}^{(k)}(x)$ is the projection of $\bar{V}(x)$ onto $\bar{H}^{(k-1)} \ominus \bar{H}^{(k)}$, and $H^{(k)}(x) = U(x)\bar{H}^{(k)}$.

3.4. Elementary vessels with zero curvature

Let dim $H = 1$ and $A_j(x)a = \lambda_j(x)a$, where $a = 1$ and $\lambda_j(x)$ $(j = 1, 2)$ are functions with values in $\mathbb{C}$. The relation $R(x)a = \left(\dfrac{\partial \lambda_2}{\partial x_1} - \dfrac{\partial \lambda_1}{\partial x_2}\right)a = 0$ implies that there exists a function $\mu(x)$ such that

$$\lambda_j = -i\frac{\partial \mu}{\partial x_j} \quad (j = 1, 2).$$

The vessel condition is:

$$(\sigma_1\nabla_2^\times - \sigma_2\nabla_1^\times)\varphi = \gamma^{in}\varphi,$$

where

$$\nabla_j^\times \varphi = \frac{1}{i}\frac{\partial\varphi}{\partial x_j} + \lambda_j^*\varphi = -i\mu^*\frac{\partial}{\partial x_j}\left(\frac{\varphi}{\mu^*}\right).$$

After some calculations we can bring the vessel conditions to the following form:

$$\sigma_2(x)\frac{\partial}{\partial x_1}\left(\frac{\varphi}{\mu^*}\right) - \sigma_1(x)\frac{\partial}{\partial x_2}\left(\frac{\varphi}{\mu^*}\right) + i\gamma^{in}(x)\left(\frac{\varphi}{\mu^*}\right) = 0\ .$$

Therefore the window $\varphi(x)$ is given by

$$\varphi(x)a = u_0(x)\mu^*(x),$$

where $u_0(x)$ is a solution of the input PDE:

$$\sigma_2(x)\frac{\partial u_0}{\partial x_1} - \sigma_1(x)\frac{\partial u_0}{\partial x_2} + i\gamma^{in}(x)u_0 = 0.$$

The colligation conditions $\frac{1}{i}(\lambda_j - \lambda_j^*) = \varphi^*\sigma_j\varphi$ can be brought to the form:

$$\frac{\partial}{\partial x_j}\left(\frac{1}{|\mu|^2}\right) = (\sigma_j(x)u_0(x), u_0(x)),$$

or, in the matrix form,

$$\frac{\partial}{\partial x_j}\left(\frac{1}{|\mu|^2}\right) = u_0^*(x)\sigma_j(x)u_0(x).$$

It is easy to check that for solutions of the input PDE the relation

$$\frac{\partial}{\partial x_1}(u_0^*\sigma_2 u_0) = \frac{\partial}{\partial x_2}(u_0^*\sigma_1 u_0)$$

holds. Therefore there exists a "potential" $M(x)$ such that

$$\frac{\partial M}{\partial x_j} = u_0^*(x)\sigma_j(x)u_0(x),$$

and $\frac{1}{|\mu|^2} = M(x) + C$, where C is a constant. If the solution $u_0(x)$ is defined in a compact domain then we can choose a constant C which satisfies the condition $M(x) + C > 0$ in the corresponding domain, and for the factor $\mu(x)$ we obtain the following formula:

$$\mu(x) = \frac{e^{i\theta(x)}}{\sqrt{M(x) + C}} \qquad (-\infty < \theta(x) < \infty).$$

Then for $\varphi(x)$ and $\lambda_j(x)$ we obtain:

$$\varphi(x) = u_0(x)\frac{e^{-i\theta(x)}}{\sqrt{M(x)+C}},$$

(39)

$$\lambda_j(x) = -i\frac{\partial\mu(x)}{\partial x_j} = \frac{i}{2}\frac{\partial M}{\partial x_j}(M+C)^{-1} + \frac{\partial\theta}{\partial x_j}.$$

Using couplings we can construct all the N-dimensional vessels with zero curvature and with given $\sigma_1(x), \sigma_2(x)$ and $\gamma^{in}(x)$:

$$V = V^{(1)} \vee V^{(2)} \vee \cdots \vee V^{(N)}.$$

To these chains of vessels there correspond the adjoint chains:

$$V^* = (V^{(N)})^* \vee \cdots \vee (V^{(2)})^* \vee (V^{(1)})^*.$$

3.5. Vessels depending only on the spatial variable

Assume that σ_1, σ_2 are constant operators and that the *curvature operator* $R(x_1, x_2)$ and the operator $\sigma_{12}(x_1, x_2)$ *vanish* in some domain of $\mathbb{R}^2$. The equalities $\operatorname{Im}\gamma^{in}(x) = \operatorname{Im}\gamma^{out}(x)$ and $\sigma_{12} = 2\operatorname{Im}\gamma^{in} + \frac{\partial\sigma_2}{\partial x_1} - \frac{\partial\sigma_1}{\partial x_2}$ imply in this case that the gyrations $\gamma^{in}(x)$, $\gamma^{out}(x)$ are selfadjoint. Assume that $\sigma_2 > 0$. Then the (strong) compatibility equations will be of the hyperbolic type, and it is convenient in this case to assume that $x_1 = t$, $x_2 = x$ ($t_0 \le t \le t_1$, $x_0 \le x \le x_1$) are the temporal and the spatial coordinates respectively. *We assume also that $A_j(x)$ ($j = 1, 2$) and $\varphi(x)$ do not depend on the time variable t.* Then the "derivatives" of the window operator are:

$$\nabla_1^\times(\varphi) = \varphi(x)A_1^*(x), \quad \nabla_2^\times\varphi = \frac{1}{i}\frac{d\varphi}{dx} + \varphi A_2^*(x).$$

The vessel conditions are:

$$\begin{aligned}\frac{1}{i}(A_j(x) - A_j^*(x)) &= \varphi^*(x)\sigma_j\varphi(x)\ ,\\ \frac{1}{i}\sigma_1\frac{d\varphi}{dx} + \sigma_1\varphi(x)A_2^*(x) - \sigma_2\varphi(x)A_1^*(x) &= \gamma^{in}(x)\varphi(x),\\ \gamma^{out}(x) &= \gamma^{in}(x) + i(\sigma_1\varphi(x)\varphi^*(x)\sigma_2 - \sigma_2\varphi(x)\varphi^*(x)\sigma_1),\\ \frac{1}{i}\sigma_1\frac{d\varphi}{dx} + \sigma_1\varphi(x)A_2(x) - \sigma_2\varphi(x)A_1(x) &= \gamma^{out}(x)\varphi(x).\end{aligned}$$

The equations of open fields will be:

$$\begin{cases} i\dfrac{\partial f}{\partial t} + A_1(x)f(t,x) = \varphi^*(x)\sigma_1 u(t,x) & (t_0 \le t \le t_1),\\ i\dfrac{\partial f}{\partial x} + A_2(x)f(t,x) = \varphi^*(x)\sigma_2 u(t,x) & (x_0 \le x \le x_1),\\ v(t,x) = u(t,x) - i\varphi(x)f(t,x), \end{cases}$$

and the condition $R_{12}(x) = 0$ has the form:

$$i\frac{dA_1}{dx} = [A_1(x), A_2(x)] \quad (x_0 \leq x \leq x_1).$$

The (strong) compatibility conditions at the input and at the output are

$$\sigma_2 \frac{\partial u}{\partial t} - \sigma_1 \frac{\partial u}{\partial x} + i\gamma^{in}(x)u = 0,$$

and

$$\sigma_2 \frac{\partial v}{\partial t} - \sigma \frac{\partial v}{\partial x} + i\gamma^{out}(x)v = 0,$$

respectively.

In the special case of separated variables when $u = e^{it\lambda}u_\lambda(x)$, $f = e^{it\lambda}f_\lambda(x)$, $v = e^{it\lambda}v_\lambda(x)$, we obtain the following

Theorem 8. *If $u_\lambda(x)$ is a solution of the equation*

$$i\sigma_1 \frac{du_\lambda}{dx} + [\gamma^{in}(x) + \lambda\sigma_2]u_\lambda = 0, \tag{41}$$

then the corresponding equations of the open field

$$(A_1(x) - \lambda I)f_\lambda(x) = \varphi^*(x)\sigma_1 u_\lambda(x), \tag{42}$$

$$(i\frac{d}{dx} + A_2(x))f_\lambda(x) = \varphi^*(x)\sigma_2 u_\lambda(x), \tag{43}$$

are compatible (we assume that λ does not belong to the spectrum of $A_1(x)$). The corresponding output $v_\lambda(x) = u_\lambda(x) - i\varphi f_\lambda(x)$ satisfies the equation

$$i\sigma_1 \frac{dv_\lambda}{dx} + [\gamma^{out}(x) + \lambda\sigma_2]v_\lambda = 0, \tag{44}$$

and the output can be represented using the characteristic function:

$$v_\lambda(x) = S(\lambda, x)u_\lambda(x),$$

where

$$S(\lambda, x) = I - i\varphi(x)(A_1(x) - \lambda I)^{-1}\varphi^*(x)\sigma_1. \tag{45}$$

Proof. Multiplying the equation (41) by φ^* (from the left) and using the adjoint vessel conditions we can write

$$i\varphi^*\sigma_1 \frac{du_\lambda}{dx} + \lambda\varphi^*\sigma_2 u_\lambda = -\varphi^*\gamma^{in}u_\lambda$$

and

$$i\frac{d\varphi^*}{dx}\sigma_1 u_\lambda + A_2\varphi^*\sigma_1 u_\lambda - A_1\varphi^*\sigma_2 u_\lambda = \varphi^*\gamma^{in}u_\lambda.$$

Adding we obtain

$$(i\frac{d}{dx} + A_2(x))\varphi^*(x)\sigma_1 u_\lambda = (A_1(x) - \lambda I)\varphi^*(x)\sigma_2 u_\lambda.$$

Using the condition $R_{12}(x) = 0$ we obtain

$$(i\frac{d}{dx} + A_2(x))(A_1(x) - \lambda I)f_\lambda(x) = (A_1(x) - \lambda I)(i\frac{d}{dx} + A_2(x))f_\lambda(x).$$

Therefore,

$$(i\frac{d}{dx} + A_2(x))(A_1(x) - \lambda I)^{-1}\varphi^*\sigma_1 u_\lambda = \varphi^*\sigma_2 u_\lambda,$$

and the solution $f_\lambda(x)$ of the equation (42) satisfies also the equation (43). □

Theorem 8 shows that there exist input-output relations between differential equations with spectral parameter λ related to the given vessel field with zero curvature.

Definition. The differential operator

$$\mathcal{L}^{out}(v) = i\sigma_1\frac{dv}{dx} + [\gamma^{out}(x) + \lambda\sigma_2]v$$

is said to be a *vascular transformation* of the differential operator

$$\mathcal{L}^{in}(u) = i\sigma_1\frac{du}{dx} + [\gamma^{in}(x) + \lambda\sigma_2]u.$$

The *inverse vascular transformation* is also a vascular transformation with the characteristic function

$$S_*(\lambda, x) = I + i\varphi^*(x)(A_1^*(x) - \lambda I)^{-1}\varphi(x)\sigma_1 = S^{-1}(\lambda, x).$$

Remark. Let $W^{in}(\lambda; x_0, x)$ and $W^{out}(\lambda; x_0, x)$ be the fundamental matrices (the Wronskians) of the input and output compatibility equations (41) and (44) respectively. Then the following relation holds

$$S(\lambda, x)W^{in}(\lambda; x_0, x) = W^{out}(\lambda, x_0, x)S(\lambda, x_0), \tag{46}$$

where $S(\lambda, x)$ is the characteristic function of the colligation

$$(A_1(x);\ H, \varphi(x), E;\ \sigma_1).$$

Indeed, let $u_\lambda(x) = W^{in}(\lambda; x_0, x)u_\lambda(x_0)$ be a solution of the input equation, satisfying the initial condition $u_\lambda(x)|_{x=x_0} = u_\lambda(x_0)$. Then $v_\lambda(x) = S(\lambda, x)u_\lambda(x)$ satisfies the output equations and therefore

$$S(\lambda, x)u_\lambda(x) = W^{out}(\lambda; x_0, x)S(\lambda, x_0)u_\lambda(x_0).$$

Analogously to the previous definition we will say that the Wronskian

$$W^{out}(\lambda; x_0, x) = S(\lambda, x)W^{in}(\lambda; x_0, x)S_*(\lambda, x_0), \tag{47}$$

where

$$S_*(\lambda, x) = I + i\varphi(x)(A_1^*(x) - \lambda I)^{-1}\varphi^*(x)\sigma_1, \tag{48}$$

is the *vascular transformation* of the Wronskian $W^{in}(\lambda; x_0, x)$:

$$W^{out} = \text{vas}(W^{in}) \qquad (x_0 \le x \le x_1).$$

Example: the Sturm-Liouville equations. Consider a special case of the input compatibility equations

$$\sigma_1 \frac{du_\lambda}{dx} + [\gamma^{in}(x) + \lambda\sigma_2]u_\lambda = 0,$$

where $u_\lambda = \begin{pmatrix} y_1 \\ y_2 \end{pmatrix}$ and

$$\sigma_1 = \begin{pmatrix} 0 & 1 \\ 1 & 0 \end{pmatrix}, \ \sigma_2 = \begin{pmatrix} 1 & 0 \\ 0 & 0 \end{pmatrix}, \ \gamma^{in}(x) = \begin{pmatrix} \gamma_{11}^{in}(x) & i\beta(x) \\ -i\beta(x) & 1 \end{pmatrix}. \tag{49}$$

Then

$$i\begin{pmatrix} 0 & 1 \\ 1 & 0 \end{pmatrix} \frac{d}{dx}\begin{pmatrix} y_1 \\ y_2 \end{pmatrix} + \begin{pmatrix} \gamma_{11}^{in} + \lambda & i\beta \\ -i\beta & 1 \end{pmatrix}\begin{pmatrix} y_1 \\ y_2 \end{pmatrix} = 0,$$

and we obtain for y_2:

$$y_2 = -i\Big(\frac{dy_1}{dx} - \beta y_1\Big),$$

where y_1 satisfies the Sturm-Liouville equation

$$-\frac{d^2 y_1}{dx^2} + q^{in}(x)y_1 = \lambda y_1,$$

with the potential

$$q^{in}(x) = (\beta^{in})^2 + \frac{d\beta^{in}}{dx} - \gamma_{11}^{in}.$$

The output equation is:

$$\sigma_1 \frac{dv_\lambda}{dx} + [\gamma^{out}(x) + \lambda\sigma_2]v_\lambda = 0,$$

where $\gamma^{out} = \gamma^{in} + i(\sigma_1\varphi\varphi^*\sigma_2 - \sigma_2\varphi\varphi^*\sigma_1)$. Let us denote $\varphi\varphi^* = \begin{pmatrix} \pi_{11} & \pi_{12} \\ \bar{\pi}_{12} & \pi_{22} \end{pmatrix}$. Then

$$\gamma^{out}(x) = \begin{pmatrix} \gamma_{11}^{in} + i(\bar{\pi}_{12} - \pi_{12}) & i(\beta - \pi_{11}) \\ -i(\beta - \pi_{11}) & 1 \end{pmatrix}$$

has the form of $\gamma^{in}(x)$ with

$$\gamma_{11}^{out} = \gamma_{11}^{in} + i(\bar{\pi}_{12} - \pi_{12}), \quad \beta^{out} = i(\beta^{in} - \pi_{11}).$$

If we denote $v_\lambda = \begin{pmatrix} \tilde{y}_1 \\ \tilde{y}_2 \end{pmatrix}$ then for $\tilde{y}_1$ we obtain the output Sturm-Liouville equation

$$-\frac{d^2\tilde{y}_1}{dx^2} + q^{out}(x)\tilde{y}_1 = \lambda\tilde{y}_1,$$

where

$$q^{out}(x) = (\beta^{out})^2 + \frac{d\beta^{out}}{dx} - \gamma_{11}^{out}.$$

Using the relation

$$W^{out}(\lambda;x_0,x)=S(\lambda,x)W^{in}(\lambda;x_0,x)S_*(\lambda,x_0)$$

we obtain a mapping of the set of solutions of the input Sturm-Liouville equation onto the set of solutions of the corresponding output Sturm-Liouville equation. Such a mapping and the connections between the two potentials $q^{out}(x)=\operatorname{vas}\big(q^{in}(x)\big)$ are uniquely defined by a vessel field $V(x)$ with σ_1, σ_2, $\gamma^{in}(x)$ of the form (49).

3.6. Elementary vessels depending only on the spatial variables

We will use the formulas for the elementary vessels ($x_1=t, x_2=x$):

$$\begin{aligned}
\varphi(t,x)&=\frac{u_0(t,x)}{(M(t,x)+C)^{1/2}}e^{-i\theta(t,x)},\\
\lambda_j(t,x)&=\frac{i}{2}\frac{\partial M/\partial x_j}{M(t,x)+C}+\frac{\partial\theta}{\partial x_j},\\
\frac{\partial M}{\partial x_j}&=[u_0(t,x)]^*\sigma_j u_0(t,x),
\end{aligned}\tag{50}$$

in the special case discussed in Section 3.5. Our goal is to find elementary vessels

$$V_0(x)=(\lambda_j(x);\ \mathbb{C}^1,\varphi(x),E;\ \sigma_j,\gamma^{in}(x),\gamma^{out}(x))$$

which do not depend on t. We will take a solution of the given input PDE of the form $u_0(t,x)=e^{itz_0}\varphi_0(x)$, where $\varphi_0(x)$ is a solution of the equation

$$i\sigma_1\frac{d\varphi_0}{dx}+[\gamma^{in}(x)+z_0\sigma_2]\varphi_0=0.\tag{51}$$

The equality $\dfrac{\partial}{\partial t}(u_0^*\sigma_2u_0)=\dfrac{\partial}{\partial x}(u_0^*\sigma_1u_0)$ implies

$$\frac{d}{dx}(\varphi_0^*\sigma_1\varphi_0)=-(2\operatorname{Im}z_0)\varphi_0^*\sigma_2\varphi_0.\tag{52}$$

We have to find a "potential" $M(t,x)$, satisfying the conditions

$$\frac{\partial M}{\partial t}=e^{-2(\operatorname{Im}z_0)t}\varphi_0^*(x)\sigma_1\varphi_0(x),\qquad \frac{\partial M}{\partial x}=\ell^{-2(\operatorname{Im}z_0)t}\varphi_0^*(x)\sigma_2\varphi_0(x),$$

and we will try to find such a potential in the form $M(t,x)=e^{-2t\operatorname{Im}z_0}F(x)$. The derivatives are then given by

$$\frac{\partial M}{\partial t}=-2(\operatorname{Im}z_0)e^{-2t\operatorname{Im}z_0}F(x),\qquad \frac{\partial M}{\partial x}=e^{-2t\operatorname{Im}z_0}\frac{dF}{dx}.$$

The condition $\dfrac{\partial M}{\partial t}=u_0^*\sigma_1u_0$ implies

$$-2(\operatorname{Im}z_0)F(x)=\varphi_0^*(x)\sigma_1\varphi_0(x).\tag{53}$$

THE CASE $\operatorname{Im} z_0 \neq 0$. In this case we must have

$$F(x) = \frac{\varphi_0^*(x)\sigma_1\varphi_0(x)}{-2\operatorname{Im} z_0}. \tag{54}$$

Using the equality (52) we obtain

$$\frac{\partial M}{\partial x} = e^{-2t\operatorname{Im} z_0}\frac{dF}{dx} = e^{-2t\operatorname{Im} z_0}\frac{\frac{d}{dx}\varphi_0^*\sigma_1\varphi_0}{-2\operatorname{Im} z_0} = e^{-2t\operatorname{Im} z_0}\varphi_0^*\sigma_2\varphi_0 = u_0^*\sigma_2 u_0.$$

Hence, the condition $\frac{\partial M}{dx} = u_0^*\sigma_2 u_0$ is also satisfied. The derivative

$$\frac{dF}{dx} = \varphi_0^*(x)\sigma_2\varphi_0(x) > 0 \qquad (x_0 \le x \le x_1)$$

(in the case $\sigma_2 > 0$). If $F(x_0) > 0$ then $F(x) > 0$ $(x_0 \le x \le x_1)$ and $M(t,x) > 0$. Taking $C = 0$ we conclude that the corresponding window is:

$$\varphi(t,x) = \frac{e^{itz_0}\varphi_0(x)e^{i\theta(t,x)}}{e^{-t\operatorname{Im} z_0}[F(x)]^{1/2}}.$$

This function does not depend on t if and only if $\theta(t,x) = t\operatorname{Re} z_0 + \theta_0(x)$. From the relations (50) it follows that

$$\begin{aligned} \varphi(x) &= \frac{\varphi_0(x)e^{-i\theta_0(x)}}{\left(\frac{\varphi_0^*(x)\sigma_1\varphi_0(x)}{-2\operatorname{Im} z_0}\right)^{1/2}}, \\ \lambda_1(x) &= z_0^*, \quad \lambda_2(x) = -i\frac{\varphi_0(x)^*\sigma_2\varphi_0(x)}{\varphi_0^*(x)\sigma_1\varphi_0}\operatorname{Im} z_0 + \frac{d\theta_0}{dx}. \end{aligned} \tag{55}$$

Hence, we obtained the general form of a one-dimensional vessel

$$V_0(x) = (\lambda_j(x);\ \mathbb{C}^1, \varphi(x), E;\ \sigma_j, \gamma^{in}(x), \gamma^{out}(x))$$

which depends only on the variable x. The corresponding "elementary" characteristic function is:

$$S_0(\lambda,x) = I - i\frac{\varphi(x)\varphi^*(x)}{\lambda_1 - \lambda}\sigma_1 = I - i\frac{2\operatorname{Im} z_0^*}{z_0^* - \lambda}\cdot\frac{\varphi_0(x)\varphi_0^*(x)}{\varphi_0^*(x)\sigma_1\varphi_0(x)}\sigma_1. \tag{56}$$

THE CASE $\operatorname{Im} z_0 = 0$. In this case

$$\frac{d}{dx}(\varphi_0^*\sigma_1\varphi_0) = -2(\operatorname{Im} z_0)\varphi_0^*\sigma_2\varphi_0 = 0 \qquad (x_0 \le x \le x_1),$$

and therefore $\varphi_0^*(x)\sigma_1\varphi_0(x) = \text{const}$. The function $M(t,x)$ has to satisfy the condition $\frac{\partial M}{\partial t} = \varphi_0^*(x)\sigma_1\varphi_0(x)$. If we choose the initial condition $\varphi_0(x_0)$ such that $\varphi_0^*(x_0)\sigma_1\varphi_0(x_0) = 0$ then $\frac{\partial M}{\partial t} \equiv 0$ $(t_0 \le t \le t_1,\ x_0 \le x \le x_1)$, and therefore

$M(t,x) = F(x)$ depends only on x. The condition $\dfrac{\partial M}{\partial x} = \varphi_0^*(x)\sigma_2\varphi_0(x)$ gives $\dfrac{dF}{dx} = \varphi_0^*(x)\sigma_2\varphi_0(x)$, and therefore $F(x)$ has the form

$$F(x) = \int_{x_0}^{x} \varphi_0^*(s)\sigma_2\varphi_0(s)\,ds + C \qquad (x_0 \leq x \leq x_1).$$

If $\sigma_2 > 0$ and $C > 0$ then $F(x) > 0$. The window $\varphi(x)$ has the form

(57)
$$\begin{aligned} \varphi(x) &= \frac{\varphi_0(x)e^{-i\theta_0(x)}}{[F(x)]^{1/2}} \qquad (x_0 \leq x \leq x_1), \\ \lambda_1(x) &= z_0 \quad (\operatorname{Im} z_0 = 0), \quad \lambda_2(x) = \frac{i}{2}\frac{F'(x)}{F(x)} + \frac{d\theta_0}{dx}. \end{aligned}$$

Using couplings of elementary vascular transformations we can construct chains (discrete or continuous) of transformations of systems of differential equations of the form

$$i\frac{du}{dx} + [\gamma(x) + \lambda\sigma]u = 0.$$

In particular, chains of potentials $q(x)$, in the case of the Sturm-Liouville equations, can be constructed.

Notes

In the zero curvature case and for constant operator colligations, the operators A_1, A_2 are commuting operators with finite-dimensional imaginary parts. For a systematic exposition of the theory of commuting nonselfadjoint operators see the monograph [3]. The generalization of this theory to the case of operator colligations on differentiable manifolds is due to H. Gauchman [1,2]. In this paper we use, in a slightly different form, some of his results.

References

[1] H. Gauchman, Operator colligations on differentiable manifolds, *Toeplitz Centennial (Tel Aviv, 1981)*, edited by I. Gohberg, Operator Theory Adv. Appl. **4**, Birkhäuser, Basel, 1982, pp. 271–302.

[2] H. Gauchman, On nonselfadjoint representations of Lie algebras, *Integral Equations Operator Theory* **6** (1983), 672–705.

[3] M. S. Livsic, N. Kravitsky, A. S. Markus, V. Vinnikov, *Theory of Commuting Nonselfadjoint Operators*, Mathematics and its Applications **332**, Kluwer Academic Publishers Group, Dordrecht, 1995.

Department of Mathematics, Ben-Gurion University of the Negev, P. O. Box 653, 84105 Beer-Sheva, Israel

Operator Theory:
Advances and Applications, Vol. 123

Realization and Factorization in Reproducing Kernel Pontryagin Spaces

D. Alpay, A. Dijksma, J. Rovnyak, and H. S. V. de Snoo

Abstract. New proofs are constructed of some realization theorems for Schur functions on arbitrary domains. A generalization of Leech's factorization theorem is obtained in which the nonnegativity condition on kernels is replaced by a hypothesis of finitely many negative squares.

1. Introduction

We are concerned with kernels on a set $\Omega \times \Omega$, where Ω is a region (open connected set) in the complex plane, of the form

$$\frac{1 - S(z)S(w)^*}{\rho(w,z)}, \qquad \frac{1 - \tilde{S}(z)\tilde{S}(w)^*}{\rho(w,z)}, \tag{1}$$

and

$$\begin{pmatrix} \dfrac{1 - S(z)S(w)^*}{\rho(w,z)} & \dfrac{S(z) - \tilde{S}(w)^*}{\sigma(w,z)} \\ \dfrac{\tilde{S}(z) - S(w)^*}{\sigma(w,z)} & \dfrac{1 - \tilde{S}(z)\tilde{S}(w)^*}{\rho(w,z)} \end{pmatrix}. \tag{2}$$

In these expressions,

$$\rho(w,z) = a(z)\overline{a(w)} - b(z)\overline{b(w)}, \tag{3}$$

$$\sigma(w,z) = b(z)\overline{a(w)} - a(z)\overline{b(w)}, \tag{4}$$

where $a(z)$ and $b(z)$ are holomorphic functions on Ω and $S(z)$ is an operator-valued function. The coefficient spaces are Kreĭn spaces, and * denotes operator adjoint. A symmetry $z \to \tilde{z}$ is defined in Ω that plays a role similar to complex conjugation (see §3), and relative to this notion we set $\tilde{S}(z) = S(\tilde{z})^*$.

1991 *Mathematics Subject Classification.* Primary 47A48, 47B50. Secondary 46C20, 46E22, 47A45.
Key words and phrases. Kreĭn space, Pontryagin space, reproducing kernel, complementation, realization, Leech theorem, Schur function, negative squares.
J. Rovnyak was supported by the National Science Foundation under DMS–9501304.

Functions $S(z)$ such that the three kernels in (1) and (2) have κ negative squares are viewed as generalizations of Schur functions (κ always denotes a nonnegative integer). Realization theorems for such functions are proved in [2, 3]. The reproducing kernel Pontryagin spaces associated with the kernels (1) and (2) serve as state spaces for the realizations. In this paper, we adapt the methods of [4] to give more transparent proofs of these results. We also use a method of Ball and Trent [6] to prove a version of Leech's factorization theorem [8] for kernels having a finite number of negative squares.

Realization theorems for the case in which Ω is the unit disk and $\rho(w,z) = 1 - z\bar{w}$ and $\sigma(w,z) = z - \bar{w}$ are models for other domains; they are stated without proof in §2. Definitions and examples of Schur functions on other domains may be found in §3. Realization theorems for arbitrary domains appear in §4, and our version of Leech's theorem is in §5.

2. Realization theorems for functions defined on the unit disk

We recall some results from [4] for functions defined on the open unit disk $\mathbf{D}$. It is natural to take the domain of a given function to be a region contained in $\mathbf{D}$. This allows us to deal only with holomorphic functions and to avoid poles that may arise in the indefinite case. Let $S(z)$ be a holomorphic function defined on a subregion $\Omega(S)$ of $\mathbf{D}$ containing the origin and having values in $\mathfrak{L}(\mathfrak{F},\mathfrak{G})$, where $\mathfrak{F}$ and $\mathfrak{G}$ are Kreĭn spaces. No essential generality is lost if we assume that $\Omega(S)$ is symmetric with respect to the real axis (see [4, Theorem 1.1.4]). Set

$$\tilde{S}(z) = S(\bar{z})^*. \tag{5}$$

We say that $S(z)$ is in the generalized Schur class $\mathbf{S}_\kappa(\mathfrak{F},\mathfrak{G})$ if each of the three kernels

$$\begin{cases} K_S(w,z) = \dfrac{1 - S(z)S(w)^*}{1 - z\bar{w}}, \\[2ex] K_{\tilde{S}}(w,z) = \dfrac{1 - \tilde{S}(z)\tilde{S}(w)^*}{1 - z\bar{w}}, \\[2ex] D_S(w,z) = \begin{pmatrix} K_S(w,z) & \dfrac{S(z) - \tilde{S}(w)^*}{z - \bar{w}} \\ \dfrac{\tilde{S}(z) - S(w)^*}{z - \bar{w}} & K_{\tilde{S}}(w,z) \end{pmatrix}. \end{cases} \tag{6}$$

has κ negative squares. The reproducing kernel Pontryagin spaces associated with the kernels (6) are denoted by $\mathfrak{H}(S)$, $\mathfrak{H}(\tilde{S})$, $\mathfrak{D}(S)$. If one of these properties holds:

(i) $\operatorname{ind}_- \mathfrak{F} = \operatorname{ind}_- \mathfrak{G} < \infty$, or
(ii) there exist fundamental decompositions

$$\mathfrak{F} = \mathfrak{F}_+ \oplus \mathfrak{F}_- \quad \text{and} \quad \mathfrak{G} = \mathfrak{G}_+ \oplus \mathfrak{G}_-$$

such that in the matrix decomposition

$$S(z) = \begin{pmatrix} S_{11}(z) & S_{12}(z) \\ S_{21}(z) & S_{22}(z) \end{pmatrix} : \begin{pmatrix} \mathfrak{F}_+ \\ \mathfrak{F}_- \end{pmatrix} \to \begin{pmatrix} \mathfrak{G}_+ \\ \mathfrak{G}_- \end{pmatrix}$$

the operator $S_{22}(0) \in \mathfrak{L}(\mathfrak{F}_-, \mathfrak{G}_-)$ is invertible,

then all three kernels (6) have κ negative squares if one does [4, Theorems 2.5.2 and 4.3.8]. The following realization theorems are known; see [4, §2.2, §3.2, §2.3, §3.3, §4.4A], where more information is given on the subregions on which the formulas are valid.

Theorem 1 (Coisometric realization). *Let $S(z) \in \mathbf{S}_\kappa(\mathfrak{F}, \mathfrak{G})$. If either* $\operatorname{ind}_- \mathfrak{F} = \operatorname{ind}_- \mathfrak{G} < \infty$ *or $S_{22}(0)$ is invertible, then on a subregion of $\mathbf{D}$ containing the origin,*

$$S(z) = H + zG(1 - zT)^{-1}F, \tag{7}$$

where the colligation

$$V = \begin{pmatrix} T & F \\ G & H \end{pmatrix} : \begin{pmatrix} \mathfrak{H}(S) \\ \mathfrak{F} \end{pmatrix} \to \begin{pmatrix} \mathfrak{H}(S) \\ \mathfrak{G} \end{pmatrix}$$

is coisometric ($VV^* = 1$) and closely outer connected ($\mathfrak{H}(S)$ is the closed span of elements $T^{*n}G^*g$, $g \in \mathfrak{G}$, $n \geq 0$). For each $h \in \mathfrak{H}(S)$ there is an $\tilde{h}(0) \in \mathfrak{F}$ such that the operators in the colligation and their adjoints are given by

$$h(z) \xrightarrow{T} \frac{h(z) - h(0)}{z}, \qquad h(z) \xrightarrow{T^*} zh(z) - S(z)\tilde{h}(0),$$

$$f \xrightarrow{F} \frac{S(z) - S(0)}{z} f, \qquad h(z) \xrightarrow{F^*} \tilde{h}(0),$$

$$h(z) \xrightarrow{G} h(0), \qquad g \xrightarrow{G^*} K_S(0, z)g,$$

$$f \xrightarrow{H} S(0)f, \qquad g \xrightarrow{H^*} \tilde{S}(0)g.$$

The coisometric and closely outer connected colligation V satisfying (7) is unique up to isomorphism.

Uniqueness in Theorem 1 means that if $\mathfrak{H}'$ is a Pontryagin space and the colligation

$$V' = \begin{pmatrix} T' & F' \\ G' & H' \end{pmatrix} : \begin{pmatrix} \mathfrak{H}' \\ \mathfrak{F} \end{pmatrix} \to \begin{pmatrix} \mathfrak{H}' \\ \mathfrak{G} \end{pmatrix}$$

is coisometric and closely outer connected and $S(z) = H' + zG'(1 - zT')^{-1}F'$ in a neighborhood of the origin, then there is a unitary operator $W : \mathfrak{H}(S) \to \mathfrak{H}'$ such that

$$\begin{pmatrix} T' & F' \\ G' & H' \end{pmatrix} = \begin{pmatrix} W & 0 \\ 0 & 1 \end{pmatrix} \begin{pmatrix} T & F \\ G & H \end{pmatrix} \begin{pmatrix} W^{-1} & 0 \\ 0 & 1 \end{pmatrix}.$$

The uniqueness parts of Theorems 2 and 3 below and the realization theorems in §4 have similar interpretations.

Theorem 2 (Isometric realization). *Let* $S(z) \in \mathbf{S}_\kappa(\mathfrak{F}, \mathfrak{G})$. *If either*

$$\operatorname{ind}_- \mathfrak{F} = \operatorname{ind}_- \mathfrak{G} < \infty$$

or $S_{22}(0)$ *is invertible, then on a subregion of* $\mathbf{D}$ *containing the origin,*

$$S(z) = \tilde{H} + z\tilde{G}(1 - z\tilde{T})^{-1}\tilde{F}, \tag{8}$$

where the colligation

$$\tilde{V} = \begin{pmatrix} \tilde{T} & \tilde{F} \\ \tilde{G} & \tilde{H} \end{pmatrix} : \begin{pmatrix} \mathfrak{H}(\tilde{S}) \\ \mathfrak{F} \end{pmatrix} \to \begin{pmatrix} \mathfrak{H}(\tilde{S}) \\ \mathfrak{G} \end{pmatrix}$$

is isometric ($\tilde{V}^*\tilde{V} = 1$) *and closely inner connected* ($\mathfrak{H}(\tilde{S})$ *is the closed span of elements* $\tilde{T}^m\tilde{F}f$, $f \in \mathfrak{F}$, $m \geq 0$). *For each* $k \in \mathfrak{H}(\tilde{S})$ *there is a* $\tilde{k}(0) \in \mathfrak{G}$ *such that the operators in the colligation and their adjoints are given by*

$$k(z) \xrightarrow{\tilde{T}} zk(z) - \tilde{S}(z)\tilde{k}(0), \qquad k(z) \xrightarrow{\tilde{T}^*} \frac{k(z) - k(0)}{z},$$

$$f \xrightarrow{\tilde{F}} K_{\tilde{S}}(0, z)f, \qquad k(z) \xrightarrow{\tilde{F}^*} k(0),$$

$$k(z) \xrightarrow{\tilde{G}} \tilde{k}(0), \qquad g \xrightarrow{\tilde{G}^*} \frac{\tilde{S}(z) - \tilde{S}(0)}{z} g,$$

$$f \xrightarrow{\tilde{H}} S(0)f, \qquad g \xrightarrow{\tilde{H}^*} \tilde{S}(0)g.$$

The isometric and closely inner connected colligation $\tilde{V}$ *satisfying* (8) *is unique up to isomorphism.*

Theorem 3 (Unitary realization). *Let* $S(z) \in \mathbf{S}_\kappa(\mathfrak{F}, \mathfrak{G})$. *If either*

$$\operatorname{ind}_- \mathfrak{F} = \operatorname{ind}_- \mathfrak{G} < \infty$$

or $S_{22}(0)$ *is invertible, then on a subregion of* $\mathbf{D}$ *containing the origin,*

$$S(z) = D + zC(1 - zA)^{-1}B, \tag{9}$$

where the colligation

$$U = \begin{pmatrix} A & B \\ C & D \end{pmatrix} : \begin{pmatrix} \mathfrak{D}(S) \\ \mathfrak{F} \end{pmatrix} \to \begin{pmatrix} \mathfrak{D}(S) \\ \mathfrak{G} \end{pmatrix}$$

is unitary ($U^*U = 1$ *and* $UU^* = 1$) *and closely connected* ($\mathfrak{D}(S)$ *is the closed span of elements* A^mBf *and* $A^{*n}C^*g$, $f \in \mathfrak{F}$, $g \in \mathfrak{G}$, $m, n \geq 0$). *The operators in the colligation and their adjoints are given by*

$$\begin{pmatrix} h(z) \\ k(z) \end{pmatrix} \xrightarrow{A} \begin{pmatrix} \dfrac{h(z) - h(0)}{z} \\ zk(z) - \tilde{S}(z)h(0) \end{pmatrix}, \qquad \begin{pmatrix} h(z) \\ k(z) \end{pmatrix} \xrightarrow{A^*} \begin{pmatrix} zh(z) - S(z)k(0) \\ \dfrac{k(z) - k(0)}{z} \end{pmatrix},$$

$$f \xrightarrow{B} \begin{pmatrix} \dfrac{S(z) - S(0)}{z} f \\ (1 - \tilde{S}(z)\tilde{S}(0)^*)f \end{pmatrix}, \qquad \begin{pmatrix} h(z) \\ k(z) \end{pmatrix} \xrightarrow{B^*} k(0),$$

$$\begin{pmatrix} h(z) \\ k(z) \end{pmatrix} \xrightarrow{C} h(0), \qquad g \xrightarrow{C^*} \begin{pmatrix} (1 - S(z)S(0)^*)g \\ \dfrac{\tilde{S}(z) - \tilde{S}(0)}{z} g \end{pmatrix},$$

$$f \xrightarrow{D} S(0)f, \qquad g \xrightarrow{D^*} \tilde{S}(0)g.$$

The unitary and closely connected colligation U satisfying (9) *is unique up to isomorphism.*

3. Function theory on arbitrary domains

The results in §2 are special not only to the unit disk $\mathbf{D}$, but also to the origin as a distinguished point of $\mathbf{D}$ and to complex conjugation $z \to \bar{z}$ as a symmetry on $\mathbf{D}$. In [4, §2.5 and §4.4B] it is shown that versions of the representations can be given with the origin replaced by an arbitrary base point τ in $\mathbf{D}$ and an appropriate symmetry $z \to \tilde{z}$ to replace complex conjugation. Since all points of $\mathbf{D}$ are then equivalent, this is called a conformally invariant view of the realization theory in [4]. Following [2, 3], we extend these results to other domains.

By a **symmetry** on a region $\mathcal{O}$ we mean a mapping $z \to \tilde{z}$ of $\mathcal{O}$ onto itself such that

$$(10) \qquad \delta(\tilde{z}) = \overline{\delta(z)}, \qquad z \in \mathcal{O},$$

for some holomorphic function $\delta(z)$ which maps $\mathcal{O}$ in a one-to-one way onto a region $\delta(\mathcal{O})$ which is closed under complex conjugation. Every symmetry is an involution and a conjugate analytic function. We call a region Ω **symmetric** if $\Omega \subseteq \mathcal{O}$ and $z \in \Omega$ implies $\tilde{z} \in \Omega$. If $S(z)$ is an operator-valued function defined on such a region, we write

$$(11) \qquad \tilde{S}(z) = S(\tilde{z})^*, \qquad z \in \Omega,$$

and call $\tilde{S}(z)$ the **reflection** of $S(z)$ with respect to the symmetry.

Let $a(z)$ and $b(z)$ be holomorphic functions on a region Ω such that both of the sets

$$(12) \qquad \Omega_+ = \{z \in \Omega : |a(z)| > |b(z)|\}, \quad \Omega_- = \{z \in \Omega : |a(z)| < |b(z)|\}$$

are nonempty. Then $\Omega_0 = \{z \in \Omega : |a(z)| = |b(z)|\}$ is also nonempty. We fix a **base point** $\tau \in \Omega_+$ and assume further that $a(\tau)$ and $b(\tau)$ are real and

$$b'(\tau)a(\tau) - a'(\tau)b(\tau) \neq 0.$$

The condition that $a(\tau)$ and $b(\tau)$ are real is not a severe restriction and can be met in examples by adjusting the functions $a(z)$ and $b(z)$ by constants of unit modulus. Write

$$(13) \qquad s(z) = \frac{b(z)}{a(z)}, \qquad \delta(z) = \frac{s(z) - s(\tau)}{1 - s(\tau)s(z)}.$$

Here $s(\tau)$ is real and $s'(\tau) \neq 0$. In particular, $\delta(z)$ is one-to-one on a subregion $\mathcal{O}_\tau$ of Ω_+ containing τ and such that $\delta(\mathcal{O}_\tau)$ is closed under complex conjugation. Let $z \to \tilde{z}$ be the symmetry on $\mathcal{O}_\tau$ which is determined by this choice of $\delta(z)$. The identity

$$s(\tilde{z}) = \overline{s(z)}, \qquad z \in \mathcal{O}_\tau. \tag{14}$$

is an equivalent form of (10) and characterizes the symmetry.

Definition 4. *Suppose that $a(z)$, $b(z)$, and τ are given as above. Define $\rho(w,z)$ and $\sigma(w,z)$ by (3) and (4). By $\mathbf{S}_\kappa(a,b,\tau;\mathfrak{F},\mathfrak{G})$ we mean the set of functions $S(z)$ which are defined and holomorphic on a symmetric subregion $\Omega(S)$ of Ω_+, with values in $\mathfrak{L}(\mathfrak{F},\mathfrak{G})$ for some Kreĭn spaces $\mathfrak{F}$ and $\mathfrak{G}$, for which each of the kernels*

$$\begin{cases} K_{S,a,b,\tau}(w,z) = \dfrac{1 - S(z)S(w)^*}{\rho(w,z)}, \\[2ex] K_{\tilde{S},a,b,\tau}(w,z) = \dfrac{1 - \tilde{S}(z)\tilde{S}(w)^*}{\rho(w,z)}, \\[2ex] D_{S,a,b,\tau}(w,z) = \begin{pmatrix} K_{S,a,b,\tau}(w,z) & \dfrac{S(z) - \tilde{S}(w)^*}{\sigma(w,z)} \\[2ex] \dfrac{\tilde{S}(z) - S(w)^*}{\sigma(w,z)} & K_{\tilde{S},a,b,\tau}(w,z) \end{pmatrix}, \end{cases} \tag{15}$$

has κ negative squares. The associated reproducing kernel Pontryagin spaces are then denoted by $\mathfrak{H}_{a,b,\tau}(S)$, $\mathfrak{H}_{a,b,\tau}(\tilde{S})$, $\mathfrak{D}_{a,b,\tau}(S)$.

There is a close relationship between the kernels (15) and their counterparts (6) in §2. To see this, suppose that $S(z)$ is an operator-valued function which is defined and holomorphic on a symmetric subregion $\Omega(S)$ of Ω_+ containing τ. Define $S_0(z)$ on a subregion of $\mathbf{D}$ which contains the origin and is closed under complex conjugation by

$$S(z) = S_0(\delta(z)). \tag{16}$$

Then

$$\tilde{S}(z) = \tilde{S}_0(\delta(z)), \tag{17}$$

where $\tilde{S}(z)$ is defined by (11) and $\tilde{S}_0(z)$ by (5). Straightforward calculations show that

$$\rho(w,z) = a(z)\overline{a(w)}\big[1 - s(z)\overline{s(w)}\big], \tag{18}$$

$$\sigma(w,z) = a(z)\overline{a(w)}\big[s(z) - \overline{s(w)}\big]. \tag{19}$$

In particular,

$$\delta(z) = \frac{\sigma(\tau,z)}{\rho(\tau,z)}. \tag{20}$$

Using the fact that $s(\tau)$ is real, we also obtain

$$\frac{1-\delta(z)\overline{\delta(w)}}{\rho(w,z)} = \frac{\rho(\tau,\tau)}{\rho(\tau,z)\overline{\rho(\tau,w)}} = \frac{\delta(z)-\overline{\delta(w)}}{\sigma(w,z)}. \tag{21}$$

By (21),

$$\begin{cases} K_{S,a,b,\tau}(w,z) = \dfrac{\rho(\tau,\tau)}{\rho(\tau,z)\overline{\rho(\tau,w)}} K_{S_0}(\delta(w),\delta(z)), \\ K_{\tilde{S},a,b,\tau}(w,z) = \dfrac{\rho(\tau,\tau)}{\rho(\tau,z)\overline{\rho(\tau,w)}} K_{\tilde{S}_0}(\delta(w),\delta(z)), \\ D_{S,a,b,\tau}(w,z) = \dfrac{\rho(\tau,\tau)}{\rho(\tau,z)\overline{\rho(\tau,w)}} D_{S_0}(\delta(w),\delta(z)). \end{cases} \tag{22}$$

Thus $S(z)$ belongs to $\mathbf{S}_\kappa(a,b,\tau;\mathfrak{F},\mathfrak{G})$ if and only if $S_0(z)$ is in the class $\mathbf{S}_\kappa(\mathfrak{F},\mathfrak{G})$ defined in §2. If one of these properties holds:

(i) $\operatorname{ind}_- \mathfrak{F} = \operatorname{ind}_- \mathfrak{G} < \infty$;

(ii) there exist fundamental decompositions $\mathfrak{F} = \mathfrak{F}_+ \oplus \mathfrak{F}_-$ and $\mathfrak{G} = \mathfrak{G}_+ \oplus \mathfrak{G}_-$ such that in the matrix decomposition

$$S(z) = \begin{pmatrix} S_{11}(z) & S_{12}(z) \\ S_{21}(z) & S_{22}(z) \end{pmatrix} : \begin{pmatrix} \mathfrak{F}_+ \\ \mathfrak{F}_- \end{pmatrix} \to \begin{pmatrix} \mathfrak{G}_+ \\ \mathfrak{G}_- \end{pmatrix}$$

the operator $S_{22}(\tau) \in \mathfrak{L}(\mathfrak{F}_-,\mathfrak{G}_-)$ is invertible,

then all three kernels (15) have κ negative squares if one does. This is immediate from the corresponding properties of the kernels (6) because of the relations (22).

Example 1. Let $a(z) = 1$ and $b(z) = z$ on $\Omega = \mathbf{C}$, and take $\tau = 0$. Then $\Omega_+ = \mathbf{D}$, and we have the classical example

$$\rho(w,z) = 1 - z\bar{w},$$
$$\sigma(w,z) = z - \bar{w}.$$

The symmetry $z \to \tilde{z}$ is complex conjugation. The three kernels (15) coincide with those of (6) in this case, and $\mathbf{S}_\kappa(a,b,\tau;\mathfrak{F},\mathfrak{G})$ is the class $\mathbf{S}_\kappa(\mathfrak{F},\mathfrak{G})$ defined in §2.

Example 2. Let $a(z) = \sqrt{\pi}\,(1-iz)$ and $b(z) = \sqrt{\pi}\,(1+iz)$ on $\Omega = \mathbf{C}$, and let $\tau = i$. Then $\Omega_\pm = \mathbf{C}_\pm$ are the open upper and lower half-planes, and

$$\rho(w,z) = 2\pi i(\bar{w} - z),$$
$$\sigma(w,z) = 2\pi i(\bar{w} + z),$$

It follows from (14) that the symmetry is reflection with respect to the imaginary axis, that is, $\tilde{z} = -\bar{z}$ and $\tilde{S}(z) = S(-\bar{z})^*$.

Example 3. Let

$$a(z) = \frac{1}{2} e^{-i\pi/4} \left[(R+z)^2 + i(R-z)^2\right],$$
$$b(z) = \frac{1}{2} e^{i\pi/4} \left[(R+z)^2 - i(R-z)^2\right],$$

on $\Omega = D(0,R)$, the open disk with center 0 and radius R ($0 < R < \infty$), and let $\tau = iR\eta$ be any point of $D(0,R)$ on the imaginary axis ($0 < \eta < 1$). Then $a(\tau)$ and $b(\tau)$ are real and $b'(\tau)a(\tau) - a'(\tau)b(\tau) \neq 0$. In this example, $\Omega_\pm = D_\pm(0,R) = D(0,R) \cap \mathbf{C}_\pm$, and

$$\rho(w,z) = 2iR(\bar{w} - z)(R^2 - z\bar{w}),$$
$$\sigma(w,z) = 2iR(\bar{w} + z)(R^2 + z\bar{w}).$$

As in Example 2, the identity (14) implies that $\tilde{z} = -\bar{z}$ and $\tilde{S}(z) = S(-\bar{z})^*$. The conformal mapping

$$s(z) = i\,\frac{(R+z)^2 - i(R-z)^2}{(R+z)^2 + i(R-z)^2}$$

of $D_+(0,R)$ onto $\mathbf{D}$ has applications in half-plane function theory. The mapping is described in [9, Chapter 6].

Additional examples are given in [5].

4. Realization theorems for functions defined on arbitrary domains

The realization theorems for the class $\mathbf{S}_\kappa(\mathfrak{F},\mathfrak{G})$ in Theorems 1–3 lead to similar results for $\mathbf{S}_\kappa(a,b,\tau;\mathfrak{F},\mathfrak{G})$. However, now we have to decompose the main operator by means of the pair $a(z), b(z)$, giving a slight extension of the notion of a colligation. Such results were first obtained in [2, 3]. The proofs of the following theorems could also be based on [4, Theorems 2.5.8, 2.5.9, 2.5.10]. Throughout this section we assume that $\mathfrak{F}$ and $\mathfrak{G}$ are Kreĭn spaces and that we are given a region Ω and functions $a(z)$, $b(z)$, and base point τ as in §3.

Theorem 5 (Coisometric realization for the base point τ). *Let*

$$S(z) \in \mathbf{S}_\kappa(a,b,\tau;\mathfrak{F},\mathfrak{G}).$$

If either $\operatorname{ind}_- \mathfrak{F} = \operatorname{ind}_- \mathfrak{G} < \infty$ *or* $S_{22}(\tau)$ *is invertible, then on a subregion of* Ω_+ *containing* τ,

$$S(z) = H_\tau + \sigma(\tau,z)G_\tau\big[a(z)T_{1\tau} - b(z)T_{2\tau}\big]^{-1}F_\tau, \tag{23}$$

where the colligation

$$\mathcal{A} = \begin{pmatrix} T_{1\tau} & 0 \\ b(\tau)G_\tau & 1 \end{pmatrix} : \begin{pmatrix} \mathfrak{H}_{a,b,\tau}(S) \\ \mathfrak{G} \end{pmatrix} \to \begin{pmatrix} \mathfrak{H}_{a,b,\tau}(S) \\ \mathfrak{G} \end{pmatrix},$$
$$\mathcal{B} = \begin{pmatrix} T_{2\tau} & F_\tau \\ a(\tau)G_\tau & H_\tau \end{pmatrix} : \begin{pmatrix} \mathfrak{H}_{a,b,\tau}(S) \\ \mathfrak{F} \end{pmatrix} \to \begin{pmatrix} \mathfrak{H}_{a,b,\tau}(S) \\ \mathfrak{G} \end{pmatrix},$$

is coisometric in the sense that $\mathcal{A}\mathcal{A}^* = \mathcal{B}\mathcal{B}^*$ *and closely outer connected in the sense that* $\mathfrak{H}_{a,b,\tau}(S)$ *is the closed span of elements* $\left[\overline{a(w)}\,T_{1\tau}^* - \overline{b(w)}\,T_{2\tau}^*\right]^{-1} G_\tau^* g$, $g \in \mathfrak{G}$, *w in a neighborhood of* τ. *For each* $h \in \mathfrak{H}_{a,b,\tau}(S)$ *there is a vector* $\tilde{h}(\tau) \in \mathfrak{F}$ *such that*

$$T_{1\tau} = \frac{a(\tau) + b(\tau)T_\tau}{\rho(\tau,\tau)}, \qquad T_{2\tau} = \frac{b(\tau) + a(\tau)T_\tau}{\rho(\tau,\tau)},$$

where

$$h(z) \xrightarrow{T_\tau} \frac{\rho(\tau,z)h(z) - \rho(\tau,\tau)h(\tau)}{\sigma(\tau,z)},$$

$$h(z) \xrightarrow{T_\tau^*} \frac{\sigma(\tau,z)h(z) - \rho(\tau,\tau)S(z)\tilde{h}(\tau)}{\rho(\tau,z)},$$

and

$$f \xrightarrow{F_\tau} \frac{S(z) - S(\tau)}{\sigma(\tau,z)}\, f, \qquad h(z) \xrightarrow{F_\tau^*} \tilde{h}(\tau),$$

$$h(z) \xrightarrow{G_\tau} h(\tau), \qquad g \xrightarrow{G_\tau^*} K_{S,a,b,\tau}(\tau,z)g,$$

$$f \xrightarrow{H_\tau} S(\tau)f, \qquad g \xrightarrow{H_\tau^*} S(\tau)^* g.$$

The coisometric and closely outer connected colligation $\mathcal{A}, \mathcal{B}$ *satisfying* (23) *is unique up to isomorphism.*

Uniqueness up to isomorphism in Theorems 5, 6, 7 is defined in a similar manner to the analogous notion used in §2.

Proof. Define $S_0(z)$ in a subregion of $\mathbf{D}$ which contains the origin and is closed under conjugation so that (16) and (17) hold. The first of the kernel function identities (22) implies (see [4, Theorem 1.5.7]) that there is a unitary operator $U : \mathfrak{H}(S_0) \to \mathfrak{H}_{a,b,\tau}(S)$ such that

$$U : h(z) \to \frac{\rho(\tau,\tau)^{\frac{1}{2}}}{\rho(\tau,z)}\, h(\delta(z)),$$

$$U^{-1} : h(z) \to \frac{\rho(\tau,\delta^{-1}(z))}{\rho(\tau,\tau)^{\frac{1}{2}}}\, h(\delta^{-1}(z)).$$

By Theorem 1 applied to $S_0(z)$,

$$\begin{aligned} S(z) &= S_0(\delta(z)) \\ &= H_0 + \delta(z)G_0(1 - \delta(z)T_0)^{-1}F_0 \\ &= H_0 + \delta(z)G_0U^{-1}(1 - \delta(z)UT_0U^{-1})^{-1}UF_0 \\ &= H + \delta(z)G(1 - \delta(z)T)^{-1}F, \end{aligned} \tag{24}$$

where T_0, F_0, G_0, H_0 are the operators associated with the realization of $S_0(z)$. The colligation

(25)
$$V = \begin{pmatrix} T & F \\ G & H \end{pmatrix} = \begin{pmatrix} U & 0 \\ 0 & 1 \end{pmatrix} \begin{pmatrix} T_0 & F_0 \\ G_0 & H_0 \end{pmatrix} \begin{pmatrix} U^{-1} & 0 \\ 0 & 1 \end{pmatrix} : \begin{pmatrix} \mathfrak{H}_{a,b,\tau}(S) \\ \mathfrak{F} \end{pmatrix} \to \begin{pmatrix} \mathfrak{H}_{a,b,\tau}(S) \\ \mathfrak{G} \end{pmatrix}$$

is coisometric in the sense that $VV^* = 1$ and closely outer connected in the sense that $\mathfrak{H}_{a,b,\tau}(S)$ is the closed span of elements $T^{*n}G^*g$, $g \in \mathfrak{G}$, $n \geq 0$. For each $h \in \mathfrak{H}_{a,b,\tau}(S)$ there is a vector $\tilde{h}(\tau) \in \mathfrak{F}$ such that the operators in (25) and their adjoints are given by

$$h(z) \xrightarrow{T} \frac{\rho(\tau,z)h(z) - \rho(\tau,\tau)h(\tau)}{\sigma(\tau,z)}, \qquad h(z) \xrightarrow{T^*} \delta(z)h(z) - \frac{\rho(\tau,\tau)}{\rho(\tau,z)} S(z)\tilde{h}(\tau),$$
$$f \xrightarrow{F} \rho(\tau,\tau)^{\frac{1}{2}} \frac{S(z) - S(\tau)}{\sigma(\tau,z)} f, \qquad h(z) \xrightarrow{F^*} \rho(\tau,\tau)^{\frac{1}{2}}\, \tilde{h}(\tau),$$
$$h(z) \xrightarrow{G} \rho(\tau,\tau)^{\frac{1}{2}}\, h(\tau), \qquad g \xrightarrow{G^*} \rho(\tau,\tau)^{\frac{1}{2}}\, K_{S,a,b,\tau}(\tau,z)g,$$
$$f \xrightarrow{H} S(\tau)f, \qquad g \xrightarrow{H^*} S(\tau)^*g.$$

Here the formula for F^* can be taken as the definition of $\tilde{h}(\tau)$. A coisometric and closely outer connected colligation V satisfying (24) is unique up to isomorphism. Recalling (20) and making a further transformation with

$$\begin{aligned} T_\tau &= T, \\ F_\tau &= \rho(\tau,\tau)^{-\frac{1}{2}} F, \\ G_\tau &= \rho(\tau,\tau)^{-\frac{1}{2}} G, \\ H_\tau &= H, \end{aligned}$$

we obtain the result by straightforward calculations. For example, from (24) and (20) we obtain

$$S(z) = H_\tau + \sigma(\tau,z) \frac{\rho(\tau,\tau)}{\rho(\tau,z)} G_\tau \big(1 - \delta(z)T_\tau\big)^{-1} F_\tau.$$

Since also

$$a(z)T_{1\tau} - b(z)T_{2\tau} = \frac{\rho(\tau,z)}{\rho(\tau,\tau)} \big(1 - \delta(z)T_\tau\big),$$

we obtain (23) and the close outer connectedness of the realization. The identities

$$\begin{aligned} T_\tau &= \overline{a(\tau)}T_{2\tau} - \overline{b(\tau)}T_{1\tau}, \\ T_{1\tau}T_{1\tau}^* - T_{2\tau}T_{2\tau}^* &= \frac{1}{\rho(\tau,\tau)} \big(1 - T_\tau T_\tau^*\big) \end{aligned}$$

are readily verified and used to check the coisometric property $\mathcal{A}\mathcal{A}^* = \mathcal{B}\mathcal{B}^*$. □

Theorem 6 (Isometric realization for the base point τ). *Let*

$$S(z) \in \mathbf{S}_\kappa(a, b, \tau; \mathfrak{F}, \mathfrak{G}).$$

If either $\operatorname{ind}_- \mathfrak{F} = \operatorname{ind}_- \mathfrak{G} < \infty$ *or* $S_{22}(\tau)$ *is invertible, then on a subregion of* Ω_+ *containing* τ,

$$S(z) = \tilde{H}_\tau + \sigma(\tau, z)\tilde{G}_\tau \left[a(z)\tilde{T}_{1\tau} - b(z)\tilde{T}_{2\tau}\right]^{-1} \tilde{F}_\tau, \tag{26}$$

where the colligation

$$\tilde{\mathcal{A}} = \begin{pmatrix} \tilde{T}_{1\tau} & b(\tau)\tilde{F}_\tau \\ 0 & 1 \end{pmatrix} : \begin{pmatrix} \mathfrak{H}_{a,b,\tau}(\tilde{S}) \\ \mathfrak{F} \end{pmatrix} \to \begin{pmatrix} \mathfrak{H}_{a,b,\tau}(\tilde{S}) \\ \mathfrak{F} \end{pmatrix},$$

$$\tilde{\mathcal{B}} = \begin{pmatrix} \tilde{T}_{2\tau} & a(\tau)\tilde{F}_\tau \\ \tilde{G}_\tau & \tilde{H}_\tau \end{pmatrix} : \begin{pmatrix} \mathfrak{H}_{a,b,\tau}(\tilde{S}) \\ \mathfrak{F} \end{pmatrix} \to \begin{pmatrix} \mathfrak{H}_{a,b,\tau}(\tilde{S}) \\ \mathfrak{G} \end{pmatrix},$$

is isometric in the sense that $\tilde{\mathcal{A}}^*\tilde{\mathcal{A}} = \tilde{\mathcal{B}}^*\tilde{\mathcal{B}}$ *and closely inner connected in the sense that* $\mathfrak{H}_{a,b,\tau}(\tilde{S})$ *is the closed span of elements* $\left[a(z)\tilde{T}_{1\tau} - b(z)\tilde{T}_{2\tau}\right]^{-1} \tilde{F}_\tau f$, $f \in \mathfrak{F}$, z *in a neighborhood of* τ. *For each* $k \in \mathfrak{H}_{a,b,\tau}(\tilde{S})$ *there is a vector* $\tilde{k}(\tau) \in \mathfrak{G}$ *such that*

$$\tilde{T}_{1\tau} = \frac{a(\tau) + b(\tau)\tilde{T}_\tau}{\rho(\tau,\tau)}, \qquad \tilde{T}_{2\tau} = \frac{b(\tau) + a(\tau)\tilde{T}_\tau}{\rho(\tau,\tau)},$$

where

$$k(z) \xrightarrow{\tilde{T}_\tau} \frac{\sigma(\tau, z)k(z) - \rho(\tau,\tau)\tilde{S}(z)\tilde{k}(\tau)}{\rho(\tau, z)},$$

$$k(z) \xrightarrow{\tilde{T}_\tau^*} \frac{\rho(\tau, z)k(z) - \rho(\tau,\tau)k(\tau)}{\sigma(\tau, z)},$$

and

$$f \xrightarrow{\tilde{F}_\tau} K_{\tilde{S},a,b,\tau}(\tau, z)f, \qquad h(z) \xrightarrow{\tilde{F}_\tau^*} k(\tau),$$

$$h(z) \xrightarrow{\tilde{G}_\tau} \tilde{k}(\tau), \qquad g \xrightarrow{\tilde{G}_\tau^*} \frac{\tilde{S}(z) - \tilde{S}(\tau)}{\sigma(\tau, z)}\, g,$$

$$f \xrightarrow{\tilde{H}_\tau} S(\tau)f, \qquad g \xrightarrow{\tilde{H}_\tau^*} S(\tau)^* g.$$

The isometric and closely inner connected colligation $\tilde{\mathcal{A}}, \tilde{\mathcal{B}}$ *satisfying* (26) *is unique up to isomorphism.*

Proof. We proceed as in the proof of Theorem 5 using the mapping $V : \mathfrak{H}(\tilde{S}_0) \to \mathfrak{H}_{a,b,\tau}(\tilde{S})$ such that

$$V : k(z) \to \frac{\rho(\tau,\tau)^{\frac{1}{2}}}{\rho(\tau, z)}\, k(\delta(z)).$$

By Theorem 2 applied to $S_0(z)$, $S(z) = \tilde{H} + \delta(z)\tilde{G}(1 - \delta(z)\tilde{T})^{-1}\tilde{F}$, where

$$\begin{pmatrix} \tilde{T} & \tilde{F} \\ \tilde{G} & \tilde{H} \end{pmatrix} = \begin{pmatrix} V & 0 \\ 0 & 1 \end{pmatrix} \begin{pmatrix} \tilde{T}_0 & \tilde{F}_0 \\ \tilde{G}_0 & \tilde{H}_0 \end{pmatrix} \begin{pmatrix} V^{-1} & 0 \\ 0 & 1 \end{pmatrix} : \begin{pmatrix} \mathfrak{H}_{a,b,\tau}(\tilde{S}) \\ \mathfrak{F} \end{pmatrix} \to \begin{pmatrix} \mathfrak{H}_{a,b,\tau}(\tilde{S}) \\ \mathfrak{G} \end{pmatrix}$$

and $\tilde{T}_0, \tilde{F}_0, \tilde{G}_0, \tilde{H}_0$ are the operators associated with the isometric realization of $S_0(z)$. For each $k \in \mathfrak{H}_{a,b,\tau}(\tilde{S})$ there is a vector $\tilde{k}(\tau) \in \mathfrak{G}$ such that

$$k(z) \xrightarrow{\tilde{T}} \delta(z)k(z) - \frac{\rho(\tau,\tau)}{\rho(\tau,z)}\, \tilde{S}(z)\tilde{k}(\tau), \qquad k(z) \xrightarrow{\tilde{T}^*} \frac{\rho(\tau,z)k(z) - \rho(\tau,\tau)k(\tau)}{\sigma(\tau,z)},$$

$$f \xrightarrow{\tilde{F}} \rho(\tau,\tau)^{\frac{1}{2}} K_{\tilde{S},a,b,\tau}(\tau,z)f, \qquad k(z) \xrightarrow{\tilde{F}^*} \rho(\tau,\tau)^{\frac{1}{2}}\, k(\tau),$$

$$k(z) \xrightarrow{\tilde{G}} \rho(\tau,\tau)^{\frac{1}{2}}\, \tilde{k}(\tau), \qquad g \xrightarrow{\tilde{G}^*} \rho(\tau,\tau)^{\frac{1}{2}}\, \frac{\tilde{S}(z) - \tilde{S}(\tau)}{\sigma(\tau,z)}\, g,$$

$$f \xrightarrow{\tilde{H}} S(\tau)f, \qquad g \xrightarrow{\tilde{H}^*} S(\tau)^* g.$$

A transformation with

$$\begin{aligned} \tilde{T}_\tau &= \tilde{T}, \\ \tilde{F}_\tau &= \rho(\tau,\tau)^{-\frac{1}{2}} \tilde{F}, \\ \tilde{G}_\tau &= \rho(\tau,\tau)^{-\frac{1}{2}} \tilde{G}, \\ \tilde{H}_\tau &= \tilde{H}, \end{aligned}$$

yields the result. (Alternatively, this result can be derived from Theorem 5; for a similar argument see [4, Theorem 2.2.2].) □

Theorem 7 (Unitary realization for the base point τ). *Let $S(z) \in \mathbf{S}_\kappa(a,b,\tau;\mathfrak{F},\mathfrak{G})$. If either $\operatorname{ind}_- \mathfrak{F} = \operatorname{ind}_- \mathfrak{G} < \infty$ or $S_{22}(\tau)$ is invertible, then on a subregion of Ω_+ containing τ,*

$$S(z) = D_\tau + \sigma(\tau,z)C_\tau\big[a(z)A_{1\tau} - b(z)A_{2\tau}\big]^{-1}B_\tau, \tag{27}$$

where the colligation $\mathcal{A}, \mathcal{B}, \hat{\mathcal{A}}, \hat{\mathcal{B}}$ defined by

$$\mathcal{A} = \begin{pmatrix} A_{1\tau} & 0 \\ b(\tau)C_\tau & 1 \end{pmatrix} : \begin{pmatrix} \mathfrak{D}_{a,b,\tau}(S) \\ \mathfrak{G} \end{pmatrix} \to \begin{pmatrix} \mathfrak{D}_{a,b,\tau}(S) \\ \mathfrak{G} \end{pmatrix},$$

$$\mathcal{B} = \begin{pmatrix} A_{2\tau} & B_\tau \\ a(\tau)C_\tau & D_\tau \end{pmatrix} : \begin{pmatrix} \mathfrak{D}_{a,b,\tau}(S) \\ \mathfrak{F} \end{pmatrix} \to \begin{pmatrix} \mathfrak{D}_{a,b,\tau}(S) \\ \mathfrak{G} \end{pmatrix},$$

$$\hat{\mathcal{A}} = \begin{pmatrix} A_{1\tau} & b(\tau)B_\tau \\ 0 & 1 \end{pmatrix} : \begin{pmatrix} \mathfrak{D}_{a,b,\tau}(S) \\ \mathfrak{F} \end{pmatrix} \to \begin{pmatrix} \mathfrak{D}_{a,b,\tau}(S) \\ \mathfrak{F} \end{pmatrix},$$

$$\hat{\mathcal{B}} = \begin{pmatrix} A_{2\tau} & a(\tau)B_\tau \\ C_\tau & D_\tau \end{pmatrix} : \begin{pmatrix} \mathfrak{D}_{a,b,\tau}(S) \\ \mathfrak{F} \end{pmatrix} \to \begin{pmatrix} \mathfrak{D}_{a,b,\tau}(S) \\ \mathfrak{G} \end{pmatrix},$$

is unitary in the sense that $\mathcal{A}\mathcal{A}^* = \mathcal{B}\mathcal{B}^*$ and $\hat{\mathcal{A}}^*\hat{\mathcal{A}} = \hat{\mathcal{B}}^*\hat{\mathcal{B}}$, and closely connected in the sense that $\mathfrak{D}_{a,b,\tau}(S)$ is the closed span of elements

$$\big[a(z)A_{1\tau} - b(z)A_{2\tau}\big]^{-1}B_\tau f, \qquad \big[\overline{a(w)}\, A_{1\tau}^* - \overline{b(w)}\, A_{2\tau}^*\big]^{-1}C_\tau^* g,$$

$f \in \mathfrak{F}$, $g \in \mathfrak{G}$, and w, z in a neighborhood of τ. The operators are given by

$$A_{1\tau} = \frac{a(\tau) + b(\tau)A_\tau}{\rho(\tau,\tau)}, \qquad A_{2\tau} = \frac{b(\tau) + a(\tau)A_\tau}{\rho(\tau,\tau)},$$

where

$$\begin{pmatrix} h(z) \\ k(z) \end{pmatrix} \xrightarrow{A_\tau} \begin{pmatrix} \dfrac{\rho(\tau,z)h(z) - \rho(\tau,\tau)h(\tau)}{\sigma(\tau,z)} \\ \dfrac{\sigma(\tau,z)k(z) - \rho(\tau,\tau)\tilde{S}(z)h(\tau)}{\rho(\tau,z)} \end{pmatrix}$$

$$\begin{pmatrix} h(z) \\ k(z) \end{pmatrix} \xrightarrow{A_\tau^*} \begin{pmatrix} \dfrac{\sigma(\tau,z)h(z) - \rho(\tau,\tau)S(z)k(\tau)}{\rho(\tau,z)} \\ \dfrac{\rho(\tau,z)k(z) - \rho(\tau,\tau)k(\tau)}{\sigma(\tau,z)} \end{pmatrix},$$

and

$$f \xrightarrow{B_\tau} \begin{pmatrix} \dfrac{S(z) - S(\tau)}{\sigma(\tau,z)} f \\ K_{\tilde{S},a,b,\tau}(\tau,z) f \end{pmatrix}, \qquad \begin{pmatrix} h(z) \\ k(z) \end{pmatrix} \xrightarrow{B_\tau^*} k(\tau),$$

$$\begin{pmatrix} h(z) \\ k(z) \end{pmatrix} \xrightarrow{C_\tau} h(\tau), \qquad g \xrightarrow{C_\tau^*} \begin{pmatrix} K_{S,a,b,\tau}(\tau,z) g \\ \dfrac{\tilde{S}(z) - \tilde{S}(\tau)}{\sigma(\tau,z)} g \end{pmatrix},$$

$$f \xrightarrow{D_\tau} S(\tau) f, \qquad g \xrightarrow{D_\tau^*} S(\tau)^* g.$$

The unitary and closely connected colligation $\mathcal{A}, \mathcal{B}, \tilde{\mathcal{A}}, \tilde{\mathcal{B}}$ *satisfying* (27) *is unique up to isomorphism.*

Proof. For this result we again proceed as in the proof of Theorem 5, but now using the mapping $W : \mathfrak{D}(S_0) \to \mathfrak{D}_{a,b,\tau}(S)$ defined by

$$W : \begin{pmatrix} h(z) \\ k(z) \end{pmatrix} \to \frac{\rho(\tau,\tau)^{\frac{1}{2}}}{\rho(\tau,z)} \begin{pmatrix} h(\delta(z)) \\ k(\delta(z)) \end{pmatrix}.$$

We apply Theorem 3 to $S_0(z)$, obtaining $S(z) = D + \delta(z)C(1-\delta(z)A)^{-1}B$, where

$$\begin{pmatrix} A & B \\ C & D \end{pmatrix} = \begin{pmatrix} W & 0 \\ 0 & 1 \end{pmatrix} \begin{pmatrix} A_0 & B_0 \\ C_0 & D_0 \end{pmatrix} \begin{pmatrix} W^{-1} & 0 \\ 0 & 1 \end{pmatrix} : \begin{pmatrix} \mathfrak{D}_{a,b,\tau}(S) \\ \mathfrak{F} \end{pmatrix} \to \begin{pmatrix} \mathfrak{D}_{a,b,\tau}(S) \\ \mathfrak{G} \end{pmatrix}$$

and A_0, B_0, C_0, D_0 are the operators associated with the realization of $S_0(z)$. We find that

$$\begin{pmatrix} h(z) \\ k(z) \end{pmatrix} \xrightarrow{A} \begin{pmatrix} \dfrac{\rho(\tau,z)h(z) - \rho(\tau,\tau)h(\tau)}{\sigma(\tau,z)} \\ \\ \delta(z)k(z) - \rho(\tau,\tau)\rho(\tau,z)\,\tilde{S}(z)h(\tau) \end{pmatrix},$$

$$\begin{pmatrix} h(z) \\ k(z) \end{pmatrix} \xrightarrow{A^*} \begin{pmatrix} \delta(z)h(z) - \dfrac{\rho(\tau,\tau)}{\rho(\tau,z)} S(z)k(\tau) \\ \dfrac{\rho(\tau,z)k(z) - \rho(\tau,\tau)k(\tau)}{\sigma(\tau,z)} \end{pmatrix},$$

$$f \xrightarrow{B} \rho(\tau,\tau)^{\frac{1}{2}} \begin{pmatrix} \dfrac{S(z)-S(\tau)}{\sigma(\tau,z)} f \\ K_{\tilde{S},a,b,\tau}(\tau,z) f \end{pmatrix}, \qquad \begin{pmatrix} h(z) \\ k(z) \end{pmatrix} \xrightarrow{B^*} \rho(\tau,\tau)^{\frac{1}{2}} k(\tau),$$

$$\begin{pmatrix} h(z) \\ k(z) \end{pmatrix} \xrightarrow{C} \rho(\tau,\tau)^{\frac{1}{2}} h(\tau), \qquad g \xrightarrow{C^*} \rho(\tau,\tau)^{\frac{1}{2}} \begin{pmatrix} K_{S,a,b,\tau}(\tau,z) g \\ \dfrac{\tilde{S}(z)-\tilde{S}(\tau)}{\sigma(\tau,z)} g \end{pmatrix},$$

$$f \xrightarrow{D} S(\tau) f, \qquad g \xrightarrow{D^*} S(\tau)^* g.$$

On setting

$$\begin{aligned} A_\tau &= A, \\ B_\tau &= \rho(\tau,\tau)^{-\frac{1}{2}} B, \\ C_\tau &= \rho(\tau,\tau)^{-\frac{1}{2}} C, \\ D_\tau &= D, \end{aligned}$$

we obtain the result. □

5. Leech's factorization theorem

A theorem of Leech gives a criterion for the existence of a factorization of operator-valued functions in terms of the nonnegativity of a certain kernel. We extend this result to kernels having a finite number of negative squares. The method used is in part an adaptation of the proof of [6, Theorem 2.2]. We assume that we are given a region Ω and functions $a(z)$, $b(z)$ and base point τ as in §3.

Theorem 8. *Let $\mathfrak{F}, \mathfrak{G}, \mathfrak{K}$ be Pontryagin spaces having the same negative index. Suppose $S(z) \in \mathbf{S}_\kappa(a,b,\tau;\mathfrak{F},\mathfrak{K})$ and $S_1(z) \in \mathbf{S}_{\kappa_1}(a,b,\tau;\mathfrak{G},\mathfrak{K})$, where $0 \le \kappa_1 \le \kappa$. Put $\kappa_2 = \kappa - \kappa_1$. The following are equivalent:*

(i) *$S(z) = S_1(z)S_2(z)$ for some $S_2(z) \in \mathbf{S}_{\kappa_2}(a,b,\tau;\mathfrak{F},\mathfrak{G})$;*
(ii) *the kernel*

$$K(w,z) = \frac{S_1(z)S_1(w)^* - S(z)S(w)^*}{\rho(w,z)}$$

has κ_2 negative squares.

It may be assumed that the functions $S(z)$, $S_1(z)$, and $S_2(z)$ in the theorem are defined on some common region $\mathcal{O}$ which contains τ and is contained in Ω_+. Since the kernel $K(w,z)$ is holomorphic, the number $\mathrm{sq}_- K$ of negative squares is independent of the choice of $\mathcal{O}$ (see [4, Theorem 1.1.4]).

Proof. Write $S(z) = S_0(\delta(z))$ and $S_1(z) = S_{10}(\delta(z))$, where $S_0(z) \in \mathbf{S}_\kappa(\mathfrak{F}, \mathfrak{K})$ and $S_{10}(z) \in \mathbf{S}_{\kappa_1}(\mathfrak{G}, \mathfrak{K})$, as in (16) and (17). By (21) we can write

$$K(w,z) = \frac{1 - \delta(z)\overline{\delta(w)}}{\rho(w,z)} K_0(\delta(w), \delta(z)) = \frac{\rho(\tau,\tau)}{\rho(\tau,z)\overline{\rho(\tau,w)}} K_0(\delta(w), \delta(z)),$$

where

$$K_0(w,z) = \frac{S_{10}(z)S_{10}(w)^* - S_0(z)S_0(w)^*}{1 - z\bar{w}}.$$

Thus $\mathrm{sq}_- K = \mathrm{sq}_- K_0$. The general case of the theorem is thus a consequence of the special case $a(z) = 1$, $b(z) = z$, and $\tau = 0$. These choices are assumed in what follows. Thus $S(z) \in \mathbf{S}_\kappa(\mathfrak{F}, \mathfrak{K})$ and $S_1(z) \in \mathbf{S}_{\kappa_1}(\mathfrak{G}, \mathfrak{K})$.

(i) $\Rightarrow$ (ii) If $S(z)$ is factored as in (i), then as in [4, §4.1A] we can write

$$(28) \qquad K_S(w,z) = K_{S_1}(w,z) + K(w,z), \qquad K(w,z) = S_1(z)K_{S_2}(w,z)S_1(w)^*.$$

Hence by [4, Theorem 1.5.5],

$$\kappa \le \kappa_1 + \mathrm{sq}_- K \le \kappa_1 + \kappa_2 = \kappa.$$

Equality holds throughout, and therefore $\mathrm{sq}_- K = \kappa_2$.

(ii) $\Rightarrow$ (i) Assume that $\mathrm{sq}_- K = \kappa_2$, and let $\mathfrak{H}$ be the Pontryagin space with reproducing kernel $K(w,z)$. The first relation in (28) holds by the form of the kernels. Since $\kappa = \kappa_1 + \kappa_2$,

$$\mathrm{sq}_- K_S = \mathrm{sq}_- K_{S_1} + \mathrm{sq}_- K.$$

By [4, Theorem 1.5.5], $\mathfrak{H}(S_1)$ and $\mathfrak{H}$ are contained contractively in $\mathfrak{H}(S)$ as complementary spaces. For each $w \in \mathcal{O}$, let $E(w) : \mathfrak{H} \to \mathfrak{K}$ be the evaluation mapping. By [4, Theorem 1.1.2],

$$(29) \qquad K(w,z) = E(z)E(w)^*.$$

Let

$$\mathfrak{M} = \overline{\mathrm{span}} \left\{ \begin{pmatrix} E(w)^*k \\ S(w)^*k \end{pmatrix} : w \in \mathcal{O},\ k \in \mathfrak{K} \right\}.$$

We show that $\mathfrak{M}$ is a regular subspace of $\mathfrak{H} \oplus \mathfrak{F}$ such that $\mathfrak{M}^\perp$ is a Hilbert space. By standard properties of closed subspaces of a Pontryagin space (for example, see [7, Chapter 1]), it is enough to show that

$$\mathfrak{M}^\perp = \left\{ \begin{pmatrix} h \\ f \end{pmatrix} : h \in \mathfrak{H},\ f \in \mathfrak{F}, \text{ and } h(z) + S(z)f \equiv 0 \right\}$$

is a Hilbert space in the inner product of $\mathfrak{H} \oplus \mathfrak{F}$. Define the operator $\Phi = V\Delta$ as the composition

$$\mathfrak{H} \oplus \mathfrak{F} \xrightarrow{\Delta} \mathfrak{H}(S) \oplus \mathfrak{F} \xrightarrow{V} \mathfrak{H}(S) \oplus \mathfrak{G},$$

where Δ is inclusion and V is the canonical coisometric colligation associated with $S(z)$ as in Theorem 1. Since the inclusion of $\mathfrak{H}$ in $\mathfrak{H}(S)$ is contractive, Δ is a contraction. Since V^* is an isometry, it is a contraction. By assumption $\mathfrak{F}$ and $\mathfrak{G}$ have the same finite negative index, and therefore $\mathfrak{H}(S) \oplus \mathfrak{F}$ and $\mathfrak{H}(S) \oplus \mathfrak{G}$, the domain and range spaces for V, are Pontryagin spaces having the same negative index. By [7, Corollary 2.5], V is also a contraction. Therefore Φ is a contraction, and so $\ker\Phi$ is a Hilbert subspace of $\mathfrak{H}\oplus\mathfrak{F}$ by [7, Theorem 2.7]. Now observe that the elements of $\mathfrak{M}^{\perp}$ have the form

$$\begin{pmatrix} -S(z)f \\ f \end{pmatrix}, \qquad f\in\mathfrak{F},$$

and as shown in [4, Theorem 3.2.3] such elements belong to the kernel of Φ. In fact $\mathfrak{M}^{\perp}$ is a closed subspace of $\ker\Phi$, so that it is a Hilbert space in the inner product of $\mathfrak{H}\oplus\mathfrak{F}$, as was to be shown. This completes the proof that $\mathfrak{M}$ is a regular subspace of $\mathfrak{H}\oplus\mathfrak{F}$ such that $\mathfrak{M}^{\perp}$ is a Hilbert space.

We consider $\mathfrak{M}$ as a Pontryagin space in its own right and define a linear relation $\mathbf{R}$ with dense domain in $\mathfrak{M}$ and range in $\mathfrak{H}\oplus\mathfrak{G}$ by

$$\mathbf{R}=\operatorname{span}\left\{\left(\begin{pmatrix} E(w)^*k \\ S(w)^*k \end{pmatrix},\begin{pmatrix} \bar{w}E(w)^*k \\ S_1(w)^*k \end{pmatrix}\right) : w\in\mathcal{O},\ k\in\mathfrak{K}\right\}.$$

It follows from (29) that

$$E(z)E(w)^*+S(z)S(w)^*=z\bar{w}E(z)E(w)^*+S_1(z)S_1(w)^*,$$

which shows that $\mathbf{R}$ is isometric. Since $\mathfrak{M}$ and $\mathfrak{H}\oplus\mathfrak{G}$ have the same negative index, by [4, Theorem 1.4.2] the closure of $\mathbf{R}$ is the graph of a continuous isometry W_0 mapping $\mathfrak{M}$ into $\mathfrak{H}\oplus\mathfrak{G}$. The range $\mathfrak{N}$ of W_0 is a regular subspace of $\mathfrak{H}\oplus\mathfrak{G}$ having the same negative index as that of $\mathfrak{M}$ and $\mathfrak{H}\oplus\mathfrak{G}$, and so $\mathfrak{N}^{\perp}$ is a Hilbert subspace of $\mathfrak{H}\oplus\mathfrak{G}$. Define a Hilbert space $\mathfrak{L}$ as follows: if $\dim\mathfrak{M}^{\perp}=\dim\mathfrak{N}^{\perp}$, put $\mathfrak{L}=\{0\}$; otherwise let $\mathfrak{L}$ be an infinite-dimensional Hilbert space. Then viewing W_0 as a mapping from $\mathfrak{M}$ into $\mathfrak{N}$, we may extend W_0 to a unitary operator

$$W=\begin{pmatrix} A & B \\ C & D \end{pmatrix} : \begin{pmatrix} \mathfrak{H}\oplus\mathfrak{L} \\ \mathfrak{F} \end{pmatrix} \to \begin{pmatrix} \mathfrak{H}\oplus\mathfrak{L} \\ \mathfrak{G} \end{pmatrix}.$$

Since W is unitary,

$$\begin{pmatrix} A^* & C^* \\ B^* & D^* \end{pmatrix}\begin{pmatrix} \bar{w}E(w)^* \\ S_1(w)^* \end{pmatrix}=\begin{pmatrix} E(w)^* \\ S(w)^* \end{pmatrix}.$$

Thus

$$\begin{aligned} \bar{w}A^*E(w)^*+C^*S_1(w)^* &= E(w)^*, \\ \bar{w}B^*E(w)^*+D^*S_1(w)^* &= S(w)^*, \end{aligned}$$

and elimination of $E(w)^*$ taking adjoints yields

$$S(w)=S_1(w)\left[D+wC(1-wA)^{-1}B\right].$$

We therefore obtain $S(z)=S_1(z)S_2(z)$, where $S_2(z)=D+zC(1-zA)^{-1}B$ is holomorphic on a neighborhood of the origin, which we can take to be $\mathcal{O}$, with values in $\mathfrak{L}(\mathfrak{F},\mathfrak{G})$. By the second relation in (28) and [4, Theorem 2.1.2(1)],

$$\kappa_2=\operatorname{sq}_-K\le\operatorname{sq}_-K_{S_2}\le\operatorname{ind}_-(\mathfrak{H}\oplus\mathfrak{L})=\kappa_2.$$

Equality holds throughout, and therefore $S_2(z)\in\mathbf{S}_{\kappa_2}(\mathfrak{F},\mathfrak{G})$. □

In Theorem 10 below we illustrate Theorem 8 with a result from [5, §4]. A preliminary theorem is needed. As before, suppose that we are given a region Ω and functions $a(z)$, $b(z)$ and base point τ as in §3. Suppose also that we are given a function

$$F(z) = V\big(a(z)A - b(z)B\big)^{-1},$$

where A and B are $n \times n$ matrices and V an $m \times n$ matrix. Assume that

(i) $\det(a(\tau)A - b(\tau)B) \neq 0$, and
(ii) the n columns of $F(z)$ are linearly independent.

The function $F(z)$ is thus meromorphic in Ω and holomorphic at the point τ. Hence,

(iii) for some $\mu \in \Omega_0$, $\det(a(\mu)A - b(\mu)B) \neq 0$.

Let $\mathbf{C}^m$ and $\mathbf{C}^n$ be the usual Euclidean spaces in their natural Hilbert space inner products. By $\mathfrak{F}$ we mean the finite-dimensional Kreĭn space which coincides with $\mathbf{C}^m$ as a vector space with the inner product $\langle x, y\rangle_{\mathfrak{F}} = y^{\times} Jx$, where J is an $m \times m$ signature matrix, that is, $J = J^{\times} = J^{-1}$. Care must be taken with adjoints. We write $M^{\times}$ for the conjugate transpose of any matrix M, that is, adjoint with respect to the Hilbert space inner products of $\mathbf{C}^m$ and $\mathbf{C}^n$. An $m \times n$ matrix M can also be viewed as an operator from $\mathbf{C}^n$ into $\mathfrak{F}$, and then its adjoint is denoted M^*. Thus $M^* = M^{\times} J$.

Theorem 9. *Let P be an invertible hermitian $n \times n$ matrix and having κ negative eigenvalues (counting multiplicities). The kernel $K(w,z) = F(z)P^{-1}F(w)^*$ is of the form $K_S(w,z)$ for some $S(z) \in \mathbf{S}_\kappa(a, b, \tau; \mathfrak{C}, \mathfrak{C})$ if and only if P satisfies the equation*

$$A^{\times} PA - B^{\times} PB = V^{\times} JV. \tag{30}$$

In this case, $S(z)$ is unique up to a constant unitary factor in $\mathfrak{L}(\mathfrak{F})$ on the right, and it may be chosen in the form $S(z) = 1_{\mathfrak{F}} - \rho(\mu, z)F(z)P^{-1}F(\mu)^*$, where μ is any point as in (iii) *above.*

See [5, Theorem 4.1]. The next result is a version of [5, Theorems 4.2 and 4.4]. It differs in that we obtain the existence of a factorization from Theorem 8 as opposed to the explicit construction which is given in [5].

Theorem 10. *Let $F(z)$ and P be given as above. Suppose that A, B, P can be decomposed in the form*

$$\begin{pmatrix} A_{11} & A_{12} \\ 0 & A_{22} \end{pmatrix}, \quad \begin{pmatrix} B_{11} & B_{12} \\ 0 & B_{22} \end{pmatrix}, \quad \begin{pmatrix} P_{11} & P_{12} \\ P_{21} & P_{22} \end{pmatrix}, \tag{31}$$

where the upper left corners are of order $k \times k$ and P_{11} is invertible. Assume that (30) *holds, and choose $S(z)$ in $\mathbf{S}_\kappa(a, b, \tau; \mathfrak{C}, \mathfrak{C})$ as in Theorem 9. In the same way, choose $S_1(z)$ in $\mathbf{S}_{\kappa_1}(a, b, \tau; \mathfrak{C}, \mathfrak{C})$ in relative to $F_1(z) = V_1\big(a(z)A_{11} - b(z)B_{11}\big)^{-1}$ and P_{11}, where $V = \begin{pmatrix} V_1 & V_2 \end{pmatrix}$ with V_1 of order $m \times k$ and κ_1 is the number of negative eigenvalues of P_{11}. Then*

$$S(z) = S_1(z)S_2(z),$$

where $S_2(z) \in \mathbf{S}_{\kappa_2}(a,b,\tau;\mathfrak{C},\mathfrak{C})$, $\kappa_2 = \kappa - \kappa_1$. The space $\mathfrak{H}_{a,b,\tau}(S_1)$ is contained isometrically in $\mathfrak{H}_{a,b,\tau}(S)$ and multiplication by $S_1(z)$ maps $\mathfrak{H}_{a,b,\tau}(S_2)$ isometrically onto the orthogonal complement of $\mathfrak{H}_{a,b,\tau}(S_1)$ in $\mathfrak{H}_{a,b,\tau}(S)$.

Proof. Notice that by (30) and (31),

$$A_{11}^{\times} P_{11} A_{11} - B_{11}^{\times} P_{11} B_{11} = V_1^{\times} J V_1.$$

Thus Theorem 9 is applicable to the function $F_1(z)$ and matrix P_{11} and hence $S_1(z)$ exists. To see that

$$\frac{S_1(z)S_1(w)^* - S(z)S(w)^*}{\rho(w,z)} = K_S(w,z) - K_{S_1}(w,z) \tag{32}$$

has κ_2 negative squares, it is sufficient to observe that $\mathfrak{H}_{a,b,\tau}(S_1)$ is contained isometrically in $\mathfrak{H}_{a,b,\tau}(S)$. In fact, following [5, p. 143] notice that $\mathfrak{H}_{a,b,\tau}(S_1)$ has reproducing kernel $K_{S_1}(w,z) = F_1(z)P_{11}^{-1}F_1(w)^*$ and hence, by (ii), coincides with the set of functions of the form $F_1(z)c$, $c \in \mathbf{C}^k$, in the inner product

$$\langle F_1(z)a, F_1(z)b\rangle_{\mathfrak{H}_{a,b,\tau}(S_1)} = \langle P_{11}a, b\rangle_{\mathbf{C}^k}.$$

In the same way, $\mathfrak{H}_{a,b,\tau}(S)$ coincides with the set of functions of the form $F(z)c$, $c \in \mathbf{C}^n$, in the inner product

$$\langle F(z)a, F(z)b\rangle_{\mathfrak{H}_{a,b,\tau}(S)} = \langle Pa, b\rangle_{\mathbf{C}^n}.$$

Writing $F(z) = \begin{pmatrix} F_1(z) & F_2(z)\end{pmatrix}$ and vectors in $\mathbf{C}^n$ in the form $\binom{c}{0}$ with $c \in \mathbf{C}^k$, we readily check the isometric inclusion of $\mathfrak{H}_{a,b,\tau}(S_1)$ in $\mathfrak{H}_{a,b,\tau}(S)$. Therefore (32) has κ_2 negative squares (for example, see [4, Theorem 1.5.6]), and we obtain the required factorization from Theorem 8.

To prove the last statement, observe that $S_1(\mu)$ is unitary by Theorem 9 and hence the values of $S_1(z)$ are invertible operators except at isolated points. Therefore the relation $S_1(z)h(z) \equiv 0$ with $h(z)$ in $\mathfrak{H}_{a,b,\tau}(S_2)$ can only hold with $h(z) \equiv 0$. The last statement is now easily proved by standard reproducing kernel arguments which we omit. □

The next result is a different version of Leech's theorem. It is stated for the case $a(z) = 1$, $b(z) = z$, and $\tau = 0$.

Theorem 11. *Let $\mathfrak{F}, \mathfrak{G}, \mathfrak{K}$ be Kreĭn spaces with* $\operatorname{ind}_- \mathfrak{F} = \operatorname{ind}_- \mathfrak{G} = \nu < \infty$. *Let $A(z)$ and $B(z)$ be holomorphic in a neighborhood of the origin with values in $\mathfrak{L}(\mathfrak{G},\mathfrak{K})$ and $\mathfrak{L}(\mathfrak{F},\mathfrak{K})$, respectively.*

(i) *Assume that*

$$K_{A,B}(w,z) = \frac{A(z)A(w)^* - B(z)B(w)^*}{1 - z\bar{w}}$$

has κ negative squares, and that

$$K_{A,zB}(w,z) = \frac{A(z)A(w)^* - z\bar{w}B(z)B(w)^*}{1 - z\bar{w}}$$

has $\kappa + \nu$ negative squares. Then $B(z) = A(z)S(z)$ for some $S(z) \in \mathbf{S}_\kappa(\mathfrak{F},\mathfrak{G})$.

(ii) *Suppose in addition that if* $A(z)g \equiv 0$ *for some* $g \in \mathfrak{G}$, *then* $g = 0$. *Then conversely, if* $B(z) = A(z)S(z)$ *where* $S(z) \in \mathbf{S}_\kappa(\mathfrak{F}, \mathfrak{G})$, $K_{A,B}(w,z)$ *has* κ *negative squares and* $K_{A,zB}(w,z)$ *has* $\kappa + \nu$ *negative squares.*

Notice that $K_{A,zB}(w,z) = K_{A,B}(w,z) + B(z)B(w)^*$ has at most $\kappa+\nu$ negative squares by [4, Theorem 1.5.5], and so the second condition in (i) is that this kernel has the largest possible number of negative squares.

Proof. (i) Let Ω be a subregion of the unit disk containing the origin on which $A(z)$ and $B(z)$ are defined. Write $\mathfrak{H}(K_{A,B})$ and $\mathfrak{H}(K_{A,zB})$ for the Pontryagin spaces with reproducing kernels $K_{A,B}(w,z)$ and $K_{A,zB}(w,z)$. Let $E(w) : \mathfrak{H}(K_{A,B}) \to \mathfrak{K}$ be evaluation at any point w of Ω. Define a relation

$$\mathbf{R} = \operatorname{span}\left\{ \left(\begin{pmatrix} E(w)^*k \\ B(w)^*k \end{pmatrix}, \begin{pmatrix} \bar{w}E(w)^*k \\ A(w)^*k \end{pmatrix} \right) : w \in \Omega,\ k \in \mathfrak{K} \right\}$$
$$\subseteq \begin{pmatrix} \mathfrak{H}(K_{A,B}) \\ \mathfrak{F} \end{pmatrix} \times \begin{pmatrix} \mathfrak{H}(K_{A,B}) \\ \mathfrak{G} \end{pmatrix}.$$

We show that the domain $\mathfrak{D}$ of $\mathbf{R}$ contains a maximal uniformly negative subspace of $\mathfrak{H}(K_{A,B}) \oplus \mathfrak{F}$. To this end, consider a Gram matrix of the form

$$M = \left(\left\langle \begin{pmatrix} E(w_j)^*k_j \\ B(w_j)^*k_j \end{pmatrix}, \begin{pmatrix} E(w_i)^*k_i \\ B(w_i)^*k_i \end{pmatrix} \right\rangle_{\mathfrak{H}(K_{A,B})\oplus\mathfrak{F}} \right)_{i,j=1}^n.$$

Thus

$$\begin{aligned} M &= \left(\langle [E(w_i)E(w_j)^* + B(w_i)B(w_j)^*]\, k_j, k_i \rangle_{\mathfrak{K}} \right)_{i,j=1}^n \\ &= \left(\langle [K_{A,B}(w_j,w_i) + B(w_i)B(w_j)^*]\, k_j, k_i \rangle_{\mathfrak{K}} \right)_{i,j=1}^n \\ &= \left(\langle K_{A,zB}(w_j,w_i)k_j, k_i \rangle_{\mathfrak{K}} \right)_{i,j=1}^n. \end{aligned}$$

Since we assume that $K_{A,zB}(w,z)$ has $\kappa+\nu$ negative squares, M has at most $\kappa+\nu$ negative eigenvalues and some such Gram matrix has exactly $\kappa+\nu$ negative eigenvalues. By [4, Lemma 1.1.1′], $\mathfrak{D}$ contains a $(\kappa+\nu)$-dimensional subspace which is the antispace of a Hilbert space in the inner product of $\mathfrak{H}(K_{A,B}) \oplus \mathfrak{F}$. Since $\operatorname{ind}_-(\mathfrak{H}(K_{A,B}) \oplus \mathfrak{F}) = \kappa + \nu$, this verifies the assertion.

It follows that the closure of $\mathfrak{D}$ in $\mathfrak{H}(K_{A,B}) \oplus \mathfrak{F}$ is a regular subspace whose orthogonal complement $\mathfrak{D}^\perp$ is a Hilbert space. By [4, Theorem 1.4.2], the closure of $\mathbf{R}$ is the graph of a continuous isometry W_0 which maps the closure of $\mathfrak{D}$ into $\mathfrak{H}(K_{A,B}) \oplus \mathfrak{G}$. As in the proof of Theorem 8, we choose a Hilbert space $\mathfrak{L}$ and a unitary operator

$$W = \begin{pmatrix} T & F \\ G & H \end{pmatrix} : \begin{pmatrix} \mathfrak{H}(K_{A,B}) \oplus \mathfrak{L} \\ \mathfrak{F} \end{pmatrix} \to \begin{pmatrix} \mathfrak{H}(K_{A,B}) \oplus \mathfrak{L} \\ \mathfrak{G} \end{pmatrix}$$

that extends W_0. We proceed as in the proof of Theorem 8 to obtain $B(z) = A(z)S(z)$, where $S(z) = H + zG(1 - zT)^{-1}F$. By [4, Theorem 2.1.2(1)], $S(z) \in$

$\mathbf{S}_{\kappa'}(\mathfrak{F},\mathfrak{G})$ for some

$$\kappa' \leq \operatorname{ind}_-\big(\mathfrak{H}(K_{A,B}) \oplus \mathfrak{L}\big) = \kappa.$$

Since

$$K_{A,B}(w,z) = A(z)\,\frac{1 - S(z)S(w)^*}{1 - z\bar{w}}\,A(w)^*$$

has κ negative squares, $\kappa' = \kappa$. This proves (i).

(ii) Assume that $A(z)g \equiv 0$ only for $g = 0$, or what is the same thing that the span of elements $A(w)^*k$, $w \in \Omega$ and $k \in \mathfrak{K}$, is dense in $\mathfrak{G}$. Then since $K_S(w,z)$ has κ negative squares and $K_{zS}(w,z)$ has $\kappa+\nu$ negative squares (see [4, Example 1 on p. 132]),

$$K_{A,B}(w,z) = A(z)K_S(w,z)A(w)^*$$

and

$$K_{A,zB}(w,z) = A(z)K_{zS}(w,z)A(w)^*$$

have κ and $\kappa+\nu$ negative squares, respectively. Thus (ii) follows. □

The next result is a variant of Theorem 2.58 of Alpay [1]. It is of the same general nature as the earlier theorems of Leech type, that is, Theorems 8 and 11, but the conditions needed to justify the constructions are verified in a different way. The result is stated for the case $a(z) = 1$, $b(z) = z$, and $\tau = 0$.

Theorem 12. *Assume $S(z) \in \mathbf{S}_\kappa(\mathfrak{F},\mathfrak{G})$ and $\sigma(z) \in \mathbf{S}_{\kappa_1}(\mathfrak{A},\mathfrak{B})$, where $\operatorname{ind}_-\mathfrak{F} = \operatorname{ind}_-\mathfrak{G} < \infty$ and $\operatorname{ind}_-\mathfrak{A} = \operatorname{ind}_-\mathfrak{B} < \infty$. Assume that $\kappa_2 = \kappa - \kappa_1 \geq 0$ and $\sigma(0) = 0$. Let $H(z)$ be holomorphic in a neighborhood of the origin with values in $\mathfrak{L}(\mathfrak{B},\mathfrak{G})$.*

(i) *Assume that there is a function*

$$\Sigma(z) = \begin{pmatrix} \Sigma_{11}(z) & \Sigma_{12}(z) \\ \Sigma_{21}(z) & \Sigma_{22}(z) \end{pmatrix} \in \mathbf{S}_{\kappa_2}(\mathfrak{F}\oplus\mathfrak{B},\mathfrak{G}\oplus\mathfrak{A}) \tag{33}$$

such that

$$\begin{cases} H(z) = \Sigma_{12}(z)\big[1_{\mathfrak{B}} - \sigma(z)\Sigma_{22}(z)\big]^{-1}, \\ S(z) = \Sigma_{11}(z) + \Sigma_{12}(z)\big[1_{\mathfrak{B}} - \sigma(z)\Sigma_{22}(z)\big]^{-1}\sigma(z)\Sigma_{21}(z), \end{cases} \tag{34}$$

and the kernel

$$\big(1_{\mathfrak{G}} \quad H(z)\sigma(z)\big)\, K_\Sigma(w,z) \begin{pmatrix} 1_{\mathfrak{G}} \\ \sigma(w)^*H(w)^* \end{pmatrix} \tag{35}$$

has κ_2 negative squares. Then the kernel

$$K(w,z) = K_S(w,z) - H(z)K_\sigma(w,z)H(w)^* \tag{36}$$

has κ_2 negative squares.

(ii) *Suppose that the only $a \in \mathfrak{A}$ such that $\sigma(z)a \equiv 0$ is $a = 0$. Then conversely, if the kernel* (36) *has κ_2 negative squares, there is a function* (33) *satisfying* (34) *such that the kernel* (35) *also has κ_2 negative squares.*

In part (ii) of the theorem, when $\mathfrak{A}$ and $\mathfrak{B}$ are Hilbert spaces we can drop the hypothesis that the only $a \in \mathfrak{A}$ such that $\sigma(z)a \equiv 0$ is $a = 0$; in the proof, where this hypothesis is invoked, we have $\langle a, a\rangle_{\mathfrak{A}} = 0$ and in the Hilbert space case this implies $a = 0$.

Proof. (i) An equivalent form of (34) is

$$\begin{pmatrix} S(z) & H(z) \end{pmatrix} = \begin{pmatrix} 1_{\mathfrak{G}} & H(z)\sigma(z) \end{pmatrix} \Sigma(z). \tag{37}$$

Thus

$$\begin{aligned}
&\begin{pmatrix} 1_{\mathfrak{G}} & H(z)\sigma(z) \end{pmatrix} K_\Sigma(w,z) \begin{pmatrix} 1_{\mathfrak{G}} \\ \sigma(w)^* H(w)^* \end{pmatrix} \\
&= \frac{1}{1 - z\bar{w}} \Bigg[1_{\mathfrak{G}} + H(z)\sigma(z)\sigma(w)^* H(w)^* \\
&\qquad - \begin{pmatrix} 1_{\mathfrak{G}} & H(z)\sigma(z) \end{pmatrix} \Sigma(z)\Sigma(w)^* \begin{pmatrix} 1_{\mathfrak{G}} \\ \sigma(w)^* H(w)^* \end{pmatrix} \Bigg] \\
&= \frac{1}{1 - z\bar{w}} \Big[1_{\mathfrak{G}} + H(z)\sigma(z)\sigma(w)^* H(w)^* - S(z)S(w)^* - H(z)H(w)^* \Big] \\
&= K_S(w,z) - H(z)K_\sigma(w,z)H(w)^*,
\end{aligned} \tag{38}$$

and (i) follows.

(ii) Write (36) in the form

$$K(w,z) = K_{A,B}(w,z) = \frac{A(z)A(w)^* - B(z)B(w)^*}{1 - z\bar{w}},$$

where

$$B(z) = \begin{pmatrix} S(z) & H(z) \end{pmatrix}, \qquad A(z) = \begin{pmatrix} 1_{\mathfrak{G}} & H(z)\sigma(z) \end{pmatrix}.$$

These functions are holomorphic in a neighborhood of the origin with values in $\mathfrak{L}(\mathfrak{F} \oplus \mathfrak{B}, \mathfrak{G})$ and $\mathfrak{L}(\mathfrak{G} \oplus \mathfrak{A}, \mathfrak{G})$, respectively. By hypothesis, $K_{A,B}(w,z)$ has κ_2 negative squares.

We proceed as in the proof of Theorem 11. Define a relation

$$\mathbf{R} \subseteq \begin{pmatrix} \mathfrak{H}(K_{A,B}) \\ \mathfrak{F} \oplus \mathfrak{B} \end{pmatrix} \times \begin{pmatrix} \mathfrak{H}(K_{A,B}) \\ \mathfrak{G} \oplus \mathfrak{A} \end{pmatrix}$$

as in that argument. Let $\mathfrak{D}$ and $\mathfrak{R}$ be the domain and range of $\mathbf{R}$. We show that $\mathfrak{D}^\perp$ and $\mathfrak{R}^\perp$ are Hilbert subspaces of $\mathfrak{H}(K_{A,B}) \oplus \mathfrak{F} \oplus \mathfrak{B}$ and $\mathfrak{H}(K_{A,B}) \oplus \mathfrak{G} \oplus \mathfrak{A}$. Since $\mathbf{R}$ is isometric, it is sufficient to do this for one of the cases (a similar situation occurs in the proof of Theorem 8), and we establish the assertion for $\mathfrak{R}^\perp$. Clearly $\mathfrak{R}^\perp$ is a closed subspace of the Pontryagin space $\mathfrak{H}(K_{A,B}) \oplus \mathfrak{G} \oplus \mathfrak{A}$. Suppose $h \oplus g \oplus a \in \mathfrak{R}^\perp$. Then by the definition of $\mathbf{R}$,

$$zh(z) + g + H(z)\sigma(z)a \equiv 0.$$

Since we assume that $\sigma(0) = 0$, $g = 0$. It follows that

$$h(z) = -H(z)\, \frac{\sigma(z) - \sigma(0)}{z}\, a \tag{39}$$

belongs to the Pontryagin space $H\mathfrak{H}(\sigma)$ with reproducing kernel

$$H(z)K_\sigma(w,z)H(w)^*, \quad \text{that is,} \quad h(z) \in \mathfrak{H}(K_{A,B}) \cap H\mathfrak{H}(\sigma).$$

By (36),

$$K_S(w,z) = K_{A,B}(w,z) + H(z)K_\sigma(w,z)H(w)^*$$

and hence

$$\kappa \le \kappa_2 + \operatorname{sq}_-\left(H(z)K_\sigma(w,z)H(w)^*\right) \le \kappa_2 + \kappa_1 = \kappa.$$

Thus $H(z)K_\sigma(w,z)H(w)^*$ has κ_1 negative squares. By [4, Theorem 1.5.5],

$$\mathfrak{M} = \mathfrak{H}(K_{A,B}) \cap H\mathfrak{H}(\sigma)$$

is a Hilbert space in the overlapping space inner product:

$$\langle k,k\rangle_{\mathfrak{M}} = \langle k,k\rangle_{\mathfrak{H}(K_{A,B})} + \langle k,k\rangle_{H\mathfrak{H}(\sigma)}, \qquad k \in \mathfrak{M}.$$

Moreover multiplication by $H(z)$ is a partial isometry from $\mathfrak{H}(\sigma)$ onto $H\mathfrak{H}(\sigma)$ whose kernel is a Hilbert space [4, Theorem 1.5.7]; in particular, multiplication by $H(z)$ acts as a contraction from $\mathfrak{H}(\sigma)$ onto $H\mathfrak{H}(\sigma)$. Let

$$V_\sigma = \begin{pmatrix} T_\sigma & F_\sigma \\ G_\sigma & H_\sigma \end{pmatrix} : \begin{pmatrix} \mathfrak{H}(\sigma) \\ \mathfrak{A} \end{pmatrix} \to \begin{pmatrix} \mathfrak{H}(\sigma) \\ \mathfrak{B} \end{pmatrix}$$

be the canonical coisometric colligation associated with $\sigma(z)$. Then V_σ is a contraction and $F_\sigma^*F_\sigma + H_\sigma^*H_\sigma \le 1_{\mathfrak{A}}$, that is,

$$F_\sigma^*F_\sigma \le 1_{\mathfrak{A}} \tag{40}$$

because $H_\sigma = \sigma(0) = 0$. By (39), $h = -HF_\sigma a$, and thus

$$\begin{aligned}
\langle h\oplus g\oplus a, h\oplus g\oplus a\rangle_{\mathfrak{H}(K_{A,B})\oplus\mathfrak{G}\oplus\mathfrak{A}} &= \langle h,h\rangle_{\mathfrak{H}(K_{A,B})} + \langle a,a\rangle_{\mathfrak{A}} \\
&= \langle h,h\rangle_{\mathfrak{H}(K_{A,B})} + \langle HF_\sigma a, HF_\sigma a\rangle_{H\mathfrak{H}(\sigma)} \\
&\qquad + \langle a,a\rangle_{\mathfrak{A}} - \langle HF_\sigma a, HF_\sigma a\rangle_{H\mathfrak{H}(\sigma)} \\
&\ge \langle h,h\rangle_{\mathfrak{M}} + \langle (1_{\mathfrak{A}} - F_\sigma^*F_\sigma)a, a\rangle_{\mathfrak{A}} \\
&\ge \langle h,h\rangle_{\mathfrak{M}}\,.
\end{aligned}$$

Hence as $\mathfrak{M}$ is a Hilbert space, the left side is always nonnegative. If the left side is equal to zero then $h = 0$ and so $\langle a,a\rangle_{\mathfrak{A}} = 0$. By (39), $HF_\sigma a = 0$ and so $F_\sigma a$ is in the Hilbert space kernel of multiplication by $H(z)$ as an operator from $\mathfrak{H}(\sigma)$ onto $H\mathfrak{H}(\sigma)$. By (40),

$$\langle F_\sigma a, F_\sigma a\rangle_{\mathfrak{H}(\sigma)} \le \langle a,a\rangle_{\mathfrak{A}} = 0,$$

and so $F_\sigma a = 0$. Since $\sigma(0) = 0$, $\sigma(z)a \equiv 0$ and hence $a = 0$ by the hypotheses of (ii). Therefore we have shown that $\mathfrak{R}^\perp$ is a closed strictly positive subspace of the Pontryagin space $\mathfrak{H}(K_{A,B}) \oplus \mathfrak{G} \oplus \mathfrak{A}$, and hence it is a Hilbert subspace (this fails for Kreĭn spaces but holds for Pontryagin spaces by the graph representation of a subspace). Hence the closures of $\mathfrak{D}$ and $\mathfrak{R}$ are regular subspaces having negative indices equal to the common negative index of the full spaces.

As in the proof of Theorem 11, we construct a function (33) such that $B(z) = A(z)\Sigma(z)$. This yields (37) and hence (34). The identity (38) shows that the kernel (36) has κ_2 negative squares. □

References

1. D. Alpay, *Algorithme de Schur, espaces à noyau reproduisant et théorie des systèmes*, Société Mathématique de France, Paris, 1998.
2. D. Alpay, V. Bolotnikov, A. Dijksma, and H. S. V. de Snoo, *On some operator colligations and associated reproducing kernel Hilbert spaces*, Operator extensions, interpolation of functions and related topics (Timişoara, 1992) (Basel), Oper. Theory Adv. Appl., vol. 61, Birkhäuser, Basel, 1993, pp. 1–27.
3. ———, *On some operator colligations and associated reproducing kernel Pontryagin spaces*, J. Funct. Anal. **136** (1996), 39–80.
4. D. Alpay, A. Dijksma, J. Rovnyak, and H. S. V. de Snoo, *Schur functions, operator colligations, and reproducing kernel Pontryagin spaces*, Oper. Theory Adv. Appl., vol. 96, Birkhäuser Verlag, Basel, 1997.
5. D. Alpay and H. Dym, *On a new class of reproducing kernel spaces and a new generalization of the Iohvidov laws*, Linear Algebra Appl. **178** (1993), 109–183.
6. J. A. Ball and T. T. Trent, *Unitary colligations, reproducing kernel Hilbert spaces, and Nevanlinna-Pick interpolation in several variables*, J. Funct. Anal. **157** (1998), 1–61.
7. M. A. Dritschel and J. Rovnyak, *Operators on indefinite inner product spaces*, Lectures on operator theory and its applications (Waterloo, ON, 1994) (Providence, RI), Fields Inst. Monogr., vol. 3, Amer. Math. Soc., Providence, RI, 1996, pp. 141–232.
8. R. B. Leech, *Factorization of analytic functions and operator inequalities*, unpublished manuscript, available at http://www.people.virginia.edu/~jlr5m/papers.html.
9. M. Rosenblum and J. Rovnyak, *Topics in Hardy classes and univalent functions*, Birkhäuser Verlag, Basel, 1994, (Birkhäuser Advanced Texts: Basler Lehrbücher).

Department of Mathematics, Ben-Gurion University of the Negev, P. O. Box 653, 84105 Beer-Sheva, Israel
E-mail address: `dany@ivory.bgu.ac.il`

Department of Mathematics, University of Groningen, P. O. Box 800, 9700 AV Groningen, The Netherlands
E-mail address: `dijksma@math.rug.nl`

Department of Mathematics, University of Virginia, Charlottesville, Virginia 22903-3199, U. S. A.
E-mail address: `rovnyak@Virginia.EDU`

Department of Mathematics, University of Groningen, P. O. Box 800, 9700 AV Groningen, The Netherlands
E-mail address: `desnoo@math.rug.nl`

Operator Theory:
Advances and Applications, Vol. 123

Models of Contractions Constructed from the Defect Function of Their Characteristic Function

S. S. Boiko, V. K. Dubovoj, B. Fritzsche, and B. Kirstein

It is well known (see [1], [2]) that a contractive holomorphic operator-valued function in the unit disc $\mathbb{D}$ can be considered as the suboperator of scattering (Heisenberg scattering matrix) of some orthogonal unitary coupling of the unilateral shifts V_- and V_+ which are associated with the input and the output of the system. The inner states of such a system are described with the aid of some contraction T acting in a Hilbert space $\mathcal{H}$ called inner space of the system. Hereby, the system can have not only outer channels, but inner channels as well. These inner channels are connected with the maximal unilateral shifts V_T and V_{T^*} contained in T and T^*, respectively. The existence of inner channels can be considered as a defect of the system because in this case that part of information which enters the inner space $\mathcal{H}$ through the outer channels is scattered through the inner channels and therefore will not come outside. For this reason, the suboperators which describe the scattering between outer and inner channels are called defect function.

The goal of this paper is to construct a model for contractions based on defect functions of a system. Note that the models for contractions considered up to now were constructed with the aid of their characteristic function, i.e., with the aid of the suboperator of scattering through the outer channels of the system (see [3]–[6]).

In Section 1 we will state basic facts from the theory of unitary colligations. For a detailed treatment of the theory of unitary colligations, we refer to [4]. Furthermore, heavily leaned on the papers [7]–[9], we will present some statements on defects functions. Following [9], in Section 2 we will introduce the essential notion of the generating shift. The main result of this section is Theorem 2.2 which contains a description of the model operator for a contraction having a generating shift.

In Section 3, we will show (see Theorem 3.1) that the model space is a Hilbert space $\mathcal{H}$ with reproducing kernel. Moreover, we will establish interrelations between the reproducing kernel and the defect functions (see Theorem 3.3 and Remarks 1 and 2).

Finally, in Section 4 we will construct and study models for a contraction T acting in $\mathcal{H}$ and having inner channels such that the smallest closed subspace which contains them and reduces T coincides with $\mathcal{H}$. In Sections 3 and 4 we will use methods worked out in [6]. In particular, we take from [6] some basic facts on reproducing kernel Hilbert spaces.

1. Contractive operators, defect functions and scattering theory

1.1. Let $\mathbb{D} := \{\zeta \in \mathbb{C} : |\zeta| < 1\}$, and let $\mathbb{T} := \{t \in \mathbb{C} : |t| = 1\}$. The symbol $\underline{\lambda}$ stands for the linear Lebesgue-Borel measure on $\mathbb{T}$. Throughout this paper, the considered Hilbert spaces are complex separable Hilbert spaces. If $\mathcal{H}$ is a Hilbert space, if A is a nonempty set, and if $\mathcal{L}_\alpha \subseteq \mathcal{H}$ for all $\alpha \in A$, then $\bigvee_{\alpha \in A} \mathcal{L}_\alpha$ designates the smallest closed linear subspace of $\mathcal{H}$ which contains each $\mathcal{L}_\alpha, \alpha \in A$. The operators we will work with are linear and bounded. If $\mathcal{H}$ and $\mathcal{G}$ are Hilbert spaces, then the notation $[\mathcal{H}, \mathcal{G}]$ stands for the space all bounded linear operators acting between $\mathcal{H}$ and $\mathcal{G}$. By a unitary colligation we will mean a tuple

$$\triangle = (\mathcal{H}, \mathcal{F}, \mathcal{G}; T, F, G, S) \tag{1.1}$$

where $\mathcal{H}, \mathcal{F}$ and $\mathcal{G}$ are Hilbert spaces and where $T : \mathcal{H} \to \mathcal{H}$, $F : \mathcal{F} \to \mathcal{H}$, $G : \mathcal{H} \to \mathcal{G}$ and $S : \mathcal{F} \to \mathcal{G}$ are operators such that the operator

$$U := \begin{pmatrix} T & F \\ G & S \end{pmatrix} : \mathcal{H} \oplus \mathcal{F} \to \mathcal{H} \oplus \mathcal{G} \tag{1.2}$$

is unitary. The space $\mathcal{H}$ is called inner whereas $\mathcal{F}$ and $\mathcal{G}$ are said to be outer. The operator T, which is called the basic operator of the unitary colligation $\triangle$, is necessarily contractive, i.e., $\|T\| \le 1$ holds true. Conversely, if a Hilbert space $\mathcal{H}$ and a contractive operator $T : \mathcal{H} \to \mathcal{H}$ are given, then there is a unitary colligation $\triangle$ such that T is the basic operator of $\triangle$ (see [4]). A unitary colligation $\triangle$ is said to be simple if $\mathcal{H} = \mathcal{H}_\mathcal{G} \bigvee \mathcal{H}_\mathcal{F}$ where

$$\mathcal{H}_\mathcal{G} := \bigvee_{n=0}^{\infty} (T^*)^n G^*(\mathcal{G}) \quad \text{and} \quad \mathcal{H}_\mathcal{F} := \bigvee_{n=0}^{\infty} T^n F(\mathcal{F}).$$

The (well-defined) function $\Theta_\triangle : \mathbb{D} \to [\mathcal{F}, \mathcal{G}]$ given by

$$\Theta_\triangle(\zeta) := S + \zeta G(I - \zeta T)^{-1} F \tag{1.3}$$

is called the characteristic (operator-valued) function of the unitary colligation $\triangle$.

If $\mathcal{F}$ and $\mathcal{G}$ are Hilbert spaces, then we will use $\mathcal{S}[\mathcal{F}, \mathcal{G}]$ for the set of all functions $\Theta : \mathbb{D} \to [\mathcal{F}, \mathcal{G}]$ which are holomorphic in $\mathbb{D}$ and the values of which are contractive operators, i.e., $I - \Theta(\zeta)^*\Theta(\zeta) \ge 0$ for each $\zeta \in \mathbb{D}$. Note that if $\Theta \in \mathcal{S}[\mathcal{F}, \mathcal{G}]$, then there is a function $\underline{\Theta} : \mathbb{T} \to [\mathcal{F}, \mathcal{G}]$ (more precisely, a unique $\underline{\lambda}$-equivalence class of functions) such that

$$\lim_{r \to 1-0} \|\Theta(rt)f - \underline{\Theta}(t)f\|_\mathcal{G} = 0$$

for all $f \in \mathcal{F}$ and for $\underline{\lambda}$-almost all $t \in \mathbb{T}$. We will continue to use $\underline{\Theta}$ to denote such a boundary function of $\Theta \in \mathcal{S}[\mathcal{H}, \mathcal{G}]$. Further, if $\mathcal{F}$ and $\mathcal{G}$ are Hilbert spaces, then let $\mathcal{L}^\infty[\mathcal{F}, \mathcal{G}]$ be the set of all functions $\sigma : \mathbb{T} \to [\mathcal{F}, \mathcal{G}]$ which are $\underline{\lambda}$-essentially bounded on $\mathbb{T}$. The corresponding space of $\underline{\lambda}$-equivalence classes of $\mathcal{L}^\infty[\mathcal{F}, \mathcal{G}]$ is designated by $L^\infty[\mathcal{F}, \mathcal{G}]$.

The following theorem (see, e.g., [4]) plays a key role in the theory of unitary colligations.

Theorem 1.1. (a) The characteristic function of an arbitrary unitary colligation (1.1) belongs to $\mathcal{S}[\mathcal{F},\mathcal{G}]$.

(b) If $\mathcal{F}$ and $\mathcal{G}$ are Hilbert spaces and if Θ belongs to $\mathcal{S}[\mathcal{F},\mathcal{G}]$, then there is a simple unitary colligation (1.1) such that Θ is its characteristic function.

Every unitary colligation generates a unitary dilation of the contraction T. In fact, if

$$\mathfrak{D}_{\mathcal{G}} := \bigoplus_{n=-\infty}^{-1} \mathcal{G} \quad , \quad \mathfrak{D}_{\mathcal{F}} = \bigoplus_{n=1}^{\infty} \mathcal{F}$$

and if $\mathfrak{K} := \mathfrak{D}_{\mathcal{G}} \oplus \mathcal{H} \oplus \mathfrak{D}_{\mathcal{F}}$, then the operator $W := \mathfrak{K} \to \mathfrak{K}$ given by the block representation $W = (W_{jk})_{j,k=-\infty}^{\infty}$ with $W_{00} = T$, $W_{-1,0} = G$, $W_{-1,1} = S$, $W_{0,1} = F, W_{j,j+1} = I_{\mathcal{H}}$ for $j \in \mathbb{Z} \setminus \{-1,0\}$ and $W_{jk} = 0$ for the other j,k, i.e.,

$$W = \begin{pmatrix} & I & 0 & & & & & \\ \ddots & 0 & I & 0 & & & & \\ & & 0 & G & S & & & \\ & & & T & F & 0 & & \\ & & & & 0 & I & 0 & \\ & & & & & 0 & I & \ddots \\ & & & & & & 0 & \ddots \end{pmatrix} \begin{matrix} \\ \\ \\ \leftarrow 0 \\ \\ \\ \\ \end{matrix} \tag{1.4}$$

$$\begin{matrix} \uparrow \\ 0 \end{matrix}$$

is a unitary dilation of T. (For a detailed treatment of unitary dilations, see [3].)

1.2. Let $\mathcal{H}$ be a Hilbert space. We will say that a unilateral shift $V : \mathcal{L} \to \mathcal{L}$ is contained in a contraction $T : \mathcal{H} \to \mathcal{H}$ if $\mathcal{L}$ is a closed subspace of $\mathcal{H}$ which is invariant which respect to T and if V is the restriction of T onto $\mathcal{L}$: $V = \mathrm{Rstr.}_{\mathcal{L}} T$. If $\tilde{\mathcal{L}}$ is a closed subspace of $\mathcal{H}$, then an operator $\tilde{V} \in [\tilde{\mathcal{L}}, \tilde{\mathcal{L}}]$ is called a unilateral coshift if $\tilde{V}^*$ is a unilateral shift. We will say that a unilateral coshift $\tilde{V}$ is contained in the contraction T if $\tilde{V}^*$ is contained in T^*. For example, the operators $V_{\mathcal{G}} := \mathrm{Rstr.}_{\mathfrak{D}_{\mathcal{G}}} W$ and $V_{\mathcal{F}} := \mathrm{Rstr.}_{\mathcal{D}_{\mathcal{F}}} W^*$ are unilateral shifts which are contained in W and W^*, respectively. Note that in the formalism of scattering theory the unilateral coshift $V_{\mathcal{F}}^*$ and the shift $V_{\mathcal{G}}$ are related to the input space and the output space, respectively. From this point of view, every unilateral coshift contained in W can be considered as an input channel operator, whereas every unilateral shift contained in W can be interpreted as an output channel operator. In the context of a scattering model with loss the inner channel operators (i.e., the unilateral shifts and coshifts contained in W and acting between subspaces of the inner space $\mathcal{H}$) play an important role. The following lemma shows that the description of the inner channel operators of the dilation W can be reduced to the description of the unilateral shifts and coshifts contained in the contraction T.

Lemma 1.2. Every unilateral shift $V : \mathcal{L} \to \mathcal{L}$ (respectively, unilateral coshift $\tilde{V} : \tilde{\mathcal{L}} \to \tilde{\mathcal{L}}$) contained in W and satisfying $\mathcal{L} \subseteq \mathcal{H}$ (respectively, $\tilde{\mathcal{L}} \subseteq \mathcal{H}$) is contained in T. Conversely, each unilateral shift $V : \mathcal{L} \to \mathcal{L}$ (respectively, unilateral coshift $\tilde{V} : \tilde{\mathcal{L}} \to \tilde{\mathcal{L}}$) contained in T is contained in W as well.

In the case of a completely nonunitary contraction T one can choose maximal elements in the set of inner channel operators (see [7]):

Lemma 1.3. Let $T : \mathcal{H} \to \mathcal{H}$ be a completely nonunitary contraction, and let $\triangle = (\mathcal{H}, \mathcal{F}, \mathcal{G}; T, F, G, S)$ be a unitary colligation such that T is the basic operator of $\triangle$. Then the subspace $\mathcal{H}_{\mathcal{G}}^{\perp} := \mathcal{H} \ominus \mathcal{H}_{\mathcal{G}}$ (respectively, $\mathcal{H}_{\mathcal{F}}^{\perp} := \mathcal{H} \ominus \mathcal{H}_{\mathcal{F}}$) is invariant with respect to T (respectively, T^*). The restriction V of T onto $\mathcal{H}_{\mathcal{G}}^{\perp}$ (respectively, the restriction V_1 of T^* onto $\mathcal{H}_{\mathcal{F}}^{\perp}$) is a unilateral shift. The shift V (respectively, V_1) is maximal in the following sense: Every unilateral shift which is contained in T (respectively, T^*) is contained in V (respectively, V_1).

The unilateral shift V given in Lemma 1.3 is said to be the maximal unilateral shift which is contained in T and will be denoted by V_T. In this sense, the operator V_1 occurring in Lemma 1.3 can be written as V_{T^*}.

Let us introduce further notations. If $\mathcal{L}$ is a closed subspace of the Hilbert space $\mathcal{H}$, then let P_U denote the orthogonal projection of $\mathcal{H}$ onto $\mathcal{L}$. If $\mathcal{E}_1$ and $\mathcal{E}_2$ are Hilbert spaces and if $Q \in [\mathcal{E}_1, \mathcal{E}_2]$, then $\mathcal{R}(Q)$ stands for the range of Q, and $\mathcal{N}(Q)$ is the kernel of Q, i.e., $\mathcal{N}(Q) := \{x \in \mathcal{E}_1 : Qx = 0\}$.

Definition. Let $\Theta \in \mathcal{S}(\mathcal{F}, \mathcal{G})$, and let $\triangle$ given by (1.1) be a simple unitary colligation such that Θ is the characteristic function of $\triangle$. Let V_T and V_{T^*} be the maximal unilateral shifts which are contained in T and T^*, respectively, and let $\mathcal{H}_0$ and $\tilde{\mathcal{H}}_0$ be the generating subspaces for V_T and V_{T^*}, respectively. Then the function $\varphi : \mathbb{D} \to [\mathcal{F}, \mathcal{H}_0]$ given by

$$\varphi(\zeta) := P_{\mathcal{H}_0}(I - \zeta T)^{-1} F \tag{1.5}$$

(respectively, $\psi : \mathbb{D} \to [\tilde{\mathcal{H}}_0, \mathcal{G}]$ given by

$$\psi(\zeta) := \mathrm{Rstr.}_{\tilde{\mathcal{H}}_0} G(I - \zeta T)^{-1} P_{\tilde{\mathcal{H}}_0} \tag{1.6}$$

is called the right (respectively, left) defect function associated with Θ (and $\triangle$).

In [7] and [9] it was shown that the function φ (respectively, the function ψ) describes the scattering connected with the inner channel which is generated by the maximal unilateral shift V_T (respectively, V_{T^*}).

1.3. In order to verify immediate interrelations between the function Θ and their defect functions φ and ψ we need the following result from [3]:

Theorem 1.4. Let $\mathcal{G}$ be a Hilbert space, and let $N : \mathbb{T} \to [\mathcal{G}, \mathcal{G}]$ be a Borel measurable function. Suppose that, for each $t \in \mathbb{T}$, the operator $N(t)$ is selfadjoint and satisfies $0 \leq N(t) \leq I_{\mathcal{G}}$. Then there exist a Hilbert space $\mathcal{K}$ and a contractive outer function $\vartheta : \mathbb{D} \to [\mathcal{G}, \mathcal{K}]$ such that the following conditions are satisfied:

(i) $\underline{\vartheta}^*\underline{\vartheta} \le N^2 \quad \underline{\lambda}$-a.e. on $\mathbb{T}$.

(ii) If $\tilde{\mathcal{K}}$ is a Hilbert space and if $\tilde{\vartheta} : \mathbb{D} \to [\mathcal{G}, \tilde{\mathcal{K}}]$ is a contractive holomorphic function satisfying $\underline{\tilde{\vartheta}}^*\underline{\tilde{\vartheta}} \le N^2 \quad \underline{\lambda}$-a.e. on $\mathbb{T}$, then $\underline{\tilde{\vartheta}}^*\underline{\tilde{\vartheta}} \le \underline{\vartheta}^*\underline{\vartheta}$ $\underline{\lambda}$-a.e. on $\mathbb{T}$.

Moreover, the function ϑ is uniquely determined up to a left constant unitary factor.

Theorem 1.5. Let $\mathcal{F}$ and $\mathcal{G}$ be Hilbert spaces, and let $\Theta \in \mathcal{S}[\mathcal{F}, \mathcal{G}]$. Then:

(a) If $N_r := \sqrt{I_\mathcal{F} - \underline{\Theta}^*\underline{\Theta}}$, then every largest minorant of N_r^2 is a right defect function associated with Θ (and some simple unitary colligation $\triangle$ given by (1.1) having Θ as characteristic function).

(b) Let $\check{\Theta} : \mathbb{D} \to [\mathcal{G}, \mathcal{F}]$ be given by $\check{\Theta}(\zeta) := \Theta(\overline{\zeta})^*$. Then $\check{\Theta} \in \mathcal{S}[\mathcal{G}, \mathcal{F}]$. If $N_l := \sqrt{I_\mathcal{G} - (\underline{\check{\Theta}})^*\underline{\check{\Theta}}}$, then every largest minorant of N_l^2 is a left defect function associated with Θ (and some simple unitary colligation $\triangle$ given by (1.1) having Θ as characteristic function).

Corollary. Let $\mathcal{F}$ and $\mathcal{G}$ be Hilbert spaces, and let $\Theta \in \mathcal{S}[\mathcal{F}, \mathcal{G}]$. Let $\triangle$ given by (1.1) be a simple unitary colligation such that Θ is its characteristic function.

(a) If φ is a right defect function associated with Θ (and $\triangle$), then φ is outer.

(b) If ψ is a left defect function associated with Θ (and $\triangle$), then ψ is *-outer.

Observe that in [7] there is a description of unitary colligations having φ (respectively, ψ) as characteristic functions.

1.4. Let $\Theta \in \mathcal{S}[\mathcal{F}, \mathcal{G}]$, and let $\varphi \in \mathcal{S}[\mathcal{F}, \mathcal{H}_0]$ (respectively, $\psi \in \mathcal{S}[\tilde{\mathcal{H}}_0, \mathcal{G}]$) be the right (respectively, left) defect function associated with Θ. For each $\eta \in \mathcal{L}^\infty[\tilde{\mathcal{H}}_0, \mathcal{H}_0]$ let

$$\alpha(\eta) := \left\| \begin{pmatrix} \eta & \underline{\varphi} \\ \underline{\psi} & \underline{\Theta} \end{pmatrix} \right\|_{L^\infty[\tilde{\mathcal{H}}_0 \oplus \mathcal{F}, \mathcal{H}_0 \oplus \mathcal{G}]} . \tag{1.7}$$

Then the following problem was studied in [8]: Determine

$$\kappa := \inf\{\alpha(\eta) : \eta \in L^\infty[\tilde{\mathcal{H}}_0, \mathcal{H}_0]\} \tag{1.8}$$

and the set of all $\chi \in L^\infty[\tilde{\mathcal{H}}_0, \mathcal{H}_0]$ such that $\alpha(\chi) = \kappa$.

An answer to the problem was given in [8]:

Theorem 1.6. Let $\Theta \in \mathcal{S}[\mathcal{F}, \mathcal{G}]$. Then there is a unique $\mathcal{X} \in L^\infty[\tilde{\mathcal{H}}_0, \mathcal{H}_0]$ such that

$$\alpha(\mathcal{X}) = 1 = \inf\left\{\alpha(\eta) : \eta \in L^\infty[\tilde{\mathcal{H}}_0 \oplus \mathcal{F}, \mathcal{H}_0 \oplus \mathcal{G}]\right\}.$$

Remark. Let Θ be the characteristic function of a simple unitary colligation $\triangle$ given by (1.1). Then it can be shown (see [8]) that the unique $\chi \in L^\infty[\tilde{\mathcal{H}}_0, \mathcal{H}_0]$ such that $\alpha(\chi) = \inf\{\alpha(\eta) : \eta \in L^\infty[\tilde{\mathcal{H}}_0, \mathcal{H}_0]\}$ is a suboperator of scattering with respect to the inner channels of a scattering system associated with the unitary colligation $\triangle$. Let $\chi^\square : \mathbb{D} \to [\tilde{\mathcal{H}}_0, \mathcal{H}_0]$ be given by

$$\chi^\square(\zeta) := \mathrm{Rstr.}_{\tilde{\mathcal{H}}_0} P_{\mathcal{H}_0} \left((\overline{\zeta}(I - \overline{\zeta}T^*)^{-1} + T(I - \zeta T)^{-1}\right) P_{\tilde{\mathcal{H}}_0}. \tag{1.9}$$

Then

$$\lim_{r\to 1-0} \|\chi^{\square}(rt)h - \chi(t)h\|_{\mathcal{H}_0} = 0$$

for all $h \in \tilde{\mathcal{H}}_0$ and $\underline{\lambda}$-almost all $t \in \mathbb{T}$. From (1.9) it can be derived that $\chi^{\square} \in \mathcal{S}[\tilde{\mathcal{H}}_0, \mathcal{H}_0]$ if and only if the subspaces where the maximal unilateral shifts V_T and V_{T^*} act are orthogonal, i.e., $\mathcal{H}_{\mathcal{G}}^{\perp} \perp \mathcal{H}_{\mathcal{F}}^{\perp}$. In this case,

$$\chi^{\square}(\zeta) = \mathrm{Rstr.}_{\tilde{\mathcal{H}}_0} P_{\mathcal{H}_0} T(I - \zeta T)^{-1} P_{\tilde{\mathcal{H}}_0}$$

for each $\zeta \in \mathbb{D}$. In connection with (1.9) we consider the holomorphic operator-valued functions $\chi_- : \mathbb{D} \to [\tilde{\mathcal{H}}_0, \mathcal{H}_0]$ and $\chi_+ : \mathbb{D} \to [\tilde{\mathcal{H}}_0, \mathcal{H}_0]$ given by

$$\chi_-(\zeta) = \mathrm{Rstr.}_{\tilde{\mathcal{H}}_0} \zeta P_{\mathcal{H}_0} (I - \zeta T^*)^{-1} P_{\tilde{\mathcal{H}}_0} \tag{1.10}$$

and

$$\chi_+(\zeta) = \mathrm{Rstr.}_{\tilde{\mathcal{H}}_0} P_{\mathcal{H}_0} T(I - \zeta T)^{-1} P_{\tilde{\mathcal{H}}_0}. \tag{1.11}$$

Obviously, then $\chi^{\square}(\zeta) = \chi_-(\overline{\zeta}) + \chi_+(\zeta)$ for all $\zeta \in \mathbb{D}$.

2. A model for a contractive operator with generating shift

2.1. Let us again assume that $\mathcal{H}$ is a Hilbert space.

Definition. (a) Let $T : \mathcal{H} \to \mathcal{H}$ be a contractive operator which contains a unilateral shift $V : \mathcal{L} \to \mathcal{L}$. Then the unilateral shift V is called generating for T if

$$\mathcal{H} = \bigvee_{n=0}^{\infty} (T^*)^n \mathcal{L}.$$

(b) Let $T : \mathcal{H} \to \mathcal{H}$ be a contraction which contains a unilateral coshift $\tilde{V} : \tilde{\mathcal{L}} \to \tilde{\mathcal{L}}$. Then the unilateral coshift $\tilde{V}$ is said to be generating for T, if $\tilde{V}^*$ is a generating shift for T^*, i.e, if

$$\mathcal{H} = \bigvee_{n=0}^{\infty} T^n \tilde{\mathcal{L}}.$$

The following theorem which is proved in [9] shows the essential role of generating unilateral shifts (coshifts).

Theorem 2.1. Let $\omega \in \mathcal{S}[\mathcal{F}, \mathcal{G}]$, and let T be a completely nonunitary contraction the characteristic function of which is ω. Then the following statements are equivalent:

(i) There is a generating unilateral shift (respectively, coshift) for T.
(ii) There is a function $\Theta \in \mathcal{S}[\mathcal{F}, \mathcal{G}]$ such that ω is exactly the right (respectively, left) defect function associated with Θ.

2.2. Let $\mathcal{E}$ be a Hilbert space, and let $h^2[\mathcal{E}]$ be the Hilbert space of all functions $f : \mathbb{D} \to \mathcal{E}$ which admit a representation

$$f(\zeta) = \sum_{k=1}^{\infty} f_{-k}\overline{\zeta}^k + \sum_{k=0}^{\infty} f_k \zeta^k \tag{2.1}$$

for each $\zeta \in \mathbb{D}$ where $(f_k)_{k=-\infty}^{\infty}$ is some sequence of vectors which belong to $\mathcal{E}$ and which satisfy

$$\sum_{k=-\infty}^{\infty} \|f_k\|_{\mathcal{E}}^2 < +\infty.$$

If f and g belong to $h^2[\mathcal{E}]$, then (with obvious notation) the scalar product $(f,g)_{h^2[\mathcal{E}]}$ is given by $(f,g)_{h^2[\mathcal{E}]} = \sum_{k=-\infty}^{\infty}(f_k, g_k)_{\mathcal{E}}$. If $f \in h^2[\mathcal{E}]$, then there is a function $\underline{f} : \mathbb{T} \to \mathcal{E}$ (more precisely, a unique $\underline{\lambda}$-equivalence class of functions) such that

$$\lim_{r\to 1-0} \|f(rt) - \underline{f}(t)\|_{\mathcal{E}} = 0$$

for $\underline{\lambda}$-almost all $t \in \mathbb{T}$ (see [3]). This function $\underline{f}$ belongs to the class $\mathcal{L}^2[\mathcal{E}]$ of all functions $v : \mathbb{T} \to \mathcal{E}$ for which $\|v\|_{\mathcal{E}}^2$ is integrable with respect $\underline{\lambda}/2\pi$ on $\mathbb{T}$.

In this way an isomorphism between $h^2[\mathcal{E}]$ and the space $L^2[\mathcal{E}]$ of all $\underline{\lambda}$-equivalence classes generated by the elements of $\mathcal{L}^2[\mathcal{E}]$ is established. The subspace of all functions $f \in h^2[\mathcal{E}]$ which admit a representation (2.1) where $f_k = 0$ for all negative integers k (respectively, where $f_k = 0$ for all nonnegative integers k) will be denoted by $H_+^2[\mathcal{E}]$ (respectively, $H_-^2[\mathcal{E}]$). We will consider a particular multiplication operator $\underline{S}_{\mathcal{E}}$ in $L^2[\mathcal{E}]$. If $v \in L^2[\mathcal{E}]$, then let $\underline{S}_{\mathcal{E}} : L^2[\mathcal{E}] \to L^2[\mathcal{E}]$ be defined by $\underline{S}_{\mathcal{E}} v := w$ where $w(t) := tv(t)$ for each $t \in \mathbb{T}$. Obviously, $\underline{S}_{\mathcal{E}}$ is a bilateral shift, and the adjoint operator $\underline{S}_{\mathcal{E}}^*$ is given by $\underline{S}_{\mathcal{E}}^* v = u$ where $u(t) = \overline{t}v(t)$ for $\underline{\lambda}$-almost all $t \in \mathbb{T}$. We will use the notation $S_{\mathcal{E}}$ for the bilateral shift in the space $h^2[\mathcal{E}]$, i.e., if $f \in h^2[\mathcal{E}]$ admits the representation (2.1), then let

$$(S_{\mathcal{E}} f)(\zeta) := \sum_{k=1}^{\infty} f_{-k-1}\overline{\zeta}^k + \sum_{k=0}^{\infty} f_{k-1}\zeta^k$$

for every choice of ζ in $\mathbb{D}$. Further, we have then

$$(S_{\mathcal{E}}^* f)(\zeta) = \sum_{k=1}^{\infty} f_{-k+1}\overline{\zeta}^k + \sum_{k=0}^{\infty} f_{k+1}\zeta^k \tag{2.2}$$

for all $\zeta \in \mathbb{D}$.

The following theorem gives a description of a model of a contraction T with shift completeness property.

Theorem 2.2. Let $T : \mathcal{H} \to \mathcal{H}$ be a contractive operator which has a generating unilateral shift $V : \mathcal{L} \to \mathcal{L}$. Let $\mathcal{E}$ be the generating subspace for the unilateral shift V, and let $\Xi : \mathbb{D} \to [\mathcal{H}, \mathcal{E}]$ be given by

$$\Xi(\zeta) := P_{\mathcal{E}}\left((I_{\mathcal{H}} - \overline{\zeta}T^*)^{-1} + (I_{\mathcal{H}} - \zeta T)^{-1} - I_{\mathcal{H}}\right). \tag{2.3}$$

Then:

(a) For each $h \in \mathcal{H}$, the function $f_h : \mathbb{D} \to \mathcal{E}$ given by

$$f_h(\zeta) := \Xi(\zeta)h \tag{2.4}$$

belongs to $h^2[\mathcal{E}]$. The mapping $Y^\square : \mathcal{H} \to h^2[\mathcal{E}]$ given by $Y^\square h := f_h$ satisfies $\mathcal{N}(Y^\square) = \{0\}$. By

$$\|Y^\square h\|_{\mathcal{M}} := \|h\|_{\mathcal{H}} \tag{2.5}$$

for each $h \in \mathcal{H}$ a (well-defined) norm $\|\cdot\|_{\mathcal{M}}$ on $\mathcal{M} := \mathcal{R}(Y^\square)$ is given, and (with respect to this norm) the operator $Y : \mathcal{H} \to \mathcal{M}$ given by $Yh := Y^\square h$ is unitary.

(b) The identity

$$S_{\mathcal{E}}^* Y = YT \tag{2.6}$$

holds true. In particular, the subspace $\mathcal{M}$ is invariant with respect to $S_{\mathcal{E}}^*$, and

$$RY = YT \tag{2.7}$$

where $R := \mathrm{Rstr.}_{\mathcal{M}} S_{\mathcal{E}}^*$.

Proof. (a) Let

$$T = \begin{pmatrix} V & r \\ 0 & T_1 \end{pmatrix}$$

be the block representation of T with respect to the orthogonal decomposition $\mathcal{H} = \mathcal{L} \oplus \mathcal{L}^\perp$ of $\mathcal{H}$. Setting $\mathcal{L}_1 := \mathcal{L}^\perp$ we have

$$r = \mathrm{Rstr.}_{\mathcal{L}_1} P_{\mathcal{L}} T \tag{2.8}$$

and

$$\begin{pmatrix} V^*V & V^*r \\ r^*V & r^*r + T_1^*T_1 \end{pmatrix} = T^*T \le I_{\mathcal{H}}.$$

In particular,

$$r^*r \le I_{\mathcal{L}_1} - T_1^*T_1 \,. \tag{2.9}$$

For each nonnegative integer k, we have

$$\mathcal{L} \subseteq \mathcal{N}(rP_{\mathcal{L}_1} T^k). \tag{2.10}$$

Now let $h \in \mathcal{H}$. Using the Neumann series representation we get that f_h defined by (2.3) and (2.4) can be represented by

$$f_h(\zeta) = \sum_{k=1}^{\infty} P_{\mathcal{E}}(T^*)^k h \overline{\zeta}^k + \sum_{k=0}^{\infty} P_{\mathcal{E}} T^k h \zeta^k \tag{2.11}$$

for each $\zeta \in \mathbb{D}$. For each positive integer k, we obtain from (2.8), (2.9) and (2.10) that

$$\begin{aligned}
\|P_{\mathcal{E}} T^K h\|_{\mathcal{H}}^2 &\le \|P_{\mathcal{L}} T^k h\|_{\mathcal{H}}^2 = \|rP_{\mathcal{L}_1} T^{k-1} h\|_{\mathcal{H}}^2 = \|rP_{\mathcal{L}_1} T^{k-1} P_{\mathcal{L}_1} h\|_{\mathcal{H}}^2 \\
&= \|rT_1^{k-1} P_{\mathcal{L}_1} h\|_{\mathcal{H}}^2 = (r^*rT_1^{k-1} P_{\mathcal{L}_1} h, T_1^{k-1} P_{\mathcal{L}_1} h)_{\mathcal{H}} \\
&\le \left((I_{\mathcal{L}_1} - T_1^*T_1) T_1^{k-1} P_{\mathcal{L}_1} h, T_1^{k-1} P_{\mathcal{L}_1} h\right)_{\mathcal{H}} \\
&= \|T_1^{k-1} P_{\mathcal{L}_1} h\|_{\mathcal{H}}^2 - \|T_1^k P_{\mathcal{L}_1} h\|_{\mathcal{H}}^2 \,.
\end{aligned}$$

Consequently, for each positive integer n,

$$\sum_{k=1}^{n} \|P_{\mathcal{E}} T^k h\|_{\mathcal{H}}^2 \le \sum_{k=1}^{n} \left(\|T_1^{k-1} P_{\mathcal{L}_1} h\|_{\mathcal{H}}^2 - \|T_1^k P_{\mathcal{L}_1} h\|_{\mathcal{H}}^2 \right)$$
$$= \|P_{\mathcal{L}_1} h\|_{\mathcal{H}}^2 - \|T_1^n P_{\mathcal{L}_1} h\|_{\mathcal{H}}^2 \le \|P_{\mathcal{L}_1} h\|_{\mathcal{H}}^2 .$$

Thus

$$\sum_{k=1}^{\infty} \|P_{\mathcal{E}} T^k h\|_{\mathcal{H}}^2 \le \|P_{\mathcal{L}_1} h\|_{\mathcal{H}}^2. \tag{2.12}$$

Since $\mathcal{E}$ is the generating subspace for the unilateral shift $V : \mathcal{L} \to \mathcal{L}$, we have

$$P_{\mathcal{L}} = \sum_{k=0}^{\infty} T^k P_{\mathcal{E}} (T^*)^k.$$

Thus it follows

$$\sum_{k=0}^{\infty} \|P_{\mathcal{E}} (T^*)^k h\|_{\mathcal{H}}^2 = \sum_{k=0}^{\infty} (T^k P_{\mathcal{E}} (T^*)^k h, h)_{\mathcal{H}} = (P_{\mathcal{L}} h, h)_{\mathcal{H}} = \|P_{\mathcal{L}} h\|_{\mathcal{H}}^2. \tag{2.13}$$

Hence we can conclude from (2.12) and (2.13) that

$$\sum_{k=0}^{\infty} \|P_{\mathcal{E}} (T^*)^k h\|_{\mathcal{H}}^2 + \sum_{k=1}^{\infty} \|P_{\mathcal{E}} T^k h\|_{\mathcal{H}}^2 \le \|P_{\mathcal{L}} h\|_{\mathcal{H}}^2 + \|P_{\mathcal{L}_1} h\|_{\mathcal{H}}^2 = \|h\|_{\mathcal{H}}^2.$$

In view of (2.11) this implies that f_h belongs to $h^2[\mathcal{E}]$ and satisfies $\|f_h\|_{h^2[\mathcal{E}]} \le \|h\|_{\mathcal{H}}$. Thus we see that $Y^{\square}$ is an operator defined on $\mathcal{H}$ the values of which belong to $h^2[\mathcal{E}]$, and that

$$\|Y^{\square} h\|_{h^2[\mathcal{E}]} \le \|h\|_{\mathcal{H}} \tag{2.14}$$

is valid for each $h \in \mathcal{H}$. Now let h belong to $\mathcal{N}(Y^{\square})$. Setting

$$T_{[k]} := \begin{cases} (T^*)^{|k|} & \text{if} \quad k \le -1 \\ I & \text{if} \quad k = 0 \\ T^k & \text{if} \quad k \ge 1. \end{cases}$$

we get from (2.11) that

$$P_{\mathcal{E}} T_{[k]} h = 0 \tag{2.15}$$

holds for each $k \in \mathbb{Z}$. Therefore

$$(T_{[k]} h, e)_{\mathcal{H}} = 0$$

for every choice of e in $\mathcal{E}$ and k in $\mathbb{Z}$. Thus

$$(h, T_{[k]} e)_{\mathcal{H}} = 0$$

is true for all $e \in \mathcal{E}$ and all $k \in \mathbb{Z}$. Hence h is orthogonal to the space $\bigvee_{k=-\infty}^{\infty} T_{[k]}\mathcal{E}$, where one can see easily that

$$\bigvee_{k=-\infty}^{\infty} T_{[k]}\mathcal{E} = \bigvee_{k=0}^{\infty} (T^*)^k \mathcal{L} = \mathcal{H}.$$

This implies $h = 0$. Hence $\mathcal{N}(Y^\square) = \{0\}$.
(b) In view of $\mathcal{N}(Y^\square) = \{0\}$, by the formula (2.5) a well-defined norm $\|.\|_{\mathcal{M}}$ on the subspace $\mathcal{M}$ of $h^2[\mathcal{E}]$ is given. Using the identity $P_{\mathcal{E}}(T^*)^k T = P_{\mathcal{E}}(T^*)^{k-1}$, which holds true for every positve integer k, and using (2.11), (2.1) and (2.2), we can conclude that

$$\begin{aligned}(YTh)(\zeta) &= \left(\sum_{k=1}^{\infty} P_{\mathcal{E}}(T^*)^k T \overline{\zeta}^k + \sum_{k=0}^{\infty} P_{\mathcal{E}} T^{k+1} \zeta^k \right)_h \\ &= \sum_{k=1}^{\infty} P_{\mathcal{E}}(T^*)^{k-1} h \overline{\zeta}^k + \sum_{k=0}^{\infty} P_{\mathcal{E}} T^{k+1} h \zeta^k = (S_{\mathcal{E}}^* Y h)(\zeta)\end{aligned}$$

for each $\zeta \in \mathbb{D}$ and each $h \in \mathcal{H}$. Thus we see that (2.6) and consequently (2.7) are verified. In view of (2.14) and (2.5) the operator $Y : \mathcal{H} \to \mathcal{M}$ is unitary. The proof is complete. □

3. Models as reproducing kernel Hilbert spaces

3.1. Let $\mathcal{E}$ be Hilbert space, the so-called coefficient space, and let Ω be a nonempty set. We will assume that $\mathcal{K}$ is a reproducing kernel Hilbert space of $\mathcal{E}$-valued functions defined on Ω, i.e., for each $\omega \in \Omega$, the mapping $e_\omega : \mathcal{K} \to \mathcal{E}$ given by $e_\omega(f) := f(\omega)$ is continuous. Then there is a so-called reproducing kernel K for $\mathcal{K}$, i.e., there exists a mapping K from $\Omega \times \Omega$ with values in the space $\mathfrak{E}$ of operators in $\mathcal{E}$ such that the following two conditions are satisfied:

(i) For each $\omega \in \Omega$, the mapping $K_{\cdot\omega} : \Omega \to \mathfrak{E}$ given by $K_{\cdot\omega}(v) := K(v, \omega)$ satisfies $K_{\cdot\omega} x \in \mathcal{K}$ for each $x \in \mathcal{E}$.

(ii) For every choice of f in $\mathcal{K}$, x in $\mathcal{E}$ and $\omega \in \Omega$, the identity

$$(f(\omega), x)_{\mathcal{E}} = (f, K_{\cdot\omega} x)_{\mathcal{K}}$$

holds true.

If Ω is a simply connected domain in the complex plane and if the functions which belong to $\mathcal{K}$ are analytic in Ω, then, for each $\omega \in \Omega$, the function $K_{\cdot\omega}$ is analytic in Ω as well, and the function $K_{\omega\cdot} : \Omega \to \mathfrak{E}$ defined by $K_{\omega\cdot}(v) := K(\omega, v)$ is conjugate analytic in Ω. Any reproducing kernel K of an arbitrary kernel Hilbert space $\mathcal{K}$ is necessarily nonnegative definite, i.e., for every positive integer N, for every sequence $(\omega_j)_{j=1}^N$ from Ω and every sequence $(x_j)_{j=1}^N$ of vectors that belong to $\mathcal{E}$, the matrix

$$[(K(\omega_j, \omega_k) x_k, x_j)_{\mathcal{E}}]_{j,k=1}^N$$

is nonnegative hermitian. Conversely, an operator-valued version of a result due to Aronszajn [10] says that if $K : \Omega \times \Omega \to \mathfrak{C}$ is an arbitrary mapping which is nonnegative definite, then K is the reproducing kernel for some reproducing kernel Hilbert space of $\mathcal{E}$-valued functions defined on Ω. The space $H_2^+[\mathcal{E}]$ with the reproducing kernel $K : \mathbb{D} \times \mathbb{D} \to \mathfrak{C}$ given by $K(\zeta, \eta) := \frac{1}{1-\zeta\overline{\eta}}$ is an example for a reproducing kernel Hilbert space. The orthogonal complement $\mathcal{B}$ of a shift-invariant subspace $\mathcal{M}$ of $H_+^2[\mathcal{E}]$ is a further example. Indeed, according to the Beurling-Lax-Halmos Theorem, any such invariant subspace $\mathcal{M}$ admits the representation $\mathcal{M} = \Theta H_+^2[\mathcal{F}]$ with some inner function $\mathcal{S}[\mathcal{F}, \mathcal{E}]$. Then $\mathcal{B} = H_+^2[\mathcal{E}] \ominus \Theta H_+^2[\mathcal{F}]$ and $K_\Theta : \mathbb{D} \times \mathbb{D} \to \mathcal{E}$ defined by

$$K_\Theta(\zeta, \eta) := \frac{1}{1 - \zeta\overline{\eta}} \left(I_\mathcal{E} - \Theta(\zeta)\Theta(\eta)^*\right)^{-1}$$

is the reproducing kernel of the reproducing kernel Hilbert space $\mathcal{B}$.

Note that a detailed discussion of models of contractive operators in reproducing kernel Hilbert spaces is given in [6].

3.2. Our next goal is to show that the space $\mathcal{M}$ described in Theorem 2.2 is a reproducing kernel Hilbert space.

Theorem 3.1. Let $T : \mathcal{H} \to \mathcal{H}$ be a contractive operator which has a generating unilateral shift V, and let $\mathcal{E}$ be the generating subspace for the unilateral shift V. Let $\Xi : \mathbb{D} \to [\mathcal{H}, \mathcal{E}]$ be given by (2.3), and let $Y : \mathcal{H} \to h^2[\mathcal{E}]$ be the operator which is defined by $Yh := f_h$ where $f_h(\zeta) := \Xi(\zeta)h$ for each $\zeta \in \mathbb{D}$. Then $\mathcal{M} := \mathcal{R}(Y)$ is a reproducing kernel Hilbert space with reproducing kernel $K : \mathbb{D} \times \mathbb{D} \to [\mathcal{E}, \mathcal{E}]$ defined by

$$K(\zeta, \eta) := \Xi(\zeta)(\Xi(\eta))^*. \tag{3.1}$$

Moreover, the kernel K admits the representation

$$K(\zeta, \eta) = \mathrm{Rstr.}_\mathcal{E} P_\mathcal{E}(I - \zeta T)^{-1}(I - \overline{\eta}T^*)^{-1} P_\mathcal{E} + \frac{1}{1 - \overline{\zeta}\eta} I_\mathcal{E} - I_\mathcal{E} \tag{3.3}$$

for every choice of ζ and η in $\mathbb{D}$.

Proof. First we verify that K given by (3.1) can be represented by (3.3). Indeed, since $\mathcal{E}$ is generating subspace for the unilateral shift $V : \mathcal{L} \to \mathcal{L}$ we have

$$\mathrm{Rstr.}_\mathcal{E} P_\mathcal{E}(T^*)^j T^k P_\mathcal{E} = \begin{cases} I_\mathcal{E} & , \quad j = k \\ 0 & , \quad j \neq k \end{cases} \tag{3.4}$$

for every choice of nonnegative integers j and k. Hence we get

(3.5)

$$\begin{aligned}
&\mathrm{Rstr.}_{\mathcal{E}} P_{\mathcal{E}}(I_{\mathcal{H}} - \overline{\zeta} T^*)^{-1}(I_{\mathcal{H}} - \eta T)^{-1} P_{\mathcal{E}} \\
&\qquad = \mathrm{Rstr.}_{\mathcal{E}} \left(\sum_{j=0}^{\infty} \sum_{k=0}^{\infty} \overline{\zeta}^j \eta^k P_{\mathcal{E}} (T^*)^j T^k P_{\mathcal{E}} \right) = \left(\sum_{j=0}^{\infty} \overline{\zeta}^j \eta^j \right) I_{\mathcal{E}} = \frac{1}{1 - \overline{\zeta}\eta} I_{\mathcal{E}}, \\
&\mathrm{Rstr.}_{\mathcal{E}} P_{\mathcal{E}}(I_{\mathcal{H}} - \overline{\zeta} T^*)^{-1}(I_{\mathcal{H}} - \overline{\eta} T^*)^{-1} P_{\mathcal{E}} \\
&\qquad = \mathrm{Rstr.}_{\mathcal{E}} \left(\sum_{j=0}^{\infty} \sum_{k=0}^{\infty} \overline{\zeta}^j \overline{\eta}^k P_{\mathcal{E}} (T^*)^{j+k} P_{\mathcal{E}} \right) = I_{\mathcal{E}}, \\
&\mathrm{Rstr.}_{\mathcal{E}} P_{\mathcal{E}}(I_{\mathcal{H}} - \overline{\zeta} T^*)^{-1} P_{\mathcal{E}} = \mathrm{Rstr.}_{\mathcal{E}} \left(\sum_{j=0}^{\infty} \overline{\zeta}^j P_{\mathcal{E}} (T^*)^j P_{\mathcal{E}} \right) = I_{\mathcal{E}}, \\
&\mathrm{Rstr.}_{\mathcal{E}} P_{\mathcal{E}}(I_{\mathcal{H}} - \zeta T)^{-1}(I_{\mathcal{H}} - \eta T)^{-1} P_{\mathcal{E}} \\
&\qquad = \mathrm{Rstr.}_{\mathcal{E}} \left(\sum_{j=0}^{\infty} \sum_{k=0}^{\infty} \zeta^j \eta^k P_{\mathcal{E}} T^{j+k} T^k P_{\mathcal{E}} \right) = I_{\mathcal{E}}, \\
&\mathrm{Rstr.}_{\mathcal{E}} P_{\mathcal{E}}(I_{\mathcal{H}} - \zeta T)^{-1} P_{\mathcal{E}} = \mathrm{Rstr.}_{\mathcal{E}} \left(\sum_{j=0}^{\infty} \zeta^j P_{\mathcal{E}} T^j P_{\mathcal{E}} \right) = I_{\mathcal{E}},
\end{aligned}$$

and consequently

$$\begin{aligned}
K(\zeta, \eta) &= \Xi(\zeta)(\Xi(\eta))^* \\
&= \mathrm{Rstr.}_{\mathcal{E}} P_{\mathcal{E}}((I_{\mathcal{H}} - \overline{\zeta} T^*)^{-1}(I_{\mathcal{H}} - \zeta T)^{-1} - I_{\mathcal{H}})((I_{\mathcal{H}} - \eta T)^{-1} \\
&\quad + (I_{\mathcal{H}} - \overline{\eta} T^*)^{-1} - I_{\mathcal{H}}) P_{\mathcal{E}} \\
&= \mathrm{Rstr.}_{\mathcal{E}} P_{\mathcal{E}}(I_{\mathcal{H}} - \zeta T)^{-1}(I_{\mathcal{H}} - \overline{\eta} T^*)^{-1} P_{\mathcal{E}} + \frac{1}{1 - \overline{\zeta}\eta} I_{\mathcal{E}} - I_{\mathcal{E}}.
\end{aligned}$$

for each $\zeta \in \mathbb{D}$ and each $\eta \in \mathbb{D}$. In order to prove that K is a reproducing kernel for $\mathcal{M}$ it remains to show that

$$(f(\eta), x)_{\mathcal{E}} = (f, K_{\cdot\eta} x)_{\mathcal{M}}$$

for every choice of f in $\mathcal{M}$, x in $\mathcal{E}$ and η in $\mathbb{D}$. Obviously, for each $f \in \mathcal{M}$, there is an $h \in \mathcal{H}$ sucht hat $f = Yh$. Hence

$$f(\eta) = (Yh)(\eta) = \Xi(\eta) h$$

is valid for all $\eta \in \mathbb{D}$. Using part (b) of Theorem 2.2 we obtain finally

$$(f(\eta), x)_{\mathcal{E}} = (\Xi(\eta)h, x)_{\mathcal{E}} = (h, \Xi(\eta)^* x)_{\mathcal{H}} = (Yh, Y(\Xi(\eta))^* x)_{\mathcal{M}} = (f, K_{\cdot\eta} x)_{\mathcal{M}}$$

for each $x \in \mathcal{E}$ and each $\eta \in \mathbb{D}$. □

Now we consider the situation given in Theorem 2.2. In particular, let $T : \mathcal{H} \to \mathcal{H}$ be a contractive operator which has a generating unilateral shift $V : \mathcal{L} \to \mathcal{L}$. We assume that T is the basic operator of the unitary colligation

$$\triangle = (\mathcal{H}, \mathcal{F}, \mathcal{G}; T, F, G, S).$$

To express the adjoint operator R^* of the operator R given in Theorem 2.2 we consider the mappings $\varphi_0 : \mathbb{D} \to [\mathcal{F}, \mathcal{E}]$ defined by

$$\varphi_0(\zeta) := P_{\mathcal{E}}(I_{\mathcal{H}} - \zeta T)^{-1} F \tag{3.6}$$

and $B : \mathcal{F} \to \mathcal{M}$ defined by $Bf := g_f$ where $g_f : \mathbb{D} \to \mathcal{E}$ is given by

$$g_f(\zeta) := \varphi_0(\zeta) f.$$

The special structure of the operator T and the definition of the unitary colligation yield $F^* T^n P_{\mathcal{E}} = 0$. Since $P_{\mathcal{E}}(T^*)^n F = 0$ holds true for each positive integer n, we see from the particular structure (1.4) of unitary dilation of T that

$$(YFf)(\zeta) = \Xi(\zeta) Ff = \varphi_0(\zeta) f = (Bf)(\zeta) \tag{3.7}$$

for all $\zeta \in \mathbb{D}$ and all $f \in \mathcal{F}$. Hence $YF = B$. This implies $\mathcal{R}(B) \subseteq \mathcal{R}(Y)$ and $B^* = F^* Y^*$.

Theorem 3.2. Let R be the operator given in Theorem 2.2. Then the adjoint operator R^* admits for all $h \in \mathcal{H}$ and all $\zeta \in \mathbb{D}$ the representation

$$(R^* h)(\zeta) = (S_{\mathcal{E}} h)(\zeta) - \zeta (BB^* h)(\zeta).$$

Proof. According to part (b) of Theorem 2.2, the operator $Y : \mathcal{H} \to \mathcal{M}$ is unitary and (2.7) holds true. This we have $R^* Y = YT^*$. Let $h \in \mathcal{H}$. Then we infer

$$\begin{aligned}
(R^* Yh)(\zeta) &= (YT^* h)(\zeta) = \Xi(\zeta) T^* h \\
&= P_{\mathcal{E}}(I_{\mathcal{H}} - \overline{\zeta} T^*)^{-1} T^* + (I_{\mathcal{H}} - \zeta T)^{-1} T^* - T^*) h \\
&= P_{\mathcal{E}} \left(\frac{1}{\overline{\zeta}} \left((I_{\mathcal{H}} - \overline{\zeta} T^*)^{-1} - I_{\mathcal{H}} \right) + \sum_{k=1}^{\infty} \zeta^k T^{k-1} T T^* \right) h
\end{aligned}$$

for all $\zeta \in \mathbb{D} \setminus \{0\}$. Since the operator U defined in (1.2) is unitary we have

$$\begin{aligned}
\sum_{k=1}^{\infty} \zeta^k T^{k-1} T T^* &= \zeta \sum_{k=1}^{\infty} \zeta^{k-1} (I_{\mathcal{H}} - FF^*) \\
&= \zeta (I_{\mathcal{H}} - \zeta T)^{-1} - \zeta (I_{\mathcal{H}} - \zeta T)^{-1} FF^*,
\end{aligned}$$

$$P_{\mathcal{E}} \left(\frac{1}{\overline{\zeta}} \left((I_{\mathcal{H}} - \overline{\zeta} T^*)^{-1} - I_{\mathcal{H}} \right) + \zeta (I_{\mathcal{H}} - \zeta T)^{-1} \right) h = S_{\mathcal{E}}(Yh)(\zeta),$$

$$P_{\mathcal{E}}(I_{\mathcal{H}} - \zeta T)^{-1} FF^* h = (BF^* h)(\zeta) = (BF^* Y^* Yh)(\zeta) = (BB^* Yh)(\zeta),$$

and consequently

$$(R^* Yh)(\zeta) = (S_{\mathcal{E}}(Yh))(\zeta) + \zeta (BB^*(Yh))(\zeta)$$

for all $\zeta \in \mathbb{D} \setminus \{0\}$. A continuity argument completes the proof. $\square$

Theorem 3.3. The kernel K given by (3.1) admits the representation

$$K(\zeta,\eta) = \frac{1}{1-\zeta\overline{\eta}} I_{\mathcal{E}} + \frac{\zeta\overline{\eta}}{1-\zeta\overline{\eta}} \left(I_{\mathcal{E}} - \varphi_0(\zeta)\varphi_0(\eta)^*\right) \tag{3.8}$$

for every choice of ζ and η in $\mathbb{D}$ where $\varphi_0 : \mathbb{D} \to [\mathcal{F},\mathcal{E}]$ is given by (3.6).

Proof. For each $\zeta \in \mathbb{D}$ and each $\eta \in \mathbb{D}$ we have

$$\begin{aligned}\varphi_0(\zeta)(\varphi_0(\eta))^* &= \mathrm{Rstr.}_{\mathcal{E}} P_{\mathcal{E}}(I_{\mathcal{H}} - \zeta T)^{-1} F F^* (I_{\mathcal{H}} - \overline{\eta}T^*)^{-1} P_{\mathcal{E}} \\ &= \mathrm{Rstr.}_{\mathcal{E}} P_{\mathcal{E}}(I_{\mathcal{H}} - \zeta T)^{-1} (I_{\mathcal{H}} - TT^*) (I_{\mathcal{H}} - \overline{\eta}T^*)^{-1} P_{\mathcal{E}} \\ &= \mathrm{Rstr.}_{\mathcal{E}} P_{\mathcal{E}}(I_{\mathcal{H}} - \zeta T)^{-1} (I_{\mathcal{H}} - \overline{\eta}T^*)^{-1} P_{\mathcal{E}} \\ &\qquad - \mathrm{Rstr.}_{\mathcal{E}} P_{\mathcal{E}}(I_{\mathcal{H}} - \zeta T)^{-1} TT^* (I_{\mathcal{H}} - \overline{\eta}T^*)^{-1} P_{\mathcal{E}}\end{aligned}$$

and, in view of (3.5)

$$\begin{aligned}&\zeta\overline{\eta}\, \mathrm{Rstr.}_{\mathcal{E}} P_{\mathcal{E}}(I_{\mathcal{H}} - \zeta T)^{-1} TT^* (I_{\mathcal{H}} - \overline{\eta}T^*)^{-1} P_{\mathcal{E}} = \\ &\qquad = \mathrm{Rstr.}_{\mathcal{E}} P_{\mathcal{E}} \left((I_{\mathcal{H}} - \zeta T)^{-1} - I_{\mathcal{H}}\right)\left((I_{\mathcal{H}} - \overline{\eta}T^*)^{-1} - I_{\mathcal{H}}\right) P_{\mathcal{E}} \\ &\qquad = \mathrm{Rstr.}_{\mathcal{E}} P_{\mathcal{E}}(I_{\mathcal{H}} - \zeta T)^{-1} (I_{\mathcal{H}} - \overline{\eta}T^*)^{-1} P_{\mathcal{E}} - I_{\mathcal{E}}\,.\end{aligned}$$

Then we see that the identity

$$\varphi_0(\zeta)\varphi_0(\eta)^* = \left(1 - \frac{1}{\zeta\overline{\eta}}\right) \mathrm{Rstr.}_{\mathcal{E}} P_{\mathcal{E}}(I_{\mathcal{H}} - \zeta T)^{-1} (I_{\mathcal{H}} - \overline{\eta}T^*)^{-1} P_{\mathcal{E}} + \frac{1}{\zeta\overline{\eta}} I_{\mathcal{E}}$$

holds true for each $\zeta \in \mathbb{D} \setminus \{0\}$ and each $\eta \in \mathbb{D} \setminus \{0\}$. Thus we can conclude that

$$\begin{aligned}&\frac{1}{1-\zeta\overline{\eta}} I_{\mathcal{E}} + \frac{\zeta\overline{\eta}}{1-\zeta\overline{\eta}} (I_{\mathcal{E}} - \varphi_0(\zeta)\varphi_0(\zeta)^*) = \\ &\qquad = \frac{1}{1-\zeta\overline{\eta}} I_{\mathcal{E}} - I_{\mathcal{E}} + \mathrm{Rstr.}_{\mathcal{E}} P_{\mathcal{E}}(I_{\mathcal{H}} - \zeta T)^{-1} (I_{\mathcal{H}} - \overline{\eta}T^*)^{-1} P_{\mathcal{E}} = K(\zeta,\eta),\end{aligned}$$

i.e., that (3.8) is valid for each $\zeta \in \mathbb{D} \setminus \{0\}$ and each $\eta \in \mathbb{D} \setminus \{0\}$. By virtue of $\Xi(0) = P_{\mathcal{E}}$ and (3.4) we see that (3.8) holds as well if $\zeta = 0$ or if $\eta = 0$. □

In view of the representation (3.8), we will denote the space $\mathcal{M}$, the mappings $Y : \mathcal{H} \to \mathcal{M}$ and $R : \mathcal{M} \to \mathcal{M}$ from Theorem 2.2 and the reproducing kernel K from Theorem 3.1 by $\mathcal{M}_{\varphi_0}, Y_{\varphi_0}, R_{\varphi_0}$ and K_{φ_0}, respectively.

Remark 1. Let $T : \mathcal{H} \to \mathcal{H}$ be a completely nonunitary contraction for which the maximal unilateral shift V_T contained in it is generating. In this case, the generating subspace $\mathcal{H}_0$ of V_T plays the role of the space $\mathcal{E}$ in Theorem 2.2. From (1.5) and (3.6) one can conclude then $\varphi_0 = \varphi$. Hence in this situation the model operator R_φ is completely determined by the defect function φ.

Remark 2. Analogously, the case that the contraction $T : \mathcal{H} \to \mathcal{H}$ has the completeness property with respect to the coshift $\tilde{V} : \tilde{\mathcal{L}} \to \tilde{\mathcal{L}}$ contained in T, where $\tilde{\mathcal{L}} \subseteq \mathcal{H}$. If T is completely nonunitary and $\tilde{V}$ coincides with the maximal coshift V_T^*, then the function $\check{\psi}$ plays the role of φ_0, where ψ is the left defect function.

4. A model of a contractive operator which contains a unilateral shift and a unilateral coshift which generate jointly the whole space

First we are going to generalize the notion of a generating shift.

Definition. Let $T : \mathcal{H} \to \mathcal{H}$ be a contractive operator which contains as well a unilateral shift $V : \mathcal{L} \to \mathcal{L}$ as a unilateral coshift $\tilde{V} : \tilde{\mathcal{L}} \to \tilde{\mathcal{L}}$. If

$$\mathcal{H} = \left(\bigvee_{n=0}^{\infty} (T^*)^n \mathcal{L}\right) \vee \left(\bigvee_{n=0}^{\infty} T^n \tilde{\mathcal{L}}\right), \tag{4.1}$$

then one says that V and $\tilde{V}$ generate jointly the space $\mathcal{H}$.

Theorem 4.1. Let $T : \mathcal{H} \to \mathcal{H}$ be a contractive operator which contains as well a unilateral shift $V : \mathcal{L} \to \mathcal{L}$ as a unilateral coshift $\tilde{V} : \tilde{\mathcal{L}} \to \tilde{\mathcal{L}}$. Suppose that V and $\tilde{V}$ generate jointly the space $\mathcal{H}$. Let $\triangle = (\mathcal{H}, \mathcal{F}, \mathcal{G}; T, F, G, S)$ be a unitary colligation the basic operator of which is T. Further, let $\mathcal{E}$ and $\tilde{\mathcal{E}}$ be the generating subspaces of V and $\tilde{V}$, respectively. For $\zeta \in \mathbb{D}$, let $\Xi(\zeta)$ and $\tilde{\Xi}(\zeta)$ be given by (2.3) and by

$$\tilde{\Xi}(\zeta) := P_{\tilde{\mathcal{E}}}\left((I_{\mathcal{H}} - \overline{\zeta} T)^{-1} + (I_{\mathcal{H}} - \zeta T^*)^{-1} - I_{\mathcal{H}}\right), \tag{4.2}$$

respectively. Then:

(a) For each $h \in \mathcal{H}$, the function $f_h^{\square} : \mathbb{D} \to \mathcal{E} \oplus \tilde{\mathcal{E}}$ defined by

$$f_h^{\square}(\zeta) := \begin{pmatrix} f_h(\zeta) \\ \tilde{f}_h(\zeta) \end{pmatrix} \tag{4.3}$$

and

$$f_h(\zeta) := \Xi(\zeta) h \quad \text{and} \quad \tilde{f}_h(\zeta) := \tilde{\Xi}(\zeta) h \tag{4.4}$$

belongs to $h^2[\mathcal{E}] \oplus h^2[\tilde{\mathcal{E}}]$. Then $Z^{\square} : \mathcal{H} \to h^2[\mathcal{E}] \oplus h^2[\tilde{\mathcal{E}}]$ given by

$$Z^{\square} h := f_h^{\square} \tag{4.5}$$

satisfies $\mathcal{N}(Z^{\square}) = \{0\}$. Let $\mathcal{M} := \mathcal{R}(Z^{\square})$. Then by

$$\|Z^{\square} h\|_{\mathcal{M}} := \|h\|_{\mathcal{H}} \tag{4.6}$$

a well-defined norm $\|\cdot\|_{\mathcal{M}}$ on $\mathcal{M}$ is given, and (with respect to this norm) the operator $Z : \mathcal{H} \to \mathcal{M}$ defined by $Zh := Z^{\square} h$ is unitary.

(b) For each $f \in h^2[\mathcal{E}]$ and each $\tilde{f} \in h^2[\tilde{\mathcal{E}}]$, the function $S^{\#}_{f,\tilde{f}} : \mathbb{D} \to \tilde{\mathcal{E}}$ defined by

$$S^{\#}_{f,\tilde{f}}(\zeta) := (S_{\tilde{\mathcal{E}}} \tilde{f})(\zeta) - \zeta (P_{\tilde{\mathcal{E}}} (I_{\mathcal{H}} - \zeta T^*)^{-1}) G^* G Z^* \begin{pmatrix} f \\ \tilde{f} \end{pmatrix} \tag{4.7}$$

belongs to $h^2[\tilde{\mathcal{E}}]$, and the operator $W : h^2[\mathcal{E}] \oplus h^2[\tilde{\mathcal{E}}] \to h^2[\mathcal{E}] \oplus h^2[\tilde{\mathcal{E}}]$ given by

$$W\left(\begin{pmatrix} f \\ \tilde{f} \end{pmatrix}\right) := \begin{pmatrix} S^*_{\mathcal{E}} f \\ S^{\#}_{f,\tilde{f}} \end{pmatrix} \tag{4.8}$$

is well defined and satisfies

$$WZ = ZT. \tag{4.9}$$

In particular, the subspace $\mathcal{M}$ is invariant with respect to W. The adjoint operator R^* of the operator $R := \text{Rstr.}_{\mathcal{M}} W$ is exactly the operator $R^\square : \mathcal{M} \to \mathcal{M}$ defined by

$$R^\square \left(\begin{pmatrix} f \\ \tilde{f} \end{pmatrix} \right) = \begin{pmatrix} R^\#_{f,\tilde{f}} \\ S^*_{\tilde{\mathcal{E}}} \tilde{f} \end{pmatrix} \tag{4.10}$$

where $R^\#_{f,\tilde{f}} : \mathbb{D} \to \mathcal{E}$ is given by

$$R^\#_{f,\tilde{f}}(\zeta) := (S_{\mathcal{E}} f)(\zeta) - \zeta P_{\mathcal{E}} (I - \zeta T)^{-1} F F^* Z^* \begin{pmatrix} f \\ \tilde{f} \end{pmatrix}. \tag{4.11}$$

Moreover, $RZ = ZT$.

Proof. Let $h \in \mathcal{H}$. The proof of Theorem 2.2 shows that f_h belongs to $h^2[\mathcal{E}]$. Analogously, $\tilde{f}_h \in h^2[\tilde{\mathcal{E}}]$ can be verified. Hence $f_h^\square$ belongs to $h^2[\mathcal{E}] \oplus h^2[\tilde{\mathcal{E}}]$. Now assume that h belongs to $\mathcal{N}(Z^\square)$. Then $h \in \mathcal{N}(Y) \cap \mathcal{N}(\tilde{Y})$ where $Y : \mathcal{H} \to h^2[\mathcal{E}]$ and $\tilde{Y} : \mathcal{H} \to h^2[\tilde{\mathcal{E}}]$ are given by $Yh := f_h$ and $\tilde{Y}h := \tilde{f}_h$, respectively. In the proof of Theorem 2.2 we already checked that $Yh = 0$ implies that h is orthogonal to $\bigvee_{n=0}^{\infty} (T^*)^n \mathcal{L}$. Analogously, it follows from $\tilde{Y}h = 0$ that h is orthogonal to $\bigvee_{n=0}^{\infty} T^n \tilde{\mathcal{L}}$. Thus we get from (4.1) that $h = 0$. Thus $\mathcal{N}(Z^\square) = \{0\}$, the formula (4.6) defines a norm $\|\cdot\|_{\mathcal{M}}$ on $\mathcal{M}$, and the operator Z is unitary. In order to verify (4.9) it is sufficient to check that

$$S^*_{\mathcal{E}} Y = YT \tag{4.12}$$

and

$$S^\#_{Yh,\tilde{Y}h}(\zeta) = (\tilde{Y}Th)(\zeta) \tag{4.13}$$

hold for every choice of h in $\mathcal{H}$ and ζ in $\mathbb{D}$. The identity (4.12) can be obtained as the identity (2.6) in the proof of Theorem 2.2. Since

$$\begin{aligned}
\tilde{Y}Th(\zeta) &= \tilde{\Xi}(\zeta)Th = P_{\tilde{\mathcal{E}}} \left((I_{\mathcal{H}} - \overline{\zeta}T)^{-1} + (I_{\mathcal{H}} - \zeta T^*)^{-1} - I_{\mathcal{H}} \right) Th \\
&= P_{\mathcal{E}} \left(\sum_{n=0}^{\infty} \overline{\zeta}^n T^{n+1} + \sum_{n=1}^{\infty} \zeta^n (T^*)^n T \right) h \\
&= P_{\mathcal{E}} \left(\sum_{n=0}^{\infty} \overline{\zeta}^n T^{n+1} + \sum_{n=1}^{\infty} \zeta^n (T^*)^{n-1} (I - G^*G) \right) h \\
&= P_{\mathcal{E}} \left(\sum_{n=0}^{\infty} \overline{\zeta}^n T^{n+1} h + \sum_{n=1}^{\infty} \zeta^n (T^*)^{n-1} h \right) - \zeta \sum_{n=0}^{\infty} \zeta^n (T^*)^n G^*Gh \\
&= (S_{\tilde{\mathcal{E}}} \tilde{Y}h)(\zeta) - \zeta P_{\mathcal{E}} (I - \zeta T^*)^{-1} G^*GZ^*Zh = S^\#_{Yh,\tilde{Y}h}(\zeta)
\end{aligned}$$

is valid for each $h \in \mathcal{H}$ and for each $\zeta \in \mathbb{D}$, the equation (4.13) is true as well. Hence (4.9) and $RZ = ZT$ are fulfilled. Since Z is unitary we infer

$$R^* Z = ZT^*. \tag{4.14}$$

Finally, similarly to the proof of (4.9), one can show that the operator $R^\square$ satisfies $R^\square Z = ZT^*$. We omit the details. $\square$

In view of the representations of the operators R and R^* stated in Theorem 4.1, we obtain representations of the operators $Q := F^* Z^*$ and $\tilde{Q} := GZ^*$. Let us continue to use the notations given above. Observe that $P_{\mathcal{E}}(T^*)^n F = 0$ and, analogously, $P_{\tilde{\mathcal{E}}} T^n F = 0$ for every positive integer n. For each $f \in \mathcal{F}$ and each $\zeta \in \mathbb{D}$, then we see

$$(YFf)(\zeta) = \Xi(\zeta) Ff = P_{\mathcal{E}}(I - \zeta T)^{-1} Ff,$$

in view of $P_{\tilde{\mathcal{E}}} F = 0$,

$$\begin{aligned}(\tilde{Y}Ff)(\zeta) &= \tilde{\Xi}(\zeta) f = P_{\tilde{\mathcal{E}}}(I - \zeta T^*)^{-1} Ff = \sum_{n=0}^{\infty} \zeta^n P_{\tilde{\mathcal{E}}}(T^*)^n Ff \\ &= \sum_{n=1}^{\infty} \zeta^n P_{\tilde{\mathcal{E}}}(T^*)^{n-1} T^* Ff = -\zeta P_{\tilde{\mathcal{E}}}(I - \zeta T^*)^{-1} G^* Sf,\end{aligned}$$

and consequently

$$(Q^* f)(\zeta) = (ZFf)(\zeta) = \begin{pmatrix} (YFf)(\zeta) \\ (\tilde{Y}Ff(\zeta) \end{pmatrix} = \begin{pmatrix} P_{\mathcal{E}}(I_{\mathcal{H}} - \zeta T)^{-1} Ff \\ -\zeta P_{\tilde{\mathcal{E}}}(I_{\mathcal{H}} - \zeta T^*)^{-1} G^* Sf \end{pmatrix}.$$

Analogously, we obtain

$$(\tilde{Q}^* g)(\zeta) = (ZG^* g)(\zeta) = \begin{pmatrix} -\zeta P_{\mathcal{E}}(I_{\mathcal{H}} - \zeta T)^{-1} FS^* g \\ P_{\tilde{\mathcal{E}}}(I_{\mathcal{H}} - \zeta T^*)^{-1} G^* g \end{pmatrix}$$

for each $\zeta \in \mathbb{D}$ and each $g \in \mathcal{G}$. Hence, if T is a completely nonunitary contraction, if the shift V and the coshift $\tilde{V}$ are maximal, i.e., $V = V_T$ and $\tilde{V}^* = V_{T^*}$, then, in view of (1.5) and (1.6), Q and $\tilde{Q}$ satisfy

$$(Q^* f)(\zeta) = \begin{pmatrix} \varphi(\zeta) f \\ -\zeta \check{\psi}(\zeta) Sf \end{pmatrix}, \quad (\tilde{Q}^* g)(\zeta) = \begin{pmatrix} \zeta \varphi(\zeta) S^* g \\ \check{\psi}(\zeta) g \end{pmatrix},$$

for each $f \in \mathcal{F}$ and all $g \in \mathcal{G}$, respectively, and therefore $Q\left(\begin{pmatrix} u \\ \tilde{u} \end{pmatrix}\right) = (\varphi(\zeta)^* u,$ $-\overline{\zeta} S^* \psi(\overline{\zeta}) u)$ and $\tilde{Q} = (-\zeta S \check{\varphi}(\zeta) u, \psi(\overline{\zeta}) u)$ for each $\begin{pmatrix} u \\ \tilde{u} \end{pmatrix} \in h^2[\mathcal{E}] \oplus h^2[\tilde{\mathcal{E}}]$. Taking into account part (b) of Theorem 4.1, the following theorem is proved.

Theorem 4.2 Let $T : \mathcal{H} \to \mathcal{H}$ be a completely nonunitary contraction for which the maximal unilateral shift V_T and the maximal unilateral coshift V_{T^*} generate jointly the space $\mathcal{H}$. Let $\triangle = (\mathcal{H}, \mathcal{F}, \mathcal{G}; T, F, G, S)$ be a unitary colligation with basic operator T. Moreover, let $\mathcal{H}_0$ and $\tilde{\mathcal{H}}_0$ be the generating subspaces for V_T

and V_{T^*}, respectively. Then the mappings $Y : \mathcal{H} \to h^2[\mathcal{H}_0], \tilde{Y} : \mathcal{H} \to h^2[\tilde{\mathcal{H}}_0]$ and $Z^\square : \mathcal{H} \to h^2[\mathcal{H}_0] \oplus h^2[\tilde{\mathcal{H}}_0]$, given by

$$(Yh)(\zeta) := P_{\mathcal{H}_0}((I - \overline{\zeta}T^*)^{-1} + (I - \zeta T^*)^{-1} - I_{\mathcal{H}})h,$$
$$(\tilde{Y}h)(\zeta) := \tilde{P}_{\mathcal{H}_0}((I - \overline{\zeta}T^*)^{-1} + (I - \zeta T^*)^{-1} - I_{\mathcal{H}})h$$

and

$$(Z^\square h)(\zeta) := \begin{pmatrix} (Yh)(\zeta) \\ (\tilde{Y}h)(\zeta) \end{pmatrix}$$

for every choice of $h \in \mathcal{H}$ and $\zeta \in \mathbb{D}$, are well defined, and $\mathcal{N}(Z^\square) = \{0\}$. If $\mathcal{M} := \mathcal{R}(Z^\square)$, then the formula

$$\|Z^\square h\|_{\mathcal{M}} := \|h\|_{\mathcal{H}}$$

defines a norm $\|\cdot\|_{\mathcal{M}}$ on $\mathcal{M}$, and by

$$\left(R\begin{pmatrix} f \\ \tilde{f} \end{pmatrix}\right)(\zeta) := \begin{pmatrix} (S^*_{\mathcal{H}_0} f)(\zeta) \\ (S_{\tilde{\mathcal{H}}_0}\tilde{f})(\zeta) - \zeta\check{\psi}(\zeta)(-\overline{\zeta}S\varphi(\zeta)^* f + \psi(\overline{\zeta})\tilde{f}) \end{pmatrix}, \zeta \in \mathbb{D},$$

an operator $R : \mathcal{M} \to \mathcal{M}$ is given, the adjoint operator R^* of which admits the representation

$$\left(R^*\begin{pmatrix} f \\ \tilde{f} \end{pmatrix}\right)(\zeta) := \begin{pmatrix} (S_{\mathcal{H}_0} f)(\zeta) - \zeta\varphi(\zeta)(\varphi(\zeta)^* f(\zeta) - \overline{\zeta}S^*\psi(\overline{\zeta})\tilde{f})) \\ (S^*_{\tilde{\mathcal{H}}_0}\tilde{f})(\zeta) \end{pmatrix}$$

for each $\begin{pmatrix} f \\ \tilde{f} \end{pmatrix} \in \mathcal{M}$ and each $\zeta \in \mathbb{D}$, where φ and ψ are the right defect function and the left defect function associated with the characteristic function Θ of $\triangle$. Moreover, $RZ = ZT$.

Similarly to Theorem 3.1 one can show the following result.

Theorem 4.3. The model space $\mathcal{M}$ in Theorem 4.1 is a reproducing kernel Hilbert space with reproducing kernel $K : \mathbb{D} \times \mathbb{D} \to [\mathcal{E} \oplus \tilde{\mathcal{E}}, \mathcal{E} \oplus \tilde{\mathcal{E}}]$ defined by

$$K(\zeta, \eta) := Z(\zeta)Z(\eta)^* \tag{4.15}$$

where $Z : \mathbb{D} \to [\mathcal{H}, \mathcal{E} \oplus \tilde{\mathcal{E}}]$ is given by

$$Z(\zeta) := \begin{pmatrix} \Xi(\zeta) \\ \tilde{\Xi}(\zeta) \end{pmatrix} \tag{4.16}$$

and where $\Xi : \mathbb{D} \to [\mathcal{H}, \mathcal{E}]$ and $\tilde{\Xi} : \mathbb{D} \to [\mathcal{H}, \tilde{\mathcal{E}}]$ are given by (2.3) and (4.2) respectively.

At the end of this paper, we are going to study the reproducing kernel in the particular case when $V = V_T$ and $\tilde{V}^* = V_{T^*}$ hold true. Note that one can consider more general situations in this way.

From (4.15) and (4.16) we get immediately $K(\zeta, \eta) = (K_{jk}(\zeta, \eta))_{j,k=1}^2$ where $K_{11}(\zeta, \eta) = \Xi(\zeta)\Xi(\eta)^*, K_{22}(\zeta, \eta) = \tilde{\Xi}(\zeta)\tilde{\Xi}(\eta)^*$ and $K_{12}(\zeta, \eta) = K_{21}(\eta, \zeta))^* =$

$\Xi(\zeta)\tilde{\Xi}(\eta)^*$ for every choice of ζ and η in $\mathbb{D}$. In view of $\mathcal{E} = \mathcal{H}_0, \tilde{\mathcal{E}} = \tilde{\mathcal{H}}_0$ and Theorem 3.3, we have

$$K_{11}(\zeta,\eta) = K_\varphi(\zeta,\eta) = \frac{1}{1-\zeta\overline{\eta}} I_{\mathcal{H}_0} + \frac{\zeta\overline{\eta}}{1-\zeta\overline{\eta}}(I_{\mathcal{H}_0} - \varphi(\zeta)\varphi(\eta)^*) \tag{4.17}$$

for each $\zeta \in \mathbb{D}$ and each $\eta \in \mathbb{D}$, where φ is the right defect function, i.e.,

$$\varphi(\zeta) = P_{\mathcal{H}_0}(I - \zeta T)^{-1} F.$$

Analogously, we obtain

$$K_{22}(\zeta,\eta) = K_{\check{\psi}}(\zeta,\eta) = \frac{1}{1-\zeta\overline{\eta}} I_{\tilde{\mathcal{H}}_0} + \frac{\zeta\overline{\eta}}{1-\zeta\overline{\eta}}(I_{\tilde{\mathcal{H}}_0} - \psi(\overline{\zeta})^*\psi(\overline{\eta})) \tag{4.18}$$

for all ζ and η that belong to $\mathbb{D}$, where ψ is the left defect function, i.e.,

$$\psi(\zeta) = G(I - \zeta T)^{-1} P_{\tilde{\mathcal{H}}_0}.$$

Further, we can conclude

$$\begin{aligned} K_{21}(\zeta,\eta) &= \tilde{\Xi}(\zeta)\Xi(\eta)^* \\ &= \mathrm{Rstr.}_{\mathcal{H}_0} P_{\tilde{\mathcal{H}}_0} \left((I_{\mathcal{H}} - \overline{\zeta} T)^{-1} + (I - \zeta T^*)^{-1} - I_{\mathcal{H}}\right) \\ &\quad \cdot \left((I_{\mathcal{H}} - \eta T)^{-1} + (I_{\mathcal{H}} - \overline{\eta} T^*)^{-1} - I_{\mathcal{H}}\right) P_{\mathcal{H}_0} \\ &= K_{21}^{[1]}(\zeta,\eta) + K_{21}^{[2]}(\zeta,\eta) + K_{21}^{[3]}(\zeta,\eta) \end{aligned}$$

where

$$\begin{aligned} K_{21}^{[1]}(\zeta,\eta) &:= \mathrm{Rstr.}_{\mathcal{H}_0} P_{\tilde{\mathcal{H}}_0} \left((I - \overline{\zeta} T)^{-1} - I\right)\left(I - \eta T^*)^{-1} - I\right) P_{\mathcal{H}_0}, \\ K_{21}^{[2]}(\zeta,\eta) &:= \mathrm{Rstr.}_{\mathcal{H}_0} P_{\tilde{\mathcal{H}}_0} \left((I - \zeta T^*)^{-1} - I\right)\left(I - \eta T)^{-1} - I\right) P_{\mathcal{H}_0} \end{aligned}$$

and

$$K_{21}^{[3]}(\zeta,\eta) := \mathrm{Rstr.}_{\mathcal{H}_0} P_{\tilde{\mathcal{H}}_0} \left((I - \overline{\zeta} T)^{-1}(I - \eta T)^{-1} + (I - \zeta T^*)^{-1}(I - \overline{\eta} T^*)^{-1} - I\right).$$

Since $\mathcal{H}_0$ and $\tilde{\mathcal{H}}_0$ are the generating subspaces for V_T and V_{T^*}, respectively, we have

$$\begin{aligned} K_{21}^{[1]}(\overline{\zeta},\overline{\eta}) &= \mathrm{Rstr.}_{\mathcal{H}_0} \sum_{j=1}^\infty \sum_{k=1}^\infty \zeta^j \eta^k P_{\tilde{\mathcal{H}}_0} T^j (T^*)^k \\ &= \mathrm{Rstr.}_{\mathcal{H}_0} \sum_{j=1}^\infty \sum_{k=1}^\infty \zeta^j \eta^k P_{\tilde{\mathcal{H}}_0} T^{j-k} + \mathrm{Rstr.}_{\mathcal{H}_0} \sum_{j=1}^\infty \sum_{k=j+1}^\infty \zeta^j \eta^k P_{\tilde{\mathcal{H}}_0} (T^*)^{k-j} \\ &= \mathrm{Rstr.}_{\mathcal{H}_0} \sum_{k=1}^\infty \sum_{j=1}^\infty \zeta^j \eta^k P_{\tilde{\mathcal{H}}_0} T^{j-k} + \mathrm{Rstr.}_{\mathcal{H}_0} \sum_{j=1}^\infty (\zeta\eta)^j \sum_{k=1}^\infty \eta^k P_{\tilde{\mathcal{H}}_0} (T^*)^k \\ &= \mathrm{Rstr.}_{\mathcal{H}_0} \sum_{k=1}^\infty (\zeta\eta)^k \sum_{j=0}^\infty \zeta^j P_{\tilde{\mathcal{H}}_0} T^j + \mathrm{Rstr.}_{\mathcal{H}_0} \frac{\zeta\eta}{1-\zeta\eta} P_{\tilde{\mathcal{H}}_0}((I_{\mathcal{H}} - \eta T^*)^{-1} - I_{\mathcal{H}}) \\ &= \frac{\zeta\eta}{1-\zeta\eta}(\omega(\zeta) + \tilde{\omega}(\eta)) \end{aligned}$$

for each $\zeta \in \mathbb{D}$ and each $\eta \in \mathbb{D}$ where $\omega : \mathcal{H} \to \tilde{\mathcal{H}}_0$ and $\tilde{\omega} : \mathcal{H}_0 \to \tilde{\mathcal{H}}_0$ are defined by

$$\omega(\zeta) := \mathrm{Rstr.}_{\mathcal{H}_0} P_{\tilde{\mathcal{H}}_0} (I_{\mathcal{H}} - \zeta T)^{-1} \tag{4.19}$$

and

$$\tilde{\omega}(\zeta) := \mathrm{Rstr.}_{\mathcal{H}_0} (\zeta P_{\tilde{\mathcal{H}}_0} T^* (I_{\mathcal{H}} - \zeta T^*)^{-1}). \tag{4.20}$$

The functions ω and $\tilde{\omega}$ can be represented by the holomorphic part and the antiholomorphic part, respectively, of a suboperator of scattering χ by the inner channels of the system. In fact, comparing (1.10) and (4.20) (respectively (1.11) and (4.19)) we infer

$$\zeta\omega = \check{\chi}_-(\zeta) \quad \text{and} \quad \tilde{\omega}(\zeta) = \zeta\check{\chi}_+(\zeta) \tag{4.21}$$

for each $\zeta \in \mathbb{D}$. Analogously, we see that

$$K_{21}^{[2]}(\zeta, \eta) = \frac{\zeta\eta}{1 - \zeta\eta} (\omega(\zeta) + \tilde{\omega}(\zeta))$$

and consequently

$$K_{21}^{[2]}(\zeta, \eta) = K_{21}^{[1]}(\overline{\zeta}, \overline{\eta})$$

hold true for every choice of ζ and η in $\mathbb{D}$. Finally, using the Hilbert Identity for resolvents we get

$$K_{21}^{[3]}(\zeta, \eta) = \frac{\overline{\zeta}\omega(\overline{\zeta}) - \eta\omega(\eta)}{\overline{\zeta} - \eta} + \frac{\zeta\tilde{\omega}(\zeta) - \overline{\eta}\tilde{\omega}(\overline{\eta})}{\zeta - \overline{\eta}}$$

for all ζ and η that belong to $\mathbb{D}$ and satisfy $\zeta \neq \overline{\eta}$. Thus the following Theorem is proved.

Theorem 4.4. Let $T : \mathcal{H} \to \mathcal{H}$ be a completely nonunitary contractive operator for which the maximal unilateral shift V_T and the maximal unilateral coshift V_{T^*} generate jointly the space $\mathcal{H}$. Then the reproducing kernel $K : \mathbb{D} \times \mathbb{D} \to [\mathcal{H}_0 \oplus \tilde{\mathcal{H}}_0, \mathcal{H}_0 \oplus \tilde{\mathcal{H}}_0]$ of the model space $\mathcal{M}$ admits, for every choice of ζ and η in $\mathbb{D}$, $\zeta \neq \overline{\eta}$, the representation

$$K(\zeta, \eta) = \begin{pmatrix} K_{11}(\zeta, \eta) & K_{12}(\zeta, \eta) \\ K_{21}(\zeta, \eta) & K_{22}(\zeta, \eta) \end{pmatrix}$$

where

$$K_{11}(\zeta, \eta) = \frac{1}{1 - \zeta\overline{\eta}} I_{\mathcal{H}_0} + \frac{\zeta\overline{\eta}}{1 - \zeta\overline{\eta}} (I_{\mathcal{H}_0} - \varphi(\zeta)\varphi(\eta)^*),$$

$$K_{12}(\zeta, \eta) = \frac{1}{1 - \zeta\overline{\eta}} I_{\tilde{\mathcal{H}}_0} + \frac{\zeta\overline{\eta}}{1 - \zeta\overline{\eta}} (I_{\tilde{\mathcal{H}}_0} - \psi(\overline{\zeta})^* \psi(\overline{\eta})),$$

$$\begin{aligned} K_{21}(\zeta, \eta) = {} & \frac{\overline{\zeta}\overline{\eta}}{1 - \overline{\zeta}\overline{\eta}} (\omega(\overline{\zeta}) + \tilde{\omega}(\overline{\eta})) + \frac{\zeta\eta}{1 - \zeta\eta} (\omega(\eta) + \tilde{\omega}(\zeta)) \\ & + \frac{\overline{\zeta}\omega(\zeta) - \eta\omega(\eta)}{\overline{\zeta} - \eta} + \frac{\zeta\tilde{\omega}(\zeta) - \overline{\eta}\tilde{\omega}(\overline{\eta})}{\zeta - \overline{\eta}} \end{aligned}$$

and

$$K_{12}(\zeta, \eta) = (K_{21}(\eta, \zeta))^*,$$

where φ and ψ are the right defect function and the left defect function, respectively, and where ω and $\tilde{\omega}$ can be represented by the holomorphic part and the antiholomorphic part, respectively, of the suboperator of scattering by the inner channels through formulas (4.21).

References

[1] V.M. Adamjan, D.Z. Arov: On the unitary couplings of isometric operators, Mat. Issled. Kisinev 1 (1966), Issue 2, 3–64.

[2] D.Z. Arov: On unitary couplings (scattering theory with losses), Funktional Anal. Appl. 8 (1974), 280–294.

[3] B. Sz.-Nagy, C. Foias: Harmonic Analysis of Operators on Hilbert Space, North-Holland, Amsterdam 1970.

[4] M.S. Brodskii: Unitary operator colligations and their characteristic functions, Russian Math. Surveys 33 (1978), Issue 4, 159–191.

[5] N.K. Nikolskii, V.I. Vasyunin: A unified approach to functional models, and the transcription problem, in: The Gohberg Anniversary Collection, Volume II: Topics in Analysis and Operator Theory (Eds.: H. Dym, S. Goldberg, M.A. Kaashoek, P. Lancaster), Operator Theory: Advances and Applications, Volume 41, Birkhäuser Basel 1989, pp. 405–434.

[6] J.A. Ball, N. Cohen: De Branges-Rovnyak operator models and systems theory: a survey, in: Topics in Matrix and Operator Theory (Eds. H. Bart, I. Gohberg, M.A. Kaashoek), Operator Theory: Advances and Applications, Volume 50, Birkhäuser, Basel 1991, pp. 93–136.

[7] V.K. Dubovoj, R.K. Mohammed: Defect functions of holomorphic contractive matrix functions, regular extensions and open systems, Math. Nachr. 160 (1993), 69–110.

[8] S.S. Boiko, V.K. Dubovoj: On some extremal problem connected with the suboperator of scattering through inner channels of a system, Dopovidi AN Ukraine 4 (1997), 7–11.

[9] S.S. Boiko, V.K. Dubovoj, B. Fritzsche, B. Kirstein: Contractive operators, defect functions and scattering theory, Ukrainian Math. J. 49 (1997), 481–489.

[10] N. Aronszajn: Theory of reproducing kernels, Trans. Amer. Math. Soc. 68 (1950), 337–404.

Department of Mathematics, Kharkov State University, Kharkov, Ukraine 310077

Department of Mathematics, Kharkov State University, Kharkov, Ukraine 310077

Mathematisches Institut, Universität Leipzig, Augustusplatz 10, D-04109 Leipzig, Germany

Mathematisches Institut, Universität Leipzig, Augustusplatz 10, D-04109 Leipzig, Germany

Operator Theory:
Advances and Applications, Vol. 123

Measure Schur Complements and Spectral Functions of Unitary Operators with Respect to Different Scales

S. S. Boiko, V. K. Dubovoy, and A. Ya. Kheifets

Dedicated to Professor Moshe Livšic on the occasion of his 80th birthday

Introduction

The main goal of the paper is computing orthogonal complements to reducing subspaces of unitary operators in terms of Schur complements to corresponding spectral measures. The results can be applied to a number of problems in analysis.

In Section 1 the Hellinger space L^σ is discussed and necessary preliminaries are given about this space.

In Section 2 we define and study measure Schur complements. The main theorems are 2.7–2.9 which give orthogonal decomposition of the Hellinger space L^σ in terms of measure Schur complements.

Section 3 deals with the special case of an operator measure σ which is of bounded variation (i.e., is absolutely continuous with respect to a scalar measure). In this case the formula for the measure Schur complement is simplified to a more classical form (Theorem 3.11).

In Section 4 we study unitary scattering systems (i.e., unitary operator together with a scale mapping) and compute orthogonal complement to the reducing subspace generated by the scale in terms of Schur complement to the corresponding spectral measure (Theorem 4.3.).

In Section 5 an application of these results is given to a completion problem motivated by the study of internal scattering channels.

Acknowledgements. The third author was partially supported by the Israel Ministry of Absorption. The second and the third author are grateful to the organizers of the conference in operator theory in honour of Moshe Livšic 80th birthday for support for participation in the conference and a wonderful opportunity to meet colleagues. It is also a pleasure for the third author to express his thanks to Victor Katsnelson and Harry Dym for inviting him to visit the Weizmann Institute in summer 1997. The authors are grateful to Miriam Abraham for her patience and a perfect typing of the manuscript.

1. Hellinger space $\mathbf{L}^\sigma$

The notion of a Hellinger space was introduced by E. Hellinger ([1]), matrix and operator generalizations are due to Yu.L. Shmulyan ([2, 3, 4]). Our exposition follows [5].

Let $\mathbb{T}$ be the unit circle of the complex plane ($t \in \mathbb{T}$ if and only if $|t| = 1$); let E be a Hilbert space; let σ be a bounded operator-valued measure on $\mathbb{T}$, that is: for any Borel subset $\Delta \subset \mathbb{T}$, $\sigma(\Delta)$ is a bounded linear nonnegative operator on E, $\sigma(\Delta) \geq 0$, and $\sigma(\Delta)$ is a strongly countably additive function of the set, i.e.,

$$\sigma\left(\bigcup_{k=1}^{\infty} \Delta_k\right) e = \sum_{k=1} \sigma(\Delta_k)e, \ \forall e \in E$$

if $\Delta_i \cap \Delta_k = \emptyset, i \neq k$.

Consider E-valued vector-chart ν on $\mathbb{T}$, that is: for any Borel $\Delta \subseteq \mathbb{T}$, $\nu(\Delta) \in E$ and ν is a countably additive function of the set (in the sense of the topology of the Hilbert space E).

We will say that the chart ν is in *Hellinger space* L^σ if there exists a scalar measure μ on the unit circle $\mathbb{T}$ so that the following operator-valued function of the set is nonnegative.

$$\begin{bmatrix} \mu(\Delta) & \nu(\Delta)^* \\ & \\ \nu(\Delta) & \sigma(\Delta) \end{bmatrix} \geq 0, \Delta \subseteq \mathbb{T} \tag{1.1}$$

Here $\nu(\Delta)^* \in E^*$ and

$$\nu(\Delta)^* e = \langle e, \nu(\Delta)\rangle_E , e \in E.$$

I.e., (1.1) is an operator-valued measure on $\mathbb{T}$,

$$\begin{bmatrix} \mu(\Delta) & \nu(\Delta)^* \\ & \\ \nu(\Delta) & \sigma(\Delta) \end{bmatrix} : \mathbb{C} \oplus E \to \mathbb{C} \oplus E.$$

It may be more natural to formulate this in terms of the corresponding quadratic form. In other words (1.1) means that for any Borel set $\Delta \subseteq \mathbb{T}$ and any $e \in E$

$$\begin{bmatrix} \mu(\Delta) & \langle e, \nu(\Delta)\rangle_E \\ & \\ \langle \nu(\Delta), e\rangle_E & \langle \sigma(\Delta)e, e\rangle_E \end{bmatrix} \geq 0. \tag{1.2}$$

A minimal measure μ_ν satisfying this exists and is unique for any $\nu \in L^\sigma$. Minimality here means that any measure μ fitting (1.2) is bigger than μ_ν on each Borel $\Delta \subseteq \mathbb{T}$

$$\mu(\Delta) \geq \mu_\nu(\Delta).$$

We will prove it and obtain a formula for μ_ν in Theorem 1.5.

In what follows the measure μ_ν is going to be referred to as the *Hellinger measure* of the chart ν with respect to (or against) the measure σ.

Lemma 1.1. Let E be a Hilbert space, $A : E \to E$ be a bounded nonnegative self-adjoint operator on $E, A \geq 0$. Then the inequality

$$\begin{bmatrix} c & \langle e, f\rangle_E \\ \langle f, e\rangle_E & \langle Ae, e\rangle \end{bmatrix} \geq 0, \ \forall e \in E \tag{1.3}$$

holds true if and only if

$$f \in \operatorname{Ran} A^{1/2}$$

and

$$\| A^{[-1/2]} f \|^2 \leq c, \tag{1.4}$$

where $A^{[-1/2]}f$ stands for the unique preimage of the smallest norm among all of the preimages of the vector f (i.e., the one which seats in $(Ker\ A^{1/2})^{\perp} = \overline{\operatorname{Ran} A^{1/2}}$). The number $\| A^{[-1/2]} f \|^2$ is the smallest constant fitting (1.3).

Proof. Let (1.4) hold true and $g = A^{[-1/2]}f$, then

$$|\langle e, f\rangle|^2 = \left|\left\langle e, A^{1/2}g\right\rangle\right|^2 = \left|\left\langle A^{1/2}e, g\right\rangle\right|^2 \leq$$

$$\leq \| A^{1/2}e \|^2 \cdot \| g \|^2 = \| g \|^2 \cdot \langle Ae, e\rangle$$

and (1.3) holds with $\| g \|^2$ as C. Let (1.3) hold true. For $h \in Ran\ A^{1/2}$ consider the linear functional

$$Q(h) = \langle e, f\rangle, h = A^{1/2}e, \ e \in E.$$

Then (1.3) implies

$$|Q(h)|^2 = |\langle e, f\rangle|^2 \leq C\langle Ae, e\rangle = C \| A^{1/2}e \|^2 = C \| h \|^2 .$$

The latter shows that Q is well defined on $\operatorname{Ran} A^{1/2}$ and Q is bounded

$$\| Q \|^2 \leq C.$$

Hence Q can be continued to $\overline{\operatorname{Ran} A^{1/2}}$ as a bounded linear functional. By means of Riesz theorem there exists a unique $g \in \operatorname{Ran} A^{1/2}$ so that

$$Q(h) = \langle h, g\rangle_E, \forall h \in \overline{\operatorname{Ran} A^{1/2}}.$$

In particular,

$$\langle e, f\rangle = \left\langle A^{1/2}e, g\right\rangle, \forall e \in E$$

which implies

$$f = A^{1/2}g \in \operatorname{Ran} A^{1/2}.$$

Since $\| Q \| = \| g \|$, we obtain

$$\| g \|^2 \leq C.$$

Meaning

$$\| A^{[-1/2]} f \|^2 \leq C,$$

for any C fitting (1.3), i.e., it is the smallest constant.

Remark. If A is an unbounded nonnegataive selfadjoint operator one has to write $\langle A^{1/2}e, A^{1/2}e \rangle$ as the South-East entry of (1.3) and to make e run over the domain of $A^{1/2}$ only. But we will use the lemma only for bounded oeprators.

The following theorem is an obvious corollary of Lemma 1.1.

Theorem 1.2. Let $\nu \in L^\sigma$ (i.e., there exists a scalar measure μ so that (1.2) holds). Then the best constant that can be substituted instead of $\mu(\Delta)$ is computed by this formula

$$C_\nu(\Delta) = \| \sigma(\Delta)^{[-1/2]} \nu(\Delta) \|^2, \tag{1.5}$$

and $C_\nu(\Delta) \leq \mu(\Delta)$ for any measure fitting (1.2).

Remark. $C_\nu(\Delta)$ is not in general an additive function of the set.

To produce the best majorizing measure μ_ν we will need some properties of the function $C_\nu(\Delta)$.

Lemma 1.3. Let $C_\nu(\Delta)$ be defined by (1.5) then

1. $C_\nu(\Delta) \geq 0$, for any Borel $\Delta \subseteq \mathbb{T}$.
2. $C_\nu(\Delta) \leq \sum_k C_\nu(\Delta_k)$, for any finite or countable disjunctive partition of Δ:

$$\Delta = \bigcup_k \Delta_k, \Delta_k \cap \Delta_j = \emptyset, \qquad k \neq j.$$

(I.e., $C_\nu(\Delta)$ is a nonnegative countably subadditive function of the set.)

Proof. Property 1) is obvious. To verify 2), observe that since ν and σ are measures then

$$\sum_k \begin{bmatrix} C_\nu(\Delta_k) & \langle e, \nu(\Delta_k) \rangle \\ \langle \nu(\Delta_k), e \rangle & \langle \sigma(\Delta_k) e, e \rangle \end{bmatrix} = \begin{bmatrix} \sum_k C_\nu(\Delta_k) & \langle e, \nu(\Delta) \rangle \\ \langle \nu(\Delta), e \rangle & \langle \sigma(\Delta) e, e \rangle \end{bmatrix}$$

for all $e \in E$. Since the left-hand side is nonnegative we can see that $\sum_k C_\nu(\Delta_k)$ is a majorant in the right-hand side, possibly not the least one. Hence

$$C_\nu(\Delta) \leq \sum_k C_\nu(\Delta_k).$$

Thus the Lemma follows.

Remark here, that C_ν need not be in general a monotone function of the set.

Lemma 1.4. Let $C(\Delta)$ be a function of the set defined on Borel subsets of $\mathbb{T}$ that enjoys properties 1) and 2) of Lemma 1.2.

$$\begin{aligned} &1.\ C(\Delta) \geq 0 \text{ for any Borel } \Delta \subseteq \mathbb{T}. \\ &2.\ C(\Delta) \leq \sum_k C(\Delta_k), \text{ for any finite or countable partition of } \Delta. \end{aligned} \tag{1.6}$$

Let

$$\mu(\Delta) \stackrel{def}{=} \sup \sum_k C(\Delta_k), \tag{1.7}$$

where supremum is taken along all *finite* disjunctive partitions of Δ, and let $\mu(\Delta) < \infty$ for any Borel $\Delta \subseteq \mathbb{T}$ then μ is a measure and it is the least measure among all the measures bigger than C $(C(\Delta) \leq \mu(\Delta))$.

Proof. $\mu(\Delta) \geq 0$ by definition. Since $C(\Delta)$ is subadditive the net

$$\sum_k C(\Delta_k), \ \Delta = \cup \Delta_k, \ \Delta_k \cap \Delta_j = \emptyset, k \neq j,$$

increases monotonically (i.e., the value $\sum_k C(\Delta_k)$ becomes bigger when we pass to a smaller partition of Δ). Hence,

$$\mu(\Delta) \stackrel{def}{=} \sup \sum_k C(\Delta_k) = \lim \sum_k C(\Delta_k) \tag{1.8}$$

(the limit is taken along the net of finite partitions of Δ). Additivity of μ is obvious now. It remains to prove its countable additivity.

Let Δ be a Borel subset of $\mathbb{T}$ and

$$\Delta = \bigcup_{k=1}^{\infty} \Delta_k, \ \Delta_k \cap \Delta_j = \emptyset, k \neq j$$

be an arbitrary countable disjunctive partition of Δ.

Additivity of μ implies the inequality

$$\mu(\Delta) \geq \sum_{k=1}^{\infty} \mu(\Delta_k).$$

Let $\varepsilon > 0$ be an arbitrary positive real. Pick up the finite partition $\Delta = \bigcup_{j=1}^{n} \tilde{\Delta}_j$ so that

$$\sum_{j=1}^{n} C(\tilde{\Delta}_j) \geq \mu(\Delta) - \varepsilon.$$

Then

$$\Delta_k = \bigcup_{j=1}^{n} (\Delta_k \cap \tilde{\Delta}_j)$$

is a finite partition of Δ_k. Hence, by definition

$$\mu(\Delta_k) \geq \sum_{j=1}^{n} C(\Delta_k \cap \tilde{\Delta}_j),$$

and, therefore,

$$\sum_{k=1}^{\infty} \mu(\Delta_k) \geq \sum_{k=1}^{\infty} \sum_{j=1}^{n} C(\Delta_k \cap \tilde{\Delta}_j) = \sum_{j=1}^{n} \sum_{k=1}^{\infty} C(\Delta_k \cap \tilde{\Delta}_j)$$

(since C is countably subadditive)

$$\geq \sum_{j=1}^{n} C(\tilde{\Delta}_j) \geq \mu(\Delta) - \varepsilon.$$

Since ε is an arbitrary positive number,

$$\sum_{k=1}^{\infty} \mu(\Delta_k) \geq \mu(\Delta).$$

Combining this inequaltiy with the converse one proved above we obtain the countable additivity of μ.

It follows from the definition that

$$C(\Delta) \leq \mu(\Delta)$$

for any Borel $\Delta \subseteq \mathbb{T}$.

Let μ' be a measure on $\mathbb{T}$ so that

$$C(\Delta) \leq \mu'(\Delta).$$

Then

$$\mu(\Delta) = \sup \sum_k C(\Delta_k) \leq \sup \sum_k \mu'(\Delta_k) = \mu'(\Delta).$$

Hence μ is the least measure bigger than C, the lemma follows.

Theorem 1.5. Let E be a Hilbert space and σ be an operator measure on $\mathbb{T}$, $\sigma(\Delta) : E \to E$. Let $\nu \in L^{\sigma}$ (i.e., ν is an E-valued vector-chart on $\mathbb{T}$ and a scalar measure μ exists so that (1.2) holds). Then the Hellinger measure of ν with respect to σ is computed by this formula:

$$\begin{aligned} \mu_\nu(\Delta) &= \sup \sum_k \| \sigma(\Delta_k)^{[-1/2]} \nu(\Delta_k) \|_E^2 = \\ &= \lim \sum_k \| \sigma(\Delta_k)^{[-1/2]} \nu(\Delta_k) \|_E^2 \,. \end{aligned} \tag{1.9}$$

Proof. By Theorem 1.2 the best constant $C_\nu(\Delta)$, fitting (1.2) is computed by formula (1.5) and $C_\nu(\Delta) \leq \mu(\Delta)$ for any measure μ fitting (1.2) on each Borel $\Delta \subseteq \mathbb{T}$. Hence, the Hellinger measure is bigger than C_ν on each Borel $\Delta \subseteq \mathbb{T}$ (it is finite since a finite measure μ exists bigger than C_ν). By Lemma 1.3 C_ν enjoys properties (1.6) and, hence, by Lemma 1.4 the least measure bigger than C_ν is computed by formula (1.9). Then the theorem follows.

The following theorems are given without proofs.

Theorem 1.6. The space L^{σ} endowed with the following norm

$$\| \nu \|_{L^\sigma}^2 \overset{def}{=} \mu_\nu(\mathbb{T}) \tag{1.10}$$

turns into a complete Hilbert space. The corresponding inner product in L^σ is computed by means of the polarization formula

$$(1.11)\qquad \langle \nu_1,\nu_2\rangle_{L^\sigma} = \lim_{\mathbb{T}=\cup\Delta_k} \sum_k \left\langle \sigma(\Delta_k)^{[-1/2]}\nu_1(\Delta_k), \sigma(\Delta_k)^{[-1/2]}\nu_2(\Delta_k)\right\rangle_E .$$

Theorem 1.7. Let $\nu \in L^\sigma$ and let μ_ν be its Hellinger measure with respect to σ. Let f be a scalar function on $\mathbb{T}$ so that

$$\int_{\mathbb{T}} |f(t)|^2 \mu_\nu(dt) < \infty.$$

Then $\nu f \in L^\sigma$ and

$$(1.12)\qquad \mu_{\nu f} = |f|^2 \mu_\nu.$$

Corollary 1.8. In particular:

1. The space L^σ is invariant under multiplication operator by independent variable t, and the latter is a unitary one.
2. If $f = \chi_\Delta$ is a characteristic function of a Borel set $\Delta \subseteq \mathbb{T}$, then

$$(1.13)\qquad \| \nu \cdot \chi_\Delta \|^2_{L^\sigma} = \mu_\nu(\Delta).$$

The space L^σ admits another realization, namely as a space of harmonic vector-functions in the unit disk $\mathbb{D} = \{\zeta : |\zeta| < 1\}$. Any chart ν on $\mathbb{T}$ admits harmonic continuation inside $\mathbb{D}$ by means of the Poisson integral:

$$(1.14)\qquad \nu(\zeta) \stackrel{def}{=} \int_{\mathbb{T}} \frac{1-|\zeta|^2}{|t-\zeta|^2} \nu(dt).$$

Let $\nu \in L^\sigma$ and let μ be a majorizing measure in the sense of (1.1). Then consider the Poisson integral

$$(1.15)\qquad \begin{bmatrix} \mu(\zeta) & \nu(\zeta)^* \\ \nu(\zeta) & \sigma(\zeta) \end{bmatrix} = \int_{\mathbb{T}} \frac{1-|\zeta|^2}{|t-\zeta|^2} \begin{bmatrix} \mu(dt) & \nu(dt)^* \\ \nu(dt) & \sigma(dt) \end{bmatrix},$$

which implies the existence of a harmonic majorant $\mu(\zeta)$ for $\nu(\zeta)$ in the following sense

$$(1.16)\qquad \begin{bmatrix} \mu(\zeta) & \nu(\zeta)^* \\ \nu(\zeta) & \sigma(\zeta) \end{bmatrix} \geq 0.$$

Conversely, if $\nu(\zeta)$ is a harmonic vector-function on $\mathbb{D}$ with a harmonic majorant $\mu(\zeta)$ in the sense of (1.16), then the left-hand side of (1.16) admits the integral representation (1.15). Hence, $\nu(\zeta)$ is the Poisson integral of a chart in L^σ.

The chart $\nu(dt)$ can be understood also as the boundary values of $\nu(\zeta)$ in the weak sense. Thus, we can formulate the following:

Theorem 1.9. The Poisson integral (1.15) gives a one-to-one correspondence between vector charts on $\mathbb{T}$ with a majorizing measure in the sense of (1.1) and harmonic vector-functions on $\mathbb{D}$ with a majorizing harmonic function in the sense of (1.16). It also gives a one-to-one correspondence between majorizing measures $\mu(dt)$ for $\nu(dt)$ and harmonic majorants $\mu(\zeta)$ for $\nu(\zeta)$, respectively. In particular the least harmonic majorant (which we will denote by $\mu_\nu(\zeta)$) corresponds to the Hellinger measure $\mu_\nu(dt)$.

We do not give here a rigorous proof. The rest of the theorems in the section are also given without proofs.

Theorem 1.10. The norm in the space L^σ (see Theorem 1.6) can be computed as the value of the least harmonic majorant $\mu_\nu(\zeta)$ at the origin

$$\| \nu \|^2_{L^\sigma} = \mu_\nu(0). \tag{1.17}$$

It can also be expressed as

$$\| \nu \|^2_{L^\sigma} = \lim_{r \nearrow 1} \int_{\mathbb{T}} \| \sigma(rt)^{[-1/2]} \nu(rt) \|^2_E \, m(dt), \tag{1.18}$$

where $m(dt)$ is the normalized Lebesgue measure on $\mathbb{T}$.

In particular, (see (1.13))

$$\mu_\nu(\Delta) = \lim_{r \nearrow 1} \int_{\mathbb{T}} \| \sigma(rt)^{[-1/2]} (\nu\chi_\Delta)(rt) \|^2_E \, m(dt), \tag{1.19}$$

for any Borel $\Delta \subseteq \mathbb{T}$, where $(\nu\chi_\Delta)(\zeta)$ is the Poisson integral of the chart $\nu\chi_\Delta$.

Remarks. 1) We use the same notation for charts and measures on $\mathbb{T}$ and the corresponding harmonic functions on $\mathbb{D}$. We think of them as the same object: the former are boundary values of the latter (in the $*$-weak sense), the latter are harmonic continuations of the former.
2) The integrand in (1.18) is the least majorant fitting (1.16) instead of $\mu(\zeta)$ (when $\zeta = rt$), but it is not a harmonic function. The formula (1.18) comes from standard arguments of producing the least harmonic majorant,

$$\| \sigma(rt)^{[-1/2]} \nu(rt) \|^2 \, m(dt) \stackrel{*-w}{\longrightarrow} \mu_\nu(dt), r \nearrow 1. \tag{1.20}$$

Theorem 1.11. The space L^σ is a *reproducing kernel Hilbert space* in the following sense

$$\left\langle \nu(dt), \frac{1-|\zeta|^2}{|t-\zeta|^2} \sigma(dt) e \right\rangle_{L^\sigma} = \langle \nu(\zeta), \; e \rangle_E \tag{1.21}$$

for all $\nu \in L^\sigma$, $\zeta \in \mathbb{D}$, $e \in E$.

The family of charts

$$\frac{1-|\zeta|^2}{|t-\zeta|^2} \sigma(dt) e, \zeta \in \mathbb{D}, e \in E \tag{1.22}$$

is called the *reproducing kernel* of the space L^σ for that reason.

Corollary 1.12. The family (1.22) is total in the space L^σ.

2. Measure Schur complement and Hellinger space orthogonal decomposition

We begin with proving several "linear algebra" lemmas which we will need in this section and which motivate the definitions below.

Lemma 2.1. Let A be a bounded selfadjoint operator on a Hilbert space M. Let M be an orthogonal sum of two subspaces

$$M = M_1 \oplus M_2. \tag{2.1}$$

Let

$$A = \begin{bmatrix} A_{11} & A_{12} \\ A_{21} & A_{22} \end{bmatrix} : M_1 \oplus M_2 \to M_1 \oplus M_2 \tag{2.2}$$

be the corresponding decomposition of A. Then A is nonnegative, $A \geq 0$, if and only if

$$\begin{aligned} &\text{i. } A_{11} \geq 0, \\ &\text{ii. } \operatorname{Ran} A_{12} \subseteq \operatorname{Ran} A_{11}^{1/2}, \\ &\text{iii. } A_{11}^{\perp} \geq 0, \end{aligned} \tag{2.3}$$

where

$$\begin{aligned} A_{11}^{\perp} &= A_{22} - \widehat{A}_{22}, \\ \widehat{A}_{22} &= (A_{11}^{[-1/2]} A_{12})^* (A_{11}^{[-1/2]} A_{12}). \end{aligned} \tag{2.4}$$

(For the explanation of the notation $A_{11}^{[-1/2]}$ see Section 1.)

Proof. Let $A \geq 0$. Then $A_{11} \geq 0$. Let $m_2 \in M_2$. Then (2.2) yields

$$\begin{bmatrix} A_{11} & A_{12} m_2 \\ (A_{12} m_2)^* & \langle A_{22} m_2, m_2 \rangle \end{bmatrix} \geq 0. \tag{2.2'}$$

Lemma 1.1 now implies ii). The following identity is obvious.

$$\begin{bmatrix} A_{11} & A_{12} \\ A_{21} & A_{22} \end{bmatrix} = \begin{bmatrix} A_{11}^{1/2} & 0 \\ 0 & \mathbf{1} \end{bmatrix} \begin{bmatrix} \mathbf{1} & A_{11}^{[-1/2]} A_{12} \\ (A_{11}^{[-1/2]} A_{12})^* & A_{22} \end{bmatrix} \begin{bmatrix} A_{11}^{1/2} & 0 \\ 0 & \mathbf{1} \end{bmatrix}. \tag{2.5}$$

Since $\operatorname{Ran} A_{11}^{1/2}$ is dense in $\overline{\operatorname{Ran} A_{11}}$, (2.5) implies

$$\begin{bmatrix} P_{\overline{\operatorname{Ran} A_{11}}} & P_{\overline{\operatorname{Ran} A_{11}}} A_{11}^{[-1/2]} A_{12} \\ (P_{\overline{\operatorname{Ran} A_{11}}} A_{11}^{[-1/2]} A_{12})^* & A_{22} \end{bmatrix} \geq 0. \tag{2.6}$$

Since $\mathbf{1} \geq P_{\overline{\mathrm{Ran}\, A_{11}}}$ and $\mathrm{Ran}\, A_{11}^{[-1/2]} A_{12} \subseteq \overline{\mathrm{Ran}\, A_{11}}$, we arrive at

$$\begin{bmatrix} \mathbf{1} & A_{11}^{[-1/2]} A_{12} \\ (A_{11}^{[-1/2]} A_{12})^* & A_{22} \end{bmatrix} \geq 0. \tag{2.7}$$

The following identity is obvious

(2.8)
$$\begin{bmatrix} \mathbf{1} & A_{11}^{[-1/2]} A_{12} \\ (A_{11}^{[-1/2]} A_{12})^* & A_{22} \end{bmatrix} = \begin{bmatrix} \mathbf{1} & 0 \\ (A_{11}^{[-1/2]} A_{12})^* & \mathbf{1} \end{bmatrix} \begin{bmatrix} \mathbf{1} & 0 \\ 0 & A_{11}^{\perp} \end{bmatrix} \begin{bmatrix} \mathbf{1} & A_{11}^{[-1/2]} A_{12} \\ 0 & \mathbf{1} \end{bmatrix}$$

which implies iii).

Conversely, let the properties (2.3) hold. Then (2.8) makes sense and (2.7) follows, which immediately implies $A \geq 0$ by means of (2.5). Remark here, that the inequality $A_{11}^{\perp} \geq 0$ implies, in particular, boundedness of $\widehat{A}_{22}$ and, hence, boundedness of $A_{11}^{[-1/2]} A_{12} : M_2 \to M_1$. The operator

$$A_{11}^{\perp} = A_{22} - (A_{11}^{[-1/2]} A_{12})^* (A_{11}^{[-1/2]} A_{12}) \tag{2.9}$$

is called the *Schur complement* of the entry A_{11}. This is named after I. Schur since he was the first who realized the particular role of $A_{11}^{\perp}$ in the subject. The operator $\widehat{A}_{22}$ is the *minimal substitution* for the entry A_{22} which does not destroy the nonnegativity of the whole block-operator A.

To formulate the next lemma, observe that a vector $m = \begin{bmatrix} m_1 \\ m_2 \end{bmatrix} \in M = M_1 \oplus M_2$ is in the range of $A^{1/2}$ if and only if there exists a number $C \geq 0$ so that

$$\begin{bmatrix} C & m_1^* & m_2^* \\ m_1 & A_{11} & A_{12} \\ m_2 & A_{21} & A_{22} \end{bmatrix} \geq 0. \tag{2.10}$$

The smallest C in (2.10) is equal to

$$C = \| A^{[-1/2]} m \|^2 . \tag{2.11}$$

It follows from (2.10) that $m_1 \in \mathrm{Ran}\, A_{11}^{1/2}$.

Lemma 2.2. The inequality (2.10) is equivalent to the following one

$$\begin{bmatrix} C & m_1^* & (m_1^{\perp})^* \\ m_1 & A_{11} & 0 \\ m_1^{\perp} & 0 & A_{11}^{\perp} \end{bmatrix} \geq 0. \tag{2.12}$$

where

$$m_1^{\perp} = m_2 - \widehat{m}_2, \qquad \widehat{m}_2 = (A_{11}^{[-1/2]}A_{12})^* A_{11}^{[-1/2]} m_1. \tag{2.13}$$

Proof. Let (2.10) hold. Since

$$\begin{bmatrix} C & m_1^* & m_2^* \\ m_1 & A_{11} & A_{12} \\ m_2 & A_{21} & A_{22} \end{bmatrix} = \begin{bmatrix} 1 & 0 & 0 \\ 0 & A_{11}^{1/2} & 0 \\ 0 & 0 & \mathbf{1} \end{bmatrix} \tag{2.14}$$

$$\times \begin{bmatrix} C & (A_{11}^{[-1/2]}m_1)^* & m_2^* \\ A_{11}^{[-1/2]}m_1 & \mathbf{1} & A_{11}^{[-1/2]}A_{12} \\ m_2 & (A_{11}^{[-1/2]}A_{12})^* & A_{22} \end{bmatrix} \begin{bmatrix} 1 & 0 & 0 \\ 0 & A_{11}^{1/2} & 0 \\ 0 & 0 & \mathbf{1} \end{bmatrix},$$

we obtain, similarly to Lemma 2.1,

$$\begin{bmatrix} C & (A_{11}^{[-1/2]}m_1)^* & m_2^* \\ A_{11}^{[-1/2]}m_1 & \mathbf{1} & A_{11}^{[-1/2]}A_{12} \\ m_2 & (A_{11}^{[-1/2]}A_{12})^* & A_{22} \end{bmatrix} \geq 0. \tag{2.16}$$

The block-operator in (2.16) can be re-expressed as

(2.17)

$$\begin{bmatrix} 1 & 0 & 0 \\ 0 & \mathbf{1} & 0 \\ 0 & (A_{11}^{[-1/2]}A_{12})^* & \mathbf{1} \end{bmatrix} \begin{bmatrix} C & (A_{11}^{[-1/2]}m_1)^* & (m_1^{\perp})^* \\ A_{11}^{[-1/2]}m_1 & \mathbf{1} & 0 \\ m_1^{\perp} & 0 & A_{11}^{\perp} \end{bmatrix} \begin{bmatrix} 1 & 0 & 0 \\ 0 & \mathbf{1} & A_{11}^{[-1/2]}A_{12} \\ 0 & 0 & \mathbf{1} \end{bmatrix}.$$

Hence

$$\begin{bmatrix} C & (A_{11}^{[-1/2]}m_1)^* & (m_1^{\perp})^* \\ A_{11}^{[-1/2]}m_1 & \mathbf{1} & 0 \\ m_1^{\perp} & 0 & A_{11}^{\perp} \end{bmatrix} \geq 0\,, \tag{2.18}$$

which implies (2.12). In a similar way we can deduce (2.11) from (2.12).

Corollary 2.3. Let A be the same operator (2.2) as above, $A : M_1 \oplus M_2 \to M_1 \oplus M_2$. Then a vector $m = \begin{bmatrix} m_1 \\ m_2 \end{bmatrix} \in M = M_1 \oplus M_2$ is in $\operatorname{Ran} A^{1/2}$ if and only if

$$\text{(2.19)} \qquad \text{i. } m_1 \in \operatorname{Ran} A_{11}^{1/2}, \qquad \text{ii. } m_1^{\perp} \in \operatorname{Ran}(A_{11}^{\perp})^{1/2}.$$

In that case

$$\text{(2.20)} \qquad \| A^{[-1/2]} m \|^2 = \| A_{11}^{[-1/2]} m_1 \|^2 + \| (A_{11}^{\perp})^{[-1/2]} m_1^{\perp} \|^2 .$$

When $m = \begin{bmatrix} m_1 \\ m_2 \end{bmatrix}$ runs over the range of $A^{1/2}$, m_1 and $m_1^{\perp}$ run independently over the ranges of $A_{11}^{1/2}$ and $(A_{11}^{\perp})^{1/2}$ respectively.

We are passing now from operators to operator-valued measures.

Let M be a Hilbert space and σ be a strong operator measure, i.e., for any Borel subset $\Delta \subseteq \mathbb{T}$, $\sigma(\Delta) : M \to M$ is a bounded linear operator on M, $\sigma(\Delta) \geq 0$, and for any $m \in M, \sigma(\Delta)m$ is a countably additive function of the set Δ. Let $M = M_1 \oplus M_2$ and let

$$\text{(2.21)} \qquad \sigma(\Delta) = \begin{bmatrix} \sigma_{11}(\Delta) & \sigma_{12}(\Delta) \\ \sigma_{21}(\Delta) & \sigma_{22}(\Delta) \end{bmatrix} \geq 0$$

be the corresponding block decomposition of the measure σ. By Lemma 2.1 the minimal substitution for $\sigma_{22}(\Delta)$ which keeps nonnegativity in (2.21) is

$$\text{(2.22)} \qquad \widehat{\sigma_{22}(\Delta)} = (\sigma_{11}(\Delta)^{[-1/2]} \sigma_{12}(\Delta))^* (\sigma_{11}(\Delta)^{[-1/2]} \sigma_{12}(\Delta)).$$

But $\widehat{\sigma_{22}(\Delta)}$ is not an additive function of the set Δ.

Lemma 2.4. $\widehat{\sigma_{22}(\Delta)}$ is a subadditive function of the set Δ, i.e.

$$\text{(2.23)} \qquad \widehat{\sigma_{22}(\Delta)} \leq \sum_{\Delta = \cup \Delta_j} \widehat{\sigma_{22}(\Delta_j)}$$

for any finite disjunctive partition of Δ.

Proof.

$$\text{(2.24)} \qquad \begin{bmatrix} \sigma_{11}(\Delta) & \sigma_{12}(\Delta) \\ \sigma_{21}(\Delta) & \widehat{\sigma_{22}(\Delta)} \end{bmatrix} \geq 0$$

and $\widehat{\sigma_{22}(\Delta)}$ is the minimal possible. Since all the entries in (2.24) except for $\widehat{\sigma_{22}(\Delta)}$ are additive functions of the set we obtain

$$\text{(2.25)} \qquad \begin{bmatrix} \sigma_{11}(\Delta) & \sigma_{12}(\Delta) \\ \sigma_{21}(\Delta) & \sum_j \widehat{\sigma_{22}(\Delta_j)} \end{bmatrix} \geq 0$$

(2.23) follows now by Lemma 2.1.

Keeping in mind the definition of the Hellinger measure, one can expect that the minimal *measure* substitution for σ_{22} in (2.21) should be

$$\widehat{\sigma}_{22}(\Delta) \overset{def}{=} \lim_{\Delta=\cup\Delta_j} \sum_j \widehat{\sigma_{22}(\Delta_j)}, \tag{2.26}$$

where the limit is taken along the net of finite partitions of the set Δ.

We emphasize here that notations $\widehat{\sigma}_{22}(\Delta)$ and $\widehat{\sigma_{22}(\Delta)}$ are different.

Lemma 2.5. The limit in (2.26) exists in the strong sense, and $\widehat{\sigma}_{22}(\Delta)$ is a strong operator-valued measure.

Proof. By Lemma 2.4, the operator $\sum_{\Delta=\cup\Delta_j} \widehat{\sigma_{22}(\Delta_j)}$ becomes bigger when we pass on to a smaller (finite) partition of Δ. I.e., the net increases monotonically. On the other hand

$$\sum_{\Delta=\cup\Delta_j} \widehat{\sigma_{22}(\Delta_j)} \leq \sum_{\Delta=\cup\Delta_j} \sigma_{22}(\Delta_j) = \sigma_{22}(\Delta), \tag{2.27}$$

meaning the net has an upper bound. Then

$$\lim \left\langle \left(\sum_{\Delta=\cup\Delta_j} \widehat{\sigma_{22}(\Delta_j)} \right) m_2, m_2 \right\rangle \tag{2.28}$$

exists for any $m_2 \in M_2$. Hence (by means of the polarization formula)

$$\lim \left\langle \left(\sum_{\Delta=\cup\Delta_j} \widehat{\sigma_{22}(\Delta_j)} \right) m_2', m_2'' \right\rangle \tag{2.29}$$

exists for any $m_2', m_2'' \in M_2$.

Obviously (2.29) is a sesquilinear form. The quadratic form (2.28) is bounded by (2.27). Hence, the sesquilinear form (2.29) is bounded. Thus, there exists a unique operator $\widehat{\sigma}_{22}(\Delta)$ such that

$$\lim \left\langle \left(\sum_{\Delta=\cup\Delta_j} \widehat{\sigma_{22}(\Delta_j)} \right) m_2', m_2'' \right\rangle = \langle \widehat{\sigma}_{22}(\Delta) m_2', m_2'' \rangle \tag{2.30}$$

Because of (2.27)

$$0 \leq \widehat{\sigma}_{22}(\Delta) \leq \sigma_{22}(\Delta). \tag{2.31}$$

So far we proved weak convergence, but again because of the monotonicity and the existence of an upper bound it implies strong convergence.

Now $\widehat{\sigma}_{22}(\Delta)$ obviously is an additive function of the set. (2.31) implies

$$\widehat{\sigma}_{22}(\Delta_n)^{1/2} m_2 \to 0, \text{as } \ \Delta_n \searrow \emptyset, \quad \forall m_2 \in M_2.$$

It also implies

$$\| \widehat{\sigma}_{22}(\Delta)^{1/2} \| \leq \| \sigma_{22}(\Delta)^{1/2} \| \leq \| \sigma_{22}(\mathbb{T})^{1/2} \|,$$

which forces

$$\widehat{\sigma}_{22}(\Delta_n)m_2 \to 0, \text{as } \ \Delta_n \searrow \emptyset, \quad \forall m_2 \in M_2,$$

meaning $\widehat{\sigma}_{22}(\Delta)$ is a strongly countably additive operator-function of the set. Thus, the lemma follows.

Remark that

$$\widehat{\sigma_{22}(\Delta)} \leq \widehat{\sigma}_{22}(\Delta) \tag{2.32}$$

which follows directly from the definition (2.26).

Definition. The measure

$$\sigma_{11}^{\perp}(\Delta) \stackrel{def}{=} \sigma_{22}(\Delta) - \widehat{\sigma}_{22}(\Delta) \tag{2.33}$$

is called the *Schur complement* for the entry σ_{11} in σ, where σ is of the form (2.21).

After introducing some notations we will formulate and prove an analogue of Lemma 2.2 for charts (Lemma 2.7).

Let σ be an operator-valued measure of the type (2.21)

$$\sigma(\Delta) = \begin{bmatrix} \sigma_{11}(\Delta) & \sigma_{12}(\Delta) \\ \sigma_{21}(\Delta) & \sigma_{22}(\Delta) \end{bmatrix} \geq 0. \tag{2.34}$$

Let $\sigma(\Delta) : M \to M, \ M = M_1 \oplus M_2, \ \Delta \subseteq \mathbb{T}$ be a Borel subset of $\mathbb{T}$. Let $\nu \in L^\sigma$ (see Section 1 for the definition of the Hellinger space L^σ). We will write ν in the form

$$\nu = \begin{bmatrix} \nu_1 \\ \nu_2 \end{bmatrix},$$

where ν_1 and ν_2 are M_1 and M_2 valued charts respectively. Since $\nu \in L^\sigma$, then

$$\begin{bmatrix} \mu_\nu(\Delta) & \nu_1(\Delta)^* & \nu_2(\Delta)^* \\ \nu_1(\Delta) & \sigma_{11}(\Delta) & \sigma_{12}(\Delta) \\ \nu_2(\Delta) & \sigma_{21}(\Delta) & \sigma_{22}(\Delta) \end{bmatrix} \geq 0, \tag{2.35}$$

where μ_ν is the Hellinger measure of the chart ν with respect to σ (see Section 1). Then, by Lemma 2.2

$$\begin{aligned} &\nu_1(\Delta) \in \operatorname{Ran} \sigma_{11}(\Delta)^{1/2} \\ &\nu_1(\Delta)^{\perp} \stackrel{def}{=} \nu_2(\Delta) - \widehat{\nu_2(\Delta)} \in \operatorname{Ran}(\sigma_{11}(\Delta)^{\perp})^{1/2}, \end{aligned} \tag{2.36}$$

where

$$\begin{aligned}
&\sigma_{11}(\Delta)^{\perp} = \sigma_{22}(\Delta) - \widehat{\sigma_{22}(\Delta)},\\
(2.37)\qquad &\widehat{\sigma_{22}(\Delta)} = (\sigma_{11}(\Delta)^{[-1/2]}\sigma_{12}(\Delta))^{*}(\sigma_{11}(\Delta)^{[-1/2]}\sigma_{12}(\Delta)),\\
&\widehat{\nu_2(\Delta)} = (\sigma_{11}(\Delta)^{[-1/2]}\sigma_{12}(\Delta))^{*}(\sigma_{11}(\Delta)^{[-1/2]}\nu_1(\Delta)).
\end{aligned}$$

But neither $\sigma_{11}(\Delta)^{\perp}$ nor $\widehat{\nu_2(\Delta)}$ are additive functions of the set Δ.

Again analogy with Hellinger measure suggests that the limit

$$(2.38)\qquad \widehat{\nu}_2(\Delta) = \lim_{\Delta=\cup\Delta_j} \sum_j \widehat{\nu_2(\Delta_j)}$$

should exist in a sense and has to be a chart.

We again distinguish here between the notations $\sigma_{11}(\Delta)^{\perp}$ and $\sigma_{11}^{\perp}(\Delta)$; between $\widehat{\nu_2(\Delta)}$ and $\widehat{\nu}_2(\Delta)$.

Lemma 2.6. If a chart $\nu = \begin{bmatrix}\nu_1\\ \nu_2\end{bmatrix} \in L^{\sigma}$, then the following weak limit

$$(2.39)\qquad \widehat{\nu}_2(\Delta) \overset{def}{=} w - \lim_{\Delta=\cup\Delta_j} \sum_j \widehat{\nu_2(\Delta_j)}$$

exists along the net of finite partitions of the set Δ, and $\widehat{\nu}_2(\Delta)$ is a vector chart ($\widehat{\nu_2(\Delta)}$ is defined by (2.37)).

Proof. Observe first, that (2.34) implies

$$(2.40)\qquad \sigma_{12}m_2 \in L^{\sigma_{11}},\ \forall m_2 \in M_2.$$

In fact, (2.34) yields

$$(2.41)\qquad \begin{bmatrix} \sigma_{11}(\Delta) & \sigma_{12}(\Delta)m_2 \\ (\sigma_{12}(\Delta)m_2)^{*} & \langle\sigma_{22}(\Delta)m_2, m_2\rangle \end{bmatrix} \geq 0,$$

meaning $\langle\sigma_{22}(\Delta)m_2, m_2\rangle$ is a majorizing measure for the chart $\sigma_{12}m_2$ with respect to the measure σ_{11}, and

$$(2.42)\qquad \mu_{\sigma_{12}m_2}(\Delta) \leq \langle\sigma_{22}(\Delta)m_2, m_2\rangle\,.$$

Hence

$$(2.43)\qquad |\mu_{\sigma_{12}m_2}(\Delta)| \leq \| \sigma_{22}(\Delta) \| \cdot \| m_2, \|^2 \leq \| \sigma_{22}(\mathbb{T}) \| \cdot \| m_2 \|^2\,.$$

Combining (1.9), (2.22), (2.26) one can see that,

$$(2.44)\qquad \mu_{\sigma_{12}m_2}(\Delta) = \langle\widehat{\sigma}_{22}(\Delta)m_2, m_2\rangle\,.$$

For any $m_2 \in M_2$ consider now

$$\left\langle \sum_{\Delta=\cup\Delta_j} \widehat{\nu_2(\Delta_j)}, m_2 \right\rangle \tag{2.45}$$

$$= \sum_{\Delta=\cup\Delta_j} \left\langle \sigma_{11}(\Delta_j)^{[-1/2]}\nu_1(\Delta_j), \sigma_{11}(\Delta_j)^{[-1/2]}\sigma_{12}(\Delta_j)m_2 \right\rangle .$$

According to formula (1.11) that converges to

$$\langle \nu_1\chi_\Delta, \sigma_{12}m_2\chi_\Delta \rangle_{L^{\sigma_{11}}} \cdot \tag{2.46}$$

The latter expression is an antilinear functional in $m_2 \in M_2$, for any Borel set $\Delta \subseteq \mathbb{T}$. Moreover, it is a bounded one.

Indeed,

$$| \langle \nu_1\chi_\Delta, \sigma_{12}m_2\chi_\Delta \rangle |^2 \leq \| \nu_1\chi_\Delta \|^2_{L^{\sigma_{11}}} \cdot \| \sigma_{12}m_2\chi_\Delta \|^2_{L^{\sigma_{11}}} \tag{2.47}$$

(by formula (1.13))

$$= \mu_{\nu_1}(\Delta) \cdot \mu_{\sigma_{12}m_2}(\Delta)$$

(by (2.43))

$$\leq \mu_{\nu_1}(\Delta) \cdot \| \sigma_{22}(\mathbb{T}) \| \cdot \| m_2 \|^2 .$$

Then by Riesz theorem, there exists a unique vector $\widehat{\nu}_2(\Delta) \in M_2$, for any Borel set $\Delta \subseteq \mathbb{T}$, so that

$$\langle \nu_1\chi_\Delta, \sigma_{12}m_2\chi_\Delta \rangle_{L^{\sigma_{11}}} = \langle \widehat{\nu}_2(\Delta), m_2 \rangle \ . \tag{2.48}$$

The desired weak convergence follows. (2.48) also shows that $\widehat{\nu}_2(\Delta)$ is a weakly countably additive vector-function of the set Δ. To prove it is a vector chart, it suffices to verify its continuity on the empty subset of $\mathbb{T}$.

Let $\Delta_k \subseteq \mathbb{T}$ and $\Delta_k \searrow \emptyset$ as $k \to \infty$, i.e., $\Delta_{k+1} \subseteq \Delta_k$, $\cap_k \Delta_k = \emptyset$. Then (2.48) and (2.47) guarantee that

$$\| \widehat{\nu}_2(\Delta_k) \| = \sup_{m_2} \frac{| \langle \widehat{\nu}_2(\Delta_k), m_2 \rangle |}{\| m_2 \|} \leq \sqrt{\mu_{\nu_1}(\Delta_k) \cdot \| \sigma_{22}(\mathbb{T}) \|}. \tag{2.49}$$

Since μ_{ν_1} is a measure, $\mu_{\nu_1}(\Delta_k) \searrow 0$ as $k \to \infty$. Hence, $\widehat{\nu}_2(\Delta_k) \to 0$, as $k \to \infty$ in norm topology of M_2. The lemma follows.

Theorem 2.7. Let σ be a strong operator valued measure on $\mathbb{T}$

$$\sigma(\Delta) = \begin{bmatrix} \sigma_{11}(\Delta) & \sigma_{12}(\Delta) \\ \sigma_{21}(\Delta) & \sigma_{22}(\Delta) \end{bmatrix} : M_1 \oplus M_2 \to M_1 \oplus M_2,$$

for a Borel set $\Delta \subseteq \mathbb{T}$. Let ν be an $M = M_1 \oplus M_2$ valued vector-chart on $\mathbb{T}$,

$$\nu(\Delta) = \begin{bmatrix} \nu_1(\Delta) \\ \nu_2(\Delta) \end{bmatrix},$$

ν_1 and ν_2 are M_1 and M_2 valued vector-charts respectively. Then $\nu \in L^\sigma$ if and only if

$$\text{i. } \nu_1 \in L^{\sigma_{11}}, \qquad \text{ii. } \nu_1^\perp \in L^{\sigma_{11}^\perp}, \tag{2.50}$$

where $\sigma_{11}^\perp$ is the Schur complement (2.33) of the entry σ_{11} and

$$\nu_1^\perp(\Delta) \stackrel{def}{=} \nu_2(\Delta) - \widehat{\nu}_2(\Delta), \tag{2.51}$$

$\widehat{\nu}_2(\Delta)$ is defined by (2.38) or (2.48) (which is the same). In that case

$$\mu_\nu = \mu_{\nu_1} + \mu_{\nu_1^\perp}, \tag{2.52}$$

where μ_ν, μ_{ν_1} and $\mu_{\nu_1^\perp}$ are the Hellinger measures of ν, ν_1 and $\nu_1^\perp$ with respect to σ, σ_{11} and $\sigma_{11}^\perp$ respectively. When $\nu = \begin{bmatrix} \nu_1 \\ \nu_2 \end{bmatrix}$ runs over the space L^σ, ν_1 and $\nu_1^\perp$ run independently over the spaces $L^{\sigma_{11}}$ and $L^{\sigma_{11}^\perp}$ respectively. In particular, (2.52) contains the formula:

$$\| \nu \|_{L^\sigma}^2 = \langle \nu_1, \nu_1' \rangle_{L^{\sigma_{11}}} + \left\langle \nu_1^\perp, (\nu_1')^\perp \right\rangle_{L^{\sigma_{11}^\perp}} . \tag{2.53}$$

Proof. I. Let $\nu \in L^\sigma$, then

$$\begin{bmatrix} \mu_\nu(\Delta) & \nu_1(\Delta)^* & \nu_2(\Delta)^* \\ \nu_1(\Delta) & \sigma_{11}(\Delta) & \sigma_{12}(\Delta) \\ \nu_2(\Delta) & \sigma_{21}(\Delta) & \sigma_{22}(\Delta) \end{bmatrix} \geq 0. \tag{2.54}$$

By Lemma 2.2, (2.54) is equivalent to

$$\begin{bmatrix} \mu_\nu(\Delta) & \nu_1(\Delta)^* & (\nu_1(\Delta)^\perp)^* \\ \nu_1(\Delta) & \sigma_{11}(\Delta) & 0 \\ \nu_1(\Delta)^\perp & 0 & \sigma_{11}(\Delta)^\perp \end{bmatrix} \geq 0. \tag{2.55}$$

Now let $\Delta = \cup \Delta_j$ be a finite disjunctive partition, then (2.55) implies

$$\begin{bmatrix} \mu_\nu(\Delta) & \nu_1(\Delta)^* & \sum_j (\nu_1(\Delta_j)^\perp)^* \\ \nu_1(\Delta) & \sigma_{11}(\Delta) & 0 \\ \sum_j \nu_1(\Delta_j)^\perp & 0 & \sum_j \sigma_{11}(\Delta_j)^\perp \end{bmatrix} \geq 0. \tag{2.56}$$

Here, we have just applied (2.55) to each one of Δ_j, summed up the inequalities and used additivity of those entries which enjoy it. Now passing to the limit (weak) in (2.56) along the net of finite partitions of Δ we will obtain (by Lemmas 2.5 and 2.6)

$$\begin{bmatrix} \mu_\nu(\Delta) & \nu_1(\Delta)^* & \nu_1^\perp(\Delta)^* \\ \nu_1(\Delta) & \sigma_{11}(\Delta) & 0 \\ \nu_1^\perp(\Delta) & 0 & \sigma_{11}^\perp(\Delta) \end{bmatrix} \geq 0. \tag{2.57}$$

Hence, $\nu_1 \in L^{\sigma_{11}}$ and $\nu_1^\perp \in L^{\sigma_{11}^\perp}$. Applying formula (1.9) for Hellinger measure we can check that the Hellinger measure of the chart $\begin{bmatrix} \nu_1 \\ \nu_1^\perp \end{bmatrix}$ with respect to the measure $\begin{bmatrix} \sigma_{11} & 0 \\ 0 & \sigma_{11}^\perp \end{bmatrix}$ is equal to $\mu_{\nu_1} + \mu_{\nu_1^\perp}$. Therefore, (2.57) yields

$$\mu_\nu \geq \mu_{\nu_1} + \mu_{\nu_1^\perp} \, . \tag{2.58}$$

II. Now let $\nu_1 \in L^{\sigma_{11}}, \nu_1^\perp \in L^{\sigma_{11}^\perp}$ be arbitrary charts. Then

$$\begin{bmatrix} \mu(\Delta) & \nu_1(\Delta)^* & \nu_1^\perp(\Delta)^* \\ \nu_1(\Delta) & \sigma_{11}(\Delta) & 0 \\ \nu_1^\perp(\Delta) & 0 & \sigma_{11}^\perp(\Delta) \end{bmatrix} \geq 0, \tag{2.59}$$

with $\mu = \mu_{\nu_1} + \mu_{\nu_1^\perp}$. Again applying Lemma 2.2 one can see that (2.59) is equivalent to

$$\begin{bmatrix} \mu(\Delta) & \nu_1(\Delta)^* & (\nu_1^\perp(\Delta) + \widehat{\nu_2(\Delta)})^* \\ \nu_1(\Delta) & \sigma_{11}(\Delta) & \sigma_{12}(\Delta) \\ \nu_1^\perp(\Delta) + \widehat{\nu_2(\Delta)} & \sigma_{21}(\Delta) & \sigma_{11}^\perp(\Delta) + \widehat{\sigma_{22}(\Delta)} \end{bmatrix} \geq 0, \tag{2.60}$$

where

$$\begin{aligned} \widehat{\sigma_{22}(\Delta)} &= (\sigma_{11}(\Delta)^{[-1/2]}\sigma_{12}(\Delta))^*(\sigma_{11}(\Delta)^{[-1/2]}\sigma_{12}(\Delta)), \\ \widehat{\nu_2(\Delta)} &= (\sigma_{11}(\Delta)^{[-1/2]}\sigma_{12}(\Delta))^*(\sigma_{11}(\Delta)^{[-1/2]}\nu_1(\Delta)) \end{aligned} . \tag{2.61}$$

Observe, that these two are the same as in the first part of the theorem. Again, considering the net of finite partitions $\Delta = \cup \Delta_j$, we can see that

$$\begin{bmatrix} \mu(\Delta) & \nu_1(\Delta)^* & \nu_1^\perp(\Delta)^* + (\sum_j \widehat{\nu_2(\Delta_j)})^* \\ \nu_1(\Delta) & \sigma_{11}(\Delta) & \sigma_{12}(\Delta) \\ \nu_1^\perp(\Delta) + \sum_j \widehat{\nu_2(\Delta_j)} & \sigma_{12}(\Delta)^* & \sigma_{11}^\perp(\Delta) + \sum_j \widehat{\sigma_{22}(\Delta_j)} \end{bmatrix} \geq 0, \tag{2.62}$$

Passing to the limit (weak) along the net we obtain

$$(2.63)\qquad \begin{bmatrix} \mu(\Delta) & \nu_1(\Delta)^* & \nu_1^\perp(\Delta)^* + \widehat{\nu}_2(\Delta)^* \\ \nu_1(\Delta) & \sigma_{11}(\Delta) & \sigma_{12}(\Delta) \\ \nu_1^\perp(\Delta) + \widehat{\nu}_2(\Delta) & \sigma_{12}(\Delta)^* & \sigma_{11}^\perp(\Delta) + \widehat{\sigma}_{22}(\Delta) \end{bmatrix} \geq 0,$$

where $\widehat{\sigma}_{22}(\Delta)$ and $\widehat{\nu}_2(\Delta)$ are the same as in the first part of the theorem. Hence

$$(2.64)\qquad \sigma_{11}^\perp(\Delta) + \widehat{\sigma}_{22}(\Delta) = \sigma_{22}(\Delta).$$

Now define

$$\nu_2(\Delta) \stackrel{def}{=} \nu_1^\perp(\Delta) + \widehat{\nu}_2(\Delta),$$

then (2.63) means that

$$\nu = \begin{bmatrix} \nu_1 \\ nu_2 \end{bmatrix} \in L^\sigma,$$

and

$$(2.65)\qquad \mu_\nu \leq \mu = \mu_{\nu_1} + \mu_{\nu_1^\perp}\ .$$

Finally observe, that the transition from $\nu = \begin{bmatrix} \nu_1 \\ \nu_2 \end{bmatrix} \in L^\sigma$ to the pair $\nu_1 \in L^{\sigma_{11}}, \nu_1^\perp \in L^{\sigma_{11}^\perp}$ in the first part of the theorem and the other way around from $\nu_1 \in L^{\sigma_{11}}, \nu_1^\perp \in L^{\sigma_{11}^\perp}$ to $\nu = \begin{bmatrix} \nu_1 \\ \nu_2 \end{bmatrix} \in L^\sigma$ in the second part of the theorem *are mutually inverse*, since $\widehat{\nu}_2(\Delta)$ appeared in both procedures is computed from ν_1 and σ by the same formula ((2.38), 2.37)). Then (2.58) and (2.65) taken together guarantee

$$(2.66)\qquad \mu_\nu = \mu_{\nu_1} + \mu_{\nu_1^\perp}\ .$$

Thus the theorem follows.

We apply the Schur complement to obtain an orthogonal decomposition of the Hellinger space L^σ in the next theorem.

Theorem 2.8. Let σ be the same as above (2.34), L^σ be the corresponding Hellinger space. Consider the subspaces in L^σ:

$$(2.67)\qquad \begin{aligned} L_1^\sigma &\stackrel{def}{=} \{\nu = \begin{bmatrix} \nu_1 \\ \nu_2 \end{bmatrix} \in L^\sigma : \| \nu \|_{L^\sigma} = \| \nu_1 \|_{L^{\sigma_{11}}} \}, \\ L_2^\sigma &\stackrel{def}{=} \{\nu = \begin{bmatrix} \nu_1 \\ \nu_2 \end{bmatrix} \in L^\sigma :\ \nu_1 = 0\}. \end{aligned}$$

Then

1. An $M = M_1 \oplus M_2$ valued vector-chart $\nu = \begin{bmatrix} \nu_1 \\ \nu_2 \end{bmatrix} \in L_1^\sigma$ if and only if

$$\nu_1 \in L^{\sigma_{11}} \quad \text{and} \quad \nu_2 = \widehat{\nu}_2, \tag{2.68}$$

where $\widehat{\nu}_2$ is defined by (2.38), (2.37) from ν_1 and σ. The correspondence $\nu_1 \to \begin{bmatrix} \nu_1 \\ \widehat{\nu}_2 \end{bmatrix}$ is unitary one to one between $L^{\sigma_{11}}$ and L_1^σ.

2. An $M = M_1 \oplus M_2$ valued vector-chart $\nu = \begin{bmatrix} \nu_1 \\ \nu_2 \end{bmatrix} \in L_2^\sigma$ if and only if

$$\nu_1 = 0 \quad \text{and} \quad \nu_2 \in L^{\sigma_{11}^\perp}, \tag{2.69}$$

where $\sigma_{11}^\perp$ is defined by (2.23) from σ. In that case

$$\| \nu \|_{L^\sigma} = \| \nu_2 \|_{L^{\sigma_{11}^\perp}} . \tag{2.70}$$

The correspondence $\nu_1^\perp \to \begin{bmatrix} 0 \\ \nu_1^\perp \end{bmatrix}$ is unitary one to one between $L^{\sigma_{11}^\perp}$ and L_2^σ.

3. There is an orthogonal decomposition

$$L^\sigma = L_1^\sigma \oplus L_2^\sigma, \tag{2.71}$$

and the orthogonal projections onto L_1^σ and L_2^σ are

$$P_1 \begin{bmatrix} \nu_1 \\ \nu_2 \end{bmatrix} = \begin{bmatrix} \nu_1 \\ \widehat{\nu}_2 \end{bmatrix} \quad \text{and} \quad P_2 \begin{bmatrix} \nu_1 \\ \nu_2 \end{bmatrix} = \begin{bmatrix} 0 \\ \nu_2 - \widehat{\nu}_2 \end{bmatrix} \tag{2.72}$$

respectively.

4. The mapping $\nu = \begin{bmatrix} \nu_1 \\ \nu_2 \end{bmatrix} \to \begin{bmatrix} \nu_1 \\ \nu_1^\perp \end{bmatrix}$ is unitary from L^σ onto $L^{\sigma_{11}} \oplus L^{\sigma_{11}^\perp}$, mapping L_1^σ onto $L^{\sigma_{11}}$ and L_2^σ onto $L^{\sigma_{11}^\perp}$.

Proof. 1. Let $\nu = \begin{bmatrix} \nu_1 \\ \nu_2 \end{bmatrix} \in L^\sigma$. Then by formula (2.53)

$$\| \nu \|_{L^\sigma}^2 = \| \nu_1 \|_{L^{\sigma_{11}}}^2 + \| \nu_2 - \widehat{\nu}_2 \|_{L^{\sigma_{11}^\perp}}^2 .$$

If $\nu \in L_1^\sigma$, then $\nu_2 - \widehat{\nu}_2 = 0$. Conversely, let $\nu_1 \in L^{\sigma_{11}}$. Define $\widehat{\nu}_2$ by formula (2.38). Then, by Theorem 2.7, $\nu = \begin{bmatrix} \nu_1 \\ \widehat{\nu}_2 \end{bmatrix} \in L^\sigma$, because $\nu_1^\perp = \widehat{\nu}_2 - \widehat{\nu}_2 = 0$. Hence

$$\| \nu \|_{L^\sigma} = \| \nu_1 \|_{L^{\sigma_{11}}} .$$

Thus, we have one to one unitary correspondence between L_1^σ and $L^{\sigma_{11}}$.

2. Let $\nu = \begin{bmatrix} \nu_1 \\ \widehat{\nu}_2 \end{bmatrix} \in L_2^\sigma$, i.e., $\nu_1 = 0$. But then $\widehat{\nu}_2 = 0$ (by (2.38), (2.37)). Therefore $\nu_1^\perp = \nu_2 \in L^{\sigma_{11}^\perp}$, and by formula (2.53)

$$\| \nu \|_{L^\sigma} = \| \nu_2 \|_{L^{\sigma_{11}^\perp}} .$$

Conversely, for any $\nu_1^\perp \in L^{\sigma_{11}^\perp}$, $\nu = \begin{bmatrix} 0 \\ \nu_1^\perp \end{bmatrix} \in L^\sigma$ by Theorem 2.7. This is one to one unitary correspondence between L_2^σ and $L^{\sigma_{11}^\perp}$.

3. For $\nu \in L_1^\sigma$ we have

$$\nu = \begin{bmatrix} \nu_1 \\ \widehat{\nu}_2 \end{bmatrix}$$

and $\nu_1^\perp = 0$, by the first statement of this theorem. For $\nu' \in L_2^\sigma$, we have

$$\nu' = \begin{bmatrix} 0 \\ \nu_2' \end{bmatrix}$$

i.e., $\nu_1' = 0$, $(\nu_1')^\perp = \nu_2' \in L^{\sigma_{11}^\perp}$, by the second statement of this theorem. Hence, by formula (2.53)

$$\begin{aligned} \langle \nu, \nu' \rangle_{L^\sigma} &= \langle \nu_1, \nu_1' \rangle_{L^{\sigma_{11}}} + \left\langle \nu_1^\perp, (\nu_1')^\perp \right\rangle_{L^{\sigma_{11}}} \\ &= \langle \nu_1, 0 \rangle_{L^{\sigma_{11}}} + \langle 0, \nu_2' \rangle_{L^{\sigma_{11}^\perp}} = 0. \end{aligned}$$

Therefore, $L_1^\sigma \perp L_2^\sigma$. For any $\nu \in L^\sigma$, we have

$$\nu = \begin{bmatrix} \nu_1 \\ \nu_2 \end{bmatrix} = \begin{bmatrix} \nu_1 \\ \widehat{\nu}_2 \end{bmatrix} + \begin{bmatrix} 0 \\ \nu_2 - \widehat{\nu}_2 \end{bmatrix}. \tag{2.73}$$

By Theorem 2.7, $\nu_1 \in L^{\sigma_{11}}$, $\nu_1^\perp = \nu_2 - \widehat{\nu}_2 \in L^{\sigma_{11}^\perp}$. By statements 1) and 2) of that theorem the summands in (2.73) belong to L_1^σ and L_2^σ respectively. The formulas for the projections P_1 and P_2 follow from (2.73) directly.

4. The unitarity of the mapping follows from Theorem 2.7. By statements 1) and 2) of that theorem, for $\nu \in L_1^\sigma$: ν_1 is arbitrary in $L^{\sigma_{11}}$, $\nu_1^\perp = 0$; for $\nu \in L_2^\sigma$: $\nu_1 = 0$, $\nu_1^\perp$ is arbitrary in $L^{\sigma_{11}^\perp}$.

Theorem 2.9. $L^\sigma = L_1^\sigma$ if and only if $\sigma_{11}^\perp = 0$, i.e., if and only if

$$\sigma_{22}(\Delta) = \underset{\Delta = \cup \Delta_j}{\text{s-lim}} \sum_j (\sigma_{11}(\Delta)^{[-1/2]} \sigma_{12}(\Delta))^* (\sigma_{11}^{[-1/2]}(\Delta) \sigma_{12}(\Delta)). \tag{2.74}$$

Proof. By Theorem 2.8

$$L^\sigma \ominus L_1^\sigma = L_2^\sigma \cong L^{\sigma_{11}^\perp}.$$

Hence, $L^\sigma = L_1^\sigma$ if and only if

$$L^{\sigma_{11}^\perp} = \{0\}.$$

But the latter is possible if and only if $\sigma_{11}^\perp = 0$ (see Section 1). The remaining part of the theorem follows from the definition of the Schur complement ((2.33), (2.26)).

We conclude this section with a remark on how the Schur complement and related objects can be defined in terms of harmonic continuations inside the disc $\mathbb{D}$, similarly to what we did defining the Hellinger measure.

Consider harmonic continuation of the measure σ:

$$\sigma(\zeta) = \begin{bmatrix} \sigma_{11}(\zeta) & \sigma_{12}(\zeta) \\ \sigma_{21}(\zeta) & \sigma_{22}(\zeta) \end{bmatrix} \geq 0, |\zeta| < 1. \tag{2.75}$$

The harmonic continuation of $\widehat{\sigma}_{22}$ is the least *harmonic* substitution for σ_{22} in (2.75) which preserves the nonnegativity of the whole matrix. Hence, it is simply the least harmonic majorant for

$$\widehat{\sigma_{22}(\zeta)} \stackrel{def}{=} (\sigma_{11}(\zeta)^{[-1/2]}\sigma_{12}(\zeta))^*(\sigma_{11}(\zeta)^{[-1/2]}\sigma_{12}(\zeta)). \tag{2.76}$$

Thus, $\widehat{\sigma}_{22}(\zeta)$ can be computed as

$$\widehat{\sigma}_{22}(\zeta) = \lim_{r\nearrow 1} \int_{\mathbb{T}} \frac{1-|\zeta|^2}{|t-\zeta|^2} \widehat{\sigma_{22}(rt)}\, m(dt), \tag{2.77}$$

and

$$\widehat{\sigma}_{22}(dt) = \lim_{r\nearrow 1} \widehat{\sigma_{22}(rt)}\, m(dt) \tag{2.78}$$

in the weak sense. Analogously (see (2.38), (2.37)),

$$\widehat{\nu}_2(\zeta) = \lim_{r\nearrow 1} \int_{\mathbb{T}} \frac{1-|\zeta|^2}{|t-\zeta|^2} \widehat{\nu_2(rt)}\, m(dt), \tag{2.79}$$

where

$$\widehat{\nu_2(\zeta)} = (\sigma_{11}(\zeta)^{[-1/2]}\sigma_{12}(\zeta))^*(\sigma_{11}(\zeta)^{[-1/2]}\nu_1(\zeta)). \tag{2.80}$$

Hence,

$$\widehat{\nu}_2(dt) = \lim_{r\nearrow 1} \widehat{\nu_2(rt)}\, m(dt) \tag{2.81}$$

in the weak sense. Then $\sigma_{11}^{\perp}$ and $\nu_1^{\perp}$ can be computed by (2.33) and (2.51) respectively.

3. Bounded variation and absolute continuity

The goal of this section is to discuss the Hellinger space L^σ in the case when σ is absolutely continuous with respect to a scalar measure τ. We will need some definitions and preliminaries first.

Let E be a Hilbert space, ν an E-valued vector-chart on $\mathbb{T}$. The value

$$\mathrm{var}_\nu(\Delta) \stackrel{def}{=} \sup_{\Delta=\cup\Delta_j} \sum_j \| \nu(\Delta_j) \| \tag{3.1}$$

is said to be the *variation of the chart* ν, where the supremum is taken over all finite partitions of Δ. We will say that ν is of *bounded variation* if

$$\mathrm{var}_\nu(\Delta) < \infty, \tag{3.2}$$

for any Borel set $\Delta \subseteq \mathbb{T}$.

Lemma 3.1. If ν is of bounded variation then the variation var_ν is a measure on $\mathbb{T}$.

Proof. Since $\nu(\Delta)$ is a countably additive function of the set, the function

$$C(\Delta) = \| \nu(\Delta) \|$$

enjoys the properties (1.6). Then var_ν is a measure by Lemma 1.4.

Remark. Obviously,

$$\| \nu(\Delta) \| \leq \mathrm{var}_\nu(\Delta), \tag{3.3}$$

for any Borel set $\Delta \subseteq \mathbb{T}$. Moreover, $\mathrm{var}_\nu(\Delta)$ is the smallest measure so that

$$\| \nu(\Delta) \| \leq \tau(\Delta), \tag{3.4}$$

as it follows from the definition (3.1). In terms of harmonic continuations $\mathrm{var}_\nu(\zeta)$ is the least harmonic majorant of $\| \nu(\zeta) \|$, $|\zeta| < 1$.

A vector-chart ν on $\mathbb{T}$ is said to be *absolutely continuous* with respect to a finite scalar measure τ on $\mathbb{T}$ if its variation var_ν is absolutely continuous with respect to τ:

$$\nu \ll \tau \overset{def}{\Longleftrightarrow} \mathrm{var}_\nu \ll \tau. \tag{3.5}$$

In particular, ν is absolutely continuous with respect to var_ν (and is of bounded variation).

Theorem 3.2 (Vector Radon-Nikodim). Let E be a Hilbert space, $\nu(\Delta)$ be an E-valued vector-chart on $\mathbb{T}$. Let ν be absolutely continuous with respect to a finite scalar measure τ on $\mathbb{T}$. Then there exists a unique measurable (in the topology of E) E-valued function $\nu'(t)$, defined τ-a.e. on $\mathbb{T}$, so that

$$\nu(\Delta) = \int_\Delta \nu'(t)\tau(dt), \tag{3.6}$$

where the integration is understood in the Bochner sense [24], and

$$\mathrm{var}_\nu(\Delta) = \int_\Delta \| \nu'(t) \| \, \tau(dt). \tag{3.7}$$

Proof. Observe first that the values $\nu(\Delta)$ run over a separable subspace $E_\nu \subseteq E$, when Δ runs over the Borel subsets of $\mathbb{T}$. One can see that, e.g., by means of harmonic continuation inside the unit disc and Fourier series expansion. Thus we can deal with the separable space E_ν only. For $e \in E_\nu$

$$| \langle \nu(\Delta), e \rangle | \leq \| \nu(\Delta) \| \cdot \| e \| \leq \mathrm{var}_\nu(\Delta) \cdot \| e \| . \tag{3.8}$$

Hence $\langle \nu(\Delta), e \rangle$ is absolutely continuous with respect to τ. By the Radon-Nikodim theorem

$$\langle \nu(\Delta), e \rangle = \int_\Delta f_e(t)\tau(dt), \tag{3.9}$$

where $f_e \in L^1_\tau$ is defined on $\mathbb{T}$ except for a set of τ-measure zero (which depends on e). Let $\{e_k\}$ be an orthonornal basis in E_ν. Fix representatives of f_{e_k} and let Δ_k be the Lebesgue subset [24] in the domain of f_{e_k} (with respect to τ), $\tau(\Delta_k) = \tau(\mathbb{T})$. Fix also $\xi \in L^1_\tau$ so that

$$\operatorname{var}_\nu(\Delta) = \int_\Delta \xi(t)\tau(dt). \tag{3.10}$$

Now let Δ^* be the intersection of all the Lebesgue sets Δ_k and the Lebesgue set of ξ, $\tau(\Delta^*) = \tau(\mathbb{T})$, since there is a countable number of them. For finite linear combinations $e = \sum c_k e_k$ define

$$f_e(t) = \sum_k \bar{c}_k f_{e_k}(t), t \in \Delta^*. \tag{3.11}$$

Obviously f is well defined on Δ^* and any point of Δ^* is a Lebesgue point for f_e (with respect to τ).

Moreover,

$$\begin{aligned} \int_\Delta f_e(t)\tau(dt) &= \sum \bar{c}_k \int_\Delta f_{e_k}(t)\tau(dt) \\ &= \sum \bar{c}_k \langle \nu(\Delta), e_k\rangle = \langle \nu(\Delta), e\rangle . \end{aligned} \tag{3.12}$$

Then, (3.12), (3.8) and (3.10) imply

$$|\int_\Delta f_e(t)\tau(dt)| \leq \int_D \xi(t)\tau(dt)\cdot \parallel e \parallel, \tag{3.13}$$

for any Borel set $\Delta \subseteq \mathbb{T}$ and e being any finite linear combination of e_k. Then

$$|f_e(t)| \leq \xi(t)\cdot \parallel e \parallel, t \in \Delta^*, \tag{3.14}$$

since points in Δ^* are Lebesgue points for all these functions. Fix now $t \in \Delta^*$, then $f_e(t)$ is an antilinear functional in e (see (3.11)) defined on a dense linear manifold in E_ν. By (3.14) it is bounded. Hence, $f_e(t)$ can be continued for any $e \in E_\nu$ as a bounded linear functional, which satisfies (3.14) for all $e \in E_\nu$. Then (3.14) guarantees (by the dominated convergence theorem) that for $f_e(t)$ defined that way

$$\int_\Delta f_e(t)\tau(dt) = \langle \nu(\Delta), e\rangle . \tag{3.15}$$

It also follows from (3.14) (by the Riesz theorem) that for any $t \in \Delta^*$ there exists unique $\nu'(t) \in E_\nu$ so that

$$f_e(t) = \langle \nu'(t), e\rangle , \ \forall e \in E_\nu, t \in \Delta^*, \tag{3.16}$$

and

$$\parallel \nu'(t) \parallel \leq \xi(t), t \in \Delta^*. \tag{3.17}$$

$\nu'(t)$ is defined on Δ^* ($\tau(\Delta^*) = \tau(\mathbb{T})$) and is weakly measurable (since $f_e(t)$ is measurable $\forall e \in E_\nu$). Since E_ν is separable $\nu'(t)$ is strongly measurable [24]. By (3.17)

$$\parallel \nu'(t) \parallel \in L^1_\tau. \tag{3.18}$$

Then [24] ν' is integrable in the sense of Bochner with respect to the measure τ and

$$\left\langle \int_{\Delta} \nu'(t)\tau(dt), e \right\rangle = \int_{D} \langle \nu'(t), e \rangle \, \tau(dt) \tag{3.19}$$

for any Borel set $\Delta \subseteq \mathbb{T}$ and $e \in E_{\nu}$. Since the integral on the right is equal to $\langle \nu(\Delta), e \rangle$, we obtain

$$\int_{D} \nu'(t)\tau(dt) = \nu(\Delta). \tag{3.20}$$

This is the desired representation. Now (3.20), (3.17), (3.10) imply

$$\| \nu(\Delta) \| \leq \int_{\Delta} \| \nu'(t) \| \, \tau(dt) \leq \int_{D} \xi(t)\tau(dt) = \mathrm{var}_{\nu}(\Delta). \tag{3.21}$$

But var_{ν} is the smallest measure which majorizes $\| \nu(\Delta) \|$. Hence,

$$\| \nu'(t) \| = \xi(t), \tau - a.e.$$

and

$$\mathrm{var}_{\nu}(\Delta) = \int_{\Delta} \| \nu'(t) \| \, \tau(dt). \tag{3.22}$$

Thus, the theorem follows.

We define the variation for operator-valued measures in terms of the operator (uniform) topology. Let E be a Hilbert space, $\sigma(\Delta) : E \to E$ be an operator-valued measure, then

$$\mathrm{var}_{\sigma}(\Delta) \stackrel{def}{=} \sup_{\Delta = \cup \Delta_j} \sum_j \| \sigma(\Delta_j) \|, \tag{3.23}$$

where the supremum is taken over all finite partitions. We say that σ is of *bounded variation* if $\mathrm{var}_{\sigma}(\Delta) < \infty$ for any Borel set $\Delta \subseteq \mathbb{T}$. If σ is of bounded variation then σ is a uniform measure (i.e., it is countably additive in the uniform topology).

Obviously,

$$\| \sigma(\Delta) \| \leq \mathrm{var}_{\sigma}(\Delta), \tag{3.24}$$

and this is the least majorizing measure. For any $e \in E$ the chart $\sigma(\Delta)e$ is of bounded variation and

$$\mathrm{var}_{\sigma e}(\Delta) \leq \mathrm{var}_{\sigma}(\Delta) \cdot \| e \| \, . \tag{3.25}$$

We say that σ is *absolutely continuous* with respect to a finite scalar measure τ if var_{σ} is absolutely continuous with respect to τ. In that case σe is also absolutely continuous with respect to τ by (3.25).

Applying Theorem 3.2 we obtain

$$\begin{aligned} \sigma(\Delta)e &= \int_{\Delta} (\sigma e)'(t)\tau(dt) \\ \mathrm{var}_{\sigma e}(\Delta) &= \int_{\Delta} \| (\sigma e)'(t) \| \, \tau(dt). \end{aligned} \tag{3.26}$$

Define $\eta(t)$ as follows

$$\operatorname{var}_\sigma(\Delta) = \int_\Delta \eta(t)\tau(dt). \tag{3.27}$$

Then (3.25), (3.26) and (3.27) imply

$$\| (\sigma e)'(t) \| \leq \eta(t) \cdot \| e \|, \tau - a.e. \tag{3.28}$$

by the Lebesgue differentiation theorem. But the exceptional set of τ-measure zero might depend on $e \in E$.

Lemma 3.3. Let E be a *separable* Hilbert space, then there exists a common set of full τ-measure where all of $(\sigma e)'$ are defined simultaneously.

$$(\sigma e)'(t) = \sigma'(t) \cdot e, \ \tau - a.e., \ \forall e \in M, \tag{3.29}$$

where $\sigma'(t)$ is an operator-function defined $\tau - a.e.$ on $\mathbb{T}$,

$$\| \sigma'(t) \| = \eta(t), \tau - a.e., \tag{3.30}$$

i.e.,

$$\operatorname{var}_\sigma(\Delta) = \int_\Delta \| \sigma'(t) \| \, \tau(dt). \tag{3.31}$$

Proof. Analogously to Theorem 3.2, pick up an orthonormal basis $\{e_k\}$ in E. There exists $\Delta^* \subseteq \mathbb{T}$ of full τ-measure, so that any $t \in \Delta^*$ is a Lebesgue point for all $(\sigma e_k)'$ and η. We define

$$\sigma'(t)e_k \stackrel{def}{=} (\sigma e_k)'(t), t \in \Delta^*.$$

Then we extend it by linearity to finite linear combinations $e = \sum c_k e_k$. Applying Lebesgue differentiation to (3.26) we can see that for those e

$$(\sigma e)'(t) = \sum_k c_k (\sigma e_k)'(t), t \in \Delta^*$$

by the linearity of $\sigma(\Delta)$. Hence,

$$\sigma'(t)e = (\sigma e)'(t), t \in \Delta^*$$

for finite linear combinations of e_k.

By means of (3.28) we can extend $\sigma'(t)$ to all e for fixed $t \in \Delta^*$ and verify (3.29). Then (3.28) implies

$$\| \sigma'(t) \| \leq \eta(t), t \in \Delta^* \tag{3.32}$$

$\| \sigma'(t) \|$ is a measurable function since E is a separable space and $\| \sigma'(t)e \|$ is measurable for all $e \in E$. Hence,

$$\| \sigma'(t) \| \in L^1_\tau \, .$$

Because of (3.29) and (3.26)

$$\sigma(\Delta)e = \int_\Delta \sigma'(t)e\tau(dt),$$

which implies

$$\| \sigma(\Delta) \| \leq \int_{\Delta} \| \sigma'(t) \| \ \tau(dt). \tag{3.33}$$

Since var_σ is the minimal measure majorizing $\| \sigma(\Delta) \|$, (3.27), (3.32) and (3.33) imply

$$\| \sigma'(t) \| = \eta(t), \tau - a.e.$$

and

$$\mathrm{var}_\sigma(\Delta) = \int_{D} \| \sigma'(t) \| \ \tau(dt).$$

Thus the lemma follows.

But if E is not separable, then it might happen that there is not one single point on $\mathbb{T}$ where all $(\sigma e)'$ are defined. It makes no sense to talk about the operator function $\sigma'(t)$ in that case. Lemmas 3.7–3.9 will help us to give it sense in the context that we need.

We turn now to discussing the Hellinger space L^σ in the case when σ is absolutely continuous.

Theorem 3.4. Let σ be an operator-valued measure of bounded variation, $\sigma(\Delta) : E \to E$, E being a Hilbert space. Let $\nu \in L^\sigma$. Then ν is of bounded variation and

$$\mathrm{var}_\nu(\Delta) \leq \mathrm{var}_\sigma(\Delta)^{1/2} \cdot \mu_\nu(\Delta)^{1/2}, \tag{3.34}$$

where μ_ν is the Hellinger measure of ν with respect to σ.

Proof. Since $\nu \in L^\sigma$, then

$$| \langle \nu(\Delta), e \rangle |^2 \leq \langle \sigma(\Delta) e, e \rangle \cdot \mu_\nu(\Delta)$$

for all $e \in E$. Hence

$$\| \nu(\Delta) \|^2 \leq \| \sigma(\Delta) \| \cdot \mu_\nu(\Delta) \leq \mathrm{var}_\sigma(\Delta) \cdot \mu_\nu(\Delta).$$

Then

$$\begin{aligned}
\mathrm{var}_\nu(\Delta) &= \sup_{\Delta = \cup \Delta_j} \sum_j \| \nu(\Delta_j) \| \\
&\leq \sup \sum_j \sqrt{\mathrm{var}_\sigma(\Delta_j) \cdot \mu_\nu(\Delta_j)} \\
&\leq \sup \{ (\sum_j \mathrm{var}_\sigma(\Delta_j))^{1/2} \cdot (\sum_j \mu_\nu(\Delta_j))^{1/2} \} \\
&= \mathrm{var}_\sigma(\Delta)^{1/2} \cdot \mu_\nu(\Delta)^{1/2}.
\end{aligned}$$

Corollary 3.5. Let σ be an operator-valued measure, $\sigma(\Delta) : E \to E$, absolutely continuous with respect to a finite scalar measure τ. Then for any $\nu \in L^\sigma, \nu$ is absolutely continuous with respect to τ.

Proof. This is immediate corollary of (3.34).

We formulate the next theorem in the general case, but we first give a proof in a special case. The lemmas following the theorem reduce the general case to the special case and give sense to the formula (3.35).

Theorem 3.6. Let σ be an operator measure on $\mathbb{T}$, $\sigma(\Delta) : E \to E, E$ being a Hilbert space. Let σ be absolutely continuous with respect to a finite scalar measure τ. Let $\nu \in L^\sigma$. Then

$$\mu_\nu(\Delta) = \int_\Delta \| \sigma'(t)^{[-1/2]}\nu'(t) \|_E^2 \ \tau(dt), \tag{3.35}$$

where σ' and ν' are the Radon-Nikodim derivatives of σ and ν respectively with respect to the measure τ (see above).

Proof (The case when E is a separable space). Consider the inequality

$$\begin{bmatrix} \mu_\nu(\Delta) & \langle e, \nu(\Delta)\rangle \\ \langle \nu(\Delta), e\rangle & \langle \sigma(\Delta)e, e\rangle \end{bmatrix} \geq 0.$$

The minimality of μ_ν forces it to be absolutely continuous with respect to τ since all the other entries are. Hence,

$$\begin{bmatrix} \mu'_\nu(t) & \langle e, \nu'(t)\rangle \\ \langle \nu'(t), e\rangle & \langle \sigma'(t)e, e\rangle \end{bmatrix} \geq 0, \ \tau - a.e.,$$

(simultaneously $\forall e \in E$, because of the separability).

The minimality of μ_ν again leads to

$$\mu'_\nu(t) = \| \sigma'(t)^{[-1/2]}\nu'(t) \|^2, \tau - a.e.$$

Thus the theorem follows.

Lemma 3.7. Let E be a Hilbert space. Let A be a countable set of bounded linear operators on E. Then for any $e \in E$ there exists a separable subspace $\tilde{E}$, so that

$$e \in \tilde{E} \subseteq E$$

which is invariant under all $a \in A$.

Proof. Consider the following sets in E:

$$E_0 = \{e\}, E_1 = \{a_1 e\}_{a_1 \in A}, E_2 = \{a_2 a_1 e\}_{\substack{a_1 \in A \\ a_2 \in A}}, \cdots .$$

Each one is obviously countable.

Let

$$E_\infty = \bigcup_{k=0}^{\infty} E_k,$$

which is also countable. The linear span $Lin\, E_\infty$ is invariant under all $a \in A$, hence $\overline{Lin\, E_\infty}$ is invariant under all $a \in A$. But $\overline{Lin\, E_\infty}$ is separable (say, orthogonalizing the system E_∞ we obtain a countable basis in $\overline{Lin\, E_\infty}$). Thus the lemma follows.

Lemma 3.8. Let E be a Hilbert space. Let $\tilde{E}$ be a separable subspace $\tilde{E} \subseteq E$. Let A be a countable set of bounded linear operators on E. Then there exists a separable subspace $\approx\!\!\!\!E$ $\overset{\approx}{E}$, so that

$$\tilde{E} \subseteq \overset{\approx}{E} \subseteq E$$

and $\overset{\approx}{E}$ is invariant under all $a \in A$.

Proof. Pick up an orthonormal basis $\{e_k\}$ in $\tilde{E}$. By Lemma 3.7 for any e_k there exists separable $\tilde{E}_k$ so that

$$e_k \in \tilde{E}_k \subseteq E$$

and $\tilde{E}_k$ is invariant under all $a \in A$. Then $\overset{\approx}{E} = \overline{Lin(\bigcup_k \tilde{E}_k)}$ contains $\tilde{E}$, is invariant under all $a \in A$, and is separable (we can choose a basis in each $\tilde{E}_k$ and then orthogonalize their union).

Lemma 3.9. Let E be a Hilbert space, $\tilde{E}$ a separable subspace, $\tilde{E} \subseteq E$. Let σ be an operator measure, $\sigma(\Delta) : E \to E$ on $\mathbb{T}$. Then there exists a separable subspace $\overset{\approx}{E}$, so that

$$\tilde{E} \subseteq \overset{\approx}{E} \subseteq E$$

and $\overset{\approx}{E}$ is invariant under $\sigma(\Delta)$ for any Borel $\Delta \subseteq \mathbb{T}$.

Proof. Let

$$\sigma(\zeta)e = \int_{\mathbb{T}} \frac{1 - |\zeta|^2}{|t - \zeta|^2} \sigma(dt)e$$

be the harmonic continuation inside the unit disc. Let σ_k be the Fourier coeffizients, they are bounded linear operators. By Lemma 3.8, there exists a separable $\overset{\approx}{E}$ so that

$$\tilde{E} \subseteq \overset{\approx}{E} \subseteq E$$

and $\overset{\approx}{E}$ is invariant under all σ_k. Then $\overset{\approx}{E}$ is invariant under $\sigma(\zeta)$, $|\zeta| < 1$. Then $\overset{\approx}{E}$ is invariant under $\sigma(\Delta)$ for any Borel set $\Delta \subseteq \mathbb{T}$. In fact, $\forall\ e \in \overset{\approx}{E},\ e' \in \overset{\approx}{E}^{\perp}$

$$\langle \sigma(\zeta)e, e' \rangle = 0.$$

Hence

$$\langle \sigma(\Delta)e, e' \rangle = 0,$$

i.e., $\sigma(\Delta)e \in \overset{\approx}{E}$. Thus the lemma follows.

Proof of Theorem 3.6 (General Case). Since the values of ν belong to a separable subspace $E_\nu \subseteq E$, by Lemma 3.9 there exists a separable subspace $\tilde{E}_\nu$, so that

$$E_\nu \subseteq \tilde{E}_\nu \subseteq E$$

and $\tilde{E}_\nu$ is invariant for $\sigma(\Delta)$ for all Borel sets $\Delta \subseteq \mathbb{T}$. $\sigma \mid_{\tilde{E}_\nu}$ is also absolutely continuous with respect to τ. Thus, we reduced the general case to a separable case: $(\sigma \mid_{\tilde{E}_\nu})'(t)$ exists $\tau - a.e.$, and

$$(\sigma \mid_{\tilde{E}_\nu})'(t)^{[-1/2]}\nu'(t) \tag{3.36}$$

is well defined $\tau - a.e.$ Moreover, the vector function (3.36) does not depend on the choice of an invariant subspace $\tilde{E}_\nu$ modulo a set of τ-measure zero.

Theorem 3.10. Let σ be an operator measure on $\mathbb{T}$, $\sigma(\Delta) : E \to E$, E being a Hilbert space. Let σ be absolutely continuous with respect to a finite scalar measure τ. Let $\nu, \tilde{\nu} \in L^\sigma$. Then

$$\langle \nu, \tilde{\nu} \rangle_{L^\sigma} = \int_{\mathbb{T}} \left\langle \sigma'(t)^{[-1/2]}\nu'(t), \sigma'(t)^{[-1/2]}\tilde{\nu}'(t) \right\rangle \tau(dt). \tag{3.37}$$

Proof. It follows from Theorem 3.6 by means of the polarization formula.

Remark. Thus in the case when σ is absolutely continuous with respect to τ, the Hellinger space can be realized as a space of vector functions $\nu'(t)$ (instead of measures ν). The inner product is defined (see (3.37)) by the scalar measure τ and the operator weight $\sigma'(t) \geq 0$, which is actually well defined $\tau - a.e.$ as an operator function on separable subspaces of E only (but the operator functions are consistent). The restrictions on the weight are

1. $\sigma'(t)e$ is strongly measurable $\forall e \in E$.
2. $\int_{\mathbb{T}} \langle \sigma'(t)e, e \rangle \, \tau(dt) < k \cdot \| e \|^2$, $\forall e \in E$.

Restrictions on vector-functions ν':

1. $\nu'(t)$ is strongly measurable (i.e., it can be strongly approximated by simple functions [24]).
2. $\int_{\mathbb{T}} \| \nu'(t) \| \, \tau(dt) < \infty$, which is equivalent to:

2′. $\| \int_\Delta \nu'(t)\tau(dt) \| < \infty$ for any Borel set $\Delta \subseteq \mathbb{T}$, the integral being understood in the Bochner sense.

3. There exists a function $C \in L^1_\tau$, so that

$$\begin{bmatrix} C(t) & \langle e, \nu'(t) \rangle \\ \langle \nu'(t), e \rangle & \langle \sigma'(t)e, e \rangle \end{bmatrix} \geq 0 \quad \tau - a.e., \quad \forall e \in E.$$

Theorem 3.11. Let $M = M_1 \oplus M_2$ be a Hilbert space. Let σ be an operator measure, $\sigma(\Delta) : M \to M$,

$$\begin{bmatrix} \sigma_{11}(\Delta) & \sigma_{12}(\Delta) \\ \sigma_{21}(\Delta) & \sigma_{22}(\Delta) \end{bmatrix} : M_1 \oplus M_2 \to M_1 \oplus M_2.$$

Let σ be absolutely continuous with respect to a finite scalar measure τ. Then $\widehat{\sigma}_{22}, \sigma_{11}^{\perp}$ (see (2.26), (2.22)) are absolutely continuous with respect to τ and

$$
\begin{aligned}
\widehat{\sigma}_{22}'(t) &= (\sigma_{11}'(t)^{[-1/2]}\sigma_{12}'(t))^*(\sigma_{11}'(t)^{[-1/2]}\sigma_{12}'(t)),\\
(\widehat{\sigma}_{11}^{\perp})'(t) &= \sigma_{22}'(t) - \widehat{\sigma}_{22}'(t).
\end{aligned} \tag{3.38}
$$

For any $\nu \in L^{\sigma}$, $\widehat{\nu}_2$ and $\nu_1^{\perp}$ (see (2.36), (2.37)) are absolutely continuous with respect to τ and

$$
\begin{aligned}
\widehat{\nu}_2'(t) &= (\sigma_{11}'(t)^{[-1/2]}\sigma_{12}'(t))^*\sigma_{11}'(t)^{[-1/2]}\nu_1'(t),\\
(\nu_1^{\perp})(t) &= \nu_2'(t) - \widehat{\nu}_2'(t).
\end{aligned} \tag{3.39}
$$

Proof. Since $\widehat{\sigma}_{22}(\Delta) \le \sigma_{22}(\Delta), \widehat{\sigma}_{22}$ is absolutely continuous with respect to τ, and hence so is $\sigma_{11}^{\perp}$. Then by Corollary 3.5 so is $\nu_1^{\perp}$ is (since $\nu_1^{\perp} \in L^{\sigma_{11}^{\perp}}$), and finally $\widehat{\nu}_2$ is absolutely continuous with respect to τ. It follows then from (3.26) that

$$
\begin{bmatrix}
\langle(\sigma_{11}m_1)'(t), m_1\rangle & \langle(\sigma_{12}m_2)'(t), m_1\rangle\\
\langle(\sigma_{21}m_1)'(t), m_2\rangle & \langle(\widehat{\sigma}_{22}m_2)'(t), m_2\rangle
\end{bmatrix} \ge 0 \tag{3.40}
$$

$\tau - a.e.$ (but the set depends on m_1 and m_2). If M is a separable space, then by Lemma 3.3, (3.40) can be rephrased as

$$
\begin{bmatrix}
\sigma_{11}'(t) & \sigma_{12}'(t)\\
\sigma_{21}'(t) & \sigma_{22}'(t)
\end{bmatrix} \ge 0, \tau - a.e. \tag{3.41}
$$

(3.38) follows now from the minimality of $\widehat{\sigma}_{22}$ (see the definition (2.22), (2.26)).

In general case (3.38) can be realized weakly. By the formula (2.44)

$$\langle\widehat{\sigma}_{22}(\Delta)m_2, m_2\rangle = \mu_{\sigma_{12}m_2}(\Delta)$$

for any Borel set $\Delta \subseteq \mathbb{T}$, $m_2 \in M_2$. Applying (3.26) and (3.35) we obtain

$$\langle(\widehat{\sigma}_{22}m_2)'(t), m_2\rangle = \| \sigma_{11}'(t)^{[-1/2]}(\sigma_{12}m_2)'(t) \|^2, \tag{3.42}$$

$\tau - a.e.$ Or, by means of polarization and linearity,

$$
\begin{aligned}
&\langle(\widehat{\sigma}_{22}m_2)'(t), \tilde{m}_2\rangle =\\
&= \left\langle \sigma_{11}'(t)^{[-1/2]}(\sigma_{12}m_2)'(t), \sigma_{11}'(t)^{[-1/2]}(\sigma_{12}\tilde{m}_2)'(t)\right\rangle
\end{aligned} \tag{3.43}
$$

$\tau - a.e.$, which is the weak form of (3.38). Analogously, by (2.48)

$$\langle\widehat{\nu}_2(\Delta), m_2\rangle_{\mu_2} = \langle\nu_1 \cdot \chi_\Delta, \sigma_{12}m_2 \cdot \chi_\Delta\rangle_{L^{\sigma_{11}}},$$

for any Borel set $\Delta \subseteq \mathbb{T}$, $m_2 \in M_2$, where χ_Δ is the characteristic function of the set Δ. By (1.13)

$$\| \nu \cdot \chi_\Delta \|_{L^{\sigma}}^2 = \mu_\nu(\Delta)$$

for any $\nu \in L^\sigma$ and a Borel set $\Delta \subseteq \mathbb{T}$. Then, by means of (3.6), (3.35) and the polarization formula

$$\langle \widehat{\nu}_2'(t), m_2 \rangle = \left\langle \sigma_{11}'(t)^{[-1/2]} \nu_1'(t), \sigma_{11}'(t)^{[-1/2]} (\sigma_{12} m_2)'(t) \right\rangle, \tag{3.44}$$

$\tau - a.e.$, which is the weak form of (3.39). In case M is separable, (3.39) makes literal sense. Thus, the theorem follows.

Remark 3.12. In fact $\widehat{\sigma}_{22}$ is absolutely continuous with respect to τ even without assuming that σ_{22} is, since $\widehat{\sigma}_{22}$ does not depend on σ_{22} (see also Corollary 3.5).

Remark 3.13. The conditions of Theorem 3.11 can be relaxed further. Let σ_{12} be absolutely continuous with respect to a scalar measure τ. Let σ_{11} be of bounded variation and let

$$\mathrm{var}_{\sigma_{11}} = \mu_a + \mu_s \ ,$$

where μ_a is absolutely continuous with respect to τ, and μ_s is mutually singular with respect to τ. Let

$$\mathbb{T} = \Delta_a \cup \Delta_s$$

where Δ_a and Δ_s are the supports of μ_a and μ_s respectively. Then

$$\mathrm{var}_{\sigma_{11}|\Delta_a} = \mu_a \ ,$$

hence $\sigma_{11} \mid \Delta_\alpha$ is absolutely continuous with respect to τ. Then

$$\widehat{\sigma}_{22}'(t) = ((\sigma_{11} \mid \Delta_a)'(t)^{[-1/2]} \sigma_{12}'(t))^* ((\sigma_{11} \mid \Delta_a)'(t)^{[-1/2]} \sigma_{12}'(t)).$$

4. Models of unitary operators

This section is devoted to discussing the role of the Hellinger spaces L^σ as functional models for unitary operators. Some statements in this section are considered as known and are given without proofs.

Let U be a unitary operator on a Hilbert space L. Let E be another Hilbert space and

$$\alpha : E \to L \tag{4.1}$$

be a bounded linear mapping of E into L. We will call the pair (E, α) a *scale* for the operator U. Or, briefly, just *the mapping α will be called a scale.* We will use the notation L_α for the *minimal subspace in L which contains $\alpha(E)$ and reduces U*, i.e.,

$$L_\alpha \stackrel{def}{=} \bigvee_{k=-\infty}^{+\infty} U^k \alpha(E), \tag{4.2}$$

where $\bigvee$ denotes the closed linear span. Let $\mathcal{P}_U(\zeta)$, $|\zeta| < 1$, be the Poisson kernel for the operator U^*:

$$\mathcal{P}_U(\zeta) \stackrel{def}{=} (\mathbf{1} - \zeta U)^{-1} + (\mathbf{1} - \bar{\zeta} U^*)^{-1} - \mathbf{1}, \quad |\zeta| < 1. \tag{4.3}$$

It is a harmonic nonnegative operator-function:

$$\mathcal{P}_U(\zeta) \geq 0, \quad |\zeta| < 1.$$

We will call the harmonic operator function

$$\sigma_\alpha(\zeta) \overset{def}{=} \alpha^* \mathcal{P}_U(\zeta)\alpha, \tag{4.4}$$

the *spectral function* of U with respect to the scale α; it is also a nonnegative harmonic function in the unit disc.

Let $\mathcal{E}_U(\Delta)$, for a Borel set $\Delta \subseteq \mathbb{T}$, be the orthogonal resolution of the identity for the operator U^*. Then,

$$\mathcal{P}_U(\zeta)\ell = \int_{\mathbb{T}} \frac{1-|\zeta|^2}{|t-\zeta|^2} \mathcal{E}_U(dt)\ell, \ell \in L, \tag{4.5}$$

and

$$\sigma_\alpha(\zeta)e = \int_{\mathbb{T}} \frac{1-|\zeta|^2}{|t-\zeta|^2} \sigma_\alpha(dt)e, \quad e \in E, \tag{4.6}$$

where

$$\sigma_\alpha(\Delta) = \alpha^* \mathcal{E}_U(\Delta)\alpha, \tag{4.7}$$

for a Borel set $\Delta \subseteq \mathbb{T}$.

The operator measure $\sigma_\alpha(\Delta)$ will be called the *spectral measure* of the operator U with respect to the scale α.

Theorem 4.1. Let U be a unitary operator on a Hilbert space L, and let $\alpha : E \to L$ be a scale. Let L_α be the corresponding reducing subspace (4.2), $L_\alpha \subseteq L$, and let $\sigma_\alpha(\Delta)$ be the spectral measure (4.7). Then the mapping Q_α

$$Q_\alpha : \ell \to \nu_\ell, \tag{4.8}$$

where $\ell \in L$, and

$$\nu_\ell(\Delta) \overset{def}{=} \alpha^* \mathcal{E}_U(\Delta)\ell, \tag{4.9}$$

for a Borel set $\Delta \subseteq \mathbb{T}$, maps L_α onto the Hellinger space L^{σ_α} (see Section 1) isometrically, Q_α vanishes on $L_\alpha^\perp = L \ominus L_\alpha$, and

$$Q_\alpha U = \bar{t} Q_\alpha, \tag{4.10}$$

where t is the independent variable on $\mathbb{T}$, $|t| = 1$;

$$Q_\alpha \alpha = \alpha', \tag{4.11}$$

where $\alpha' : E \to L^{\sigma_\alpha}$ is defined by the formula

$$\alpha' e = \sigma_\alpha e. \tag{4.12}$$

In particular, the operator

$$U_\alpha \overset{def}{=} U \mid L_\alpha \tag{4.13}$$

is unitarily equivalent to the multiplication operator

$$\bar{t} : L^{\sigma_\alpha} \to L^{\sigma_\alpha}, \tag{4.14}$$

which is called the *model operator* for U_α.

The mapping Q_α is said to be the *Fourier representation* of the operator U with respect to the scale α.

The triplet consisting of a Hilbert space L, a unitary operator U, and a scale α is said to be a *unitary scattering system.*

Let L, U, $\alpha : E \to L$ and L', U', $\alpha' : E \to L'$ be two unitary scattering systems. They are said to be *unitarily equivalent* if there exists a unitary mapping $Q : L \to L'$, such that

$$\begin{aligned} QU &= U'Q, \\ Q\alpha &= \alpha'. \end{aligned} \tag{4.14$'$}$$

Theorem 4.1$'$. If two unitary scattering systems are unitarily equivalent, then $\sigma'_{\alpha'} = \sigma_\alpha$. Conversely, if $\sigma'_{\alpha'} = \sigma_\alpha$ and $L = L_\alpha$, $L' = L'_{\alpha'}$, then the systems are unitarily equivalent and the equivalence Q is unique.

Theorem 4.2. Let E be a Hilbert space. An arbitrary strong operator measure σ, $\sigma(\Delta) : E \to E$ for a Borel set $\Delta \subseteq \mathbb{T}$, is the spectral measure of a unitary scattering system.

Proof. The multiplication operator

$$\bar{t} : L^\sigma \to L^\sigma$$

and the mapping $\alpha : E \to L^\sigma$, defined as

$$\alpha : e \to \sigma e$$

may serve as such a unitary operator and a scale, respectively. Thus the theorem follows.

Now let $\alpha : E \to L$ and $\beta : N \to L$ be two scales for the same operator U. We define the *sum of the two scales* as follows

$$\rho \stackrel{def}{=} \alpha + \beta,$$

if

$$\begin{aligned} \rho : E \oplus N &\to L, \\ \rho \mid E &= \alpha, \\ \rho \mid N &= \beta. \end{aligned} \tag{4.15}$$

We emphasize here that the scale space of ρ is the orthogonal sum of E and N *by definition*, but the ranges of E and N under ρ need not be orthogonal in L.

Theorem 4.3 (Main). Let U be a unitary operator on a Hilbert space L. Let $\alpha : E \to L$ and $\beta : N \to L$ be two scales and $\rho : M = E \oplus N \to L$ be their sum $\rho = \alpha + \beta$. Let L_α, L_β and L_ρ be the reducing subspaces for U in L generated by these scales. Let σ_α, σ_β and σ_ρ be the spectral functions of U with respect to the scales α, β and ρ respectively. Then

$$\sigma_\rho = \begin{bmatrix} \sigma_\alpha & \sigma_{\alpha\beta} \\ \sigma_{\beta\alpha} & \sigma_\beta \end{bmatrix},$$

where $\sigma_{\alpha\beta}(\Delta) = \alpha^* \mathcal{E}_U(\Delta)\beta$, $\sigma_{\beta\alpha} = \sigma_{\alpha\beta}^*$. The Fourier representation Q_ρ maps unitarily L_ρ onto L^{σ_ρ}, so that L_α goes onto $L_1^{\sigma_\rho}$ and $L_\rho \ominus L_\alpha$ goes onto $L_2^{\sigma_\rho}$ (see the definitions (2.67)). $L_\rho = L_\alpha$ if and only if the Schur complement (see Section 2) of σ_α in σ_ρ is zero

$$\sigma_\alpha^\perp = 0, \tag{4.16}$$

i.e.,

$$\sigma_\beta(\Delta) = \underset{\Delta = \cup \Delta_j}{\text{s-lim}} \sum_j (\sigma_\alpha(\Delta_j)^{[-1/2]} \sigma_{\alpha\beta}(\Delta_j))^* (\sigma_\alpha(\Delta_j)^{[-1/2]} \sigma_{\alpha\beta}(\Delta_j)) \tag{4.17}$$

along the net of finite partitions.

Proof. By Theorem 4.1, Q_ρ maps L_ρ onto L^{σ_ρ} unitarily and Q_α maps $L_\alpha \subseteq L_\rho$ onto L^{σ_α} unitarily. But

$$(Q_\rho \ell)(\Delta) = \begin{bmatrix} \alpha^* \\ \beta^* \end{bmatrix} \mathcal{E}_U(\Delta)\ell = \begin{bmatrix} Q_\alpha(\Delta)\ell \\ Q_\beta(\Delta)\ell \end{bmatrix}.$$

Hence, for $\ell \in L_\alpha$ we have

$$\| Q_\rho \ell \|_{L^{\sigma_\rho}} = \| \ell \| = \| Q_\alpha \ell \|_{L^{\sigma_\alpha}}.$$

I.e., for $\nu = \begin{bmatrix} \nu_1 \\ \nu_2 \end{bmatrix} \in Q_\rho L_\alpha$ we have

$$\| \nu \|_{L^{\sigma_\rho}} = \| \nu_1 \|_{L^{\sigma_\alpha}}.$$

Hence,

$$Q_\rho L_\alpha \subseteq L_1^{\sigma_\rho}. \tag{4.18}$$

Since Q_α vanishes on $L_\alpha^\perp = L_\rho \ominus L_\alpha$, for $\nu = \begin{bmatrix} \nu_1 \\ \nu_2 \end{bmatrix} \in Q_\rho L_\alpha^\perp$ we have

$$\nu_1 = 0.$$

Hence,

$$Q_\rho L_\alpha^\perp \subseteq L_2^{\sigma_\rho}. \tag{4.19}$$

By Theorem 2.8

$$L^{\sigma_\rho} = L_1^{\sigma_\rho} \oplus L_2^{\sigma_\rho}.$$

Since Q_ρ maps $L_\rho = L_\alpha \oplus L_\alpha^\perp$ onto $L^{\sigma_\rho} = L_1^{\sigma_\rho} \oplus L_2^{\sigma_\rho}$, then (4.18) and (4.19) imply

$$Q_\rho L_\alpha = L_1^{\sigma_\rho},$$

$$Q_\rho L_\alpha^\perp = L_2^{\sigma_\rho}.$$

In particular, $L_\rho = L_\alpha$ if and only if $L^{\sigma_\rho} = L_1^{\sigma_\rho}$. The rest of the theorem follows from Theorem 2.9.

We discuss now several frequently encountered special cases of unitary scattering systems.

4.1. Cotlar-Sadosky scattering systems [23]

Let U be a unitary operator on a Hilbert space L, let $\alpha : E \to L$ be a scale, and let $E = E_1 \oplus E_2$. Then $\alpha = \alpha_1 + \alpha_2$, where

$$\alpha_1 = \alpha \mid E_1, \alpha_2 = \alpha \mid E_2.$$

In that case the spectral measure is of the form

$$\sigma_\alpha = \begin{bmatrix} \sigma_{\alpha_1} & \sigma_{\alpha_1\alpha_2} \\ \sigma_{\alpha_2\alpha_1} & \sigma_{\alpha_2} \end{bmatrix}, \tag{4.20}$$

where the entries are of the form $\alpha_i^* \mathcal{E}_U \alpha_j$.

4.2. Adamjan-Arov scattering systems [8, 9]

In the previous case we assume now in addition that α_1 and α_2 are isometries, and the subspaces $\alpha_1(E_1) \subseteq L$, $\alpha_2(E_2) \subseteq L$ are *wandering* for the operator U. (A subspace $\tilde{L} \subseteq L$ is said to be wandering for an operator U if $U^k\tilde{L} \perp U^j\tilde{L}$, where k, j are integers, $k \neq j$). In that case

$$\sigma_{\alpha_1}(\zeta) = \mathbf{1}_{E_1}, \sigma_{\alpha_2}(\zeta) = \mathbf{1}_{E_2},$$

which follows directly from the definition (4.3) by the wandering properties and the isometricity of the scales α_1 and α_2. Hence,

$$\sigma_{\alpha_1}(\Delta) = m(\Delta) \cdot \mathbf{1}_{E_1}, \sigma_{\alpha_2}(\Delta) = m(\Delta) \cdot \mathbf{1}_{E_2},$$

where $m(\Delta)$ is the normalized Lebesgue measure on $\mathbb{T}$. Since $\sigma_\alpha \geq 0$, (4.20) implies

$$\mid \langle \sigma_{\alpha_2\alpha_1}(\Delta)e_1, e_2 \rangle \mid \leq m(\Delta) \cdot \parallel e_1 \parallel \cdot \parallel e_2 \parallel . \tag{4.21}$$

Hence,

$$\parallel \sigma_{\alpha_2\alpha_1}(\Delta) \parallel \leq m(\Delta), \tag{4.22}$$

and $\sigma_{\alpha_2\alpha_1}$ is also absolutely continuous with respect to the Lebesgue measure. Then (see (3.26))

$$\sigma_{\alpha_2\alpha_2}(\Delta)e_1 = \int_\Delta (we_1)(t)m(dt) \tag{4.23}$$

and (by (3.28))

$$\parallel (we_1)(t) \parallel \leq \parallel e_1 \parallel, a.e. \text{ on } \mathbb{T} \tag{4.24}$$

(in general the set depends on e_1). The harmonic continuation $w(\zeta) = \sigma_{\alpha_2\alpha_1}(\zeta)$ is well defined as a contractive valued operator function on $\mathbb{D}$, $\parallel w(\zeta) \parallel \leq 1, |\zeta| < 1$. In this case w is said to be the *scattering function* of the scattering system.

By the wandering properties and the isometricity of the scales α_i the subspaces L_{α_i} have the following structure

$$L_{\alpha_i} = \cdots \oplus E_i^{(-1)} \oplus E_i^{(0)} \oplus E_i^{(1)} \oplus \cdots , \tag{4.25}$$

where $E_i^{(0)} = \alpha_i(E_i)$, $E_i^{(k)} = U^{-k}E_i^{(0)}, k \in \mathbb{Z}$, and $U_{\alpha_i} = U \mid_{L_{\alpha_i}}$ are backward shifts. The subspaces L_{α_i} are called in this case the *scattering channels.*

The results of Section 3 are applicable in this case, which leads to the following functional model of U_α: L^w is a space of (strongly) measurable vector functions on $\mathbb{T}$, where L^w is a brief notation for $L^{\left[\begin{smallmatrix}\mathbf{1} & w^* \\ w & \mathbf{1}\end{smallmatrix}\right]}$:

$$L^w = \{f = \begin{bmatrix} f_1 \\ f_2 \end{bmatrix} : \int_{\mathbb{T}} \| \begin{bmatrix} \mathbf{1}_{E_1} & w(t)^* \\ w(t) & \mathbf{1}_{E_2} \end{bmatrix}^{[-1/2]} \begin{bmatrix} f_1 \\ f_2 \end{bmatrix} \|^2 m(dt) < \infty\}. \tag{4.26}$$

The integral is the square of the norm on L^w and U_α acts as the multiplication by $\bar{t}$. Now the scattering channels are

$$\begin{aligned} L_{\alpha_1} &\longrightarrow \begin{bmatrix} \mathbf{1} \\ w \end{bmatrix} L^2(E_1), \\ L_{\alpha_2} &\longrightarrow \begin{bmatrix} w^* \\ \mathbf{1}_{E_2} \end{bmatrix} L^2(E_2), \end{aligned} \tag{4.27}$$

where $L^2(E_i)$ are vector L^2 spaces of E_i valued functions with respect to the Lebesgue measure on $\mathbb{T}$.

An arbitrary harmonic contractive valued operator function $w(\zeta) : E_1 \to E_2$, $\| w(\zeta) \| \leq 1$, $|\zeta| < 1$, may serve as the scattering function.

4.3. Lax-Phillips (orthogonal) scattering systems. Unitary colligations [10, 8]

This is a special case of the previous case under an additional assumption

$$\bigvee_{n\geq 0} (U^n)^*\alpha_1(E_1) \perp \bigvee_{n\geq 1} U^n\alpha_2(E_2). \tag{4.28}$$

In this case one can see from (4.3) that $w(\zeta)$ is an analytic contractive operator function. The space L_α can be decomposed as

$$L_\alpha = \cdots \oplus E_1^{(1)} \oplus E_1^{(0)} \oplus H_\alpha \oplus E_2^{(-1)} \oplus E_2^{(-2)} \oplus \cdots \tag{4.29}$$

due to the orthogonality assumption (4.28). Since $U \mid_{L_{\alpha_1}}$ are backward shifts,

$$U : E_1^{(0)} \oplus H_\alpha \to H_\alpha \oplus E_2^{(-1)}. \tag{4.30}$$

This restriction is called the *unitary colligation*, H_α is called the *state space*, $E_1^{(0)}$ and $E_2^{(-1)}$ are called the *input* and the *output* spaces respectively. U_α is called the *unitary dilation* of the unitary colligation (4.30).

The characteristic function of the unitary colligation coincides with the scattering function of its unitary dilation [9]. The space L^w has the following structure inherited from the structure of L_α (4.29)

$$L^w = \begin{bmatrix} \mathbf{1} \\ w \end{bmatrix} H^2_+(E_1) \oplus H^w \oplus \begin{bmatrix} w^* \\ \mathbf{1} \end{bmatrix} H^2_-(E_2), \tag{4.31}$$

where $H^2_+(E_1)$ and $H^2_-(E_2)$ are vector Hardy spaces, and H^w is the de Branges-Rovnyak functional space. H^w is characterized as follows:

$$\begin{bmatrix} f_1 \\ f_2 \end{bmatrix} \in H^w \text{ if and only if } \begin{bmatrix} f_1 \\ f_2 \end{bmatrix} \in L^w, f_1 \in H^2_-(E_1), f_2 \in H^2_+(E_2).$$

If we fix the scale

$$\tilde{\alpha} : E_1 \oplus E_2 \to L^w,$$

$$\tilde{\alpha} \begin{bmatrix} e_1 \\ e_2 \end{bmatrix} \to \begin{bmatrix} \mathbf{1} & w^* \\ w & \mathbf{1} \end{bmatrix} \begin{bmatrix} e_1 \\ e_2 \end{bmatrix}$$

then the unique (see Theorem 4.1) unitary equivalence sends H_α onto H^w and "tails" onto "tails", respectively.

An arbitrary analytic contractive-valued operator function $w(\zeta) : E_1 \to E_2$, $\| w(\zeta) \| \leq 1$, $|\zeta| < 1$, may serve as the scattering function for such a system.

5. Completion problems related to internal scattering channels

As an application, consider the following *completion problem*: we want to complete the following matrix

$$\begin{bmatrix} \sigma_{11} & \sigma_{12} & \sigma_{13} \\ \sigma_{21} & \sigma_{22} & ? \\ \sigma_{31} & ? & \sigma_{33} \end{bmatrix} \geq 0, \tag{5.1}$$

where the entries are operator-valued measures on $\mathbb{T}$ (or the corresponding harmonic operator-functions on $\mathbb{D}$, see Section 1), $\sigma_{21} = \sigma_{12}^*$, $\sigma_{31} = \sigma_{13}^*$, $\sigma_{32} = \sigma_{23}^*$. Denote the space where σ acts by $M = M_1 \oplus M_2 \oplus M_3$.

Theorem 5.1. The completion problem (5.1) has a solution if and only if

$$\sigma_1 \stackrel{def}{=} \begin{bmatrix} \sigma_{11} & \sigma_{12} \\ \sigma_{21} & \sigma_{22} \end{bmatrix} \geq 0 \tag{5.2}$$

and

$$\sigma_2 \stackrel{def}{=} \begin{bmatrix} \sigma_{11} & \sigma_{13} \\ \sigma_{31} & \sigma_{33} \end{bmatrix} \geq 0. \tag{5.3}$$

Proof. The conditions are obviously necessary. Now we will prove the sufficiency. (5.2) implies that there exists

$$\widehat{\sigma}_{22}(\Delta) \stackrel{\text{def}}{=} \underset{\Delta=\cup\Delta_j}{\text{s-lim}} \sum_j (\sigma_{11}(\Delta_j)^{[-1/2]}\sigma_{12}(\Delta_j))^*(\sigma_{11}(\Delta_j)^{[-1/2]}\sigma_{12}(\Delta_j)) \tag{5.4}$$

(see Lemma 2.5), $\widehat{\sigma}_{22}(\Delta)$ is a strong operator measure, and

$$\widehat{\sigma}_{22}(\Delta) \le \sigma_{22}(\Delta). \tag{5.5}$$

In the same way, (5.3) implies the existence of the limit

$$\widehat{\sigma}_{33}(\Delta) \stackrel{\text{def}}{=} \underset{\Delta=\cup\Delta_j}{\text{s-lim}} \sum_j (\sigma_{11}(\Delta_j)^{[-1/2]}\sigma_{13}(\Delta_j))^*(\sigma_{11}(\Delta_j)^{[-1/2]}\sigma_{13}(\Delta_j)), \tag{5.6}$$

and the inequality

$$\widehat{\sigma}_{33}(\Delta) \le \sigma_{33}(\Delta). \tag{5.7}$$

Then, by means of the polarization formula, one can see that the limit

$$\widehat{\sigma}_{23}(\Delta) \stackrel{\text{def}}{=} \lim_{\Delta=\cup\Delta_j} \sum_j (\sigma_{11}(\Delta_j)^{[-1/2]}\sigma_{12}(\Delta_j))^*(\sigma_{11}(\Delta_j)^{[-1/2]}\sigma_{13}(\Delta_j)), \tag{5.8}$$

exists at least in weak sense. In fact, (5.2) guarantees that

$$\sigma_{12}e_2 \in L^{\sigma_{11}}, \ \forall e_2 \in E_2 \tag{5.9}$$

(see Section 1), (5.3) guarantees that

$$\sigma_{13}e_3 \in L^{\sigma_{11}}, \ \forall e_3 \in E_3 \ . \tag{5.10}$$

(5.9) and (5.10) imply

$$c_1\sigma_{12}e_2 + c_2\sigma_{13}e_3 \in L^{\sigma_{11}}, \ \forall e_2 \in E_2, \ e_3 \in E_3,$$

$c_1, \ c_2$ are arbitrary numbers, which imply (5.8). Moreover,

$$\begin{aligned}\begin{bmatrix} \widehat{\sigma}_{22}(\Delta) & \widehat{\sigma}_{23}(\Delta) \\ \widehat{\sigma}_{32}(\Delta) & \widehat{\sigma}_{33}(\Delta) \end{bmatrix} &= \lim_{\Delta=\cup\Delta_j} \sum_j (\sigma_{11}(\Delta_j)^{[-1/2]}[\sigma_{12}(\Delta_j), \sigma_{13}(\Delta_j)])^* \\ &\quad \cdot (\sigma_{11}(\Delta_j)^{[-1/2]}[\sigma_{12}(\Delta_j), \sigma_{13}(\Delta_j)]) \ge 0,\end{aligned} \tag{5.11}$$

which, in particular, implies that $\widehat{\sigma}_{23}(\Delta)$ is also a strong operator measure and

$$\begin{bmatrix} \sigma_{11}(\Delta) & \sigma_{12}(\Delta) & \sigma_{13}(\Delta) \\ \sigma_{21}(\Delta) & \widehat{\sigma}_{22}(\Delta) & \widehat{\sigma}_{23}(\Delta) \\ \widehat{\sigma}_{31}(\Delta) & \widehat{\sigma}_{32}(\Delta) & \widehat{\sigma}_{33}(\Delta) \end{bmatrix} \ge 0. \tag{5.12}$$

(5.12) together with (5.5) and (5.7) guarantee that $\widehat{\sigma}_{23}(\Delta)$ is a solution of the completion problem (5.1). Thus the theorem follows.

Theorem 5.2. Using the above notations, if $\sigma_{22} = \widehat{\sigma}_{22}$ and $\sigma_{33} = \widehat{\sigma}_{33}$ then $\widehat{\sigma}_{23}$ (see (5.8)) is the unique solution to the completion problem (5.1).

Proof. Let σ_{23} be a solution, i.e.,

$$\begin{bmatrix} \sigma_{11} & \sigma_{12} & \sigma_{13} \\ \sigma_{21} & \sigma_{22} & \sigma_{23} \\ \sigma_{31} & \sigma_{32} & \sigma_{33} \end{bmatrix} \geq 0. \tag{5.13}$$

Then (see Lemma 2.5)

$$\begin{bmatrix} \sigma_{22} & \sigma_{23} \\ \sigma_{32} & \sigma_{33} \end{bmatrix} - \begin{bmatrix} \widehat{\sigma}_{22} & \widehat{\sigma}_{23} \\ \widehat{\sigma}_{32} & \widehat{\sigma}_{33} \end{bmatrix} \geq 0,$$

i.e.

$$\begin{bmatrix} \sigma_{22} - \widehat{\sigma}_{22} & \sigma_{23} - \widehat{\sigma}_{23} \\ \sigma_{32} - \widehat{\sigma}_{32} & \sigma_{33} - \widehat{\sigma}_{33} \end{bmatrix} \geq 0.$$

Diagonal entries are zero by assumption. Then all the other entries have to be zero. Thus the theorem follows.

Remark 5.3. If in (5.1) σ_{11} is of bounded variation, and σ_{12}, σ_{13} are absolutely continuous with respect to a scalar measure τ (see Section 3), then by Theorem 3.11, Remarks 3.12, 3.13 (see also Theorem 3.6)

$$\begin{bmatrix} \widehat{\sigma}'_{22}(t) & \widehat{\sigma}'_{12}(t) \\ \widehat{\sigma}'_{21}(t) & \widehat{\sigma}'_{33}(t) \end{bmatrix} = ((\sigma_{11} \mid \Delta_a)'(t)^{[-1/2]}[\sigma'_{12}(t), \sigma'_{13}(t)])^* \cdot ((\sigma_{11} \mid \Delta_a)'(t)^{[-1/2]}[\sigma'_{12}(t).\sigma'_{13}(t)]) \tag{5.14}$$

(at least in the weak sense).

Remark. The general solution to the completion problem (5.1) is obviously of the form

$$\begin{bmatrix} \sigma_{22} & \sigma_{23} \\ \sigma_{32} & \sigma_{33} \end{bmatrix} \geq \begin{bmatrix} \widehat{\sigma}_{22} & \widehat{\sigma}_{23} \\ \widehat{\sigma}_{32} & \widehat{\sigma}_{33} \end{bmatrix}.$$

Or, equivalently,

$$\begin{bmatrix} \sigma_{22} - \widehat{\sigma}_{22} & \sigma_{23} - \widehat{\sigma}_{23} \\ \sigma_{32} - \widehat{\sigma}_{32} & \sigma_{33} - \widehat{\sigma}_{33} \end{bmatrix} \geq 0.$$

Everything here (except for σ_{23}) is either given or computed from the data.

We turn now to special cases. Let σ be an operator-measure on $\mathbb{T}$, $\sigma(\Delta) : E \to E$ for any Borel set $\Delta \subseteq \mathbb{T}$, and let $\sigma(\zeta)$ be its harmonic continuation inside the unit disc $\mathbb{D}$ (see Section 1).

Let

$$\varphi(\zeta) : E \to M_1 \tag{5.15}$$

and

$$\psi(\zeta) : M_2 \to E \tag{5.16}$$

be *harmonic* operator-functions on $\mathbb{D}$, so that

$$\sigma(\zeta) \geq \varphi(\zeta)^* \varphi(\zeta), \tag{5.17}$$

$$\sigma(\zeta) \geq \psi(\zeta)\psi(\zeta)^*, \tag{5.18}$$

or, equivalently, so that

$$\begin{bmatrix} \sigma(\zeta) & \varphi(\zeta)^* \\ \varphi(\zeta) & \mathbf{1}_{M_1} \end{bmatrix} \geq 0 \tag{5.19}$$

and

$$\begin{bmatrix} \sigma(\zeta) & \psi(\zeta) \\ \psi(\zeta)^* & \mathbf{1}_{M_2} \end{bmatrix} \geq 0. \tag{5.20}$$

Denote the measures corresponding to (5.19) and (5.20) by

$$\begin{bmatrix} \sigma_{11}(\Delta) & \sigma_{12}(\Delta) \\ \sigma_{21}(\Delta) & \sigma_{22}(\Delta) \end{bmatrix} \geq 0 \tag{5.21}$$

and

$$\begin{bmatrix} \sigma_{11}(\Delta) & \sigma_{13}(\Delta) \\ \sigma_{31}(\Delta) & \sigma_{33}(\Delta) \end{bmatrix} \geq 0 \tag{5.22}$$

where $\sigma_{12}(\zeta) = \varphi(\zeta)^*, \sigma_{13}(\zeta) = \psi(\zeta), |\zeta| < 1$,

$$\sigma_{11}(\Delta) = \sigma(\Delta), \quad \sigma_{22}(\Delta) = \mathbf{1}_{M_1} \cdot m(\Delta), \quad \sigma_{33}(\Delta) = \mathbf{1}_{M_2} \cdot m(\Delta), \tag{5.23}$$

and m is the Lebesgue measure.

Theorem 5.4. Using the notations (5.15)–(5.23), there exists σ_{23} so that

$$\begin{bmatrix} \sigma_{11}(\Delta) & \sigma_{12}(\Delta) & \sigma_{13}(\Delta) \\ \sigma_{21}(\Delta) & \sigma_{22}(\Delta) & \sigma_{23}(\Delta) \\ \sigma_{31}(\Delta) & \sigma_{32}(\Delta) & \sigma_{33}(\Delta) \end{bmatrix} \geq 0. \tag{5.24}$$

If

$$\sigma_{22} = \widehat{\sigma}_{22} \quad \text{and} \quad \sigma_{33} = \widehat{\sigma}_{33}, \tag{5.25}$$

then σ_{23} is unique and $\sigma_{23} = \widehat{\sigma}_{23}$, which is computed by the formula (5.8).

Proof. This is a special case of Theorems 5.1 and 5.2.

Theorem 5.5. Using the notations (5.15)–(5.23), σ_{12} and σ_{13} are absolutely continuous with respect to the Lebesgue measure m on $\mathbb{T}$ (see Section 3), and m – *a.e.* on $\mathbb{T}$

$$(\sigma_{21}e)'(t) = (\varphi e)(t) \text{ and } (\sigma_{31}e)'(t) = (\psi^* e)(t), \tag{5.26}$$

where the right-hand sides in (5.26) are boundary values of harmonic vector-functions $\varphi(\zeta)e$ and $\psi(\zeta)^*e$, respectively, for any $e \in E$. Let σ be of bounded variation (see Section 3), then $\widehat{\sigma}_{23}$ is absolutely continuous with respect to the Lebesgue measure, and its density is

$$\widehat{\sigma}'_{23}(t) \equiv \widehat{\chi}(t) = ((\sigma \mid \Delta_a)'(t)^{[-1/2]}\varphi(t)^*)^*((\sigma \mid \Delta_a)'(t)^{[-1/2]}\psi(t)) \ , \tag{5.27}$$

which is understood at least in the weak sense.

Proof. By (5.19)–(5.22)

$$\sigma_{21}e \in L^{\mathbf{1}}{}_{M_1 \cdot m}, \ \sigma_{31}e \in L^{\mathbf{1}}{}_{M_2 \cdot m}. \tag{5.28}$$

Hence they are absolutely continuous with respect to the Lebesgue measure by Corollary 3.5. By Theorem 3.2 the densities exist with respect to the Lebesgue measure m (see (3.6)). Hence,

$$\varphi(\zeta)e \equiv \sigma_{21}(\zeta)e = \int_{\mathbb{T}} \frac{1-|\zeta|^2}{|t-\zeta|^2}(\sigma_{21}e)'(t)m(dt)$$

and

$$\psi(\zeta)^*e \equiv \sigma_{31}(\zeta)e = \int_{\mathbb{T}} \frac{1-|\zeta|^2}{|t-\zeta|^2}(\sigma_{31}e)'(t)m(dt), \tag{5.29}$$

which follows from (3.6) if one approximates Poisson kernel by simple functions. Then the Fatou Lemma leads to (5.26) and (5.27) follows from Remark 5.3.

We turn now to a special important case of φ and ψ in (5.17), (5.18), when (5.25) is satisfied.

Theorem 5.6. Among all the *analytic* functions φ and ψ in $\mathbb{D}$, that satisfy (5.17) and (5.18) respectively, there exist maximal ones (in the sense of [11]); the function φ is outer, and the function ψ is $*$-outer (see also [11] for all the definitions).

Proof. It is essentially the same as the one given in [11] for a slightly less general case.

Theorem 5.7. The maximal analytic functions φ and ψ of Theorem 5.6 satisfy (5.25).

Proof. By the proof of Theorem 5.6 (see [11]), the mapping

$$\sigma p \mapsto \varphi p, \tag{5.30}$$

where p are trigonometric polynomials with coeffizients in E, is a partial isometry from L^σ onto $L_2(M_1)$, where $L_2(M_1)$ is standard vector L_2 space with coeffizients M_1. Hence, the adjoint mapping

$$q \to \varphi^* q \in L^\sigma, \tag{5.31}$$

where q are trigonometric polynomials with coeffizients M_1, is an isometry from $L_2(M_1)$ into L^σ, i.e.,

$$\langle \varphi^* q, \varphi^* \tilde{q} \rangle_{L^{\sigma_{11}}} = \langle q, \tilde{q} \rangle_{L_2(M_2)} . \tag{5.32}$$

Now picking up q and $\tilde{q}$ to be of the form $g \cdot m_2$, $\tilde{g} \cdot m_2$, where g and $\tilde{g}$ are scalar trigonometric polynomials, we obtain (see Theorem 1.7)

$$\langle \varphi^* g m_2, \varphi^* \tilde{g} \cdot m_2 \rangle = \int_{\mathbb{T}} \bar{\tilde{g}} g \mu_{\varphi^* m_2}, \tag{5.33}$$

where $\mu_{\varphi^* m_2}$ is the Hellinger measure of $\varphi^* m_2$. Now (5.32) and (5.33) mean that

$$\mu_{\varphi^* m_2}(\Delta) = m(\Delta) \cdot \| m_2 \|_{M_2}^2, \tag{5.34}$$

where m is the Lebesgue measure on $\mathbb{T}$. By formulas (1.9) and (2.29), (2.26), the equality (5.34) means that

$$\widehat{\sigma}_{22}(\Delta) = m(\Delta) \cdot \mathbf{1}_{M_2}, \tag{5.35}$$

which is equivalent to the first equality in (5.25). The second one is proved analogously. Thus the theorem follows.

Corollary 5.8. For the maximal analytic φ and ψ the completion σ_{23} in (5.24) exists, is unique and can be computed by formula (5.27).

The following completion problem was considered in [16]. We give it here in a slightly more general setting than it was done in [16]. In a sense, it is a special case of the one considered in Theorems 5.4 and 5.5 but with stronger constraints. In this problem $E = E_1 \oplus E_2$ and $\sigma(\zeta) : E \to E$

$$\sigma(\zeta) = \begin{bmatrix} \mathbf{1}_{E_1} & w(\zeta)^* \\ w(\zeta) & \mathbf{1}_{E_2} \end{bmatrix} \geq 0, |\zeta| < 1, \tag{5.36}$$

w is harmonic. Instead of (5.17), (5.18), the factorizations considered are of the special form

$$\sigma(\zeta) \geq \begin{bmatrix} \varphi(\zeta)^* \\ 0 \end{bmatrix} \begin{bmatrix} \varphi(\zeta) & 0 \end{bmatrix}, \tag{5.37}$$

$$\sigma(\zeta) \geq \begin{bmatrix} 0 \\ \psi(\zeta) \end{bmatrix} \begin{bmatrix} \psi(\zeta)^* & 0 \end{bmatrix}, \tag{5.38}$$

where φ and ψ are harmonic. The splitting here reflects the splitting of $E = E_1 \oplus E_2$. Then (5.37) and (5.38) are equivalent to

$$\mathbf{1} - w(\zeta)^* w(\zeta) \geq \varphi(\zeta)^* \varphi(\zeta), \tag{5.39}$$

$$\mathbf{1} - w(\zeta) w(\zeta)^* \geq \psi(\zeta) \psi(\zeta)^*, \tag{5.40}$$

respectively, because of (5.36). On the other hand (5.37) and (5.38) are equivalent to

$$\begin{bmatrix} \mathbf{1}_{E_1} & w^* & \varphi^* \\ w & \mathbf{1}_{E_2} & 0 \\ \varphi & 0 & \mathbf{1}_{M_1} \end{bmatrix} \geq 0, \tag{5.41}$$

$$\begin{bmatrix} \mathbf{1}_{E_1} & w^* & 0 \\ w & \mathbf{1}_{E_2} & \psi \\ 0 & \psi^* & \mathbf{1}_{M_2} \end{bmatrix} \geq 0, \tag{5.42}$$

respectively.

The problem is to find a harmonic function $\chi(\zeta)$ in $\mathbb{D}$ so that

$$\begin{bmatrix} \chi(\zeta) & \varphi(\zeta) \\ \psi(\zeta) & w(\zeta) \end{bmatrix}, \quad |\zeta| < 1, \tag{5.43}$$

is a contraction.

Lemma 5.9. The contractivity of (5.43) is equivalent to the positivity

$$\begin{bmatrix} \mathbf{1}_{E_1} & w^* & \varphi^* & 0 \\ w & \mathbf{1}_{E_2} & 0 & \psi \\ \varphi & 0 & \mathbf{1}_{M_1} & \chi \\ 0 & \psi^* & \chi^* & \mathbf{1}_{M_2} \end{bmatrix} \geq 0. \tag{5.44}$$

Proof. The contractivity of (5.43) is equivalent to the positivity

$$\begin{bmatrix} \mathbf{1} & 0 & \chi & \varphi \\ 0 & \mathbf{1} & \psi & w \\ \chi^* & \psi^* & \mathbf{1} & 0 \\ \varphi^* & w^* & 0 & \mathbf{1} \end{bmatrix} \geq 0. \tag{5.45}$$

Thus (5.44) differs from (5.45) by permutations of certain rows and columns. Then the lemma follows.

Now denote

$$\sigma_{11}(\zeta) = \begin{bmatrix} \mathbf{1}_{E_1} & w(\zeta)^* \\ w(\zeta) & \mathbf{1}_{E_2} \end{bmatrix},$$

$$\sigma_{12}(\zeta) = \begin{bmatrix} \varphi(\zeta)^* \\ 0 \end{bmatrix}, \quad \sigma_{13}(\zeta) = \begin{bmatrix} 0 \\ \psi(\zeta) \end{bmatrix},$$

$$\sigma_{22}(\zeta) = \mathbf{1}_{M_1}, \ \sigma_{33}(\zeta) = \mathbf{1}_{M_2},$$

$$\sigma_{23}(\zeta) = \chi(\zeta).$$

All that we did for the factorizations (5.17), (5.18) is still valid for the factorizations (5.39), (5.40). But some simplifications can be made in formula (5.14).

Theorem 5.10. For the special form (5.36) of σ and the factorizations (5.37), (5.38) the formula (5.14) is simplified to (see also (5.27))

$$(5.47) \qquad \widehat{\sigma}_{22}(t) = ((\mathbf{1} - w(t)^*w(t))^{[-1/2]}\varphi(t)^*)^*((\mathbf{1} - w(t)^*w(t))^{[-1/2]}\varphi(t)^*),$$

$$(5.48) \qquad \widehat{\sigma}_{33}(t) = ((\mathbf{1} - w(t)w(t)^*)^{[-1/2]}\psi(t))((\mathbf{1} - w(t)w(t))^*)^{[-1/2]}\psi(t)),$$

(5.49)

$$\widehat{\sigma}'_{23}(t) = \left(\begin{bmatrix} \mathbf{1} & w(t)^* \\ w(t) & \mathbf{1} \end{bmatrix}^{[-1/2]} \begin{bmatrix} \varphi(t)^* \\ 0 \end{bmatrix}\right)^* \left(\begin{bmatrix} \mathbf{1} & w(t)^* \\ w(t) & \mathbf{1} \end{bmatrix}^{[-1/2]} \begin{bmatrix} 0 \\ \psi(t) \end{bmatrix}\right),$$

$m - a.e.$ on $\mathbb{T}$. All the densities are with respect to the Lebesgue measure m on $\mathbb{T}$.

Proof is straightforward.

Thus for the completion problem (5.43) (which is equivalent to (5.44)) conditions (5.25) that guarantee the uniqueness are

$$(5.50) \qquad ((\mathbf{1} - w(t)^*w(t))^{[-1/2]}\varphi(t)^*)^*((\mathbf{1} - w(t)^*w(t))^{[-1/2]}\varphi(t)^*) = \mathbf{1}_{M_1}$$

and

$$((\mathbf{1} - w(t)w(t))^*)^{[-1/2]}\psi(t))^*((\mathbf{1} - w(t)w(t)^*)^{[-1/2]}\psi(t)) = \mathbf{1}_{M_2}$$

$m - a.e.$ on $\mathbb{T}$, respectively. In this case the unique completion is

$$(5.51) \quad \widehat{\chi}(t) = \widehat{\sigma}'_{23}(t)$$

$$= \left(\begin{bmatrix} \mathbf{1} & w(t)^* \\ w(t) & \mathbf{1} \end{bmatrix}^{[-1/2]} \begin{bmatrix} \varphi(t)^* \\ 0 \end{bmatrix}\right)^* \left(\begin{bmatrix} \mathbf{1} & w(t)^* \\ w(t) & \mathbf{1} \end{bmatrix}^{[-1/2]} \begin{bmatrix} 0 \\ \psi(t) \end{bmatrix}\right),$$

$m - a.e.$ on $\mathbb{T}$.

Remark. Equalities (5.50) imply (5.39) and (5.40), respectively.

For the completion problem (5.43) there is also an important special case when properties (5.50) are enjoyed. It is similar to the one of Theorems 5.6, 5.7. We formulate the corresponding theorems without proofs since the proofs are completely similar to the proofs of Theorems 5.6, 5.7.

Theorem 5.11. There exist maximal *analytic* functions φ and ψ that satisfy (5.39) and (5.40), respectively.

Theorem 5.12. The maximal analytic functions φ and ψ satisfy the properties (5.50) (i.e., (5.25)).

Corollary 5.13. For the maximal analytic φ and ψ a completion χ in (5.43) exists, is unique and can be computed by formula (5.51).

Remark 5.14. It was shown in [16] that under the assumptions (5.50) (we already know that the completion χ in (5.43) exists and is unique in this case) the contractivity of (5.43) implies

$$\left\| \begin{bmatrix} \chi(t) & \varphi(t) \\ \psi(t) & w(t) \end{bmatrix} \right\| = 1 \text{ a.e. on } \mathbb{T}. \tag{5.52}$$

In fact, if one assumes that

$$\begin{bmatrix} \chi(t) & \varphi(t) \\ \psi(t) & w(t) \end{bmatrix}^* \begin{bmatrix} \chi(t) & \varphi(t) \\ \psi(t) & w(t) \end{bmatrix} \le C < 1$$

on a set of positive measure, then, in particular,

$$\varphi^*\varphi + w^*w \le C < 1.$$

Since

$$w^*w \ge C \cdot w^*w,$$

then

$$\varphi^*\varphi \le C(1 - w^*w),$$

which implies

$$((\mathbf{1} - w^*w)^{[-1/2]}\varphi^*)^*((\mathbf{1} - w^*w)^{[-1/2]}\varphi^*) \le C < 1.$$

But the latter contradicts the first of the properties (5.50)

Remark 5.15. The equality $\widehat{\sigma}_{22} = \mathbf{1}$ in Theorem 5.4 means that the subspace

$$\varphi^* L^2(M_1) \subseteq L^\sigma \tag{5.53}$$

is a channel, with a wandering subspace

$$\varphi^* M_1 \subseteq L^\sigma \tag{5.54}$$

(under the multiplication by $\bar{t}$). The analyticity of φ means that the semichannel

$$\varphi^* H^2_-(M_1) \subseteq L^\sigma \tag{5.55}$$

is orthogonal to the subspace

$$\overline{\{\sigma p\}} \subseteq L^\sigma, \tag{5.56}$$

where p's are analytic polynomials in t with coeffizients in E. The maximality of an analytic factorization φ means that (5.54) is the maximal subspace in L^σ with these properties.

Analogously, for ψ, $\widehat{\sigma}_{33} = \mathbf{1}$ means that the subspace $\psi M_2 \subseteq L^\sigma$ is wandering, the analyticity of ψ means that the semichannel

$$\psi H^2_+(M_2) \subseteq L^\sigma \tag{5.57}$$

is orthogonal to the subspace

$$\overline{\{\sigma p\}} \subseteq L^\sigma, \tag{5.58}$$

where p's are antianalytic polynomials in t vanishing at the origin. The maximality of an analytic factorization ψ means that ψM_2 is the maximal subspace with these properties in L^σ.

If in addition, $E = E_1 \oplus E_2$ and

$$\sigma = \begin{bmatrix} \mathbf{1} & w^* \\ w & \mathbf{1} \end{bmatrix}, \tag{5.59}$$

then $L^\sigma \cong L^w$ (see Section 4) and L^w is generated by two scattering channels

$$\begin{bmatrix} \mathbf{1} \\ w \end{bmatrix} L^2(E_1) \quad \text{and} \quad \begin{bmatrix} w^* \\ \mathbf{1} \end{bmatrix} L^2(E_2). \tag{5.60}$$

The factorizations (5.37), (5.38) under the assumptions (5.25) (or equivalently (5.50)) generate new scattering channels in L^w:

$$\begin{bmatrix} \varphi^* \\ 0 \end{bmatrix} L^2(M_1) \quad \text{and} \quad \begin{bmatrix} 0 \\ \psi \end{bmatrix} L^2(M_2). \tag{5.61}$$

Obviously

$$\begin{bmatrix} \varphi^* \\ 0 \end{bmatrix} L^2(M_1) \perp \begin{bmatrix} w^* \\ \mathbf{1} \end{bmatrix} L^2(E_2) \tag{5.62}$$

in L^w (see Section 4), and

$$\begin{bmatrix} 0 \\ \psi \end{bmatrix} L^2(M_2) \perp \begin{bmatrix} \mathbf{1} \\ w \end{bmatrix} L^2(E_1). \tag{5.63}$$

If φ is analytic then we have the additional orthogonality of semichannels

$$\begin{bmatrix} \varphi^* \\ 0 \end{bmatrix} H^2_-(M_1) \perp \begin{bmatrix} \mathbf{1} \\ w \end{bmatrix} H^2_+(E_1). \tag{5.64}$$

If ψ is analytic then

$$\begin{bmatrix} 0 \\ \psi \end{bmatrix} H^2_+(M_2) \perp \begin{bmatrix} w^* \\ \mathbf{1} \end{bmatrix} H^2_-(E_2). \tag{5.65}$$

If φ and ψ are maximal analytic in (5.39), (5.40), respectively, then the corresponding channels are the maximal ones with these properties.

If in addition w is analytic then (see Section 4)

$$L^w = \begin{bmatrix} \mathbf{1} \\ w \end{bmatrix} H^2_+(E_1) \oplus H^w \oplus \begin{bmatrix} w^* \\ \mathbf{1} \end{bmatrix} H^2_-(E_2), \tag{5.66}$$

and (5.62)–(5.65) mean that semichannels

$$\begin{bmatrix} \varphi^* \\ 0 \end{bmatrix} H^2_-(M_1) \text{and} \begin{bmatrix} 0 \\ \psi \end{bmatrix} H^2_+(M_2) \tag{5.67}$$

are contained in the state space H^w.

The study of these internal semichannels was initiated by D. Z. Arov [18]. An extensive study was carried out in [13, 16]. The completion problem (5.43) appeared in the context of that study. All the results related to the completion problem (5.43), except for the explicit formula (5.51), were obtained there in different terms.

References

[1] U. Grenander and G. Szegö, *Toeplitz Forms and their Applications*, University of California Press, Berkeley and Los Angeles, 1958.

[2] Yu.L. Schmulyan, *Operator Hellinger integral*, Mat. Sbornik. **49** (91) (1958), 381–430 [in Russian].

[3] Yu.L. Schmulyan, *On reduction of operator Hellinger integral to Lebesgue integral*, Izv. Vuzov (Matem.) **4** (23) (1961), 159–160 [in Russian].

[4] Yu.L. Schmulyan, *Topics on theory of operators with finite nonhermitian rank*, Mat. Sbornik **57** (99) (1962), 105–136 [in Russian].

[5] A.Ya. Kheifets, *Hamburger moment problem: Parseval equality and A-singularity*, J. Funct. Anal. **141** (2) (1996), 374–420.

[6] N.I. Akhiezer and I.M. Glazman, *Theory of Linear Operators on Hilbert Space*, Nauka, Moscow, 1966 [in Russian]. English translation: Monographs and Studies in Mathematics, Vol. 9, 10, Pitman Advanced Publishing Program, Boston–London–Melbourne, 1981.

[7] I. Schur, *Über Potenzreihen, die im Innern des Einheitskreises beschränkt sind* I, II. J. für die reine und angewandte Math. **147** (1916), 205–232; **148** (1917), 122–145. English translation in: Oper. Th. Adv. Appl. 18 (1986), Birkhäuser Verlag, Basel.

[8] V.M. Adamjan and D.Z. Arov, *On unitary couplings of semiunitary operators*, Mat. Issled. Kishinev. **1** (2) (1966), 3–64 [in Russian]. English translation: Amer. Math. Soc. Transl. Ser. 2 **95** (1970), 75–129.

[9] D.Z. Arov, *A class of scattering operators and of characteristic operator-functions of contractions*, Dokl. Acad. Nauk SSSR **160**, 1965, 9–12. English translation: Soviet Math. Dokl. **6** (1965), 1–5.

[10] P. Lax and R. Phillips, *Scattering Theory*, Academic Press, New York, 1967.

[11] B. Sz. Nagy and C. Foias, *Harmonic Analysis of Operators on Hilbert Space*, North Holland Publ. Co., Amsterdam–Budapest, 1970.

[12] A. Ya. Kheifets, *Nevanlinna–Adamjan–Arov–Krein Theorem in semideterminate case*, Teor. Funk., Funk. Anal. i Pril. (56) (1991), 128–137. English translation: J. Math. Sci. **76** (4) (1995), 2542–2549.

[13] S.S. Boiko and V.K. Dubovoy, *Unitary couplings and regular factorizations of operator functions in L^∞*, Dopovidi Acad. Nauk. Ukr. **1** (1997), 41–44.

[14] M.S. Brodskii, *Unitary operator colligations and characteristic functions*, Uspekhi Math. Nauk. **33** (4) (1978), 141–168. English translation: Russian Math. Surveys **33** (4) (1978), 159–191.

[15] V.K. Dubovoy and R.K. Mohammed, *Defect functions of holomorphic contractive matrix functions, regular extensions and open systems*, Math. Nachr. **160** (1993), 69–110.

[16] S.S. Boiko and V.K. Dubovoy, *On an external problem related to scattering operator function between internal channels of scattering system*, Dopovidi Acad. Nauk. Ukr., (1997), (4), 7–11.

[17] V.K. Dubovoy, *Indefinite metric in Schur interpolation problem for analytic functions*, Teor. Funk., Funk. Anal. i Pril. 1982 (37), 14–26; 1982 (38), 32–39; 1984 (41), 55–64; 1984 (42), 46–57; 1986 (45), 16–26; 1987 (47), 112–119. [in Russian].

[18] D.Z. Arov, *Stable dissipative linear stationary dinamical scattering systems*, J. Oper. Th. **2** (1979), 95–126.

[19] S.S. Boiko, V.K. Dubovoy, B. Kirstein, and B. Fritzsche, *Contractive operators, defect functions and scattering theory*, Ukr. Mat. J. **49** (4) (1997), 481–489.

[20] V. Katsnelson, *Integral representations of positive–semidefinite kernels of mixed type and generalized Nehari Theorem* I. Teor. Funk., Funk. Anal. i Pril. **43** (1985), 54–70.

[21] B. Fritzsche and B. Kirstein, *A Nehari type problem for matrix–functions of the Schur class*, Zeitschr. Anal. **9** (1990), 73–84.

[22] V.K. Dubovoy, B. Fritzsche, and B. Kirstein, *On spectrally associated Schur functions, Arov–inner functions and Nehari–type completion problems for Schur class functions*, Integr. Eq. Oper. Theory **17** (1993), 247–276.

[23] M. Cotlar and C. Sadosky, *Integral representations of bounded Hankel forms defined in scattering systems with a multiparametric evolution group*, Contributions to operator theory and its applications (Mesa, AZ, 1987), Oper. Th. Adv. Appl. 35, Birkhäuser, Basel and Boston, 1988, 357–375.

[24] D.L. Cohn, *Measure Theory*, Birkhäuser, Boston, 1980.

Department of Mathematics, Kharkov State University, Kharkov, Ukraine 310077

Department of Mathematics, Kharkov State University, Kharkov, Ukraine 310077

College of Judea and Samaria, Ariel, Israel 44837

Operator Theory:
Advances and Applications, Vol. 123

On the Second Order Interpolation for Rational Vector Functions

V. Bolotnikov

Abstract. All rational solutions of an interpolation problem for Hardy functions with prescribed first m Fourier coefficients of the autocorrelation function are described. Necessary and sufficient conditions for the existence and uniqueness of a solution are established.

1. Introduction

The questions considered in the present paper originate with the following problem formulated and studied in [14]. In the statement, $\mathbf{H}_2^p$ denotes the space of $\mathbb{C}^p$-valued functions with entries in the Hardy space $\mathbf{H}_2$ of the unit disc $\mathbb{D}$.

Problem 1.1. *Given vectors $h_0, \ldots, h_n \in \mathbb{C}^p$ and numbers $r_0, \ldots, r_m \in \mathbb{C}$, find necessary and sufficient conditions for the existence of a function $H \in \mathbf{H}_2^p$ satisfying the following conditions:*

1. *For $j = 0, \ldots, n$, it holds that*

$$\frac{H^{(j)}(0)}{j!} = h_j. \tag{1.1}$$

2. *For $k = 0, \ldots, m$, it holds that*

$$\frac{1}{2\pi}\int_0^{2\pi} e^{-ikt} H(e^{it})^* H(e^{it}) dt = r_k. \tag{1.2}$$

In the setting of [14] the problem consists of finding of an impulse response sequence $\{h_k\}_{k=0}^{\infty} \in \ell_2^p$ with prescribed first $n+1$ impulse parameters $h_0, \ldots, h_n$ and with prescribed first $m+1$ correlation parameters which are defined as

$$r_k = \sum_{j=0}^{\infty} h_j^* h_{j+k} = r_{-k}^* \quad (k \geq 0).$$

Setting $H(z)=\sum_{j=0}^{\infty} h_j z^j$ it is readily checked that

$$H(e^{it})^*H(e^{it})=\sum_{k=-\infty}^{\infty} r_k e^{ikt}$$

and thus, r_k is the k-th Fourier coefficient of $H(e^{it})^*H(e^{it})$, which is frequently termed as the *autocorrelation function* of the impulse response sequence $\{h_k\}_{k=0}^{\infty}$.

Problem 1.1, initially considered for $p=1$, has been extended to the multivariable case by Inouye [11]. It was shown that the necessary condition for Problem 1.1 to be solvable is that

$$\mathbf{K}_{m,n} := \mathbf{R}_m - \mathbf{H}_{m,n}^*\mathbf{H}_{m,n} \geq 0. \tag{1.3}$$

where

$$\mathbf{R}_m = \begin{pmatrix} r_0 & r_1^* & \cdots & r_m^* \\ r_1 & r_0 & \cdots & r_{m-1}^* \\ \cdot & \cdot & \cdot & \cdot \\ r_m & r_{m-1} & \cdots & r_0 \end{pmatrix} \tag{1.4}$$

and $\mathbf{H}_{m,n} \in \mathbb{C}^{(n+1)p\times(m+1)q}$ is the lower triangular Toeplitz block matrix which is defined as

$$\mathbf{H}_{m,n} = \begin{pmatrix} h_0 & 0 & \cdots & 0 \\ \vdots & \ddots & \ddots & \vdots \\ h_n & \cdots & h_0 & 0 \end{pmatrix} \quad \text{or as} \quad \mathbf{H}_{m,n} = \begin{pmatrix} h_0 & & 0 \\ \vdots & \ddots & \\ h_m & & h_0 \\ \vdots & & \vdots \\ h_n & \cdots & h_{n-m} \end{pmatrix} \tag{1.5}$$

depending on $n<m$ or $n\geq m$, respectively.

Following the traditional terminology we shall call $\mathbf{K}_{m,n}$ *the Pick matrix* of Problem 1.1. It was also shown in [11, Example 1] that the condition (1.3) is not sufficient. However, the inequality (1.3) is necessary and sufficient for the following modified problem to be solvable:

Problem 1.2. *Given vectors $h_0,\dots,h_n \in \mathbb{C}^p$ and $r_0,\dots,r_m \in \mathbb{C}$ as above, find all functions $H \in \mathbf{H}_2^p$ satisfying* (1.1) *and the inequality*

$$\frac{1}{2\pi}\int_0^{2\pi} \begin{pmatrix} 1 \\ e^{-it} \\ \vdots \\ e^{-imt} \end{pmatrix} H(e^{it})^*H(e^{it})\left(1,\ e^{it},\ \dots,\ e^{imt}\right) dt \leq \mathbf{R}_m. \tag{1.6}$$

The latter condition can be interpreted as a norm constraint on the interpolating function H and for the case of the equality sign is equivalent to conditions (1.2).

In [12] the set of all minimal degree solutions of Problem 1.2 was characterized in the case when $n = m + 1$ and $K_{m,m+1} > 0$ and the same method turns to be applicable to arbitrary m and n. A description of the set of all the solutions of Problem 1.2 (for matrix-valued functions H) can be found in [3].

It is the usual case for classical boundary interpolation problems that a part of interpolation conditions are given in the inequality form. We mention the truncated Hamburger moment problem (see e.g., [1]) which consists of finding of all nonnegative measures $d\sigma$ on the real line such that

$$\int_{-\infty}^{\infty} \lambda^k d\sigma(\lambda) = s_k \quad (k = 0, \dots, 2n-1) \quad \text{and} \quad \int_{-\infty}^{\infty} \lambda^{2n} d\sigma(\lambda) \le s_{2n} \tag{1.7}$$

for prescribed numbers (or square matrices of the same size) $s_0, \dots, s_{2n}$. Another example is the Nevanlinna-Pick boundary problem (see e.g., [6]): find all Schur functions S (analytic and bounded in modulus by one in the unit disk $\mathbb{D}$) which take prescribed boundary limit values S_j at prescribed points ζ_j on the unit circle $\mathbb{T}$ and whose boundary derivatives are bounded by preassigned numbers γ_j:

$$\lim_{r\to 1} S(r\zeta_j) = S_j \quad \text{and} \quad \lim_{r\to 1} \frac{1 - S(r\zeta_j)S(r\zeta_j)^*}{1-r^2} \le \gamma_j \quad (\zeta_j \in \mathbb{T}). \tag{1.8}$$

The sets of all the solutions for the abovementioned problems are parametrized by special linear fractional transformations. However, the question about all (and even the existence of) "exact" solutions for which equality sign prevails in all the interpolation conditions, is more delicate. We refer to [7] and [16] for some results concerning exact solutions to the problems (1.7) and (1.8), respectively.

From this point of view Problem 1.1 can be interpreted as the "exact" version of Problem 1.2. Note that unlike (1.7), the inequality (1.6) involves all the given correlation parameters r_j and unlike (1.8), this inequality can not be reduced to simpler inequalities for each parameter r_j separately.

The necessary and sufficient condition for Problem 1.1 to be solvable will be established in Section 6. We shall show also that the condition $\mathbf{K}_{m,n} > 0$ is necessary and sufficient for Problem 1.1 to have infinitely many solutions and describe all rational solutions to Problem 1.1 for this case. In [3] we have considered the following

Problem 1.3. *Given numbers $r_0, \dots, r_n \in \mathbb{C}$ and given matrices $W \in \mathbb{C}^{N\times N}$ such that* spec $W \subset \mathbb{D}$ *and given $B \in \mathbb{C}^{N\times p}$, $C \in \mathbb{C}^{N\times 1}$ such that*

$$\bigcup_{j\ge 0} W^j B = \mathbb{C}^{N\times 1} \quad \textit{and} \quad \operatorname{spec} W \subset \mathbb{D}, \tag{1.9}$$

find all functions $H \in \mathbf{H}_2^p$ which satisfy the inequality (1.6) *and the left sided residue interpolation condition*

$$\sum_{z\in\mathbb{D}} \operatorname{Res}\, (zI_N - W)^{-1} BH(z) = C. \tag{1.10}$$

The first condition in (1.9) means that $(W,\ B)$ is a *full-range pair* and is not restrictive [5]. The interpolation conditions (1.1) may be represented in the residue form (1.10) for a special choice of

$$W = \begin{pmatrix} 0 & & \cdots & & 0 \\ I_p & 0 & & & \\ 0 & I_p & & & \\ \vdots & \ddots & \ddots & \ddots & \vdots \\ 0 & \cdots & 0 & I_p & 0 \end{pmatrix}, \quad B = \begin{pmatrix} I_p \\ 0 \\ \vdots \\ 0 \end{pmatrix} \quad \text{and} \quad C = \begin{pmatrix} h_0 \\ h_1 \\ \vdots \\ h_n \end{pmatrix} \tag{1.11}$$

and thus, Problem 1.3 contains Problem 1.2 as a particular case. The set of all solutions of Problem 1.3 has been established in [3] using reproducing kernel spaces methods and orthogonal polynomials assocoiated to the Toeplitz matrix $\mathbf{R}_m$ defined in (1.4). These methods turned to be suitable to study the second order interpolation problem (see [4]) in a more general framework for Hardy functions in a certain general class of domains. We refer also to [2, Section 6.6] for further discussions of the second order interpolation problem in the framework of reproducing kernel Hilbert spaces.

In this paper we shall focus on the equality analogue of Problem 1.3 for rational Hardy functions. In the statement, $\mathcal{R}^p$ denotes the set of all rational $\mathbb{C}^p$-valued functions with poles off the closed unit disk.

Problem 1.4. *Given numbers $r_0, \ \dots, \ r_n \in \mathbb{C}$ and matrices W, B and C as in Problem 1.3, find all functions $H \in \mathcal{R}^p$ which satisfy* (1.2) *and the left sided interpolation condition* (1.10).

2. On an interpolation problem for Schur functions

In this section we reduce Problem 1.4 to a left sided interpolation problem for rational inner functions in the unit disk. We recall that a $\mathbb{C}^{p\times q}$-valued function S is said to belong to the Schur class $\mathcal{S}^{p\times q}$ if it is analytic and takes contractive values in $\mathbb{D}$. If moreover,

$$S(e^{it})^*S(e^{it}) = I_q \quad \text{or} \quad S(e^{it})S(e^{it})^* = I_p$$

almost everywhere on $\mathbb{T}$, then S is called *inner* or *coinner*, respectively.

We also denote by $\mathcal{C}$ the Carathéodory class of functions which are analytic and have the positive semidefinite real part in $\mathbb{D}$. Every function $\Phi \in \mathcal{C}$ admits a Riesz-Herglotz representation

$$\Phi(z) = i\Im\,\Phi(0) + \frac{1}{2\pi}\int_0^{2\pi} \frac{e^{it}+z}{e^{it}-z} d\sigma(t) \tag{2.1}$$

for some nonnegative finite measure $d\sigma$ on the unit circle $\mathbb{T}$. By Fatou's lemma, the boundary values $\Phi(e^{it})$ exist almost everywhere on $\mathbb{T}$ and the absolutely continuous part of $d\sigma$ in (2.1) equals to

$$d\sigma_{ac} = \frac{\Phi(e^{it}) + \Phi(e^{it})^*}{2}. \tag{2.2}$$

Lemma 2.1. *Let $H \in \mathbf{H}_2^p$ and let Φ be defined by*

$$\Phi(z) = \frac{1}{2\pi}\int_0^{2\pi} \frac{e^{it}+z}{e^{it}-z} H(e^{it})^* H(e^{it}) dt. \tag{2.3}$$

Then $\Phi \in \mathcal{C}$ and satisfies

$$\Phi(e^{it}) + \Phi(e^{it})^* = 2H(e^{it})^* H(e^{it}) \quad \text{almost everywhere on } \mathbb{T}, \tag{2.4}$$

$$\Phi(z) + \Phi(z)^* \geq 2H(z)^* H(z) \quad (|z| < 1). \tag{2.5}$$

Proof. The function Φ given by (2.3) belongs to $\mathcal{C}$ by Riesz-Herglotz theorem, the corresponding measure $d\sigma(t) = H(e^{it})^* H(e^{it}) dt$ is absolutely continuous and (2.4) holds in view of (2.2). Furthermore, in view of (2.3),

$$\frac{\Phi(z) + \Phi(z)^*}{2(1-|z|^2)} = \frac{1}{2\pi}\int_0^{2\pi} \frac{H(e^{it})^* H(e^{it})}{|e^{it}-z|^2} dt \tag{2.6}$$

and changing variable

$$e^{it} = \frac{z - e^{i\tau}}{1 - e^{i\tau}\bar{z}}, \qquad \frac{dt}{|e^{it}-z|^2} = \frac{d\tau}{1-|z|^2}, \qquad \widetilde{H}(e^{i\tau}) := H(e^{it})$$

in the right-hand side integral, we get

$$\begin{aligned} \frac{1}{2\pi}\int_0^{2\pi} \frac{H(e^{it})^* H(e^{it})}{|e^{it}-z|^2} dt &= \frac{1}{2\pi(1-|z|^2)}\int_0^{2\pi} \widetilde{H}(e^{i\tau})^* \widetilde{H}(e^{i\tau}) d\tau \\ &\geq \frac{\widetilde{H}(0)^* \widetilde{H}(0)}{1-|z|^2} = \frac{H(z)^* H(z)}{1-|z|^2}, \end{aligned}$$

which together with (2.6) implies (2.5). □

Substituting the expansion of the Schwartz kernel $\frac{e^{it}+z}{e^{it}-z} = 1 + 2\sum_{k=1}^{\infty} e^{-ikt} z^k$ into (2.3) we get

$$\Phi(z) = \frac{1}{2\pi}\int_0^{2\pi} H(e^{it})^* H(e^{it}) dt + 2\sum_{k=1}^{\infty} z^k \frac{1}{2\pi}\int_0^{2\pi} e^{-ikt} H(e^{it})^* H(e^{it}) dt \tag{2.7}$$

which reduces the second order conditions (1.2) for H to the Carathéodory-Fejér interpolation conditions

$$\Phi(0) = r_0 \quad \text{and} \quad \Phi^{(k)}(0) = 2k! r_k \quad (k = 1, \dots, m) \tag{2.8}$$

for the associated function Φ. For rational H, the associated function Φ is characterized by relations (2.4) and (2.5).

Lemma 2.2. *Let $H \in \mathcal{R}^p$ and let*

$$H(z) = G\left(I_d - zT\right)^{-1} F \tag{2.9}$$

be its minimal realization of such form. Then the associated function Φ defined by (2.3) belongs to $\mathcal{C} \bigcap \mathcal{R}$, satisfies conditions (2.4), (2.5) and admits a representation

$$\Phi(z) = F^* \mathbb{P}\left(I + zT\right)\left(I - zT\right)^{-1} F \tag{2.10}$$

where $\mathbb{P} \in \mathbb{C}^{d\times d}$ is a unique solution of the Stein equation

$$\mathbb{P} - T^* \mathbb{P} T = G^* G. \tag{2.11}$$

Conversely, let $\Phi \in \mathcal{R}$ satisfy (2.4) and (2.5) for some $H \in \mathcal{R}^p$. Then Φ belongs to $\mathcal{C}$ and admits a representation (2.3).

Proof. Since the realization (2.9) is minimal and H has no poles in $\overline{\mathbb{D}}$, all eigenvalues of the matrix T lie inside the open unit disk and the pair $(G,\ T)$ is observable. Therefore, the Stein equation (2.11) has a unique solution $\mathbb{P}$ which is strictly positive and equal to

$$\mathbb{P} = \sum_{k=0}^{\infty} T^{*k} G^* G T^k > 0. \tag{2.12}$$

It follows from (2.9) that

$$H(z) = \sum_{j=0}^{\infty} H_j z^j, \quad \text{where} \quad H_j = GT^j F,$$

which together with (2.12) implies

$$\begin{aligned} \frac{1}{2\pi}\int_0^{2\pi} e^{-ikt} H(e^{it})^* H(e^{it}) dt &= \sum_{j=0}^{\infty} H_j^* H_{j+k} \\ &= \sum_{j=1}^{\infty} F^* T^{*j} G^* G T^{j+k} F = F^* \mathbb{P} T^k F \end{aligned}$$

for $k \geq 0$. Substituting the last equalities into (2.7) we obtain

$$\Phi(z) = F^* \mathbb{P} F + 2\sum_{k=1}^{\infty} z^k F^* \mathbb{P} T^k F = F^* \mathbb{P} F + 2zF^* \mathbb{P} T\left(I - zT\right)^{-1} F,$$

which coincides with (2.10). In particular, (2.10) implies that $\Phi \in \mathcal{R}$. Relations (2.4) and (2.5) are in force by Lemma 2.1 and since Φ is analytic in $\overline{\mathbb{D}}$, (2.4) holds for all t.
Conversely, let $\Phi \in \mathcal{R}$ satisfy (2.4) and (2.5). By (2.5), $\Re\, \Phi \geq 0$ and since $\Im\, \Phi(0) = 0$,

$$\Phi(z) = \frac{1}{2\pi}\int_0^{2\pi} \frac{e^{it} + z}{e^{it} - z} d\sigma(t) \tag{2.13}$$

for a nonnegetive measure $d\sigma$ on $\mathbb{T}$ with the absolutely continuous part

$$d\sigma_{ac}(t) = \frac{\Phi(e^{it}) + \Phi(e^{it})^*}{2} = H(e^{it})^* H(e^{it}) dt. \tag{2.14}$$

Since the function $\Phi(z) + \Phi(z)^*$ is harmonic and continuous in $\overline{\mathbb{D}}$,

$$\Phi(0) = \frac{\Phi(0) + \Phi(0)^*}{2} = \frac{1}{2\pi} \int_0^{2\pi} \frac{\Phi(e^{it}) + \Phi(e^{it})^*}{2} dt = \frac{1}{2\pi} \int_0^{2\pi} d\sigma_{ac}(t).$$

On the other hand, (2.13) implies $\Phi(0) = \frac{1}{2\pi} \int_0^{2\pi} d\sigma(t)$ and hence $d\sigma$ has no singular part. In view of (2.14),

$$d\sigma(t) = d\sigma_{ac}(t) = H(e^{it})^* H(e^{it}) dt$$

which together with (2.13) implies (2.3). □

Lemma 2.3. *The functions $H \in \mathcal{R}^p$ and $\Phi \in \mathcal{R}$ satisfy* (2.4) *and* (2.5) *if and only if they admit representations*

$$H(z) = S_2(z) \left(1 - S_1(z)\right)^{-1} \quad (|z| \le 1) \tag{2.15}$$

and

$$\Phi(z) = \left(1 + S_1(z)\right) \left(1 - S_1(z)\right)^{-1} \quad (|z| \le 1) \tag{2.16}$$

for some inner rational function

$$S(z) = \begin{pmatrix} S_1(z) \\ S_2(z) \end{pmatrix} \in \mathcal{R}^{p+1} \tag{2.17}$$

where S_1 takes scalar values and is subject to

$$S_1(e^{it}) \neq 1 \quad \text{for all } t. \tag{2.18}$$

Proof. Let $H \in \mathcal{R}^p$ and $\Phi \in \mathcal{R}$ satisfy (2.4) and (2.5). Since $\Re\, \Phi(z) \ge 0$ in $\overline{\mathbb{D}}$,

$$\Phi(z) + 1 \neq 0 \qquad (z \in \overline{\mathbb{D}}) \tag{2.19}$$

and the function

$$S(z) = \begin{pmatrix} S_1(z) \\ S_2(z) \end{pmatrix} = \begin{pmatrix} \Phi(z) - 1 \\ H(z) \end{pmatrix} \left(\Phi(z) + 1\right)^{-1} \tag{2.20}$$

belongs to $\mathcal{R}^{p+1}$. Furthermore, it follows readily from (2.20) that

$$\Phi(z) + \Phi(z)^* - 2H(z)^* H(z) = \frac{1}{2} (\Phi(z)^* + 1) \left(1 - S(z)^* S(z)\right) (\Phi(z) + 1), \tag{2.21}$$

which together with (2.4), (2.5) and (2.19) implies that S is inner. Finally, in view of (2.20) and (2.19),

$$1 - S_1(z) = 2(\Phi(z) + 1)^{-1} \neq 0 \qquad (z \in \overline{\mathbb{D}}). \tag{2.22}$$

Conversely, let S be a $\mathbb{C}^{p+1}$-valued rational inner function with the top entry S_1 satisfying (2.18). Then by the maximum modulus principle, $S_1 \neq 1$ in $\overline{\mathbb{D}}$ and the functions H and Φ defined via (2.15) and (2.16), respectively, are rational and with poles off $\overline{\mathbb{D}}$. Then relations (2.4) and (2.5) follow from (2.21) in view of (2.19). □

The following proposition follows immediately from (2.21).

Remark 2.4. *The functions $H \in \mathbf{H}_2^p$ and $\Phi \in \mathcal{C}$ satisfy (2.5) if and only if they admit representations (2.15) and (2.16) for some Schur function $S = \begin{pmatrix} S_1 \\ S_2 \end{pmatrix} \in \mathcal{S}^{(p+1)\times 1}$.*

Let

$$F = \begin{pmatrix} 0 & & \cdots & & 0 \\ 1 & 0 & & & \\ 0 & 1 & & & \\ \vdots & \ddots & \ddots & \ddots & \vdots \\ 0 & \cdots & 0 & 1 & 0 \end{pmatrix} \in \mathbb{C}^{(m+1)\times(m+1)},$$

$$E = \begin{pmatrix} 1 \\ 0 \\ \vdots \\ 0 \end{pmatrix} \in \mathbb{C}^{m+1} \quad \text{and} \quad C_R = \begin{pmatrix} \frac{1}{2}r_0 \\ r_1 \\ \vdots \\ r_m \end{pmatrix}. \tag{2.23}$$

It is easily seen that the interpolation conditions (2.8) can be written in the residue form as

$$\operatorname{Res}_{z=0} (zI - F)^{-1} E\Phi(z) = 2C_R. \tag{2.24}$$

To rewrite the latter condition in terms of the function S_1 we need the following auxiliary lemma (for the proof see [4, Lemma 7.2]).

Lemma 2.5. *Let W be any square matrix with $\operatorname{spec} W \subset \mathbb{D}$, and let Ψ and Υ be matix-valued functions analytic on $\operatorname{spec} W$. Then*

$$\sum_{z\in\mathbb{D}} \operatorname{Res}(zI - W)^{-1}\Psi(z)\Upsilon(z) = \sum_{z\in\mathbb{D}} \operatorname{Res}(zI - W)^{-1}L\Upsilon(z) \tag{2.25}$$

where

$$L = \sum_{z\in\mathbb{D}} \operatorname{Res}(zI - W)^{-1}\Psi(z).$$

Lemma 2.6. *Let $H \in \mathcal{R}^p$ and $\Phi \in \mathcal{R}$ satisfy (2.4) and (2.5), and consider their representations (2.15) and (2.15) with an inner function $S \in \mathcal{R}^{p+1}$ satisfying (2.18). Then, Φ and H satisfy interpolation conditions (1.10) and (2.24) (i.e., H is a solution of Problem 1.4) if and only if S satisfies the condition*

$$\operatorname{Res}_{z=0} \left(zI_M - \mathbf{W}\right)^{-1} \mathbf{B}S(z) = \mathbf{C} \tag{2.26}$$

where $M = N + m + 1$,

$$\mathbf{W} = \begin{pmatrix} F & 0 \\ 0 & W \end{pmatrix}, \quad \mathbf{B} = \begin{pmatrix} V_+ & 0 \\ C & B \end{pmatrix}, \quad \mathbf{C} = \begin{pmatrix} V_- \\ C \end{pmatrix}. \tag{2.27}$$

and

$$V_+ = C_R + \frac{1}{2}E = \begin{pmatrix} \frac{1}{2}(r_0+1) \\ r_1 \\ \vdots \\ r_m \end{pmatrix}, \quad V_- = C_R - \frac{1}{2}E = \begin{pmatrix} \frac{1}{2}(r_0-1) \\ r_1 \\ \vdots \\ r_m \end{pmatrix}. \tag{2.28}$$

Proof. Taking into account partitions (2.17) and (2.27) we conclude that the condition (2.26) is reduced to the following two conditions

$$\sum_{z\in\mathbb{D}} \operatorname{Res}\,(zI_N - W)^{-1}\,(C,\ B)S(z) = C, \tag{2.29}$$

$$\operatorname{Res}_{z=0}\,(zI_{m+1} - F)^{-1}\,V_+ S_1(z) = V_-. \tag{2.30}$$

We show that (1.10) and (2.24) are equivalent to (2.29) and (2.30), respectively. Let H satisfy (1.10) and be of the form (2.15). Applying Lemma 2.5 to $\Psi(z) = BH(z)$ and $\Upsilon(z) = 1 - S_1(z)$, we get

$$\begin{aligned} \sum_{z\in\mathbb{D}} \operatorname{Res}\,(zI_N - W)^{-1}\,BS_2(z) &= \sum_{z\in\mathbb{D}} \operatorname{Res}\,(zI_N - W)^{-1}\,BH(z)\,(1 - S_1(z)) \\ &= \sum_{z\in\mathbb{D}} \operatorname{Res}\,(zI_N - W)^{-1}\,C\,(1 - S_1(z))\,, \end{aligned} \tag{2.31}$$

which can be rewritten as

$$\sum_{z\in\mathbb{D}} \operatorname{Res}\,(zI_N - W)^{-1}\,(C,\ B)\begin{pmatrix} S_1(z) \\ S_2(z) \end{pmatrix} = \sum_{z\in\mathbb{D}} \operatorname{Res}\,(zI_N - W)^{-1}\,C = C.$$

Conversely, let S of the form (2.17) and let H be defined by (2.15). Then (2.31) holds and applying Lemma 2.5 to $\Psi(z) = C\,(1 - S_1(z))\,,\ \Upsilon(z) = (1 - S_1(z))^{-1}$ we get

$$\begin{aligned} C &= \sum_{z\in\mathbb{D}} \operatorname{Res}\,(zI - W)^{-1}\,C\,(1 - S_1(z))\,(1 - S_1(z))^{-1} \\ &= \sum_{z\in\mathbb{D}} \operatorname{Res}\,(zI_N - W)^{-1}\,BS_2(z)\,(1 - S_1(z))^{-1} \\ &= \sum_{z\in\mathbb{D}} \operatorname{Res}\,(zI_N - W)^{-1}\,BH(z). \end{aligned}$$

Thus, (1.10) is equivalent to (2.29). Furthermore, it follows from (2.24) and (2.28) that

$$\begin{aligned} &\operatorname{Res}_{z=0}\,(zI_{m+1} - F)^{-1}\,E(\Phi(z) + 1) = V_+, \\ &\operatorname{Res}_{z=0}\,(zI - F_{m+1})^{-1}\,E(\Phi(z) - 1) = V_- \end{aligned}$$

and applying Lemma 2.5 to $\Psi(z)=E(\Phi(z)+1)$, $\Upsilon=S_1$ we get

$$\begin{aligned} V_- &= \operatorname{Res}_{z=0}\left(zI_{m+1}-F\right)^{-1}E(\Phi(z)-1)\\ &= \operatorname{Res}_{z=0}\left(zI_{m+1}-F\right)^{-1}E(\Phi(z)+1)S_1(z)\\ &= \operatorname{Res}_{z=0}\left(zI_{m+1}-F\right)^{-1}V_+S_1(z). \end{aligned}$$

The converse is checked quite similarly. □

3. The inequality case

In this section we shall show that every Schur function $S\in\mathcal{S}^{(p+1)\times 1}$ satisfying the interpolation condition (2.26) gives rise via (2.15) to some solution H of Problem 1.3.

Lemma 3.1. *A function $H\in\mathbf{H}_2^p$ is subject to* (1.6) *if and only if there exists a Carathéodory function $\Phi\in\mathcal{C}$ which satisfies interpolation conditions* (2.8) *and the inequality* (2.5).

Proof. Let (2.8) and (2.5) hold for some $\Phi\in\mathcal{C}$. Since Φ is analytic in $\mathbb{D}$, it is readily checked by the residue theorem that

$$\frac{1}{2\pi}\int_0^{2\pi}\begin{pmatrix}1\\ \vdots\\ \rho^{-m}e^{-imt}\end{pmatrix}\frac{\Phi(\rho e^{it})+\Phi(\rho e^{it})^*}{2}\left(1,\ \dots\ ,\ \rho^m e^{imt}\right)dt=\mathbf{R}_m \tag{3.1}$$

for $0<\rho<1$. Futhermore, the inequality (2.5) may be extended to almost all points on $\mathbb{T}$ at which H and Φ have nontangential boundary limits. Therefore, by Fatou's lemma and on account of (3.1),

$$\begin{aligned} &\frac{1}{2\pi}\int_0^{2\pi}\begin{pmatrix}1\\ \vdots\\ e^{-imt}\end{pmatrix}H(e^{it})^*H(e^{it})\left(1,\ \dots\ ,\ e^{imt}\right)dt\\ &\le\frac{1}{2\pi}\int_0^{2\pi}\begin{pmatrix}1\\ \vdots\\ e^{-imt}\end{pmatrix}\frac{\Phi(e^{it})+\Phi(e^{it})^*}{2}\left(1,\ \dots\ ,\ e^{imt}\right)dt\\ &\le\lim_{\rho\to 1}\frac{1}{2\pi}\int_0^{2\pi}\begin{pmatrix}1\\ \vdots\\ \rho^{-m}e^{-imt}\end{pmatrix}\frac{\Phi(\rho e^{it})+\Phi(\rho e^{it})^*}{2}\left(1,\ \dots\ ,\ \rho^m e^{imt}\right)dt=\mathbf{R}_m. \end{aligned}$$

Conversely, let $H\in\mathbf{H}_2^p$ satisfy (1.6). Then the function Φ defined via (2.3) satisfies (2.8) and (2.5), by Lemma 2.1. □

Lemma 3.2. *The function $H\in\mathbf{H}_2^p$ is a solution of Problem 1.3 if it admits the representation* (2.15) *for some Schur function $S\in\mathcal{S}^{(p+1)\times 1}$ of the form* (2.17), *which satisfies the interpolation condition* (2.26).

Proof. By Lemma 3.1, H satisfies the second order condition (1.6) if and only if there exists a Carathéodory function $\Phi \in \mathcal{C}$ which satisfies interpolation conditions (2.8) and the inequality (2.5). By Remark 2.4, the inequality (2.5) is equivalent to representations (2.15) and (2.16) for some Schur function $S = \begin{pmatrix} S_1 \\ S_2 \end{pmatrix} \in \mathcal{S}^{(p+1)\times 1}$. It remains to note that by Lemma 2.6, conditions (2.8) and (1.10) are equivalent to (2.30) and (2.29), respectively, which in turn, are equivalent to (2.26). □

4. Description of all solutions

Since $\operatorname{spec} W \subset \mathbb{D}$ and $\operatorname{spec} F = \{0\}$, it follows from (2.27) that $\operatorname{spec} \mathbf{W} \subset \mathbb{D}$ and therefore, that the Stein equation

$$\mathbf{K} - \mathbf{W}\mathbf{K}\mathbf{W}^* = \mathbf{B}\mathbf{B}^* - \mathbf{C}\mathbf{C}^* \tag{4.1}$$

has a unique solution $\mathbf{K}$, which is given by the converging series

$$\mathbf{K} = \sum_{j=0}^{\infty} \mathbf{W}^j (\mathbf{B}\mathbf{B}^* - \mathbf{C}\mathbf{C}^*) \mathbf{W}^{*j}. \tag{4.2}$$

Making use of (2.27) and (2.28) we get

$$\mathbf{B}\mathbf{B}^* - \mathbf{C}\mathbf{C}^* = \begin{pmatrix} C_R E^* + E C_R^* & E C^* \\ C E^* & B B^* \end{pmatrix}, \tag{4.3}$$

which being substituted into (4.2) leads to the decomposition

$$\mathbf{K} = \begin{pmatrix} \mathbf{R}_m & K_1 \\ K_1^* & K_2 \end{pmatrix}, \tag{4.4}$$

where $\mathbf{R}_m = \sum_{j=0}^{m} F^j (C_R E^* + E C_R^*) F^{*j}$ is the matrix defined in (1.4),

$$K_1 = \sum_{j=0}^{m} F^j E C^* W^{*j} \quad \text{and} \quad K_2 = \sum_{j=0}^{\infty} W^j B B^* W^{*j}. \tag{4.5}$$

The necessary and sufficient condition for the existence of a Schur function S satisfying (2.26) is that $\mathbf{K} \geq 0$ (see e.g., [5]). Then, by Lemma 3.2, this condition is necessary and sufficient for Problem 1.3 to be solvable. In particular, the same condition is necessary (and not sufficient, due to [11, Example 1]) for Problem 1.3 to have a solution. We show that the strict positivity of $\mathbf{K}$ ensures the existence of infinitely many inner rational functions $S \in \mathcal{R}^{p+1}$ of the form (2.17) satisfying (2.26) and (2.18) and it is therefore the sufficient condition for Problem 1.4 to be solvable.

The following theorem is the immediate adaptation of the well-known results on tangential interpolation for matrix-valued Schur functions (see e.g., [2], [5], [10], [15]) to the vector-valued case.

Theorem 4.1. *Let* $\mathbf{K}$ *be the strictly positive solution of the Stein equation* (4.1) *and let* J *be the signature matrix defined by*

$$J = \begin{pmatrix} -I_{p+1} & 0 \\ 0 & 1 \end{pmatrix}. \tag{4.6}$$

1. The $\mathbb{C}^{(p+2)\times(p+2)}$*-valued function*

$$\Theta(z) = I_{p+2} + (1-z)\begin{pmatrix} \mathbf{B}^* \\ \mathbf{C}^* \end{pmatrix}(I - z\mathbf{W}^*)^{-1}\mathbf{K}^{-1}(I-\mathbf{W})^{-1}(\mathbf{B},\, \mathbf{C})\,J \tag{4.7}$$

is J*-expansive in* $\mathbb{D}$ *and* J*-unitary on* $\mathbb{T}$ *and moreover, for every point* $z \in \mathbb{C}$ *at which* Θ *is analytic,*

(4.8)
$$\Theta(z)J\Theta(z)^* - J = (1-|z|^2)\begin{pmatrix} \mathbf{B}^* \\ \mathbf{C}^* \end{pmatrix}(I - z\mathbf{W}^*)^{-1}\mathbf{K}^{-1}(I-\bar{z}\mathbf{W})^{-1}(\mathbf{B},\, \mathbf{C})$$

2. Let Θ *be partitioned by*

$$\Theta = \begin{pmatrix} \theta_{11} & \theta_{12} \\ \theta_{21} & \theta_{22} \\ \theta_{31} & \theta_{32} \end{pmatrix} \; : \; \begin{pmatrix} \mathbb{C}^{p+1} \\ \mathbb{C} \end{pmatrix} \to \begin{pmatrix} \mathbb{C} \\ \mathbb{C}^p \\ \mathbb{C} \end{pmatrix}. \tag{4.9}$$

Then the linear fractional transformation

$$S(z) = \begin{pmatrix} S_1(z) \\ S_2(z) \end{pmatrix} = \begin{pmatrix} \theta_{11}(z)\mathcal{E}(z) + \theta_{12}(z) \\ \theta_{21}(z)\mathcal{E}(z) + \theta_{22}(z) \end{pmatrix}(\theta_{31}(z)\mathcal{E}(z) + \theta_{32}(z))^{-1} \tag{4.10}$$

parametrizes the set of all Schur functions S *satisfying* (2.26) *when the parameter* $\mathcal{E}$ *varies on the set* $\mathcal{S}^{p+1}$.

Remark 4.2. Since the spectrum of $\mathbf{W}$ lies inside $\mathbb{D}$, the function Θ defined by (4.7) is rational and therefore, the function S of the form (4.10) belongs to $\mathcal{R}^{p+1}$ if and only if the corresponding parameter $\mathcal{E}$ belongs to $\mathcal{R}^{p+1}$.

It follows from (4.8) that $\Theta(e^{it})J\Theta(e^{it})^* = J$ for all t, and in particular, on account of (4.6) and (4.9), that

$$\theta_{32}(e^{it})\theta_{32}(e^{it})^* - \theta_{31}(e^{it})\theta_{31}(e^{it})^* = 1.$$

Therefore, $\theta_{32}(e^{it}) \neq 0$, the function $\theta_{32}(e^{it})^{-1}\theta_{31}(e^{it})$ is strictly contractive for all t and hence, for each $\mathcal{E} \in \mathcal{S}^{p+1}$,

$$\theta_{31}(e^{it})\mathcal{E}(e^{it}) + \theta_{32}(e^{it}) \neq 0.$$

Remark 4.3. Since Θ is J-unitary on $\mathbb{T}$, it follows from (4.10) and (4.6) that

$$(\theta_{31}(e^{it})\mathcal{E}(e^{it})+\theta_{32}(e^{it}))^*\left(1-S(e^{it})^*S(e^{it})\right)(\theta_{31}(e^{it})\mathcal{E}(e^{it})+\theta_{32}(e^{it}))$$
$$=(\theta_{31}(e^{it})\mathcal{E}(e^{it})+\theta_{32}(e^{it}))^*\left(S(e^{it})^*,\ 1\right)J\begin{pmatrix} S(e^{it}) \\ 1 \end{pmatrix}(\theta_{31}(e^{it})\mathcal{E}(e^{it})+\theta_{32}(e^{it}))$$
$$=\left(\mathcal{E}(e^{it})^*,\ 1\right)\Theta(e^{it})^*J\Theta(e^{it})\begin{pmatrix} \mathcal{E}(e^{it}) \\ 1 \end{pmatrix}$$
$$=\left(\mathcal{E}(e^{it})^*,\ 1\right)J\begin{pmatrix} \mathcal{E}(e^{it}) \\ 1 \end{pmatrix}=1-\mathcal{E}(e^{it})^*\mathcal{E}(e^{it})$$

and therefore, S is inner if and only if $\mathcal{E}$ is inner.

Theorem 4.4. *Let* $\mathbf{K}$ *be the strictly positive solution of the Stein equation* (4.1) *and let the* $\mathbb{C}^{(p+1)\times(p+2)}$*-valued function*

$$\begin{aligned}\Psi(z) &= \begin{pmatrix} 0 & I_p & 0 \\ -1 & 0 & 1 \end{pmatrix} \\ &+(1-z)\begin{pmatrix} 0 & B^* \\ -E^* & 0 \end{pmatrix}(I-z\mathbf{W}^*)^{-1}\mathbf{K}^{-1}(I-\mathbf{W})^{-1}(\mathbf{B},\ \mathbf{C})J \end{aligned} \tag{4.11}$$

be partitioned by

$$\Psi=\begin{pmatrix} \psi_{11} & \psi_{12} \\ \psi_{21} & \psi_{22} \end{pmatrix}\ :\ \begin{pmatrix} \mathbb{C}^{p+1} \\ \mathbb{C} \end{pmatrix}\to\begin{pmatrix} \mathbb{C}^{p} \\ \mathbb{C} \end{pmatrix}. \tag{4.12}$$

Then the linear fractional transformation

$$H(z)=(\psi_{11}(z)\mathcal{E}(z)+\psi_{12}(z))\,(\psi_{21}(z)\mathcal{E}(z)+\psi_{22}(z))^{-1} \tag{4.13}$$

parametrizes the set of all solutions H of Problem 1.4 when the parameter $\mathcal{E}$ varies on $\mathcal{R}^{p+1}$, is inner and satisfies

$$\psi_{21}(e^{it})\mathcal{E}(e^{it})+\psi_{22}(e^{it})\neq 0. \tag{4.14}$$

Moreover, the formula (4.13 *gives all the solutions of Problem 1.3 when the parameter $\mathcal{E}$ varies on the set $\mathcal{S}^{p+1}$ of all $\mathbb{C}^{p+1}$-valued Schur functions.*

Proof. By Lemmas 2.3 and 2.6 all solutions H of Problem 1.4 are parametrized by formula (2.15) for the inner parameter $S\in\mathcal{R}^{p+1}$ which is of the form (2.17) and satisfies (2.18) and (2.26). By Theorem 4.1 and Remarks 4.2 and 4.3, all inner vector-valued functions $S\in\mathcal{R}^{p+1}$ satisfying conditions (2.18) and (2.26) are described by the transformation (4.10) when the parameter $\mathcal{E}$ varies on $\mathcal{R}^{p+1}$, is inner and satisfies (4.14). Substituting fractional expressions for S_1 and S_2 from (2.26) into (2.15) we get

$$\begin{aligned} H(z) &= S_2(z)\,(1-S_1(z))^{-1} \\ &= (\theta_{21}(z)\mathcal{E}(z)+\theta_{22}(z))\,((\theta_{31}(z)-\theta_{11}(z))\,\mathcal{E}(z)+\theta_{32}(z)-\theta_{12}(z))^{-1} \\ &= (\psi_{11}(z)\mathcal{E}(z)+\psi_{12}(z))\,(\psi_{21}(z)\mathcal{E}(z)+\psi_{22}(z))^{-1} \end{aligned}$$

where

$$\Psi(z) = \begin{pmatrix} \psi_{11}(z) & \psi_{12}(z) \\ \psi_{21}(z) & \psi_{22}(z) \end{pmatrix} = \begin{pmatrix} 0 & I_p & 0 \\ -1 & 0 & 1 \end{pmatrix} \Theta(z). \tag{4.15}$$

In view of (2.27) and (2.28),

$$\begin{pmatrix} 0 & I_p & 0 \\ -1 & 0 & 1 \end{pmatrix} \begin{pmatrix} \mathbf{B}^* \\ \mathbf{C}^* \end{pmatrix} = \begin{pmatrix} 0 & B^* \\ -E^* & 0 \end{pmatrix} \tag{4.16}$$

and now the formula (4.11) follows from (4.7) and (4.15). Finally, on account of (4.10) and (4.15),

$$\begin{aligned} 1 - S_1(z) &= ((\theta_{31}(z) - \theta_{11}(z))\mathcal{E}(z) + \theta_{32}(z) - \theta_{12}(z)) (\theta_{31}(z)\mathcal{E}(z) + \theta_{32}(z))^{-1} \\ &= (\psi_{21}(z)\mathcal{E}(z) + \psi_{22}(z)) (\theta_{31}(z)\mathcal{E}(z) + \theta_{32}(z))^{-1} \end{aligned}$$

and therefore, (2.18) is equivalent to (4.14).

The second assertion of the theorem is deduced in much the same way Lemmas 3.2 and 2.6. □

Theorem 4.5. *Let* $\mathbf{K}$ *be strictly positive. Then Problem 1.4 has infinitely many solutions.*

Proof. By Theorem 4.4, it suffices to show that there exist infinitely many inner functions $\mathcal{E} \in \mathcal{R}^{p+1}$ which satisfy the condition (4.14). The rest is broken into steps.

Step 1: $\psi_{22}(z) \neq 0$ at every point z from the closed unit disk and therefore, the condition (4.14) is equivalent to

$$\sigma(e^{it})\mathcal{E}(e^{it}) \neq 1, \quad \text{where} \quad \sigma(z) = -\psi_{22}^{-1}(z)\psi_{21}(z). \tag{4.17}$$

Proof of Step 1. It follows from (4.15), (4.16) and (4.8) that

$$\begin{aligned} &\Psi(z)J\Psi(z)^* - \begin{pmatrix} -I_p & 0 \\ 0 & 0 \end{pmatrix} \\ &= (1-|z|^2) \begin{pmatrix} 0 & B^* \\ -E^* & 0 \end{pmatrix} (I - z\mathbf{W}^*)^{-1} \mathbf{K}^{-1} (I - \bar{z}\mathbf{W})^{-1} \begin{pmatrix} 0 & -E \\ B & 0 \end{pmatrix}. \end{aligned} \tag{4.18}$$

Comparing the right bottom entries in the last equality and making use of decompositions (4.12) and (4.6) we conclude that

$$\psi_{22}(z)\psi_{22}(z)^* > \psi_{21}(z)\psi_{21}(z)^* \quad \text{for } |z| < 1 \tag{4.19}$$

and

$$\psi_{21}(e^{it})\psi_{21}(e^{it})^* = \psi_{22}(e^{it})\psi_{22}(e^{it})^*. \tag{4.20}$$

Since the function Θ defined in (4.7) is J-unitary and has no poles on $\mathbb{T}$, it is invertible everywhere on $\mathbb{T}$. Furthermore, it follows from (4.19) that $\psi_{22}(z) \neq 0$ for every $z \in \mathbb{D}$. Assuming that $\psi_{22}(\zeta) = 0$ for some $\zeta \in \mathbb{T}$ we conclude from (4.20) that $\psi_{21}(\zeta) = 0$ and therefore, on account of (4.15), that $(-1,\ 0,\ 1)\Theta(\zeta) = 0$. which is impossible, since $\Theta(\zeta)$ is invertible.

Step 2: There exists a unit vector $\mathcal{E} \in \mathbb{C}^{p+1}$ such that

$$\sigma(e^{it})\mathcal{E} \neq 1, \quad \text{where} \quad \sigma(z) = -\psi_{22}^{-1}(z)\psi_{21}(z). \tag{4.21}$$

Proof of Step 2. It follows from (4.19) and (4.20) that the rational function $\sigma \in \mathcal{R}^{1\times(p+1)}$ defined as in (4.21) is coinner. Let σ be partitioned by

$$\sigma(z) = (\sigma_1(z),\, \sigma_2(z)), \quad \text{where } \sigma_1 \in \mathcal{R} \quad \text{and} \quad \sigma_2 \in \mathcal{R}^{1\times p}.$$

Since σ is coinner, $\sigma_2(\zeta) = 0$ at every point $\zeta \in \mathbb{T}$ at which $|\sigma_1(\zeta)| = 1$. Let the latter hold for infinitely many points on $\mathbb{T}$. Since σ_2 is rational, then it is equal to zero identically and therefore, every unit vector of the form $\mathcal{E} = (0, \dots, 0, \xi)^*$ ($|\xi| = 1$) satisfies (4.21).

Let now $|\sigma_1(\zeta)|$ take the value one at finitely many points and let for the definitness

$$\sigma_1(\zeta_j) = \xi_j, \quad \zeta_j,\, \xi_j \in \mathbb{T} \quad (j = 1, \dots, \ell).$$

Let ξ be any point on $\mathbb{T}$ different from all the ξ_j. Then the unit vector $\mathcal{E} = (\zeta,\, 0, \dots, 0,)^*$ satisfies (4.21). To complete the proof, it remains to note that the function $\mathcal{E}(z) \equiv \mathcal{E}$ is inner and satisfies (4.17). □

5. The degenerate case

In this section we consider the case when the Pick matrix $\mathbf{K}$ is positive semidefinite. By Lemma 3.2, Problem 1.3 is reduced to the interpolation problem (2.26) for vector-valued Schur functions. Now this problem is degenerate and the formula (4.7) (and therefore, the description (4.10)) makes no sense. The degenerate bitangential interpolation problem for matrix-valued Schur functions has been studied in [8] (we refer also to [9] for the degenerate classical Schur problem). Below we recall some results from [8] suitably adapted to the vector-valued case. As in Theorem 4.1, the special structure (2.27) of the matrices $\mathbf{W}$, $\mathbf{B}$ and $\mathbf{C}$ plays no role.

Let us introduce the subspace

$$\mathcal{K} = \operatorname{Ker}\mathbf{K} \,\cap\, \operatorname{Ker}\mathbf{C}^* \subset \mathbb{C}^M \tag{5.1}$$

and let Q be an isometric matrix whose columns span $\mathcal{K}^\perp$, the orthogonal complement of $\mathcal{K}$ in $\mathbb{C}^M$ with respect to the standard inner product. Let us consider the matrix-valued function

$$\Delta(z) = \mathbf{K}(I - z\mathbf{W}^*) + \mathbf{C}\mathbf{C}^* \tag{5.2}$$

It was shown in [8] that if $\operatorname{spec}\mathbf{W} \subset \mathbb{D}$, then

$$\operatorname{Ker}\Delta(z) = \operatorname{Ker}\Delta(z)^* = \mathcal{K} \quad \text{for every} \quad z \in \overline{\mathbb{D}}$$

and the rational matrix-valued function

$$\Delta^{[-1]}(z) := Q\left(Q^*\Delta(z)Q\right)^{-1}Q^* \tag{5.3}$$

has no poles in $\overline{\mathbb{D}}$ and is the Moore-Penrose pseudoinverse (see e.g. [13, Section 12.8] for the definition) of $\Delta(z)$ for every point z at which $\Delta^{[-1]}(z)$ is analytic.

Theorem 5.1. *Let* spec$\mathbf{W} \subset \mathbb{D}$, *let* $\mathbf{K}$ *be a nonnegative solution of the Stein equation* (4.1) *and let* ν *be the integer defined as*

$$\nu = \text{rank}\,(\mathbf{K} + \mathbf{C}\mathbf{C}^*) - \text{rank}\,\mathbf{K} = \text{rank}\,(\mathbf{W}\mathbf{K}\mathbf{W}^* + \mathbf{B}\mathbf{B}^*) - \text{rank}\,\mathbf{K}. \tag{5.4}$$

1. If $\nu = 0$, *then the problem* (2.26) *has ifinitely many solutions.*
2. If $\nu = 1$, *then the problem has a unique solution*

$$S(z) = \mathbf{B}^*\Delta^{[-1]}(z)\mathbf{C}, \tag{5.5}$$

which is rational and inner.

Note that there are only two possible values which ν may take: zero and one. Indeed, it is readily seen from the definition (5.4) that $\nu \geq 0$ and since rank $\mathbf{C}\mathbf{C}^* = 1$, it follows that $\nu \leq 1$. Note also that $\nu = 0$ if $\mathbf{K}$ is strictly positive.

Lemma 5.2. *Let* $\mathbf{K}$ *be a nonnegative singular solution of the Stein equation* (4.1) *and let* $(\mathbf{W},\ \mathbf{B})$ *be a full-range pair. Then* $\nu = 1$.

Proof. Let $\mathbf{K} \geq 0$ be singular and let us assume that $n \neq 1$. By the above remark, $\nu = 0$ and therefore,

$$\text{rank}\,(\mathbf{K} + \mathbf{C}\mathbf{C}^*) = \text{rank}\,\mathbf{K}.$$

Since Ker $(\mathbf{K} + \mathbf{C}\mathbf{C}^*) \subseteq \text{Ker}\,\mathbf{K}$, the last equality implies that for every $f \in \text{Ker}\,\mathbf{K}$ we have also $\mathbf{C}^*f = 0$. Then we conclude from the Stein equation (4.1) that

$$\mathbf{K}\mathbf{W}^*f = 0 \quad \text{and} \quad \mathbf{B}^*f = 0 \quad \text{for every} \quad f \in \text{Ker}\,\mathbf{K}. \tag{5.6}$$

Take any vector $f \neq 0$ from $\text{Ker}\,\mathbf{K}$. Then we obtain from (5.6) iteratively that $\mathbf{W}^{*j}f \in \text{Ker}\,\mathbf{K}$ for all integers $j \geq 0$. Then, again by (5.6),

$$\mathbf{B}^*\mathbf{W}^{*j}f = 0 \quad \text{for} \quad j = 0, 1, \ldots$$

Thus, f belongs to $\bigcap\limits_{j\geq 0}\text{Ker}\mathbf{B}^*\mathbf{W}^{*j}$ and therefore, is orthogonal to the set $\bigcup\limits_{j\geq 0}\mathbf{W}^j\mathbf{B}$, which contradicts the assumption that $(\mathbf{W},\ \mathbf{B})$ is the full-range pair. □

Now we shall use the special structure (2.27) of the matrices $\mathbf{W}$, $\mathbf{B}$ and $\mathbf{C}$.

Lemma 5.3. *The subspace* $\mathcal{K}$ *defined in* (5.1) *admits a representation*

$$\mathcal{K} = \text{Ker}\,\mathbf{K}\ \cap\ \text{Ker}\left(\begin{array}{cc} E^* & 0 \\ 0 & B^* \end{array}\right). \tag{5.7}$$

Proof. Let a vector f belongs to $\text{Ker}\,\mathbf{K}\ \cap\ \text{Ker}\,\mathbf{C}^*$. Then by the Stein equation (4.1), $\mathbf{B}^*f = 0$ and therefore,

$$\left(\begin{array}{cc} E^* & 0 \\ 0 & B^* \end{array}\right) f = \left(\begin{array}{cc} V_+^* - V_-^* & C^* - C^* \\ 0 & B^* \end{array}\right) f = \mathbf{B}^*f - \left(\begin{array}{c} \mathbf{C}^* \\ 0 \end{array}\right) f = 0,$$

which means that f belongs to the set indicated on the right-hand side in (5.7). Conversely, let f belongs to this last set. Then upon applying the Stein equation (4.1) to f and making use of (4.3) we get

$$-\mathbf{WKW}^* f = \begin{pmatrix} E \\ 0 \end{pmatrix} (C_R^*,\, C^*)\, f. \tag{5.8}$$

Multiplying the latter equality by f^* on the left we obtain $f^*\mathbf{WKW}^*f = 0$ and therefore, $\mathbf{KW}^*f = 0$. Now it follows from (5.8) that $(C_R^*,\, C^*)\, f = 0$. Therefore,

$$\mathbf{C}^* f = \left(V_-^*,\, C^*\right) f = (C_R^*,\, C^*)\, f + \frac{1}{2}\,(E^*,\, 0)\, f = 0$$

and f belongs to the set indicated on the right-hand side in (5.7). Thus, the sets on the right-hand sides in (5.1) and (5.7) coincide. □

Let us define the matrix-valued function

$$\Delta_1(z) = \mathbf{K}(I - z\mathbf{W}^*) - \mathbf{C}\,(E^*,\, 0)\,, \tag{5.9}$$

where E is the unit vector defined in (2.23). It follows readily from Lemma 5.3 that

$$\mathcal{K} \subseteq \operatorname{Ker} \Delta_1(z) \bigcap \operatorname{Ker} \Delta_1(z)^* \quad \text{for all } z \in \mathbb{C}. \tag{5.10}$$

Lemma 5.4. *The function $\Delta_1(z)$ has the nonnegative real part in $\mathbb{D}$ and*

$$\operatorname{Ker} \Delta_1(z) = \operatorname{Ker} \Delta_1(z)^* = \mathcal{K} \tag{5.11}$$

for every point $z \in \mathbb{D}$. Moreover, for every point $\zeta \in \mathbb{T}$,

$$\operatorname{Ker} \Delta_1(\zeta) = \operatorname{Ker} \mathbf{K}(I_M - \zeta\mathbf{W}^*) \bigcap \operatorname{Ker} \begin{pmatrix} E^* & 0 \\ 0 & B^* \end{pmatrix}. \tag{5.12}$$

Proof. Using the Stein equation (4.1) and block decompositions (2.27) and (2.28), we rewrite (5.9) as

$$\begin{aligned} \Delta_1(z) &= (\mathbf{W} - zI)\mathbf{KW}^* - \mathbf{C}\,(E^*,\, 0) + \mathbf{BB}^* - \mathbf{CC}^* \\ &= (\mathbf{W} - zI)\mathbf{KW}^* + \begin{pmatrix} EV_+^* & EC^* \\ 0 & BB^* \end{pmatrix} \end{aligned}$$

Taking adjoints in the last equality and making use of (5.9) we get

(5.13)

$$\begin{aligned} &\Delta_1(z) + \Delta_1(z)^* \\ &\quad = \mathbf{K}(I - z\mathbf{W}^*) + \mathbf{WK}(\mathbf{W}^* - \bar{z}I) + \begin{pmatrix} V_+E^* & 0 \\ CE^* & BB^* \end{pmatrix} - \begin{pmatrix} V_- \\ C \end{pmatrix} (E^*,\, 0) \\ &\quad = (1 - |z|^2)\mathbf{K} + (zI - \mathbf{W})K(\bar{z}I - \mathbf{W}^*) + \begin{pmatrix} EE^* & 0 \\ 0 & BB^* \end{pmatrix}, \end{aligned}$$

which implies the first assertion of lemma. Let $z \in \mathbb{D}$ and $f \in \operatorname{Ker} \Delta_1(z) \cup \operatorname{Ker} \Delta_1(z)^*$. Then

$$f^*\left\{\Delta_1(z) + \Delta_1(z)^*\right\} f = 0,$$

which in view of (5.13) and (5.7) implies that $f \in \mathcal{K}$. Thus,

$$\operatorname{Ker} \Delta_1(z) \bigcup \operatorname{Ker} \Delta_1(z)^* \subseteq \mathcal{K},$$

which together with (5.10) implies (5.11). Let now $\zeta \in \mathbb{T}$ and $f \in \operatorname{Ker} \Delta_1(\zeta)$. Then we conclude from (5.13) that $f \in \operatorname{Ker} \begin{pmatrix} EE^* & 0 \\ 0 & BB^* \end{pmatrix}$ and then it follows from (5.9) that f belongs also to $\operatorname{Ker} \mathbf{K}(I_M - \zeta \mathbf{W}^*)$. This shows that

$$\operatorname{Ker} \Delta_1(\zeta) \subseteq \operatorname{Ker} \mathbf{K}(I_M - \zeta \mathbf{W}^*) \bigcap \operatorname{Ker} \begin{pmatrix} E^* & 0 \\ 0 & B^* \end{pmatrix}.$$

The inverse inclusion is obvious. □

Let Q be the isometric matrix from (5.3). Since the columns of Q span $\mathcal{K}^\perp$, the rational matrix-valued function

$$\Delta_1^{[-1]}(z) := Q \left(Q^* \Delta_1(z) Q\right)^{-1} Q^* \tag{5.14}$$

has no poles in $\mathbb{D}$, by (5.11). Let $\mathbf{P}_\mathcal{K}$ denote the orthogonal projection of $\mathbb{C}^M$ onto $\mathcal{K}$. It follows readily from (5.11) and (5.14) that

$$\begin{aligned} &(1)\ \ \Delta_1^{[-1]}(z)\Delta_1(z)\Delta_1^{[-1]}(z) = \Delta_1^{[-1]}(z), \\ &(2)\ \ \Delta_1(z)\Delta_1^{[-1]}(z)\Delta_1(z) = \Delta_1(z), \\ &(3)\ \ \Delta_\omega(z)\Delta_1^{[-1]}(z) = \Delta_1^{[-1]}(z)\Delta_1(z) = I_M - \mathbf{P}_\mathcal{K} \end{aligned}$$

for every point z at which $\Delta_1^{[-1]}(z)$ is analytic and therefore, $\Delta_1^{[-1]}(z)$ is the Moore-Penrose pseudoinverse of $\Delta_1(z)$.

Lemma 5.5. *If $\mathbf{K} \geq 0$ is a solution of the Stein equation (4.1) and $\omega \in \mathbb{D}$, then the resolventlike identity*

$$\Delta_1^{[-1]}(z) - \Delta^{[-1]}(z) = \Delta^{[-1]}(z)\mathbf{C}\left(V_+^*,\ C\right)\Delta_1^{[-1]}(z) \tag{5.15}$$

hold for every point z at which $\Delta^{[-1]}(z)$ and $\Delta_1^{[-1]}(z)$ are analytic.

Proof. In view of (5.2) and (5.9),

$$\Delta(z) - \Delta_1(z) = \mathbf{C}\mathbf{C}^* + \mathbf{C}(E^*,\ 0) = \mathbf{C}\left(V_+^*,\ C\right),$$

and therefore, on account of (5.3) and (5.14),

$$\begin{aligned} \Delta_1^{[-1]}(z) - \Delta^{[-1]}(z) &= Q\left(Q^*\Delta_1(z)Q\right)^{-1}Q^* - Q\left(Q^*\Delta(z)Q\right)^{-1}Q^* \\ &= Q\left(Q^*\Delta(z)Q\right)^{-1}Q^*\left(\Delta_1(z) - \Delta(z)\right)Q\left(Q^*\Delta_1(z)Q\right)^{-1}Q^* \\ &= \Delta_1^{[-1]}(z)\mathbf{C}\left(V_+^*,\ C\right)\Delta^{[-1]}(z). \end{aligned}$$

□

Lemma 5.6. *Let $Y \in \mathbb{C}^{k\times M}$, $Y \in \mathbb{C}^{M\times k}$ and let $X \in \mathbb{C}^{M\times M}$ be invertible. Then the matrix $I_k - YX^{-1}Z$ is invertible if and only if $X - ZY$ is invertible and*

$$\left(I_k - YX^{-1}Z\right)^{-1} = I_M + Y\left(X - ZY\right)^{-1}Z. \tag{5.16}$$

Proof. Let $U = \begin{pmatrix} X & Z \\ Y & I_k \end{pmatrix}$. The matrices $I_k - YX^{-1}Z$ and $X - ZY$ are Schur complements of the entries X and I_k of U, respectively. Therefore, the invertibility of each one of them is equivalent to the invertibility of U. The equality (5.16) is easily verified by straightforward calculation. □

Theorem 5.7. *Let conditions* (1.9) *be in force, let* $\mathbf{K}$ *be a nonnegative (and singular) solution of the Stein equation* (4.1) *and let* $\mathcal{K}$ *be the subspace defined via* (5.1) *or* (5.7). *Then*

1. *Problem 1.3 has a unique solution which is given by*

$$H(z) = (0,\ B^*)\Delta_1^{[-1]}(z)\mathbf{C}, \tag{5.17}$$

   where Δ_1 *is the function defined in* (5.9).
2. *This function* H *is a (unique) solution of Problem 1.4 if and only if the following condition holds:*

$$\operatorname{Ker} \mathbf{K}(I_M - \zeta \mathbf{W}^*) \cap \operatorname{Ker} \begin{pmatrix} E^* & 0 \\ 0 & B^* \end{pmatrix} = \mathcal{K} \quad \text{for every } \zeta \in \mathbb{T}. \tag{5.18}$$

Proof. It is readily seen that $(F,\ E)$ is the full-range pair, which together with the first condition in (1.9) and decomposition (2.27) of $\mathbf{W}$ implies that $(\mathbf{W},\ \mathbf{B})$ is the full-range pair. Then by Lemma 5.2, the integer ν defined in (5.4) is equal to one. By Theorem 5.1, there exists the only Schur function S given by (5.5) which satisfies the interpolation condition (2.26). Using the structure (2.27) of $\mathbf{B}$ we get

$$S_1(z) = (0,\ B^*)\Delta^{[-1]}(z)\mathbf{C} \quad \text{and} \quad S_2(z) = (V_+^*,\ C^*)\Delta^{[-1]}(z)\mathbf{C}.$$

Applying identity (5.16) to the matrices

$$X = Q^*\Delta(z)Q, \quad Y = (V_+^*,\ C^*)Q, \quad Z = Q\mathbf{C}$$

and taking (5.2), (5.9) and (5.14) into account, we obtain for $z \in \mathbb{D}$,

$$\begin{aligned} (1 - S_1(z))^{-1} &= \left(1 - (V_+^*,\ C^*)\Delta^{[-1]}(z)\mathbf{C}\right)^{-1} \\ &= 1 + (V_+^*,\ C^*)Q\left(Q^*\Delta(z)Q - Q\mathbf{C}(V_+^*,\ C^*)Q\right)^{-1}Q\mathbf{C} \\ &= 1 + (V_+^*,\ C^*)Q\left(Q^*\Delta_1(z)Q\right)^{-1}Q\mathbf{C} \\ &= 1 + (V_+^*,\ C^*)\Delta_1^{[-1]}(z)\mathbf{C}. \end{aligned} \tag{5.19}$$

Upon substituting the latter equality into (2.15) and making use of (5.15) we come to (5.17):

$$H(z) = (0,\ B^*)\Delta^{[-1]}(z)\mathbf{C}\left(1 + (V_+^*,\ C^*)\Delta_1^{[-1]}(z)\mathbf{C}\right) = (0,\ B^*)\Delta_1^{[-1]}(z)\mathbf{C}.$$

The obtained function is a solution of Problem 1.3, by Lemma 3.2.

Since every solution of Problem 1.4 is a solution of Problem 1.3, the degenerate Problem 1.4 has at most one solution H defined in (5.17). The corresponding Schur function S from (5.5) from is rational and inner by Theorem 5.1. Thus, H is a solution of Problem 1.4 if and only if the condition (2.18) is fulfilled. By (5.19) and Lemma 5.6, $1 - S_1(\zeta)$ is invertible if and only if $Q^*\Delta_1(\zeta)Q$ is invertible. By definition of Q and (5.12), $Q^*\Delta_1(\zeta)Q$ is invertible everywhere on $\mathbb{T}$ if and only if (5.18) holds. □

Remark 5.8. Since spec $\mathbf{W} \in \mathbb{D}$, it follows that for strictly positive $\mathbf{K}$ condition (5.18) is fulfilled automatically and therefore, it is necessary and sufficient for Problem 1.4 to have a solution.

6. Concluding remarks

In conclusion we apply the obtained results to Problems 1.1 and 1.2, which are particular cases of Problems 1.4 and 1.3 for the special choice (1.11) of W, B and C. Substituting (1.11) into (4.5) we get

$$K_1 = \mathbf{H}_{m,n}, \quad K_2 = I_{p(n+1)} \quad \text{and} \quad \mathbf{K} = \begin{pmatrix} \mathbf{R}_m & \mathbf{H}^*_{m,n} \\ \mathbf{H}_{m,n} & I_{p(n+1)} \end{pmatrix}.$$

The matrix $\mathbf{K}$ is nonnegative (positive) if and only if $\mathbf{K}_{m,n} = \mathbf{R}_m - \mathbf{H}^*_{m,n}\mathbf{H}_{m,n}$, the Schur complement of $I_{p(n+1)}$ in $\mathbf{K}$ is nonnegative (positive). Thus, by Theorem 5.7 we obtain that Problem 1.2 has a unique solution if and only if the matrix $\mathbf{K}_{m,n}$ is positive semidefinite (and singular). Upon substituting (1.11), condition (5.18) takes the form

$$\operatorname{Ker} \mathbf{K}_{m,n}(I - \zeta F^*) \cap \operatorname{Ker} E^* = \operatorname{Ker} \mathbf{K}_{m,n} \cap \operatorname{Ker} E^* \quad \text{for all } \zeta \in \mathbb{T} \tag{6.1}$$

and is together with (1.3) necessary and sufficient for Problem 1.1 to be solvable. We illustrate the last condition by two examples. The first example is taken from [11].

Example 6.1. Let $r_0 = 2$, $r_1 = 1$, $h_0 = 1$, $h_1 = 0$. Then $E^* = (1,\ 0)$,

$$F = \begin{pmatrix} 0 & 0 \\ 1 & 0 \end{pmatrix}, \quad \mathbf{K}_{1,1} = \begin{pmatrix} 1 & 1 \\ 1 & 1 \end{pmatrix} \quad \text{and} \quad \mathbf{K}_{1,1}(I - \zeta F^*) = \begin{pmatrix} 1 & 1-\zeta \\ 1 & 1-\zeta \end{pmatrix}.$$

It is easily checked that condition (6.1) is broken at the point $\zeta = 1$ and thus, the corresponding Problem 1.1 has no solutions.

Example 6.2. Let $r_0 = 1$, $r_1 = 0$, $h_0 = 0$, $h_2 = 1$. Then

$$\mathbf{K}_{1,1} = \mathbf{K}_{1,1}(I - \zeta F^*) = \begin{pmatrix} 0 & 0 \\ 0 & 1 \end{pmatrix}$$

and therefore, (6.1) holds for all $\zeta \in \mathbb{T}$. The corresponding Problem 1.1 has a unique solution $H(z) = z$.

References

[1] N.I. Akhiezer. *The classical moment problem.* Hafner, New-York, 1965.

[2] D. Alpay. *Algorithme de Schur, espaces à noyau reproduisant et théorie des systèmes,* volume 6 of *Panoramas et Synthèses.* Société Mathématique de France, 1998.

[3] D. Alpay, V. Bolotnikov, and Ph. Loubaton. On tangential H_2 interpolation with second order norm constraints. *Integral Equations and Operator Theory,* 24:156–178, 1996.

[4] D. Alpay, V. Bolotnikov, and Ph. Loubaton. On interpolation for Hardy functions in a certain class of domains under moment type constraints. *Houston Journal of Mathematics,* 23:539–571, 1997.

[5] J. Ball, I. Gohberg, and L. Rodman. *Interpolation of rational matrix functions.* Birkhäuser Verlag, Basel, 1990.

[6] J. Ball and J.W. Helton. Interpolation problems of Pick-Nevanlinna and Loewner types for meromorphic matrix-functions: parametrization of the set of all solutions. *Integral Equations and Operator Theory,* 9:155–203, 1986.

[7] V. Bolotnikov. Extensions of nonnegative Hankel matrices and degenerate Hamburger moment problem. *Integral Equations and Operator Theory,* 25:253–276, 1996.

[8] V. Bolotnikov and H. Dym. On degenerate interpolation maximum entropy and extremal problems for matrix Schur functions. To appear in Integral Equations and Operator Theory.

[9] V. Dubovoy. Indefinite metric in the interpolation problem of Schur for analytic matrix functions IV. *Theor. Funktsii, Func. Anal. i Prilozen,* 42:46–57, 1984.

[10] H. Dym. A basic interpolation problem. In D. Sarason, S. Axler, J. McCarthy, editors, *Holomorphic spaces.* Cambridge University press, 1998.

[11] Y. Inouye. Approximation of multivariable linear systems with impulse response and autocorrelation sequences. *Automatica,* 19:265–277, 1983.

[12] A.M. King, U.B. Desai, and R.E. Skelton. A generalized approach to q-Markov covariance equivalent realizations for discrete systems. *Automatica,* 24:507–515, 1988.

[13] P. Lancaster and M. Tismenetsky. *The theory of matrices.* Academic Press, Orlando, 1985.

[14] C.T. Mullis and R.A. Roberts. The use of second-order information in the approximation of discrete-time linear systems. *IEEE Transactions on Acoustics, Speech and Signal Processing,* 24:226–238, 1976.

[15] A.A. Nudelman. On a new problem of moment problem type. *Sov. Math. Dokl.,* 18:507–510, 1977.

[16] D. Sarason. Nevanlinna-Pick interpolation with boundary data. *Integral Equations and Operator Theory,* 30:231–250, 1998.

Department of Mathematics, The College of William and Mary, Williamsburg, VA 23187-8795, USA

Operator Theory:
Advances and Applications, Vol. 123

Generalized Moments, Center-Focus Conditions, and Compositions of Polynomials

M. Briskin, J.-P. Françoise, and Y. Yomdin

To M.S. Livšic on his 80th birthday

Preface

In his paper [25], which has started a long line of investigations, M.S. Livšic introduced generalized moments and gave conditions for solvability of a generalized moment problem.

In the authors' recent work on Center-Focus conditions for plane vector fields and Abel equations, a special type of generalized moments appeared, as one of the main analytic structures, underlying the problem. In particular, the problem of vanishing of all the moments (or simply of orthogonality of a polynomial to all the powers of another polynomial on a given interval) turned out to be equivalent to the so-called "tangential Center-Focus problem" for a polynomial Abel equation (see below).

In the present paper we give a solution of this last problem. It involves another important structure — the algebra of univariate polynomials under composition.

In M.S. Livšic's theorem algebra of compositions does not appear explicitly. However, there are some indications of possible connections between these two problems, which we discuss below.

The authors would like to thank M.S. Livšic, Y. Kannai, V. Katsnelson, C. Christopher and S. Yakovenko for inspiring discussions, and the referee for suggesting several important improvements. The third author would like to thank M.S. Livšic for an exciting opportunity to follow his research and to enjoy his advice for twenty years.

The paper is organized as follows: In the introduction we give a short overview of the main results presented.

In Section 2 equivalence of the tangential Center-Focus problem and the problem of vanishing of the generalized moments is shown.

In Section 3 the algebra of univariate polynomials under composition is discussed.

The research of the first and the third author was supported by the Israel Science Foundation, Grant No. 101/95-1, and by the Minerva Foundation.

In Section 4 the main theorem, which gives necessary and sufficient conditions for vanishing of generalized polynomial moments is stated and detailed sketches of three different approaches to its proof are given.

In Section 5, Livšic's theorem is stated and its possible relation to the Moments vanishing problem is discussed.

Finally, in Section 6, we discuss the status of the general Center-Focus problem for Abel equations. In particular, we investigate some of its simplified versions.

1. Introduction

This paper is a continuation of [6–11].

Consider an Abel equation

$$y' = p(x)y^2 + q(x)y^3, \quad y(0) = y_0, \tag{1.1}$$

with $p(x), q(x)$-polynomials in x of the degrees d_1, d_2 respectively. We will write $p(x) = \lambda_{d_1}x^{d_1} + \cdots + \lambda_0$, $q(x) = \mu_{d_2}x^{d_2} + \cdots + \mu_0$, $(\lambda_{d_1}, \ldots, \lambda_0, \mu_{d_2}, \ldots, \mu_0) = (\lambda, \mu) \in \mathbb{C}^{d_1+d_2+2}$.

The following two problems for Equation (1.1) have been intensively studied (see [2–4], [13–23] and references there):

a. Is it possible to bound the number of solutions $y(x)$ of (1.1), satisfying $y(0) = y(1)$, in terms of the degrees d_1, d_2 only?
b. Is it possible to give explicit conditions on p and q for $y(0) \equiv y(1)$ for all solutions of (1.1)?

These two problems are well known to be closely related to the classical Hilbert's 16th problem (second part) and Poincaré's center-focus problem for polynomial vector fields on the plane.

Adopting the standard terminology in these problems, we shall call solutions of (1.1), satisfying $y(0) = y(1)$, the closed ones, and the conditions for $y(0) \equiv y(1)$ the center conditions.

Now consider a parametric form of the above equation:

$$y' = p(x)y^2 + \varepsilon q(x)y^3, \; y(0) = y_0, \tag{1.2}$$

with $\varepsilon \in \mathbb{C}$, p and q as above.

Consider a family $y(\varepsilon, x)$ of solutions of (1.2) with $y(\varepsilon, 0) = y_0$ for any ε. We call it a parametrically closed solution of (1.2), if $y(\varepsilon, 1) = y_0$ for any ε.

In a similar way we introduce "tangentially closed trajectories" in Section 2 (just differentiating with respect to ε).

Let $P(x) = \int_0^x p(t)dt$, and let the "moments" $m_k(x)$ be defined by $m_k(x) = \int_0^x P^k(t)q(t)dt$, $k = 0, 1, \ldots$. In particular, $m_0(x) = \int_0^x q(t)dt = Q(x)$.

It turns out that the distribution of parametrically closed trajectories of (1.2) is governed by the generating function $M(x, y) = y \cdot \sum_{k=0}^{\infty} m_k(x)y^k$ of the sequence $m_k(x)$. Namely, we show in Section 2 below that if $y(\varepsilon, x)$ with $y(\varepsilon, 0) = y(\varepsilon, 1) = y_0$ is a parametrically (or a tangentially) closed solution, then $P(1) = 0$ and

$M(1, y_0) = 0$. In the same way, if (1.2) has a parametric (or a tangential) center, then $P(1) = 0$ and $m_k(1) = 0$ for $k = 0, 1, \ldots$.

Let us introduce the following question, which we call "the Moments Center problem":

Find conditions on p *and* q*, under which 1 is a common zero of* $P(x)$ *and of all the moments* $m_k(x)$*.*

An equivalent statement of this problem is:

Find conditions on $P(x)$ *(with* $P(0) = P(1) = 0$*) and* $q(x)$ *for* q *to be orthogonal to each power* P^k *of* P *on* [0,1].

Below we shortly denote the center problem (or the center conditions) for the equations (1.1), (1.2) and for the moments $m_k(x)$, by C, PC and MC, respectively (and the tangential center problem, as defined in Section 2, we denote by TC).

Let us introduce (following [2], [8]) the following "composition conditions" (which will be denoted CC):

$P(x) = \int_0^x p(t)dt$ *and* $Q(x) = \int_0^x q(t)dt$ *can be represented as*

$$P(x) = \tilde{P}(W(x)), \; Q(x) = \tilde{Q}(W(x)),$$

with $\tilde{P}$, $\tilde{Q}$, W *polynomials, such that* $\tilde{P}(0) = \tilde{Q}(0) = 0, \; W(0) = W(1) = 0$.

The following "Composition conjecture" has been stated and discussed in various forms in [8–11].

Conjecture 1.1. *The conditions* C, TC, PC, MC *and* CC *are equivalent.*

Some implications are quite immediate. Clearly, PC implies TC and C. It is easy to show (see [2], [8]) that CC implies each of C, TC, PC and MC. In Section 2 below we show that TC implies MC.

As one of the main results of this paper we prove the equivalence of CC and MC (i.e., the composition conjecture for the moments center problem), and settle a closely related problem of bounding the "stabilization moment":

Given degrees d_1, d_2 *of* p *and* q*, how many equations* $m_0(1) = 0, \ldots, m_k(1) = 0$ *do we need to conclude that composition conditions are satisfied? (We denote the minimal such* k *by* $N(d_1, d_2)$*.)*

In particular, this implies equivalence of the tangential and parametric Center problems TC and PC and their equivalence to the Moment Center problem MC and thus provides explicit conditions for the tangential (= parametric) center.

From the point of view of the Moment theory, the moments center problem appears as the most classical and natural question of description of the orthogonal subspace to all the functions of the form $f(x) = \tilde{f}(P(x))$ on [0,1] for a given polynomial P.

However, we could not find any reference to this specific question. Also in M.S. Livšic's paper [25], which was the starting point for a long line of investigations of the generalized Moment problem, it is not addressed explicitly. On the other hand, one of the conditions for solvability of a generalized Moment problem in Livšic theorem is in terms of the divided difference $\frac{P(x)-P(y)}{x-y}$. As we shall see below, composition representability of P and Q as assumed in Composition Conditions

CC, can be also expressed in terms of such divided differences. In Section 5 below we state Livšic's theorem and reformulate our result in a form, convenient for comparison.

Let us return now to the original Center problem C. The Composition Conjecture has been verified here for small degrees of p and q in [6, 8]. We consider the verification of the Composition Conjecture for the Moment, Tangential and Parametric Center problems, given in this paper, as an important evidence, supporting it for the original problem C.

One can consider other simplified versions of the Center problem, where verification of Composition Conjecture is quite simple.

For example, let us say that (1.1) has a *continuous center*, if for any x_0 near zero there is x_1 near 1 (with $x_1 \to 1$ as $x_0 \to 0$) such that $y(x_0) \equiv y(x_1)$ for any solution of (1.1).

Using the characterization of a "Composition GCD" of P and Q, given in Section 3, one can easily show that Continuous Center and Composition Conditions are equivalent.

One can consider also a local version of the above property: (1.1) is said to have a local continuous center, if for any x_0 near zero there is x_1 near 0, $x_1 \neq x_0$, with $y(x_0) \equiv y(x_1)$. Here one can prove equivalence to a corresponding local Composition representability by elementary considerations, not referring to global composition factorization. This local version of the Center problem turns out to be closely related to the Center problem for the so-called Lienard equation (see [13–18]).

It is also easy to show that in these continuous versions of Center Conditions, one can assume existence of x_1 only for a finite number of x_0, if this number exceeds a certain constant, depending only on the degree of p and q.

2. Parametric center implies moments center

The parametric version of the center problem for (1.1) is defined above by the equation (1.2) with the parameter ε, multiplied by the second term. Apparently, more natural setting is to consider equation

$$y' = \varepsilon_1(p(x)y^2 + \varepsilon_2 q(x)y^3 \tag{2.1}$$

with two free parameters. However, the following simple remark shows that essentially, these settings are equivalent, unless when bifurcating from $\varepsilon_1 = \varepsilon_2 = 0$. Indeed, rescaling $y = \alpha y_1$, which clearly preserves center conditions, transforms (2.1) into

$$y_1' = \alpha\varepsilon_1 p(x)y_1^2 + \alpha^2\varepsilon_2 q(x)y^3. \tag{2.2}$$

Thus, for $\varepsilon_1 \neq 0$ we can put $\alpha = 1/\varepsilon_1$ and to obtain (1.2). Respectively, for $\varepsilon_2 \neq 0$, putting $\alpha^2 = 1/\varepsilon_2$, we get deformation of the first term only.

In fact, MC is implied by (and equivalent to) a much weaker condition, than PC.

We say that (1.2) *satisfies a tangential (or an infinitesimal) center condition at* $\varepsilon = 0$ *(TC below) if* (1.2) *has a center for* $\varepsilon = 0$ *and for any solution* $y(\varepsilon, x)$ *of* (1.2), $\frac{d}{d\varepsilon}\, y(\varepsilon, 1) \mid_{\varepsilon=0} = 0$.

(Respectively, we call a solution $y(0, x)$ *with* $y(0, 0) = y_0$ *tangentially (or infinitesimally) closed, if* $y(0, 1) = y_0$ *and* $\frac{d}{d\varepsilon}\, y(\varepsilon, 1) \mid_{\varepsilon=0} = 0$ *for this solution.)*

Theorem 2.1. $y(0, x)$ *with* $y(0, 0) = y(0, 1) = y_0$ *is tangential closed if and only if* $P(1) = 0$ *and* y_0 *is a zero of* $M(1, y)$.

(1.2) *satisfies a tangential center condition at* $\varepsilon = 0$ *if and only if* $P(1) = 0$ *and* $m_k(1) = 0$ *for any* k.

Proof. For $\varepsilon = 0$ (1.2) takes a form

$$y' = p(x)y^2, \; y(0) = y_0. \tag{2.3}$$

Thus $y(0, x) = \frac{y_0}{1 - y_0 P(x)}$. If $y(0, 1) = y_0$, then $P(1) = 0$.

Now dividing (1.2) by y^2 we get

$$-\left(\frac{1}{y}\right)' = p(x) + \varepsilon y q(x). \tag{2.4}$$

Integration of (2.4) from zero to one gives $\frac{1}{y_0} - \frac{1}{y(\varepsilon,1)} = P(1) + \varepsilon \int_0^1 q(t) \cdot y(\varepsilon, t)dt$. Taking into account that $P(1) = 0$ and computing up to the first order in ε, we obtain

$$-\frac{d}{d\varepsilon}\left(\frac{1}{y(\varepsilon, 1)}\right) \mid_{\varepsilon=0} = \int_0^1 q(t)y(0, t)dt. \tag{2.5}$$

Substituting $y_0(x) = \frac{y_0}{1 - y_0 P(x)}$, we conclude that

(2.6)

$$-\frac{d}{d\varepsilon}\left(\frac{1}{y(\varepsilon, 1)}\right) \mid_{\varepsilon=0} = y_0 \sum_{k=0}^{\infty} (\int_0^1 P^k(t)q(t)dt)y_0^k = y_0 \sum_{k=0}^{\infty} m_k(1)y_0^k = M(1, y_0).$$

$y \equiv 0$ is always a closed trajectory of (1.2). Now from (2.6) we obtain that for $y_0 \neq 0$ the trajectory $y(0, x)$ with $y(0, 0) = y_0$ is infinitesimally closed if and only if $M(1, y_0) = 0$. Finally, (1.2) has an infinitesimal center at $\varepsilon = 0$ if and only if all the solutions are infinitesimally closed, i.e., $M(1, y_0) \equiv 0$ or all $m_k(1) = 0$. Theorem 2.1 is proved.

Remark 1. The above result is completely similar to the results of an infinitesimal analysis of the perturbations of a Hamiltonian vector field on the plane (see, e.g., [5]). Here the infinitesimally closed trajectories correspond to zeroes of a certain Abelian integral (the perturbation form is integrated along the level curves of the Hamiltonian). In our case the function $M(1, y)$ play this role. There is a well-developed theory, describing distribution of zeroes of Abelian integrals. It would be important to have similar results for the functions of the form $M(1, y)$. Notice, that the calculation above gives an explicit formula for $M(x, y) : M(x, y) = y \int_0^x \frac{q(t)y}{1 - P(t)y} dt$.

Remark 2. The choice of the point $x = 1$ above has no special reasons. Any point $x = a$ can serve equally well. Below we take (following [8]) a more invariant point of view, considering all the "periods", i.e., the values of x, for which $y(0) \equiv y(x)$. Theorem 2.1 can be reformulated accordingly:

All the tangential periods of (1.2) at $\varepsilon = 0$ are exactly all the common zeroes of $m_0(x), m_1(x), \dots$.

Remark 3. Considering the equation (1.1) with ε multiplied by the first term, we obtain a symmetric moment sequence $\tilde{m}_k(x) = \int_0^x Q^k(t)p(t)dt$. Thus the corresponding tangential center conditions are equivalent to vanishing $Q(1)$ and all the moments $\tilde{m}_k(1)$. In particular, since the parametric center conditions imply both the infinitesimal ones, we get:

TC implies both MC and $\tilde{MC}$ (where $\tilde{MC}$ denotes the symmetric Moments center problem for the moments $\tilde{m}_k(x)$).

As we shall see below, PC, TC, MC and $\tilde{MC}$ are in fact equivalent.

3. Composition algebra

From an abstract algebraic viewpoint, the algebra of univariant polynomials under composition is an associative monoid; the linear polynomials are the units. One defines primes (= indecomposable polynomials), prime factorizations (= maximal decompositions) and associated primes and equivalent decompositions just as in any noncommutative monoids.

In 1922 J.F. Ritt [29] proved fundamental results about such polynomial decompositions. We state below some of them. We also need more recent results [24, 26, 27], which give effective criteria for a decomposability of a given polynomial and for an existence of a nontrivial common divisor, with respect to composition, of two given polynomials.

One can expect that an algebraic structure of polynomial decomposition is reflected in the integrability structure of Abel equations and plane vector fields, via the connection described above. However, only the very initial steps in this direction have been done. (See [2].)

For a polynomial $w = f(x)$ of degree n the monodromy group G_f of f^{-1} is the group of permutations of the elements $x_1, \dots, x_n$ of the fiber $f^{-1}(w_0)$, w_0 a regular value of f, induced by following all the closed loops in w-plane, starting at w_0 and avoiding critical values of f.

A group G of permutations of $\{x_1, \dots, x_n\}$ is called imprimitive, if for a certain subset $S = \{x_{i_1}, \dots, x_{i_\ell}\}$ of $\{x_1, \dots, x_n\}$, $\ell \neq 1, n$, $g(S) = S$ or $g(S) \cap S = \phi$.

Theorem 3.1. (Ritt, [29].) *A polynomial $f(x)$ is composite (i.e., allows representation $f(x) \equiv f_1(f_2(x))$, with $\deg f_1 > 1, \deg f_2 > 1$) if and only if the monodromy group G_f is imprimitive.*

In [29] Ritt describes all the maximal decompositions of a given polynomial, and, in particular, all the prime polynomials, which commute in composition algebra. These last turn out to be only powers of x and Chebyshev polynomials.

An important tool in study of a composition representability of polynomials turns out to be the following construction: to any polynomial $f(x)$ we associate the curve

$$f(x) - f(y) = 0 \qquad in \qquad \mathbb{C}^2$$

(or better, the same curve without the diagonal $\Delta = \{x - y = 0\}$, given by $\frac{f(x)-f(y)}{x-y} = 0$, which we denote by Γ_f).

The following result has been communicated to us by C. Christopher ([19]), with a reference to P. Mueller ([27]):

Theorem 3.2. *A polynomial $f(x)$, which is not a power of x or a Chebyshev polynomial, is composite if and only if the curve Γ_f is reducible.*

In fact, by Theorem 3.1, f is composite if and only if its monodromy group G_f is imprimitive, and one can show ([27]) that Γ_f is irreducible if and only if G_f is 2-transitive. Now it follows from the classification of the monodromy groups of indecomposible polynomials, given in [26], that such a group can be primitive but not 2-transitive, only for f either a cyclic polynomial x^p, or a Chebyshev polynomial T, which can be defined implicitly by $T(x + 1/x) = x^p + 1/x^p$.

Also the composition conditions CC, given above, can be checked in terms of the curves Γ_f.

Theorem 3.3. *Two polynomials $f_1(x)$ and $f_2(x)$ have a nontrivial common right divisor in composition algebra if and only if $\Gamma_{f_1} \cap \Gamma_{f_2}$ has one-dimensional components.*

This theorem can be proved in several ways. In [24] it is obtained in a much more general form via Grauert's theorem on graphs of entire functions. In algebraic setting it is essentially equivalent to Luroth theorem. The composition GCD from the right of f_1 and f_2 is the generator of the field $\mathbb{C}(f_1, f_2)$. (This proof has been communicated to us by C. Christopher [20].) Also a direct application of the Ritt approach allows one to prove Theorem 3.3. Since it is very important for our computations, we give below two other formulations of essentially the same result.

Proposition 3.4. (N. Roytvarf, [30].) *f_1 and f_2 have a nontrivial common right divisor if and only if in a neighborhood of some point w_0, f_2 glues together two different branches of f_1^{-1} (i.e. for two different branches $f_{1i}^{-1}(w)$ and $f_{1j}^{-1}(w)$, w near w_0, $f_2(f_{1i}^{-1}(w)) \equiv f_2(f_{1j}^{-1}(w))$).*

Notice that apparently this condition is not symmetric. However, reducing it to the existence of a one-dimensionalal component of $\Gamma_{f_1} \cap \Gamma_{f_2}$, one restores the symmetry.

Another version of the same result is especially convenient for the proof of Composition conjecture for the Moments center problem, given in Section 4.3.

Proposition 3.5. *f_1 and f_2 have a common right divisor W, with $W(0) = W(1) = 0$, if and only if there is a germ of an analytic transformation (possibly ramified) $U(x)$ at zero, $U(0) = 1$, such that $f_1(U(x)) \equiv f_1(x)$ and $f_2(U(x)) \equiv f_2(x)$.*

Proof. If such a U exists, $(x,\ U(x))$ locally parametrize a one-dimensionalal component of $\Gamma_{f_1} \cap \Gamma_{f_2}$. Since this component contains a point (0,1), one can easily check that the corresponding common right divisor W, provided by Theorem 3.3, satisfies $W(0) = W(1) = 0$.

Conversely, if $f_1(x) = \tilde{f}_1(W(x))$ and $f_2(x) = \tilde{f}_2(W(x))$, with $W(0) = W(1) = 0$, we can define $U(x)$ by an algebraic equation $W(x) = W(U(x))$, $U(x)$ near 1.

A direct constructive proof of the first part of this proposition (without reference to Theorem 3.3) can be given.

In conclusion of this section notice that the subset of decomposable polynomials in the space of all polynomials of a given degree (or a subset of pairs of polynomials with a nontrivial GCD from the right) can be described explicitly. A parametric representation of these subspaces is obtained by explicitly writing all the "combinatorial" possibilities of a decomposition. Ritt's results provide important information concerning the ramification structure of this parametrization.

4. Moments center problem

This section is devoted to our main result: equivalence of the Moments Center conditions and Composition conditions.

We formulate the result in a slightly more general form than in the introduction. Indeed, the choice of point 1 as a possible zero of moments sequence has no invariant meaning. It is more natural to consider all the common zeroes of the moments m_k.

Let two polynomials $P(x)$ and $Q(x)$ be given, $P(0) = Q(0) = 0$. As above, denote by $q(x)$ the derivative $Q'(x)$ (and by $p(x)$ the derivative $P'(x)$). Define the polynomial $m_k(x)$, $k = 0, 1, \dots$, by $m_k(x) = \int_0^x P^k(t)q(t)dt$.

Theorem 4.1. *Common zeroes of all the moments $m_k(x)$, $k = 0, 1, \dots$, are exactly the zeroes of the GCD from the right W of P and Q in the composition algebra of polynomials without free term. (In particular, they are contained among common zeroes of P and Q.)*

As an immediate corollary we obtain:

Theorem 4.2. *$a \in \mathbb{C}$ is a common zero of $m_k(x)$ if and only if P and Q satisfy Composition conditions CC : $P(x) = \tilde{P}(W(x))$, $Q(x) = \tilde{Q}(W(x))$, $W(0) = W(a) = 0$.*

Theorem 4.3. *The conditions TC and PC for the Abel equation (1.1) are equivalent, and they both are equivalent to the Moments Center condition MC.*

Proof. It was shown above that $PC \Longrightarrow TC \Longrightarrow MC$. Now by Theorem 4.2, MC is equivalent to Composition conditions CC, which, in turn, imply PC.

Theorem 4.4. *There exists an explicitly computable function $N(\deg P, \deg Q)$, such that if the moments $m_k(x)$ vanish at a certain point a for $k = 0, 1, \dots, N$, then $m_k(a) = 0$ for any k.*

Theorems 4.1–4.4 solve completely the Moments, Parametric and Tangential Center problems. Extended sketch of three different approaches to these results is given in subsections 4.1, 4.2 and 4.3 below. Our complete proofs are based on the approach of 4.3; they will appear separately.

4.1. Canonical representation of the moment sequence

In this subsection we present an algebraic approach to the Moments Center problem. It is based on a certain recursive representation of the moments $m_k(x)$ (under assumption that these moments vanish at given points), which allows to replace the sequence $m_k(x)$ by another sequence $S_\nu(x)$, with lower degrees. Ultimately (assuming that the given zeroes of m_k survive long enough) the algebraic structure created can be "read backward", producing the required representation of P and Q as functions of a certain polynomial R, comprising the survived zeroes of m_k.

The presentation below contains rather complicated computations. We believe that besides their importance for understanding the composition algebra hidden in the generalized moments, they unfold subtle additional algebraic structures. In particular, the Canonical representation below turns out to be closely related to certain constructions in D-modules (recall that the problem of analytic continuation of the generalized moments m_k, as a function in k, was a starting point in the development of D-modules; we plan to present these important relations separately).

The reader interested in a simplified version of the Canonical representation can skip Theorems 4.1.1 to 4.1.3 and pass directly to Theorems 4.1.1′ to 4.1.3′.

As usual, $P(x) = \int_0^x p(t)dt$, $Q(x) = \int_0^x q(t)dt$. We introduce the following notations:

a. $Y(P) \subset \mathbb{C}$ is a set of distinct zeroes of a polynomial $P(x)$.

b. Let $\Sigma = \{a_1 = 0, \ a_2, \ldots, a_m\}$ be a subset of $Y(P)$.

We denote by $n_0, n_1, \ldots, n_k$ the different multiplicities of points in Σ as zeroes of P, and by Σ_s, $0 \leq s \leq k$, the disjoint subsets of Σ, consisting of all the zeroes with the same multiplicity n_s.

Let $U_s = \prod_{a_j \in \Sigma_s} (x - a_j)$ and $W = \prod_{i=0}^k U_i$. By definition all the zeroes of U_i and W are simple and $Y(W) = \Sigma$. Below we denote by W_s the product of all the U_i except U_s, and by $W_{s,k}$ the product of all the U_i except U_s and U_k.

c. We now write $P(x)$ in the form

$$P(x) = P_0(x) \cdot P_1(x), \tag{4.1.1}$$

where $P_0(x) = \prod_{i=0}^k U_i^{n_i}(x)$ vanishes at each point of Σ, and $P_1(a_i) \neq 0$ for each $a_i \in \Sigma$.

d. For given $\Sigma \subset Y(P)$ we denote by $\kappa(Q, P, \Sigma, \ell)$ a maximal length of a chain of consecutive moments $m_\ell, m_{\ell+1}, \ldots$ vanishing on Σ:

$$\kappa(Q, P, \Sigma, \ell) = \begin{cases} 0, \text{ if } m_\ell(a_j) \neq 0 \text{ for some } a_j \in \Sigma \\ \sup\{\nu \geq 1 \mid m_{\ell+i-1}(a_j) = 0 \text{ for } a_j \in \Sigma \text{ and } 1 \leq i \leq \nu\}. \end{cases}$$

The following theorem presents what we call a "canonical representation" of the moments $m_k(x)$. It involves, in fact, some "combinatorial" choices, which are important for further applications.

Theorem 4.1.1. *For given $\Sigma \subset Y(P)$ and $\ell \geq 0$ we denote $\kappa(Q, P, \Sigma, \ell)$ by $d+1$. Choose one of the polynomials U_s, defined above. Then the moments $m_{\ell+i}(x)$, $0 \leq i \leq d$, can be represented in the form*

$$m_{\ell+i} = P_0^{\ell+i} \sum_{\nu=1}^{i+1} \alpha_\nu^i U_s^\nu P_1^{i-\nu+1} S_\nu, \tag{4.1.2}$$

where α_ν^i are positive constants, depending on ℓ and n_s only, and $\{S_\nu\}, 1 \leq \nu \leq d+1$, is a sequence of new polynomials, satisfying recurrence relations

$$m_\ell = P_0^\ell \cdot U_s \cdot S_1,$$

$$n_s P_1 W_s S_\nu' - \nu P_1 \sum_{i=0, i\neq s}^{k} n_i U_i' W_{s,i} S_\nu - (n_0(\ell+\nu-1)+\nu) P_1' W_s S_\nu = \tag{4.1.3}$$

$$= [(n_0\ell + \nu + 1) U_s' W_s + \nu \sum_{i=0}^{k} n_i U_i' W_i] S_{\nu+1} + W_s S_{\nu+1}'.$$

Remark 1. The representation (4.1.2) holds for every choice of U_s, but constants α_ν^i and polynomials $S_\nu(x)$ will be different for different choices.

Remark 2. It can be easily seen that $\deg S_{\nu+1} = \deg S_\nu + \deg P_1 - \deg U_s$. Therefore, polynomials S_ν have decreasing degrees for $\deg P_1 < \deg U_s$. In this case, as the following Theorems 4.1.2, 4.1.3 show, the composition conjecture is completely settled and the explicit bound on the "stabilization moment" is received.

Theorem 4.1.2. *Assume that $\Sigma \neq P(Y)$, i.e. the polynomial $P_1(x)$ is at least of degree 1. Assume also that for some U_s we have $n_s = \deg U_s > \deg P_1$. Let $m_s = \deg U_s - \deg P_1$. Then*

$$\kappa(Q, P, \Sigma, \ell) \leq \left[\frac{\deg Q - n_s}{m_s}\right] + 1.$$

Theorem 4.1.3. *Assume that $\Sigma = Y(P)$, i.e. $\deg P_1 = 0$. Then either*

$$\kappa(Q, P, \Sigma, \ell) \leq \min_{0\leq s\leq k} \left[\frac{\deg Q}{\deg U_s}\right],$$

or $\kappa(Q, P, \Sigma, \ell) = \infty$. In the latter case the composition representation

$$P = R^n, \quad Q = \tilde{Q}(R) \tag{4.1.4}$$

holds, where $n \in \mathbb{N}$, $\tilde{Q}(0) = 0$ and $R(S)$ is a polynomial with $Y(R) = \Sigma$.

The proof of Theorem 4.1.1 includes rather long and complicated computations based on consecutive integrations by parts, organized in a special way. We prefer to give here the proofs of Theorems 4.1.1–4.1.3 in the simplest case, when all the zeroes of a polynomial $P(x)$ have the same multiplicity n_0 and Σ includes all of them, i.e, $\Sigma = \Sigma_0 = Y(P)$, $W = U_0$, $P_1 = const$ and $P = W^{n_0}$. In this special case the bound of Theorem 4.1.3 can be improved by 1. We will also use below the notation $\kappa(q, p, \ell)$ for $\kappa(Q, P, \Sigma, \ell)$.

Theorem 4.1.1′. *Let $P = W^{n_0}$ and $\kappa(q, P, \ell) \geq d+1$. Then $m_{\ell+i}(x)$ for $i = 0, \dots, d$ may be represented in the form*

$$m_{\ell+i} = \sum_{\nu=1}^{i+1} \alpha_\nu^i W^{n_0(\ell+i)+\nu} S_\nu \tag{4.1.5}$$

where α_ν^i are positive constants not depending on q and P, and $\{S_\nu, 1 \leq \nu \leq d+1\}$ is a new sequence of polynomials, with $\deg S_\nu = \deg Q - \nu k$, satisfying

$$n_0 S'_\nu = (n_0(\ell+\nu) + \nu + 1) W' S_{\nu+1} + W S'_{\nu+1} \tag{4.1.6}$$

(S'_ν is a derivative of S_ν) and

$$\int_0^{a_i} W^{n_0(\ell+\nu)+\nu} S'_\nu dt = 0, \ \nu = 1, \dots, d \tag{4.1.7}$$

(α_ν^i and S_ν depend on ℓ and n_0, but we omit it from the notations).

Proof. We will prove the formula (4.1.5) by induction. For $d = 0$ we have $m_\ell(a_j) = \int_0^{a_j} P^\ell(t) q(t) dt = \int_0^{a_j} W^{n_0 \ell}(t) q(t) dt = 0$ for $j = 1, \dots, k$, i.e., $m_\ell(x)$ has each zero a_j with the multiplicity $n_0 \ell + 1$ and so it can be written as $m_\ell = W^{n_0 \ell + 1} \cdot S_1$, where S_1 is a polynomial with $\deg S_1 = \deg Q - \deg W$. Hence the formula (4.1.5) is true with $\alpha_1^1 = 1$.

Assume that (4.1.5) is true for $m_\ell, \dots, m_{\ell+i}$. Then by integrating by parts we have the following:

$$\begin{aligned}
m_{\ell+i+1}(x) &= \int_0^x P^{\ell+i+1}(t) q(t) dt \\
&= \int_0^x W^{n_0(\ell+i+1)}(t) q(t) dt \\
&= m_{\ell+i}(x) \cdot W^{n_0} - \int_0^x m_{\ell+i}(t) \cdot (W^{n_0}(t))' dt \\
&= \left(\sum_{\nu=1}^{i+1} \alpha_\nu^i W^{n_0(\ell+i)+\nu} S_\nu \right) W^{n_0} \\
&\quad - \int_0^x \left(\sum_{\nu=1}^{i+1} \alpha_\nu^i W^{n_0(\ell+i)+\nu} S_\nu \right) n_0 W^{n_0 - 1} W' dt \\
&= \sum_{\nu=1}^{i+1} \alpha_\nu^i W^{n_0(\ell+i+1)+\nu} S_\nu - \sum_{\nu=1}^{i+1} \alpha_\nu^i \frac{W^{n_0(\ell+i+1)+\nu}}{n_0(\ell+i+1)+\nu} n_0 S_\nu \\
&\quad + \int_0^x \sum_{\nu=1}^{i+1} \frac{\alpha_\nu^i}{n_0(\ell+i+1)+\nu} W^{n_0(\ell+i+1)+\nu} \cdot n_0 S'_\nu dt.
\end{aligned}$$

Substitute (4.1.6) in the last sum for $\nu = 1, \dots, i$.

Then

$$\begin{aligned} m_{\ell+i+1}(x) = &\sum_{\nu=1}^{i+1} \alpha_\nu^i \left(1 - \frac{n_0}{n_0(\ell+i+1)+\nu}\right) W^{n_0(\ell+i+1)+\nu} S_\nu \\ &+ \sum_{\nu=1}^{i} \frac{\alpha_\nu^i}{n_0(\ell+i+1)+\nu} \int_0^x W^{n_0(\ell+i+1)+\nu} [(n_0(\ell+\nu)+\nu+1) W' S_{\nu+1} \\ &+ W S'_{\nu+1}] dt + \frac{\alpha_{i+1}^i}{n_0(\ell+i+1)+i+1} \int_0^a W^{n_0(\ell+i+1)+i+1} \cdot n_0 S'_{i+1} \\ = &\sum_{\nu=1}^{i+1} \alpha_\nu^i \left(1 - \frac{n_0}{n_0(\ell+i+1)+\nu}\right) W^{n_0(\ell+i+1)+\nu} S_\nu \\ &+ \sum_{\nu=1}^{i} \alpha_\nu^i \frac{n_0(\ell+\nu)+\nu+1}{n_0(\ell+i+1)+\nu} \int_0^x W^{n_0(\ell+i+1)+\nu} \cdot W' S_{\nu+1} \\ &+ \sum_{\nu=1}^{i} \frac{\alpha_\nu^i}{n_0(\ell+i+1)+\nu} \int_0^x W^{n_0(\ell+i+1)+\nu+1} \cdot S'_{\nu+1} dt \\ &+ \frac{\alpha_{i+1}^i}{n_0(\ell+i+1)+i+1} \int_0^x W^{n_0(\ell+i+1)+i+1} S'_{i+1} dt. \end{aligned}$$

After integrating by parts once more in the second term, we have:

$$\begin{aligned} m_{\ell+i+1}(x) = &\sum_{\nu=1}^{i+1} \tilde{\alpha}_\nu^i W^{n_0(\ell+i+1)+\nu} S_\nu + \sum_{\nu=1}^{i-1} \beta_\nu^i \int_0^x W^{n_0(\ell+i+1)+\nu+1} \cdot S'_{\nu+1} + \\ &+ \beta_{i+1}^i \int_0^x W^{n_0(\ell+i+1)+i+1} \cdot n_0 S'_{i+1}, \end{aligned}$$

where $\tilde{\alpha}_\nu^i$, β_ν^i are linear combinations of α_ν^i with positive coeffizients, therefore they are also positive constants.

The second sum contains S'_j for $2 \le j \le i$ only (and doesn't contain S'_1). Substituting once more for S'_ν the expression (4.1.6) and integrating by parts, we will have the similar formula, in which the second sum contains $S'_3, \ldots, S'_i$ only. Repeating the same procedure till the second sum disappears completely, we finally have:

$$m_{\ell+i+1}(x) = \sum_{\nu=1}^{i+1} \alpha_\nu^{i+1} W^{n_0(\ell+i+1)+\nu} \cdot S_\nu + \alpha_{i+2}^{i+1} \int_0^x W^{n_0(\ell+i+1)+i+1} \cdot n_0 S'_{i+1} dt,$$

where α_ν^{i+1} are positive.

Then from $m_{\ell+i+1}(a_j) = 0$ follows that

$$\int_0^{a_j} W^{n_0(\ell+i+1)+i+1} \cdot n_0 S'_{i+1} dt = 0, \quad 1 \le j \le k. \tag{4.1.8}$$

Therefore the integral $\int_0^x W^{n_0(\ell+i+1)+i+1} \cdot S'_{i+1} dt$ has each zero a_j with the multiplicity $n_0(\ell+i+1)+i+2$ and

$$\int_0^x W^{n_0(\ell+i+1)+i+1} S'_{i+1} dt = W^{n_0(\ell+i+1)+i+2} \cdot S_{i+2}(x), \tag{4.1.9}$$

where $S_{i+2}(x)$ is a polynomial with $\deg S_{i+2} = \deg S_{i+1} - k$.

By differentiating (4.1.9) we have

$$n_0 S'_{i+1} = [n_0(\ell+i+1)+i+2] W' S_{i+2} + W S'_{i+2},$$

which is the formula (4.1.6) for $i+1$.

We introduced the next polynomial $S_{i+2}(x)$ according to the formulas (4.1.6)–(4.1.7). Finally we have

$$m_{\ell+i+1} = \sum_{\nu=1}^{i+2} \alpha_\nu^{i+1} W^{n_0(\ell+i+1)+\nu} S_\nu$$

Theorem 4.1.1′ is proved.

As an immediate consequence of Theorem 4.1.1′ we have the following two theorems, describing the behavior of the set of zeroes of m_k in the special case considered.

Theorem 4.1.2′. *Let $W, P = W^{n_0}$ be as above and $Q(x)$ be a polynomial of degree $n = dk + \alpha$, where $0 < \alpha \le k-1$.*

Then $\kappa(Q,P,\ell) \le d$ for each ℓ.

Proof. If $\kappa(Q,P,\ell) \ge d+1$ for some ℓ, i.e., the zeroes of W survive for $(d+1)$ consecutive moments $m_\ell, \dots, m_{\ell+d}$, then by Theorem 4.1.1′ we can construct polynomials $S_1, \dots, S_{d+1}$ satisfying (4.1.5)–(4.1.6). For S_d we have $\deg S_d = \deg Q - dk = \alpha > 0$, therefore $S'_\alpha \not\equiv 0$ and $\int_0^{a_i} W^{n_0(\ell+d)+d} S'_d dt = 0$, $i = 1, \dots, k$ yields $\deg \left(\int_0^x W^{n_0(\ell+d)+d} S'_d dt\right) \ge [n_0(\ell+d)+d+1]k$ (because the multiplicity of each a_i increases by 1).

On the other hand, $\deg S_d = \alpha \le k-1$ and after integration we have

$$\deg \left(\int_0^x W^{n_0(\ell+d)+d} S'_d dt \right) = k[n_0(\ell+d)+d] + \alpha.$$

Then $\alpha \ge k$ in contradiction to the choice of α.

Theorem 4.1.3′. *Let W, P be as above and $Q(x)$ be a polynomial of degree $n = d \cdot k$. Then for each ℓ either $\kappa(Q,P,\ell) \le d-1$ or $\kappa(Q,P,\ell) = \infty$. In the second case $Q(x)$ can be represented as $Q = T(W)$ for a certain polynomial T of degree d.*

Proof. Assume that $\kappa(Q,P,\ell) \ge d$ for some ℓ, i.e., the zeroes of W survive for d consecutive moments $m_\ell, \dots, m_{\ell+d-1}$.

Let $S_1, \dots, S_d$ be polynomials constructed according to Theorem 4.1.1′. In the last step we have $\deg S_d = 0$. But then $S_d \equiv s_0 = \text{const}$ and by (4.1.6)

$$n_0 S'_{d-1} = [n_0(\ell+d-1)+d] W' s_0,$$

which yields $S_{d-1} = \tilde{S}_{d-1}(W)$, $\deg \tilde{S}_{d-1} = 1$.

Now we can use the formula (4.1.6) to construct backwards all the polynomials $S_{d-2}, \dots, S_1$ in the form $S_\nu(x) = \tilde{S}'_\nu(W)$. Then $m_\ell(x) = \int_0^x W^{n_0 \ell} q dt = W^{n_0 \ell + 1} \cdot \tilde{S}_1(W)$ and by differentiating we finally have $Q(x) = T(W)$.

Theorems 4.1.2′ and 4.1.3′ completely solve the model problem in a special case, where $P = W^{n_0}$, as defined above and all the zeroes of W survive: this happens if and only if $Q = T(W)$, and the number of equations $m_j = 0$ required to get this conclusion, is at most $[\deg Q/k] + 1$. (M. Blinov has shown that this bound is sharp.)

4.2. Moment techniques

In this section we first of all obtain, using the classical moments techniques, an effective necessary and sufficient condition for the vanishing of the moment sequence

$$m_j(a) = \int_0^a P^j(t) q(t) dt,$$

and an explicit bound for the stabilization moment $N(d_1, d_2), \deg p = d_1, \deg q = d_2$.

However this condition, being given in terms of a certain semialgebraic function, does not provide any immediate information for the rest of the center problems C, TC and PC (in contrast with the composition conditions).

Consequently, the second part of this section is devoted to the analysis, from the point of view of composition, of the above vanishing condition and to the proof of a composition conjecture (in some special cases).

Assume that f and g are two real piecewise differentiable and piecewise monotone functions. (In this section, in contrast with Sections 2 and 3, we assume that all the functions and polynomials considered are real.)

Let the moments $m_j(x)$ be defined by

$$m_j(x) = \int_0^x F^j(t) g(t) dt, \tag{4.2.1}$$

where $F(t) = \int_0^t f(\tau) d\tau$.

For a given $a \in \mathbb{R}$ define $\psi(f, g, a, s)$ (or, shortly, $\psi(s)$) for s between $s_0 = \min_{[0,a]} F$ and $s_1 = \max_{[0,a]} F$, as follows:

Let $x_1(s), x_2(s), \dots, x_m(s)$ be the solutions of the equation $F(x) = s$ on the interval $[0, a]$. Then $\psi(s) = \sum_{i=1}^m sgn(f(x_i(s))(g(x_i(s))/f(x_i(s)))$. Thus $\psi(s)$ is the sum over all the points $F^{-1}(s)$ between 0 and a, of $\pm g/f$, where the sign at each point is chosen according to the increasing or decreasing of F at this point.

Theorem 4.2.1. *The moments $m_j(a)$ vanish for all $j = 0, 1, \dots,$ if and only if $\psi(s) \equiv 0$. If $\psi(s)$ has a finite number N of sign changes, vanishing of the first N moments $m_0(a), \dots, m_N(a)$ implies vanishing of all the moments.*

Proof. Take $s = F(x)$ as a new integration variable. We get

$$dx = ds/F'(x) = ds/f(x), \tag{4.2.2}$$

$$m_j(a) = \int_0^a F^j(t)g(t)dt = \int_{s_0}^{s_1} s^j \psi(s)ds.$$

By one of the main results of the moment theory, vanishing of all the moments $\int_{s_0}^{s_1} s^j \psi(s)ds$ implies $\psi(s) \equiv 0$. (See [1], [31].)

In the case where $\psi(s)$ has a finite number N of sign changes, we can sharpen this conclusion as follows: assume that $m_0(a), \dots, m_N(a)$ vanish.

As a linear combination of $1, s, \dots, s^N$ we can construct a polynomial $\omega(s)$, which changes sign exactly at the same points as $\psi(s)$. Hence $\int_{s_0}^{s_1} \omega(s)\psi(s)$ (which is zero as a linear combination of $m_0(a), \dots, m_N(a)$) would be positive, unless $\psi(s) \equiv 0$. Theorem 4.2.1 is proved.

Now let us return to our initial situation, where $f = p$ and $g = q$ are polynomials (with real coeffizients). Here the function $\psi(s)$ is semialgebraic, and it is well known that the property $\psi \equiv 0$ can be verified effectively. We get the following result:

Theorem 4.2.2. *For any given degrees d_1, d_2 of p and q one can write effectively semialgebraic conditions on the coeffizients of p and q and on a, which are necessary and sufficient for all the moments $m_j(a)$ to vanish.*

The stabilization moment $N(d_1, d_2)$ of the sequence $m_j(a)$ can also be computed explicitly.

Proof. It follows from a general principle in real algebraic geometry — Tarski-Zeidenberg theorem, which allows, in principle, to write explicitly the required conditions. (See e.g. [12].) We do not give here this construction, since on one side it produces a complicated answer, and, on the other side, the resulting expressions have no direct consequences for our initial problems C and PC.

Remark 1. Theorem 4.2.2 shows that in contrast with the original center problem, where the mere existence of an effective answer is unknown, the Moments Center problem is "algorithmic".

Remark 2. Theorem 4.2.1 can be interpreted as an orthogonality condition: $g(x)$ *is orthogonal to all the functions $F^j(x)$ on $[0, a]$ if and only if $\psi(f, g, a, s) \equiv 0$ on $[s_0, s_1]$.*

Remark 3. For f, g-polynomials, composition conjecture implies

$$\psi(f, g, a, s) \equiv 0 \Longleftrightarrow \psi(g, f, a, s) \equiv 0.$$

It would be interesting to investigate this symmetry property by itself.

In the rest of this section we produce some consequences of the condition $\psi(s) \equiv 0$, obtaining the required composition representability of P and Q under some restrictions and indicating the steps of the proof in a general case. Although part of these results is valid for general f and g as above, we shall always restrict ourselves to the case of real polynomials p and q.

Proposition 4.2.1. *If $\psi(p, q, a, s) \equiv 0$, then $P(a) = 0$.*

Proof. Assume that $P(a) \neq 0$. By definition, $P(0) = 0$. Hence, as s passes the value $\bar{s} = P(a)$, one nonzero summand in $\psi(s)$ disappears. Assuming that $\bar{s}$ is not a critical value of P, we obtain a contradiction: by analytic continuation, a regular function $\psi(s)$ near $\bar{s}$ cannot be identically zero on both sides of $\bar{s}$.

Thus, vanishing of $P(a)$ is implied by vanishing of all the moments $m_j(a)$.

Proposition 4.2.2. *Let $x_0 \in (0, a)$ be a nondegenerated critical point of P with a simple critical value $\hat{s}$ (i.e., $\hat{s} = P(x_0) \neq P(x_j)$ for any critical point $x_j \neq x_0$ of P on $[0, a]$).*

If $\psi(p, q, a, s) \equiv 0$, then $q(x_0) = 0$ and locally $Q(x)$ is an analytic function of $P(x) = s$.

Proof. As s passes $\hat{s}$, exactly two summands of ψ disappear. By the same argument as above, the sum of these two summands alone must be zero. We get

$$q(x_1(s))/p(x_1(s)) \equiv q(x_2(s))/p(x_2(s)) \tag{4.2.3}$$

for s near $\hat{s}$.

In particular, since $p(x_1)$ and $p(x_2)$ have different signs, the same must be true for q. Consequently, $q(x_0) = 0$, and q/p is a regular function near x_0.

Locally near $(x_0, \hat{s})$, the covering $s = P(x)$ has exactly two lists, branching at $\bar{s}$. (4.3) means that q/p takes the same values on the real part of these two lists. By analytic continuation we conclude that q/p is a univalued regular function of s near $\bar{s}$.

Now $q/p = \frac{dQ}{dx} \,/\, \frac{dP}{dx} = \frac{dQ}{ds}$, and hence Q is locally a regular function of s.

Theorem 4.2.3. *Let for real polynomials p and q the moments*

$$m_j(a) = \int_0^a P^j(t)q(t)dt$$

vanish for all j. Assume that all the critical points of P between 0 and a are nondegenerate and critical values at these points distinct.

Then $m_j(b)$ vanish for each zero b of $P(x)$ between 0 and a.

Proof. By Proposition 4.2.1, $P(a) = 0$. Applying Proposition 4.2.2 to each of the critical points of P in $[0, a]$, we conclude, by analytic continuation, that for any s between s_0 and s_1, the values of q/p are the same at each of the points $x_i(s)$.

Let b be a zero of P between 0 and a. Define $\psi(p, q, a, b, s)$ as a signed sum of q/p at those of $x_i(s)$, which lie between a and b. By the above, all these values cancel, and we get $\psi(p, q, a, b, s) \equiv 0$. By virtue of Theorem 4.2.1, $m_j(b) = 0$ for any j.

Theorem 4.2.4. *Assume that all the zeroes of P are real and distinct and all the critical values distinct. Let a be the biggest in absolute value zero of P.*

If the moments $m_j(a)$ vanish for $j = 0, 1, \ldots, N(d_1, d_2)$, then $Q = \tilde{Q}(P)$ for some polynomial $\tilde{Q}$, $\tilde{Q}(0) = 0$.

Proof. By Theorem 4.2.1, vanishing of $m_j(a)$, $j = 0, \ldots, N(d_1, d_2)$, implies that all the moments vanish, and $\psi(p, q, a, s) \equiv 0$. By Theorem 4.2.2 $m_j(b)$ vanish for any zero of P, and the values q/p are the same for any x_1, x_2 with $P(x_1) = P(x_2)$.

Since by conditions all the branches of $P^{-1}(s)$ have real parts, q/p takes the same values at all the complex preimages $P^{-1}(s)$ for any s. Hence, q/p is a univalued function of s. Finally, by Proposition 4.3.2, q/p is a regular function of s at all the critical values of P (which are the only points, where q/p could have singularities). Hence, q/p, as a univalued regular algebraic function of s, is a polynomial. Finally, $q/p = dQ/ds$, and hence $Q(s)$ is a polynomial of s.

Theorem 4.2.5. *Let all the zeroes $a_1 = 0, a_2, \dots, a_k$ of P be real and distinct. If $m_j(a_i) = 0$ for $j = 0, \dots, N(d_1, d_2)$, $i = 1, \dots, k$, then Q is a polynomial of P.*

Proof. Our assumption implies that all the critical points of P are nondegenerated. We do not assume, that the critical values are distinct. But since $m_j(a_i) = 0$ for any i, we can apply the above results to each of the consecutive zeroes of P.

By Proposition 4.2.2 we conclude that q/p, as a multivalued function of s, has no critical points. The same argument as above shows, that then it is a polynomial of s, as well as the function Q itself.

Theorems 4.2.3–4.2.5 illustrate the way in which the condition $\psi(s) \equiv 0$ implies the composition conditions for P and Q, under some special (also generic) assumptions on P.

Let us now outline the way in which a general case is settled. First of all we prove (still under some genericity assumptions) the property, which is valid essentially also in general.

Proposition 4.2.3. *Let 0 be a regular value of P. If $\psi(p, q, a, s) \equiv 0$ in s, then there is a germ of a regular analytic transformation $U(x)$, with $U(0) = a$, and $U(x)$ real for each x, such that $P(x) = P(U(x))$ and $(q/p)(x) = (q/p)(U(x))$.*

Proof. As s passes 0, either exactly two summands of ψ disappear, or one is replaced by another (according to the sign of $p(a) = P'(a)$). These summands correspond to the solutions $x_1(s)$ and $x_2(s)$ of $P(x) = s$ near 0 and a. By the same argument as above, in the first case the sum of these two summands alone must be zero, and in the second case they are equal. We get $(q/p)(x_1(s)) \equiv (q/p)(x_2(s))$ for s near 0.

Defining U by $U(x_1(s)) = x_2(s)$ we get the required equalities (U is regular since 0 is a regular value of P).

Now the general result is the following:

Theorem 4.2.6. *For any p and q, if $\psi(p, q, a, s) \equiv 0$, then there exists a germ of an analytic transformation $U(x)$, $U(0) = a$ (possibly, non-real and ramified) such that $P(x) \equiv P(U(x))$, $Q(x) \equiv Q(U(x))$.*

The proof comprises the following main ingredients:

i. The equality $(q/p)(x_1(s)) \equiv (q/p)(x_2(s))$ can be extended, by analytic continuation, to any pair $x_{\sigma(1)}(s)$, $x_{\sigma(2)}(s)$, where σ is a permutation of $x_i(s)$ from the monodromy group G_P of P^{-1}.
ii. For each critical value of P we can conclude (as in Proposition 4.2.2) that the sum of the contributions to ψ of all the critical points with this critical value, is zero. Once more, this is true for any permutation of the corresponding indices by the monodromy group.

iii. If 0 is not a regular value of P, some other summands, except those corresponding to $x_1(s)$ and $x_2(s)$, can disappear, as s passes 0. However, a detailed analysis of these additional summand (for s complex) allows still to conclude that the contributions of $x_1(s)$ and $x_2(s)$ cancel.

A combination of i, ii and iii, and the analysis of the group G_p in the spirit of Ritt's theorem, allows us finally to "integrate" the equality $(q/p)(x_1(s)) \equiv (q/p)(x_2(s))$ (or $\frac{dQ}{ds}(x_1(s)) \equiv \frac{dQ}{ds}(x_2(s))$), and to conclude that

$$Q(x_1(s)) \equiv Q(x_2(s)).$$

Now the main results of Section 4 follow immediately from Theorem 4.2.6.

Indeed, if $m_k(a) = 0$ for $k = 0, 1, \ldots, N(\deg p, \deg q)$, then, by Theorem 4.2.2, $\psi(p, q, a, s) \equiv 0$. By Theorem 4.2.6 and by Proposition 3.5, P and Q have a common right divisor W, such that $W(0) = W(a) = 0$. This proves Theorems 4.2 and 4.4.

To prove Theorem 4.1 we apply Theorem 4.2 to one of the common zeroes a of the moments $m_k(x)$. We get $P(x) = \tilde{P}(W(x))$, $Q(x) = \tilde{Q}(W(x))$, with $\tilde{P}(0) = \tilde{Q}(0) = 0$, $W(0) = W(a) = 0$. We get $m_k(x) = \int_0^x P^k(t)q(t)dt = \int_0^x \tilde{P}^k(W(t)) \cdot \tilde{Q}'(W(t))W'(t)dt = \int_0^{W(x)} \tilde{P}^k(w)\tilde{Q}'(w)dw = \tilde{m}_k(w)$. Now we perform a change of variables $W(x) = w$, and get a new sequence of moments $\tilde{m}_k(w)$. If it has nontrivial common zeroes, we repeat the procedure. The composition of all the polynomials W produced, gives us a right common divisor $\hat{W}$ of P and Q, which vanishes at all the common zeroes of $m_k(x)$. On the other hand, all the zeroes of $\hat{W}$ survive as common zeroes of $m_k(x)$. Hence the zeroes of $\hat{W}$ are exactly all the common zeroes of $m_k(x)$.

Now let $\tilde{W}$ be the *GCD* from the right of P and Q in the composition algebra of polynomials without free term. Since $\hat{W}$ divides $\tilde{W}$ from the right, $\tilde{W}$ vanishes at all the zeroes of $\hat{W}$, and since zeroes of $\tilde{W}$ survive in m_k, in fact zeroes of $\tilde{W}$ and $\hat{W}$ coincide. (One can easily show that $\tilde{W}$ is a power of $\hat{W}$.) Theorem 4.1 is proved.

4.3. Puiseux series technique

A different proof of Theorem 4.1 has been obtained by N. Roytvarf [30]. It comprises the following main steps:

1. First of all we perform a change of variables (ramified) $P(x) = s$. In what follows, we consider algebraic functions (in particular, of the form $R(P^{-1}(s))$ with R polynomials or rational functions) which are univalued on the Riemann surface Z of $P^{-1}(s)$.

In particular, the moments $m_k(x)$ take the form $\tilde{m}_k(s) = \int s^k (q/p)(P^{-1}(s))ds$. Integrating by parts one can show that the functions $I_k(s) = \underbrace{\int \cdots \int}_{k\ times} Q(P^{-1}(s))ds$ have the same common zeroes as $\tilde{m}_k$. (As polynomials in x, $\hat{I}_k(x)$ are given by the recurrence relation $\hat{I}_k' = p\hat{I}_{k-1}$, $I_0 = Q$; see [8], Section 6 and Theorem 1.3.)

2. Let us illustrate the general idea of N. Roytvarf's proof in a special case, where all the zeroes $0 = a_1, \dots, a_d$ of P are simple and they all survive as zeroes of $m_k(x)$. In this case $\hat{I}_k(x)$ has at a_i zeroes of a multiplicity $k+1$ and the same is true for I_k at the corresponding points on the Riemann surface Z of $P^{-1}(s)$.

Now let us consider an algebraic function $y(s)$, which is the difference of two different branches of $I_k(s)$. y is first defined in a neighborhood of a regular value w_0 of P and then extended to its Riemann surface V by analytic continuation. Notice that y is not any more a composition of a polynomial in x with P^{-1}. Respectively, V is generally a "bigger" Riemann surface than Z: it consists of all the pairs of points in the fibers of $P^{-1}(s)$, which can be obtained from the initial pair by continuation. Thus V is one of the irreducible components of the curve Γ_P, introduced in Section 3 (compare Theorem 3.2).

Assume that the monodromy group of P acts transitively on the pairs of points in the fibers of $P^{-1}(s)$ (is 2-transitive in notations of Theorem 3.2).

One can easily see that y has zeroes of multiplicity at least k at each point (a_i, a_j) on $V, i \neq j$. But it has at least $d-1$ additional zeroes on V: at each branching point, corresponding to a certain critical value of P, two branches of $I_k(s)$ come together and hence their difference is zero. By the transitivity assumption this corresponds to a certain point on V. (One can easily see this for P real and for y corresponding to the difference of two "neighboring" branches of I_k.) Since y is a k-fold integral of the difference of the corresponding branches of I_0, also the multiplicity of these additional zeroes is at least k. Hence the total multiplicity of zeroes of y on V is at least $k[d(d-1)+(d-1)]$.

On the other hand, one can easily show that this multiplicity is bounded from above by $(d-1)(m+kd)$, where $m = \deg Q$ and d, as above is the degree of P ($m+kd$ is simply the degree of $\hat{I}_k(x)$ as a polynomial in x, and y has $d-1$ branches of this multiplicity at infinity).

Comparing these two bounds we see that for $k > m$, y has more zeroes on V than it is allowed by the Bezout theorem. This is possible only if $y \equiv 0$. But then also the difference of the corresponding branches of $I_0(s) = Q(P^{-1}(s))$ is zero. By Proposition 3.4, P and Q have a nontrivial right common divisor, and we can proceed exactly as in the proof of Theorem 4.1 given in Section 4.2.

(In the above situation the difference of any two branches of $Q(P^{-1}(s))$ is zero and hence Q is a function of P. The same proof works in the case, where zeroes of P are not necessarily simple, but have equal multiplicities.)

3. Now we "localize" considerations of step 2. Consider a certain critical value s_0 of P and consider one branch of an algebraic function P^{-1} at s_0.

Using some techniques from local algebraic geometry of curves (in particular, Puiseux expansions) one can show that the derivatives of any prescribed order of $Q(P^{-1}(s))$ can be expressed near s_0 as polynomials (without free term) of I_k. The coeffizients $\psi_j^k(s)$ of these polynomials are univalued algebraic functions of s near s_0.

At any point s_1, where ψ_j^k are finite (don't have poles), vanishing of I_k implies vanishing of the corresponding derivatives of $Q(P^{-1}(s))$.

Now the counting of multiplicities as in p.2, can be applied to ψ_j. It turns out that for N big enough, $\psi_j^k(s_1)$ cannot have poles for $k = 0, 1, \dots, N$ at any s_1.

We conclude that if all the integrals I_k take the same value at two different points of $P^{-1}(s_1)$, all the derivatives of the difference of the two corresponding branches of $Q(P^{-1}(s))$ at s_1 vanish. Hence these two branches coincide, and we can split a common right divisor of P and Q.

4. Finally, we show that the assumption of s_1 being "close enough" to s_0 is not essential, and all the constructions of p.3 can be performed globally. This completes the proof.

Remark. A different approach to the proof of Theorem 4.1 has been proposed by C. Christopher ([21]). It is based on an investigation of the generating function of the moments $m_k(x) : M(x,y) = \sum_{k=0}^{\infty} m_k(x) y^k = \int_0^x \frac{q(t)dt}{1-yP(t)}$.

Yet another approach, based on extension of the moment techniques above to the complex domain, has been suggested by S. Yakovenko [32].

5. Comparison with M.S. Livšic's theorem

M.S. Livšic's construction of a generalized moment problem and his theorem, giving sufficient conditions of solvability of such a problem ([25]), are based on a spectral theory of operators. Since in this paper we give only a very preliminary indication of possible connections of Livšic's approach with our results, presented above, we give only a short description of Livšic's generalized moments problem and a statement of his theorem.

The following is a quotation from [1]:

"We take as our starting point a certain linear aggregate G of complex valued continuous functions $f(s)(-\infty < s < \infty)$. We consider all possible products $f(s)\overline{g(s)}$, where $f(s), g(s) \in G$, and we denote by $\mathfrak{M}$ the linear hull of all such products. This linear hull will just be the domain of definition of the functionals. A functional $\mathfrak{P}$, linear (in the algebraic sense), given on $\mathfrak{M}$ will be called *positive* (and correspondingly non-negative) if for any element $\phi \in \mathfrak{M}$ of the form $\phi = f(s)\overline{f(s)}$, where $f(s) \in G$ and $f(s) \not\equiv 0$, the inequality $\mathfrak{P}(\phi) > 0$ (or $\mathfrak{P}(\phi) \geqq 0$) holds.

Thus this definition of the positive character of a functional involves not all functions $\phi = \phi(s)$ belonging to $\mathfrak{M}$ for which $\phi(s) \geqq 0(-\infty < s < \infty)$, but only all squares. This distinction is sometimes only apparent, for instance in the case when G is the set of all polynomials.

The formulation of the general moment problem is then the following: *find the conditions under which a given non-negative functional $\mathfrak{P}$ defined over $\mathfrak{M}$ permits the integral representation*

$$(*) \qquad \mathfrak{P}\{\phi\} = \int_{-\infty}^{\infty} \phi(t) d\sigma(t) \qquad (\phi \in M)$$

with a non-decreasing function $\sigma(t)$ $(-\infty < t < \infty)$. Evidently it is sufficient to obtain the representation $(*)$ for functions $\phi(s) = f(s)\overline{f(s)}$, where $f(s) \in G$. We shall call this problem the generalized Livšic moment problem. It becomes the ordinary power moment problem if G is the aggregate of all polynomials in s.

Theorem 5.1. (Livšic [25].) *The moment problem* $(*)$ *is certainly soluble if*

a) $1 \in G$

b) *The set* D *of all functions* $g(s) \in G$, *for which* $sg(s) \in G$ *is dense in* G *with the norm*

$$\| f \| = \sqrt{\mathfrak{P}\{f(s)\overline{f(s)}\}}[f(s) \in G],$$

c) *From* $f(s) \in G$ *it follows that for any real* t *the function of* s

$$\frac{f(s) - f(t)}{s - t}$$

belongs to G".

In our case of the moments $m_k(s) = \int_0^s P^k(t)q(t)dt$, G consists of all the polynomials $R(s)$, which have the form $R(s) = \tilde{R}(P(s))$. For $\deg P > 1$ the conditions of Theorem 5.1 clearly are not satisfied for this G.

One can conjecture that condition c) in this theorem is closely related to the question of completeness of G. Indeed, one can reformulate our Theorem 4.2 in the following way (using Theorem 3.3):

Theorem 5.2. $q(s)$ *is orthogonal to all the powers* $P^k(s)$ *on* $[0, 1]$ *if and only if for any* $t \in [0, 1]$ *the functions of* s

$$\frac{P(s) - P(t)}{s - t}$$

and

$$\frac{Q(s) - Q(t)}{s - t}$$

have a common zero $s_0(t)$, $s(0) = 1$, $s(1) = 0$.

We plan to further investigate the role of the composition algebra (and, in particular, of the curves $\Gamma_f = \{\frac{f(s)-f(t)}{s-t} = 0\}$) in Livšic's generalized moment problem in separate publications.

6. Continuous and local center problems

We shall give all the definitions and results for the case of the Abel equation (1.1):

$$y' = p(x)y^2 + q(x)y^3, \qquad y(0, x_0) = y_0.$$

The case of the Moments Center conditions can be considered in a completely similar way and we describe shortly the corresponding results at the end of the section.

Let us call two points $x_1, x_2 \in \mathbb{C}$ *conjugate* with respect to (1.1), if $y(x_1) \equiv y(x_2)$ for any solution $y(x)$ of (1.1). As above, the points a_i, conjugate to 0, are called periods of (1.1).

Definition 6.1. (1.1) *has a continuous center with the periods*

$$a_1 = 0,\ a_2,\ \ldots,\ a_n$$

if for any x_1 near 0, there are points $x_2, \ldots, x_n$, x_j near a_j, conjugate to x_1.

It would seem that continuous center conditions are much stronger than the original ones. However the following results show that modulo the Composition Conjecture, these conditions are equivalent.

Theorem 6.1. *(1.1) has a continuous center with the periods $a_1 = 0,\ a_2, \ldots, a_n$ if and only if $a_i,\ i = 1, \ldots, n$ are exactly all the zeroes of the GCD from the right of P and Q in the composition algebra of polynomials without free term.*

Proof. Let W be a GCD of P and Q, and let $a_1 = 0,\ a_1, \ldots, a_n$ be all the zeroes of W. Then all the solutions of (1.1) are in fact functions of $W : y(x) = \tilde{y}(W(x))$.

Now for each x_1 near zero define $x_j,\ j = 2, \ldots, n$, as (one of) the solutions of $W(x_1) = W(x_j), x_j$ near a_j. Since $y(x_j) = \tilde{y}(W(x_j)) = \tilde{y}(W(x_1)) = y(x_1)$, all the points $x_1, \ldots, x_n$ are conjugate.

Conversely, if (1.1) has a continuous center at $0, a_2, \ldots, a_n$, then, for x_1 near zero, $(x_1, x_2),\ (x_1, x_3), \ldots, (x_1, x_n)$ define parametrizations of $n-1$ neighborhoods U_j on the curve $\Gamma_P \cap \Gamma_Q$ (as defined in Theorem 3.3). Indeed, the first two center equations (see e.g. [8]) give us $P(x_1) = P(x_2) = \cdots P(x_n),\ Q(x_1) = Q(x_2) = \cdots = Q(x_n)$ for any x_1 near 0. Clearly, U_j contain the points $(0, a_j), j = 2, \ldots, n$. Hence by Theorem 3.3, the GCD W of P and Q vanishes at $a_1 = 0, a_2, \ldots, a_n$.

Thus we have shown that the zeroes of W are the periods of a continuous center for (1.1) and all such periods are zeroes of W. Theorem 6.1 is proved.

In fact, one does not need to have a continuous family of conjugate points to conclude that P and Q have a nontrivial factorization.

Theorem 6.2. *If there are more than $\frac{1}{2} \deg P \cdot \deg Q$ different (unordered) pairs of conjugate points (x_1, x_2) for* (1.1), *then P and Q have a nontrivial GCD W from the right. In particular, zeroes of W, different from 0, are nontrivial periods of* (1.1).

Proof. As above, each conjugate pair (x_1, x_2) satisfies $P(x_1) = P(x_2),\ Q(x_1) = Q(x_2)$, and hence belongs to $\Gamma_P \cap \Gamma_Q$. But by the Bezout theorem, $\Gamma_P \cap \Gamma_Q$ cannot contain more than $\deg P \cdot \deg Q$ different points, unless it has a one-dimensionalal component. Application of Theorem 3.3 completes the proof.

Remark 1. Specifying the positions of the conjugate pairs in Theorem 6.2, one can get more specific information about W. See, for example, the Local Center conditions below.

Remark 2. The equations of the curves Γ_P have a very special structure. One can hope to get more precise information on $\Gamma_P \cap \Gamma_Q$ than the bound of Bezout theorem. In particular, some bounds for the local multiplicity of $\Gamma_P \cap \Gamma_Q$ have been obtained in [17, 18], in relation with the bounding of the number of limit cycles in the Lienard equation.

We can “localize” the situation, described in definition 6.1:

Definition 6.2. (1.1) *has a local continuous center at* x_0, *if for any* x_1, *close to* x_0, *there is* x_2 *close to* x_0, x_2 *conjugate to* x_1 *and different from it.*

Theorem 6.3. (1.1) *has a local continuous center at* x_0 *if and only if* x_0 *is a critical point of the GCD from the right* W *of* P *and* Q.

Proof. The same as for Theorem 6.1. We notice in addition that the conjugate points x_1 and x_2 are the solutions of $W(x_1) = W(x_2)$, both near x_0, and hence x_0 must be a critical point of W.

Remark 1. One can give a direct proof of Theorem 6.3, without reference to Theorem 3.3, as follows: by the first two center equations, $P(x_1) = P(x_2)$ and $Q(x_1) = Q(x_2)$. After local algebraic change of variables we can assume $P(x) = (x - x_0)^m$. Hence $x_2 - x_0 = \varepsilon(x_1 - x_0)$, $\varepsilon^m = 1$. Substituting this expression into the power series for $Q(x)$ we easily obtain that it can contain only the powers $(x - x_0)^{k\ell}$, with ℓ-one of the divisors of m.

Remark 2. Local continuous center conditions for the Abel equation (1.1) are closely related to the center conditions for the Lienard Equation (see, e.g. [13, 14], [15–17]). Indeed, the transformatiom $Y = 1/y$ brings one into another.

Different "localization" of Theorem 6.2 are possible. One of the simplest is the following:

Theorem 6.4. *If the Composition conjecture is true, then in the convex hull of any complete set of conjugate points of (1.1) there is a continuous local center.*

Proof. Conjugate points of (1.1) are exactly the solution of $W = const$, and local continuous centers are zeroes of W'. The result follows by the Gauss theorem.

The conjugate points can be defined for the Moments Center problem in a similar way: x_1 is conjugate to x_2 if $m_k(x_1) = m_k(x_2)$ for any k (or, equivalently, $M(x_1, y) \equiv M(x_2, y)$ in y). Local centers are defined exactly as above. Since the Composition Conjecture is true for the Moments Center problem, we have the following:

Theorem 6.5. *The sets of conjugate points for the Moments Center problem are exactly the solutions of* $W = const$, *where* W *is the GCD of* P *and* Q. *Local continuous centers are exactly the critical points of* W.

References

[1] N.I. Akhiezer, The classical moment problem, Oliver and Boyd, Edinburgh, London, 1965.

[2] M.A.M. Alwash and N.G. Lloyd, Non-autonomous equations related to polynomial two-dimensional systems, *Proceedings of the Royal Society of Edinburgh*, **105A** (1987), 129–152.

[3] M.A.M. Alwash and N.G. Lloyd, Periodic solutions of a quartic nonautonomous equation, *Nonlinear Analysis. Theory, Methods and Applications*, **11**, 7, (1987), 809–820.

[4] M.A.M. Alwash, On a condition for a centre of cubic non-autonomous equations, *Proceedings of the Royal Society of Edinburgh*, **113A** (1989), 289–291.

[5] V.I. Arnold, Yu. Il'yashenko, Ordinary differential equations, Encyclopedia of Mathematical Sciences 1 (Dynamical Systems - I), Springer, Berlin, 1988.

[6] M. Blinov and Y. Yomdin, Generalized center conditions and multiplicities for polynomial Abel equations of small degree, to appear. Nonlinearity, 1999.

[7] M. Briskin, J.-P. Francoise and Y. Yomdin, The Bautin ideal of the Abel equation, *Nonlinearity* **11**, (1998), 431–443.

[8] M. Briskin, J.-P. Francoise and Y. Yomdin, Center conditions, composition of polynomials and moments on algebraic curves, to appear. Erg. Theory and Dyn. Systems, 1999.

[9] M. Briskin, J.-P. Francoise and Y. Yomdin, Center conditions II: Parametric and model center problems, to appear.

[10] M. Briskin, J.-P. Francoise and Y. Yomdin, Center conditions III: More on parametric and model center problems, to appear.

[11] M. Briskin, J.-P. Francoise and Y. Yomdin, Une approche au problème de centre-foyer de Poincaré, to appear, C.R.Acad.Sci.

[12] R. Benedetti, J.-J. Risler, Real algebraic and semi-algebraic sets, Actualities Mathematiques, Hermann, Paris 1990.

[13] L.A. Cherkas, Conditions for a Lienard equation to have a center, *Differ. Eq.* **12**, (1976), 201–202.

[14] L.A. Cherkas, Conditions for the equation $yy' = p(x, y)$ to have a center, *Differ. Eq.* **14**, (1978), 1133–1137.

[15] C.J. Christopher, N.G. Lloyd and J.M. Pearson, On Cherkas's method for center conditions, *Nonlin. World* **2** (1995), 459–469.

[16] C.J. Christopher and N.G. Lloyd, On the paper of Jun and Wang concerning the conditions for a centre in certain cubic systems, *Bull. London Math. Soc.* **22** (1990). 5–12.

[17] C.J. Christopher and N.G. Lloyd, Small-amplitude limit cycles in Lienard Systems. *Non-linear Diff. Equations and Applications*, **3** (1996), 183–190.

[18] C.J. Christopher, An algebraic approach to the classification of Centres in Polynomial Lienard Systems, to appear.

[19] C.J. Christopher, private communication, I.

[20] C.J. Christopher, private communication, II.

[21] C.J. Christopher, private communication, III.

[22] J. Devlin, Word problem related to periodic solutions of a non-autonomous system, *Math. Proc. Camb. Phil. Soc.* **108** (1990), 127–151.

[23] J. Devlin, Word problems related to derivatives of the displacement map, *Math. Proc. Camb. Phil. Soc.* **110** (1991), 569–579.

[24] A. Eremenko and L.A. Rubel, The arithmetic of entire functions under composition, Advances in Math., **124**, 1996, 334–354.

[25] M.S. Livšic, On an application of the theory of Hermitian Operators to Generalized Moment problem, DAN SSSR, 1944.

[26] P. Mueller, Primitive Monodromy Groups of Polynomials, In: M. Fried (ed), Recent developments in the inverse Galois problem, Contemp. Math. **186**, 385–401.

[27] P. Mueller, Private communication.

[28] A. Lins Neto, On the number of solutions of the equation $x' = P(x,t)$ for which $x(0) = x(1)$, Inventiones Math., **59** (1980), 67–76.

[29] J.F. Ritt, Prime and composite polynomials, Trans. A.M.S. **23** (1922), 51–66.

[30] N. Roytvarf, A proof of Composition conjecture for the Moments Center Problem, to appear.

[31] D.V. Widder, The Laplace transform, Princeton University Press, N.J., 1946.

[32] S. Yakovenko, private communication.

Ort Academic College for Teachers in Technology, Givat Ram, Jerusalem 95435, Israel

Université de Paris VI, U.F.R. 920, 46–56, B.P. 172, Département de Mathématiques, 4 Place Jussieu, 75252 Paris, France

Department of Mathematics, The Weizmann Institute of Science, Rehovot 76100, Israel

Operator Theory:
Advances and Applications, Vol. 123

Inverse Spectral problem for the Sturm-Liouville Operator with Eigenvalue Parameter Dependent Boundary Conditions

M. V. Chugunova

Abstract. To solve the inverse problem for the Sturm-Liouville operator with eigenvalue parameter dependent boundary conditions we reconstruct the spectral distribution function from two spectra of the boundary-value problems with equal $\Theta(\lambda)$ and different real constants in the boundary conditions. The well-known results of A.V. Strauss [5] concerning the connection between the eigenvalue problems with the spectral parameter in the boundary conditions and the theory of generalized resolvents is used.

I. Let us consider the regular differential equation

$$-y'' + q(x)y = \lambda y, \tag{1}$$

where $q(x)$ $(0 \le x \le l)$ is a real-valued continuous function and two boundary-value problems:

$$y'(0) - h_1 y(0) = 0, \quad y(l) = \Theta(\lambda) y'(l), \tag{2}$$

$$y'(0) - h_2 y(0) = 0, \quad y(l) = \Theta(\lambda) y'(l), \tag{3}$$

where h_1 and h_2 are finite real numbers, $h_1 \ne h_2$ and $\Theta(\lambda)$ is a rational function for which $\operatorname{Im}\Theta(\lambda)\operatorname{Im}\lambda \le 0$. We write $\Theta(\lambda) = \frac{\Theta_1(\lambda)}{\Theta_2(\lambda)}$, where $\Theta_1(\lambda)$ and $\Theta_2(\lambda)$ are relatively prime polynomials. If the boundary conditions (2), (3) do not contain the eigenvalue parameter λ, then the solution of the inverse problem from two spectra is given by M.G. Krein, B.M. Levitan, I.M. Gelfand [1,2,3].

Let $u_1(x,\lambda)$ and $u_2(x,\lambda)$ be the solutions of (1) satisfying the following initial conditions:

$$u_1(0,\lambda) = 1, \quad u_1'(0,\lambda) = h_1,$$

$$u_2(0,\lambda) = 1, \quad u_2'(0,\lambda) = h_2.$$

Then the spectra of the boundary-value problems (1), (2) and (1), (3) are the zero-sequences $\{\lambda_n\}_{n=0}^{\infty}$ and $\{\mu_n\}_{n=0}^{\infty}$ of the entire functions:

$$\Phi_1(\lambda) = \Theta_2(\lambda) u_1(l,\lambda) - \Theta_1(\lambda) u_1'(l,\lambda),$$

$$\Phi_2(\lambda) = \Theta_2(\lambda) u_2(l,\lambda) - \Theta_1(\lambda) u_2'(l,\lambda).$$

1991 *Mathematics Subject Classification.* 47A10, 47A20, 47A45, 47A67, 47B25.

As the spectra of these problems are bounded from below we can number the eigenvalues so that $\lambda_0 < \lambda_1 < \cdots < \lambda_n < \cdots$; $\mu_0 < \mu_1 < \cdots < \mu_n < \cdots$.

The spectral distribution function $\rho(\lambda)$ of the problem (1), (2) is a jump function [4,5] and it is determined by equality

$$\rho(\lambda) = \sum_{\lambda_n < \lambda} \frac{1}{\alpha_n}.$$

The numbers α_n are known as the normalizing numbers [3]. In the classical case ($\Theta(\lambda) = const$) the equality $\alpha_n = \int_0^l |u_1(x, \lambda_n)|^2 dx$ takes place. Next we will show how to use $\{\lambda_n\}_{n=0}^{\infty}$ and $\{\mu_n\}_{n=0}^{\infty}$ to obtain α_n, the normalizing numbers of eigenfunctions of the first boundary-value problem.

Let A_1 be the differential operator of the second order acting in $L^2(0, l)$ which is defined by the expression

$$l[f] = -f'' + q(x)f$$

and the boundary conditions

$$f_1'(0) = h_1 f_1(0), \quad f_1'(l) = f_1(l) = 0.$$

Let m be the number of poles $\Theta(\lambda)$. The selfadjoint extension $\tilde{A}$ of the differential operator is in some sense a coupling of the operators A_1 and A_2 (A_2 is a difference operator) acting in the orthogonal spaces $L^2(0, l)$ and C^m, respectively [4,6]. Eigenfunctions of the selfadjoint operator $\tilde{A}$ are discretely extended into the space C^m. Therefore we can write

$$\alpha_n = \alpha_{n1} + \alpha_{n2},$$

where

$$\alpha_{n1} = \int_0^l |u_1(x, \lambda_n)|^2 dx, \quad \alpha_{n2} = -\Theta'(\lambda_n)|u_1'(l, \lambda_n)|^2.$$

Example 1. *Let us consider the boundary-value problem*

$$-y'' + q(x)y = \lambda y,$$
$$y(0) = \lambda y'(0), \quad y'(l) = hy(l).$$

Let $\tilde{A}$ be the selfadjoint operator

$$D(\tilde{A}) = \{\langle f, f'(0)\rangle : f(x) \in L^2(0, l),\ f'(l) = hf(l)\},$$

$$\tilde{A}\langle f(x), f'(0)\rangle = \langle -f''(x) + q(x)f(x), f(0)\rangle.$$

The eigenfunction system $u(x, \lambda_n)$ is complete in $L^2(0, l)$, but not minimal.

$$\tilde{A}u = \lambda u, \quad u(x, \lambda_n) = \langle u_2(x, \lambda_n) + \alpha(\lambda)u_1(x, \lambda_n), -1\rangle.$$

Denote by

$$f(x,\lambda) = u_2(x,\lambda) + m(\lambda)u_1(x,\lambda),$$

where the Weyl function $m(\lambda)$ is defined by the condition

$$f(l,\lambda) = \Theta(\lambda)f'(l,\lambda)$$

and hence

$$m(\lambda) = \frac{\Theta_2(\lambda)u_2(l,\lambda) - \Theta_1(\lambda)u_2'(l,\lambda)}{\Theta_1(\lambda)u_1'(l,\lambda) - \Theta_2(\lambda)u_1(l,\lambda)} = -\frac{\Phi_1(\lambda)}{\Phi_2(\lambda)}.$$

The formula shows that $m(\lambda)$ is a meromorphic function, whose poles and zero-sequences are the spectra of the appropriate boundary-value problems (1), (2) and (1), (3) respectively. In addition, by the Green's formula, we have

$$(\lambda-\overline{\mu})\int_0^l f(x,\lambda)\overline{f(x,\mu)}dx = (h_1-h_2)(m(\lambda)-\overline{m(\mu)})+(\Theta(\lambda)-\overline{\Theta(\mu)})f'(l,\lambda)\overline{f'(l,\mu)}.$$

If $\mu = \lambda$, then, because $\operatorname{Im}\Theta(\lambda)\operatorname{Im}\lambda \le 0$,

$$(h_1-h_2)\frac{\operatorname{Im} m(\lambda)}{\operatorname{Im}\lambda} = \int_0^l |f(x,\lambda)|^2dx - \frac{\operatorname{Im}\Theta(\lambda)}{\operatorname{Im}\lambda}|f'(l,\lambda)|^2 \ge 0.$$

Hence, in the case $h_1 > h_2$, $m(\lambda)$ is a Nevanlinna function and in the case $h_1 < h_2$, $-m(\lambda)$ is a Nevanlinna function [8]. We have thus proved

Theorem 1. *The spectra of boundary-value problems with equal $\Theta(\lambda)$ and different real constants in boundary conditions are alternating.*

Let us use Green's formula once more

$$(\lambda-\lambda_n)\int_0^l f(x,\lambda)u_1(x,\lambda_n)dx = h_2 - h_1 + (\Theta(\lambda)-\Theta(\lambda_n))f'(l,\lambda)u_1'(l,\lambda_n),$$

$$\begin{aligned}(\lambda-\lambda_n)&\int_0^l f(x,\lambda)u_1(x,\lambda_n)dx\\ &= (\lambda-\lambda_n)\int_0^l u_2(x,\lambda)u_1(x,\lambda_n)dx - (\lambda-\lambda_n)\frac{\Phi_2(\lambda)}{\Phi_1(\lambda)}\int_0^l u_1(x,\lambda)u_1(x,\lambda_n)dx\\ &= h_2 - h_1 + (\Theta(\lambda)-\Theta(\lambda_n))u_2'(l,\lambda)u_1'(l,\lambda_n)\\ &\quad - (\Theta(\lambda)-\Theta(\lambda_n))\left(\frac{\Phi_2(\lambda)}{\Phi_1(\lambda)}\right)u_1'(l,\lambda)u_1'(l,\lambda_n).\end{aligned}$$

Change the formula in the form of

$$\int_0^l u_1(x,\lambda)u_1(x,\lambda_n)dx - \frac{\Theta(\lambda)-\Theta(\lambda_n)}{\lambda-\lambda_n}u_1'(l,\lambda)u_1'(l,\lambda_n)$$
$$= \frac{\Phi_1(\lambda)}{\Phi_2(\lambda)}\int_0^l u_2(x,\lambda)u_1(x,\lambda_n)dx - (h_2-h_1)\frac{\Phi_1(\lambda_n)}{\Phi_2(\lambda_n)(\lambda-\lambda_n)}$$
$$- \frac{\Theta(\lambda)-\Theta(\lambda_n)}{\lambda-\lambda_n}\frac{\Phi_1(\lambda_n)}{\Phi_2(\lambda_n)}u_2'(l,\lambda)u_1'(l,\lambda_n).$$

But now as $\lambda \to \lambda_n$ we see that

$$\int_0^l |u_1(x,\lambda_n)|^2 dx - \Theta'(\lambda_n)|u_1'(l,\lambda_n)|^2 = -(h_2-h_1)\frac{\Phi_1'(\lambda_n)}{\Phi_2(\lambda_n)},$$

since $\Phi_1(\lambda_n) = 0$.

$$\alpha_n = -(h_2-h_1)\frac{\Phi_1'(\lambda)}{\Phi_2(\lambda)}. \tag{4}$$

Using (4) we find that

$$\alpha_n = \frac{h_2-h_1}{\mu_n-\lambda_n}\prod_{k=0}^{\infty}{}' \frac{\lambda_k-\lambda_n}{\mu_k-\lambda_n}. \tag{5}$$

$\Phi_1(\lambda)$ and $\Phi_2(\lambda)$ are entire functions which behave as $O(1/|\lambda|^{1/2})$ [1,2]. We can write the infinite product

$$\Phi_1(\lambda) = c_1\prod_{k=0}^{\infty}\left(1-\frac{\lambda}{\lambda_k}\right), \quad \Phi_2(\lambda) = c_2\prod_{k=0}^{\infty}\left(1-\frac{\lambda}{\mu_k}\right),$$

where c_1 and c_2 are constants. It follows from the statement above and (4) that

$$\alpha_n = (h_1-h_2)\frac{c_1}{c_2}\frac{1}{\lambda_n}\frac{\prod_{k=0}^{\prime\infty}\left(1-\frac{\lambda_n}{\lambda_k}\right)}{\prod_{k=0}^{\infty}\left(1-\frac{\lambda_n}{\mu_k}\right)} = \frac{h_2-h_1}{\mu_n-\lambda_n}\frac{c_1}{c_2}\prod_{k=0}^{\infty}\frac{\mu_k}{\lambda_k}\prod_{k=0}^{\infty}{}'\frac{\lambda_k-\lambda_n}{\mu_k-\lambda_n}.$$

Therefore, to obtain (5) we must show that

$$\frac{c_1}{c_2}\prod_{k=0}^{\infty}\frac{\mu_k}{\lambda_k} = 1.$$

Let us consider two cases.

(a) In this case $\deg\Theta_1(\lambda) \geq \deg\Theta_2(\lambda)$ [5]. Using the classical asymptotic formulas for the solutions of Sturm-Liouville equation to obtain the limit

$$\lim_{\lambda\to-\infty}\frac{\Phi_1(\lambda)}{\Phi_2(\lambda)} = \lim_{\lambda\to-\infty}\frac{\Theta_2(\lambda)u_1(l,\lambda)-\Theta_1(\lambda)u_1'(l,\lambda)}{\Theta_2(\lambda)u_2(l,\lambda)-\Theta_1(\lambda)u_2'(l,\lambda)}$$
$$= \lim_{\lambda\to-\infty}\frac{Hu_1(l,\lambda)-u_1'(l,\lambda)}{Hu_2(l,\lambda)-u_2'(l,\lambda)},$$

where $H = \lim_{\lambda\to-\infty} \frac{\Theta_2(\lambda)}{\Theta_1(\lambda)}$.

$$\lim_{\lambda\to-\infty} \frac{c_1}{c_2} \prod_{k=0}^{\infty} \left(1-\frac{\lambda}{\lambda_k}\right)\left(1-\frac{\lambda}{\mu_k}\right)^{-1} = \frac{c_1}{c_2} \prod_{k=0}^{\infty} \frac{\mu_k}{\lambda_k} \lim_{\lambda\to-\infty} \prod_{k=0}^{\infty} \frac{\lambda_k-\lambda}{\mu_k-\lambda} = 1.$$

We must show that

$$\lim_{\lambda\to-\infty} \prod_{k=0}^{\infty} \frac{\lambda_k-\lambda}{\mu_k-\lambda} = \lim_{\lambda\to-\infty} \prod_{k=0}^{\infty} \left(1+\frac{\lambda_k-\mu_k}{\mu_k-\lambda}\right) = 1. \tag{6}$$

The asymptotic behavior of eigenvalues (2), (3) is given by formulas [6,7]

$$\lambda_k = (k-m)^2 + O(1), \quad \mu_k = (k-m)^2 + O(1).$$

Therefore $\lambda_k - \mu_k = O(1)$, i.e., the spectra of boundary-value problems are asymptotically close and the series $\sum_{k=0}^{\infty} \left|\frac{\lambda_k-\mu_k}{\mu_k-\lambda}\right|$ is uniformly convergent in the neighborhood of the point $\lambda = -\infty$. This allows us to write

$$\lim_{\lambda\to-\infty} \prod_{k=0}^{\infty} \left(1+\frac{\lambda_k-\mu_k}{\mu_k-\lambda}\right) = 1.$$

This shows that (5) holds.

(b) In this case $\deg\Theta_1(\lambda) < \deg\Theta_2(\lambda)$.

$$\lim_{\lambda\to-\infty} \frac{\Phi_1(\lambda)}{\Phi_2(\lambda)} = \lim_{\lambda\to-\infty} \frac{u_1(l,\lambda) - \frac{\Theta_1(\lambda)}{\Theta_2(\lambda)} u_1'(l,\lambda)}{u_2(l,\lambda) - \frac{\Theta_1(\lambda)}{\Theta_2(\lambda)} u_2'(l,\lambda)} = 1.$$

Moreover, $\lambda_k = (k-m+1/2) + O(1)$ and $\mu_k = (k-m+1/2) + O(1)$, therefore $\lambda_k - \mu_k = O(1)$, i.e., the linear term of $\Theta(\lambda)$ does not affect the asymptotic closeness of the spectra of boundary-value problems.

Using formula (5) we can reconstruct the spectral distribution function. If $q(x)$ is a sufficiently differentiable (twice differentiable) function then asymptotic formulas take place

$$\sqrt{\lambda_{n+m}} = n + \frac{a_0}{n} + \frac{a_1}{n^3} + o\left(\frac{1}{n^3}\right), \tag{7}$$

$$\sqrt{\mu_{n+m}} = n + \frac{a_0'}{n} + \frac{a_1'}{n^3} + o\left(\frac{1}{n^3}\right), \tag{8}$$

where

$$a_0 = (h_1 + H)/l + (2l)^{-1} \int_0^l q(x)dx,$$

$$a_0' = (h_2 + H)/l + (2l)^{-1} \int_0^l q(x)dx,$$

and therefore

$$a_0 - a_0' = (h_2 - h_1)/l.$$

Theorem 2. *Let two spectra $\{\lambda_n\}_{n=0}^{\infty}$ and $\{\mu_n\}_{n=0}^{\infty}$ be given so that $\lambda_0 < \mu_0 < \lambda_1 < \mu_1 < \cdots$, equalities (7), (8) take place, moreover $a_0 \neq a_0'$. Then there exist an absolutely continuous function $q(x)$, real numbers h_1, h_2 and a rational function $\Theta(\lambda)$ for which $\operatorname{Im}\Theta(\lambda)\operatorname{Im}\lambda \leq 0$, such that λ_n is the spectrum of the problem (2), μ_n is the spectrum of the problem (3).*

II. Let us consider the regular differential equation

$$-y'' + q(x)y = \lambda y, \tag{9}$$

where $q(x)$ $(0 \leq x \leq l)$ is a real-valued continuous function and two boundary-value problems:

$$y(0) = \Theta(\lambda)y'(0), \quad y'(l) + H\,y(l) = 0, \tag{10}$$

$$y(0) = (\Theta(\lambda) + h)y'(0), \quad y'(l) + H\,y(l) = 0, \tag{11}$$

where h and H are finite real numbers, $h \neq 0$ and $\Theta(\lambda)$ is a rational function for which $\operatorname{Im}\Theta(\lambda)\operatorname{Im}\lambda \geq 0$. Denote by $\lambda_0 < \lambda_1 < \lambda_2 < \cdots$; $\mu_0 < \mu_1 < \mu_2 < \cdots$, the eigenvalues of boundary-value problems (9), (10) and (9), (11), respectively. Let $u_1(x,\lambda)$ and $u_2(x,\lambda)$ be the solutions of (9) satisfying the following initial conditions:

$$u_1(0,\lambda) = 1, \quad u_1'(0,\lambda) = 0,$$
$$u_2(0,\lambda) = 0, \quad u_2'(0,\lambda) = -1.$$

Then the solutions of the differential equation (9)

$$\varphi(x,\lambda) = \Theta_1(\lambda)u_1(x,\lambda) - \Theta_2(\lambda)u_1'(x,\lambda),$$
$$\psi(x,\lambda) = (\Theta_1(\lambda) + h\,\Theta_2(\lambda))u_1(x,\lambda) - \Theta_2(\lambda)u_1'(x,\lambda)$$

satisfy right boundary condition (10) and right boundary condition (11), respectively.

Then the spectra of the boundary-value problems are the zero-sequences of the entire functions [5]:

$$\Phi_1(\lambda) = \varphi'(l,\lambda) + H\varphi(l,\lambda),$$
$$\Phi_2(\lambda) = \psi'(l,\lambda) + H\psi(l,\lambda).$$

Next we will show how to use $\{\lambda_n\}_{n=0}^{\infty}$ and $\{\mu_n\}_{n=0}^{\infty}$ to obtain α_n the normalizing numbers of eigenfunctions of the first boundary-value problem. Following the work [3], we shall determine

$$\alpha_n = \alpha_{n1} + \alpha_{n2}, \quad \text{where} \quad \alpha_{n1} = \int_0^l |\varphi(x,\lambda_n)|^2 dx, \quad \alpha_{n2} = \Theta'(\lambda_n)|\Theta_2(\lambda_n)|^2.$$

Let us define

$$f(x,\lambda) = \psi(x,\lambda) + m(\lambda)\varphi(x,\lambda).$$

The Weyl function $m(\lambda)$ is defined by the condition $f'(l,\lambda) + H\,f(l,\lambda) = 0$. Hence we obtain $m(\lambda) = -\frac{\Phi_1(\lambda)}{\Phi_2(\lambda)}$.

The formula shows that $m(\lambda)$ is a meromorphic function, whose poles and zero-sequences are spectra of appropriate boundary-value problems (9), (10) and (9),

(11) respectively. In addition, by Green's formula, we have

$$
\begin{aligned}
(\lambda-\overline{\mu})\int_0^l f(x,\lambda)\overline{f(x,\mu)}dx &= f'(0,\lambda)\overline{f(0,\mu)} - f(0,\lambda)\overline{f'(0,\mu)} \\
&= (\Theta_2(\lambda)+m(\lambda)\Theta_2(\lambda))(\overline{\Theta_1(\mu)}+h\overline{\Theta_2(\mu)}+\overline{m(\mu)}\,\overline{\Theta_1(\mu)}) \\
&- (\Theta_1(\lambda)+h\Theta_2(\lambda)+m(\lambda)\Theta_1(\lambda))(\overline{\Theta_2(\mu)}+\overline{m(\mu)}\,\overline{\Theta_2(\mu)}) \\
&= (\overline{\Theta(\mu)}-\Theta(\lambda))\Theta_2(\lambda)\,\overline{\Theta_2(\mu)}+h\Theta_2(\lambda)\overline{\Theta_2(\mu)}(m(\lambda)-\overline{m(\mu)}) \\
&+ m(\lambda)\overline{m(\mu)}(\overline{\Theta(\mu)}-\Theta(\lambda))\Theta_2(\lambda)\overline{\Theta_2(\mu)} \\
&+ (\overline{m(\mu)}+m(\lambda))(\Theta_2(\lambda)\overline{\Theta_1(\mu)}-\Theta_1(\lambda)\overline{\Theta_2(\mu)}).
\end{aligned}
$$

On the other hand, if $\mu=\lambda$, then

$$
\begin{aligned}
(12)\quad \int_0^l |f(x,\lambda)|^2dx &= -\frac{\operatorname{Im}\Theta(\lambda)}{\operatorname{Im}\lambda}|\Theta_2(\lambda)|^2+h\,\frac{\operatorname{Im}\, m(\lambda)}{\operatorname{Im}\lambda}|\Theta_2(\lambda)|^2 \\
&- \frac{\operatorname{Im}\Theta(\lambda)}{\operatorname{Im}\lambda}|\Theta_2(\lambda)|^2|m(\lambda)|^2-(\overline{m(\lambda)}+m(\lambda))\frac{\operatorname{Im}\Theta(\lambda)}{\operatorname{Im}\lambda}|\Theta_2(\lambda)|^2.
\end{aligned}
$$

It is obvious that $1+|m(\lambda)|^2+2\operatorname{Re}m(\lambda)\geq 0$. Hence we obtain

$$
\int_0^l |f(x,\lambda)|^2dx+\frac{\operatorname{Im}\Theta(\lambda)}{\operatorname{Im}\lambda}|\Theta_2(\lambda)|^2(\quad 1+|m(\lambda)|^2+2\operatorname{Re}m(\lambda))=h\,\frac{\operatorname{Im}\, m(\lambda)}{\operatorname{Im}\lambda}|\Theta_2(\lambda)|^2
$$

in the case $h>0$, $m(\lambda)$ is a Nevanlinna function (in the case $h<0$, $-m(\lambda)$ is a Nevanlinna function). We have thus proved

Theorem 3. *The spectra of the boundary-value problems (10) and (11) are alternating.*

Let us use Green's formula once more

$$
\begin{aligned}
(13)\quad (\lambda-\lambda_n)\int_0^l f(x,\lambda)\varphi(x,\lambda_n)dx &= f'(0,\lambda)\varphi(0,\lambda_n)-f(0,\lambda)\varphi'(0,\lambda_n) \\
&= (\Theta_2(\lambda)+m(\lambda)\Theta_2(\lambda))(\Theta_1(\lambda_n)) \\
&- (\Theta_1(\lambda)+h\,\Theta_2(\lambda)+m(\lambda)\Theta_1(\lambda))(\Theta_2(\lambda_n)),
\end{aligned}
$$

where $m(\lambda)=-\frac{\Phi_2(\lambda)}{\Phi_1(\lambda)}$. On the other hand,

$$
\begin{aligned}
(14)\quad & (\lambda-\lambda_n)\int_0^l f(x,\lambda)\varphi(x,\lambda_n)dx \\
&= (\lambda-\lambda_n)\int_0^l \psi(x,\lambda)\varphi(x,\lambda_n)dx-(\lambda-\lambda_n)\frac{\Phi_2(\lambda)}{\Phi_1(\lambda)}\int_0^l \varphi(x,\lambda)\varphi(x,\lambda_n)dx.
\end{aligned}
$$

But now as $\lambda \to \lambda_n$ using the statement (13) we see that

$$\alpha_{n1} = -\Theta'(\lambda_n)|\Theta_2(\lambda_n)|^2 - h\,|\Theta_2(\lambda_n)|^2 \frac{\Phi_1'(\lambda_n)}{\Phi_2(\lambda_n)}$$

From this it follows that

$$\alpha_n = \frac{h|\Theta_2(\lambda_n)|^2}{\mu_n - \lambda_n} \prod_{k=0}^{\infty}{}' \frac{\lambda_k - \lambda_n}{\mu_k - \lambda_n}.$$

Theorem 4. *Let two spectra $\{\lambda_n\}_{n=0}^{\infty}$, $\{\mu_n\}_{n=0}^{\infty}$ and $\Theta_2(\lambda)$ be given so that $\lambda_0 < \mu_0 < \lambda_1 < \mu_1 < \cdots$, equalities (7), (8) take place, moreover $a_0 \neq a_0'$. Then there exist an absolutely continuous function $q(x)$, a real number h and a rational function $\Theta(\lambda)$ for which $\operatorname{Im}\Theta(\lambda)\operatorname{Im}\lambda \geq 0$, such that λ_n is the spectrum of the problem (10), μ_n is the spectrum of the problem (11).*

References

[1] Krein M.G. On some cases of effective determination of the density of a nonhomogeneous string by its spectral function. Dokl.Akad.Nauk SSSR, (1953), **93**, N4, 617–620.

[2] M.G. Krein, On inverse problems for inhomogeneous strings, Dokl. Akad. Nauk SSSR, 82 (1952), 669–672.

[3] I.M. Gel'fand and B.M. Levitan, On a simple identity for the characteristic values of a differential operator of the second order, Dokl. Akad. Nauk SSSR, 88 (1953), 593–596.

[4] Strauss A.V. On selfadjoint operators acting in the orthogonal sum of Hilbert spaces. Dokl. Akad. Nauk SSSR, (1962), **144**, N3, 512–515.

[5] A.V.Strauss, On the spectral functions of differential operators, Izv. Akad. Nauk SSSR, Ser. Mat., 19 (1955), 201–220.

[6] M.V. Chugunova, On the inverse spectral problem for the finite interval, Funct. Analysis, Ulyanovsk, 35 (1994), 113–123.

[7] M.V. Chugunova, On the effective methods for the solution of some inverse problems, Funct. Analysis, Ulyanovsk, 36 (1997), 66–75.

[8] N. Aronszajn, W.F. Donoghue, On the exponential representations of analytic functions in the upper half-plane with positive imaginary part, J. d'Analyse Math., 5 (1956), 321–388.

Department of Physics and Mathematics, Ulyanovsk State Pedagogical University, Ulyanovsk, 432700 Russia

E-mail address: `smike@POP.ul.ru`

Operator Theory:
Advances and Applications, Vol. 123

A Module Approach to Commutant Lifting on Kreĭn Spaces

Michael A. Dritschel

For Moshe Livšic, in honor of his eightieth birthday.

Abstract. We formulate and prove the commutant lifting theorem for contractions on Kreĭn spaces in the language of categories of modules. Our method is inspired by a proof of the commutant lifting theorem on Hilbert spaces due to Arocena, and provides a possible framework for other versions of the commutant lifting theorem. Additionally, the projective modules in our category are described, and it is shown that certain indices related to the adjoint of the intertwining operator are preserved during lifting.

Introduction

The commutant lifting theorem for contractions on Kreĭn spaces was first proved by the author in his dissertation, the results being published in [12]. Since that time, other proofs have been given (eg., [8]), and various generalizations have been found ([2], [3], [15]).

The present work draws its inspiration from several sources. The author has felt for some time that there should be a Kreĭn space version of the more algebraic ways of considering extension and lifting problems akin to those which can be found in the Hilbert space context, primary sources for which include [9], [10], [4], and [16]. The recent paper by Robert Clancy and Scott McCullough [6], as well as discussions with these authors and Sarah Ferguson at Purdue University led in part to the methods used in this work. In addition, a paper by Arocena [1] contains an ingenious proof of the commutant lifting theorem on Hilbert spaces, which inspired the proof found here. The author feels that the algebraic methods shed light on Arocena's ideas and indicates a possible way of considering other commutant lifting problems. Even if the reader is not interested in Kreĭn spaces, the proof makes sense in the context of Hilbert modules, and gives a new and simpler method.

The standard approach to the algebraization of such problems is to reword them in terms of categories of modules over some algebra. The standard commutant lifting theorem focuses on operators which are (similar to) contractions over Hilbert

---

1991 *Mathematics Subject Classification.* Primary 47A15, 47A55, 47B50; Secondary 46C20.

spaces, and so in this context, the algebra of choice is the disk algebra. On the other hand, no set other than the whole complex plane is a spectral set for all Kreĭn space contractions, so with these one is reduced to working with the algebra polynomials in a single complex variable. There appears to be two quite different viewpoints which can be taken, at least when dealing with contractions over Hilbert spaces. One is to only ask that modules correspond to operators which are similar to contractions and that module maps be bounded. The other is to require that modules correspond to operators which are actually contractions and module maps be contractions. The latter allows for greater control in lifting, while the former makes available the power of Ext groups (see, for example, [14]). We take the latter approach in this paper.

The setup

For background information on Kreĭn spaces and operator theory on them, using notation similar to that adapted here, the reader is referred to [13]. A good reference for the categorical notions employed is [17].

For a Kreĭn space $\mathcal{H}$, let $\mathbf{L}(\mathcal{H})$ denote the bounded operators on $\mathcal{H}$. Every $T \in \mathbf{L}(\mathcal{H})$ determines a module over $\mathbb{C}[z]$ via

$$p \cdot f = p(T)f, \qquad p \in \mathbb{C}[z],\ f \in \mathcal{H}.$$

We denote this module by $\{\mathcal{H}_T, T\}$, or when there is no ambiguity, simply $\mathcal{H}_T$. In this case, we use $\mathcal{H}_T$ to refer to the Kreĭn space as well. We say that this module is a *contractive* Kreĭn module if T is a continuous contraction on $\mathcal{H}_T$ (that is, $1-T^*T \geq 0$, or equivalently, $\operatorname{ind}_-(1-T^*T)$ is zero). A contractive module is said to be *isometric* if $1-T^*T = 0$.

We shall assume that all module maps are continuous and contractive; that is, if $\alpha : \mathcal{H}_A \to \mathcal{K}_B$ is a module map, then as an operator from the Kreĭn space $\mathcal{H}_A$ to the Kreĭn space $\mathcal{K}_B$, α is a contraction (with respect to the indefinite inner products). Note that if α and β are module maps, then

$$1-(\alpha\beta)^*(\alpha\beta) = 1-\alpha^*\alpha+\alpha^*(1-\beta^*\beta)\alpha.$$

Hence it is clear that composition of module maps is contractive, and so is also a module map according to our definition. Thus the contractive Kreĭn modules with contractive module maps as morphisms form a well-defined category.

This category has a shortcoming. There is of course a zero module 0, but if $\mathcal{H}_T$, $\mathcal{K}_R$ are any other Kreĭn modules, the requirement that all module maps be contractive means that the zero morphism $0_{\mathcal{K}_R,\mathcal{H}_T}$ is defined only if $\mathcal{K}_R$ is a Hilbert space. This would seem at first to present problems in defining kernels and cokernels, module extensions, and exact sequences. Recall that a morphism κ in a category with zero morphisms is called a *kernel* of a morphism α if $\alpha\kappa = 0$, and if $\alpha\psi = 0$, then there is a unique morphism ψ' such that $\psi = \kappa\psi'$. On the other hand, κ is a *cokernel* if $\kappa\alpha = 0$, and if $\psi\alpha = 0$, then there is a unique morphism ψ' such that $\psi = \psi'\kappa$.

We can carry the construction of kernels over to the present setting by noting that the kernel of a contraction operator between two Kreĭn spaces is a Hilbert space. Thus if $\alpha : \mathcal{H}_T \to \mathcal{G}_S$ is a contractive module map (and so a contraction operator between $\mathcal{H}_T$ and $\mathcal{G}_S$), and if $\mathcal{K}_R$ is a contractive module with $\mathcal{K}_R$ isometrically isomorphic as a Kreĭn space to the kernel of α, then any isometry $\kappa : \mathcal{K}_R \to \mathcal{H}_T$ mapping $\mathcal{K}_R$ onto the kernel of α is a kernel in the categorical sense, and all kernels have this form. For suppose $\psi : \mathcal{F}_Q \to^H_T$ is another contractive module map with $\alpha\psi = 0$. The fact that ψ is contractive and the range of ψ is a contained in a Hilbert space means that $\mathcal{F}_Q$ is a Hilbert space. The range of ψ is contained in the range of κ, so by Douglas' lemma, there is an operator ψ' such that $\psi = \kappa\psi'$. The kernel of κ being zero is a necessary and sufficient condition for ψ' to be uniquely defined. Also since κ is isometric and the range of κ is a Hilbert space, $\psi' = \kappa^*\psi$ is a contraction. Observe that choosing κ to be an isometry was necessitated by the desire that ψ' be a contraction. The last thing we must verify is that ψ' is a morphism (that is, $\psi' Q = R\psi'$). But

$$\kappa\psi' Q = \psi Q = T\psi = T\kappa\psi' = \kappa R\psi',$$

which since the $\ker \kappa = \{0\}$, implies $\psi' Q = R\psi'$, as wanted.

The construction of cokernels is not generally possible under the present circumstances. This is because if $\alpha : \mathcal{H}_T \to \mathcal{G}_S$ is a contractive module map and $\mathcal{H}_T$ is not a Hilbert space, then the range of α contains a nontrivial negative subspace, which would be in the kernel of any cokernel. However as was mentioned above, the kernel of any contractive map is a Hilbert space. In any case we may define a cokernel in the special situation that the closure of the range of α is a Hilbert space (which implies that $\mathcal{H}_T$ is a Hilbert space). An important example where this happens is when α is a kernel. By arguments like those given in the last paragraph, one shows that $\kappa : \mathcal{G}_S \to \mathcal{C}_R$ may be chosen as a contractive coisometry. To see that all cokernels have this form, first note that if κ is a cokernel, then for all ψ such that $\psi\alpha = 0$, the uniqueness of ψ' such that $\psi = \psi'\kappa$ means that the closure of the range of κ is all of $\mathcal{C}_R$. In particular, if ψ itself is a coisometric cokernel, then the ranges of ψ^* and κ^* must both be equal to the orthogonal complement of the range of α, and ψ' is one-to-one from the range of κ onto the range of ψ. Also $1 - \psi^*\psi = P$, where P is the orthogonal projection onto the orthogonal complement of the range of α. But we also have

$$P = 1 - (\psi'\kappa)^*(\psi'\kappa) = (1 - \kappa^*\kappa) + \kappa^*(1 - \psi'^*\psi')\kappa.$$

Both of the terms on the right side are positive by assumption, and the range of κ^* is orthogonal to the range of P, so as a consequence, $\kappa^*(1 - \psi'^*\psi')\kappa = 0$, and hence $1 - \kappa^*\kappa = P$. In other words, κ is coisometric.

We will want to consider extensions of modules. In an abelian category these are formulated in terms of short exact sequences; that is a sequence of the form

$$B \xrightarrow{\kappa} C \xrightarrow{\nu} A,$$

where κ is a kernel and ν is a cokernel. We say that C is an *extension* of A by B.

This notion holds in a restricted sense in the present context. In this case we have a diagram of the form

$$\mathcal{L}_X \xrightarrow{\kappa} \mathcal{K}_A \xrightarrow{\nu} \mathcal{H}_T\ , \tag{1}$$

where $\mathcal{L}_X$ is a Hilbert space, κ is an isometry, ν is a contractive coisometry, and the range of κ is the kernel of ν. In other words, A is isometrically isomorphic to a contractive operator of the form

$$A_0 = \begin{pmatrix} T & 0 \\ Q & X \end{pmatrix} : \mathcal{H}_T \oplus \mathcal{L}_X \to \mathcal{H}_T \oplus \mathcal{L}_X, \tag{2}$$

where Q is chosen so that the resulting operator is a contraction. (In [5] it is shown that $Q = \tilde{D}_T^* Y D_X$, where $\tilde{D}_T$ and D_X are defect operators for T and X^*, and Y is a Hilbert space contraction and that all Q of this form make the above operator matrix into a contraction.)

We will be most interested in the case where $\mathcal{K}_A$ is isometric. In this case, it is said to be a *minimal isometric extension* of $\mathcal{H}_T$ if $\bigvee_0^n A_0^n \mathcal{H}_T = \mathcal{H}_T \oplus \mathcal{L}_X$. It is a standard result (see [12]) that all minimal isometric extensions are isometrically isomorphic. Furthermore, any contractive module has a minimal isometric extension. It is not difficult to see that if $\mathcal{K}_A$ and $\mathcal{H}_T$ are both isometric, then $Q = 0$ in (2). As a consequence, in this situation the sequence in (1) splits: that is, there is a contractive module map $\nu' : \mathcal{H}_T \to \mathcal{K}_A$ with $\nu\nu' = 1_{\mathcal{H}_T}$. Indeed $\nu' = \nu^*$ is an isometry.

According to the next lemma, any two isometric extensions are in a sense equivalent.

Lemma. *Let $\tilde{\mathcal{H}}_V$ and $\tilde{\mathcal{H}}_W$ be isometric extensions of $\mathcal{H}_T$ (that is, there are coisometric contractive module maps π_V and π_W from $\tilde{\mathcal{H}}_V$ and $\tilde{\mathcal{H}}_W$ to $\mathcal{H}_T$). Then there exists a partially isometric contractive module map $\beta : \tilde{\mathcal{H}}_V \to \tilde{\mathcal{H}}_W$ such that the adjoint $\beta^* : \tilde{\mathcal{H}}_W \to \tilde{\mathcal{H}}_V$ is a module map and the following diagram commutes:*

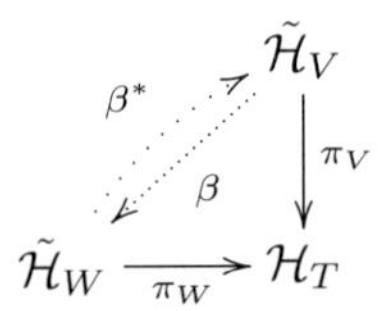

If $\tilde{\mathcal{H}}_V$ (resp., $\tilde{\mathcal{H}}_W$) is a minimal extension, then β (resp., β^*) is an isometric module map.

Proof. It suffices to consider the case where $\tilde{\mathcal{H}}_V$ is a minimal isometric extension and show that β is an isometric module map, since the more general statement follows from applying this special case twice.

So take $\tilde{\mathcal{H}}_V$ to be a minimal isometric extension. Without loss of generality, we may assume that π_W is a projection. Then W has the form

$$W = \begin{pmatrix} T & 0 \\ * & * \end{pmatrix} : \mathcal{H}_T \oplus (\tilde{\mathcal{H}}_W \ominus \mathcal{H}_T) \to \mathcal{H}_T \oplus (\tilde{\mathcal{H}}_W \ominus \mathcal{H}_T). \tag{2.5}$$

Let $\tilde{\mathcal{H}}_U = \bigvee_0^\infty W^n \mathcal{H}_T$. Then $\tilde{\mathcal{H}}_U$ is invariant under W and $U = W|\tilde{\mathcal{H}}_T$ is an isometry. We need to verify that $\tilde{\mathcal{H}}_U$ is a Kreĭn space. But the orthogonal complement of $\tilde{\mathcal{H}}_U$ in $\tilde{\mathcal{H}}_W$ is contained in the orthogonal complement of $\mathcal{H}_T$ in $\tilde{\mathcal{H}}_W$, which is a Hilbert space; and so is itself a Hilbert space (a particular kind of Kreĭn space). It follows that $\tilde{\mathcal{H}}_U$ is a Kreĭn space since it is the orthogonal complement of a Kreĭn space.

Furthermore, it is clear that $\tilde{\mathcal{H}}_U$ extends $\mathcal{H}_T$, and by construction is a minimal isometric extension. By the invariance of $\tilde{\mathcal{H}}_U$ under W, the map β^* defined as the orthogonal projection from $\tilde{\mathcal{H}}_W$ onto $\tilde{\mathcal{H}}_U$ is a coisometric contractive module map. It is contractive since $\tilde{\mathcal{H}}_W \ominus \tilde{\mathcal{H}}_U$ is a Hilbert space. In order to show that β is a module map, it suffices to show that $\tilde{\mathcal{H}}_U$ is invariant under W^*. By the definition of $\tilde{\mathcal{H}}_U$ and the fact that W is isometric, this reduces to showing that $W^*\mathcal{H}_T \subseteq \mathcal{H}_T$. But this is obvious upon examining (2.5). □

Commutant lifting

The commutant lifting problem may be abstractly formulated as follows. Suppose $\{\mathcal{H}_1, T_1\}$, $\{\mathcal{H}_2, T_2\}$, $\{\tilde{\mathcal{H}}_1, V_1\}$, and $\{\tilde{\mathcal{H}}_2, V_2\}$ are modules in our category and that we have the following diagram, where π_1 and π_2 are cokernels (that is, contractive coisometric module maps), and α is a contractive module map:

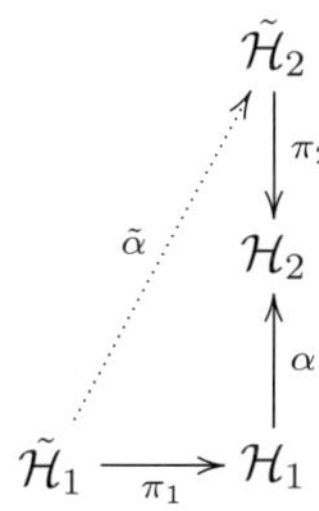

We wish to find $\tilde{\alpha}$ so that the diagram commutes. This can be changed to an equivalent problem where we have

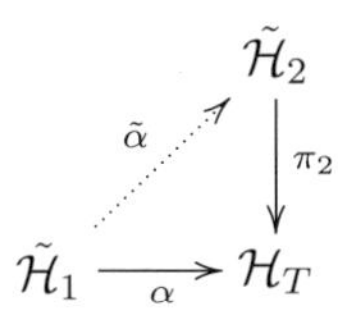

with π a cokernel, and α, $\tilde{\alpha}$ contractive module maps. (The transcription between the two diagrams in one direction is found by replacing $\mathcal{H}_2$ by $\mathcal{H}_T$ and $\alpha\pi_1$ by α, and in the other by taking $\mathcal{H}_1 = \tilde{\mathcal{H}}_1$, $\mathcal{H}_2 = \mathcal{H}_T$, and $\pi_1 = \text{id}$, the identity map.)

The key ingredient in proving that the map $\tilde{\alpha}$ exists is the following factorization theorem. It is roughly akin to the notion from abelian categories of the *analysis* of a morphism. It might also be viewed as a representation theorem. In our setting, it is a consequence of the existence of a solution to an operator matrix completion problem.

Factorization Theorem. *Let $\{\mathcal{H}_1, T_1\}$ and $\{\mathcal{H}_2, T_2\}$ be contractive Kreĭn modules, $\alpha : \mathcal{H}_1 \to \mathcal{H}_2$ a contractive module map.*

(1) *There exists a contractive Kreĭn module $\mathcal{H}_C$, an isometric module map $\phi : \mathcal{H}_1 \to \mathcal{H}_C$, and a contractive coisometric module map $\pi : \mathcal{H}_C \to \mathcal{H}_2$ such that the following diagram commutes:*

(3)

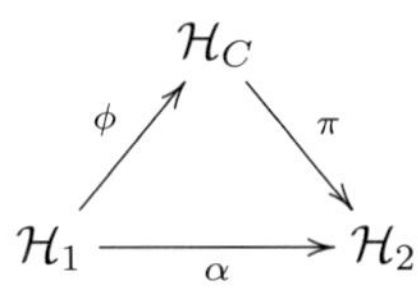

(2) *If, in addition, the module $\mathcal{H}_1$ is isometric, then $\mathcal{H}_C$ can be chosen isometric.*

Proof. We begin with (1). As a Kreĭn space we set $\mathcal{H}_C = \mathcal{H}_2 \oplus \tilde{\mathcal{D}}_\alpha$, where $\tilde{\mathcal{D}}_\alpha$ is a defect space for α. That is, there is an operator $\tilde{D}_\alpha : \tilde{\mathcal{D}}_\alpha \to \mathcal{H}_1$ with zero kernel such that $\tilde{D}_\alpha \tilde{D}_\alpha^* = 1 - \alpha^*\alpha$. Note that since α is a contraction, $\tilde{\mathcal{D}}_\alpha$ is a Hilbert space.

An obvious candidate for the map π is the projection from $\mathcal{H}_2 \oplus \tilde{\mathcal{D}}_\alpha$ onto $\mathcal{H}_2$. Since π is to be a module map, it should be contractive. But since $\ker \pi = \tilde{\mathcal{D}}_\alpha$ is a Hilbert space, this is automatic. Whatever C is, it must satisfy $\pi C = T_2 \pi$. That is, C has the form

$$\begin{pmatrix} T_2 & 0 \\ * & * \end{pmatrix}.$$

Since we want the diagram (3) to commute, ϕ should have the form $\begin{pmatrix} \alpha \\ * \end{pmatrix}$. We also want ϕ to be isometric, so the simplest choice is to set

$$\phi = \begin{pmatrix} \alpha \\ \tilde{D}_\alpha^* \end{pmatrix}.$$

For ϕ to be a module map, we need $\phi T_1 = C\phi$. This fixes the form of C on the range of ϕ. It also agrees with the assumption that α and π are module maps, since

(4) $$\alpha T_1 = \pi \phi T_1 = \pi C \phi = T_2 \pi \phi = T_2 \alpha.$$

Viewing C as a 2×2 operator matrix from $\mathcal{H}_2 \oplus \tilde{\mathcal{D}}_\alpha$ to $\operatorname{ran} \phi \oplus (\operatorname{ran} \phi)^\perp$, we see that we have so far specified the top row and left column, and (4) ensures that these agree at the (1,1) entry. The only thing left unspecified is the (2,2) entry, and we are free to choose it as we like so long as the resulting operator is a contraction. By Theorem 2.4.4 of [12], this can be done.

We next turn to (2). From part (1), we know there exists some contractive module $\mathcal{H}_C$ and isometric and coisometric module maps ϕ and π making the diagram in (3) commutes. Suppose that $\tilde{\mathcal{H}}_V$ is an isometric extension of $\mathcal{H}_C$.

Without loss of generality, we may take the module map $\pi_C : \tilde{\mathcal{H}}_V \to \mathcal{H}_C$ to be a projection. Clearly $\tilde{\pi} = \pi_C \pi$ is a module map from $\tilde{\mathcal{H}}_V$ to $\mathcal{H}_C$.

Since T_1 is assumed to be isometric and $\phi T_1 = C\phi$ with ϕ isometric, we see that $C|\operatorname{ran}\phi$ is isometric. Now V_C has the form

$$V_C = \begin{pmatrix} C|\operatorname{ran}\phi & * \\ V_1 & * \end{pmatrix}.$$

Observe that $\operatorname{ran} V_1$ is contained in a Hilbert space and the first column of V_C is contractive, hence

$$1_{\operatorname{ran}\phi} - \begin{pmatrix} (C|\operatorname{ran}\phi)^* & V_1^* \end{pmatrix} \begin{pmatrix} C|\operatorname{ran}\phi \\ V_1 \end{pmatrix} = -V_1^* V_1 \geq 0,$$

and so $V_1 = 0$. It follows that the isometry

$$\tilde{\phi} = \begin{pmatrix} \phi \\ 0 \end{pmatrix} : \mathcal{H}_1 \to \mathcal{H}_C \oplus (\tilde{\mathcal{H}}_V \ominus \mathcal{H}_C)$$

is a module map from $\mathcal{H}_1$ to $\tilde{\mathcal{H}}_V$, and the diagram

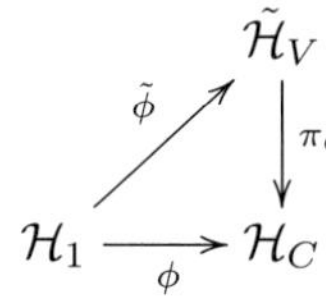

commutes. Consequently the diagram

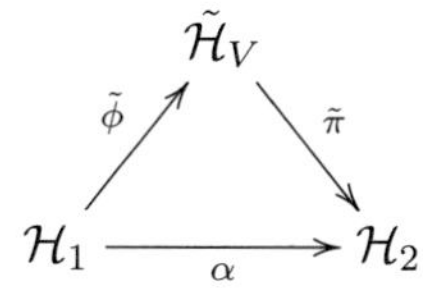

commutes. □

We are now in a position to solve the commutant lifting problem.

Commutant Lifting Theorem. *Suppose that $\mathcal{H}_T$ is a contractive Kreĭn module, $\{\tilde{\mathcal{H}}_1, V_1\}$, and $\{\tilde{\mathcal{H}}_2, V_2\}$ are isometric Kreĭn modules, and that we have the following diagram, where π_2 is a coisometric contractive module map, α a contractive module map:*

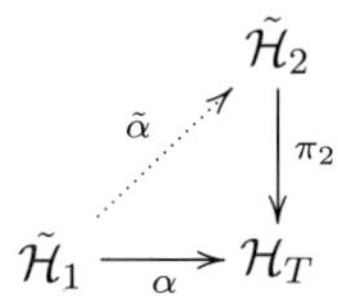

Then there exists a contractive module map $\tilde{\alpha}$ so that the diagram commutes.

Proof. By part (2) of the Factorization Theorem, we may factor the map α via

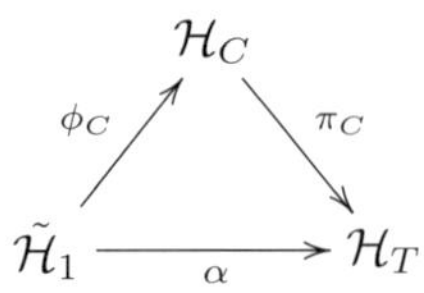

where ϕ_C and π_C are isometric and coisometric module maps, and $\mathcal{H}_C$ is an isometric module. Now applying the lemma, we find a partially isometric module map β so that the following diagram commutes:

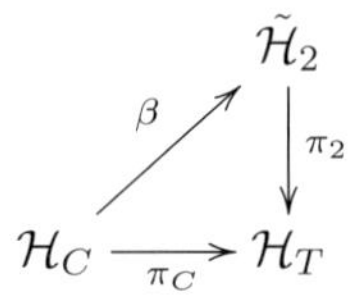

Combining the two diagrams and letting $\tilde{\alpha} = \beta\phi_C$ finishes the proof. □

Projective modules

We say that a module $\mathcal{H}_T$ is *projective* if for any module $\mathcal{H}_X$, module map $\alpha : \mathcal{H}_T \to \mathcal{H}_X$, and module $\mathcal{H}_C$ and cokernel $\pi : \mathcal{H}_C \to \mathcal{H}_X$, there exists a module map $\tilde{\alpha}$ making the following diagram commute:

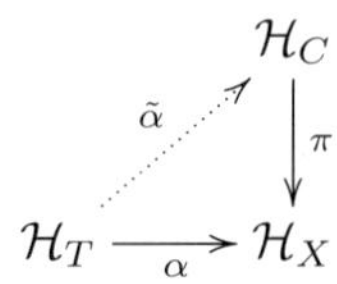

The next theorem describes all of the projective modules in the category of contractive Kreĭn modules. For similar results over Hilbert modules, see for example, [4], [7], [6], [10], and [16]. Our approach follows that of [6].

Theorem 2. *The module $\mathcal{H}_T$ is projective module in the category of contractive Kreĭn modules if and only if it is isometric.*

Proof. We first show that an isometric module is projective. To this end, suppose $\mathcal{H}_T$ is an isometric module, $\mathcal{H}_X$ a contractive module, and $\alpha : \mathcal{H}_T \to \mathcal{H}_X$ a module map. Furthermore, assume $\mathcal{H}_C$ is a contractive module and $\pi_C : \mathcal{H}_C \to \mathcal{H}_X$ is a coisometric module map. Let $\tilde{\mathcal{H}}_W$ be an isometric extension of $\mathcal{H}_C$, with $\pi_W : \tilde{\mathcal{H}}_W \to \mathcal{H}_C$ the corresponding cokernel. Then $\pi_C\pi_W$ is a contractive coisometric module map, and by the commutant lifting theorem, there exists a map α_W making the following diagram commute:

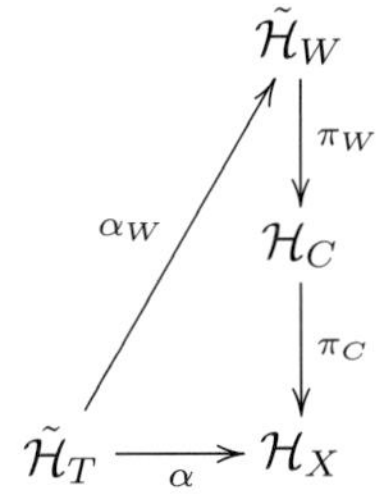

By letting $\tilde{\alpha} = \pi_W \alpha_W$, we see that $\mathcal{H}_T$ is projective.

On the other hand, suppose $\mathcal{H}_T$ is projective. Let $\tilde{\mathcal{H}}_V$ be an isometric extension of $\mathcal{H}_T$. Then there is a module map $\tilde{\alpha}$ so that the following diagram commutes:

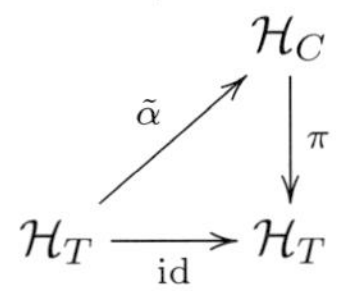

where id is the identity map on $\mathcal{H}_T$. So id $= \pi\tilde{\alpha}$, and consequently, $\tilde{\alpha}^* | \operatorname{ran} \pi^* = \pi$. Since $\tilde{\alpha}$ is a contraction and π is a coisometry,

$$0 \leq 1 - \tilde{\alpha}^*\tilde{\alpha} = 1 - \pi\pi^* - \tilde{\alpha}^* P_{(\operatorname{ran} \pi_V)^\perp} \tilde{\alpha} = -\tilde{\alpha}^* P_{(\operatorname{ran} \pi_V)^\perp} \tilde{\alpha},$$

and since the range of $P_{(\operatorname{ran} \pi_V)^\perp} \tilde{\alpha}$ is a Hilbert space, this operator is zero. Hence V is isometrically isomorphic to an operator containing T as a direct summand, and so T is an isometry. □

Defect indices for adjoints of maps

Let $A \in \mathbf{L}(\mathcal{H})$ be a selfadjoint operator on a Kreĭn space. Recall that $\operatorname{ind}_\pm(A)$ can be defined in terms of a factorization of A. Namely, every selfadjoint operator A may be factored as $A = DD^*$, where $D \in \mathbf{L}(\mathcal{D}, \mathcal{H})$ with $\ker D = \{0\}$, $\mathcal{D}$ a Kreĭn space. The space $\mathcal{D}$ has a fundamental decomposition $\mathcal{D} = \mathcal{D}_- \oplus \mathcal{D}_+$, where $\mathcal{D}_+$ is a Hilbert space and $\mathcal{D}_-$ is a Hilbert space with the norm replaced by its negative. Then $\operatorname{ind}_\pm(A) = \dim \mathcal{D}_\pm$. We also use $\operatorname{ind}_\pm(\mathcal{D})$ for the dimensions of $\mathcal{D}_\pm$ in the fundamental decomposition of a Kreĭn space $\mathcal{D}$. The operator A being positive corresponds to $\operatorname{ind}_-(A) = 0$. As we noted earlier, a contraction T on a Kreĭn space is characterized by $\operatorname{ind}_-(1 - T^*T) = 0$. Such an operator need not have a contractive adjoint. When it does, it is termed a bicontraction.

In [11], it was first shown that the commutant lifting theorem in Kreĭn spaces is valid if the intertwining operator is a bicontraction, in which case the operator lifts to a bicontraction. In the notation of the lifting theorem given above, this says that if $\operatorname{ind}_-(1 - \alpha\alpha^*) = 0$, then $\operatorname{ind}_-(1 - \tilde{\alpha}\tilde{\alpha}^*) = 0$. As we shall see, this is part of a more general result on the preservation of negative indices of maps during liftings.

Theorem 3. *If $A = V^*B$, with A, B, and V continuous Kreĭn space operators and V^* a contractive coisometry. Furthermore, assume*

$$\operatorname{ind}_-(1 - A^*A) = \operatorname{ind}_-(1 - B^*B).$$

Then

$$\operatorname{ind}_-(1 - AA^*) = \operatorname{ind}_-(1 - BB^*).$$

Proof. By multiplying through by V in $A = V^*B$, the equation becomes $C = PB$, where P is a contractive projection, and $C = VA$ satisfying $\operatorname{ind}_-(1 - C^*C) = \operatorname{ind}_-(1 - A^*A)$ and $\operatorname{ind}_-(1 - CC^*) = \operatorname{ind}_-(1 - AA^*)$ (this is readily verified by using the fact that V is an isometry and the kernel of V^* is a Hilbert space). Hence it suffices to prove the theorem in the case that V^* is a contractive projection. In this case $B \in \mathbf{L}(\mathcal{H}, \mathcal{K})$ decomposes as

$$B = \begin{pmatrix} A \\ R \end{pmatrix} : \mathcal{H} \longrightarrow (\ker V^*)^\perp \oplus \ker V^*.$$

Form the isometric extension

$$\begin{aligned} \tilde{B} &= \begin{pmatrix} B \\ \tilde{D}_B^* \end{pmatrix} \\ &= \begin{pmatrix} A \\ \begin{pmatrix} R \\ \tilde{D}_B^* \end{pmatrix} \end{pmatrix} \\ &= \begin{pmatrix} A \\ W\tilde{D}_A^* \end{pmatrix}, \end{aligned}$$

where we view $\tilde{B}$ as an isometric extension of A, and then use Theorem 2.12 of [13] to express the operator by which A is being extended in terms of a defect operator $\tilde{D}_A^*$ of A (that is, $\tilde{D}_A\tilde{D}_A^* = 1 - A^*A$, $\tilde{D}_A : \tilde{\mathcal{D}}_A \to \mathcal{H}$ with zero kernel) and an operator W which is a bicontractive isometry. Also by this same theorem, we have that $\operatorname{ind}_-(1 - \tilde{B}\tilde{B}^*) = \operatorname{ind}_-(1 - BB^*)$ when viewing $\tilde{B}$ as an extension of B, and $\operatorname{ind}_-(1 - \tilde{B}\tilde{B}^*) = \operatorname{ind}_-(1 - AA^*) + \operatorname{ind}_-(1 - WW^*) = \operatorname{ind}_-(1 - AA^*)$, viewing $\tilde{B}$ as an extension of A. The result then follows. □

As a consequence of this theorem, any time we have a commutative diagram as in the commutant lifting theorem

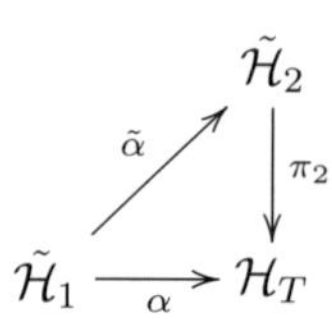

where π_2 is a coisometric contractive module map and $\operatorname{ind}_-(1 - \tilde{\alpha}^*\tilde{\alpha}) = \operatorname{ind}_-(1 - \alpha^*\alpha)$ (a particular case being when this index is zero, as in the version of the commutant lifting theorem proved above), then $\mathrm{ind}_-(1 - \tilde{\alpha}\tilde{\alpha}^*) = \mathrm{ind}_-(1 - \alpha\alpha^*)$. Thus we have the following corollary.

Commutant Lifting Theorem, II. *Suppose that $\mathcal{H}_T$ is a contractive Kreĭn module, $\{\tilde{\mathcal{H}}_1, V_1\}$, and $\{\tilde{\mathcal{H}}_2, V_2\}$ are isometric Kreĭn modules, and that we have the following diagram, where π_2 is a coisometric contractive module map, α a contractive module map:*

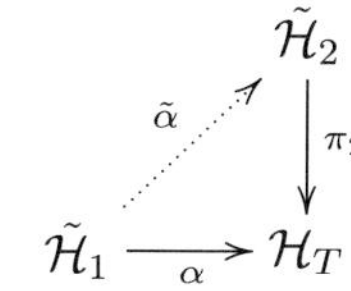

Then there exists a contractive module map $\tilde{\alpha}$ so that the diagram commutes and $\mathrm{ind}_-(1 - \tilde{\alpha}\tilde{\alpha}^*) = \mathrm{ind}_-(1 - \alpha\alpha^*)$. *In particular, if α is bicontractive, then so is $\tilde{\alpha}$.*

References

[1] R. Arocena, *Unitary extensions of isometries and contractive intertwining dilations*, Oper. Theory: Adv. Appl., vol. 41, Birkhäuser, Basel-Boston, 1989, pp. 13–23; MR 91c:47009.

[2] R. Arocena, T. Ya. Azizov, A. Dijksma, and S. Marcantognini, *On commutant lifting with finite defect*, J. Operator Theory **35** (1996), 117–132; MR 97d:47008.

[3] R. Arocena, T. Ya. Azizov, A. Dijksma, and S. A. M. Marcantognini, *On commutant lifting with finite defect. II*, J. Funct. Anal. **144** (1997), 105–116.

[4] J. F. Carlson, D. N. Clark, C. Foias,J. P. Williams, *Projective Hilbert* **A(D)***-modules*, New York J. Math. **1** (1994), 26–38; MR 95g:46136.

[5] G. Christner, Kin Y. Li, and J. Rovnyak, *Julia Operators and coefficient problems*, Oper. Theory Adv. Appl., vol. 73, Birkhäuser, Basel-Boston, 1994, pp. 138–181; MR 95m:47058.

[6] R. S. Clancy and S. McCullough, *Projective modules and Hilbert spaces with a Nevalinna-Pick kernel*, Proc. Amer. Math. Soc. (to appear).

[7] D. N. Clark, *Projectivity and lifting of Hilbert module maps*, Ann. Polon. Math. **66** (1997), 43–48.

[8] A. Dijksma, M. A. Dritschel, S. Marcantognini, and H. S. V. de Snoo, *The commutant lifting theorem for contractions on Krein spaces*, Oper. Theory: Adv. Appl., vol. 61, Birkhäuser, Basel-Boston, 1993, pp. 65–83; MR 94m:47068.

[9] R. G. Douglas and C. Foias, *A homological view in dilation theory* (1976); (unpublished manuscript).

[10] R. G. Douglas and V. I. Paulsen, *Hilbert modules over function algebras*, Pitman Research Notes in Mathematics, 217. Longman Scientific & Technical, Harlow, 1989; MR 91g:46084.

[11] M. A. Dritschel, *A lifting theorem for bicontractions on Kreĭn spaces*, J. Funct. Anal. **89** (1990), 61–89; MR 91c:47066.

[12] M. A. Dritschel and J. Rovnyak, *Extension theorems for contraction operators on Kreĭn spaces*, Oper. Theory: Adv. Appl., vol. 47, Birkhäuser, Basel-Boston, 1990, pp. 221–305; MR 92m:47068.

[13] M. A. Dritschel and J. Rovnyak, *Operators on indefinite inner product spaces*, Lectures on operator theory and its applications (Waterloo, ON, 1994), Fields Inst. Monogr. vol. 3, Amer. Math. Soc., Providence, RI, 1996, pp 141–232; MR 96k:47001.

[14] S. H. Ferguson, *Polynomially bounded operators and* Ext *groups*, Proc. Amer. Math. Soc. **124** (1996), 2779–2785; MR 96k:47012.

[15] A. Gheondea, *Contractive intertwining dilations of quasi-contractions*, Z. Anal. Anwendungen **15** (1996), 31–44; MR 97e:47030.

[16] P. S. Muhly and B. Solel, *Hilbert modules over operator algebras*, Mem. Amer. Math. Soc. **117** (1995); MR 96c:47060.

[17] P. J. Hilton and U. Stammbach, *A course in homological algebra, 2nd ed.*, Graduate Texts in Mathematics, vol. 4, Springer-Verlag, New York, 1997; MR 97k:18001.

Department of Statistics, Purdue University, West Lafayette, Indiana 47907-1399
Current address: Department of Mathematics, The University of Newcastle, Newcastle upon Tyne, England NE1 7RU
E-mail address: `m.a.dritschel@ncl.ac.uk`

Operator Theory:
Advances and Applications, Vol. 123

On Dual Canonical Systems and Dual Matrix String Equations

Harry Dym and Lev A. Sakhnovich

Dedicated to M. S. Livšic with respect, admiration and affection

Abstract. A class of canonical systems and dual canonical systems which is equivalent to a class of matrix string equations and dual matrix string equations, respectively, under appropriate smoothness conditions is introduced. This serves to generalize the notion of dual string equations which was introduced by Kac and Krein for scalar strings some thirty years ago. A complete description of the set of spectral functions $\tau(\mu)$ on $[0, \infty)$ for the original matrix string equation such that $\widetilde{\tau}(\mu) = \int_0^\mu \lambda d\tau(\lambda)$ is a spectral function for the dual matrix string is furnished.

1. Introduction

The string equation

$$-\frac{d^2\varphi}{dx^2}(x,\lambda) = \lambda\rho^2(x)\varphi(x,\lambda) \ , \ 0 \le x \le \ell \ , \tag{1.1}$$

with smooth mass distribution

$$M(x) = \int_0^x \rho^2(t)dt \ , \ \rho(t) \ge 0 \ , \tag{1.2}$$

can be written in the form

$$-\frac{d^2\varphi}{dx^2}(x,\lambda) = \lambda\frac{dM}{dx}\varphi(x,\lambda) \ . \tag{1.3}$$

If $\rho(t) > 0$ for a.e. $t \in (0, \ell)$, then x can be viewed as a function of M and the equation

$$-\frac{d^2\widetilde{\varphi}}{dM^2}(M,\lambda) = \lambda\frac{dx}{dM}\widetilde{\varphi}(M,\lambda) \ , \ 0 \le M \le M(\ell) \ , \tag{1.4}$$

1991 *Mathematics Subject Classification.* 34B20, 34B24, 34L05, 47A48.

The authors would like to thank the Israel-Ukraine Exchange Program for supporting Lev A. Sakhnovich's visit to the Weizmann Institute, where the work described in this paper was carried out. H. Dym would also like to thank Renee and Jay Weiss for endowing the chair which supports his research.

is said to be dual to equation (1.3). The notion of a dual string was introduced and investigated by I. S. Kac and M. G. Krein [KaKr2] for the general string equation with a general mass distribution function $M(x)$; see also Sections 6.8–6.9 of [DMcK]. For applications of dual strings to prediction theory and oscillation, see the discussion and cited references in Section 12.6 of [KaKr2].

Kac and Krein obtained the dual string equation (1.4) from the original string equation (1.3) by interchanging the variables x and $M(x)$. They proved that there exists a monotone nondecreasing function $\tau(\mu)$ on the halfaxis $[0, \infty)$ such that:

(1) $\tau(\mu)$ is a spectral function for the string equation (1.3) subject to the initial conditions

$$\varphi(0, \lambda) = 1 \text{ and } \varphi'(0, \lambda) = 0 \,. \tag{1.5}$$

(2) $\widetilde{\tau}(\mu) = \int_0^\mu \lambda d\tau(\lambda)$ is a spectral function for the dual string equation (1.4) subject to the initial conditions

$$\widetilde{\varphi}(0, \lambda) = 0 \text{ and } \widetilde{\varphi}'(0, \lambda) = 1 \,. \tag{1.6}$$

In this paper we shall generalize the notion of dual equations to a class of canonical systems. From these canonical systems we shall obtain a matrix string equation and a dual matrix string equation. In particular, we shall show that if $p(x)$ and $q(x)$ are continuous, invertible $m \times m$ mvf's on the interval $[0, \ell)$ with square summable entries, then the differential system

$$-(p^{-1}\frac{d}{dx}q^{-1})(q^{-*}\frac{d}{dx}p^{-*})\varphi = \lambda\varphi \tag{1.7}$$

$$(p^{-*}\varphi)(0, \lambda) = I_m \,, \; (p^{-*}\varphi)'(0, \lambda) = 0 \tag{1.8}$$

plays the role of the original string equation (1.3) subject to the initial conditions (1.5), whereas the differential system

$$-(q^{-*}\frac{d}{dx}p^{-*})(p^{-1}\frac{d}{dx}q^{-1})\widetilde{\varphi} = \lambda\widetilde{\varphi} \tag{1.9}$$

$$(q^{-1}\widetilde{\varphi})(0, \lambda) = 0 \,, \; (q^{-1}\widetilde{\varphi})'(0, \lambda) = (pp^*)(0) \tag{1.10}$$

plays the role of the dual string equation (1.4) subject to the initial conditions (1.6).[1] Moreover, we shall give a complete description of the set of $m \times m$ monotone mvf's $\tau(\mu)$ on the halfaxis $[0, \infty)$ such that $\tau(\mu)$ is a spectral mvf for the system (1.7)–(1.8) and $\widetilde{\tau}(\mu) = \int_0^\mu \lambda d\tau(\lambda)$ is a spectral mvf for the system (1.9)–(1.10). To the best of our knowledge, this result is new even in the scalar case.

Notice that the original differential equation (1.7) may be obtained from the dual differential equation (1.9) by interchanging $p(t)$ and $q(t)^*$, and conversely. However, the stated conclusions depend also on the change in initial conditions in going from one system to the other.

---

[1]For additional discussion of this pair of equations, see Remark 7.1.

The analysis is based on the construction of a pair of operator identities which is then used to generate a pair of canonical systems, the original and its dual. These are then shown to be equivalent to the pair of matrix string equations (1.7)–(1.8) and (1.9)–(1.10), under appropriate assumptions.

The analysis referred to above is a particular application of a general abstract scheme which can be formulated in terms of a pair of complex separable Hilbert spaces G and H and two sets of bounded linear operators $B, C : H \longrightarrow H$ and $\Pi_1, \Pi_2 : G \longrightarrow H$ which satisfy the operator identity

$$B^* - C = \Pi_2 \Pi_1^* \ . \tag{1.11}$$

It is then readily checked that

$$CB - B^*C^* = B^*\Pi_1\Pi_2^* - \Pi_2\Pi_1^*B \tag{1.12}$$

and

$$BC - C^*B^* = \Pi_1\Pi_2^*B^* - B\Pi_2\Pi_1^* \ . \tag{1.13}$$

In the sequel we shall fix $G = \mathbb{C}^m$ and $H = L_2^m(0, \ell)$, the set of $m \times 1$ vector valued functions f which are measurable and square integrable on the interval $(0, \ell)$. The string equation and its dual will then emerge from (1.12) and (1.13), respectively, upon making appropriate choices of the operators B, C, Π_1 and Π_2. The analysis depends upon some abstract interpolation theorems, which are summarized in the next section. We then turn to a class of canonical systems for which it is possible to define dual canonical systems and some associated operator identities in Section 3. The passage from dual canonical systems to dual strings is discussed in Section 5. Section 4 summarizes some general information on spectral mvf's for canonical systems of the kind treated here. It is based largely on [Sa1], which in turn incorporates a number of generalizations of the pioneering work of M. S. Livšic; see e.g. [BrLi] for a good introduction to the latter. The main conclusion of this paper is Theorem 6.1, which gives a full description of the set of spectral mvf's $\tau(\mu)$ on $[0, \infty)$ such that $\widetilde{\tau}(\mu)$ (as defined above) is a spectral mvf for the dual string. The proof of this theorem depends essentially upon the analysis of a related interpolation problem in [BoSa]. Some examples are sketched in Section 7.

We remark that the operator identities (1.12) and (1.13) can be reexpressed as

$$A - A^* = i(\Psi_1\Psi_2^* + \Psi_2\Psi_1^*) \tag{1.14}$$

and

$$\widetilde{A} - \widetilde{A}^* = i(\widetilde{\Psi}_1\widetilde{\Psi}_2^* + \widetilde{\Psi}_2\widetilde{\Psi}_1^*) \tag{1.15}$$

respectively, upon setting

$$A = CB \ , \ \ \Psi_1 = B^*\Pi_1 \ , \ \ \Psi_2 = i\Pi_2 \tag{1.16}$$

$$\widetilde{A} = BC \ , \ \ \widetilde{\Psi}_1 = \Pi_1 \ , \ \ \widetilde{\Psi}_2 = iB\Pi_2 \ . \tag{1.17}$$

This choice of operators is not unique. Thus, for example, the identity (1.14) still holds if A is replaced by $A + A_1$ when $A_1 = A_1^*$ and/or if

$$\begin{bmatrix}\Psi_1^* \\ \Psi_2^*\end{bmatrix} \text{ is replaced by } \begin{bmatrix}\Phi_1^* \\ \Phi_2^*\end{bmatrix} = K^* \begin{bmatrix}\Psi_1^* \\ \Psi_2^*\end{bmatrix},$$

when K is a constant J-unitary matrix with respect to the signature matrix

$$J = \begin{bmatrix} 0 & I_m \\ I_m & 0 \end{bmatrix}, \tag{1.18}$$

i.e., $K^*JK = J$. We shall in fact make such a substitution in Section 3 and discuss the connection with the set of operators given above in Section 4.

The notation is pretty standard except perhaps for the symbols

$$f^{\#}(\lambda) = f(\bar{\lambda})^*, \quad \mathbb{R}_+ = [0, \infty)$$

and $\mathbb{C}_+$ [$\mathbb{C}_-$] for the open upper [lower] half plane. We shall also use the abbreviation mvf for matrix valued function and shall let $\{H_1; H_2\}$ denote the set of bounded linear operators from the Hilbert space H_1 into the Hilbert space H_2.

2. Three abstract interpolation theorems

In this section we shall formulate three abstract theorems on interpolation for future use. The first two theorems are adapted from the papers [IvSa] and [BoSa], respectively, both of which in fact contain more general results. However, the presented formulations suffice for the needs at hand and have the added advantage of simplicity. The third theorem is a consequence of the other two.

For the rest of this section we shall fix a complex separable Hilbert space H and shall let $B, C \in \{H; H\}$ be bounded linear operators from H into H and $\Pi_1, \Pi_2 \in \{\mathbb{C}^m; H\}$ be bounded linear operators from $\mathbb{C}^m$ (endowed with the standard inner product) into H such that the basic identity

$$B^* - C = \Pi_2 \Pi_1^* \tag{2.1}$$

holds.

Recall that a pair of $m \times m$ mvf's $\{\mathcal{P}(\lambda), \mathcal{Q}(\lambda)\}$ is said to be a Nevanlinna pair (with respect to $\mathbb{C}_+$) if $\mathcal{P}$ and $\mathcal{Q}$ are meromorphic in $\mathbb{C}_+$ and the two conditions

(1) $\mathcal{P}(\lambda)^*\mathcal{P}(\lambda) + \mathcal{Q}(\lambda)^*\mathcal{Q}(\lambda) > 0$

(2) $\mathcal{P}(\lambda)^*\mathcal{Q}(\lambda) + \mathcal{Q}(\lambda)^*\mathcal{P}(\lambda) \geq 0$

hold at every point $\lambda \in \mathbb{C}_+$ at which both $\mathcal{P}$ and $\mathcal{Q}$ are holomorphic.

Now, for any $2m \times 2m$ mvf

$$\mathfrak{A}(\lambda) = \begin{bmatrix} a(\lambda) & b(\lambda) \\ c(\lambda) & d(\lambda) \end{bmatrix}$$

with blocks of size $m \times m$, define the linear fractional transformation

$$T_{\mathfrak{A}}[\{\mathcal{P}, \mathcal{Q}\}] = \{a(\lambda)\mathcal{P}(\lambda) + b(\lambda)\mathcal{Q}(\lambda)\}\{c(\lambda)\mathcal{P}(\lambda) + d(\lambda)\mathcal{Q}(\lambda)\}^{-1} \tag{2.2}$$

for all points λ at which $\mathfrak{A}$, $\mathcal{P}$ $\mathcal{Q}$ are holomorphic and $c\mathcal{P} + d\mathcal{Q}$ is invertible.

Theorem 2.1. *Assume that the bounded linear operators B, $C \in \{H; H\}$ and Π_1, $\Pi_2 \in \{\mathbb{C}^m; H\}$ satisfy the basic identity (2.1) and that furthermore:*

(1) *The spectrum of the operator CB is concentrated at the origin.*

(2) $\ker\ \Pi_2 = \{0\}$.

(3) $\ker(CB) = \{0\}$.

(4) $\mathrm{range}(CB) \cap \mathrm{range}\ \Pi_2 = \{0\}$.

Let

$$\mathfrak{A}(\lambda) = I_{2m} + \lambda \begin{bmatrix} \Pi_1^* B \\ -i\Pi_2^* \end{bmatrix} (I - \lambda B^* C^*)^{-1} [\Pi_2 \quad -iB^*\Pi_1] \,. \tag{2.3}$$

Then

$$\det\{c(\lambda)\mathcal{P}(\lambda) + d(\lambda)\mathcal{Q}(\lambda)\} \not\equiv 0 \text{ in } \mathbb{C}_+$$

for every Nevanlinna pair $\{\mathcal{P}, \mathcal{Q}\}$ and $iT_{\mathfrak{A}}[\{\mathcal{P}, \mathcal{Q}\}]$ maps the set of Nevanlinna pairs into the class of $m \times m$ mvf's of the form

$$v(\lambda) = \alpha + \int_{-\infty}^{\infty} \left\{ \frac{1}{\mu - \lambda} - \frac{\mu}{1 + \mu^2} \right\} d\tau(\mu) \tag{2.4}$$

where $\alpha = \alpha^*$ and τ is a nondecreasing mvf on $\mathbb{R}$ such that

$$\int_{-\infty}^{\infty} (1 + \mu^2)^{-1} d\tau(\mu) \text{ is finite.} \tag{2.5}$$

Moreover, the following two statements are equivalent for any such α and τ:

(1) $$\alpha + \int_{-\infty}^{\infty} \left\{ \frac{1}{\mu - \lambda} - \frac{\mu}{1 + \mu^2} \right\} d\tau(\mu) = iT_{\mathfrak{A}}[\{\mathcal{P}, \mathcal{Q}\}] \tag{2.6}$$

for some Nevalinna pair $\{\mathcal{P}, \mathcal{Q}\}$.

(2) *The identities*

$$I = \int_{-\infty}^{\infty} (I - \mu CB)^{-1} \Pi_2 d\tau(\mu) \Pi_2^* (I - \mu B^* C^*)^{-1} \tag{2.7}$$

and

$$B^*\Pi_1 = \int_{-\infty}^{\infty} \left\{ CB(I - \mu CB)^{-1} + \frac{\mu I}{1 + \mu^2} \right\} \Pi_2 d\tau(\mu) - \Pi_2 \alpha \tag{2.8}$$

hold.

We remark that the condition

$$\limsup_{r \to \infty} \frac{\ln \left[\max\limits_{0 \le \theta < 2\pi} \|(I - re^{i\theta} CB)^{-1}\| \right]}{r^\rho} < \infty$$

which appears in the formulation of Theorem 4.2 in [IvSa] is automatically fulfilled when $CB - B^*C^*$ has finite dimensional range and the spectrum of CB is concentrated at the origin. The requisite estimates follow from the triangular model of Livšic; see Section 8 of [BrLi] and, for additional discussion of Theorem 2.1, [Sa3].

We further remark that the form of identity (2.8) is a consequence of the two identities (2.1) and (2.7) and assumption (2): If Λ denotes the integral on the right-hand side of (2.8), then, with the aid of (2.7), it is readily checked that

$$\Lambda\Pi_2^* - \Pi_2\Lambda^* = CB - B^*C^* \ .$$

Therefore, by (1.12),

$$\Lambda\Pi_2^* - \Pi_2\Lambda^* = B^*\Pi_1\Pi_2^* - \Pi_2\Pi_1^*B \ .$$

But this in turn implies that

$$\Pi_2(\Pi_1^*B - \Lambda^*) = (B^*\Pi_1 - \Lambda)\Pi_2^*$$

and hence, in view of assumption (2), that

$$B^*\Pi_1 - \Lambda = \Pi_2\alpha_1$$

for some $m \times m$ matrix α_1. Thus, upon substituting this information in the preceding identity, we obtain

$$\Pi_2\alpha_1^*\Pi_2^* = \Pi_2\alpha_1\Pi_2^* \ ,$$

which, in view of assumption (2), leads easily to the conclusion that

$$\alpha_1^* = \alpha_1$$

and hence that

$$B^*\Pi_1 = \Lambda + \Pi_2\alpha_1$$

for some selfadjoint $m \times m$ matrix α_1.

We shall say that a pair $\{\mathcal{P}, \mathcal{Q}\}$ of $m \times m$ mvf's is a Stieltjes pair if $\mathcal{P}(\lambda)$ and $\mathcal{Q}(\lambda)$ are meromorphic in $\mathbb{C}\backslash\mathbb{R}_+$ and if the following three inequalities are fulfilled at every point $\lambda \in \mathbb{C}\backslash\mathbb{R}_+$ at which $\mathcal{P}(\lambda)$ and $\mathcal{Q}(\lambda)$ are holomorphic:

(1) $\mathcal{P}(\lambda)^*\mathcal{P}(\lambda) + \mathcal{Q}(\lambda)^*\mathcal{Q}(\lambda) > 0$.

(2) $\dfrac{\mathcal{P}(\lambda)^*\mathcal{Q}(\lambda) + \mathcal{Q}(\lambda)^*\mathcal{P}(\lambda)}{i(\bar{\lambda} - \lambda)} \geq 0$.

(3) $\dfrac{\bar{\lambda}\mathcal{P}(\lambda)^*\mathcal{Q}(\lambda) + \mathcal{Q}(\lambda)^*\lambda\mathcal{P}(\lambda)}{i(\bar{\lambda} - \lambda)} \geq 0$.

This definition has been changed a little from the definition used in [BoSa] in order to ensure that the Stieltjes pairs are a subclass of the Nevanlinna pairs. For additional information on Stieltjes functions and Stieltjes pairs, see [KaKr1] and [DvKa].

Let

$$\mathfrak{B}(\lambda) = \left\{ I_{2m} + \lambda \begin{bmatrix} \Pi_1^*C^* \\ -\Pi_2^* \end{bmatrix} (I - \lambda B^*C^*)^{-1}[\Pi_2 \ C\Pi_1] \right\} K_1 \ , \tag{2.9}$$

where

$$K_1 = \begin{bmatrix} I_m & \Pi_1^*\Pi_1 \\ 0 & I_m \end{bmatrix} . \tag{2.10}$$

Then, upon combining the implications of Lemma 6.9 and Theorems 2.5, 4.1 and 7.3 in [BoSa], we obtain the following result:

Theorem 2.2. *Assume that the bounded linear operators* $B,\ C \in \{H; H\}$ *and* Π_1, $\Pi_2 \in \{\mathbb{C}^m; H\}$ *satisfy the basic identity (2.1) and that furthermore:*

(1) $I - \mu CB$ *has a bounded inverse for every point* $\mu \in (0, \infty)$.

(2) $\ker\ \Pi_2 = \{0\}$.

(3) $\ker(BC) = \{0\} = \ker(CB)$.

(4) $\mathrm{range}(CB) \cap \mathrm{range}\ \Pi_2 = \{0\}$.

(5) $\mathrm{range}(BC) \cap \mathrm{range}(B\Pi_2) = \{0\}$.

Then for every Stieltjes pair $\{\mathcal{P}, \mathcal{Q}\}$ *the mvf*

$$v(\lambda) = T_{\mathfrak{B}}[\{\mathcal{P}, -i\mathcal{Q}\}] \tag{2.11}$$

admits a representation of the form

$$v(\lambda) = \gamma + \int_0^\infty \frac{1}{\lambda - \mu} d\tau(\mu) , \tag{2.12}$$

where $\gamma \geq 0$ *and* τ *is a nondecreasing* $m \times m$ *mvf on* $[0, \infty)$ *such that*

$$\int_0^\infty (1+\mu)^{-1} d\tau(\mu) \text{ is finite.} \tag{2.13}$$

Moreover, the following two statements are equivalent for any such γ *and* τ:

(1)
$$\gamma + \int_0^\infty \frac{1}{1+\mu} d\tau(\mu) = T_{\mathfrak{B}}[\{\mathcal{P}, -i\mathcal{Q}\}] \tag{2.14}$$

for some Stieltjes pair $\{\mathcal{P}, \mathcal{Q}\}$.

(2) *The following three identities hold:*

$$I = \int_0^\infty (I - \mu CB)^{-1}\Pi_2 d\tau(\mu)\Pi_2^*(I - \mu B^*C^*)^{-1} . \tag{2.15}$$

$$I = B\int_0^\infty (I - \mu CB)^{-1}\Pi_2 \mu d\tau(\mu)\Pi_2^*(I - \mu B^*C^*)^{-1}B^* . \tag{2.16}$$

$$\Pi_1 = \int_0^\infty (I - \mu BC)^{-1} B\Pi_2 d\tau(\mu) . \tag{2.17}$$

In view of the identity (2.1), the formula for $\mathfrak{B}(\lambda)$ can be reexpressed as

$$\mathfrak{B}(\lambda) = K_1 \mathfrak{A}_1(\lambda) ,$$

where

$$\mathfrak{A}_1(\lambda) = I_{2m} + \lambda \begin{bmatrix} \Pi_1^* B \\ -\Pi_2^* \end{bmatrix} (I - \lambda B^*C^*)^{-1}[\Pi_2\ B^*\Pi_1]$$

which agrees with the formula (2.3) for $\mathfrak{A}(\lambda)$ except for some multiples of $\pm i$ in the off-diagonal blocks $b(\lambda)$ and $c(\lambda)$ of $\mathfrak{A}(\lambda)$.

In particular, it turns out that

$$iT_{\mathfrak{A}}[\{\mathcal{P},\mathcal{Q}\}] = T_{\mathfrak{A}_1}[\{\mathcal{P},-i\mathcal{Q}\}] \ ,$$

which is as it should be, since $\{\mathcal{P},\mathcal{Q}\}$ is a Stieltjes pair in the sense used here if and only if $\{\mathcal{P},-i\mathcal{Q}\}$ is a Stieltjes pair in the sense used in [BoSa]. Moreover, if both Theorems 2.1 and 2.2 are applicable, then

$$\begin{aligned} T_{\mathfrak{B}}[\{\mathcal{P},-i\mathcal{Q}\}] &= T_{K_1\mathfrak{A}_1}[\{\mathcal{P},-i\mathcal{Q}\}] \\ &= \Pi_1^*\Pi_1 + T_{\mathfrak{A}_1}[\{\mathcal{P},-i\mathcal{Q}\}] \\ &= \Pi_1^*\Pi_1 + iT_{\mathfrak{A}}[\{\mathcal{P},\mathcal{Q}\}] \\ &= \Pi_1^*\Pi_1 + \alpha + \int_0^\infty \left\{\frac{1}{\mu-\lambda} - \frac{\mu}{\mu^2+1}\right\} d\tau(\mu) \ . \end{aligned}$$

It then follows from Theorem 2.2 that

$$\tau(\mu) = 0 \text{ for } \mu < 0 \ , \tag{2.18}$$

$$\int_0^\infty \frac{1}{1+\mu} d\tau(\mu) \text{ is finite} \ , \tag{2.19}$$

and

$$\gamma = \Pi_1^*\Pi_1 + \alpha - \int_0^\infty \frac{\mu}{1+\mu^2} d\tau(\mu) \ .$$

Next, upon combining the implications of (2.8) and (2.17), we see that

$$\begin{aligned} B^*\Pi_1 &= \int_0^\infty CB(I-\mu CB)^{-1}\Pi_2 d\tau(\mu) + \Pi_2 \left\{\int_0^\infty \frac{\mu}{1+\mu^2} d\tau(\mu) - \alpha\right\} \\ &= C\Pi_1 + \Pi_2 \left\{\int_0^\infty \frac{\mu}{1+\mu^2} d\tau(\mu) - \alpha\right\} \ . \end{aligned}$$

Therefore, by (2.1),

$$\Pi_2\Pi_1^*\Pi_1 = \Pi_2 \left\{\int_0^\infty \frac{\mu}{1+\mu^2} d\tau(\mu) - \alpha\right\} \ .$$

Thus, as $\ker\ \Pi_2 = \{0\}$, we must have

$$\Pi_1^*\Pi_1 = \int_0^\infty \frac{\mu}{1+\mu^2} d\tau(\mu) - \alpha \ , \tag{2.20}$$

i.e., $\gamma = 0$.

In our applications, both CB and BC will be Volterra operators, i.e., compact operators with a single point of spectrum at the origin. In this setting, both condition (1) of Theorem 2.1 and condition (1) of Theorem 2.2 are automatically met. Therefore, the preceding analysis leads directly to the following result:

Theorem 2.3. *Assume that the bounded linear operators $B,\ C \in \{H;H\}$ and Π_1, $\Pi_2 \in \{\mathbb{C}^m;H\}$ satisfy the basic identity (2.1) and that furthermore:*

(1) *CB and BC are Volterra operators.*

(2) $\ker\ \Pi_2 = \{0\}$.

(3) $\ker(BC) = \{0\} = \ker(CB)$.

(4) $\mathrm{range}(CB) \cap \mathrm{range}\ \Pi_2 = \{0\}$.

(5) $\mathrm{range}(BC) \cap \mathrm{range}(B\Pi_2) = \{0\}$.

Then the following conclusions hold for every Stieltjes pair $\{\mathcal{P},\mathcal{Q}\}$:

(1) $\det\{c(\lambda)\mathcal{P}(\lambda)+d(\lambda)\mathcal{Q}(\lambda)\} \not\equiv 0$ *in $\mathbb{C}_+$ or in $\mathbb{C}_-$.*

(2)
$$v(\lambda) = iT_{\mathfrak{A}}[\{\mathcal{P},\mathcal{Q}\}] \tag{2.21}$$

admits a representation of the form

$$v(\lambda) = \alpha + \int_{-\infty}^{\infty} \left\{ \frac{1}{\mu-\lambda} - \frac{\mu}{1+\mu^2} \right\} d\tau(\mu) \ , \tag{2.22}$$

where

$$\tau(\mu) = 0 \ for \ \mu < 0 \ , \tag{2.23}$$

$$\int_0^{\infty} \frac{(1+\mu)d\tau(\mu)}{1+\mu^2} \ is \ finite \tag{2.24}$$

and

$$\alpha = \int_0^{\infty} \frac{\mu d\tau(\mu)}{1+\mu^2} - \Pi_1^*\Pi_1 \ . \tag{2.25}$$

Moreover, the three identities (2.15)–(2.17) hold.

Conversely, if $\tau(\mu)$ is any monotone nondecreasing mvf on $\mathbb{R}$ such that (2.23), (2.24) and the three identities (2.15)–(2.17) hold, then

$$v(\lambda) = \int_0^{\infty} \frac{1}{\mu-\lambda} d\tau(\mu) - \Pi_1^*\Pi_1 \tag{2.26}$$

is generated by a Stieltjes pair $\{\mathcal{P},\mathcal{Q}\}$ via formula (2.21).

3. Canonical systems

In this paper we shall focus on canonical systems of the form

$$\frac{dU}{dx}(x,\lambda) = i\lambda J\mathcal{H}(x)U(x,\lambda)\ ,\ 0 \leq x \leq \ell\ , \tag{3.1}$$

$$U(0,\lambda) = I_{2m}\ , \tag{3.2}$$

where J is given by (1.18) and the Hamiltonian $\mathcal{H}(x)$ is a $2m \times 2m$ summable mvf on $(0,\ell)$ which can be expressed in the form

$$\mathcal{H}(x) = R(x)R(x)^*\ , \tag{3.3}$$

where

$$R \in L_2^{2m\times m}(0,\ell) \text{ and } R(x)^*JR(x) = 0\ . \tag{3.4}$$

Dual canonical systems

Let $p(x)$ and $q(x)$ be a pair of $m \times m$ mvf's in $L_2^{m\times m}(0,\ell)$ and let

$$M(x) = \int_0^x p(t)p(t)^*dt \text{ and } N(x) = \int_0^x q(t)^*q(t)dt\ . \tag{3.5}$$

Then, for canonical systems with Hamiltonians $\mathcal{H}(x)$ of the form (3.3) with

$$R(x) = \begin{bmatrix} r_1(x) \\ r_2(x) \end{bmatrix} = \begin{bmatrix} -iN(x) \\ I_m \end{bmatrix} p(x)\ , \tag{3.6}$$

we define the dual canonical system

$$\frac{d\widetilde{U}}{dx}(x,\lambda) = i\lambda J\widetilde{\mathcal{H}}(x)\widetilde{U}(x,\lambda)\ ,\ 0 \leq x \leq \ell\ , \tag{3.7}$$

$$\widetilde{U}(0,\lambda) = I_{2m} \tag{3.8}$$

with Hamiltonian

$$\widetilde{\mathcal{H}}(x) = \widetilde{R}(x)\widetilde{R}(x)^*\ , \tag{3.9}$$

where

$$\widetilde{R}(x) = \begin{bmatrix} \widetilde{r}_1(x) \\ \widetilde{r}_2(x) \end{bmatrix} = iJ\begin{bmatrix} -iM(x) \\ I_m \end{bmatrix} q(x)^*\ . \tag{3.10}$$

Thus, $-i\widetilde{R}(x)$ is obtained from $R(x)$ by first interchanging $p(x)$ and $q(x)^*$ and then interchanging the positions of the block entries. The latter interchange is what causes the principal change in the initial conditions when passing from the original string (1.7)–(1.8) to the dual string (1.9)–(1.10). The constant i in (3.10) is just a convenience, which is introduced in order to end up with real initial conditions.

We remark that $R(x)$ can always be expressed in the form exhibited in formula (3.6) if r_2 is invertible, $r_1r_2^{-1}$ is differentiable and $i(r_1r_2^{-1})' \geq 0$ on $[0,\ell)$. In this case,

$$\widetilde{r}_1\widetilde{r}_1^* = i(r_1r_2^{-1})' \text{ and } \widetilde{r}_2(x) = \left\{-i\int_0^x r_2(t)r_2(t)^*dt\right\}\widetilde{r}_1(x)\ .$$

Associated operator identities

Let

$$(Bf)(x) = q(x)\int_0^x p(t)f(t)dt \tag{3.11}$$

and

$$(Cf)(x) = -p(x)^*\int_0^x q(t)^*f(t)dt \tag{3.12}$$

be operators from the Hilbert space $H = L_2^m(0,\ell)$ into itself and let

$$\Pi_1 g = q(x)g \text{ and } \Pi_2 g = p(x)^* g \tag{3.13}$$

be operators from the Hilbert space $G = \mathbb{C}^m$ into H. It is then readily checked that:

$$(B^* - C)f(x) = p(x)^*\int_0^\ell q(t)^*f(t)dt = \Pi_2\Pi_1^* f(x)\ , \tag{3.14}$$

$$(CBf)(x) = i\int_0^x R(x)^*JR(t)f(t)dt\ , \tag{3.15}$$

$$(BCf)(x) = i\int_0^x \widetilde{R}(x)^*J\widetilde{R}(t)f(t)dt\ , \tag{3.16}$$

$$\Pi_2^* f = \int_0^\ell p(t)f(t)dt \tag{3.17}$$

and

$$\Pi_1^* Bf = \int_0^\ell \{N(\ell) - N(t)\}p(t)f(t)dt\ . \tag{3.18}$$

Moreover, in view of (3.14), it is easily seen that

$$CB - B^*C^* = B^*\Pi_1\Pi_2^* - \Pi_2\Pi_1^* B \tag{3.19}$$

and

$$BC - C^*B^* = \Pi_1\Pi_2^* B^* - B\Pi_2\Pi_1^*\ . \tag{3.20}$$

The identities (3.19) and (3.20) can of course be expressed in the general forms (1.14) and (1.15), respectively, by invoking the definitions given in (1.16) and (1.17). The form of these identities is maintained if we replace Ψ_j by Φ_j and $\widetilde{\Psi}_j$ by $\widetilde{\Phi}_j$, where

$$\Phi_j g = r_j(x)^* g \text{ and } \widetilde{\Phi}_j g = \widetilde{r}_j(x)^* g \tag{3.21}$$

for $j = 1,2$ and $g \in \mathbb{C}^m$.

4. Spectral functions

A nondecreasing $m \times m$ mvf $\tau(\mu)$ on $\mathbb{R}$ is said to be a spectral mvf for the canonical system (3.1), (3.2) if the operator

$$V : f \longrightarrow \int_0^\ell [0 \; I_m] U^\#(x,\lambda)\mathcal{H}(x)f(x)dx \tag{4.1}$$

is an isometric mapping from

$$L_2(\mathcal{H}) = \left\{ 2m \times 1 \text{ measurable functions } f : \int_0^\ell f(x)^*\mathcal{H}(x)f(x)dx < \infty \right\}$$

into

$$L_2(\tau) = \left\{ m \times 1 \text{ measurable functions } h : \int_{-\infty}^\infty h(\mu)^* d\tau(\mu) h(\mu) < \infty \right\} .$$

In this section we shall give a description of the set of all spectral mvf's $\tau(\mu)$ for canonical systems of the form (3.1), (3.2) by identifying $U^\#(\ell,\lambda)$ as the characteristic function of a Livšic-Brodskii node and invoking Theorems 3.2–3.4 of [Sa1]. The same analysis is of course applicable to the dual canonical system (3.7), (3.8).

To begin, we define the operators

$$\Phi_j : g \in \mathbb{C}^m \longrightarrow r_j(x)^* g \in L_2^m(0,\ell)$$

in terms of the components of

$$R(x)^* = [r_1(x)^* \; r_2(x)^*] ,$$

both of which belong to $L_2^{m\times m}(0,\ell)$, just as in (3.21). Then

$$\Phi_j^* : f \in L_2^m(0,\ell) \longrightarrow \int_0^\ell r_j(x)f(x)dx$$

and

$$i(\Phi_1\Phi_2^* + \Phi_2\Phi_1^*)f = i\int_0^\ell R(x)^* JR(t)f(t)dt = (A - A^*)f ,$$

where $A = CB$ is given by formula (3.15). Then $U(\ell,\lambda)$ admits a realization of the form

$$U(\ell,\lambda) = I_{2m} + i\lambda J \begin{bmatrix} \Phi_1^* \\ \Phi_2^* \end{bmatrix} (I - \lambda A)^{-1} [\Phi_1 \; \Phi_2] . \tag{4.2}$$

Moreover, since

$$\Pi_1^*\Pi_1 = N(\ell) , \tag{4.3}$$

it follows that

$$[\Phi_1 \; \Phi_2] = [-iB^*\Pi_1 \; \Pi_2]K , \tag{4.4}$$

where

$$K = \begin{bmatrix} I_m & 0 \\ i\Pi_1^*\Pi_1 & I_m \end{bmatrix} \tag{4.5}$$

and

$$K^*JK = J\ .$$

Therefore, $U^\#(\ell,\lambda)$ can be expressed in terms of B, C, Π_1 and Π_2 as

$$U^\#(\ell,\lambda) = K^*\left\{I_{2m} + \lambda\begin{bmatrix}\Pi_1^*B\\ -i\Pi_2^*\end{bmatrix}(I-\lambda B^*C^*)^{-1}[\Pi_2\quad -iB^*\Pi_1]\right\}K^{-*} \tag{4.6}$$

Lemma 4.1. *The operators $A = CB$ and $\widetilde{A} = BC$ which are defined by formulas (3.15) and (3.16), respectively, are Volterra operators.*

Proof. It is readily seen that A is a Hilbert-Schmidt operator since

$$\int_0^\ell\int_0^\ell \|R(x)^*JR(t)\|^2dxdt \le \left\{\int_0^\ell \|R(x)\|^2dx\right\}^2 .$$

Therefore A is compact. The estimate

$$\|A^nf\|^2 \le \left\{\int_0^\ell \|R(x)\|^2dx\right\}^n \|f\|^2/n!\ ,$$

which is obtained by standard arguments, then serves to guarantee that A has no nonzero points of spectrum and hence that A is a Volterra operator, as claimed. The proof for $\widetilde{A}$ is similar. □

Lemma 4.2. *Let the operators $A = CB, \widetilde{A} = BC$, Φ_j and $\widetilde{\Phi}_j$, $j = 1, 2$, be defined by formulas (3.15), (3.16) and (3.21), respectively. Assume further that the entries $r_j(x)$ and $\widetilde{r}_j(x)$ are given by formulas (3.6) and (3.10) and that the mvf's $p(x)$ and $q(x)$ are invertible for a.e. point $x \in (0,\ell)$ (in addition to belonging to $L_2^{m\times m}(0,\ell)$). Then:*

(1) $\ker\ A = \{0\} = \ker\ \widetilde{A}$.

(2) $\ker\ \Phi_2 = \{0\} = \ker\ \widetilde{\Phi}_2$.

(3) $\text{range}\ A \cap \text{range}\ \Phi_2 = \{0\} = \text{range}\ \widetilde{A} \cap \text{range}\ \widetilde{\Phi}_2$.

Proof. Suppose first that $f \in \ker\ A$. Then, since $p(x)$ is invertible a.e.,

$$\int_0^x \{N(x) - N(t)\}p(t)f(t)dt = 0$$

for every point $x \in (0,\ell)$ and hence

$$N'(x)\int_0^x p(t)f(t)dt = 0$$

for every point $x \in (0,\ell)$. But this implies that

$$\int_0^x p(t)f(t)dt = 0$$

for every point $x \in (0,\ell)$, which in turn leads easily to the conclusion that $f = 0$.

The proof that ker $\Phi_2 = \{0\}$ is selfevident. Let us therefore suppose that

$$Af = \Phi_2 g$$

for some $f \in L_2^{2m}(0,\ell)$ and some $g \in \mathbb{C}^m$. This forces the equality

$$\int_0^x \{N(t) - N(x)\}p(t)f(t)dt = g \in \mathbb{C}^m$$

for all $x \in (0,\ell)$. Thus, upon letting $x \to 0$, we see that $g = 0$ and therefore $Af = 0$, i.e., $f \in$ ker A and so f must also be equal to zero by the first part of the proof.

The stated assertions for $\widetilde{A}$ and $\widetilde{\Phi}_2$ are proved in much the same way. □

The next theorem is a special case of Theorems 3.2–3.4 of [Sa1], which is applicable thanks to the preceding two lemmas. In the present setting, the operator S which is mentioned in that reference is equal to the identity operator. This means that the node in question is really a Livšic-Brodskii node; see [BrLi] for an introduction and additional references to the pioneering work on the subject.

Theorem 4.1. *Let $A = CB$ and Π_j, $j = 1, 2$, be given by formulas (3.15) and (3.13), respectively, where the $r_j(x)$ are specified in (3.6). Assume that the mvf's $p(x)$ and $q(x)$ are invertible for a.e. point $x \in (0,\ell)$ (in addition to belonging to $L_2^{m\times m}(0,\ell)$) and let*

$$\mathfrak{A}(\lambda) = K^{-*}U^{\#}(\ell,\lambda)K^* = I_{2m} + \lambda \begin{bmatrix} \Pi_1^* B \\ -i\Pi_2^* \end{bmatrix} (I - \lambda B^*C^*)^{-1}[\Pi_2 \ \ -iB^*\Pi_1] \ . \qquad (4.7)$$

Then for every Nevanlinna pair $\{\mathcal{P},\mathcal{Q}\}$ the following statements are in force:

(1) $\det\{c(\lambda)\mathcal{P}(\lambda) + d(\lambda)\mathcal{Q}(\lambda)\} \not\equiv 0$ *in* $\mathbb{C}_+$.

(2) $v(\lambda) = iT_{\mathfrak{A}}[\{\mathcal{P},\mathcal{Q}\}]$ *admits a representation of the form*

$$v(\lambda) = \alpha + \int_{-\infty}^{\infty} \left\{\frac{1}{\mu - \lambda} - \frac{\mu}{1+\mu^2}\right\} d\tau(\mu) \qquad (4.8)$$

where $\alpha = \alpha^*$ and $\int_{-\infty}^{\infty} (1+\mu^2)^{-1} d\tau(\mu)$ is finite.

(3) *The monotone mvf $\tau(\mu)$ in (4.8) is a spectral mvf for the canonical system with Hamiltonian $H(x) = R(x)R(x)^*$. Moreover, every spectral mvf $\tau(\mu)$ of this canonical system arises this way.*

(4) *The identity*

$$I = \int_{-\infty}^{\infty} (I - \mu A)^{-1}\Phi_2[d\tau(\mu)]\Phi_2^*(I - \mu A^*)^{-1}$$

holds and every solution $\tau(\mu)$ of this interpolation problem arises this way.

5. Dual matrix string equations

Let $U(x,\lambda)$ be the solution of the canonical system (3.1), (3.2) with a Hamiltonian $\mathcal{H}(x)$ of the form (3.3) and let

$$\varphi(x,\lambda) = R(x)^* U(x,\lambda) \begin{bmatrix} 0 \\ I_m \end{bmatrix} . \tag{5.1}$$

Then, since

$$U(x,\lambda) = I_{2m} + i\lambda J \int_0^x R(t)R(t)^* U(t,\lambda)dt ,$$

it is readily seen that

$$\varphi(x,\lambda) = r_2(x)^* + i\lambda \int_0^x R(x)^* J R(t)\varphi(t,\lambda)dt \tag{5.2}$$

and hence that the operator V defined by (4.1) can be expressed in terms of φ as

$$(Vf)(\lambda) = \int_0^\ell \varphi^\#(x,\lambda)R(x)^* f(x)dx .$$

Therefore, since

$$\int_0^\ell f(x)^* \mathcal{H}(x) f(x)dx = \int_0^\ell (R^* f)(x)^* (R^* f)(x)dx$$

and

$$\{(R^* f)(x) : f \in L_2(\mathcal{H})\} = L_2^m(0,\ell) ,$$

$\tau(\mu)$ is a spectral mvf for the canonical system (3.1) if and only if the mapping

$$g \longrightarrow \int_0^\ell \varphi^\#(x,\lambda)g(x)dx$$

is an isometry from $L_2^m(0,\ell)$ into $L_2(\tau)$.

Lemma 5.1. *If* $R(x) = \begin{bmatrix} -iN(x) \\ I_m \end{bmatrix} p(x)$, *and if* $p(x)$ *and* $q(x)$ *are invertible for a.e. point* $x \in (0,\ell)$, *then* $\varphi(x,\lambda)$ *is a solution of the integral equation (5.2) if and only if it is a solution of the differential equation (1.7) with initial conditions (1.8).*

Proof. Upon inserting the given choice of $R(x)$ into equation (5.2) it is readily checked that

$$p(x)^{-*}\varphi(x,\lambda) = I_m + \lambda \int_0^x \{N(t) - N(x)\} p(t)\varphi(t,\lambda)dt$$

and

$$\frac{d}{dx}\{p(x)^{-*}\varphi(x,\lambda)\} = -\lambda q(x)^* q(x) \int_0^x p(t)\varphi(t,\lambda)dt .$$

But that leads easily to the stated conclusions. □

Lemma 5.2. *If* $p(x)$ *and* $q(x)$ *are invertible for a.e. point* $x \in (0,\ell)$, *then* $\widetilde{\varphi}(x,\lambda)$ *is a solution of the integral equation (5.2) with* $R(x)$ *replaced by* $\widetilde{R}(x)$, *as specified* *in formula (3.10), if and only if it is a solution of the differential equation (1.9) with initial conditions (1.10).*

Proof. The integral equation for $\widetilde{\varphi}(x,\lambda)$ is

$$\widetilde{\varphi}(x,\lambda) = q(x)M(x) - \lambda \int_0^x q(x)\{M(x) - M(t)\}q(t)^*\widetilde{\varphi}(t,\lambda)dt \ .$$

Therefore,

$$q(x)^{-1}\widetilde{\varphi}(x,\lambda) = M(x) - \lambda \int_0^x \{M(x) - M(t)\}q(t)^*\varphi(t,\lambda)dt$$

and

$$\frac{d}{dx}q(x)^{-1}\widetilde{\varphi}(x,\lambda) = M'(x) - \lambda M'(x) \int_0^x q(t)\widetilde{\varphi}(t,\lambda)dt \ .$$

The rest is straightforward. □

6. The main theorem

Theorem 6.1. *Let $p(x)$ and $q(x)$ be continuous invertible $m \times m$ mvf's on a finite interval $[0,\ell]$. Let B, C, Π_1, Π_2 and $\mathfrak{A}(\lambda)$ be defined by formulas (2.11)–(2.13) and (3.8), respectively.*

Then the following conclusions hold for every Stieltjes pair $\{\mathcal{P},\mathcal{Q}\}$:

(1) $\det\{c(\lambda)\mathcal{P}(\lambda) + d(\lambda)\mathcal{Q}(\lambda)\} \not\equiv 0$ *in* $\mathbb{C}_+$ *or in* $\mathbb{C}_-$.

(2)
$$v(\lambda) = i\mathcal{T}_{\mathfrak{A}}[\{\mathcal{P},\mathcal{Q}\}] \tag{6.1}$$

admits a representation of the form

$$v(\lambda) = \alpha + \int_{-\infty}^{\infty} \left\{ \frac{1}{\mu - \lambda} - \frac{\mu}{1+\mu^2} \right\} d\tau(\mu) \ , \tag{6.2}$$

where

$$\tau(\mu) = 0 \ for \ \mu < 0 \ , \tag{6.3}$$

$$\int_0^{\infty} \frac{(1+\mu)d\tau(\mu)}{1+\mu^2} \ is \ finite \tag{6.4}$$

and

$$\alpha = \int_0^{\infty} \frac{\mu d\tau(\mu)}{1+\mu^2} - \Pi_1^*\Pi_1 \ . \tag{6.5}$$

(3) $\tau(\mu)$ *is a spectral mvf for the original generalized string equation (1.7), (1.8).*

(4) $\widetilde{\tau}(\mu) = \int_0^{\mu} \lambda d\tau(\lambda)$ *is a spectral mvf for the generalized dual string equation (1.9), (1.10).*

Conversely, every monotone nondecreasing mvf $\tau(\mu)$ on $[0,\infty)$ for which condition (6.3) and statements (3) and (4) hold, is obtained from the linear fractional transformation of some Stieltjes pair as in (2).

Proof. By Lemmas 4.1 and 4.2, the hypotheses of Theorem 2.3 are satisfied for the given choice of the operators B, C, Π_1 and Π_2. Thus, (1) and (2) of Theorem 2.3 and the three identities (2.15)–(2.17) all hold. Therefore, by Theorem 3.2 in Chapter III of [Sa1], $\tau(\mu)$ is a spectral mvf for the canonical system (3.1) based on (3.6) and $\widetilde{\tau}(\mu) = \int_0^\mu \lambda d\tau(\lambda)$ is a spectral mvf for the dual canonical system (3.7) based on (3.10). But this is equivalent to assertions (3) and (4).

Now suppose conversely that conditions (6.3) and (6.4) and assertions (3) and (4) hold. Then $\tau(\mu)$ is a spectral mvf for the canonical system based on (3.6) and $\widetilde{\tau}(\mu)$ is a spectral mvf for the dual canonical system based on (3.10). Therefore, by Theorem 3.3 in Chapter III of [Sa1], the identities (2.15) and (2.16) hold. Moreover, it follows from (2.16) that

$$\begin{aligned} C &= \int_0^\infty \{CB\mu - I + I\}\{I - \mu CB\}^{-1}\Pi_2[d\tau(\mu)]\Pi_2^*(I - \mu B^*C^*)^{-1}B^* \\ &= -\int_0^\infty \Pi_2[d\tau(\mu)]\Pi_2^*(I - \mu B^*C^*)^{-1}B^* \\ &\quad + \int_0^\infty (I - \mu CB)^{-1}\Pi_2[d\tau(\mu)]\Pi_2^*(I - \mu B^*C^*)^{-1}B^* \\ &= -\Pi_2\Pi_{1,\tau}^* + B^* , \end{aligned}$$

by (2.15), where $\Pi_{1,\tau}$ denotes the right-hand side of (2.17). Consequently, in view of (1.11), it follows that

$$\Pi_2\{\Pi_1^* - \Pi_{1,\tau}^*\} = 0$$

and hence, since ker $\Pi_2 = \{0\}$, that (2.17) holds. Thus, by Theorem 2.3, $\tau(\mu)$ is generated by a Stieltjes pair. □

7. Examples

In this section we shall present a few examples in order to illustrate the main theorem.

Example 7.1. Let $U(x, \lambda)$ denote the solution of the integral equation

$$U(x, \lambda) = I_{2m} + i\lambda J \int_0^x dF(t)U(t, \lambda) , \quad 0 \le x \le \ell , \tag{7.1}$$

where $F(t)$ is a continuous monotone nondecreasing $2m \times 2m$ mvf such that

$$\rho(t) = \text{trace } F(t)$$

is a continuous strictly increasing function of t on $[0, \ell]$. Then, since $F(t)$ is absolutely continuous with respect to $\rho(t)$, we can reexpress equation (7.1) in the form

$$U(x, \lambda) = I_{2m} + i\lambda J \int_0^x \mathcal{H}(t)U(t, \lambda)d\rho(t) , \tag{7.2}$$

where

$$\mathcal{H}(t) = \frac{dF(t)}{d\rho(t)}$$

is subject to the bounds $0 \le \mathcal{H}(t) \le I_{2m}$. Moreover, upon setting $y = \rho(x)$, $x = \psi(y)$, $U_1(y, \lambda) = U(\psi(y), \lambda)$ and $\mathcal{H}_1(y) = \mathcal{H}(\psi(y))$, it follows from the last equation that

$$U_1(y, \lambda) = I_{2,m} + i\lambda J \int_0^y \mathcal{H}_1(s) U_1(s, \lambda) ds \ , \ 0 \le y \le \rho(\ell) \ , \tag{7.3}$$

which is of the general form considered earlier.

Example 7.2. Let $m = 1$, $q(x) = 1$ and $p(x) > 0$. Then, upon setting $\varphi_1(x, \lambda) = p(x)^{-1}\varphi(x, \lambda)$, the system (1.7)–(1.8) reduces to the differential equation

$$-\frac{d^2}{dx^2}\varphi_1(x, \lambda) = \lambda p(x)^2 \varphi_1(x, \lambda) \tag{7.4}$$

with initial conditions

$$\varphi_1(0, \lambda) = 1 \text{ and } \varphi_1'(0, \lambda) = 0 \ . \tag{7.5}$$

Moreover, the dual system (1.9)–(1.10) reduces to the differential equation

$$-\frac{d}{dx} p^{-2} \frac{d}{dx} \widetilde{\varphi}_1(x, \lambda) = \lambda \widetilde{\varphi}_1(x, \lambda) \tag{7.6}$$

with initial conditions

$$\widetilde{\varphi}_1(0, \lambda) = 0 \text{ and } \widetilde{\varphi}_1'(0, \lambda) = p(0)^2 \tag{7.7}$$

for $\widetilde{\varphi}_1(x, \lambda) = \widetilde{\varphi}(x, \lambda)$. Clearly, equation (7.4) is of the form (1.3), whereas the dual equation (7.6) is of the form (1.4), with $M(x) = \int_0^x p(t)^2 dt$ in both cases. Thus Theorem 6.1 gives a full description for the set of spectral functions $\tau(u)$ on $[0, \infty)$ for the string equation (1.3) subject to (1.5) such that $\widetilde{\tau}(\mu) = \int_0^\mu \lambda d\tau(\lambda)$ is a spectral function for the dual string (1.4) subject to (1.6).

To amplify this example just a little, we recall that conventional Sturm-Liouville theory implies that

$$\int_0^\infty \left| \int_0^\ell \varphi_1^\#(s, \lambda) g(s) p(s)^2 ds \right|^2 d\tau(\lambda) = \int_0^\ell |g(s)|^2 p(s)^2 ds \ ,$$

which is the same as

$$\int_0^\infty \left| \int_0^\ell \varphi^\#(s, \lambda) g(s) p(s) ds \right|^2 d\tau(\lambda) = \int_0^\ell |g(s) p(s)|^2 ds \ .$$

Example 7.3. Let $m = 1$, $p(x) = q(x) = 1$ and $0 < \ell < \infty$. Then

$$\varphi(x, \lambda) = \cos \sqrt{\lambda} x \text{ and } \widetilde{\varphi}(x, \lambda) = \frac{\sin \sqrt{\lambda} x}{\sqrt{\lambda}}$$

are the solutions of the systems (1.7)–(1.8) and (1.9)–(1.10), respectively. The operators B and C that are defined by (3.11) and (3.12) are Volterra operators, as is $BC = CB$ and

$$(-CB)^n f = \int_0^x \frac{(x-t)^{2n-1}}{(2n-1)!} f(t)dt \ , \ n = 1, 2, \dots \ .$$

Therefore,

$$(I - \lambda CB)^{-1} f = f(x) - \int_0^x \sqrt{\lambda} \sin \sqrt{\lambda}(x-t) f(t)dt \ ,$$

and

$$(I - \lambda B^* C^*)^{-1} f = f(x) - \int_x^\ell \sqrt{\lambda} \sin \sqrt{\lambda}(t-x) f(t)dt \ .$$

For constant vectors $g \in \mathbb{C}^m$,

$$\Pi_1 g = \Pi_2 g = g \ ,$$

$$B^* \Pi_1 g = (\ell - x) g \ ,$$

$$(I - \lambda B^* C^*)^{-1} \Pi_2 g = \cos \sqrt{\lambda}(\ell - x) g \ ,$$

$$(I - \lambda B^* C^*)^{-1} B^* \Pi_1 g = \frac{\sin \sqrt{\lambda}(\ell - x)}{\sqrt{\lambda}} g \ ,$$

$$\Pi_1^* B (I - \lambda B^* C^*)^{-1} \Pi_2 g = \frac{\cos \sqrt{\lambda}\ell - 1 + \ell\sqrt{\lambda} \sin \sqrt{\lambda}\ell}{\sqrt{\lambda}} g \ ,$$

$$\Pi_1^* B (I - \lambda B^* C^*)^{-1} B^* \Pi_1 g = \frac{\lambda^{-\frac{1}{2}} \sin \sqrt{\lambda}\ell - \ell \cos \sqrt{\lambda}\ell}{\sqrt{\lambda}} g \ ,$$

$$\Pi_2^* (I - \lambda B^* C^*)^{-1} \Pi_2 g = \frac{\sin \sqrt{\lambda}\ell}{\sqrt{\lambda}} g$$

and

$$\Pi_2^* (I - \lambda B^* C^*)^{-1} B^* \Pi_1 g = \frac{1 - \cos \sqrt{\lambda}\ell}{\lambda} g \ .$$

Thus,

$$\mathfrak{A}(\lambda) = \left[\begin{array}{c|c} \cos \sqrt{\lambda}\ell + \sqrt{\lambda}\ell \sin \sqrt{\lambda}\ell & \dfrac{-i \sin \sqrt{\lambda}\ell}{\sqrt{\lambda}} + i\ell \cos \sqrt{\lambda}\ell \\ & \\ -i\sqrt{\lambda} \sin \sqrt{\lambda}\ell & \cos \sqrt{\lambda}\ell \end{array} \right]$$

and

$$\mathfrak{B}(\lambda) = \begin{bmatrix} \cos\sqrt{\lambda}\ell & \sin\sqrt{\lambda}\ell/\sqrt{\lambda} \\ -\sqrt{\lambda}\sin\sqrt{\lambda}\ell & \cos\sqrt{\lambda}\ell \end{bmatrix} .$$

Hence, for the particular choice $\mathcal{P}(\lambda) = \frac{1}{\sqrt{\lambda}}, \mathcal{Q}(\lambda) = 1$, we obtain

$$iT_{\mathfrak{A}}\left[\left\{\frac{1}{\sqrt{\lambda}}, 1\right\}\right] = \frac{i}{\sqrt{\lambda}} - \ell = \int_0^\infty \frac{1}{\mu - \lambda} d\tau(\mu) - \ell ,$$

where

$$\tau(\mu) = \begin{cases} 0 & \text{if } \mu < 0 \\ \frac{2}{\pi}\mu^{\frac{1}{2}} & \text{if } \mu \geq 0 \end{cases}$$

and

$$\ell = \Pi_1^* \Pi_1 .$$

The spectral function for the corresponding dual system is given by the formula

$$\widetilde{\tau}(\mu) = \begin{cases} 0 & \text{if } \mu < 0 \\ \frac{2}{\pi}\mu^{3/2} & \text{if } \mu \geq 0 . \end{cases}$$

Example 7.4. Let $M_1(t)$ and $N_1(t)$ be monotonically increasing continuous $m \times m$ mvf's, let

$$dF(t) = \begin{bmatrix} N_1(t)[dM_1(t)]N_1(t) & -iN_1(t)dM_1(t) \\ i[dM_1(t)]N_1(t) & dM_1(t) \end{bmatrix}$$

and let

$$\rho(t) = \text{trace } F(t) .$$

Then, just as in Example 7.1, the solution $U(x, \lambda)$ of the integral equation can now be reexpressed as the solution of equation (7.2) in terms of the Hamiltonian

$$\mathcal{H}(t) = \begin{bmatrix} N_1(t)L_1(t)N_1(t) & -iN_1(t)L_1(t) \\ -iL_1(t)N_1(t) & L_1(t) \end{bmatrix} ,$$

where

$$L_1(t) = \frac{dM_1(t)}{d\rho(t)} , \ 0 \leq L_1(t) \leq I_m .$$

The Hamiltonian $\mathcal{H}(x)$ in formula (3.6) is of this form with $N_1(x) = N(x)$ and $M_1(x) = M(x)$. Indeed, in this case, we may write

$$\mathcal{H}(x) = \begin{bmatrix} N(x)M'(x)N(x) & -iN(x)M'(x) \\ iM'(x)N(x) & M'(x) \end{bmatrix} ;$$

similarly the Hamiltonian $\widetilde{\mathcal{H}}(x)$ for the dual canonical system that is given by (3.10) can be written as

$$\widetilde{\mathcal{H}}(x) = \begin{bmatrix} N'(x) & iN'(x)M(x) \\ -iM(x)N'(x) & M(x)N'(x)M(x) \end{bmatrix} .$$

Remark 7.1. Upon letting $a(x) = p(x)p(x)^*$, $b(x) = q(x)^*q(x)$, $y(x, \lambda) = p(x)^{-*}\varphi(x, \lambda)$ and $\widetilde{y}(x, \lambda) = q(x)^{-1}\varphi(x, \lambda)$, it is readily seen that the system (1.7), (1.8) can be rewritten as

$$-\frac{d}{dx}b^{-1}\frac{dy}{dx} = \lambda ay$$

$$y(0, \lambda) = I_m \text{ and } y'(0, \lambda) = 0 \ ,$$

whereas the dual system (1.9), (1.10) takes the form

$$-\frac{d}{dx}a^{-1}\frac{d\widetilde{y}}{dx} = \lambda b\widetilde{y}$$

$$\widetilde{y}(0, \lambda) = 0 \text{ and } \widetilde{y}\,'(0, \lambda) = a(0) \ .$$

A number of concrete calculations of the spectral function for the case $a(x) = b(x)^{-1}$ may be found in [D] and in the paper of Krein [Kr] for $m = 1$.

References

[BoSa] V. Bolotnikov and L. A. Sakhnovich, On an operator approach to interpolation problems for Stieltjes functions, *Integral Equations Operator Theory*, in press.

[BrLi] M. S. Brodskii and M. S. Livšic, Spectral analysis of non-selfadjoint operators and intermediate systems, *Uspehi Mat. Nauk* **13**:1 (1958), 3–85; English transl. *Amer. Math. Soc. Transl.* (2) **13** (1960), 265–346.

[D] H. Dym, An introduction to de Branges spaces of entire functions with applications to differential equations of the Sturm-Liouville type, *Adv. Math.* **5** (1970), 395–471.

[DI] H. Dym and A. Iacob, Positive definite extensions, canonical equations and inverse problems, in: *Topics in Operator Theory, Systems and Networks*, (H. Dym and I. Gohberg, eds.), *Oper. Theory Adv. Appl.* **OT12**, Birkhäuser, Basel, 1984, pp.141–240.

[DMcK] H. Dym and H.P. McKean, *Gaussian Processes, Function Theory and the Inverse Spectral Problem*, Academic Press, New York, 1976.

[DvKa] Yu. M. Dyukarev and V. E. Katsnelson, Multiplicative and additive classes of Stieltjes analytic matrix-valued functions and interpolation problems associated with them. I, *Teor. Funktsii, Funktsional. Anal. i Prilozhen* **36** (1981), 13–27; English transl. *Amer. Math. Soc. Transl.* (2) **131** (1986), 55–70.

[Kr] M. G. Krein, On a method of effective solution of an inverse boundary problem, *Dokl. Akad. Nauk SSSR* **94**:6 (1954), 987–990.

[KaKr1] I. S. Kac and M. G. Krein, *R*-functions–analytic functions mapping the upper half plane into itself, Supplement I to the Russian translation of F. V. Atkinson, *Discrete and Continuous Boundary Problems*, Mir, Moscow, 1968, pp.629–647; English transl. *Amer. Math. Soc. Transl.* (2) **103** (1974), 1–18.

[KaKr2] I. S. Kac and M. G. Krein, On the spectral functions of the string, Supplement II to the Russian translation of F. V. Atkinson, *Discrete and Continuous Boundary Problems*, Mir, Moscow, 1968, pp.648–737; English transl. *Amer. Math. Soc. Transl.* (2) **103** (1974), 19–102.

[IvSa] T. S. Ivanchenko and L. A. Sakhnovich, An operator approach to the Potapov scheme for the solution of interpolation problems, in: *Matrix and Operator Valued Functions* (I. Gohberg and L. A. Sakhnovich, eds.), *Oper. Theory Adv. Appl.* **OT72**, Birkhäuser, Basel, 1994, pp. 48–86.

[Sa1] L. A. Sakhnovich, Factorization problems and operator identities, *Russian Math. Surveys* **41**:1 (1986), 1–64.

[Sa2] L. A. Sakhnovich, Method of operator identities and problems of analysis, *St. Petersburg Math. J.* **5**:1 (1994), 1–69.

[Sa3] L. A. Sakhnovich, *Interpolation Theory and its Applications*, Kluwer, Dordrecht, 1997.

[Sa4] L. A. Sakhnovich, Classical Spectral Problems, in: *Differential and Integral Operators* (I. Gohberg, R. Mennicken and C. Tretter, eds.), *Oper. Theory Adv. Appl.* **OT102**, Birkhäuser, Basel, 1998, pp.243–255.

Department of Theoretical Mathematics, The Weizmann Institute of Science, Rehovot 76100, Israel

Department of Higher Mathematics, Ukranian State Academy of Communications, 270021 Odessa, Ukraine

Operator Theory:
Advances and Applications, Vol. 123

Angle Between Subspaces of Analytic and Antianalytic Functions in Weighted L_2 Space on the Boundary of a Multiply Connected Domain

Sergei Fedorov

Abstract. The problem of describing those positive weights on the boundary Γ of a finitely connected domain Ω for which the angle in the corresponding weighted L_2 space on Γ between the linear space $\mathcal{R}(\Omega)$ of all rational functions on $\overline{\mathbf{C}}$ with poles outside of $\operatorname{Clos}\Omega$ and the linear space $\mathcal{R}(\Omega)_- = \{\overline{f} | f \in \mathcal{R}(\Omega)\}$ of antianalytic rational functions, is a natural analog of the problem solved in the famous Helson-Szegö theorem. In this paper we give a complete description (in terms of necessary and sufficient conditions) of those positive weights w on Γ for which the angle in $L_2(\Gamma, w)$ between the spaces of analytic and antianalytic character-automorphic (non only single-valued) functions is nonzero. The given description is similar to the one in the Helson-Sarason Theorem for the unit disk: the "modified" weight obtained by factoring out finitely many zeros on Γ must satisfy the Muckenhoupt condition, provided an additional condition on the divisor of these zeros and the character of the space under consideration holds.

1. Introduction

The famous Helson-Szegö [1, 2] and Hunt-Muckenhoupt-Wheeden [3] theorems give descriptions of those finite positive measures $d\mu$ on the unit circle $\mathbf{T}$, for which the angle in $L_2(d\mu, \mathbf{T})$ between $\bigvee_{n<0}\{z^n\}$ and $\bigvee_{n\geq 0}\{z^n\}$ is nonzero. This result turned out to be very important for the solution of the problem of Riesz basis properties of the reproducing kernels in the co-invariant subspaces of the shift operator, and in particular for the solution of the classical problem of harmonic analysis on bases of exponentials in the L_2 space on a segment (see [2, 4, 5]).

1991 *Mathematics Subject Classification.* Primary 46J15, 30D55, 30F99, 47B25. Secondary 30F15, 30F30, 14K20, 46E22, 30H05.

Key words and phrases. Angle between subspaces, Hardy spaces, Riemann surface, Muckenhoupt condition.

The research was supported by Marsden Fund grant 96-UOA-MIS-0098.

In the present paper we deal with a similar problem in the weighted L_2 spaces on the boundary of a finitely ($g+1$) connected domain Ω.

Let us fix a point $a \in \Omega$ and denote by $d\eta_a$ the harmonic measure of $\Gamma = \partial\Omega$ with respect to the point a, i.e., the measure which gives rise to the continuous linear functional: $u \mapsto u(a)$, where $u \in C(\Gamma)$ and $u(a)$ is the value of the harmonic extension of u into Ω. Let us also denote by $\mathcal{R}(\Omega)$ the linear space of all rational functions on $\overline{\mathbf{C}}$ with poles outside of $\operatorname{Clos}\Omega$. The closure of $\mathcal{R}(\Omega)$ in $L_2(\Gamma, d\eta_a) = L_2(\Gamma) = L_2$ gives us the usual Hardy space $H^2_+(\Omega)$ (see for example [6]). Let us denote by $(\mathcal{R}(\Omega))^\perp$ the orthogonal complement of $\mathcal{R}(\Omega)$ in $L_2(\Gamma, d\eta_a)$ and by $\mathcal{R}(\Omega)_-$ the linear space $\{\overline{f} | f \in \mathcal{R}(\Omega), f(a) = 0\}$. Obviously $\mathcal{R}(\Omega)_- \subset (\mathcal{R}(\Omega))^\perp$. The natural analogs of the problem in the unit disk are to describe those finite positive measures $d\mu$ on Γ for which the angle in $L_2(\Gamma, d\mu)$ between $\mathcal{R}(\Omega)$ and either $(\mathcal{R}(\Omega))^\perp \cap L_2(\Gamma, d\mu)$ or $\mathcal{R}(\Omega)_-$ is nonzero.

The first and essential difference in the case under consideration from that in the unit disk is that the mentioned above two options of posing the problem are essentially different. Namely, as opposed to the case of the unit disk, where $L_2(\mathbf{T}) = H^2_+ \oplus H^2_-$, $H^2_- = \{\overline{f} \mid f \in H^2_+, f(0) = 0\}$, (and in general the sum of the spaces of analytic and antianalytic trigonometric polynomials is dense in $L_2(d\mu, \mathbf{T})$ for any $d\mu$), in our case there exists a g-dimensional space of measures which are orthogonal to $\mathcal{R}(\Omega) + \mathcal{R}(\Omega)_-$, and consequently

$$L_2(\Gamma, d\eta_a) = H^2_+ \oplus H^2_- \oplus \mathcal{M},$$

where $H^2_- = \{\overline{f} \mid f \in H^2_+, f(a) = 0\}$ and $\mathcal{M}$ is a so-called defect space. This means that the sum of the spaces $H^2_+ \cap L_2(d\mu)$ and $H^2_- \cap L_2(d\mu)$ is not necessarily dense in $L_2(d\mu)$. Another difference is connected with the fact that we actually have a whole g-dimensional torus of similar decompositions: $L_2(\Gamma) = H^2_{+,\kappa} \oplus H^2_{-,\kappa} \oplus \mathcal{M}_\kappa$, where $H^2_{+,\kappa}, H^2_{-,\kappa}$ are the so-called character-automorphic Hardy spaces on Ω, corresponding to a character $\kappa \in \mathbb{R}^g/\mathbb{Z}^g$, and $\mathcal{M}_\kappa$ is a g-dimensional κ-automorphic defect space. All these spaces are of equal importance and for each particular character the problem on the angle in $L_2(d\mu)$ between the spaces of analytic and antianalytic character-automorphic functions can be posed.

Having in mind as a final goal the study of problems similar to the basis problem for the reproducing kernels, it is natural to suspect that, just as in the case of the disk, results on the angles between subspaces in weighted L_2 spaces on Γ will play an essential role.

The first steps towards the solution of this problem were given by the author in [7], where the positive weights w on Γ for which the angle in $L_2(\Gamma, w d\eta_a)$ between $L_2(\Gamma, w d\eta_a) \cap \{H^2_{-,\kappa} + \mathcal{M}_\kappa\}$ and $L_2(\Gamma, w d\eta_a) \cap H^2_{+,\kappa}$ is nonzero were described. These weights should satisfy the Muckenhoupt condition, which in this case has the same form as the classical one and the corresponding angles are nonzero for all the characters simultaneously. As a particular case this result contains the solution of the above mentioned problem on non-zero angle in $L_2(\Gamma, d\mu)$ between $\mathcal{R}(\Omega)$ and $(\mathcal{R}(\Omega))^\perp \cap L_2(\Gamma, d\mu)$.

However, this description is not enough for the solution of the problem on the basis properties of reproducing kernels. Namely, the problem can be reduced (this reduction and the solution of the problem even in a more general setting is a matter of a forthcoming paper) to the problem of the description of those weights w on Γ for which the angle in $L_2(\Gamma, wd\eta_a)$ between $L_2(\Gamma, wd\eta_a) \cap H^2_{-,\kappa}$ and $L_2(\Gamma, wd\eta_a) \cap H^2_{+,\kappa}$ is nonzero. Note that the co-dimension of the sum of these subspaces in ordinary $L_2(\Gamma)$ is equal to g and one could expect that the situation here is similar to that in the Helson-Sarason theorem for the unit disk [15].

The present paper contains such a description and the answer is, in fact, similar to the one in the Helson-Sarason Theorem (the Muckenhoupt condition for a modified weight), but depends essentially on a character. Namely, it can happen that for one and the same weight w on Γ, for one character the corresponding angle is zero and for another character it is nonzero. The solution is essentially based on the results of the papers [7, 8].

2. Preliminaries

In what follows we consider a planar domain as a half of its Schottky double (a compact Riemann surface), and make essential use of function theory on this surface. Though the description of the basic objects and constructions on these doubles is well known (see for example [9, 10]) and is already present in the previous papers [7, 8, 11] by the author, it is worthwhile to repeat this material here once more. We start this section with a brief discussion of Schottky doubles, meromorphic differentials, divisors, the Riemann-Roch theorem, the Abel-Jacobi transform and multiplicative meromorphic functions. Then we mention the definition and basic properties of character-automorphic Hardy spaces, and finish the section with recalling the notion of an angle between subspaces in a Hilbert space.

2.1.

We consider our $g+1$ connected domain $\Omega = \Omega_+$ with nondegenerate analytic or piecewise analytic boundary components $\Gamma_0, \ldots, \Gamma_g$ as a bordered Riemann surface and denote by $\hat{\Omega}$ its Schottky double – the compact Riemann surface of genus g, which is obtained topologically from two copies Ω_+ and Ω_- of the domain Ω by the identification of the points of $\partial\Omega_\pm \stackrel{def}{=} \Gamma = \bigcup_{j=0}^{g} \Gamma_j$ (gluing a second copy Ω_- of Ω to Ω_+ along $\partial\Omega_+$) with the complex structure obtained by "reflecting" the complex structure from $\Omega_+ \cup \partial\Omega_+$ to Ω_-. The involution $J : \hat{\Omega} \longrightarrow \hat{\Omega}$ which fixes the points of Γ and interchanges the same points on Ω_+ and Ω_- is antiholomorphic.

Fix a point $P_0 \in \Gamma_0$ and let $\gamma_1, \ldots, \gamma_g$ be oriented crosscuts from P_0 to the boundary components $\Gamma_1, \ldots, \Gamma_g$ respectively, which except for their end points lie in Ω_+ and intersect each other only in the point P_0. We define a homology basis on the compact Riemann surface $\hat{\Omega}$ of genus g in the following way: the b-cycles are the boundary components, $b_j = \Gamma_j$, $j = 1, \ldots, g$, and the a-cycles are constructed from the crosscuts γ_j and their antiholomorphic reflections $J(\gamma_j)$ on Ω_-, $a_j = \gamma_j - J(\gamma_j)$, $j = 1, \dots, g$.

We consider the g-dimensional vector space $\mathcal{N}$ of holomorphic differentials on $\hat{\Omega}$ and the normalized basis $\omega_1, \dots, \omega_g$ in this space that is dual to the homology basis

$$\int_{a_j} \omega_k = \delta_{j,k}$$

where $\delta_{j,k}$ is the Kronecker symbol. The matrix of b-periods $\tau = \left\{ \int_{b_j} \omega_k \right\}_{j,k=1}^{g}$ has the form $\tau = iP$ where P is a real positive definite matrix.

We denote by $d\eta_{P,Q}$, $P, Q \in \hat{\Omega}$, $P \neq Q$, the normalized differential of the third kind with only simple poles at the points P, Q with residues $\frac{1}{2\pi i}$ and $-\frac{1}{2\pi i}$ respectively and $\int_{a_j} d\eta_{P,Q} = 0$, $j = 1, \dots, g$. The bilinear relations between the differentials of the first and the third kind give us the following expressions for the b-periods of $d\eta_{P,Q}$

$$\int_{b_j} d\eta_{P,Q} = \int_Q^P \omega_j, \; j = 1, \dots, g \tag{2.1}$$

where the path of integration does not intersect the a- and b-cycles. Note that for the point $P \in \Omega_+$ the restriction of the differential $d\eta_{P,J(P)}$ on Γ determines exactly the harmonic measure $d\eta_P$.

There is a close relation between the harmonic measure of Ω_+ with respect to the point P and the Green function $G(z, P)$ of the domain Ω_+ with singularity at the point P,

$$2\pi \operatorname{Im} \int_{P_0}^{z} d\eta_{P,J(P)} = G(z, P) \quad \text{or} \quad 2\pi d\eta_{P,J(P)} = 2i \frac{\partial G(z, P)}{\partial z} dz.$$

2.2.

We will write the divisors on $\hat{\Omega}$ in the additive form, $\mathcal{A} = \sum_{j=1}^{k} n_j P_j$, $n_j \in \mathbb{Z}$, $P_j \in \hat{\Omega}$, using the standard notations: $\mathcal{A}$ is positive ($\mathcal{A} \geq 0$) if $n_j \geq 0$ for all j, $ord(\mathcal{A}) \stackrel{def}{=} \sum_{j=1}^{k} n_j$. The action of the antiholomorphic involution J extends obviously to the divisors.

The divisor of a meromorphic function f on $\hat{\Omega}$ is called a principal divisor and is denoted by (f). Obviously $ord(f) = 0$. Let us denote by $L(\hat{\Omega})$ the field of meromorphic functions on $\hat{\Omega}$ and by $L(\mathcal{A})$ the vector space $\left\{ f \in L(\hat{\Omega}) : (f) \geq \mathcal{A} \right\}, r(\mathcal{A}) \stackrel{def}{=} \dim L(\mathcal{A})$.

The divisors $\mathcal{A}, \mathcal{B}$ are equivalent ($\mathcal{A} \sim \mathcal{B}$) if their difference is a principal divisor.

The divisor of a meromorphic differential ω on $\hat{\Omega}$ is called a canonical divisor and is also denoted by (ω). The corresponding vector space of meromorphic differentials on $\hat{\Omega}$ is denoted by $\mathcal{N}(\hat{\Omega})$ and $\mathcal{N}(\mathcal{A}) \stackrel{def}{=} \left\{\omega \in \mathcal{N}(\hat{\Omega}) : (\omega) \geq \mathcal{A}\right\}, i(\mathcal{A}) \stackrel{def}{=} \dim \mathcal{N}(\mathcal{A})$. Any two canonical divisors are equivalent and the order of any canonical divisor is equal to $2g-2$.

The Riemann-Roch theorem asserts that

$$r(-\mathcal{A}) = ord(\mathcal{A}) - g + 1 + i(\mathcal{A}).$$

We make use of Abel-Jacoby mapping, which is defined by

$$\varphi(P) = \int_{P0}^{P} \vec{\omega},\ P \in \hat{\Omega},$$

where $\vec{\omega} = (\omega_1, \ldots, \omega_g)^t$. Note that $\mathbb{Z}^g + \tau\mathbb{Z}^g$ is the period lattice of the holomorphic differentials in $\mathbb{C}^g$ and hence the mapping $\varphi : \hat{\Omega} \longrightarrow \mathbb{C}^g$, being multivalued, is a correctly defined mapping from $\hat{\Omega}$ to the complex torus $Jac(\hat{\Omega}) = \mathbb{C}^g/(\mathbb{Z}^g + \tau\mathbb{Z}^g)$, which is called the Jacobian variety of the Riemann surface $\hat{\Omega}$.

The action of the Abel-Jacoby mapping can obviously be extended to $D_0(\hat{\Omega})$ - the divisors of order 0 on $\hat{\Omega}$ (for details see [9, 10, 12, 13] or Preliminaries in [8])

The divisor of the zeros of $\eta_{P,J(P)}$ has order 2g;it is symmetric with respect to J and has the form $Z^* + J(Z^*)$ where $Z^* \subset \Omega_+$, $Z^* = z_1^* + \cdots + z_g^*$, and $z_1^*, \ldots, z_g^*$ are the critical points of the Green function $G(z,P)$. This is why the divisor $Z^* = Z^*(P)$ is often referred to as the critical Green's divisor. The φ-image of the divisor $Z^* - gP_0$ in $Jac(\hat{\Omega})$ can be described in terms of Riemann theta functions on $Jac(\hat{\Omega})$ (see [9, 12]).

2.3.

For any divisor $\mathcal{A}$ of order 0, $\mathcal{A} = \sum_{j=1}^k Q_j - \sum_{j=1}^k P_j$, and any vector $\vec{c} = (c_1, \ldots, c_g) \in \mathbb{C}^g$, the function

$$f_{\mathcal{A},\vec{c}}(P) = e^{2\pi i \int_{P0}^{P} \omega(\mathcal{A},\vec{c})}, \quad \omega(\mathcal{A},\vec{c}) = \sum_{j=1}^{k} \eta_{Q_j,P_j} + \sum_{l=1}^{g} c_l\omega_l\ , \tag{2.2}$$

is a so-called multiplicative meromorphic function on $\hat{\Omega}$ corresponding to the divisor $\mathcal{A}$ and vector $\vec{c}$. That is, $f_{\mathcal{A},\vec{c}}(P)$ is a locally meromorphic multivalued function on $\hat{\Omega}$, whose multivalued behavior is determined by the $2g$-dimensional vector

$$\chi = (\int_{a_1} \omega(\mathcal{A},\vec{c}), \ldots, \int_{a_g} \omega(\mathcal{A},\vec{c}) : \int_{b_1} \omega(\mathcal{A},\vec{c}), \ldots, \int_{b_g} \omega(\mathcal{A},\vec{c}))$$

of a and b periods of the differential $\omega(\mathcal{A},\vec{c})$ which due to (2.1) is equal to $(\vec{c} : \tau\vec{c} + \varphi(\mathcal{A}))$ modulo $\mathbb{Z}^{2g}$. Two values f_1, f_2 of $f_{\mathcal{A},\vec{c}}$ at the point P corresponding to two different paths of integration l_1, l_2 from P_0 to P, for which the cycle $l =$

$l_2 - l_1$ is homologous to $\sum_{j=1}^{g} n_j a_j + \sum_{j=1}^{g} m_j b_j$, are related by the equality $f_2 = e^{2\pi i \langle \chi, \vec{\mathbf{n}}_\mathbf{l} \rangle} f_1$, where $\langle \chi, \vec{\mathbf{n}}_\mathbf{l} \rangle = \sum_{j=1}^{g} n_j \chi_j + \sum_{j=1}^{g} m_j \chi_{g+j}$. The converse statement is also true (see for example [13]): any meromorphic multiplicative function on $\hat{\Omega}$ has the form (2.2).

2.4.

We will consider the vector space $\mathcal{H}_\kappa$, $\kappa \in \mathbb{R}^g/\mathbb{Z}^g$, of functions f, locally analytic on Ω_+, with single-valued modulus, and such that the analytic continuation of any functional element of f along a closed curve homologous to $\sum_{j=1}^{g} m_j b_j$ leads to multiplication of the initial value by the unimodular factor (character)

$$e^{2\pi i \sum_{j=1}^{g} m_j \kappa_j}.$$

From now on we reserve the term character for the elements $\kappa \in \mathbb{R}^g/\mathbb{Z}^g$; this means that in place of the usual multiplicative representation of the fundamental group of Ω_+ in $\mathbf{T}^g$ we use the (equivalent) additive one.

The functions from the space $\mathcal{H}_\kappa$ are called modulus automorphic or character-automorphic functions corresponding to the character κ, or simply κ-automorphic functions. They can be considered as single-valued analytic functions f in the simply connected domain $\Omega'_+ = \Omega_+ \setminus \bigcup_{j=1}^{g} \gamma_j$, with $|f|$ continuous in Ω_+, and such that the limits $f(P\pm) = \lim_{z\to P\pm} f(z)$ exist on $\bigcup_{j=1}^{g} \gamma_j \cap \Omega_+$ and satisfy $f(P+) = e^{2\pi i \lambda_j} f(P-)$, $P \in \gamma_j$, $j = 1, \dots, g$. Here the limits as $z \to P+$, $z \to P-$ on γ_j are respectively from the left and from the right side of γ_j.

It should be noted that the unit disk D is the universal covering of Ω_+, and let us denote by $\pi, \pi : D \longrightarrow \Omega_+$ the corresponding universal covering map. The group of covering transformations Σ (a Fuchsian group of the second kind) is isomorphic to the fundamental group $\pi_1(\Omega_+, P_0)$ of the domain. Any element $f \in \mathcal{H}_\kappa$ can be lifted to a single-valued holomorphic function $\tilde{f}$ on D such that $|\tilde{f} \circ \sigma| = |\tilde{f}|$, $\sigma \in \Sigma$, and $\tilde{f} \circ \sigma_j = e^{2\pi i \lambda_j} \tilde{f}$, where $\sigma_j \in \Sigma$ corresponds to the element ${\gamma_j}^{-1} b_j \gamma_j$ in $\pi_1(\Omega_+, P_0)$. The corresponding lifted space $\mathcal{H}_\kappa = \mathcal{H}_\kappa(D)$ will also be called a space of character automorphic functions.

From now on we fix the universal covering mapping $\pi : D \longrightarrow \Omega_+$ by the condition $\pi(0) = a$ for some fixed point $a \in \Omega_+$ and consider the harmonic measure $d\eta_a$ as the basic measure on Γ. By $L_p(\Gamma), p \geq 1$, we denote the usual L_p space on Γ with respect to measure $d\eta_a$.

The spaces $H^p_{+,\kappa}$, $1 \leq p \leq \infty$, mentioned in the previous section are the spaces of functions f from $\mathcal{H}_\kappa$ for which $|f|^p$ have harmonic majorant on Ω_+. Consequently (see for example [6]) any such function f has non-tangential limits on Γ a.e. $d\eta_a$ belonging to $L_p(\Gamma)$. A function f in $H^p_{+,\kappa}$ can be lifted to the universal covering as a function $\tilde{f}$ in $\mathcal{H}_\kappa(D) \cap H^p_+(D)$, where $H^p_+(D)$ is the usual Hardy space in the unit disk. For the lifted space we will use the same notation as for the space in the domain.

The multiplicative meromorphic functions on $\hat{\Omega}$ of the form (2.2), with $P_j = J(Q_j)$, $Q_j \in \Omega_+, \operatorname{Re} c_j = 0$, $j = 1, \ldots, g$, for the point P and the path of integration from P_0 to P contained in $\overline{\Omega_+}$, give important examples of character-automorphic functions. Thus, the function

$$b_Q \overset{def}{=} f_{Q-J(Q),\vec{0}}, \quad Q \in \Omega_+$$

is a character-automorphic function with $\kappa = \kappa(Q) = \operatorname{Re} \varphi(Q - J(Q))$, contractive in Ω_+, unimodular on Γ, with only one simple zero at the point Q (and only one simple pole at $J(Q)$). We will call this function an elementary Blaschke factor. By a character-automorphic Blaschke product we will mean the product (finite or infinite) of elementary Blaschke factors (for conditions on the zero set of Blaschke factors that are necessary and sufficient for the local uniform convergence of a Blaschke product see for example [11]). The lifting of b_Q to the universal covering is an ordinary Blaschke product in the unit disk with the zero set $\pi^{-1}(Q)$.

In what follows we will denote by B_Z the Blaschke product with the divisor of zeros Z and the corresponding character by $\kappa(Z)$. Thus B_{Z^*} is the finite Blaschke product corresponding to the critical Green's divisor Z^* (with the zeros at the critical points of the Green function $G(z, a)$) with character $\kappa^* = \kappa_{Z^*} = \operatorname{Re} \varphi(Z^* - J(Z^*)) = 2 \operatorname{Re} \varphi(Z^*)$.

By analogy with the theory in the unit disk we also consider the space

$$H^2_{-,\kappa} = \left\{ \overline{f} : f \in H^2_{+,-\kappa}, f(a) = 0 \right\}.$$

The spaces $H^2_{+,\kappa}$ and $H^2_{-,\kappa}$ are orthogonal since $\langle f, \overline{g} \rangle = \int_\Gamma f g d\eta_a = f(a) g(a)$ for $f \in H^2_{+,\kappa}$, $g \in H^2_{+,-\kappa}$. The space $H^2_{-,\kappa}$ can also be treated as a Hardy space on the second "sheet" Ω_- of the double $\hat{\Omega}$.

The space $H^2_{+,\kappa}$ is a functional Hilbert space and for any $t \in \Omega_+$ there exists the reproducing kernel $k(\cdot, t, \kappa) = k_t^\kappa : \langle f, k_t^\kappa \rangle = f(t)$, for any $f \in H^2_{+,\kappa}$ (for an expression of k_t^κ in terms of Riemann theta-functions on $Jac(\hat{\Omega})$ see [10]).

For the critical Green's divisor $Z^* = z_1^* + \cdots + z_g^*$ we consider the g-dimensional space $K_{Z^*}^\kappa = H^2_{+,\kappa} \ominus B_{Z^*} H^2_{+,\kappa-\kappa^*}$, $\kappa^* = \kappa(Z^*) = \operatorname{Re} \varphi(Z^* - J(Z^*))$, which is spanned by the reproducing kernels at the points z_j^*, $j = 1, \ldots, g$. Then by [8, Sec.3]

$$L^2(\Gamma) = H^2_{-,\kappa} \oplus \mathcal{M}_\kappa \oplus H^2_{+,\kappa}, \quad \mathcal{M}_\kappa = \frac{1}{B_{Z^*}} K_{Z^*}^{\kappa+\kappa^*},$$

where $\mathcal{M}_\kappa$ is the so-called κ-automorphic defect space, $\dim \mathcal{M}_\kappa = g$.

2.5.

By a character-automorphic meromorphic function on $\hat{\Omega}$ we mean a function of the form (2.2) with $\vec{c} \in \mathbb{Z}^g$ and $\varphi(\mathcal{A}) = \kappa \in \mathbb{R}^g / \mathbb{Z}^g$ modulo $\mathbb{Z}^g + \tau \mathbb{Z}^g$. Clearly such a function corresponds exactly to the character κ. The analog of Abel's theorem (see for example [13]) asserts that a divisor $\mathcal{A}$ of order zero is the divisor of a character-automorphic function if and only if $\varphi(\mathcal{A}) = \kappa \in \mathbb{R}^g / \mathbb{Z}^g$, and κ is exactly the character of this function. The divisor of a character-automorphic function f will be also denoted by (f) (clearly $ord(f) = 0$). We denote by $L_\kappa(\hat{\Omega})$ the vector space of κ-automorphic meromorphic functions on $\hat{\Omega}$ and by $L_\kappa(\mathcal{A})$ the vector space $\left\{f \in L_\kappa(\hat{\Omega}) : (f) \geq \mathcal{A}\right\}$, $r^\kappa[\mathcal{A}] = \dim L_\kappa(\mathcal{A})$. We can also consider character-automorphic meromorphic differentials on $\hat{\Omega}$. Actually all such differentials are obtained by multiplication of meromorphic differentials by character-automorphic functions on $\hat{\Omega}$. The vector space of κ-automorphic meromorphic differentials will be denoted by $\mathcal{N}_\kappa$ and $\mathcal{N}_\kappa(\mathcal{A}) = \{\omega \in \mathcal{N}_\kappa : (\omega) \geq \mathcal{A}\}$, $i^\kappa[\mathcal{A}] = \dim \mathcal{N}_\kappa(\mathcal{A})$, where by (ω) we again denote the divisor of ω. The version of Riemann-Roch theorem for character-automorphic functions and differentials asserts that for any divisor $\mathcal{A}$ on $\hat{\Omega}$

$$r^\kappa[\mathcal{A}] = -ord(\mathcal{A}) - g + 1 + i^{-\kappa}[-\mathcal{A}].$$

In these notations we clearly have

$$\mathcal{M}_\kappa = L_\kappa(J(a) - Z^* - J(Z^*)),$$

and consequently $r^\kappa[J(a) - J(Z^*)] = 0$, since any nonzero function from $L_\kappa(J(a) - J(Z^*))$ would be in both $\mathcal{M}_\kappa$ and $H^2_{+,\kappa}$ simultaneously.

2.6.

Let L, M be subspaces of the Hilbert space H with $L \cap M = \{0\}$. Then on the sum $L + M$ we can define the operator

$$\mathcal{P}_{L\|M} : l + m \mapsto l, \ l \in L, \ m \in M,$$

the so-called skew projection onto L parallel to M. By the closed graph theorem the boundedness of this operator is equivalent to the equality $L + M = \text{Clos}(L + M)$. The angle between L and M is defined as follows:

$$\angle(L, M) \in [0, \pi/2], \ \cos\angle(L, M) = \sup_{x \in L, y \in M} \frac{|\langle x, y\rangle|}{\|x\|\|y\|}.$$

It is easy to see that

$$\cos\angle(L, M) = \sup_{x \in L} \frac{\|P_M x\|}{\|x\|} = |P_M P_L|,$$

$$\sin\angle(L, M) = \inf_{x \in L} \frac{\|(I - P_M)x\|}{\|x\|} = |\mathcal{P}_{L\|M}|^{-1},$$

where P_L, P_M are the orthogonal projections onto L and M respectively.

From the above definition it is easy to see that to define an angle between L and M, we do not need to suppose that L and M are closed. They should simply be the linear spaces, contained in H. Then, by definition, $\angle(L, M) = \angle(\text{Clos}_H L, \text{Clos}_H M)$. Note also that if $L \cap M = \{0\}$ but $\text{Clos}_H L \cap \text{Clos}_H M \neq \{0\}$ we can still consider the corresponding angle and this angle is obviously equal to zero. But in this case we could not define the skew projection $\mathcal{P}_{\text{Clos}\, L\|\,\text{Clos}\, M}$. Nevertheless, we can still define the skew projection $\mathcal{P}_{L\|M} : l + m \mapsto l, \ l \in L, \ m \in M$. Then the angle between L and M is nonzero if and only if the operator $\mathcal{P}_{L\|M}$ can be extended from $L + M$ to $\text{Clos}_H(L + M)$ as a bounded operator.

3. Two technical lemmas on divisors of character-automorphic functions on $\hat{\Omega}$

Let $Z = z_1 + z_2 + \cdots + z_g$ be a positive divisor of order g on $\hat{\Omega}$. By Abel's theorem, for any $\kappa \neq 0 \in \mathbb{R}^g/\mathbb{Z}^g$ we can find another positive divisor $T = t_1 + \cdots + t_g$ of order g such that $T - Z$ is a divisor of some κ-automorphic function on $\hat{\Omega}$. Unfortunately in general we can say nothing about the positions of the points $t_1, \dots, t_g$ on $\hat{\Omega}$. But if the character κ is not fixed and we are looking for a divisor T such that $T - Z$ is a divisor of a character-automorphic function for some character, then it is possible to choose T in two (positionally) different ways: T is totally contained inside Ω_+ and (or) T is totally contained in Γ, and in the last case T may be chosen in such a way that $t_1 \in \Gamma_1, \dots, t_g \in \Gamma_g$.

Probably the above statements are already known, but they are important in what follows and we prove these facts in the two following lemmas. The proofs are essentially based on the results from [12, Chapter VI].

Lemma 3.1. *For any positive divisor Z of order g on $\hat{\Omega}$, $\varphi(Z) = \mu + \tau\nu$, $\mu, \nu \in \mathbb{R}^g/\mathbb{Z}^g$, there exists a positive divisor T of order g, $T = t_1 + \cdots + t_g$, $t_1 \in \Gamma_1, \dots, t_g \in \Gamma_g$, such that $\varphi(T) = (1, 1, \dots, 1)^t + \tau\nu$. That is, $\operatorname{Im}\varphi(T - Z) = 0$ and, by Abel's theorem for character-automorphic functions, there exists a $((1, \dots, 1) - \mu)$-automorphic function with the divisor $T - Z$.*

Proof. Due to [12, Propositions 6.2 and 6.3] the subvariety S of $Jac(\hat{\Omega})$, defined by $S = \{s \in Jac(\hat{\Omega}) : [s] = -[s]\}$, where $[s]$ denotes the equivalence class of s in $Jac(\hat{\Omega})$, is a disjoint union of 2^g real g-dimensional torii $S_\mu = \frac{1}{2}\mu + \tau\nu$, $\mu \in (\mathbb{Z}/2\mathbb{Z})^g$, $\nu \in \mathbb{R}^g/\mathbb{Z}^g$, and each torus $S_\mu, \mu = (\mu_1, \dots, \mu_g)^t$, consists of the φ-images of all divisors of the form $T - gb_0$, where $b_0 \in \Gamma_0$ is a base point and T is a positive symmetric ($J(T) = T$) divisor of order g, containing, modulo 2, μ_k points on Γ_k, $k = 1, \dots, g$ [1].

Now to prove the Lemma we simply take that positive symmetric divisor T of order g for which $\varphi(T - gb_0) = (1, \dots, 1)^t + \tau\nu$, where $\nu = \operatorname{Im}\varphi(Z)$. Then, due to Fay's description, T contains exactly one point on each boundary component Γ_k, $k = 1, \dots, g$. □

Lemma 3.2. *For any positive divisor Z of order g on $\hat{\Omega}$, $\varphi(Z) = \mu_Z + \tau\nu_Z$, $\mu_Z, \nu_Z \in \mathbb{R}^g/\mathbb{Z}^g$, there exists a positive divisor T of order g that is contained totally inside Ω_+ and for which $\varphi(T) = \mu_T + \tau\nu_Z$. That is $\operatorname{Im}\varphi(T - Z) = 0$ and, by Abel's theorem for character-automorphic function, there exists a $(\mu_T - \mu_Z)$-automorphic function with the divisor $T - Z$.*

Proof. First we note that by the previous Lemma it is enough to prove this only for positive divisors of order g that contain exactly one point from each of the boundary components $\Gamma_1, \dots, \Gamma_g$.

[1] Since $Jac(\hat{\Omega})$ can be identified with D_0 – the group of divisors on $\hat{\Omega}$ of order 0 modulo principal divisors, it is often convenient to say that the divisor $\mathcal{A} \in D_0$ belongs to S_μ, $\mu \in (\mathbb{Z}/2\mathbb{Z})^g$, if $\varphi(\mathcal{A}) \in S_\mu$.

Thus, let $Z = z_1 + \cdots + z_g$ with $z_k \in \Gamma_k$, $k = 1, \dots, g$. It is very easy to see (see for example [11]) that $i[Z] = 0$, i.e., there are no holomorphic differentials on $\hat{\Omega}$ which have zeros at all points $z_1, \dots, z_g$. This means that the matrix $\{\varphi_j(z_k)\}_{k,j=1}^g$ is nondegenerate. But the matrix $\{\varphi_j(u_k)\}_{k,j=1}^g$, $u_k \in \hat{\Omega}$, $k = 1, \dots, g$, is exactly equal to the Jacobian of the analytic mapping $\Phi(u_1, \dots, u_g) = \varphi(u_1 + \cdots + u_g - gb_0)$ from a (sufficiently small) neighborhood U_Z of the point $(z_1, \dots, z_g) \in \hat{\Omega} \times \cdots \times \hat{\Omega}$ into $\mathbb{C}^g$. So the Jacobian of this mapping is nondegenerate at $(z_1, \dots, z_g)$ and there exists the inverse analytic mapping Φ^{-1} from a small neighborhood $V_{\varphi(Z)}$ of the point $\mu_Z + \tau\nu_Z$ onto a neighborhood of the point $(z_1, \dots, z_g)$. We consider all the points $(u_1, \dots, u_g)$ close to $(z_1, \dots, z_g)$ such that $\operatorname{Im} \varphi(u_1 + \cdots + u_g - gb_0) = \operatorname{Im} \varphi(Z - gb_0)$. These points form a real manifold of dimension g which is exactly the image of the g-dimensional real subspace $i \operatorname{Im} \mathbb{C}^g$ of $\mathbb{C}^g$ under the mapping Φ^{-1} (here $i \operatorname{Im}(w_1, w_2, \dots, w_g) = (i \operatorname{Im} w_1, i \operatorname{Im} w_2, \dots, i \operatorname{Im} w_g)$). Now since we already know that the Jacobian of this mapping at the point $\mu_Z + \tau\nu_Z$ is nondegenerate and all its elements are purely imaginary (see Sec. 2.1 for the properties of the differentials $\varphi_1, \dots, \varphi_g$), we conclude that the tangent space of this manifold $\Phi^{-1}(i \operatorname{Im} \mathbb{C}^g)$ at the point $(z_1, \dots, z_g)$ is exactly $\operatorname{Re} \mathbb{C}^g$. This analysis shows that for any fixed l, $1 \le l \le g$, there is a point $(u_1, \dots, u_g)$ in the neighborhood U_Z with $\operatorname{Im} \varphi(u_1 + \cdots + u_g) = \tau\nu_Z$ and $u_l \ne z_l$. Indeed, otherwise, say for $l = 1$, $\Phi^{-1}(i \operatorname{Im} \mathbb{C}^g) \cap U_Z \subset \{(u_1, \dots, u_g) \in U_Z : u_1 = z_1\}$. But the last inclusion means that the above mentioned tangent space is $(g-1)$-dimensional, since all the vectors of the form $(x + iy, 0, \dots, 0), \forall x, y \in \mathbb{R}$, are orthogonal to the manifold $\Phi^{-1}(i \operatorname{Im} \mathbb{C}^g) \cap U_Z$ itself and in particular to its tangent space at $(z_1, \dots, z_g)$.

Thus, for each l, $1 \le l \le g$, there exists a character κ_l such that the space $L_{\kappa_l}(-Z)$ contains a function f_l with a nondegenerate pole at the point z_l. Then the meromorphic functions $f_l(z)\overline{f_l(J(z))}$ are single-valued, positive on Γ and belong to $L(-2Z)$. Their sum is also positive on Γ and has nondegenerate double poles at the points of Z. Hence, for a positive constant C, the divisor of the function $f = \sum_{l=1}^g f_l(z)\overline{f_l(J(z))} + C$, which is strictly positive on Γ, is symmetric $(J((f)) = (f))$ and has the form $(f) = T + J(T) - 2Z$, where T is positive divisor of order g totally contained inside Ω_+.

We claim that T is the desired divisor and $\sqrt{1/(fb_T)}$ is the desired function, where as above b_T is a finite character-automorphic Blaschke product, $(b_T) = T - J(T)$. To show this we must prove that $\operatorname{Im}(\varphi(Z - T)) = \tau(\nu_Z - \nu_T) = 0$. Actually, since f is a single-valued meromorphic function on $\hat{\Omega}$, we already know that $2(\nu_Z - \nu_T) \in \mathbb{Z}^g$, and to see that $(\nu_Z - \nu_T) \in \mathbb{Z}^g$ we simply note that, by construction of the function f, $\omega = (1/f)d\eta_a$ is a symmetric meromorphic differential on $\hat{\Omega}$, nonnegative on Γ, with divisor $(\omega) = 2Z - T - J(T) + Z^* + J(Z^*) - a - J(a)$. Hence, by [12, Proposition 6.8], $\varphi(Z - T + Z^* - a) + [\Delta_0] \in T_0 = \{\mu + \tau\nu : \mu \in \mathbb{R}^g/\mathbb{Z}^g, \nu \in \mathbb{Z}^g\}$, where Δ_0 is a vector of Riemann constants, which depends only on the choice of homology basis and base point b_0. But $d\eta_a$ itself is a symmetric meromorphic differential on $\hat{\Omega}$ which is positive on Γ. Hence, by the same Proposition from [12] $\varphi(Z^* - a) + [\Delta_0] \in T_0$ and, finally, $\operatorname{Im} \varphi(Z - T) \in \mathbb{Z}^g$. □

4. The weight might be bad

4.1.

Let $\mathcal{B}_+^\kappa$ be the space of all functions in $H_\kappa(\Omega_+)$ which are continuous up to the boundary Γ. It is known (see for example [7, Sec.3]) that $\mathcal{B}_+^\kappa$ is dense in $H^2_{+,\kappa}$. By $\mathcal{B}_-^\kappa$ we denote the corresponding dense subset of $H^2_{-,\kappa}$, $\mathcal{B}_-^\kappa = \{\overline{f \circ J} : f \in \mathcal{B}_+^{-\kappa}, f(a) = 0\}$. Note that the sum $\mathcal{B}_+^\kappa + \mathcal{B}_-^\kappa + \mathcal{M}_\kappa$ is obviously dense in $L_2(d\eta_a)$ and moreover (see [7, Sec.3]) is dense in $L_2(d\mu)$ for any finite positive measure $d\mu$ as well.

The question we are interested in is: What are necessary and sufficient conditions on a finite positive measure $d\mu$ on Γ for the angle between the subspaces $\mathrm{Clos}_{L_2(d\mu)}\,\mathcal{B}_-^\kappa$ and $\mathrm{Clos}_{L_2(d\mu)}\,\mathcal{B}_+^\kappa$ to be nonzero? An equivalent question asks when (for what measures $d\mu$) the orthogonal projection P_+^κ in $L_2(\Gamma, d\eta_a) = L_2(\Gamma)$ onto $H^2_{+,\kappa}$ can be extended from $\mathcal{B}_-^\kappa + \mathcal{B}_+^\kappa$ to $\mathrm{Clos}_{L_2(d\mu)}\{\mathcal{B}_-^\kappa + \mathcal{B}_+^\kappa\}$ as a bounded operator?

Proposition 4.1. *If the angle in $L_2(d\mu)$ between $\mathrm{Clos}_{L_2(d\mu)}\,\mathcal{B}_-^\kappa$ and $\mathrm{Clos}_{L_2(d\mu)}\,\mathcal{B}_+^\kappa$ is nonzero, then $d\mu$ is absolutely continuous with respect to $d\eta_a$, $d\mu = w d\eta_a$, and $\log w \in L_1(\Gamma)$.*

Proof. Let $d\mu = w d\eta_a + d\mu_s$, where $w \in L_1(d\eta_a)$ and $d\mu_s$ is the singular component of $d\mu$. The statement about w follows immediately from the character-automorphic analog of Szegö's theorem ([7, Theorem 3.1, Corollary 3.1]), which in particular asserts that if $\log w \notin L_1(d\eta_a)$ then $\mathrm{Clos}_{L_2(d\mu)}\,\mathcal{B}_+^\kappa = L_2(d\mu)$. The latter is not true in our case since the closures of $\mathcal{B}_+^\kappa$ and $\mathcal{B}_-^\kappa$ in $L_2(d\mu)$ are at nonzero angle. As for the absence of the singular component $d\mu_s$, this assertion was actually proved in [7, Proposition 4.1], which contains the same statements on the measures in the case where $\mathrm{Clos}_{L_2(d\mu)}\{\mathcal{B}_-^\kappa + \mathcal{M}_\kappa\}$ and $\mathrm{Clos}_{L_2(d\mu)}\,\mathcal{B}_+^\kappa$ are at nonzero angle. □

From now on we denote by h a character-automorphic outer function in H^2_{+,κ_0}, with a certain character κ_0, such that $|h|^2\,|_\Gamma = w$, $d\mu = |h|^2\, d\eta_a$.

4.2.

As in the case of the unit disk for $\lambda \in \Omega_+$ and $f \in L_1(\Gamma)$ we denote by $f(\lambda)$ the value at the point λ of the harmonic extension of f in Ω_+, $f(\lambda) = \int_\Gamma f\, d\eta_\lambda$.

Remark 4.2. Since we know that the measure $d\mu = w d\eta_a = |h|^2 d\eta_a$ is absolutely continuous with respect to harmonic measure $d\eta_a$, it is easy to see that in place of the spaces $\mathcal{B}_+^\kappa, \mathcal{B}_-^\kappa$ we can use the spaces $H^\infty_{+,\kappa}, H^\infty_{-,\kappa}$. Indeed, we have $\mathcal{B}_+^\kappa \subset H^\infty_{+,\kappa}$, $\mathcal{B}_-^\kappa \subset H^\infty_{-,\kappa}$ and

$$\mathcal{D}_+^\kappa = \mathrm{Clos}_{L_2(d\mu)}\, H^\infty_{+,\kappa} = \mathrm{Clos}_{L_2(d\mu)}\, \mathcal{B}_+^\kappa = \frac{1}{h} H^2_{+,\kappa+\kappa_0},$$

$$\mathcal{D}_-^\kappa = \mathrm{Clos}_{L_2(d\mu)}\, H^\infty_{-,\kappa} = \mathrm{Clos}_{L_2(d\mu)}\, \mathcal{B}_-^\kappa = \frac{1}{h} H^2_{-,\kappa-\kappa_0}.$$

To see that, for example, the first of these chains of equalities is valid one should simply note that due to the properties of outer functions, $h\mathcal{B}^\kappa$ is dense in $H^2_{+,\kappa+\kappa_0}$, and that the convergence of the sequence of functions $\{f_n\}_{n=1}^\infty$ to the function f in $L_2(d\mu)$ is equivalent to the convergence of the sequence $\{hf_n\}_{n=1}^\infty$ to the function hf in L_2.

Proposition 4.3. *If the "Riesz" projection P_+^κ can be extended from $\mathcal{B}_-^\kappa + \mathcal{B}_+^\kappa$ to $\mathrm{Clos}_{L_2(d\mu)}\{\mathcal{B}_-^\kappa + \mathcal{B}_+^\kappa\}$ as a bounded operator then the following alternative takes place:*

1. *$1/h$ is a $(-\kappa_0)$-automorphic outer function and $|h|^2$ satisfies the $A_2'(\Gamma)$ Muckenhoupt condition (see [7]):*

$$\sup_{\lambda\in\Omega_+} (|h|^2)(\lambda)\cdot(1/|h|^2)(\lambda) < \infty.$$

2. *There exists a positive divisor $U_h = u_1 + \cdots + u_k$ of order $k \le g$ contained totally in Γ, such that $1/h$ is locally square summable (from L_2^{loc}) at any point from $\Gamma \setminus U_h$, is not square summable at any of the points u_j, $1 \le j \le k$, but at each of these points the function $(z-u_j)^{\nu_j}/h$ is locally square summable, where $\nu_j = ordU_h|_{u_j}$, the order of divisor U_h at the point u_j (note that some of the points u_j may coincide). Moreover U_h is a minimal divisor possessing these properties in the sense that if U is any other divisor with the same properties, then $U_h \le U$.*

Proof. As we know the nonzero angle condition means that

$$H^2_{-,\kappa+\kappa_0}/\bar{h} \cap H^2_{+,\kappa+\kappa_0}/h = 0$$

and $H^2_{-,\kappa+\kappa_0}/\bar{h} + H^2_{-,\kappa+\kappa_0}/h$ is closed. Note that if $(H^2_{-,\kappa+\kappa_0} + \mathcal{M}_{\kappa+\kappa_0})/\bar{h} \cap H^2_{+\kappa+\kappa_0}/h = 0$ then the angle between the last two subspaces is nonzero and by [7] the weight satisfies the Muckenhoupt condition. This case exactly corresponds to the condition 1 of the Proposition. But if the dimension of the intersection of $(H^2_{-,\kappa+\kappa_0} + \mathcal{M}_{\kappa+\kappa_0})/\bar{h}$ and $H^2_{+,\kappa+\kappa_0}/h = 0$ is greater than zero then $1/h \notin L_2(\Gamma)$ and respectively $1/w \notin L_1(\Gamma)$. Indeed, if $1/h$ is square summable on Γ then it is an outer function from $H^2_{+,-\kappa_0}$ and for the nonzero function $f \in \mathcal{D}_- \cap \mathcal{D}_+$: $f = f_-/\bar{h} = f_+/h$, where $f_- \in H^2_{-,\kappa-\kappa_0} + \mathcal{M}_{\kappa-\kappa_0}$ and $f_+ \in H^2_{+,\kappa+\kappa_0}$. But then $f \in \{H^1_{-,\kappa} + \mathcal{M}_\kappa\} \cap H^1_{+,\kappa}$, which is impossible since the last intersection is zero by character-automorphic analog of Riesz brothers theorem (see [8]).

Therefore we are interested exactly in a case when

$$\int_\Gamma \frac{1}{w} d\eta_a = \infty. \tag{4.1}$$

On the other hand, since $\log w \in L_1$, we know (see [7, Sec.3] that for any fixed $\lambda \in \Omega_+$ the distance in $L_2(wd\eta_a)$ between the reproducing kernel k_λ^κ in $H^2_{+,\kappa}$ at the point λ and $b_\lambda \mathcal{B}_+^{\kappa-\kappa(\lambda)}$ is nonzero and, hence the orthogonal projection $P^+_{k_\lambda^\kappa}$ in

$H^2_{+,\kappa}$ onto the one-dimensional space spanned by k^κ_λ can be extended from $\mathcal{B}^\kappa_+$ as a bounded operator to $\mathrm{Clos}_{L_2(d\mu)}\mathcal{B}^\kappa_+$:

$$|P^+_{k^\kappa_\lambda}|_{L_2(d\mu)} = \frac{\|k^\kappa_\lambda\|_{L_2(d\mu)}}{dist_{L_2(d\mu)}(k^\kappa_\lambda, b_\lambda \mathcal{B}^{\kappa-\kappa(\lambda)}_+)} < \infty.$$

The orthogonal projection $P_{k^\kappa_\lambda}$ in $L_2(\Gamma)$ onto the same one-dimensional space spanned by k^κ_λ can be extended from $\mathcal{B}^\kappa_- + \mathcal{B}^\kappa_+$ as a bounded operator as well to $\mathrm{Clos}_{L_2(d\mu)}\{\mathcal{B}^\kappa_- + \mathcal{B}^\kappa_+\}$. Indeed, this fact immediately follows from the expression $P_{k^\kappa_\lambda} = P^+_{k^\kappa_\lambda} P^\kappa_+$.Therefore

$$|P_{\{k^\kappa_\lambda\}}|_{L_2(d\mu)} = \frac{\|k^\kappa_\lambda\|_{L_2(d\mu)}}{dist_{L_2(d\mu)}(k^\kappa_\lambda, \mathcal{B}^\kappa_- + b_\lambda \mathcal{B}^{\kappa-\kappa(\lambda)}_+)} < \infty. \tag{4.2}$$

Let us put for any positive ε $w_\varepsilon = w + \varepsilon$ and $d\mu_\varepsilon = w_\varepsilon d\eta_a$. Then for any $f \in \mathcal{B}^\kappa_- + \mathcal{M}_\kappa + \mathcal{B}^\kappa_+$

$$\lim_{\varepsilon \to 0+} \int_\Gamma |f|^2 d\mu_\varepsilon = \int_\Gamma |f|^2 d\mu$$

by Lebesgue's theorem on monotone convergence.

Now using the measures $d\mu_\varepsilon$ and taking into account that in $L_2(\Gamma)$

$$\left(\sqrt{w_\varepsilon}\{\mathcal{B}^\kappa_- + b_\lambda \mathcal{B}^{\kappa-\kappa_\lambda}_+\}\right)^\perp = \frac{1}{\sqrt{w_\varepsilon}}(\{\mathbb{C}k^\kappa_\lambda\} + \mathcal{M}_\kappa),$$

we get

$$dist^2_{L_2(d\mu_\varepsilon)}(k^\kappa_\lambda, \mathcal{B}^\kappa_- + b_\lambda \mathcal{B}^{\kappa-\kappa(\lambda)}_+) = dist^2_{L_2}(\sqrt{w_\varepsilon}k^\kappa_\lambda, \sqrt{w_\varepsilon}(\mathcal{B}^\kappa_- + b_\lambda \mathcal{B}^{\kappa-\kappa(\lambda)}_+))$$

$$= \sup_{\substack{\varphi \in \left\{\sqrt{w_\varepsilon}\left(\mathcal{B}^\kappa_- + b_\lambda \mathcal{B}^{\kappa-\kappa(\lambda)}_+\right)\right\}^\perp \\ \|\varphi\|_2 = 1}} |\langle \sqrt{w_\varepsilon}k^\kappa_\lambda, \varphi\rangle|^2 = \sup_{\varphi \in \{\mathbb{C}k^\kappa_\lambda\} + \mathcal{M}_\kappa} \frac{|\langle k^\kappa_\lambda, \varphi\rangle|^2}{\int_\Gamma |\varphi|^2 \frac{1}{w_\varepsilon} d\eta_a}$$

$$= (k^\kappa_\lambda(\lambda))^2 \left(\inf_{\varphi \in \mathcal{M}_\kappa} \int_\Gamma |k^\kappa_\lambda + \varphi|^2 \tfrac{1}{w_\varepsilon} d\eta_a\right)^{-1},$$

which, by compactness arguments in the finite dimensional space $\mathcal{M}_\kappa$, means that there exists $\varphi_\varepsilon \in \mathcal{M}_\kappa$, such that

$$dist^2_{L_2(d\mu_\varepsilon)}(k^\kappa_\lambda, \mathcal{B}^\kappa_- + b_\lambda \mathcal{B}^{\kappa-\kappa(\lambda)}_+) = (k^\kappa_\lambda(\lambda))^2 \left(\int_\Gamma |k^\kappa_\lambda + \varphi_\varepsilon|^2 \frac{1}{w_\varepsilon} d\eta_a\right)^{-1},$$

Since $\lim_{\varepsilon\to 0+} dist_{L_2(d\mu_\varepsilon)}(k^\kappa_\lambda, \mathcal{B}^\kappa_- + b_\lambda \mathcal{B}^{\kappa-\kappa(\lambda)}_+) = dist^2_{L_2(d\mu)}(k^\kappa_\lambda, \mathcal{B}^\kappa_- + b_\lambda \mathcal{B}^{\kappa-\kappa(\lambda)}_+)$, taking into account (4.2) we see that $\lim_{\varepsilon\to 0+} \int_\Gamma |k^\kappa_\lambda + \varphi_\varepsilon|^2 \frac{1}{w_\varepsilon} d\eta_a$ exists and is finite. Now if we assume that the norms $\|k^\kappa_\lambda + \varphi_\varepsilon\|_\infty$ are not bounded for all positive ε, then there exists a subsequence $\{\varepsilon_n\}$ of positive numbers with a zero limit such that $\|k^\kappa_\lambda + \varphi_{\varepsilon_n}\|_\infty \longrightarrow \infty$ as n tends to ∞. The functions $\varphi'_n = (k^\kappa_\lambda + \varphi_{\varepsilon_n})/\|k^\kappa_\lambda + \varphi_{\varepsilon_n}\|_\infty$ belong to $(g+1)$-dimensional space $\mathbb{C}k^\kappa_\lambda + \mathcal{M}_\kappa$ and are of unit L_∞-norm. Hence, without loss of generality, by compactness arguments, we can assume that this sequence itself is convergent. Let let $\varphi' = \lim_{n\to\infty} \varphi'_n$. Clearly $\varphi' \in \mathbb{C}k^\kappa_\lambda +$

$\mathcal{M}_\kappa$, $\|\varphi'\|_\infty = 1$. But, by the definition of the functions φ'_n, $|\varphi'_n|^2/w_{\varepsilon_n} \longrightarrow |\varphi'|^2/w$ as $n \to \infty$ almost everywhere, and by Fatou's Lemma

$$\lim_{n\to\infty} \int_\Gamma \frac{|\varphi'_n|^2}{w_{\varepsilon_n}} d\eta_a = \int_\Gamma \frac{|\varphi'|^2}{w} d\eta_a = 0.$$

But this is impossible, since φ' is a nonzero function, locally analytic on Γ, and hence the equality $\int_\Gamma \frac{|\varphi'|^2}{w} d\eta_a = 0$ implies that $w = |h|^2$ is equal to infinity on a set of positive Lebesgue measure on Γ.

This contradiction shows that the norms $\|k^\kappa_\lambda + \varphi_\varepsilon\|_\infty$ are bounded for all positive ε, and again, by compactness arguments, we can find a sequence of positive numbers ε_n, decreasing to zero, such that $(k^\kappa_\lambda + \varphi_{\varepsilon_n})$ tends to $k^\kappa_\lambda + \varphi_\lambda$, $\varphi_\lambda \in \mathcal{M}_\kappa$, uniformly on Γ as n tends to infinity. Finally,

$$\int_\Gamma |k^\kappa_\lambda + \varphi_\lambda|^2 \frac{1}{w} d\eta_a = \lim_{\varepsilon\to 0+} \int_\Gamma |k^\kappa_\lambda + \varphi_\varepsilon|^2 \frac{1}{w_\varepsilon} d\eta_a < \infty. \tag{4.3}$$

Now let us compare the expressions (4.1) and (4.3). On the one hand expression (4.1) tells us that $1/w \notin L_1(\Gamma)$ but on the another hand (4.3) tells us that for any $\lambda \in \Omega_+$ it is possible to find $\varphi_\lambda \in \mathcal{M}_\kappa$, such that $|k^\kappa_\lambda + \varphi_\lambda|^2/w \in L_1(\Gamma)$. Since $k^\kappa_\lambda + \varphi_\lambda$ is locally analytic on Γ, it means that $1/w$ may be not locally summable only at the zeros of $k^\kappa_\lambda + \varphi_\lambda$ on Γ. Since $k^\kappa_\lambda + \varphi_\lambda \in L_\kappa(J(a) - J(\lambda) - Z^* - J(Z^*))$ (see for example [8, Proposition 3.2]) it may have at most $2g$ zeros on Γ. Thus the number of points on Γ at which $1/w$ is not locally summable is also finite. So $1/h$ is not locally square summable on Γ only at a finite number of points.

Let us denote by $U = U_h$ the positive divisor of these points on Γ. Note that due to (4.3) the order of U_h at a zero of $k^\kappa_\lambda + \varphi_\lambda$ is less or equal to the order of this zero. Therefore if $\nu = ordU_h|_u$, $u \in \Gamma$ then $(z-u)^\nu/h$ is locally square summable at u.

To finish the proof of Proposition 4.3 we must show that $ordU_h \le g$. Note that since (4.3) is valid for all $\lambda \in \Omega_+$, $k^\kappa_\lambda + \varphi_\lambda \in L^\kappa(U_h + J(a) - J(\lambda) - Z^* - J(Z^*))$ for all $\lambda \in \Omega_+$. Hence $r^\kappa[U_h + J(a) - J(\lambda) - Z^* - J(Z^*)] \ge 1$ for all $\lambda \in \Omega_+$. An application of the Riemann-Roch theorem gives us

$$\begin{aligned} 1 &\le r^\kappa[U_h + J(a) - J(\lambda) - Z^* - J(Z^*)] \\ &= g + 1 - ord(U_h) + i^{-\kappa}[J(\lambda) + Z^* + J(Z^*) - U_h - J(a)], \end{aligned}$$

$$r^\kappa[U_h + J(a) - Z^* - J(Z^*)] = g - ord(U_h) + i^{-\kappa}[Z^* + J(Z^*) - U_h - J(a)]. \tag{4.4}$$

Clearly $r^\kappa[U_h + J(a) - J(\lambda) - Z^* - J(Z^*)] \ge r^\kappa[U_h + J(a) - Z^* - J(Z^*)]$ and $i^{-\kappa}[Z^* + J(Z^*) - U_h - J(a)] \ge i^{-\kappa}[J(\lambda) + Z^* + J(Z^*) - U_h - J(a)]$. We will show that

$$i^{-\kappa}[Z^* + J(Z^*) - U_h - J(a)] = 0. \tag{4.5}$$

This will imply that $i^{-\kappa}[J(\lambda) + Z^* + J(Z^*) - U_h - J(a)] = 0$ and, by (4.4), that $ordU_h \le g$. Suppose for the moment that $i^{-\kappa}[Z^* + J(Z^*) - U_h - J(a)] > 0$, then the equality $i^{-\kappa}[Z^* + J(Z^*) - U_h - J(a)] = i^{-\kappa}[J(\lambda) + Z^* + J(Z^*) - U_h - J(a)]$

can be valid only for a finite number of points $\lambda \in \Omega_+$. Indeed, this equality means that all differentials from $\mathcal{N}_\kappa(Z^* + J(Z^*) - U_h - J(a))$ have a zero at the point λ, which is not true if λ does not belong to the set of zeros of one certain differential from this space. Now for such λ $i^{-\kappa}[Z^* + J(Z^*) - U_h - J(a)] = i^{-\kappa}[J(\lambda) + Z^* + J(Z^*) - U_h - J(a)] + 1$ and, by (4.4)

$$\begin{aligned} & r^\kappa[U_h + J(a) - Z^* - J(Z^*)] \\ = \; & g + 1 - ord(U_h) + i^{-\kappa}[J(\lambda) + Z^* + J(Z^*) - U_h - J(a)] \\ = \; & r^\kappa[U_h + J(a) - J(\lambda) - Z^* - J(Z^*)]. \end{aligned}$$

But due to (4.3) $k_\lambda^\kappa + \varphi_\lambda \in L_\kappa(U_h + J(a) - J(\lambda) - Z^* - J(Z^*))$ and $k_\lambda^\kappa + \varphi_\lambda \notin L_\kappa(U_h + J(a) - Z^* - J(Z^*))$. This contradiction proves (4.5) and finishes the proof of the Proposition 4.3. □

Corollary 4.4. *For any $\lambda \in \Omega_+$ and function $k_\lambda^\kappa + \varphi_\lambda$ from (4.3),*

$$\frac{k_\lambda^\kappa + \varphi_\lambda}{h} \in \mathcal{M}_{\kappa-\kappa_0} + H^2_{+,\kappa-\kappa_0}.$$

This fact immediately follows from the observation that this function is square summable on Γ and meromorphic on Ω_+ with the divisor $(k_\lambda^\kappa + \varphi_\lambda) \geq -J(Z^*)$ on Ω_+.

Corollary 4.5. 1. *We have*

$$r^\kappa[U_h + J(a) - Z^* - J(Z^*)] = g - ordU_h,$$

and for any function $\varphi_{\mathcal{M}_\kappa} \in L_\kappa(U_h + J(a) - Z^* - J(Z^*))$

$$\frac{\varphi_{\mathcal{M}_\kappa}}{h} \in \mathcal{M}_{\kappa-\kappa_0} + H^2_{+,\kappa-\kappa_0}, \quad \frac{\varphi_{\mathcal{M}_\kappa}}{\bar{h}} \in \mathcal{M}_{\kappa+\kappa_0} + H^2_{+,\kappa+\kappa_0}.$$

2. *Any function $\varphi_{\mathcal{M}_\kappa} \in \mathcal{M}_\kappa$ for which*

$$\frac{\varphi_{\mathcal{M}_\kappa}}{h} \in \mathcal{M}_{\kappa-\kappa_0} + H^2_{+,\kappa-\kappa_0}, \quad \frac{\varphi_{\mathcal{M}_\kappa}}{\bar{h}} \in \mathcal{M}_{\kappa+\kappa_0} + H^2_{+,\kappa+\kappa_0},$$

belongs to $L_\kappa(U_h + J(a) - Z^* - J(Z^*))$

The proof immediately follows from (4.4), (4.5) and the same observation as for Corollary 4.4.

Remark 4.6. It follows from the proof of Proposition 4.3, that for any character $\nu \in \mathbb{R}^g/\mathbb{Z}^g$ and any $\lambda \in \Omega_+$

$$r^\nu[U_h + J(a) - J(\lambda) - Z^* - J(Z^*)] \geq 1,$$

and for any $\psi \in L_\nu(U_h + J(a) - J(\lambda) - Z^* - J(Z^*))$ the function ψ/h is square summable on Γ.

The converse statement is also true: If $\psi/h \in L^2(\Gamma)$, $\psi \in k_\lambda^\nu + \mathcal{M}_\nu$, then $\psi \in L_\nu(U_h + J(a) - J(\lambda) - Z^* - J(Z^*))$.

5. The weight can be improved

5.1.

Only case (2) of the alternative from the Proposition 4.3 is of interest for us, so in what follows we suppose that $0 < ordU_h \leq g$. Our first aim is to study the function h and the divisor U_h corresponding to its "bad behavior" on Γ. To do this we will use a technique, similar to that in [14, 2, 3], which involves Hankel and Toeplitz operators.

As above P_+^ν, $\nu \in \mathbb{R}^g/\mathbb{Z}^g$, is the "Riesz" orthogonal projection from $L_2(\Gamma)$ onto $H^2_{+,\nu}$. Let $P_-^\nu = I - P_+^\nu$ be the orthogonal projection from $L_2(\Gamma)$ onto $H^2_{-,\nu} + \mathcal{M}_\nu$. Recall that (see for example [8]) for any $f \in L_\infty(\Gamma)$ and $\mu, \nu \in \mathbb{R}^g/\mathbb{Z}^g$ the Hankel operator $\mathbf{H}_f^{\nu,\mu}$ with symbol f is the operator acting from $H^2_{+,\nu}$ into $H^2_{-,\nu+\mu} + \mathcal{M}_{\nu+\mu}$ according to the formula

$$\mathbf{H}_f^{\nu,\mu} g = P_-^{\nu+\mu} fg, \quad \forall g \in H^2_{+,\nu}.$$

The character-automorphic Toeplitz operator $\mathbf{T}_f^{\nu,\mu}$ with symbol f acts from $H^2_{+,\nu}$ into $H^2_{+,\nu+\mu}$:

$$\mathbf{T}_f^{\nu,\mu} g = fg - \mathbf{H}_f^{\nu,\mu} g = P_+^{\nu+\mu} fg, \quad \forall g \in H^2_{+,\nu}.$$

Here the character ν describes the Hardy space on which the operators act, and the character μ prescribes the character of the symbol $f \in L_\infty(\Gamma)$ [2].

If f is a unimodular function on Γ then clearly, for any $g \in H^2_{+,\nu}$,

$$\|\mathbf{H}_f^{\nu,\mu} g\|_2^2 + \|\mathbf{T}_f^{\nu,\mu} g\|_2^2 = \|g\|_2^2, \tag{5.1}$$

and the adjoint operator $\left(\mathbf{T}_f^{\nu,\mu}\right)^*$ is the Toeplitz operator $\mathbf{T}_{\overline{f}}^{\nu+\mu,-\mu}$.

Just as in the case of the unit disk, the estimation of the angles between subspaces in weighted L_2 on Γ is equivalent to the estimation of the norms of corresponding Hankel operators:

$$\cos\angle(\mathrm{Clos}_{L_2(d\mu)} \mathcal{B}_-^\kappa, \mathrm{Clos}_{L_2(d\mu)} \mathcal{B}_+^\kappa) = \sup_{f\in\mathcal{B}_+^\kappa, g\in\mathcal{B}_-^\kappa} \frac{|\langle fh, gh\rangle|}{\|fh\|_2 \cdot \|g\overline{h}\|_2}$$

$$= \sup_{\substack{f\in H^2_{+,\kappa+\kappa_0} \\ g\in H^2_{-,\kappa-\kappa_0}}} \frac{|\langle f\frac{\overline{h}}{h}, g\rangle|}{\|f\|_2 \cdot \|g\|_2} = \sup_{\substack{f\in H^2_{+,\kappa+\kappa_0} \\ g\in\left(H^2_{+,\kappa-\kappa_0+\kappa^*}\right)^\perp}} \frac{|\langle B_{Z^*}\frac{\overline{h}}{h} f, g\rangle|}{\|f\|_2 \cdot \|g\|_2}$$

$$= \left|\mathbf{H}^{\kappa+\kappa_0,-2\kappa_0+\kappa^*}_{B_{Z^*}\frac{\overline{h}}{h}}\right|.$$

Therefore case (1) of the alternative from Proposition 4.2 corresponds to the case $\left|\mathbf{H}^{\nu,2\kappa_0}_{\frac{\overline{h}}{h}}\right| < 1$ for all $\nu \in \mathbb{R}^g/\mathbb{Z}^g$ (see [7]), which, in particular, implies that $\left|\mathbf{H}^{\kappa+\kappa_0,-2\kappa_0+\kappa^*}_{B_{Z^*}\frac{\overline{h}}{h}}\right| < 1$, and case (2) corresponds exactly to the case $\left|\mathbf{H}^{\nu,2\kappa_0}_{\frac{\overline{h}}{h}}\right| = 1$ for

[2]Note that (see [8]) $L^\infty(\Gamma)$ can be identified with $L^\infty_\mu(\Gamma)$ for all characters μ.

all $\nu \in \mathbb{R}^g/\mathbb{Z}^g$ (see again [7]) and $\left|\mathbf{H}^{\kappa+\kappa_0,-2\kappa_0+\kappa^*}_{B_{Z^*}\frac{\bar h}{h}}\right| < 1$. In what follows, to simplify the formulas, we will omit the upper indices for Hankel and Toeplitz operators, and the reader should keep in mind that the characters of the symbols, which are the products or quotients of character-automorphic functions in our case, could be restored from these expressions, and the characters of the Hardy spaces could be restored from the context. For example, as above, the character of the symbol $\bar h/h$ is the difference of the characters of $\bar h$ and h, i.e., is equal to $-2\kappa_0$, and the character of the symbol $B_{Z^*}\bar h/h$ in the same manner is equal to $\kappa^* - 2\kappa_0$.

5.2.

Proposition 5.1. *Let the Toeplitz operator* $\mathbf{T}_{B_{Z^*}\frac{\bar h}{h}}$ *act on the space* $H^2_{+,\kappa+\kappa_0}$ *and* $\mathbf{T}_{\overline{B_{Z^*}}\frac{h}{\bar h}}$ *be its adjoint acting on the space* $H^2_{+,\kappa-\kappa_0+\kappa^*}$. *Then*

$$\mathcal{M} = \mathbf{T}_{B_{Z^*}\frac{\bar h}{h}}\left(H^2_{+,\kappa+\kappa_0}\right)\bigcap K^{\kappa-\kappa_0+\kappa^*}_{Z^*} \neq \{0\}, \quad \dim \mathcal{M} = ordU_h$$

$$\dim Ker\mathbf{T}_{\overline{B_{Z^*}}\frac{h}{\bar h}} = r^{\kappa}[U_h + J(a) - Z^* - J(Z^*)] = g - ordU_h,$$

$$Ker\mathbf{T}_{\overline{B_{Z^*}}\frac{h}{\bar h}} = \frac{B_{Z^*}}{h} L_{\kappa}(U_h + J(a) - Z^* - J(Z^*)).$$

Proof. Using the definition of Toeplitz operators it is easy to see that $\mathbf{T}_{\frac{\bar h}{h}} = T_{\overline{B_{Z^*}}} T_{B_{Z^*}\frac{\bar h}{h}}$. Note also that for $g \in H^2_{+,\nu}$ represented in the form $g = k_{Z^*} + B_{Z^*} f$, where $k_{Z^*} \in K^{\nu}_{Z^*}$, $f \in H^2_{+,\nu-\kappa^*}$,

$$\|T_{\overline{B_{Z^*}}} g\|_2 = \|P^{\nu-\kappa^*}_+ \frac{1}{B_{Z^*}}(k_{Z^*} + B_{Z^*} f)\|_2 = \|f\|_2 = dist(g, K^{\nu}_{Z^*})$$

Now, since $|\mathbf{H}_{\frac{\bar h}{h}}| = 1$, due to (5.1)

$$\inf_{\substack{x\in H^2_{+,\kappa+\kappa_0}\\ \|x\|_2=1}} \|\mathbf{T}_{B_{Z^*}\frac{\bar h}{h}} x - P_{K^{\kappa-\kappa_0+\kappa^*}_{Z^*}} \mathbf{T}_{B_{Z^*}\frac{\bar h}{h}} x\|_2 = 0.$$

But since $\inf\limits_{\substack{x\in H^2_{+,\kappa+\kappa_0}\\ \|x\|_2=1}} \|\mathbf{T}_{B_{Z^*}\frac{\bar h}{h}} x\|_2 > 0$, the image of the unit "sphere" $S_{+,\kappa+\kappa_0} \overset{def}{=} \{x \in H^2_{+,\kappa+\kappa_0} : \|x\|_2 = 1\}$ is separated from the origin and

$$dist(\mathbf{T}_{B_{Z^*}\frac{\bar h}{h}}\left(S_{+,\kappa+\kappa_0}\right), K^{\kappa-\kappa_0+\kappa^*}_{Z^*}) = 0.$$

Therefore, since $\dim K^{\kappa-\kappa_0+\kappa^*}_{Z^*} = g$,

$$\mathcal{M} = \mathbf{T}_{B_{Z^*}\frac{\bar h}{h}}\left(H^2_{+,\kappa+\kappa_0}\right)\bigcap K^{\kappa-\kappa_0+\kappa^*}_{Z^*} \neq \{0\}.$$

Let us find the dimension of the subspace $\mathcal{M}$. To do this note that, by (5.1), $\mathbf{T}_{B_{Z^*}\frac{\bar h}{h}}$ is an isomorphism of $H^2_{+,\kappa+\kappa_0}$ onto its image, and therefore

$$\mathbf{T}_{B_{Z^*}\frac{\bar h}{h}}\left(H^2_{+,\kappa+\kappa_0}\right) \oplus Ker\mathbf{T}_{\overline{B_{Z^*}}\frac{h}{\bar h}} = H^2_{+,\kappa-\kappa_0+\kappa^*}.$$

Let us first find the dimension of $Ker\mathbf{T}_{\overline{B}_{Z^*}\frac{h}{\overline{h}}}$ in $H^2_{+,\kappa-\kappa_0+\kappa^*}$. If $q \in Ker\mathbf{T}_{\overline{B}_{Z^*}\frac{h}{\overline{h}}}$ then, by the inner-outer factorization for character-automorphic functions, $q = kv$, where k is a character-automorphic outer function and v is a character-automorphic inner function. Now, since $\overline{B}_{Z^*}\frac{h}{\overline{h}}q \in \left(H^2_{+,\kappa-\kappa_0+\kappa^*}\right)^\perp = H^2_{-,\kappa-\kappa_0+\kappa^*} \oplus \mathcal{M}_{\kappa-\kappa_0+\kappa^*} = B_{Z^*}H^2_{-,\kappa-\kappa_0}$ and the symbol $\overline{B}_{Z^*}\frac{h}{\overline{h}}$ is unimodular on Γ, there exists a character-automorphic inner function v' such that

$$\frac{B_{Z^*}}{b_a}\overline{kv'} = \overline{B}_{Z^*}\frac{h}{\overline{h}}kv, \quad \text{or equivalently} \quad \frac{B_{Z^*}}{b_a}\overline{hkv'} = \overline{B}_{Z^*}hkv. \tag{5.2}$$

Note that both sides of this equality are summable on Γ and, moreover $\overline{B}_{Z^*}hkv \in H^1_{+,\kappa} + \mathcal{M}_\kappa$, $\frac{B_{Z^*}}{b_a}\overline{hkv'} \in H^1_{-,\kappa} + \mathcal{M}_\kappa$. Now, since this function is "orthogonal" to all functions from $H^\infty_{+,\kappa} + H^\infty_{-,\kappa}$, by an analog of the Riesz brothers theorem (see for example [8]), it belongs to $\mathcal{M}_\kappa$. If we denote this function by $\varphi_{\mathcal{M}_\kappa}$, then, by (5.2),

$$\frac{\varphi_{\mathcal{M}_\kappa}}{h} \in \mathcal{M}_{\kappa-\kappa_0} + H^2_{+,\kappa-\kappa_0}, \quad \frac{\varphi_{\mathcal{M}_\kappa}}{\overline{h}} \in \mathcal{M}_{\kappa+\kappa_0} + H^2_{-,\kappa+\kappa_0},$$

and, by part (2) of Corollary 4.5 $\varphi_{\mathcal{M}_\kappa} \in L_\kappa(U_h + J(a) - Z^* - J(Z^*))$.

Conversely, if $\varphi_{\mathcal{M}_\kappa} \in L_\kappa(U_h + J(a) - Z^* - J(Z^*))$ then, by part (1) of Corollary 4.5, $q = B_{Z^*}\frac{\varphi_{\mathcal{M}_\kappa}}{h} \in H^2_{\kappa-\kappa_0+\kappa^*}$ and

$$\overline{B}_{Z^*}\frac{h}{\overline{h}}q = \frac{\varphi_{\mathcal{M}_\kappa}}{\overline{h}} \in \mathcal{M}_{\kappa+\kappa_0} + H^2_{-,\kappa+\kappa_0}$$

and $q \in Ker\mathbf{T}_{\overline{B}_{Z^*}\frac{h}{\overline{h}}}$. Thus we have proved that

$$\dim Ker\mathbf{T}_{\overline{B}_{Z^*}\frac{h}{\overline{h}}} = r^\kappa[U_h + J(a) - Z^* - J(Z^*)] = g - ordU_h,$$

$$Ker\mathbf{T}_{\overline{B}_{Z^*}\frac{h}{\overline{h}}} = \frac{B_{Z^*}}{h}L_\kappa(U_h + J(a) - Z^* - J(Z^*)). \tag{5.3}$$

Now for any $q \in Ker\mathbf{T}_{\overline{B}_{Z^*}\frac{h}{\overline{h}}}$ write $q = k_q + B_{Z^*}f_q$, where $k_q = P_{K^{\kappa-\kappa_0+\kappa^*}_{Z^*}}q$, $f \in H^2_{+,\kappa-\kappa_0}$; then $\mathcal{M} = \{g \in K^{\kappa-\kappa_0+\kappa^*}_{Z^*} : 0 = \langle g, q\rangle = \langle k_q, g\rangle, \forall q \in Ker\mathbf{T}_{\overline{B}_{Z^*}\frac{h}{\overline{h}}}\}$. Note that for any such q, $k_q \neq 0$. This last fact follows from (5.3), since otherwise the expression

$$q = B_{Z^*}\frac{\varphi_{\mathcal{M}_\kappa}}{h} = B_{Z^*}f_q, \quad \varphi_{\mathcal{M}_\kappa} \in L_\kappa(U_h + J(a) - Z^* - J(Z^*)) \subset \mathcal{M}_\kappa$$

would imply that $\varphi_{\mathcal{M}_\kappa} = hf_q \in H^1_{+,\kappa}$.

This means that

$$\dim P_{K^{\kappa-\kappa_0+\kappa^*}_{Z^*}} Ker\mathbf{T}_{\overline{B}_{Z^*}\frac{h}{\overline{h}}} = \dim \bigvee_{q \in Ker\mathbf{T}_{\overline{B}_{Z^*}\frac{h}{\overline{h}}}} \{k_q\} = g - ordU_h.$$

Since $\mathcal{M} = K^{\kappa-\kappa_0+\kappa^*}_{Z^*} \ominus Ker\mathbf{T}_{\overline{B}_{Z^*}\frac{h}{\overline{h}}}$, we have proved that $\dim \mathcal{M} = ord(U_h)$. $\square$

Now let us consider all possible combinations of $ord(U_h)$ distinct points from the set $\{z_1^*, \ldots, z_g^*\}$ or, equivalently, the set $\mathcal{Z}$ of all positive divisors Z of order

$ordU_h$, such that $Z < Z^*$. For each such divisor Z let us consider the finite Blaschke product B_Z for which Z is the divisor of its zeros.

Proposition 5.2. *There exists at least one divisor $Z \in \mathcal{Z}$, for which the norm of the Hankel operator on $H^2_{+,\kappa+\kappa_0}$ with symbol $B_Z \frac{\overline{h}}{h}$ is strictly less than 1, i.e., $|\mathbf{H}_{B_Z \frac{\overline{h}}{h}}| < 1$.*

Proof. Suppose on the contrary that for all such divisors Z, $|\mathbf{H}_{B_Z \frac{\overline{h}}{h}}| = 1$. Then the same reasoning as used above in the proof of Proposition 5.1 tells us that for the complementary divisor $Z^* - Z$ of order $g - ordU_h$ (for which $B_{Z^*} = B_Z B_{Z^*-Z}$), $\mathbf{T}_{B_Z \frac{\overline{h}}{h}} = \mathbf{T}_{\overline{B}_{Z^*-Z}} \mathbf{T}_{B_{Z^*} \frac{\overline{h}}{h}}$ and, hence

$$\mathbf{T}_{B_{Z^*} \frac{\overline{h}}{h}} \left(H^2_{+,\kappa+\kappa_0}\right) \bigcap K^{\kappa-\kappa_0+\kappa^*}_{Z^*-Z} \neq \{0\},$$

where $K^{\kappa-\kappa_0+\kappa^*}_{Z^*-Z} \subset K^{\kappa-\kappa_0+\kappa^*}_{Z^*}$, $\dim K^{\kappa-\kappa_0+\kappa^*}_{Z^*-Z} = g - ordU_h$, is spanned by the reproducing kernels at the points of $Z^* - Z$. But we already know that

$$\mathbf{T}_{B_{Z^*} \frac{\overline{h}}{h}} \left(H^2_{+,\kappa+\kappa_0}\right) \bigcap K^{\kappa-\kappa_0+\kappa^*}_{Z^*} = \mathcal{M}, \ \dim \mathcal{M} = ordU_h.$$

Therefore our assumption implies that for all divisors Z, $ordZ = ordU_h, 0 < Z < Z^*$, $K^{\kappa-\kappa_0+\kappa^*}_{Z^*-Z} \cap \mathcal{M} \neq \{0\}$.

But this is impossible due to the following simple Lemma.

Lemma 5.3. *Let $\mathcal{H}$ be an n-dimensional vector space with basis $B = \{v_1, \dots, v_n\}$ and let $\mathbf{K_k}$, $k < n$, be the family of all k-dimensional subspaces of $\mathcal{H}$, each of which is spanned by some k vectors from B. Then for any $n - k$-dimensional subspace $\mathcal{M} \subset \mathcal{H}$, there exists at least one subspace $\mathcal{K} \in \mathbf{K_k}$, such that $\mathcal{M} \cap \mathcal{K} = \{0\}$.*

Proof. Without loss of generality we can assume that $\mathcal{H} = \mathbb{R}^n$ and B is the standard basis of $\mathbb{R}^n$. Let $A = \{a_{jl}\}^{n \quad ,n-k}_{j=1,l=1}$ be $n \times (n-k)$ matrix, whose columns form the basis in $\mathcal{M}$. Clearly $rankA = n - k$. Let $1 \leq j_1 < j_2 < \cdots < j_{n-k} \leq n$ be the indices of $n - k$ linear independent rows of A. Then the k-dimensional space $\mathcal{K}$ from $\mathbf{K_k}$ which is defined by the condition that all its vectors have zeros at the $j_1, j_2, \dots, j_{n-k}$ coordinate positions is the desired one. Indeed, if $\mathcal{M} \cap \mathcal{K} \neq \{0\}$ then there exists a vector $\vec{c} \in \mathbb{R}^{n-k}$ such that for nondegenerate $(n-k) \times (n-k)$ matrix $A_{n-k} = \{a_{j_s,l}\}^{n-k}_{s,l=1}$, $A_{n-k}\vec{c} = \mathbf{0}$. □

All we need now is to apply this Lemma to the case $\mathcal{H} = K^{\kappa-\kappa_0+\kappa^*}_{Z^*}$, $n = g$, $k = g - ordU_h$ and B is the basis of reproducing kernels at the points of Z^*. □

Renumbering, if necessary, the zeroes on Ω_+ of the harmonic measure $d\eta_a$, we can assume that for the divisor $Z_h = z_1^* + z_2^* + \cdots + z^*_{ordU_h}$ the norm of Hankel operator on $H^2_{+,\kappa+\kappa_0}$ with symbol $B_{Z_h} \frac{\overline{h}}{h}$ is less than 1,

$$|\mathbf{H}_{B_{Z_h} \frac{\overline{h}}{h}}| < 1 \quad \text{and} \quad \mathcal{M} \cap K^{\kappa-\kappa_0+\kappa^*}_{Z^*-Z_h} = \{0\}. \tag{5.4}$$

Let $\kappa_h = \mathrm{Re}\varphi(Z_h - (ordU_h)b_0)$ be the character of the finite Blaschke product B_{Z_h}.

Proposition 5.4. *For the divisor Z_h of Proposition 5.2 the norm of the Hankel operator $\mathbf{H}_{\overline{B_{Z_h}}\frac{h}{\bar h}}$, acting on the space $H^2_{+,\kappa-\kappa_0+\kappa_h}$ is strictly less than one. Here $\kappa_h = \operatorname{Re}\varphi(Z_h - (ordU_h)b_0)$ is the character of the finite Blaschke product B_{Z_h}.*

Proof. Together with (5.1), (5.4) immediately implies that $Ker\mathbf{T}_{B_{Z_h}\frac{\bar h}{h}} = \{0\}$ and moreover that $\mathbf{T}_{B_{Z_h}\frac{\bar h}{h}}$ is an isomorphism of $H^2_{+,\kappa+\kappa_0}$ onto its image in $H^2_{+,\kappa-\kappa_0+\kappa_h}$. Hence $\mathbf{T}_{B_{Z_h}\frac{\bar h}{h}}(H^2_{+,\kappa+\kappa_0})$ is closed and, therefore

$$\mathbf{T}_{B_{Z_h}\frac{\bar h}{h}}(H^2_{+,\kappa+\kappa_0}) \oplus Ker\mathbf{T}_{\overline{B_{Z_h}}\frac{h}{\bar h}} = H^2_{+,\kappa-\kappa_0+\kappa_h}. \tag{5.5}$$

Let us show that $Ker\mathbf{T}_{\overline{B_{Z_h}}\frac{h}{\bar h}} = 0$ on $H^2_{+,\kappa-\kappa_0+\kappa_h}$.

Note that $K^{\kappa-\kappa_0+\kappa^*}_{Z^*} = K^{\kappa-\kappa_0+\kappa^*}_{Z^*-Z_h} \oplus B_{Z^*-Z_h}K^{\kappa-\kappa_0+\kappa_h}_{Z_h}$. Here $K^{\kappa-\kappa_0+\kappa_h}_{Z_h} \subset K^{\kappa-\kappa_0+\kappa_h}_{Z^*}$, $\dim K^{\kappa-\kappa_0+\kappa_h}_{Z_h} = ordU_h$, is spanned by the reproducing kernels at the points of Z_h. Now since $\dim \mathcal{M} = \dim K^{\kappa-\kappa_0+\kappa_h}_{Z_h} = ordU_h$, using the second part of (5.4) we see that

$$P_{B_{Z^*-Z_h}K^{\kappa-\kappa_0+\kappa_h}_{Z_h}}\mathcal{M} = B_{Z^*-Z_h}K^{\kappa-\kappa_0+\kappa_h}_{Z_h}.$$

Indeed, otherwise the projection on the left-hand side of above equality has a nontrivial kernel in $\mathcal{M}$ and therefore we can find a nonzero element from $\mathcal{M} \cap K^{\kappa-\kappa_0+\kappa^*}_{Z^*-Z_h}$. Using the representation $\mathbf{T}_{B_{Z_h}\frac{\bar h}{h}} = \mathbf{T}_{\overline{B_{Z^*-Z_h}}}\mathbf{T}_{B_{Z^*}\frac{\bar h}{h}}$ we see that

$$P^{\kappa-\kappa_0+\kappa_h}_{+}\overline{B}_{Z^*-Z_h}\mathcal{M} = P_{K^{\kappa-\kappa_0+\kappa_h}_{Z_h}}\overline{B}_{Z^*-Z_h}\mathcal{M} = K^{\kappa-\kappa_0+\kappa_h}_{Z_h} \subset \mathbf{T}_{B_{Z_h}\frac{\bar h}{h}}\left(H^2_{+,\kappa+\kappa_0}\right).$$

Hence $Ker\mathbf{T}_{\overline{B_{Z_h}}\frac{h}{\bar h}} \perp K^{\kappa-\kappa_0+\kappa_h}_{Z_h}$. Therefore any element from $Ker\mathbf{T}_{\overline{B_{Z_h}}\frac{h}{\bar h}}$ has the form $B_{Z_h}f$, with $f \in H^2_{+,\kappa-\kappa_0}$. But this means that $f \in Ker\mathbf{T}_{\frac{h}{\bar h}}$ and so $fh/\bar h \in B_{Z^*}H^2_{-,\kappa+\kappa_0-\kappa^*}$. But then, multiplying both sides of this inclusion by $\bar h$, we obtain that fh simultaneously belongs to $H^1_{+,\kappa}$ and $H^1_{-,\kappa} + \mathcal{M}_\kappa$. Thus $f = 0$ and $Ker\mathbf{T}_{\overline{B_{Z_h}}\frac{h}{\bar h}} = \{0\}$.

Now, by (5.5), $\mathbf{T}_{B_{Z_h}\frac{\bar h}{h}}$ is an isomorphism of $H^2_{+,\kappa+\kappa_0}$ onto $H^2_{+,\kappa-\kappa_0+\kappa_h}$ and so it has a bounded inverse operator, defined on $H^2_{+,\kappa-\kappa_0+\kappa_h}$. But then its adjoint operator $\mathbf{T}_{\overline{B_{Z_h}}\frac{h}{\bar h}}$ also has a bounded inverse operator, defined on $H^2_{+,\kappa+\kappa_0}$.

Therefore

$$\inf_{\substack{f\in H^2_{+,\kappa-\kappa_0+\kappa_h}\\ \|f\|_2=1}} \|\mathbf{T}_{\overline{B_{Z_h}}\frac{h}{\bar h}}f\|_2 > 0,$$

and

$$\mathbf{T}_{\overline{B_{Z_h}}\frac{h}{\bar h}}\left(H^2_{+,\kappa-\kappa_0+\kappa_h}\right) = H^2_{+,\kappa+\kappa_0} \quad \textit{and} \quad Ker\mathbf{T}_{\overline{B_{Z_h}}\frac{h}{\bar h}} = \{0\}, \tag{5.6}$$

Finally, using (5.1) we see that

$$|\mathbf{H}_{\overline{B_{Z_h}}\frac{h}{\bar h}}| < 1. \tag{5.7}$$

□

5.3.

Proposition 5.5. *Let U_h be the divisor from Proposition 4.3. Then for $k = ordU_h|_u$, $u \in \Gamma$, the expression $h/(z-u)^k$ is locally square summable at u.*

Proof. Due to (5.6) $Ker\mathbf{T}_{\overline{b}_a \overline{B}_{Z_h} \frac{h}{\overline{h}}} \neq \{0\}$. Let $q \in H^2_{+,\kappa-\kappa_0+\kappa_h}$ be a nonzero function from this kernel with inner-outer factorization $q = kv$, where k is an outer function and v is an inner function. Now repeating the above arguments used to obtain (5.2), we have

$$\frac{B_{Z^*}}{b_a}\overline{kv'} = \frac{1}{b_a B_{Z_h}}\frac{h}{\overline{h}}kv,$$

where v' is an inner function. This expression implies that

$$B_{Z^*}\overline{hkv'} = \frac{hkv}{B_{Z_h}}. \tag{5.8}$$

In this case the left-hand side of this equality belongs to $H^1_{-,\kappa} + \mathcal{M}_\kappa + k_a^\kappa$ and the right-hand side belongs to $H^1_{+,\kappa} + \mathcal{M}_\kappa$. Again an application of an analog of the Riesz brother's theorem gives us that this function belongs to $\mathcal{M}_\kappa + k_a^\kappa$. From (5.8) we observe that

$$\psi = \frac{hkv}{B_{Z_h}} \in L_\kappa((v)_+ + J((v')_+) + U_h - Z_h - J(Z^*)),$$

here $(v)_+$ and $(v')_+$ are the divisors of the zeros of the inner functions v and v'. Clearly $ord(v)_+ + ord(v')_+ \leq g$ and all other zeros of ψ lie on Γ. Thus $(\psi) = (v)_+ + J((v')_+) + U_h - Z_h - J(Z^*) + U_\psi$, where U_ψ is a nonnegative divisor contained completely in Γ, $ordU_\psi + ord(v)_+ + ord(v')_+ = g$. By Lemma 3.2 we can find a positive divisor U^* of order g contained completely in Γ, such that there exists a character-automorphic function f with divisor $(f) = U^* - \{U_\psi + (v)_+ + J((v')_+)\}$ on $\hat{\Omega}$. Then since $1/h$ is locally square summable at all points on $\Gamma \setminus U_h$, the function $\tilde{\psi} = B_{Z_h} f\psi$ belongs to the space $L_\nu(U_h + U^* - J(Z_h) - J(Z^*))$ for some character ν, $\overline{\tilde{\psi}} = B_{Z^*}B_{Z_h}\tilde{\psi}$. Therefore, by (5.8) and Proposition 4.3, the function $h^* = \tilde{\psi}/h = B_{Z_h} f\psi/h = kvf = \frac{B_{Z_h}\psi}{h} f$ is square summable on Γ and belongs to $H^2_{+,\nu-\kappa_0}$. The function h^* is outer, since it has no zeros in Ω_+ and its product with the outer function h is an outer function. Thus

$$\frac{\overline{h^*}}{h^*}B_{Z^*} = \overline{B}_{Z_h}\frac{h}{\overline{h}}, \tag{5.9}$$

and by (5.2)

$$|\mathbf{H}_{\frac{\overline{h^*}}{h^*}B_{Z^*}}| < 1. \tag{5.10}$$

The last inequality, as we already checked, is equivalent to the nonzero angle condition in $L^2(\Gamma, |h^*|^2 d\eta_a)$ between the closures of the subspaces $\mathcal{B}_-^{\kappa-\nu+\kappa_h}$ and $\mathcal{B}_+^{\kappa-\nu+\kappa_h}$. Using Proposition 4.3 we find that there exists a positive divisor $U = U_{h^*}$ of order less or equal to g, contained totally in Γ, such that $1/h^*$ is locally square summable at all points of $\Gamma \setminus U_{h^*}$, $1/h^*$ is not square summable at the points of U_{h^*}, but the expression $(z-u)^k/h^*$ is square summable for $k = ordU_{h^*}|_u$, and k is the minimal nonnegative integer with this property. Now, since $\tilde{\psi}/h^* = h$ is square summable on Γ, by Remark 4.6, $0 \le U_{h^*} \le U_h + U^*$.

Let us show that $U_{h^*} = U^*$. Since by the definition of the divisors U_h and U_{h^*} the numbers $k_{h^*} = ordU_{h^*}|_u$ and $k_h = ordU_h|_u$ are the smallest nonnegative integers for which the respective expressions $(z-u)^{k_{h^*}}/h^*$ and $(z-u)^{k_h}/h$ are locally square summable at u, and since, by the definitions of the function $\tilde{\psi} = hh^*$ and the divisor U^*, $k_{h^*}, k_h \le k_u = ord(U_h + U^*)|_u = ord(\tilde{\psi})|_u$, using Holder's inequality for a sufficiently small interval X_u containing the point u, we have

$$\infty = \int\limits_{X_u} \frac{1}{|z-u|} d\eta_a(z) = \int\limits_{X_u} \frac{h}{|z-u|^{k_u - k_{h^*}}} \frac{|z-u|^{k_u - k_{h^*} - 1}}{h} d\eta_a(z)$$

$$\le C_u \left\{ \int\limits_{X_u} \left| \frac{(z-u)^{k_{h^*}}}{h^*} \right|^2 d\eta_a(z) \right\}^{1/2} \left\{ \int\limits_{X_u} \left| \frac{(z-u)^{k_u - k_{h^*}}}{h(z-u)} \right|^2 d\eta_a(z) \right\}^{1/2},$$

where C_u is a constant depending on the derivative of $\tilde{\psi}$ of order k_u at u. This means that $(z-u)^{k_u - k_{h^*} - 1}/h$ is not locally square summable at u, and therefore $k_h \ge k_u - k_{h^*}$ or, equivalently, $k_h + k_{h^*} \ge k_u$. Summing up the last inequality over all $u \in \Gamma, ord(\tilde{\psi})|_u > 0$, we get

$$ordU_h + ordU_{h^*} \ge ordU_h + g.$$

But we know that $ordU_{h^*} \le g$ and so $ordU_{h^*} = g$, which means that we have the equality $ordU_h + ordU_{h^*} = ordU_h + g$ and, therefore, we also have equality $k_h + k_{h^*} = k_u$ at every $u \in \Gamma$. But the last equality is nothing but

$$ordU_h|_u + ordU_{h^*}|_u = ordU_h|_u + ordU^*|_u,$$

and thus $U^* = U_{h^*}$.

Moreover the above reasoning has shown us that in addition to the statement of Proposition 4.3 not only is $(z-u)^{k_h}/h$ locally square summable at $u \in \Gamma$, but also $h/(z-u)^{k_h} \sim C_u(z-u)^{k_{h^*}}/h^*$ is locally square summable at u. □

Corollary 5.6. *For any character-automorphic meromorphic function f on $\hat{\Omega}$, whose divisor on Γ is exactly equal to U_h, both of the functions h/f and f/h are square summable on Γ.*

6. Necessary and sufficient conditions on the weight

6.1.

Now we are ready to obtain a necessary condition on the weight $w = |h|^2$. Let U_h be the divisor corresponding to the "bad behaviour" of the function h and Z_h the corresponding subdivisor of Z^* of order equal to $ordU_h$, discussed in Propositions 5.2 and 5.4. Then, by Lemma 3.2 we can find a divisor Q of order g, contained totally inside Ω_+, such that for a certain $\nu \in \mathbb{R}^g/\mathbb{Z}^g$ there exists a ν-automorphic meromorphic function ϕ on $\hat{\Omega}$ with the divisor $(\phi) = U_h + J(Z^* - Z_h) - J(Q)$. Note that ϕ itself is an outer function from $H^\infty_{+,\nu}$ and, by Corollary 5.6, h/ϕ and ϕ/h are outer functions from $H^2_{+,\kappa_0-\nu}$ and $H^2_{+,\nu-\kappa_0}$ respectively. Recall that by B_Q we denote the finite character-automorphic Blaschke product whose divisor of zeroes equals to Q; its corresponding character is $\kappa_Q = 2\operatorname{Re}\varphi(q - gb_0)$.

Proposition 6.1. *If the angle in $L_2(|h|^2 d\eta_a)$ between D^κ_- and D^κ_+ is nonzero then for any character-automorphic function ψ meromorphic on $\hat{\Omega}$, with divisor on Γ exactly equal to U_h, the weight $|h/\psi|^2$ satisfies the $A_2'(\Gamma)$ Muckenhoupt condition, i.e.,*

$$\sup_{\lambda \in \Omega_+} \int_\Gamma \left|\frac{h}{\psi}\right|^2 d\eta_\lambda \cdot \int_\Gamma \left|\frac{\psi}{h}\right|^2 d\eta_\lambda < \infty. \tag{6.1}$$

Proof. First we note that if the $A_2'(\Gamma)$ condition holds for a certain ψ with $(\psi)|_\Gamma = U_h$, then it holds for all such functions. Indeed, by [7, Theorem 4.2] the $A_2'(\Gamma)$ condition on the weight w is equivalent to the $A_2(\Gamma)$ condition on w, which is the direct analog of the classical Muckenhoupt condition for the unit circle:

$$\sup_I \frac{1}{m(I)} \int_I w\, d\eta_a \cdot \frac{1}{m(I)} \int_I \frac{1}{w} d\eta_a < \infty,$$

where I is an interval on Γ of harmonic measure $\int_I d\eta_a = m(I)$. Now if ψ and ψ' are two character-automorphic meromorphic functions on $\hat{\Omega}$, $(\psi)|_\Gamma = (\psi')|_\Gamma = U_h$, then $|\psi'/\psi|$ is a continuous function on Γ, bounded away from zero and infinity on Γ, and thus the weights $|h/\psi|^2$ and $|h/\psi'|^2 = |h/\psi|^2 |\psi/\psi'|^2$ satisfy or do not satisfy the $A_2(\Gamma)$ condition simultaneously. Therefore we will show that (6.1) holds for the weight $|h/\phi|^2$, where ϕ is the function introduced at the beginning of Sec. 6.1. We already know that if the angle under consideration is nonzero then (5.4) holds. In the same way as was done in Sec. 5, it can be checked easily that (5.4) is equivalent to the nonzero angle condition in $L_2(w d\eta_a)$, $w = |h|^2$, between the subspaces $\operatorname{Clos}_{L_2(wd\eta_a)} H^\infty_{+,\kappa}$ and $\operatorname{Clos}_{L_2(wd\eta_a)} B_{Z^*-Z_h} H^\infty_{-,\kappa-\kappa^*+\kappa_h}$. Indeed,

$$1 > |\mathbf{H}_{\overline{B_{Z_h}} \frac{h}{\bar{h}}}| = \sup_{\substack{f_+ \in H^2_{+,\kappa+\kappa_0} \\ f_- \in H^2_{-,\kappa-\kappa_0+\kappa_h-\kappa^*}}} \frac{|\langle B_{Z_h} \frac{\bar{h}}{h} f_+, B_{Z^*} f_- \rangle|}{\|f_+\|_2 \cdot \|f_-\|_2}$$

$$= \sup_{\substack{f_+ \in H^2_{+,\kappa+\kappa_0} \\ f_- \in H^2_{-,\kappa-\kappa_0+\kappa_h-\kappa^*}}} \frac{|\langle B_{Z_h} \frac{f_+}{h} h, B_{Z^*} \frac{f_-}{\bar{h}} h \rangle|}{\|f_+\|_2 \cdot \|f_-\|_2} = \sup_{\substack{f_+ \in H^\infty_{+,\kappa} \\ f_- \in H^\infty_{-,\kappa+\kappa_h-\kappa^*}}} \frac{|\langle f_+ h, B_{Z^*-Z_h} f_- h \rangle|}{\|f_+ h\|_2 \cdot \|f_- h\|_2}$$

$$= \cos\angle(\operatorname{Clos}_{L_2(wd\eta_a)} H^\infty_{+,\kappa}, \operatorname{Clos}_{L_2(wd\eta_a)} B_{Z^*-Z_h} H^\infty_{-,\kappa-\kappa^*+\kappa_h}).$$

Now note that the divisor of the character-automorphic meromorphic function $\bar{\phi}/\phi = \overline{\phi \circ J}/\phi$ on $\hat{\Omega}$ is exactly equal to $(Z^* - Z_h) + J(Q) - J(Z^* - Z_h) - Q$, which in its turn is exactly the divisor of the function $B_{Z^*-Z_h}/B_Q$. Hence $\bar{\phi}/\phi =$

$B_{Z^*-Z_h}/B_Q$. Clearly $-2\nu = \kappa^* - \kappa_h - \kappa_Q$. Since h/ϕ is an outer function from $H^2_{+,\kappa_0-\nu}$, for any character μ the linear spaces $H^\infty_{+,\mu+\nu}/\phi$ and $H^\infty_{-,\mu-\nu}/\overline{\phi}$ are contained in $L^2(wd\eta_a)$ and, moreover, are dense in $\mathcal{D}^\mu_+$ and $\mathcal{D}^\mu_-$ respectively. Therefore, we can continue the previous chain of equalities in the following way

$$1 > \cos\angle(\mathrm{Clos}_{wd\eta_a} H^\infty_{+,\kappa}, \mathrm{Clos}_{wd\eta_a} B_{Z^*-Z_h} H^\infty_{-,\kappa-\kappa^*+\kappa_h})$$

$$= \sup_{\substack{f_+\in\mathcal{D}^\kappa_+ \\ f_-\in\mathcal{D}^{\kappa+\kappa_h-\kappa^*}_-}} \frac{|\langle f_+h, B_{Z^*-Z_h} f_-h\rangle|}{\|f_+h\|_2\cdot\|f_-h\|_2} = \sup_{\substack{f_+\in H^\infty_{+,\kappa+\nu} \\ f_-\in H^\infty_{-,\kappa+\kappa_h-\kappa^*-\nu}}} \frac{|\langle \frac{f_+}{\phi}h, B_{Z^*-Z_h}\frac{f_-}{\overline{\phi}}h\rangle|}{\|\frac{f_+}{\phi}h\|_2\cdot\|\frac{f_-}{\overline{\phi}}h\|_2}$$

$$= \sup_{\substack{f_+\in H^\infty_{+,\kappa+\nu} \\ f_-\in H^\infty_{-,\kappa+\kappa_h-\kappa^*-\nu}}} \frac{|\langle \frac{h}{\phi}f_+, B_{Z^*-Z_h}\frac{\phi}{\overline{\phi}}f_-\frac{h}{\phi}\rangle|}{\|f_+\frac{h}{\phi}\|_2\cdot\|f_-\frac{h}{\phi}\|_2} = \sup_{\substack{f_+\in H^\infty_{+,\kappa+\nu} \\ f_-\in H^\infty_{-,\kappa+\kappa_h-\kappa^*-\nu}}} \frac{|\langle \frac{h}{\phi}f_+, B_Q f_-\frac{h}{\phi}\rangle|}{\|f_+\frac{h}{\phi}\|_2\cdot\|f_-\frac{h}{\phi}\|_2}$$

$$\geq \sup_{\substack{f_+\in H^\infty_{+,\kappa+\nu} \\ f_-\subset \overline{B}_Q H^\infty_{-,\kappa+\nu}}} \frac{|\langle \frac{h}{\phi}f_+, B_Q f_-\frac{h}{\phi}\rangle|}{\|f_+\frac{h}{\phi}\|_2\cdot\|f_-\frac{h}{\phi}\|_2} = \sup_{\substack{f_+\in H^\infty_{+,\kappa+\nu} \\ f_-\in H^\infty_{-,\kappa+\nu}}} \frac{|\langle \frac{h}{\phi}f_+, f_-\frac{h}{\phi}\rangle|}{\|f_+\frac{h}{\phi}\|_2\cdot\|f_-\frac{h}{\phi}\|_2}$$

But the right-hand side of the last expression, which, as we now see, is strictly less than 1, is exactly equal to the cosine of the angle in $L_2(|\frac{h}{\phi}|^2 d\eta_a)$ between the subspaces $\mathrm{Clos}_{L_2(|\frac{h}{\phi}|^2 d\eta_a)} H^\infty_{-,\kappa+\nu}$ and $\mathrm{Clos}_{L_2(|\frac{h}{\phi}|^2 d\eta_a)} H^\infty_{+,\kappa+\nu}$. Thus this angle is nonzero. Now we can apply Proposition 4.3 to the weight $|\frac{h}{\phi}|^2$. Since, as we remarked at the beginning of this section, by Corollary 5.6, both functions h/ϕ and ϕ/h are square summable, the divisor of "bad behaviour" for this weight is zero. Therefore case (1) of the alternative from Proposition 4.3 takes place and this weight satisfies the $A_2'(\Gamma)$ Muckenhoupt condition. □

6.2.

Note that (see (4.5) and statement (1) of corollary 4.5) if the weight $|h|^2$ provides a nonzero angle between subspaces $\mathcal{D}^\kappa_-$ and $\mathcal{D}^\kappa_+$, then necessarily (4.5) holds. This condition is equivalent to the following one

$$r^{-\kappa}[a - U_h] = r^\kappa[J(a) - U_h] = 0, \tag{6.2}$$

since after dividing any differential from $\mathcal{N}_{-\kappa}(Z^* + J(Z^*) - U_h - J(a))$ by the harmonic measure $d\eta_a$, which is a meromorphic (**0**-automorphic) differential on $\hat{\Omega}_+$ with the divisor $(d\eta_a) = Z^* + J(Z^*) - a - J(a)$, we will get a function from $L_{-\kappa}(a - U_h)$. If condition (6.2) does not hold then, by the properties of the divisor U_h, $L_\kappa(J(a) - U_h) \subset H^2_{-,\kappa-\kappa_0}/\overline{h} \cap H^2_{+,\kappa+\kappa_0}/h$ and the angle is obviously zero.

We will show that conditions (6.1) and (6.2) together are also sufficient for our problem.

Proposition 6.2. *If an outer function $h \in H^2_{+,\kappa_0}$ is such that there exists a character-automorphic meromorphic function ψ on $\hat{\Omega}$ with a nonnegative divisor $U =$*

$(\psi)|_\Gamma$ *on* Γ, $0 \le ordU \le g$, *and*

$$r^{-\kappa}[a - U] = 0, \quad \kappa \in \mathbb{R}^g/\mathbb{Z}^g, \tag{6.3}$$

$$\sup_{\lambda\in\Omega_+} \int_\Gamma \left|\frac{h}{\psi}\right|^2 d\eta_\lambda \cdot \int_\Gamma \left|\frac{\psi}{h}\right|^2 d\eta_\lambda < \infty,$$

then the angle in $L_2(wd\eta_a)$, $w = |h|^2$, *between the subspaces* $\mathcal{D}_-^\kappa = \mathrm{Clos}_{L_2(wd\eta_a)} H^\infty_{-,\kappa}$ *and* $\mathcal{D}_+^\kappa = \mathrm{Clos}_{L_2(wd\eta_a)} H^\infty_{+,\kappa}$ *is nonzero.*

Proof. Note first that, by the definition of a Hankel operator, the nonzero angle condition we want to prove is also equivalent to the nonzero angle in $L_2(d\eta_a)$ between the subspaces $H^2_{-,\kappa-\kappa_0}$ and $\frac{\bar h}{h} H^2_{+,\kappa+\kappa_0}$. Using Lemma 3.2, in the same way as was done in the proof of Proposition 6.1 we can find a ν-automorphic meromorphic function ϕ on $\hat\Omega$, such that $(\phi) = U + J(Z) - J(Q)$, where Z is any fixed nonnegative subdivisor of the critical Green's divisor Z^* of order $ordZ = g - ordU$ and Q is a positive divisor of order g contained totally in Ω_+. Clearly ϕ is an outer function from $H^2_{+,\nu}$, $\bar\phi/\phi = B_Z/B_Q$, $(\phi)|_\Gamma = (\psi)|_\Gamma = U$ and again, by the reasoning from the proof of Proposition 6.1, the weight $|h/\phi|^2$ satisfies the $A_2'(\Gamma)$ Muckenhoupt condition. Let $\tilde h = h/\phi$, then $\tilde h$ is an outer function from $H^2_{+,\kappa_0-\nu}$ and the norm of the Hankel operator $\mathbf{H}_{\frac{\bar{\tilde h}}{\tilde h}}$ is strictly less than one on $H^2_{+,\mu}$ for any character μ. In particular $|\mathbf{H}_{\frac{\bar{\tilde h}}{\tilde h}}| < 1$ on $H^2_{+,\kappa+\kappa_0}$, which means that the angle in $L_2(d\eta_a)$ between the subspaces $H^2_{-,\kappa-\kappa_0} + \mathcal{M}_{\kappa-\kappa_0}$ and $\frac{\bar{\tilde h}}{\tilde h} H^2_{+,\kappa+\kappa_0}$ is nonzero. But $\bar h/h = \{\bar{\tilde h}/\tilde h\}\{\bar\phi/\phi\} = \{B_Z\bar{\tilde h}\}/\{B_Q\tilde h\}$. Thus, in terms of $\tilde h$, we want to show that the angle in $L_2(d\eta_a)$ between the subspaces $H^2_{-,\kappa-\kappa_0}$ and $\frac{B_Z}{B_Q}\frac{\bar{\tilde h}}{\tilde h} H^2_{+,\kappa+\kappa_0-\kappa_Z+\kappa_Q}$ is nonzero. But $H^2_{+,\kappa+\kappa_0-\kappa_Z+\kappa_Q} = K_Q^{\kappa+\kappa_0-\kappa_Z+\kappa_Q} \oplus B_Q H^2_{+,\kappa+\kappa_0-\kappa_Z}$ and therefore $\frac{B_Z}{B_Q} H^2_{+,\kappa+\kappa_0-\kappa_Z+\kappa_Q} = \frac{B_Z}{B_Q} K_Q^{\kappa+\kappa_0-\kappa_Z+\kappa_Q} \oplus B_Z H^2_{+,\kappa+\kappa_0-\kappa_Z}$. But we already know that the angle between $H^2_{-,\kappa-\kappa_0}$ and $B_Z \frac{\bar{\tilde h}}{\tilde h} H^2_{+,\kappa+\kappa_0-\kappa_Z}$ is nonzero and the sum of these subspaces $H^2_{-,\kappa-\kappa_0} \dot{+} B_Z \frac{\bar{\tilde h}}{\tilde h} H^2_{+,\kappa+\kappa_0-\kappa_Z}$ is direct. Thus, if the angle we are interested in is zero, due to the finite dimension of the subspace $\frac{B_Z}{B_Q} K_Q^{\kappa+\kappa_0-\kappa_Z+\kappa_Q}$, there exist functions $f_- \in H^2_{-,\kappa-\kappa_0}$ and $f_+ \in H^2_{+,\kappa+\kappa_0-\kappa_Z+\kappa_Q}$, such that $f_- = \frac{B_Z}{B_Q}\frac{\bar{\tilde h}}{\tilde h} f_+$. Therefore

$$\frac{f_-}{\bar{\tilde h}} = \frac{B_Z f_+}{\tilde h B_Q}. \tag{6.4}$$

But since the weight $|\tilde{h}|^2$ satisfies the Muckenhoupt condition for any character μ the "Riesz" projection P_+^μ is a bounded operator in $L_2(|\tilde{h}|^2 d\eta_a)$ (can be extended as a bounded operator from the dense subset in $L_2(d\eta_a) \cap L_2(|\tilde{h}|^2 d\eta_a)$ to the whole space $L_2(|\tilde{h}|^2 d\eta_a)$). Therefore the operator $P_{K_Q}^\mu = P_+^\mu - B_Q P_+^{\mu-\kappa_Q} \overline{B}_Q$ is also bounded in $L_2(|\tilde{h}|^2 d\eta_a)$). Thus for any $\mu \in \mathbb{R}^g/\mathbb{Z}^g$,

$$\operatorname{Clos}_{L_2(|\tilde{h}|^2 d\eta_a)} H_{+,\mu}^\infty = \frac{H^2_{+,\mu+\kappa_0-\nu}}{\tilde{h}} = K_Q^\mu \dot{+} B_Q \frac{H^2_{+,\mu+\kappa_0-\nu-\kappa_Q}}{\tilde{h}}, \tag{6.5}$$

where the right-hand sum in (6.5) is direct.

Now applying the decomposition (6.5) to $\mu = \kappa - \kappa_Z + \kappa_Q + \nu$ and the function $f_+/\tilde{h}$ from (6.4), we get that

$$\frac{f_+}{\tilde{h}} = k_Q + B_Q \frac{\tilde{f}_+}{\tilde{h}},$$

where $k_Q \in K_Q^{\kappa+\nu-\kappa_Z+\kappa_Q}$ and $\tilde{f}_+ \in H^2_{+,\kappa+\kappa_0-\kappa_Z}$. We can rewrite (6.4) in the form

$$\frac{B_Z}{B_Q} k_Q = \frac{f_-}{\overline{\tilde{h}}} - B_Z \frac{\tilde{f}_+}{\tilde{h}}. \tag{6.6}$$

Note that $\frac{f_-}{\overline{\tilde{h}}} \in H^1_{-,\kappa+\nu}$, $\frac{\tilde{f}_+}{\tilde{h}} \in H^1_{+,\kappa+\nu-\kappa_Z}$ and $\frac{B_Z}{B_Q} k_Q \in H^2_{-,\kappa+\nu} + \mathcal{M}_{\kappa+\nu} + K_Z^{\kappa+\nu}$. Hence $\frac{B_Z}{B_Q} k_Q$ is "orthogonal" to $B_Z H^\infty_{+,\kappa+\nu-\kappa_Z}$ and, by the analog of the Riesz brothers' theorem, it follows from (6.6) that $\tilde{f}_+ = 0$. This means that $\frac{B_Z}{B_Q} k_Q = \frac{f_-}{\overline{\tilde{h}}} \in H^2_{-,\kappa+\nu}$. But from [8, Sec.3] we know that k_Q is a character-automorphic meromorphic function on $\hat{\Omega}$ with divisor $(k_Q) \geq J(a) - J(Q) - J(Z^*)$. Hence $(\frac{B_Z}{B_Q} k_Q) \geq J(a) + Z - Q - J(Z^*) - J(Z)$ and, moreover, since $\frac{B_Z}{B_Q} k_Q \in H^2_{-,\kappa+\nu}$,

$$(\frac{B_Z}{B_Q} k_Q) \geq J(a) + Z - Q.$$

Hence, multiplying $\frac{B_Z}{B_Q} k_Q$ by $\frac{1}{\phi}$, we will get a function χ from $L^\kappa(J(a) - U)$; but then $\overline{\chi \circ J} \in L^{-\kappa}(a - U)$, which contradicts the first condition from (6.3). This contradiction proves that the angle under consideration is nonzero. Reversing the above reasoning we see that the fulfilment of the first condition from (6.3) is essential for the nonzero angle, i.e., if this condition does not hold, even if the weight $|h/\phi|^2$ satisfies the $A_2'(\Gamma)$ condition, the angle is zero. □

Now combining Propositions 6.1 and 6.2 we get the following necessary and sufficient condition on the weight:

Theorem 6.3. *Let $d\mu$ be a finite positive measure on Γ. Then the subspaces $\mathcal{D}_-^\kappa = \mathrm{Clos}_{L_2(d\mu)} B_-^\kappa$ and $\mathcal{D}_+^\kappa = \mathrm{Clos}_{L_2(d\mu)} B_+^\kappa$ are at nonzero angle in $L_2(d\mu)$ if and only if*

1. *The measure $d\mu$ is absolutely continuous with respect to the harmonic measure $d\eta_a$, $d\mu = wd\eta_a$, $w \in L_1(\Gamma)$, $\log w \in L_1(\Gamma)$, and*
2. *There exists a character-automorphic meromorphic function ψ on $\hat{\Omega}$ with a nonnegative divisor on Γ, $(\psi)|_\Gamma = U_w$, $0 \le ordU_w \le g$, such that the conditions (6.3) hold for $U = U_w$ and the character-automorphic outer function h, $|h|_\Gamma^2 = w$.*

Since the $A_2'(\Gamma)$ condition is equivalent to the $A_2(\Gamma)$ condition (see [7, Theorem 4.2]) and the $A_2(\Gamma)$ condition is local, i.e., it must be fulfilled at any point of Γ for sufficiently small intervals of Γ containing this point, we can reformulate the previous theorem in the following way:

Theorem 6.4. *Let $d\mu$ be a finite positive measure on Γ. Then the subspaces $\mathcal{D}_-^\kappa = \mathrm{Clos}_{L_2(d\mu)} B_-^\kappa$ and $\mathcal{D}_+^\kappa = \mathrm{Clos}_{L_2(d\mu)} B_+^\kappa$ are at nonzero angle in $L_2(d\mu)$ if and only if*

1. *The measure $d\mu$ is absolutely continuous with respect to the harmonic measure $d\eta_a$, $d\mu = wd\eta_a$, $w \in L_1(\Gamma)$, $\log w \in L_1(\Gamma)$, and*
2. *There exists a nonnegative divisor U_w on Γ, $0 \le ordU_w \le g$, such that at any point $u \in \Gamma$ the weight $w/|z-u|^{2k}$, $k = ordU_w|_u$, is locally a Muckenhoupt weight, and*
3. *The first condition from (6.3) holds for $U = U_w$.*

References

[1] H. Helson and G. Szegö: A problem in prediction theory, Ann. Mat. Pura Appl. **51**, 1960, 107–138

[2] N. K. Nikol'skii: Treatise on the shift operator, "Nauka", Moscow, 1980; English translation: Springer-Verlag, 1986.

[3] R. Hunt, B. Muckenhoupt and R. Wheeden: Weighted norm inequalities for conjugate function and Hilbert transform, Trans. Amer. Math. Soc. **176**, 1973, 227–251.

[4] B. S. Pavlov: Basicity of an exponential system and Muckenhoupt's condition, Soviet Math. Dokl. **20**, no. 4, 1979, 655–659.

[5] S. V. Khruschev, N. K. Nikolskii, and B. S. Pavlov: Unconditional bases of exponentials and of reproducing kernels, Lecture Notes in Math. **864** (Complex analysis and spectral theory), Springer Verlag, 1981, 214–335.

[6] S. D. Fisher: Function theory on planar domains, Wiley, 1983.

[7] S. Fedorov: Weighted Norm Inequalities and the Muckenhoupt condition in a multiply connected domain, Ind. Univ. Math. J. **46**, no. 4, 1997, 1221–1253.

[8] S. Fedorov: On harmonic analysis in a multiply connected domain and character automorphic Hardy spaces, St. Petersbg. Math. J. **9**, no. 2, 1997, 192–240.

[9] K. F. Clancey: Representing measures on multiply connected domains, Illinois J. of Math. **35**, no. 2, 1991, 286–311.

[10] J. A. Ball and K. F. Clancey: Reproducing kernels for Hardy spaces on multiply connected domains, Integral Equations Operator Theory **25**, no. 1996, 35–57.

[11] S. I. Fedorov: Harmonic analysis in a multiply connected domain I, II, Matem. Sbornik **181**, no. 6, 7, 1990, 863–864, 867–909; English translation in Math. USSR Sbornik **70**, no. 1, 2, 1991, 263–296, 297–339.

[12] J. D. Fay: Theta functions on Riemann surfaces,Lecture Notes in Math. **352**, Springer-Verlag, 1973.

[13] I. Kra: Automorphic forms and Kleinian groups, Math. Lecture Note Series, W. A. Benjamin, Inc., Reading, Massachussets, 1972.

[14] D. Sarason: Function theory on the unit circle, Notes for lectures at a conference at Virginia Polytechnic Inst. and State Univ., Blacksburg, Virginia, June 19-23, 1978.

[15] H. Helson and D. Sarason: Past and Future, Math. Scand. **21**, 1967, 5–16.

Department of Mathematics, University of Auckland, Private Bag 92019, Auckland, New Zealand
E-mail address: `f.serguei@auckland.ac.nz`

Operator Theory:
Advances and Applications, Vol. 123

Bitangential Interpolation for Upper Triangular Operators with Infinitely Many Interpolation Nodes

Boris Freydin

Abstract. This paper deals with bitangential interpolation problems of the Nevanlinna-Pick type in the case when the number of the interpolation conditions is infinite. This extends the case when the number of the interpolation conditions is finite which was considered in [DF].

1. Introduction

In this paper we consider bitangential interpolation problems of the Nevanlinna-Pick type in the general setting of upper triangular operators in the case when the number of the interpolation nodes is infinite. These problems in the case when the number of the interpolation nodes is finite were considered in [DF]. Similar problems (with finite number of interpolation nodes) have also been considered by Ball, Gohberg and Kaashoek [BGK2]. See also [BGK1], [DD], [K1], [vV] where different tangential and the new book [DvV] where also bitangential interpolation problems with finite number of interpolation nodes were treated.

We remark that an interpolation problem with infinitely many interpolation nodes V_j was considered in the proof of Theorem 5.2 of [DF] under the added assumption that $r_{sp}(ZV_j^*) \leq q < 1$, i.e., when all the interpolation nodes belong to a smaller "disc". In this paper we shall relax this constraint. As in [DF] we shall follow the general strategy which was introduced by Katsnelson, Kheifets and Yuditskii [KKY] for solving the "Abstract Interpolation Problem"; for a good expository account and additional references see [KY]. We remark that in the classical setting the problem which we consider here is a particular case of Schur-Nevanlinna-Pick (SNP) problem (see [Ar] for so-called non-degenerate case and [K] for the general case).

The basic setting, which is described in more detail in [DD] and [DF], is the space $\mathcal{X}(\ell^2_{\mathcal{N}}; \ell^2_{\mathcal{M}})$ of bounded linear operators from the Hilbert space $\ell^2_{\mathcal{N}} = \bigoplus_{i=-\infty}^{\infty} \mathcal{N}_i$ into the Hilbert space $\ell^2_{\mathcal{M}} = \bigoplus_{i=-\infty}^{\infty} \mathcal{M}_i$, where each of the coordinate spaces $\mathcal{N}_i$ and $\mathcal{M}_i$, $i = 0, \pm 1, \ldots$, is itself a separable Hilbert spaces. Every

---

1991 *Mathematics Subject Classification.* Primary 47A57, Secondary 47A48.

$A \in \mathcal{X}(\ell^2_{\mathcal{N}}; \ell^2_{\mathcal{M}})$ has a natural block matrix representation with $A_{ij} : \mathcal{N}_j \to \mathcal{M}_i$. Let $\mathcal{U}, \mathcal{L}$ and $\mathcal{D}$ denote the subspaces of $\mathcal{X}$ consisting of upper triangular, lower triangular and diagonal operators, respectively, in the indicated matrix representation, and let $Z \in \mathcal{U}(\ell^2_{\mathcal{B}}; \ell^2_{\mathcal{B}})$ denote the shift operator which is specified by the rule

$$(Zf)_j = f_{j+1} \text{ for } f \in \ell^2_{\mathcal{B}} = \bigoplus_{i=-\infty}^{\infty} \mathcal{B}_i \ ,$$

for the case when $\mathcal{B}_i = \mathcal{B}_0$ for $i = \pm 1, \pm 2, \ldots$ (and in the more general setting in which the coordinate spaces $\mathcal{B}_i, \mathcal{M}_i, \mathcal{N}_i$ of $\ell^2_{\mathcal{B}}, \ell^2_{\mathcal{M}}, \ell^2_{\mathcal{N}}$ respectively are allowed to vary with the "time" index i, the shift operator Z applied to the space

$$\ell^2_{\mathcal{B}(t)} = \cdots \oplus \mathcal{B}_{t-1} \oplus \boxed{\mathcal{B}_t} \oplus \mathcal{B}_{t+1} \oplus \cdots$$

maps it onto the "displaced" space $\ell^2_{\mathcal{B}(t+1)}$; here the box designates the zeroth coordinate space, and $\ell^2_{\mathcal{B}(0)} = \ell^2_{\mathcal{B}}$, i.e., $Z_{\mathcal{B}(t)} \in \mathcal{U}(\ell^2_{\mathcal{B}(t-1)}, \ell^2_{\mathcal{B}(t)})$).

To keep the presentation simple we shall not consider the latter more general case except for a few places where this is mentioned explicitly. In this setting of constant coordinate spaces, the embedding operators

$$\pi_i : u \in \mathcal{B}_i \longmapsto f \in \ell^2_{\mathcal{B}} \ , \text{ where } \begin{cases} f_i = u \\ f_j = 0 & \text{for } j \neq i \end{cases}$$

and their adjoints

$$\pi_i^* : f \in \ell^2_{\mathcal{B}} \longmapsto f_i \in \mathcal{B}_i$$

can all be expressed in terms of π_0 and the shift operator Z:

$$\pi_i = Z^{*i} \pi_0 \ .$$

We shall also use the letter Z (symbols π_i and π_i^*, $i \in \mathbb{Z}$) for the forward shift (for the embedding operators and their adjoints) in different spaces and the dependence of Z (π_i and π_i^*) on the space in question will not be indicated explicitly. We denote by

$$\mathcal{U}' = Z\mathcal{U} \text{ and } \mathcal{L}' = Z^*\mathcal{L}$$

the spaces of strictly upper and strictly lower triangular operators, respectively.

Now let $\mathcal{X}_2(\ell^2_{\mathcal{N}}; \ell^2_{\mathcal{M}})$ denote the set of operators in $\mathcal{X}(\ell^2_{\mathcal{N}}; \ell^2_{\mathcal{M}})$ which are Hilbert-Schmidt and let

$$\mathcal{U}_2 = \mathcal{U} \cap \mathcal{X}_2 \ , \ \mathcal{U}_2' = \mathcal{U}' \cap \mathcal{X}_2 \ , \ \mathcal{D}_2 = \mathcal{D} \cap \mathcal{X}_2, \ , \ \mathcal{L}_2 = \mathcal{L} \cap \mathcal{X}_2 \text{ and } \mathcal{L}_2' = \mathcal{L}' \cap \mathcal{X}_2.$$

All of these spaces are Hilbert spaces with respect to the inner product

$$\langle F, G \rangle_{HS} = \text{trace } G^* F \ .$$

It is readily checked that

$$\langle Z^j D, Z^k E \rangle_{HS} = \begin{cases} \text{trace } E^* D & \text{if } j = k \\ 0 & \text{if } j \neq k \ , \end{cases}$$

for every choice of D and E in $\mathcal{D}_2(\ell^2_{\mathcal{N}}, \ell^2_{\mathcal{M}})$, and hence that the spaces $\mathcal{L}'_2, \mathcal{D}_2$ and $\mathcal{U}'_2$ are orthogonal with respect to this inner product:

$$\mathcal{X}_2 = \mathcal{U}'_2 \oplus \mathcal{D}_2 \oplus \mathcal{L}'_2 = \mathcal{U}_2 \oplus \mathcal{L}'_2 = \mathcal{U}'_2 \oplus \mathcal{L}_2 .$$

Let

$$\underline{\underline{p}} = \text{the orthogonal projection of } \mathcal{X}_2 \text{ onto } \mathcal{U}_2,$$
$$\underline{\underline{q}}' = \text{the orthogonal projection of } \mathcal{X}_2 \text{ onto } \mathcal{L}'_2,$$

and let $\mathcal{S}(\ell^2_{\mathcal{N}}; \ell^2_{\mathcal{M}})$ denote the set of $S \in \mathcal{U}(\ell^2_{\mathcal{N}}; \ell^2_{\mathcal{M}})$ with operator norm $\|S\| \leq 1$.

We start by presenting the BIP (basic interpolation problem) for this more general case:

We say that $S \in \mathcal{S}(\ell^2_{\mathcal{N}}, \ell^2_{\mathcal{M}})$ is a solution of the BIP based on a given set of operators

$$f_i = \begin{bmatrix} g_i \\ h_i \end{bmatrix}, \quad i \in \mathbb{Z},$$

from $l^2_{\mathcal{B}}$ to $l^2_{\mathcal{M}} \bigoplus l^2_{\mathcal{N}}$ with

$$g_i \in \mathcal{U}(\ell^2_{\mathcal{B}}, \ell^2_{\mathcal{M}}) \text{ and } h_i \in \mathcal{U}(\ell^2_{\mathcal{B}}, \ell^2_{\mathcal{N}}), \text{ for } i \leq 0$$

and

$$g_i \in \mathcal{L}'(\ell^2_{\mathcal{B}}, \ell^2_{\mathcal{M}}) \text{ and } h_i \in \mathcal{L}'(\ell^2_{\mathcal{B}}, \ell^2_{\mathcal{N}}), \text{ for } i > 0$$

if, for every choice of $E \in \mathcal{D}_2(\ell^2_{\mathcal{B}}, \ell^2_{\mathcal{B}})$

$$\underline{\underline{p}} S^* g_i E = h_i E \text{ for } i \leq 0 \tag{1.1}$$

and

$$\underline{\underline{q}}' S h_i E = g_i E \text{ for } i > 0. \tag{1.2}$$

In the present setting, upper (lower) triangular operators play the role of functions which are analytic inside (outside) the unit disc in the classical case and diagonal operators play the role of scalars. The classical case emerges upon choosing all the operators to be constant along diagonals, i.e., to be block Toeplitz.

The special choice

$$\begin{aligned} g_i &= \xi_i (I - ZV_i^*)^{-1} \text{ and } h_i = \eta_i (I - ZV_i^*)^{-1} \text{ for } i \leq 0 , \\ g_i &= \xi_i (Z - V_i)^{-1} \text{ and } h_i = \eta_i (Z - V_i)^{-1} \text{ for } i > 0 \end{aligned} \tag{1.3}$$

with $\xi_i \in \mathcal{D}(\ell^2_{\mathcal{B}}; \ell^2_{\mathcal{M}})$, $\eta_i \in \mathcal{D}(\ell^2_{\mathcal{B}}; \ell^2_{\mathcal{N}})$, $r_{sp}(ZV_i^*) < 1$ for $i \in \mathbb{Z}$, corresponds to the two-sided Nevanlinna-Pick problem in the present setting. Here r_{sp} designates the spectral radius of the indicated operator.

For this choice of g_i and h_i, the interpolation conditions (1.1) and (1.2) can be reformulated in terms of a pair of diagonal transforms $(\xi_i^* S)^\wedge(V_i)$ and $(S\eta_i)^\Delta(V_i)$ as follows:

$$(\xi_i^* S)^\wedge (V_i) = \eta_i^* \text{ for } i \leq 0 \tag{1.4}$$

and

$$(S\eta_i)^\Delta (V_i) = \xi_i \text{ for } i > 0 . \tag{1.5}$$

Here, for $V \in \mathcal{D}(\ell^2_{\mathcal{B}}; \ell^2_{\mathcal{B}})$ with $r_{sp}(ZV^*) < 1$, $F \in \mathcal{U}(\ell^2_{\mathcal{N}}; \ell^2_{\mathcal{B}})$ and $G \in \mathcal{U}(\ell^2_{\mathcal{B}}; \ell^2_{\mathcal{M}})$, the diagonal transform $F^\wedge(V)$ may be characterized as the unique element in $\mathcal{D}(\ell^2_{\mathcal{N}}; \ell^2_{\mathcal{B}})$ such that $(Z-V)^{-1}\{F - F^\wedge(V)\} \in \mathcal{U}(\ell^2_{\mathcal{N}}; \ell^2_{\mathcal{B}})$, whereas the diagonal transform $G^\Delta(V)$ may be characterized as the unique element in $\mathcal{D}(\ell^2_{\mathcal{B}}; \ell^2_{\mathcal{M}})$ such that $\{G - G^\Delta(V)\}(Z-V)^{-1} \in \mathcal{U}(\ell^2_{\mathcal{B}}; \ell^2_{\mathcal{M}})$; see Theorems 3.3 and 3.4 of [ADD].

Other choices of g_i and h_i lead to assorted analogues of classical interpolation problems (see for example Section 1 of [DD]). See also [D1] for more detail discussion on the relation of the BIP and different analogues of classical interpolation problems.

2. BIP and bitangential interpolation problems

In this section we present necessary and sufficient conditions for the existence of a solution to the BIP.

We will explicitly use the fact that the problem we consider is of a "local" type, i.e., the property to be a solution of the initial problem (1.1) and (1.2) can be checked separately for each index $i \in \mathbb{Z}$.

Following the treatment of [DF] we define the form

$$\Lambda^S_n(D,E) = \sum_{i,j=-n}^{n} \Lambda^S_{ij}(D_j, E_i)$$

on $\{\mathcal{D}_2(\ell^2_{\mathcal{B}}; \ell^2_{\mathcal{B}})\}^n \times \{\mathcal{D}_2(\ell^2_{\mathcal{B}}; \ell^2_{\mathcal{B}})\}^n$ with components

$$(\Lambda^S_n)_{ij}(D_j, E_i) = \left\langle \begin{bmatrix} I & -S \\ -S^* & I \end{bmatrix} f_j D_j, f_i E_i \right\rangle_{HS}$$

for $S \in \mathcal{S}(\ell^2_{\mathcal{N}}; \ell^2_{\mathcal{M}})$ and $D_j, E_i \in \mathcal{D}_2(\ell^2_{\mathcal{B}}; \ell^2_{\mathcal{B}})$.

Since

$$\begin{bmatrix} I & -S \\ -S^* & I \end{bmatrix} \geq 0$$

for $S \in \mathcal{S}(\ell^2_{\mathcal{N}}; \ell^2_{\mathcal{M}})$, it is selfevident that

$$\Delta^S_n(E,E) = \sum_{i,j=-n}^{n} (\Delta^S_n)_{ij}(E_j, E_i) \geq 0$$

for every choice of $E_{-n}, \dots, E_n$ in $\mathcal{D}_2(\ell^2_{\mathcal{B}}; \ell^2_{\mathcal{B}})$.

Theorem 2.1. *If $S \in \mathcal{S}(\ell^2_{\mathcal{N}}; \ell^2_{\mathcal{M}})$ is a solution of the BIP (1.1) and (1.2) with data based on the set $f_i, i \in \mathbb{Z}$, then for every $n \in \mathbb{N}$ the form Λ^S_n defined just above is positive semidefinite, where Λ^S_n is given as follows:*

$$\Delta^S_n(E_j, E_i) = \begin{cases} \langle Jf_jE_j, f_iE_i\rangle_{HS} & \text{for} \quad -n \leq i,j \leq 0, \\ -\langle Sh_jE_j, g_iE_i\rangle_{HS} & \text{for} \quad -n \leq i \leq 0, 0 < j \leq n, \\ -\langle Jf_jE_j, f_iE_i\rangle_{HS} & \text{for} \quad 0 \leq i,j \leq n \ , \end{cases} \tag{2.1}$$

where

$$J = \begin{bmatrix} I_{\ell^2_{\mathcal{M}}} & 0 \\ 0 & -I_{\ell^2_{\mathcal{N}}} \end{bmatrix}$$

and $E_i \in \mathcal{D}_2(\ell^2_{\mathcal{B}}; \ell^2_{\mathcal{B}})$, *for* $-n \le i \le n$.

Proof. This is an immediate consequence of Theorem 2.1 of [DF]. □

This is as far as one can go unless additional structure is imposed upon the data. As in [DF] we restrict the operators $f_{-n} \dots, f_1, \dots, f_n$ to the following form (when the coordinate spaces do not vary with the index i):

$$[f_{-n}, \dots, f_0, \dots, f_n] = \begin{bmatrix} C_{1,n} \\ C_{2,n} \end{bmatrix} (M_n - \mathbf{Z} N_n)^{-1} , \tag{2.2}$$

where

$$C_{1,n} = [C_{11,n}\ C_{12,n}] = [\xi_{-n} \cdots \xi_0\ \xi_1 \cdots \xi_n],\ (\xi_i \in \mathcal{D}(\ell^2_{\mathcal{B}}; \ell^2_{\mathcal{M}}))$$
$$C_{2,n} = [C_{21,n}\ C_{22,n}] = [\eta_{-n} \cdots \eta_0\ \eta_1 \cdots \eta_n],\ (\eta_i \in \mathcal{D}(\ell^2_{\mathcal{B}}; \ell^2_{\mathcal{N}}))$$
$$M_n = \begin{bmatrix} I & 0 \\ 0 & -A_{2,n} \end{bmatrix},\ N_n = \begin{bmatrix} A_{1,n} & 0 \\ 0 & -I \end{bmatrix},$$

$$A_{1,n} = \begin{bmatrix} (A_{1,n})_{11} & \cdots & (A_{1,n})_{1,n+1} \\ \vdots & & \vdots \\ (A_{1,n})_{n+1,1} & \cdots & (A_{1,n})_{n+1,n+1} \end{bmatrix},$$
$$A_{2,n} = \begin{bmatrix} (A_{2,n})_{11} & \cdots & (A_{2,n})_{1n} \\ \vdots & & \vdots \\ (A_{2,n})_{n1} & \cdots & (A_{2,n})_{nn} \end{bmatrix},$$

where $(A_{i,n})_{st} \in \mathcal{D}(\ell^2_{\mathcal{B}}; \ell^2_{\mathcal{B}})$, $i = 1, 2$.

In addition we assume that for every $n \in \mathbb{N}$ operator matrices $A_{1,n}$ and $A_{2,n}$ are upper triangular

$$(A_{i,n})_{st} = 0,\ s < t \tag{2.3a}$$

and satisfy some consistency conditions (which mean that operator-matrices $A_{1,n}$ and $A_{2,n}$ are truncations of some infinite upper triangular operator-matrices A_1 and A_2):

$$(A_{i,n})_{st} = (A_{i,m})_{st},\ s, t \le \min(n, m). \tag{2.3b}$$

Operator $\mathbf{Z}$ is defined as follows:

$$\mathbf{Z} = Z_1 \oplus Z_2 ,$$
$$Z_1 = \operatorname{diag}(Z, \dots, Z)\ (n+1 \text{ times}),$$
$$Z_2 = \operatorname{diag}(Z, \dots, Z)\ (n \text{ times}),$$

and it is assumed that

$$r_{sp}(Z_1 A_{1,n}) < 1 \text{ and } r_{sp}(A_{2,n} Z_2^*) < 1 . \tag{2.4}$$

This ensures that

$$F_{11,n} = C_{11,n}(I - Z_1 A_{1,n})^{-1} \text{ and } F_{21,n} = C_{21,n}(I - Z_1 A_{1,n})^{-1}$$

are bounded upper triangular operators, whereas

$$F_{12,n} = C_{12,n} Z_2^*(I - A_{2,n} Z_2^*)^{-1} \text{ and } F_{22,n} = C_{22,n} Z_2^*(I - A_{2,n} Z_2^*)^{-1}$$

are bounded strictly lower triangular operators.

In particular for the bitangential NP problem with the data specified in (1.3) the operator $[f_{-n}, \dots, f_0, \dots, f_n]$ for every $n \in \mathbb{N}$ can be expressed in the following form:

$$[f_{-n}, \dots, f_0, \dots, f_n] = \begin{bmatrix} C_{1,n} \\ C_{2,n} \end{bmatrix} (M_n - \mathbf{Z} N_n)^{-1},$$

with corresponding operators $A_{1,n} = \mathrm{diag}(V^*_{-n}, \dots, V^*_0)$ and $A_{2,n} = \mathrm{diag}(V_1, \dots, V_n)$.

We remark that condition (2.3), i.e., the condition that operator-matrices $A_{1,n}$ and $A_{2,n}$ are truncations of some infinite upper triangular operator-matrices A_1 and A_2, is not really necessary and is added for the sake of simplicity of presentation. This condition ensures that the data of our problems has some chain structure that will play an important role in our construction in the next section. Besides, condition (2.3) is not too restrictive, since the NP problem (as we saw just above with $A_1, A_2 \in \mathcal{D}$), the Carathéodory-Fejér problem and mixture of the two (with $A_1, A_2 \in \mathcal{U}$, see [DD]) admit the prescribed representation.

Next, for every n we express the form Λ_n^S in terms of the operator $P(n)$ which is expressed in block form (in much the same way as in Section 2 of [DF]):

$$P(n) = \begin{bmatrix} P_{11}(n) & P_{12}(n) \\ P_{21}(n) & P_{22}(n) \end{bmatrix},$$

corresponding to the indicated decomposition of M_n and N_n, i.e., $P(n)$ maps $(\ell^2_{\mathcal{B}})^{n+1} \oplus (\ell^2_{\mathcal{B}})^n$ into itself with components $P_{11}(n) \in \{\mathcal{D}(\ell^2_{\mathcal{B}}; \ell^2_{\mathcal{B}})\}^{(n+1)\times(n+1)}$, $P_{12}(n) \in \{\mathcal{D}(\ell^2_{\mathcal{B}}; \ell^2_{\mathcal{B}})\}^{(n+1)\times n}$, $P_{21}(n) = P_{12}^*(n)$ and $P_{22}(n) \in \{\mathcal{D}(\ell^2_{\mathcal{B}}; \ell^2_{\mathcal{B}})\}^{n\times n}$. Here $P(n)$ is defined by the rule

$$\Delta_n^S(D, E) = \langle P(n) D, E \rangle_{HS}$$

for every choice of D and E in $\{\mathcal{D}_2(\ell^2_{\mathcal{B}}; \ell^2_{\mathcal{B}})\}^{(2n+1)\times 1}$.

The following is a simple consequence of Theorem 2.2 of [DF]:

Theorem 2.2. *If $S \in \mathcal{S}(\ell^2_{\mathcal{N}}; \ell^2_{\mathcal{M}})$ is a solution of the BIP with data of the form (2.2), then for every $n \geq 0$ the operator $P(n)$ is a positive semidefinite solution of the Lyapunov-Stein equation*

$$M_n^* P(n) M_n - N_n^* \mathbf{Z}^* P(n) \mathbf{Z} N_n = C_{1,n}^* C_{1,n} - C_{2,n}^* C_{2,n}\,. \tag{2.5}$$

Moreover, for every $m \geq n > 0$, the following condition holds:

$$\langle P(n) x_i, x_j \rangle_{\{l^2_{\mathcal{B}}\}^{(2n+1)\times 1}} = \langle P(m) \hat{x}_i, \hat{x}_j \rangle_{\{l^2_{\mathcal{B}}\}^{(2n+1)\times 1}} \tag{2.6}$$

Where $x_i, x_j \in \{l^2_{\mathcal{B}}\}^{(2n+1)\times 1}$ *and* $\hat{x}_i, \hat{x}_j \in \{l^2_{\mathcal{B}}\}^{(2m+1)\times 1}$ *are such that the latter are images of the former under the natural embedding of* $\{l^2_{\mathcal{B}}\}^{(2n+1)\times 1}$ *into* $\{l^2_{\mathcal{B}}\}^{(2m+1)\times 1}$: $(\hat{x}_i)_k = (x_i)_k, |k| \le n$ *and* $(\hat{x}_i)_k = 0, |k| > n$.

Theorem 2.2 implies that every solution $S \in \mathcal{S}(\ell^2_{\mathcal{N}}; \ell^2_{\mathcal{M}})$ of the BIP generates a sequence of solutions

$$P(n) = \begin{bmatrix} Q_{-n,-n} & \cdots & Q_{-n,n} \\ \vdots & & \vdots \\ Q_{n,-n} & \cdots & Q_{nn} \end{bmatrix} = \begin{bmatrix} P_{11}(n) & P_{12}(n) \\ & \\ P_{21}(n) & P_{22}(n) \end{bmatrix} \ge 0$$

of the equation (2.5), for every $n \in \mathbb{N}$, subject to (2.6) with unique diagonal blocks $P_{11}(n) \in \{\mathcal{D}(\ell^2_{\mathcal{B}}; \ell^2_{\mathcal{B}})\}^{(n+1)\times(n+1)}$ and $P_{22}(n) \in \{\mathcal{D}(\ell^2_{\mathcal{B}}; \ell^2_{\mathcal{B}})\}^{n\times n}$.

In general, the off diagonal block $P_{12}(n)$ is not uniquely determined by the equation. As in [DF], we shall say that S is a solution of the augmented BIP based on each sequence of solutions $P(n) \ge 0$ of (2.5) which satisfies condition (2.6) as well as on the data, if, in addition to conditions (1.1) and (1.2), the condition

$$-\langle Sh_j E_j, g_i E_i \rangle_{HS} = \langle Q_{ij} E_j, E_i \rangle_{HS}$$

is satisfied for every choice of E_i and E_j in $\mathcal{D}_2(\ell^2_{\mathcal{B}}; \ell^2_{\mathcal{B}})$ with $i = -n, \dots, 0$, $j = 1, \dots, n$. The preceding analysis shows that every solution S of the BIP based on $f_{-n}, \dots, f_n$ is a solution of the augmented BIP based on $f_{-n}, \dots, f_n$ and the particular sequence of solutions $P(n) = P^S(n)$ of (2.5) corresponding to the form Λ^S.

In terms of the notation introduced between formula (2.2) and the constraints (2.4), the solutions of the augmented BIP can be characterized as the set of operators $S \in \mathcal{S}(\ell^2_{\mathcal{N}}; \ell^2_{\mathcal{M}})$ such that for every $n \in \mathbb{N}$ the following three conditions hold true:

(1) $(C_{1,n} - SC_{2,n})(M_n - \mathbf{Z}N_n)^{-1}E \in \mathcal{U}_2(\ell^2_{\mathcal{B}}; \ell^2_{\mathcal{M}})$

(2) $(-S^*C_{1,n} + C_{2,n})(M_n - \mathbf{Z}N_n)^{-1}E \in \mathcal{L}_2'(\ell^2_{\mathcal{B}}; \ell^2_{\mathcal{N}})$

(3) $$-\left\langle SC_{2,n}(M_n - \mathbf{Z}N_n)^{-1}\begin{bmatrix} 0 \\ E_2 \end{bmatrix}, C_{1,n}(M_n - \mathbf{Z}N_n)^{-1}\begin{bmatrix} E_1 \\ 0 \end{bmatrix}\right\rangle_{HS} = \left\langle P(n)\begin{bmatrix} 0 \\ E_2 \end{bmatrix}, \begin{bmatrix} E_1 \\ 0 \end{bmatrix}\right\rangle_{HS}$$

for every choice of

$$E \in \{\mathcal{D}_2(\ell^2_{\mathcal{B}}; \ell^2_{\mathcal{B}})\}^{(2n+1)\times 1}, \ E_1 \in \{\mathcal{D}_2(\ell^2_{\mathcal{B}}; \ell^2_{\mathcal{B}})\}^{(n+1)\times 1}, \text{ and } E_2 \in \{\mathcal{D}_2(\ell^2_{\mathcal{B}}; \ell^2_{\mathcal{B}})\}^{n\times 1}.$$

At the end of this section we prove (using the sufficiency conditions for the existence of a solution to the BIP with finite number of interpolation conditions established in [DF]) that the necessary conditions obtained in Theorem 2.1 are also sufficient.

Theorem 2.3. *Let* $P(n)$ *be a sequence of positive semidefinite operators that satisfy conditions (2.5) and (2.6). Then there exists a solution of the augmented BIP (1), (2) and (3).*

Proof. From Theorem 3.2 of [DF] and conditions of this theorem, it follows that for every $n \in \mathbb{N}$ there exists a solution of the augmented BIP based on the data $\{f_i\}$, $-n \leq i \leq n$. Since a solution of the augmented BIP based on the data $\{f_i\}$, $-m \leq i \leq m$, is a solution of the augmented BIP based on the data $\{f_i\}$, $-n \leq i \leq n$, for $m \geq n$, the statement of this theorem follows from the compactness of the unit ball in $\mathcal{U}$ in the weak operator topology (see the proof of Theorem 7.3 of [DD] where similar arguments were used). □

3. Representation of all the solutions

Our next objective is to give a description of all the solutions to the augmented BIP (1), (2) and (3). We recall that in [DF] a "finite" version of the augmented BIP (1), (2) and (3) was treated. The description was formulated in terms of the characteristic functions of two coupled unitary colligations, the first of which is defined in terms of the data of the problem whereas the second is arbitrary. This leads to a description of all solutions in terms of a linear fractional transformation of the Redheffer type.

We shall give here a description of the same kind of all the solutions to the augmented BIP in this more general case. To do this we shall slightly change the construction used in [DF]. More precisely we will use the sequence of semi-positive operators $\{P(n)\}$ that satisfy conditions (2.5) and (2.6) to define a new norm on a vector space (which will be defined below)

$$\mathcal{H} \subset X = \bigoplus_{i=-\infty}^{\infty} \mathcal{X}_i,$$

where $\mathcal{X}_i = l^2_{\mathcal{B}}, i \in \mathbb{Z}$. We shall see that if $r_{sp}(ZV_j^*) \leq q < 1$ for every $j \in \mathbb{Z}$, then this new norm is equivalent to the original norm on $\bigoplus_{i=-\infty}^{\infty} l^2_{\mathcal{B}}$, since in this case the sequence $P(n)$ is a sequence of uniformly bounded operators on $\mathcal{X}_2$ which is generically not true if the sequence $r_{sp}(ZV_j^*)$, $j \in \mathbb{Z}$, is not separated from 1.

Following the scheme used in [DF], first we need to define the isometry V corresponding to the augmented BIP (1), (2) and (3). To this end we will use the Lyapunov-Stein equation (2.5), associated with the BIP, written in the equivalent form:

$$\begin{aligned}(3.1)\quad \langle P(n)M_n x_{j,n}, M_n x_{j,n}\rangle &+ \langle C_{2,n}x_{j,n}, C_{2,n}x_{j,n}\rangle \\ &= \langle \mathbf{Z}^* P(n)\mathbf{Z}N_n x_n, N_n x_{j,n}\rangle + \langle C_{1,n}x_{j,n}, C_{1,n}x_{j,n}\rangle,\end{aligned}$$

for every choice of $x_{j,n} \in \{l^2_{\mathcal{B}}\}^{(2n+1)\times 1}$, such that all coordinates except the j-th one are equal to 0, $|j| \leq n$ (in what follows, making use of the natural embedding specified in the statement of Theorem 2.2, we shall treat vectors $x_{j,n} \in \{l^2_{\mathcal{B}}\}^{(2n+1)\times 1}$ with $n \geq |j|$ as one vector x_j even though they belong to the different spaces $\{l^2_{\mathcal{B}}\}^{(2n+1)\times 1}$ with $n \geq |j|$, i.e., we will consider these vectors, as well as their linear combinations, as elements of the space X).

We remark that, since $P(n)$ satisfies (2.6) and $M_n, N_n, C_{2,n}$ and $C_{1,n}$ are given by (2.2)–(2.4) and therefore are diagonal operators, all the terms in the formula (3.1) are independent of n, as long as n is chosen such that $n \geq |j|$.

Let X and $X^{(-1)}$ be the spaces of all doubly infinite sequences with finite support and with coordinate spaces $l^2_{\mathcal{B}}$ and $l^2_{\mathcal{B}^{(-1)}}$, respectively. With help of the sequence $\{P(n)\}$ we define the form Λ on $X \times X$ ($\Lambda^{(1)}$ on $X^{(-1)} \times X^{(-1)}$) by the rule: $\Lambda(x,x) = \langle P(n)x, x\rangle$ for $x \in X$ such that support$(x) \subset [-n,n]$ ($\Lambda^{(1)}(y,y) = \langle P(n)Zy, Zy\rangle$ for $y \in X^{(-1)}$ such that support$(y) \subset [-n,n]$) . This definition makes sense since $\{P(n)\}$ satisfy the compatibility condition (2.6).

By assumption, $P(n)$ is positive semidefinite for every $n \in \mathbb{N}$, and hence

$$\Lambda(x,x) \geq 0$$

for every $x \in X$. Let $\mathcal{H}$ ($\mathcal{H}^{(-1)}$) denote the Hilbert space obtained by completing X ($X^{(-1)}$) with respect to the metric Λ ($\Lambda^{(1)}$). It is easily seen that $\mathcal{H}$ ($\mathcal{H}^{(-1)}$) is an l^2-space, since, for every $n \in \mathbb{N}$, $P(n)$ is a diagonal operator.

Now, making use of the equation (3.1), we define an isometry V: for every $n \in \mathbb{N}$

$$V(M_n x \oplus C_{2,n}x) = N_n x \oplus C_{1,n}x, \tag{3.2}$$

where $x \in X$, with support$(x) \subset [-n,n]$.

We remark that relations (3.2) define an isometric operator

$$V : d_V \longrightarrow \Delta_V,$$

where

$$d_V = \text{closure}\{M_n \oplus C_{2,n}x : \ x \in X, \text{ with support}(x) \subset [-n,n], \ n \in \mathbb{N}\} \subseteq \mathcal{H} \oplus \mathcal{E}_1$$

and

$$\Delta_V = \text{closure}\{N_n x \oplus C_{1,n}x : \ x \in X, \text{ with support}(x) \subset [-n,n], \ n \in \mathbb{N}\} \subseteq \mathcal{H}^{(-1)} \oplus \mathcal{E}_2 \ .$$

Symbols $\mathcal{E}_1$ and $\mathcal{E}_2$ denote the spaces $l^2_{\mathcal{N}}$ and $l^2_{\mathcal{M}}$ respectively.

Observe that, since $M_n, N_n, C_{1,n}$ and $C_{2,n}$, $n \in \mathbb{N}$, are diagonal operators, d_V and Δ_V are l^2-spaces, though, generally speaking, the underlining coordinates of these two spaces may vary with the index i (and therefore the symbol $\mathcal{H}^{(-1)}$ in the definition of the Δ_V). The last observation applies equally to the spaces $d_V^{\perp}$ and $\Delta_V^{\perp}$,where

$$d_V^{\perp} = (\mathcal{H} \oplus \mathcal{E}_1) \ominus d_V$$

and

$$\Delta_V^{\perp} = (\mathcal{H}^{(-1)} \oplus \mathcal{E}_2) \ominus \Delta_V.$$

Moreover, $V \in \mathcal{D}(d_V; \Delta_V)$. This follows easily from the definition of V (3.2) and the form of the spaces d_V and Δ_V.

We shall refer to $\mathcal{H}$ as the state space of the isometry V and to the spaces $\mathcal{E}_1$ and $\mathcal{E}_2$ as the input space and the output space, respectively.

Let $\mathcal{K} = \{\ell^2_{\mathcal{C}}\}^n$ be a complex separable Hilbert space which contains $\mathcal{H}$, i.e., assume that $\mathcal{C}_i \supset \mathcal{B}_i$, where the spaces $\mathcal{C}_i$ are allowed to vary with i, and let $U \in \mathcal{D}(\mathcal{K}\oplus\mathcal{E}_1;\mathcal{K}^{(-1)}\oplus\mathcal{E}_2)$ be a unitary extension of V from $\mathcal{K}\oplus\mathcal{E}_1$ onto $\mathcal{K}^{(-1)}\oplus\mathcal{E}_2$:

$$U|_{d_V} = V\ .$$

The entries in the block decomposition

$$U = \begin{bmatrix} U_{11} & U_{12} \\ & \\ U_{21} & U_{22} \end{bmatrix} : \begin{matrix} \mathcal{K} \\ \oplus \\ \mathcal{E}_1 \end{matrix} \longrightarrow \begin{matrix} \mathcal{K}^{(-1)} \\ \oplus \\ \mathcal{E}_2 \end{matrix}$$

of U with respect to the given spaces taken together with the spaces themselves:

$$\left\{U_{11}, U_{12}, U_{21}, U_{22}; \mathcal{K}, \mathcal{E}_1, \mathcal{K}^{(-1)}, \mathcal{E}_2\right\}$$

is usually termed a colligation. It is said to be a unitary colligation if, as in the present case, U is unitary. It is important to bear in mind that because of the prevailing assumptions on U, each of the components U_{ij}, $i, j = 1, 2$, is a diagonal operator.

We shall refer to the operator valued function

$$W(r) = U_{22} + U_{21}(I - rZU_{11})^{-1} rZU_{12}\ ,\ \ 0 \le r < 1\ , \tag{3.3}$$

which maps $\mathcal{E}_1$ into $\mathcal{E}_2$ as the characteristic (scattering or transfer) function of the unitary colligation which was described above.

It is readily checked (see Lemma 3.3 of [DF]) that the operator $W(r)$ belongs to $\mathcal{U}(\mathcal{E}_1;\mathcal{E}_2)$ and is contractive: $\|W(r)\| \le 1$, for every choice of $r \in [0,1)$.

We recall that there exists an operator

$$W_U x = \lim_{r\uparrow 1} U_{22} + U_{21}(I - rZU_{11})^{-1} rZU_{12}x,\ \ x \in \mathcal{E}_1 \tag{3.3'}$$

which belongs to $\mathcal{U}(\mathcal{E}_1;\mathcal{E}_2)$ and is contractive (see Lemma 3.6 of [DF] and discussion after it if needed). We shall refer to the operator W_U as the characteristic operator of the unitary colligation U.

We remark that the latter definition implies that the "k-th diagonal" of $W(r)$ up to the multiplier r^k is equal to the "k-th diagonal" of W_U, which makes it natural to consider $W(r)$, $0 \le r < 1$, and W_U as the same object. To clarify this further we recall the following definition used in [DF] and which will be needed in the sequel:

The symbol $F(r)$ which is defined for each upper triangular operator $F \in \mathcal{U}(\ell^2_{\mathcal{M}};\ell^2_{\mathcal{N}})$ stands for the operator in $\mathcal{U}(\ell^2_{\mathcal{M}};\ell^2_{\mathcal{N}})$ with blocks

$$F(r)_{ij} = r^{j-i} F_{ij} \tag{3.4}$$

for $j \ge i$.

It is easily seen that this notation is consistent with the notation introduced in formula (3.3) because $W_U(r)$ in this new sense (for the operator W_U which is defined by (3.3′)) is equal to $W(r)$ as defined by (3.3), thanks to Lemma 3.6 and formula (3.10) of [DF]. We would also like to mention that thanks to Theorem

5.3 of [DF] every upper triangular operator of norm less than one can be represented as the characteristic operator of some unitary colligation, i.e., for every $F \in \mathcal{U}(\ell^2_{\mathcal{M}}; \ell^2_{\mathcal{N}})$, $||F|| \leq 1$, there exists a unitary colligation U such that

$$F(r) = W(r), \ 0 \leq r < 1,$$

and consequently $F = W_U$.

Theorem 3.1. *Let $\{P(n)\}, n \in \mathbb{N}$, be a given sequence of positive semidefinite operators that satisfy conditions (2.5) and (2.6) and let*

$$U = \begin{bmatrix} U_{11} & U_{12} \\ U_{21} & U_{22} \end{bmatrix}$$

be any unitary colligation with components

$$U_{11} \in \{\mathcal{D}(\mathcal{H}; \mathcal{H}^{(-1)})\}, \ U_{12} \in \{\mathcal{D}(\ell^2_{\mathcal{N}}; \mathcal{H}^{(-1)})\},$$
$$U_{21} \in \{\mathcal{D}(\mathcal{H}; \ell^2_{\mathcal{M}})\}, \ \textit{and} \ U_{22} \in \mathcal{D}(\ell^2_{\mathcal{N}}; \ell^2_{\mathcal{M}}),$$

which extends the isometry V which is defined by (3.2). Then the characteristic operator of U, W_U, is a solution of the augmented BIP (1), (2) and (3).

Proof. Observe first that for every $n \in \mathbb{N}$, (3.2) defines an isometry V_n which can be rewritten in the equivalent form:

$$V_n : \begin{bmatrix} P^{\frac{1}{2}}(n)M_n \\ C_{2,n} \end{bmatrix} x \longrightarrow \begin{bmatrix} (P^{\frac{1}{2}}(n))^{(1)}N_n \\ C_{1,n} \end{bmatrix} x \tag{3.5}$$

But this V_n is the isometry operator associated with the augmented BIP (1), (2) and (3) based on the set $\{f_i\}, -n \leq i \leq n$, given by (2.2)–(2.4) and the positive semidefinite operator

$$P(n) : (l^2_{\mathcal{B}})^{2n+1} \longrightarrow (l^2_{\mathcal{B}})^{2n+1}.$$

Then, from Theorem 3.1 and Theorem 5.2 of [DF], one can conclude that for every $n \in \mathbb{N}$ the characteristic operator W_U of the unitary colligation U is a solution of the partial BIP described just above. This completes the proof. □

One can check that the inverse statement also holds true:

Theorem 3.2. *Every solution of the augmented BIP (1), (2) and (3) is the characteristic operator of a unitary colligation U described in Theorem 3.1.*

The proof of this theorem can be carried out just in the same way as the proof of the first assertion of Theorem 5.2 of [DF] and therefore is omitted.

Our next objective is to obtain a representation of all the solutions of the BIP via a linear fractional transformation of the Redheffer type.

In view of Theorem 3.1 and Theorem 3.2, it is enough to give this representation to the set of characteristic operators of the unitary colligations described in Theorem 3.1. To do that we will need the description of this class of unitary colligations given in Section 4 of [DF], which we restate here for the convenience of the reader.

First we define the "standard" unitary operator A corresponding to the augmented BIP (cf. (4.2) of [DF]):

$$A : \mathcal{H} \oplus \mathcal{E}_1 \oplus \mathcal{N}_2 \longrightarrow \mathcal{H}^{(-1)} \oplus \mathcal{E}_2 \oplus \mathcal{N}_1,$$

$$\text{with } \mathcal{E}_1 = l^2_{\mathcal{N}} \text{ and } \mathcal{E}_2 = l^2_{\mathcal{M}}, \tag{3.6}$$

$$A|_{d_V} = V \ , \ A|_{d_V^\perp} = \alpha \ , \text{ and } A|_{\mathcal{N}_2} = \beta^* \ ,$$

where $\alpha \in \mathcal{D}(d_V^\perp; \mathcal{N}_1)$ and $\beta \in \mathcal{D}(\Delta_V^\perp; \mathcal{N}_2)$ are diagonal unitary operators (i.e., the operators α_{ii} and $\beta_{ii}, i \in \mathbb{Z}$ (the diagonal entries of operators α and β), are unitary, and $(\mathcal{N}_1)_i = \operatorname{Im}\{\alpha_{ii}\}$ and $(\mathcal{N}_2)_i = \operatorname{Im}\{\beta_{ii}\}$).

We shall refer to A as a unitary colligation with state space $\mathcal{H}$, input space $\mathcal{E}_1 \oplus \mathcal{N}_2$ and output space $\mathcal{E}_2 \oplus \mathcal{N}_1$, even though "$A_{11}$" maps $\mathcal{H}$ into $\mathcal{H}^{(-1)}$.

Now let $B \in \mathcal{D}(\mathcal{G} \oplus \mathcal{N}_1; \mathcal{G}^{(-1)} \oplus \mathcal{N}_2)$ be a second unitary colligation with state space $\mathcal{G} = \ell^2_{\mathcal{R}}$, input space $\mathcal{N}_1$ and output space $\mathcal{N}_2$. Then

$$A(h + e_1 + n_2) = h' + e_2 + n_1 \tag{3.7}$$

and

$$B(g + n_1) = g' + n_2 \ , \tag{3.8}$$

where h and h' stand for elements in $\mathcal{H}$, g and g' for elements in $\mathcal{G}$, $e_j \in \mathcal{E}_j$ and $n_j \in \mathcal{N}_j$ for $j = 1, 2$, embedded as needed for the formulas to make sense.

This pair of relations (considered as a system of equations, i.e., $n_j \in \mathcal{N}_j$ for $j = 1, 2$, in (3.6) and (3.7) are considered to be the same) serves to define (see Lemma 4.2 of [DF]) a unitary colligation $U \in \mathcal{D}(\mathcal{G} \oplus \mathcal{H} \oplus \mathcal{E}_1; \mathcal{G}^{(-1)} \oplus \mathcal{H}^{(-1)} \oplus \mathcal{E}_2)$ from $\mathcal{G} \oplus \mathcal{H} \oplus \mathcal{E}_1$ onto $\mathcal{G}^{(-1)} \oplus \mathcal{H}^{(-1)} \oplus \mathcal{E}_2$ such that

$$U(g + h + e_1) = g' + h' + e_2 \tag{3.9}$$

and

$$U|_{d_V} = V \ .$$

We shall refer to such a U as the coupling of (the "standard" colligation) A with (the arbitrary colligation) B.

This coupling exhibits a method of constructing a unitary colligation $U \in \mathcal{D}(\mathcal{G} \oplus \mathcal{H} \oplus \mathcal{E}_1; \mathcal{G}^{(-1)} \oplus \mathcal{H}^{(-1)} \oplus \mathcal{E}_2)$ with state space $\mathcal{G} \oplus \mathcal{H}$, input space $\mathcal{E}_1$ and output space $\mathcal{E}_2$ such that

$$U|_{d_V} = V \ .$$

The following theorem says that every unitary colligation $U \in \mathcal{D}(\mathcal{G} \oplus \mathcal{H} \oplus \mathcal{E}_1; \mathcal{G}^{(-1)} \oplus \mathcal{H}^{(-1)} \oplus \mathcal{E}_2)$ such that $U|_{d_V} = V$ arises this way:

Theorem 4.1 [DF]. *Let $U \in \mathcal{D}(\mathcal{G} \oplus \mathcal{H} \oplus \mathcal{E}_1; \mathcal{G}^{(-1)} \oplus \mathcal{H}^{(-1)} \oplus \mathcal{E}_2)$ be a unitary colligation (with state space $\mathcal{G} \oplus \mathcal{H}$, input space $\mathcal{E}_1$ and output space $\mathcal{E}_2$) such that $U|_{d_V} = V$. Let $\mathcal{N}_1 = \alpha d_V^\perp$ and let $\mathcal{N}_2 = \beta \Delta_V^\perp$, where α and β are unitary diagonal operators (i.e., $\alpha \in \mathcal{D}(d_V^\perp; \mathcal{N}_1)$ and $\beta \in \mathcal{D}(\Delta_V^\perp; N_2)$). Then U can be expressed as the coupling of a standard unitary colligation $A \in \mathcal{D}(\mathcal{H} \oplus \mathcal{E}_1 \oplus \mathcal{N}_2; \mathcal{H}^{(-1)} \oplus \mathcal{E}_2 \oplus \mathcal{N}_1)$ (with state space $\mathcal{H}$, input space $\mathcal{E}_1 \oplus \mathcal{N}_2$ and output space $\mathcal{E}_2 \oplus \mathcal{N}_1$) which is defined* *as in (3.6) and a unitary colligation* $B \in \mathcal{D}(\mathcal{G} \oplus \mathcal{N}_1; \mathcal{G}^{(-1)} \oplus \mathcal{N}_2)$ *(with state space* $\mathcal{G}$*, input space* $\mathcal{N}_1$ *and output space* $\mathcal{N}_2$*).*

Moreover, one can express the characteristic function of the coupling of the unitary colligations A and B through their characteristic functions:

Theorem 4.2 [DF]. *Let*

$$\Sigma(r) = W_A(r) : \ \mathcal{E}_1 \oplus \mathcal{N}_2 \longrightarrow \mathcal{E}_2 \oplus \mathcal{N}_1 \ ,$$

$$\Omega(r) = W_B(r) : \ \mathcal{N}_1 \longrightarrow \mathcal{N}_2 \ ,$$

and

$$S(r) = W_U(r) : \ \mathcal{E}_1 \longrightarrow \mathcal{E}_2$$

denote the characteristic functions of the unitary colligations $A \in \mathcal{D}(\mathcal{H} \oplus \mathcal{E}_1 \oplus \mathcal{N}_2; \mathcal{H}^{(-1)} \oplus \mathcal{E}_2 \oplus \mathcal{N}_1)$ *which is defined by (3.6),* $B \in \mathcal{D}(\mathcal{G} \oplus \mathcal{N}_1; \mathcal{G}^{(-1)} \oplus \mathcal{N}_2)$ *and their coupling* $U \in \mathcal{D}(\mathcal{G} \oplus \mathcal{H} \oplus \mathcal{E}_1; \mathcal{G}^{(-1)} \oplus \mathcal{H}^{(-1)} \oplus \mathcal{E}_2)$*, respectively. Then*

$$\|\Sigma_{22}(r)\| \ \leq \ r$$

and

$$S(r) = \Sigma_{11}(r) + \Sigma_{12}(r)\Omega(r)\{I - \Sigma_{22}(r)\Omega(r)\}^{-1}\Sigma_{21}(r) \tag{3.10}$$

for every choice of $r \in [0,1)$.

Hence, in view of Theorem 4.1 and Theorem 4.2 of [DF] together with the discussion just before Theorem 3.1, it follows from Theorem 3.1 and Theorem 3.2 that all the solutions to the augmented BIP (1), (2) and (3) can be represented via a linear fractional formula of the Redheffer type (the same one which was described in the statement of Theorem 5.4 of [DF]):

Theorem 3.3. *An operator* $S \in \mathcal{S}(\mathcal{E}_1; \mathcal{E}_2)$ *is a solution of the augmented BIP if and only if it can be expressed in the form*

$$S(r) = \Sigma_{11}(r) + \Sigma_{12}(r)\left\{I - \Omega(r)\Sigma_{22}(r)\right\}^{-1}\Omega(r)\Sigma_{21}(r) \ ,$$

for $0 \leq r < 1$*, where*

$$\Sigma(r) \ = \ \begin{bmatrix} \Sigma_{11}(r) & \Sigma_{12}(r) \\ \Sigma_{21}(r) & \Sigma_{22}(r) \end{bmatrix}$$

is the characteristic function of the standard unitary colligation A *which was defined by (3.6) and* Ω *is any contractive upper triangular operator in the class* $\mathcal{S}(\mathcal{N}_1; \mathcal{N}_2)$.

References

[ADD] D. Alpay, P. Dewilde and H. Dym, Lossless inverse scattering and reproducing kernels for upper triangular operators, in: *Extension and Interpolation of Linear Operators and Matrix Functions* (I. Gohberg, ed.), *Operator Theory: Advances and Applications*, **OT47**, Birkhäuser Verlag, Basel, 1990, pp. 61–135.

[Ar] D. Arov, Regular J-inner matrix-functions and related continuation problems, *Operator Theory: Advances and Applications*, **OT43**, Birkhäuser Verlag, Basel, 1990, pp. 63–87.

[BGK1] J.A. Ball, I. Gohberg, and M.A. Kaashoek, Nevanlinna-Pick interpolation for time-varying input-output maps: The discrete case, in: *Time-Variant Systems and Interpolation* (I. Gohberg, ed.), *Operator Theory: Advances and Applications*, **OT56**, Birkhäuser Verlag, Basel, 1992, pp. 1–51.

[BGK2] J.A. Ball, I. Gohberg, and M.A. Kaashoek, Two-sided Nudelman interpolation for input-output operators of discrete time-varying systems, *Integral Equations Operator Theory* **21** (1994), pp. 174-211.

[DD] P. Dewilde and H. Dym, Interpolation for upper triangular operators, in: *Time-Variant systems and Interpolation* (I. Gohberg, ed.), *Operator Theory: Advances and Applications*, **OT56**, Birkhäuser Verlag, Basel, 1992, pp. 153–260.

[DvV] P. Dewilde and A. van der Veen, *Time-Varying Systems and Computations*, Kluwer, 1998.

[D1] H. Dym, Remarks on interpolation for upper triangular operators, in: *Challenges of a Generalized System Theory*, (P. Dewilde, M. A. Kaashoek and M. Verhaegen, eds.), North-Holland, Amsterdam, 1993, pp. 9–24.

[DF] H. Dym and B. Freydin, Bitangential interpolation for upper triangular operators, in: *Topics in Interpolation Theory* (H. Dym et al., eds.), *Operator Theory: Advances and Applications*, **OT95**, Birkhäuser Verlag, Basel , 1997, pp. 105–142.

[KKY] V.E. Katsnelson, A.Ya. Kheifets and P.M. Yuditskii, An abstract interpolation problem and the theory of extensions of isometric operators, in: *Operators in Function Spaces and Problems in Function Theory* (V.A. Marchenko, ed.), "Naukova Dumka", Kiev, 1987, pp. 83–96, 146; English translation in: *Topics in Interpolation Theory* (H. Dym et al., eds.), *Operator Theory: Advances and Applications*, **OT95**, Birkhäuser Verlag, Basel , 1997, pp. 283–298.

[K] A.Ya. Kheifets, The generalized bitangential Schur-Nevanlinna-Pick problem and the related Parseval equality, *Journal of Soviet Mathematic* **58**, No. 4 (1992), 358–364.

[KY] A.Ya. Kheifets and P.M. Yuditskii, An analysis and extension of V.P. Potapov's approach to interpolation problems with applications to the generalized bitangential Schur-Nevanlinna-Pick problem and J-inner-outer factorization, in: *Matrix and Operator Valued Functions* (I. Gohberg and L.A. Sakhnovich, eds.), *Operator Theory: Advances and Application*, **OT72** (1994), Birkhäuser Verlag, Basel, pp. 133–161.

[Ko1] J. Kos, Higher order time-varying Nevanlinna-Pick interpolation, in: *Challenges of a Generalized System Theory,* (P. Dewilde, M. A. Kaashoek, and M. Verhaegen, eds.), North Holand, Amsterdam, 1993, pp. 59–72.

[vV] A. van der Veen, *Time-varying Theory and Computational Modeling*, Ph.D. Thesis, Delft (1993).

Department of Mathematics, Ben-Gurion University, Beer-Sheva 84105, Israel
E-mail address: fred@indigo.cs.bgu.ac.il

Operator Theory:
Advances and Applications, Vol. 123

Some Applications of Operator-valued Herglotz Functions

Fritz Gesztesy, Nigel J. Kalton, Konstantin A. Makarov, and Eduard Tsekanovskii

Dedicated to Moshe Livšic on the occasion of his 80th birthday

Abstract. We consider operator-valued Herglotz functions and their applications to self-adjoint perturbations of self-adjoint operators and self-adjoint extensions of densely defined closed symmetric operators. Our applications include model operators for both situations, linear fractional transformations for Herglotz operators, results on Friedrichs and Krein extensions, and realization theorems for classes of Herglotz operators. Moreover, we study the concrete case of Schrödinger operators on a half-line and provide two illustrations of Livšic's result [44] on quasi-hermitian extensions in the special case of densely defined symmetric operators with deficiency indices $(1,1)$.

1. Introduction

The principal purpose of this paper is to extend some of our recent results on matrix-valued Herglotz functions in [30] to the infinite-dimensional context.

Given a complex Hilbert space $\mathcal{K}$, a map $M : \mathbb{C}_+ \to \mathcal{B}(\mathcal{K})$ is called a $\mathcal{K}$-valued Herglotz function (or simply a Herglotz operator) if M is analytic on $\mathbb{C}_+$ and $\mathrm{Im}(M(z)) \geq 0$ for all $z \in \mathbb{C}_+$. (We refer to the end of this introduction for a glossary on the notation used in this paper.) $\mathcal{B}(\mathcal{K})$-valued Herglotz functions admit the celebrated Nevanlinna-Riesz-Herglotz representation studied, for instance, by Brodskii [17], Sect. I.4, Krein and Ovcharenko [40], [41], and Shmulyan [62] in the infinite-dimensional context,

$$M(z) = C + Dz + \int_{\mathbb{R}} d\Omega(\lambda)((\lambda - z)^{-1} - \lambda(1+\lambda^2)^{-1}), \quad z \in \mathbb{C}_+, \tag{1.1}$$

where,

$$C = C^* \in \mathcal{B}(\mathcal{K}), \quad 0 \leq D \in \mathcal{B}(\mathcal{K}), \tag{1.2}$$

1991 *Mathematics Subject Classification.* Primary 30D50, 30E20, 47A10; Secondary 47A45.
Research supported by the US National Science Foundation under Grant Nos. DMS-9623121, DMS-9500125, and DMS-9870027.

and Ω is a $\mathcal{B}(\mathcal{K})$-valued measure satisfying

$$\int_{\mathbb{R}} d(\xi, \Omega(\lambda)\xi)_{\mathcal{K}}(1+\lambda^2)^{-1} < \infty \text{ for all } \xi \in \mathcal{K}. \tag{1.3}$$

In this paper we study a subclass of $\mathcal{B}(\mathcal{K})$-valued Herglotz functions where $D = 0$ and the Stieltjes integral in (1.1) is either understood in the norm (cf. Section 3) or the strong operator topology (cf. Section 4) in $\mathcal{K}$. For detailed discussions of operator-valued Herglotz functions and their boundary value behavior, see, for instance, [19], [57], Ch. 4, [63], Ch. V, [68]. Throughout this paper we will adhere to the usual convention

$$M(\bar{z}) = M(z)^*, \quad z \in \mathbb{C}_+ \tag{1.4}$$

(see, however, Lemma 4.13).

As discussed in some detail in [30], our notion of Herglotz functions is not without controversy. In fact, the names Pick, Nevanlinna, Nevanlinna-Pick, and R-functions (depending on whether the open upper half-plane $\mathbb{C}_+$ or the open unit disk D are involved, as well as depending on the geographical origin of authors) are also frequently in use. Here we follow a tradition in mathematical physics which appears to favor the terminology of Herglotz functions.

A crucial role in our analysis is played by linear fractional transformations of the type

$$M(z) \longrightarrow M_A(z) = (A_{2,1} + A_{2,2}M(z))(A_{1,1} + A_{1,2}M(z))^{-1}, \quad z \in \mathbb{C}_+, \tag{1.5}$$

where

$$A = \left(A_{p,q}\right)_{1\leq p,q\leq 2} \in \mathcal{A}(\mathcal{K}\oplus\mathcal{K}),$$
$$\mathcal{A}(\mathcal{K}\oplus\mathcal{K}) = \{A \in \mathcal{B}(\mathcal{K}\oplus\mathcal{K}) \,|\, A^*JA = J\}, \quad J = \begin{pmatrix} 0 & -I_{\mathcal{K}} \\ I_{\mathcal{K}} & 0 \end{pmatrix}. \tag{1.6}$$

M_A is a Herglotz operator in $\mathcal{K}$ whenever M is one and we refer to Krein and Shmulyan [42] for a detailed study in connection with (1.5), (1.6).

Section 2 provides a detailed study of a model Hilbert space, variants of which are used in Sections 3 and 4. This construction appears to be of independent interest.

In Section 3 we consider self-adjoint perturbations H_L of a self-adjoint (possibly unbounded) operator H_0 in some separable complex Hilbert space $\mathcal{H}$

$$H_L = H_0 + KLK^*, \quad \operatorname{dom}(H_L) = \operatorname{dom}(H_0), \tag{1.7}$$

where $L = L^* \in \mathcal{B}(\mathcal{K})$ and $K \in \mathcal{B}(\mathcal{K},\mathcal{H})$, with $\mathcal{K}$ another separable complex Hilbert space. We introduce a model operator $\widehat{H}_L$ in $\widehat{\mathcal{H}} = L^2(\mathbb{R},\mathcal{K}; d\Omega_L)$ for H_L in $\mathcal{H}$, define the Herglotz operator

$$M_L(z) = K^*(H_L - z)^{-1}K = \int_{\mathbb{R}} d\Omega_L(\lambda)(\lambda - z)^{-1}, \quad z \in \mathbb{C}\backslash\mathbb{R}, \tag{1.8}$$

where

$$\Omega_L(\lambda) = K^*E_L(\lambda)K, \tag{1.9}$$

with $\{E_L(\lambda)\}_{\lambda\in\mathbb{R}}$ the family of orthogonal spectral projections of H_L, and study the pair (H_L, H_0) in terms of $(M_L(z), M_0(z))$ following Donoghue's treatment [25] of rank-one perturbations of H_0. Moreover, we prove a realization theorem for the class of Herglotz operators exemplified by (1.8).

In Section 4 we consider self-adjoint extensions H of a densely defined closed symmetric operator $\dot{H}$ with deficiency indices (k,k), $k \in \mathbb{N}\cup\{\infty\}$ in some separable complex Hilbert space $\mathcal{H}$. We review our recent note [28] on Krein's formula relating self-adjoint extensions of $\dot{H}$ and introduce the corresponding Weyl operators $M_{H,\mathcal{N}}(z)$

$$M_{H,\mathcal{N}}(z) = zI_{\mathcal{N}} + (1+z^2)P_{\mathcal{N}}(H-z)^{-1}P_{\mathcal{N}}\big|_{\mathcal{N}} \tag{1.10}$$

$$= \int_{\mathbb{R}} d\Omega_{H,\mathcal{N}}(\lambda)((\lambda-z)^{-1} - \lambda(1+\lambda^2)^{-1}), \quad z \in \mathbb{C}\backslash\mathbb{R}, \tag{1.11}$$

where $\mathcal{N}$ is a closed linear subspace of the deficiency subspace $\mathcal{N}_+ = \ker(\dot{H}^* - i)$, $P_{\mathcal{N}}$ the orthogonal projection onto $\mathcal{N}$, and

$$\Omega_{H,\mathcal{N}}(\lambda) = (1+\lambda^2)(P_{\mathcal{N}}E_H(\lambda)P_{\mathcal{N}}\big|_{\mathcal{N}}), \tag{1.12}$$

with $\{E_H(\lambda)\}_{\lambda\in\mathbb{R}}$ the family of orthogonal spectral projections of H. Following [28] we study linear fractional transformation of $M_{H,\mathcal{N}_+}(z)$ involving different self-adjoint extensions H of $\dot{H}$. Moreover, following Donoghue [25] in the special case $\dim_{\mathbb{C}}(\mathcal{N}_+) = 1$, we consider a model $(\widehat{\dot{H}}, \widehat{H})$ in $\widehat{\mathcal{H}} = L^2(\mathbb{R}, \mathcal{N}_+; d\Omega_{H,\mathcal{N}_+})$ for the pair $(\dot{H}, H)$ in $\mathcal{H}$, and discuss Friedrichs and Krein extensions of $\dot{H}$ assuming $\dot{H}$ to be bounded from below. We conclude Section 4 with realization theorems for various classes of Weyl operators of the type (1.11).

Section 5 provides concrete applications of the formalism of Section 4 specialized to the case $\dim_{\mathbb{C}}(\mathcal{N}_+) = 1$. We study Schrödinger operators on a half-line and provide two illustrations of Livšic's result [44] on quasi-hermitian extensions in the special case of densely defined closed prime symmetric operators with deficiency indices $(1,1)$.

Finally, we briefly introduce some of the notation used in this paper. $\mathbb{C}_\pm = \{z \in \mathbb{C} \mid \mathrm{Im}(z) \gtrless 0\}$ denote the open upper/lower half-plane, $\bar{z}$ the complex conjugate of $z \in \mathbb{C}$. Complex Hilbert spaces are denoted by $\mathcal{H}$ or $\mathcal{K}$, the scalar product in $\mathcal{H}$ (linear in the second factor) by $(\cdot,\cdot)_{\mathcal{H}}$, with $I_{\mathcal{H}}$ the identity operator in $\mathcal{H}$. Direct sums of linear subspaces are indicated by $\dot{+}$, orthogonal direct sums by $\oplus$ (or $\oplus_{\mathcal{H}}$, if necessary). The Banach space of bounded linear operators from $\mathcal{K}$ into $\mathcal{H}$ is denoted by $\mathcal{B}(\mathcal{K},\mathcal{H})$ (and simply by $\mathcal{B}(\mathcal{H})$ if $\mathcal{K} = \mathcal{H}$). The domain, range, and kernel (null space) of a linear operator T are denoted by $\mathrm{dom}(T)$, $\mathrm{ran}(T)$ and $\ker(T)$, respectively; the resolvent set and spectrum of T by $\rho(T)$ and $\mathrm{spec}(T)$. The adjoint of T is denoted by T^*, $\mathrm{Re}(T) = (T+T^*)/2$ and $\mathrm{Im}(T) = (T-T^*)/(2i)$ (assuming $\mathrm{dom}(T) = \mathrm{dom}(T^*)$) abbreviate the real and imaginary part of T, respectively. The symbol χ_B denotes the characteristic function of $B \subset \mathbb{R}$; Σ denotes the Borel σ-algebra on $\mathbb{R}$.

2. Construction of a model Hilbert space

This section describes in some detail the construction of a model Hilbert space, variants of which will be of crucial importance in Sections 3 and 4. Rather than referring to the theory of direct integrals of Hilbert spaces (see, e.g., [11], Ch. 4, [12], Ch. 7) we briefly develop the necessary machinery from scratch and hint at the construction of related Banach spaces as well.

Let μ denote a σ-finite Borel measure on $\mathbb{R}$, Σ the Borel σ-algebra on $\mathbb{R}$, and suppose for each $\lambda \in \mathbb{R}$ we are given a separable complex Hilbert space $\mathcal{K}_\lambda$. Let $\mathcal{S}(\{\mathcal{K}_\lambda\}_{\lambda\in\mathbb{R}})$ be the vector space associated with the product space $\prod_{\lambda\in\mathbb{R}} \mathcal{K}_\lambda$ equipped with the obvious linear structure. Elements f of $\mathcal{S}(\{\mathcal{K}_\lambda\}_{\lambda\in\mathbb{R}})$ are maps

$$\mathbb{R} \ni \lambda \to f = \{f(\lambda) \in \mathcal{K}_\lambda\}_{\lambda\in\mathbb{R}} \in \prod_{\lambda\in\mathbb{R}} \mathcal{K}_\lambda. \tag{2.1}$$

Definition 2.1. A *measurable family of Hilbert spaces* $\mathcal{M}$ modelled on μ and $\{\mathcal{K}_\lambda\}_{\lambda\in\mathbb{R}}$ is a linear subspace $\mathcal{M} \subset \mathcal{S}(\{\mathcal{K}_\lambda\}_{\lambda\in\mathbb{R}})$ such that $f \in \mathcal{M}$ if and only if the map $\mathbb{R} \ni \lambda \to (f(\lambda), g(\lambda))_{\mathcal{K}_\lambda} \in \mathbb{C}$ is μ-measurable for all $g \in \mathcal{M}$.
Moreover, $\mathcal{M}$ is said to be generated by some subset $\mathcal{F}$, $\mathcal{F} \subset \mathcal{M}$, if for every $g \in \mathcal{M}$ we can find a sequence of functions $h_n \in \text{lin.span}\{\chi_B f \in \mathcal{S}(\{\mathcal{K}_\lambda\} \mid B \in \Sigma, f \in \mathcal{F}\}$ with $\lim_{n\to\infty} \|g(\lambda) - h_n(\lambda)\|_{\mathcal{K}_\lambda} = 0$ μ-a.e.

The definition of $\mathcal{M}$ was chosen with its maximality in mind and we refer to Lemma 2.3 and Theorem 2.6 for more details in this respect. An explicit construction of an example of $\mathcal{M}$ will be given in Theorem 2.5.

Remark 2.2. The following properties are proved in a standard manner:
(i) If $f \in \mathcal{M}$, $g \in \mathcal{S}(\{\mathcal{K}_\lambda\}_{\lambda\in\mathbb{R}})$ and $g = f$ μ-a.e. then $g \in \mathcal{M}$.
(ii) If $\{f_n\}_{n\in\mathbb{N}} \in \mathcal{M}$, $g \in \mathcal{S}(\{\mathcal{K}_\lambda\}_{\lambda\in\mathbb{R}})$ and $f_n(\lambda) \to g(\lambda)$ as $n \to \infty$ μ-a.e. (i.e., $\lim_{n\to\infty} \|f_n(\lambda) - g(\lambda)\|_{\mathcal{K}_\lambda} = 0$ μ-a.e.) then $g \in \mathcal{M}$.
(iii) If ϕ is a scalar-valued μ–measurable function and $f \in \mathcal{M}$ then $\phi f \in \mathcal{M}$.
(iv) If $f \in \mathcal{M}$ then $\mathbb{R} \ni \lambda \to \|f(\lambda)\|_{\mathcal{K}_\lambda} \in [0,\infty)$ is μ–measurable.

Let us remark that we shall identify functions in $\mathcal{M}$ which coincide μ-a.e.; thus $\mathcal{M}$ is more precisely a set of equivalence classes of functions.

Lemma 2.3. *Let* $\{f_n\}_{n\in\mathbb{N}} \subset \mathcal{S}(\{\mathcal{K}_\lambda\}_{\lambda\in\mathbb{R}})$ *such that*
(α) $\mathbb{R} \ni \lambda \to (f_m(\lambda), f_n(\lambda)_{\mathcal{K}_\lambda} \in \mathbb{C}$ *is μ–measurable for all* $m, n \in \mathbb{N}$.
(β) For μ-a.e. $\lambda \in \mathbb{R}$, $\overline{\text{lin.span}\{f_n(\lambda)\}} = \mathcal{K}_\lambda$.
Then setting

$$\mathcal{M} = \{g \in \mathcal{S}(\{\mathcal{K}_\lambda\}_{\lambda\in\mathbb{R}}) \mid (f_n(\lambda), g(\lambda))_{\mathcal{K}_\lambda} \text{ is } \mu\text{–measurable for all } n \in \mathbb{N}\}, \tag{2.2}$$

one infers
(i) $\mathcal{M}$ is a measurable family of Hilbert spaces.
(ii) $\mathcal{M}$ is generated by $\{f_n\}_{n\in\mathbb{N}}$.
(iii) $\mathcal{M}$ is the unique measurable family of Hilbert spaces containing the sequence $\{f_n\}_{n\in\mathbb{N}}$.
(iv) If $\{g_n\}$ is any sequence satisfying (β) then $\mathcal{M}$ is generated by $\{g_n\}$.

Sketch of proof. (i) Without loss of generality, we may assume $\{f_n\}_{n\in\mathbb{N}}$ contains all rational linear combinations, that is, all elements of the type $\sum_{n=1}^{N}\alpha_n f_n$, with $\alpha_n\in\mathbb{Q}$, $n=1,\dots,N$, $N\in\mathbb{N}$. For $f\in\mathcal{S}(\{\mathcal{K}_\lambda\}_{\lambda\in\mathbb{R}})$,

$$\|f(\lambda)\|_{\mathcal{K}_\lambda}=\sup_{n\in\mathbb{N}}|(f(\lambda),\chi_{B_n}(\lambda)f_n(\lambda))_{\mathcal{K}_\lambda}|, \tag{2.3}$$

where $B_n=\{\lambda\in\mathbb{R}\,|\,\|f_n(\lambda)\|_{\mathcal{K}_\lambda}\le 1\}$. Hence, if $f\in\mathcal{M}$ then the map $\mathbb{R}\ni\lambda\to\|f(\lambda)\|_{\mathcal{K}_\lambda}\in[0,\infty)$ is μ–measurable. It then follows easily that $\mathcal{M}$ is a measurable family of Hilbert spaces.
(ii) If $g\in\mathcal{M}$ then

$$\inf_{n\in\mathbb{N}}\|g(\lambda)-f_n(\lambda)\|_{\mathcal{K}_\lambda}=0\quad \mu-\text{a.e.} \tag{2.4}$$

It follows that if $\varepsilon(\lambda)$ is any measurable function with $\varepsilon>0$ on $\mathbb{R}$, then one can find a measurable partition $\{B_n\}_{n\in\mathbb{N}}$ of $\mathbb{R}$ so that

$$\|g(\lambda)-\sum_{n\in\mathbb{N}}\chi_{B_n}(\lambda)f_n(\lambda)\|_{\mathcal{K}_\lambda}\le\epsilon(\lambda). \tag{2.5}$$

Indeed, for each $\lambda\in\mathbb{R}$ let $N(\lambda)$ be the first n such that

$$\|g(\lambda)-f_{N(\lambda)}(\lambda)\|_{\mathcal{K}_\lambda}<\varepsilon(\lambda). \tag{2.6}$$

Then $\mathbb{R}\ni\lambda\to N(\lambda)\in\mathbb{N}$ is μ-measurable and $B_n=\{\lambda\in\mathbb{R}\,|\,N(\lambda)=n\}$ is the desired partition. This implies (ii).
(iii) If $\mathcal{M}'\subset\mathcal{S}(\{\mathcal{K}_\lambda\}_{\lambda\in\mathbb{R}})$ is a measurable family of Hilbert spaces containing each $\{f_n\}_{n\in\mathbb{N}}$, then $\mathcal{M}'\subseteq\mathcal{M}$. However, $\mathcal{M}\subseteq\mathcal{M}'$ by (ii) then completes the argument.
(iv)This follows immediately from (iii), since we can define $\mathcal{M}'$ in a similar way, that is,

$$\mathcal{M}'=\{h\in\mathcal{S}(\{\mathcal{K}_\lambda\}_{\lambda\in\mathbb{R}})\,|\,(g_n(\lambda),h(\lambda))_{\mathcal{K}_\lambda}\text{ is }\mu\text{–measurable for all }n\in\mathbb{N}\}, \tag{2.7}$$

and then $\mathcal{M}=\mathcal{M}'$ is clear from (iii). □

Next, let w be a μ-measurable function, $w>0$ μ-a.e., and consider the space

$$\dot{L}^2(\mathcal{M};wd\mu)=\{f\in\mathcal{M}\,|\int_{\mathbb{R}}w(\lambda)d\mu(\lambda)\|f(\lambda)\|^2_{\mathcal{K}_\lambda}<\infty\} \tag{2.8}$$

with its obvious linear structure. On $\dot{L}^2(\mathcal{M};wd\mu)$ one defines a semi-inner product $(\cdot,\cdot)_{\dot{L}^2(\mathcal{M};wd\mu)}$ (and hence a semi-norm $\|\cdot\|_{\dot{L}^2(\mathcal{M};wd\mu)}$) by

$$(f,g)_{\dot{L}^2(\mathcal{M};wd\mu)}=\int_{\mathbb{R}}w(\lambda)d\mu(\lambda)(f(\lambda),g(\lambda))_{\mathcal{K}_\lambda},\quad f,g\in\dot{L}^2(\mathcal{M};wd\mu). \tag{2.9}$$

That (2.9) defines a semi-inner product immediately follows from the corresponding properties of $(\cdot,\cdot)_{\mathcal{K}_\lambda}$ and the linearity of the integral. Hence $\dot{L}^2(\mathcal{M};wd\mu)$ represents a pre-Hilbert space and one can complete it in a standard manner as follows. One defines the equivalence relation $\sim$, for elements $f,g\in\dot{L}^2(\mathcal{M};wd\mu)$ by

$$f\sim g\text{ if and only if }f=g\quad\mu-\text{a.e.} \tag{2.10}$$

and hence introduces the set of equivalence classes of $\dot{L}^2(\mathcal{M}; wd\mu)$ denoted by

$$L^2(\mathcal{M}; wd\mu) = \dot{L}^2(\mathcal{M}; wd\mu)/\sim. \tag{2.11}$$

In particular, introducing the subspace of null functions

$$\begin{aligned} \mathcal{N}(\mathcal{M}; wd\mu) &= \{f \in \dot{L}^2(\mathcal{M}; wd\mu) \,|\, \|f(\lambda)\|_{\mathcal{K}_\lambda} = 0 \text{ for } \mu - \text{a.e. } \lambda \in \mathbb{R}\} \\ &= \{f \in \dot{L}^2(\mathcal{M}; wd\mu) \,|\, \|f\|_{\dot{L}^2(\mathcal{M}; wd\mu)} = 0\}, \end{aligned} \tag{2.12}$$

$L^2(\mathcal{M}; wd\mu)$ is precisely the quotient space $\dot{L}^2(\mathcal{M}; wd\mu)/\mathcal{N}(\mathcal{M}; wd\mu)$. Denoting the equivalence class of $f \in \dot{L}^2(\mathcal{M}; wd\mu)$ temporarily by $[f]$, the semi-inner product on $L^2(\mathcal{M}; wd\mu)$

$$([f],[g])_{L^2(\mathcal{M}; wd\mu)} = \int_{\mathbb{R}} w(\lambda) d\mu(\lambda) (f(\lambda), g(\lambda))_{\mathcal{K}_\lambda} \tag{2.13}$$

is well defined (i.e., independent of the chosen representatives of the equivalence classes) and actually an inner product. Thus $L^2(\mathcal{M}; wd\mu)$ is a normed space and by the usual abuse of notation we denote its elements in the following again by f, g, etc. The fundamental fact that $L^2(\mathcal{M}; wd\mu)$ is also complete is discussed next.

Theorem 2.4. *$L^2(\mathcal{M}; wd\mu)$ is complete and hence a Hilbert space.*

Proof. It suffices to prove the following fact: For each $\{f_n\}_{n\in\mathbb{N}} \in L^2(\mathcal{M}; wd\mu)$ with $\sum_{n\in\mathbb{N}} \|f_n\|_{L^2(\mathcal{M}; wd\mu)} < \infty$, there is an $f \in L^2(\mathcal{M}; wd\mu)$ such that $\sum_{n\in\mathbb{N}} f_n = f$. Given such a sequence $\{f_n\}_{n\in\mathbb{N}}$ with $\sum_{n\in\mathbb{N}} \|f_n\|_{L^2(\mathcal{M}; wd\mu)} = A$ define

$$G(\lambda) = \bigg(\sum_{n\in\mathbb{N}} \|f_n(\lambda)\|_{\mathcal{K}_\lambda} \bigg)^2. \tag{2.14}$$

Then G is μ-measurable. From $\sum_{n=1}^N \|f_n\|_{L^2(\mathcal{M}; wd\mu)} \leq A$ one computes using Minkowski's inequality,

$$\begin{aligned} \bigg(\int_{\mathbb{R}} w(\lambda) d\mu(\lambda) \bigg(\sum_{n=1}^N \|f_n(\lambda)\|_{\mathcal{K}_\lambda} \bigg)^2 \bigg)^{1/2} &\leq \sum_{n=1}^N \bigg(\int_{\mathbb{R}} w(\lambda) d\mu(\lambda) \|f_n(\lambda)\|^2_{\mathcal{K}_\lambda} \bigg)^{1/2} \\ &= \sum_{n=1}^N \|f_n\|_{L^2(\mathcal{M}; wd\mu)} \leq A, \end{aligned} \tag{2.15}$$

that is,

$$\int_{\mathbb{R}} w(\lambda) d\mu(\lambda) \bigg(\sum_{n=1}^N \|f_n(\lambda)\|_{\mathcal{K}_\lambda} \bigg)^2 \leq A^2. \tag{2.16}$$

Applying the Monotone Convergence Theorem one then concludes

$$\int_{\mathbb{R}} w(\lambda) d\mu(\lambda) G(\lambda) \leq A^2. \tag{2.17}$$

Thus G is integrable and hence μ-a.e. finite. Consequently, we may define

$$f(\lambda) = \begin{cases} \sum_{n\in\mathbb{N}} f_n(\lambda), & \text{if } \sum_{n\in\mathbb{N}} \|f_n(\lambda)\|_{\mathcal{K}_\lambda} < \infty, \\ 0, & \text{otherwise.} \end{cases} \tag{2.18}$$

Then $\|f(\lambda)\|^2_{\mathcal{K}_\lambda} \leq G(\lambda)$ for μ-a.e. $\lambda \in \mathbb{R}$ and

$$\sum_{n\in\mathbb{N}} f_n(\lambda) = f(\lambda) \quad \mu - \text{a.e.} \tag{2.19}$$

In particular, $f \in L^2(\mathcal{M}; wd\mu)$. Finally, since

$$\left\| \sum_{n=1}^{N} f_n(\lambda) - f(\lambda) \right\|_{\mathcal{K}_\lambda} \to 0 \text{ as } N \to \infty \quad \mu - \text{a.e.} \tag{2.20}$$

and

$$\left\| f(\lambda) - \sum_{n=1}^{N} f_n(\lambda) \right\|^2_{\mathcal{K}_\lambda} = \left\| \sum_{n=N+1}^{\infty} f_n(\lambda) \right\|^2_{\mathcal{K}_\lambda} \leq G(\lambda) \quad \mu - \text{a.e.}, \tag{2.21}$$

the Lebesgue Dominated Convergence theorem yields

$$\lim_{N\to\infty} \left\| f - \sum_{n=1}^{N} f_n \right\|_{L^2(\mathcal{M}; wd\mu)} = 0. \tag{2.22}$$

□

Clearly, the analogous construction defines the Banach spaces $L^p(\mathcal{M}; wd\mu)$, $p \geq 1$. The case $p = 2$ corresponds precisely to the direct integral of the Hilbert spaces $\mathcal{K}_\lambda$ with respect to the measure $wd\mu$ (see, e.g., [11], Ch. 4, [12], Ch. 7).

Next, suppose $\mathcal{K}$ is a separable complex Hilbert space and $\Omega : \Sigma \to \mathcal{B}(\mathcal{K})$ is a positive measure (i.e., countably additive with respect to the strong operator topology in $\mathcal{K}$). Assume

$$\Omega(\mathbb{R}) = T \geq 0, \quad T \in \mathcal{B}(\mathcal{K}). \tag{2.23}$$

Moreover, let μ be a control measure for Ω, that is,

$$\mu(B) = 0 \text{ if and only if } \Omega(B) = 0 \text{ for all } B \in \Sigma. \tag{2.24}$$

(E.g., $\mu(B) = \sum_{n\in\mathcal{I}} 2^{-n}(e_n, \Omega(B)e_n)_{\mathcal{K}}$, with $\{e_n\}_{n\in\mathcal{I}}$ a complete orthonormal system in $\mathcal{K}$, $\mathcal{I} \subseteq \mathbb{N}$ an appropriate index set.)

Theorem 2.5. *There are separable complex Hilbert spaces $\mathcal{K}_\lambda$, $\lambda \in \mathbb{R}$, a measurable family of Hilbert spaces $\mathcal{M}_\Omega(\mu)$ modelled on μ and $\{\mathcal{K}_\lambda\}_{\lambda\in\mathbb{R}}$, and a bounded linear map $\underline{\Lambda} \in \mathcal{B}(\mathcal{K}, L^2(\mathcal{M}_\Omega(\mu); d\mu))$ so that*
(i) For all $B \in \Sigma$, $\xi, \eta \in \mathcal{K}$,

$$(\xi, \Omega(B)\eta)_{\mathcal{K}} = \int_B d\mu(\lambda)((\underline{\Lambda}\xi)(\lambda), (\underline{\Lambda}\eta)(\lambda))_{\mathcal{K}_\lambda}. \tag{2.25}$$

(ii) $\underline{\Lambda}(\{e_n\}_{n\in\mathcal{I}})$ *generates* $\mathcal{M}_\Omega(\mu)$*, where* $\{e_n\}_{n\in\mathcal{I}}$ *denotes any sequence of linearly independent elements in* $\mathcal{K}$ *with the property* $\overline{lin.span\{e_n\}_{n\in\mathcal{I}}} = \mathcal{K}$*,* $\mathcal{I} \subseteq \mathbb{N}$*. In particular,* $\underline{\Lambda}(\mathcal{K})$ *generates* $\mathcal{M}_\Omega(\mu)$*.*

(iii) For all $\xi \in \mathcal{K}$*,*

$$\underline{\Lambda}(\Omega(B)\xi) = \chi_B \underline{\Lambda}\xi \quad \mu - a.e. \tag{2.26}$$

Proof. Denote $\mathcal{V} = \text{lin.span}\{e_n\}_{n\in\mathcal{I}}$. By the Radon-Nikodym theorem, there exist μ–measurable $\phi_{m,n}$ such that

$$\int_B d\mu(\lambda)\phi_{m,n}(\lambda) = (e_m, \Omega(B)e_n)_\mathcal{K}. \tag{2.27}$$

Next, suppose $v = \sum_{n=1}^N \alpha_n e_n \in \mathcal{V}$, $\alpha_n \in \mathbb{C}$, $n = 1, \dots, N$, $N \in \mathcal{I}$. Then

$$(v, \Omega(B)v)_\mathcal{K} = \int_B d\mu(\lambda) \sum_{m,n=1}^N \phi_{m,n}(\lambda)\overline{\alpha_m}\alpha_n. \tag{2.28}$$

By considering only rational linear combinations we can deduce that for μ-a.e. $\lambda \in \mathbb{R}$,

$$\sum_{m,n} \phi_{m,n}(\lambda)\overline{\alpha_m}\alpha_n \geq 0 \text{ for all finite sequences } \{\alpha_n\} \subset \mathbb{C}. \tag{2.29}$$

Hence we can define a semi-inner product $(\cdot,\cdot)_\lambda$ on $\mathcal{V}$ such that

$$(v, w)_\lambda = \sum_{m,n} \phi_{m,n}(\lambda)\overline{\alpha_m}\beta_n \quad \mu - \text{a.e} \tag{2.30}$$

if $v = \sum_n \alpha_n e_n$, $w = \sum_n \beta_n e_n$.
Next, let $\mathcal{K}_\lambda$ be the completion of $\mathcal{V}$ with respect $\|\cdot\|_\lambda$ (or, more precisely the completion of $\mathcal{V}/\mathcal{N}_\lambda$ where $\mathcal{N}_\lambda = \{\xi \in \mathcal{V} \,|\, (\xi,\xi)_\lambda = 0\}$) and consider $\mathcal{S}(\{\mathcal{K}_\lambda\}_{\lambda\in\mathbb{R}})$. Each $v \in \mathcal{V}$ defines an element $\underline{v} = \{\underline{v}(\lambda)\}_{\lambda\in\mathbb{R}} \in \mathcal{S}(\{\mathcal{K}_\lambda\}_{\lambda\in\mathbb{R}})$ by

$$\underline{v}(\lambda) = v \text{ for all } \lambda \in \mathbb{R}. \tag{2.31}$$

Again we identify an element $v \in \mathcal{V}$ with an element in $\mathcal{V}/\mathcal{N}_\lambda \subseteq \mathcal{K}_\lambda$. Applying Lemma 2.3, the collection $\{\underline{e_n}\}_{n\in\mathcal{I}}$ then generates a measurable family of Hilbert spaces $\mathcal{M}_\Omega(\mu)$. If $v \in \mathcal{V}$ then

$$\|\underline{v}\|^2_{L^2(\mathcal{M}_\Omega(\mu);d\mu)} = \int_\mathbb{R} d\mu(\lambda)(\underline{v}(\lambda), \underline{v}(\lambda))_\lambda = (v, Tv)_\mathcal{K} = \|T^{1/2}v\|^2_\mathcal{K}. \tag{2.32}$$

Hence we can define

$$\dot{\underline{\Lambda}} : \mathcal{V} \to L^2(\mathcal{M}_\Omega(\mu); d\mu), \quad v \to \dot{\underline{\Lambda}}v = \underline{v} = \{\underline{v}(\lambda) = v\}_{\lambda\in\mathbb{R}} \tag{2.33}$$

and denote by $\underline{\Lambda} \in \mathcal{B}(\mathcal{K}, L^2(\mathcal{M}_\Omega(\mu); d\mu))$, $\|\underline{\Lambda}\|_{\mathcal{B}(\mathcal{K},L^2(\mathcal{M}_\Omega(\mu);d\mu))} = \|T^{1/2}\|_{\mathcal{B}(\mathcal{K})}$, the closure of $\dot{\underline{\Lambda}}$. Then properties (i)–(iii) hold. □

We now show that this construction is essentially unique.

Theorem 2.6. *Suppose $\mathcal{K}'_\lambda$, $\lambda \in \mathbb{R}$ is a family of separable complex Hilbert spaces, $\mathcal{M}'$ is a measurable family of Hilbert spaces modelled on μ and $\{\mathcal{K}'_\lambda\}$, and $\underline{\Lambda}' \in \mathcal{B}(\mathcal{K}, L^2(\mathcal{M}'; d\mu))$ is a map satisfying (i),(ii), and (iii) of the preceding theorem. Then for μ-a.e. $\lambda \in \mathbb{R}$ there is a unitary operator $U_\lambda : \mathcal{K}_\lambda \to \mathcal{K}'_\lambda$ such that $f = \{f(\lambda)\}_{\lambda\in\mathbb{R}} \in \mathcal{M}_\Omega(\mu)$ if and only if $U_\lambda f(\lambda) \in \mathcal{M}'$ and for all $\xi \in \mathcal{K}$,*

$$(\underline{\Lambda}'\xi)(\lambda) = U_\lambda(\underline{\Lambda}\xi)(\lambda) \quad \mu - a.e. \tag{2.34}$$

Proof. We use the notation of the preceding theorem. We select representatives $f'_n \in \mathcal{M}'$ of $\underline{\Lambda}'e_n$. It follows from condition (i) that for μ-a.e. $\lambda \in \mathbb{R}$ and every $m, n \in \mathcal{I}$ we have

$$(f'_m(\lambda), f'_n(\lambda))_{\mathcal{K}'_\lambda} = (e_m, e_n)_\lambda = (\underline{e}_m(\lambda), \underline{e}_n(\lambda))_{\mathcal{K}_\lambda}. \tag{2.35}$$

Hence we can induce an isometry $U_\lambda : \mathcal{K}_\lambda \to \mathcal{K}'_\lambda$ such that $U_\lambda \underline{e}_n(\lambda) = f'_n(\lambda)$.

It is easy to see that if $v \in \mathcal{V}$ we must have $U_\lambda \underline{v}(\lambda) = (\underline{\Lambda}'v)(\lambda)$ μ-a.e. From the L_2-continuity of both $\underline{\Lambda}$ and $\underline{\Lambda}'$ it follows that for every $\xi \in \mathcal{K}$ we have

$$(\underline{\Lambda}'\xi)(\lambda) = U_\lambda(\underline{\Lambda}\xi)(\lambda) \quad \mu - \text{a.e.} \tag{2.36}$$

We next observe that if $\underline{\Lambda}'(\mathcal{K})$ generates $\mathcal{M}'$ then by a density argument it must also be true that $\{f'_n\}_{n\in\mathcal{I}}$ generates $\mathcal{M}'$. It is then immediate that the linear span of $\{f'_n(\lambda)\}_{n\in\mathcal{I}}$ must be dense for μ-a.e. $\lambda \in \mathbb{R}$. Thus U_λ is actually surjective μ-a.e. and so is unitary.

Finally, if $\xi \in \mathcal{K}$ and $B \in \Sigma$ then $U_\lambda(\chi_B(\lambda)(\underline{\Lambda}\xi)(\lambda)) = \chi_B(\lambda)(\underline{\Lambda}'\xi)(\lambda)$ μ-a.e. Thus it follows by approximation that if $f \in \mathcal{M}_\Omega(\mu)$ then $U_\lambda f(\lambda) \in \mathcal{M}'$. Conversely, a similar argument shows that if $f \in \mathcal{M}'$ then $U_\lambda^{-1} f(\lambda) \in \mathcal{M}_\Omega(\mu)$. □

Without going into further details, we note that $\mathcal{M}_\Omega(\mu)$ depends of course on μ. However, a change in μ merely effects a change in density and so $\mathcal{M}_\Omega(\mu)$ can essentially be viewed as μ-independent.

Next, using the notation employed in the proof of Theorem 2.4 we recall

$$\mathcal{V} = \text{lin.span}\{e_n \in \mathcal{K} \,|\, n \in \mathcal{I}\} \tag{2.37}$$

and define

$$\underline{\mathcal{V}}_\Omega = \text{lin.span}\{\chi_B \underline{e}_n \in L^2(\mathcal{M}_\Omega(\mu); d\mu) \,|\, B \in \Sigma,\, n \in \mathcal{I}\}. \tag{2.38}$$

The fact that $\{\underline{e}_n\}_{n\in\mathcal{I}}$ generates $\mathcal{M}_\Omega(\mu)$ implies that $\underline{\mathcal{V}}_\Omega$ is dense in $L^2(\mathcal{M}_\Omega(\mu); d\mu)$, that is,

$$\overline{\underline{\mathcal{V}}_\Omega} = L^2(\mathcal{M}_\Omega(\mu); d\mu). \tag{2.39}$$

The following result will be used in Section 3.

Lemma 2.7. *Suppose $\mathcal{K}$, $\mathcal{H}$ are separable complex Hilbert spaces, $K \in \mathcal{B}(\mathcal{K}, \mathcal{H})$, $\{E(B)\}_{B\in\Sigma}$ is a family of orthogonal projections in $\mathcal{H}$, and assume*

$$\overline{\textit{lin.span}\{E(B)Ke_n \in \mathcal{H} \,|\, B \in \Sigma,\, n \in \mathcal{I}\}} = \mathcal{H}, \tag{2.40}$$

with $\{e_n\}_{n\in\mathcal{I}}$, $\mathcal{I} \subset \mathbb{N}$ a complete orthonormal system in $\mathcal{K}$. Define

$$\Omega : \Sigma \to \mathcal{B}(\mathcal{K}), \quad \Omega(B) = K^* E(B) K, \tag{2.41}$$

and introduce

$$\dot{U} : \underline{\mathcal{V}}_\Omega \to \mathcal{H},$$

$$\underline{\mathcal{V}}_\Omega \ni \sum_{m=1}^{M}\sum_{n=1}^{N} \alpha_{m,n}\chi_{B_m}\underline{e}_n \to \dot{U}\Big(\sum_{m=1}^{M}\sum_{n=1}^{N} \alpha_{m,n}\chi_{B_m}\underline{e}_n\Big) = \sum_{m=1}^{M}\sum_{n=1}^{N} \alpha_{m,n}E(B_m)Ke_n \in \mathcal{H}, \tag{2.42}$$

$$\alpha_{m,n} \in \mathbb{C},\ m = 1,\dots,M,\ n = 1,\dots,N,\ M,N \in \mathcal{I}.$$

Then $\dot{U}$ extends to a unitary operator $U : L^2(\mathcal{M}_\Omega(\mu); d\mu) \to \mathcal{H}$.

Proof. One computes

$$\begin{aligned}
&\Big\|\dot{U}\Big(\sum_{m=1}^{M}\sum_{n=1}^{N} \alpha_{m,n}\chi_{B_m}\underline{e}_n\Big)\Big\|_{\mathcal{H}}^2\\
&= \sum_{m_1,m_2=1}^{M}\sum_{n_1,n_2=1}^{N} \overline{\alpha_{m_1,n_1}}\alpha_{m_2,n_2}(e_{n_1}, K^*E(B_{m_1}\cap B_{m_2})Ke_{n_2})_{\mathcal{K}}\\
&= \sum_{m_1,m_2=1}^{M}\sum_{n_1,n_2=1}^{N} \overline{\alpha_{m_1,n_1}}\alpha_{m_2,n_2}(e_{n_1}, \Omega(B_{m_1}\cap B_{m_2})e_{n_2})_{\mathcal{K}}\\
&= \sum_{m_1,m_2=1}^{M}\sum_{n_1,n_2}^{N} \overline{\alpha_{m_1,n_1}}\alpha_{m_2,n_2}\int_{B_{m_1}\cap B_{m_2}} d\mu(\lambda)(\underline{e}_{n_1}(\lambda), \underline{e}_{n_2}(\lambda))_{\mathcal{K}_\lambda}\\
&= \Big\|\sum_{m=1}^{M}\sum_{n=1}^{N} \alpha_{m,n}\chi_{B_m}\underline{e}_n\Big\|_{L^2(\mathcal{M}_\Omega(\mu);d\mu)}^2.
\end{aligned} \tag{2.43}$$

By (2.39), $\dot{U}$ is densely defined and thus extends to an isometry U of $L^2(\mathcal{M}_\Omega(\mu); d\mu)$ into $\mathcal{H}$. In particular, $\operatorname{ran}(U)$ is closed in $\mathcal{H}$. Thus,

$$\operatorname{ran}(U) \supseteq \overline{\operatorname{lin.span}\{E(B)Ke_n \in \mathcal{H} \,|\, B \in \Sigma,\ n \in \mathcal{I}\}} = \mathcal{H} \tag{2.44}$$

by hypothesis (2.41) and hence $U : L^2(\mathcal{M}_\Omega(\mu); d\mu) \to \mathcal{H}$ is a unitary operator. □

In view of our comment following Theorem 2.6, concerning the mild dependence on the control measure μ of $\mathcal{M}_\Omega(\mu)$, we will put more emphasis on the operator-valued measure Ω and hence use the notation $L^2(\mathbb{R}, \mathcal{K}; wd\Omega)$ instead of the more precise $L^2(\mathcal{M}_\Omega(\mu); wd\mu)$ in Section 3.

Finally we adapt Lemma 2.7 to the content of Section 4.

Suppose $\mathcal{N}$ is a separable complex Hilbert space and $\widetilde{\Omega} : \Sigma \to \mathcal{B}(\mathcal{N})$ a positive measure. Assume

$$\widetilde{\Omega}(\mathbb{R}) = \widetilde{T} \geq 0, \quad \widetilde{T} \in \mathcal{B}(\mathcal{N}) \tag{2.45}$$

and let $\tilde{\mu}$ be a control measure for $\widetilde{\Omega}$. Moreover, let $\{u_n\}_{n\in\mathcal{I}}$, $\mathcal{I}\subseteq\mathbb{N}$ be a sequence of linearly independent elements in $\mathcal{N}$ with the property $\overline{\text{lin.span}\{u_n\}_{n\in\mathcal{I}}}=\mathcal{N}$. As discussed in Theorem 2.5, this yields a measurable family of Hilbert spaces $\mathcal{M}_{\widetilde{\Omega}}(\tilde{\mu})$ modelled on $\tilde{\mu}$ and $\{\mathcal{N}_\lambda\}_{\lambda\in\mathbb{R}}$ and a bounded map $\underline{\Lambda}\in\mathcal{B}(\mathcal{N},L^2(\mathcal{M}_{\widetilde{\Omega}}(\tilde{\mu});d\tilde{\mu}))$, $\|\underline{\Lambda}\|_{\mathcal{B}(\mathcal{N},L^2(\mathcal{M}_{\widetilde{\Omega}}(\tilde{\mu});d\tilde{\mu}))}=\|\widetilde{T}^{1/2}\|_{\mathcal{B}(\mathcal{N})}$, such that $\underline{\Lambda}(\{u_n\}_{n\in\mathcal{I}})$ generates $\mathcal{M}_{\widetilde{\Omega}}(\tilde{\mu})$ and

$$\underline{\Lambda}:\mathcal{V}\to L^2(\mathcal{M}_{\widetilde{\Omega}}(\tilde{\mu});d\tilde{\mu})),\quad v\to\underline{\Lambda}v=\underline{v}=\{\underline{v}(\lambda)=v\}_{\lambda\in\mathbb{R}},\tag{2.46}$$

where

$$\mathcal{V}=\text{lin.span}\{u_n\}_{n\in\mathcal{I}}.\tag{2.47}$$

Each $v\in\mathcal{V}$ defines an element

$$\underline{\underline{v}}=\{\underline{\underline{v}}(\lambda)=(\lambda-i)^{-1}v\}_{\lambda\in\mathbb{R}}\in\mathcal{S}(\{\mathcal{N}_\lambda\}_{\lambda\in\mathbb{R}})\tag{2.48}$$

and introducing the weight function

$$w_1(\lambda)=1+\lambda^2,\quad \lambda\in\mathbb{R}\tag{2.49}$$

and Hilbert space $L^2(\mathcal{M}_{\widetilde{\Omega}}(\tilde{\mu});w_1d\tilde{\mu})$ one computes

$$\|\underline{v}\|^2_{L^2(\mathcal{M}_{\widetilde{\Omega}}(\tilde{\mu});d\tilde{\mu})}=\int_{\mathbb{R}}d\tilde{\mu}(\lambda)\|\underline{v}(\lambda)\|^2_{\mathcal{N}_\lambda}=(v,\widetilde{T}v)_{\mathcal{N}}=\|\widetilde{T}^{1/2}v\|^2_{\mathcal{N}}.\tag{2.50}$$

Thus, the linear map

$$\dot{\underline{\underline{\Lambda}}}:\mathcal{V}\to L^2(\mathcal{M}_{\widetilde{\Omega}}(\tilde{\mu});w_1d\tilde{\mu}),\quad v\to\dot{\underline{\underline{\Lambda}}}v=\underline{\underline{v}}=\{\underline{\underline{v}}(\lambda)=(\lambda-i)^{-1}v\}_{\lambda\in\mathbb{R}}\tag{2.51}$$

extends to $\underline{\underline{\Lambda}}\in\mathcal{B}(\mathcal{N},L^2(\mathcal{M}_{\widetilde{\Omega}}(\tilde{\mu});w_1d\tilde{\mu}))$, $\|\underline{\underline{\Lambda}}\|_{\mathcal{B}(\mathcal{N},L^2(\mathcal{M}_{\widetilde{\Omega}}(\tilde{\mu});w_1d\tilde{\mu}))}=\|\widetilde{T}^{1/2}\|_{\mathcal{B}(\mathcal{N})}$. Introducing

$$\underline{\underline{\mathcal{V}}}_{\widetilde{\Omega}}=\text{lin.span}\{\chi_B\underline{\underline{v}}\in L^2(\mathcal{M}_{\widetilde{\Omega}}(\tilde{\mu});w_1d\tilde{\mu})\,|\,B\in\Sigma,\,n\in\mathcal{I}\}\tag{2.52}$$

one infers that $\underline{\underline{\mathcal{V}}}_{\widetilde{\Omega}}$ is dense in $L^2(\mathcal{M}_{\widetilde{\Omega}}(\tilde{\mu});w_1d\tilde{\mu})$, that is,

$$\overline{\underline{\underline{\mathcal{V}}}_{\widetilde{\Omega}}}=L^2(\mathcal{M}_{\widetilde{\Omega}}(\tilde{\mu});w_1d\tilde{\mu}).\tag{2.53}$$

Given these preliminaries we can state the following result.

Lemma 2.8. *Suppose $\mathcal{H}$ is a separable complex Hilbert space, $\mathcal{N}$ a closed linear subspace of $\mathcal{H}$, $P_{\mathcal{N}}$ the orthogonal projection in $\mathcal{H}$ onto $\mathcal{N}$, $\{E(B)\}$, $B\in\Sigma$ a family of orthogonal projections in $\mathcal{H}$, and assume*

$$\overline{\textit{lin.span}\{E(B)u_n\in\mathcal{H}\,|\,B\in\Sigma,\,n\in\mathcal{I}\}}=\mathcal{H},\tag{2.54}$$

with $\{u_n\}_{n\in\mathcal{I}}$, $\mathcal{I}\subseteq\mathbb{N}$ a complete orthonormal system in $\mathcal{N}$. Define

$$\widetilde{\Omega}:\Sigma\to\mathcal{B}(\mathcal{N}),\quad \widetilde{\Omega}(B)=P_{\mathcal{N}}E(B)P_{\mathcal{N}}\big|_{\mathcal{N}},\tag{2.55}$$

and introducc

$$\begin{aligned}&\dot{\widetilde{U}}:\underline{\underline{\mathcal{V}}}_{\Omega}\to\mathcal{H},\\ &\underline{\underline{\mathcal{V}}}_{\Omega}\ni\sum_{m=1}^{M}\sum_{n=1}^{N}\alpha_{m,n}\chi_{B_m}\underline{\underline{u}}_n\to\dot{\widetilde{U}}\bigg(\sum_{m=1}^{M}\sum_{n=1}^{N}\alpha_{m,n}\chi_{B_m}\underline{\underline{u}}_n\bigg)\end{aligned}\tag{2.56}$$

$$= \sum_{m=1}^{M} \sum_{n=1}^{N} \alpha_{m,n} E(B_m) u_n \in \mathcal{H},$$
$$\alpha_{m,n} \in \mathbb{C},\ m = 1, \dots, M,\ n = 1, \dots, N,\ M, N \in \mathcal{I}.$$

Then $\dot{\widetilde{U}}$ extends to a unitary operator $\widetilde{U} : L^2(\mathcal{M}_{\widetilde{\Omega}}(\tilde{\mu}); w_1 d\mu) \to \mathcal{H}$.

Proof. One computes

$$\begin{aligned}
&\bigg\| \dot{\widetilde{U}} \bigg(\sum_{m=1}^{M} \sum_{n=1}^{N} \alpha_{m,n} \chi_{B_m} \underline{u}_n \bigg) \bigg\|_{\mathcal{H}}^2 \\
&= \sum_{m_1, m_2 = 1}^{M} \sum_{n_1, n_2 = 1}^{N} \overline{\alpha_{m_1, n_1}} \alpha_{m_2, n_2} (u_{n_1}, E(B_{m_1} \cap B_{m_2}) u_{n_2})_{\mathcal{N}} \\
&= \sum_{m_1, m_2 = 1}^{M} \sum_{n_1, n_2 = 1}^{N} \overline{\alpha_{m_1, n_1}} \alpha_{m_2, n_2} (u_{n_1}, \widetilde{\Omega}(B_{m_1} \cap B_{m_2}) u_{n_2})_{\mathcal{N}} \\
&= \sum_{m_1, m_2 = 1}^{M} \sum_{n_1, n_2}^{N} \overline{\alpha_{m_1, n_1}} \alpha_{m_2, n_2} \int_{B_{m_1} \cap B_{m_2}} d\tilde{\mu}(\lambda) (\underline{u}_{n_1}(\lambda), \underline{u}_{n_2}(\lambda))_{\mathcal{N}_\lambda}
\end{aligned}$$

(2.57) $$= \bigg\| \sum_{m=1}^{M} \sum_{n=1}^{N} \alpha_{m,n} \chi_{B_m} \underline{u}_n \bigg\|_{L^2(\mathcal{M}_{\widetilde{\Omega}}(\tilde{\mu}); w_1 d\tilde{\mu})}^2 .$$

By (2.53), $\dot{\widetilde{U}}$ is densely defined and extends to an isometry $\widetilde{U}$ of $L^2(\mathcal{M}_{\widetilde{\Omega}}(\tilde{\mu}); w_1 d\tilde{\mu})$ into $\mathcal{H}$. In particular, $\operatorname{ran}(\widetilde{U})$ is closed in $\mathcal{H}$. Thus,

(2.58) $$\operatorname{ran}(\widetilde{U}) \supseteq \overline{\operatorname{lin.span}\{E(B) u_n \in \mathcal{H} \,|\, B \in \Sigma,\ n \in \mathcal{I}\}} = \mathcal{H}$$

by hypothesis (2.54) and hence $U : L^2(\mathcal{M}_{\widetilde{\Omega}}(\tilde{\mu}); w_1 d\tilde{\mu}) \to \mathcal{H}$ is a unitary operator. □

Analogous to our comments following Lemma 2.7, in Section 4 we will emphasize the role of $\widetilde{\Omega}$ and hence use the somewhat imprecise notation $L^2(\mathbb{R}, \mathcal{N}; w d\widetilde{\Omega})$, with various weight functions w, as opposed to the precise notation $L^2(\mathcal{M}_{\widetilde{\Omega}}(\tilde{\mu}); w d\tilde{\mu})$.

3. On self-adjoint perturbations of self-adjoint operators

In this section we will focus on the following perturbation problem. Assuming

Hypothesis 3.1. *Let $\mathcal{H}$ and $\mathcal{K}$ be separable complex Hilbert spaces, H_0 a self-adjoint (possibly unbounded) operator in $\mathcal{H}$, L a bounded self-adjoint operator in $\mathcal{K}$, and $K : \mathcal{K} \to \mathcal{H}$ a bounded operator,*

we define the self-adjoint operator H_L in $\mathcal{H}$,

(3.1) $$H_L = H_0 + KLK^*, \quad \operatorname{dom}(H_L) = \operatorname{dom}(H_0).$$

Given the perturbation H_L of H_0, we introduce the associated operator-valued Herglotz function in $\mathcal{K}$,

$$M_L(z) = K^*(H_L - z)^{-1}K, \quad z \in \mathbb{C}\backslash\mathbb{R}, \tag{3.2}$$

$$\frac{1}{\mathrm{Im}(z)}\mathrm{Im}(M_L(z)) = ((H_L - z)^{-1}K)^*(H_L - z)^{-1}K \geq 0, \quad z \in \mathbb{C}\backslash\mathbb{R}, \tag{3.3}$$

and study the pair (H_L, H_0) in terms of the corresponding pair $(M_L(z), M_0(z))$. In the special case where $\dim_{\mathbb{C}}(\mathcal{K}) = 1$, this perturbation problem has been studied in detail by Donoghue [25] and later by Simon and Wolf [60] (see also [59]). The case $\dim_{\mathbb{C}}(\mathcal{K}) = n \in \mathbb{N}$, has recently been treated in depth in [30]. In this section we treat the general case $\dim_{\mathbb{C}}(\mathcal{K}) \in \mathbb{N} \cup \{\infty\}$.

Next, let $\{E_0(\lambda)\}_{\lambda\in\mathbb{R}}$ be the family of strongly right-continuous orthogonal spectral projections of H_0 in $\mathcal{H}$ and suppose that $K\mathcal{K} \subseteq \mathcal{H}$ is a generating subspace for H_0, that is, one of the following (equivalent) equations holds in

Hypothesis 3.2.

$$\mathcal{H} = \overline{lin.span\{(H_0 - z)^{-1}Ke_n \in \mathcal{H} \,|\, n \in \mathcal{I},\, z \in \mathbb{C}\backslash\mathbb{R}\}} \tag{3.4a}$$

$$= \overline{lin.span\{E_0(\lambda)Ke_n \in \mathcal{H} \,|\, n \in \mathcal{I},\, \lambda \in \mathbb{R}\}}, \tag{3.4b}$$

where $\{e_n\}_{n\in\mathcal{I}}$, $\mathcal{I} \subseteq \mathbb{N}$ an appropriate index set, represents a complete orthonormal system in $\mathcal{K}$.

Denoting by $\{E_L(\lambda)\}_{\lambda\in\mathbb{R}}$ the family of strongly right-continuous orthogonal spectral projections of H_L in $\mathcal{H}$ one introduces

$$\Omega_L(\lambda) = K^*E_L(\lambda)K, \quad \lambda \in \mathbb{R} \tag{3.5}$$

and hence verifies

$$\begin{aligned} M_L(z) &= K^*(H_L - z)^{-1}K = K^* \int_{\mathbb{R}} dE_L(\lambda)(\lambda - z)^{-1}K \\ &= \int_{\mathbb{R}} d\Omega_L(\lambda)(\lambda - z)^{-1}, \quad z \in \mathbb{C}\backslash\mathbb{R}, \end{aligned} \tag{3.6}$$

where the operator Stieltjes integral (3.6) converges in the norm of $\mathcal{B}(\mathcal{K})$ (cf. Theorems I.4.2 and I.4.8 in [17]). Since $\text{s-lim}_{z\to i\infty} z(H_L - z)^{-1} = -I_{\mathcal{H}}$, (3.5) implies

$$\Omega_L(\mathbb{R}) = K^*K. \tag{3.7}$$

Moreover, since $\text{s-lim}_{\lambda\downarrow-\infty} E_L(\lambda) = 0$, $\text{s-lim}_{\lambda\uparrow\infty} E_L(\lambda) = I_{\mathcal{H}}$, one infers

$$\operatorname*{s-lim}_{\lambda\downarrow-\infty} \Omega_L(\lambda) = 0, \quad \operatorname*{s-lim}_{\lambda\uparrow\infty} \Omega_L(\lambda) = K^*K \tag{3.8}$$

and $\{\Omega_L(\lambda)\}_{\lambda\in\mathbb{R}} \subset \mathcal{B}(\mathcal{K})$ is a family of uniformly bounded, nonnegative, nondecreasing, strongly right-continuous operators from $\mathcal{K}$ into itself. Let μ_L be a σ-finite control measure on $\mathbb{R}$ defined, for instance, by

$$\mu_L(\lambda) = \sum_{n\in\mathcal{I}} 2^{-n}(e_n, \Omega_L(\lambda)e_n)_{\mathcal{K}}, \quad \lambda \in \mathbb{R}, \tag{3.9}$$

where $\{e_n\}_{n\in\mathcal{I}}$ denotes a complete orthonormal system in $\mathcal{K}$, and then introduce $L^2(\mathcal{M}_{\Omega_L}(\mu_L); d\mu_L)$ as in Section 3, replacing the pair (Ω, μ) by (Ω_L, μ_L), etc. As noted in Section 2, we will actually use the more suggestive notation $L^2(\mathbb{R}, \mathcal{K}; w d\Omega_L)$ instead of the more precise $L^2(\mathcal{M}_{\Omega_L}(\mu_L); w d\mu_L)$ ($w > 0$ a weight function), for the remainder of this section. Abbreviating $\widehat{\mathcal{H}}_L = L^2(\mathbb{R}, \mathcal{K}; d\Omega_L)$, we introduce the unitary operator $U_L : \widehat{\mathcal{H}}_L \to \mathcal{H}$, as the operator U in Lemma 2.7 and define $\widehat{H}_L$ in $\widehat{\mathcal{H}}_L$ by

$$(\widehat{H}_L \hat{f})(\lambda) = \lambda \hat{f}(\lambda), \quad \hat{f} \in \operatorname{dom}(\widehat{H}_L) = L^2(\mathbb{R}, \mathcal{K}; (1+\lambda^2) d\Omega_L). \tag{3.10}$$

Theorem 3.3. *Assume Hypotheses 3.1 and 3.2. Then H_L in $\mathcal{H}$ is unitarily equivalent to $\widehat{H}_L$ in $\widehat{\mathcal{H}}_L$,*

$$H_L = U_L \widehat{H}_L U_L^{-1}. \tag{3.11}$$

The family of strongly right-continuous orthogonal spectral projections $\{\widehat{E}_L(\lambda)\}_{\lambda\in\mathbb{R}}$ of $\widehat{H}_L$ in $\widehat{\mathcal{H}}_L$ is given by

(3.12)

$$(\widehat{E}_L(\lambda)\hat{f})(\nu) = \theta(\lambda - \nu)\hat{f}(\nu) \textit{ for } \Omega_L - \textit{a.e. } \nu \in \mathbb{R}, \quad \hat{f} \in \widehat{\mathcal{H}}_L, \ \theta(x) = \begin{cases} 1, & x \geq 0, \\ 0, & x < 0. \end{cases}$$

Proof. Consider

$$\underline{e}_n = \{\underline{e}_n(\lambda) = e_n\}_{\lambda\in\mathbb{R}} \in \widehat{\mathcal{H}}_L, \quad n \in \mathcal{I}, \tag{3.13}$$

then

$$U_L \underline{e}_n = \int_{\mathbb{R}} dE_L(\lambda) K e_n = K e_n, \quad n \in \mathcal{I} \tag{3.14}$$

and

$$((\widehat{H}_L - z)^{-1}\underline{e}_n)(\lambda) = (\lambda - z)^{-1}\underline{e}_n(\lambda) = (\lambda - z)^{-1} e_n, \quad n \in \mathcal{I}, \ z \in \mathbb{C}\backslash\mathbb{R} \tag{3.15}$$

yield

(3.16)

$$U_L(\widehat{H}_L - z)^{-1}\underline{e}_n = \int_{\mathbb{R}} dE_L(\lambda)(\lambda - z)^{-1} K e_n = (H_L - z)^{-1} K e_n, \ n \in \mathcal{I}, \ z \in \mathbb{C}\backslash\mathbb{R}.$$

Using the resolvent equation for H_L and H_0,

$$(H_L - z)^{-1} = (H_0 - z)^{-1} - (H_L - z)^{-1} K L K^* (H_0 - z)^{-1} \tag{3.17a}$$

$$= (H_0 - z)^{-1} - (H_0 - z)^{-1} K L K^* (H_L - z)^{-1}, \quad z \in \mathbb{C}\backslash\mathbb{R}, \tag{3.17b}$$

one verifies

$$(I_{\mathcal{K}} + L K^*(H_0 - z)^{-1} K)(I_{\mathcal{K}} - L K^*(H_L - z)^{-1} K) \tag{3.18}$$

$$= (I_{\mathcal{K}} - L K^*(H_L - z)^{-1} K)(I_{\mathcal{K}} + L K^*(H_0 - z)^{-1} K) = I_{\mathcal{K}}, \quad z \in \mathbb{C}\backslash\mathbb{R} \tag{3.19}$$

and

$$(H_L - z)^{-1} K = (H_0 - z)^{-1} K (I_{\mathcal{K}} + L K^*(H_0 - z)^{-1} K)^{-1}, \quad z \in \mathbb{C}\backslash\mathbb{R}. \tag{3.20}$$

Since

$$(I_{\mathcal{K}} + LK^*(H_0 - z)^{-1}K)^{-1} \in \mathcal{B}(\mathcal{K}), \quad z \in \mathbb{C}\backslash\mathbb{R} \tag{3.21}$$

by (3.18), one infers

$$\operatorname{ran}((I_{\mathcal{K}} + LK^*(H_0 - z)^{-1}K)^{-1}) = \mathcal{K}, \quad z \in \mathbb{C}\backslash\mathbb{R}. \tag{3.22}$$

Since by our assumption (3.4), finite linear combinations of $(H_0 - z)^{-1}Ke_n$, $n \in \mathcal{I}$, $z \in \mathbb{C}\backslash\mathbb{R}$ are dense in $\mathcal{H}$, (3.20) and (3.22) then yield the same assertion for $(H_L - z)^{-1}Ke_n$. (I.e., (3.4) is valid with H_0 replaced by any H_L.) Since U_L is unitary by Lemma 2.7, finite linear combinations of vectors of the form $(\widehat{H}_L - z)^{-1}\underline{e}_n$ (cf. (3.16)) are also dense in $\widehat{\mathcal{H}}$. This fact, (3.16), and the first resolvent equation for $\widehat{H}_L$ yield

$$\begin{aligned} U_L(\widehat{H}_L - z)^{-1}U_L^{-1}U_L(\widehat{H}_L - z')^{-1}\underline{e}_n &= U_L(\widehat{H}_L - z)^{-1}U_L^{-1}(H_L - z')^{-1}Ke_n \\ &= (H_L - z)^{-1}(H_L - z')^{-1}Ke_n, \quad n \in \mathcal{I},\ z, z' \in \mathbb{C}\backslash\mathbb{R}. \end{aligned} \tag{3.23}$$

Since finite linear combinations of $(H_L - z')^{-1}Ke_n$, $n \in \mathcal{I}$ are dense in $\mathcal{H}$ we get

$$U_L(\widehat{H}_L - z)^{-1}U_L^{-1} = (H_L - z)^{-1}, \quad z \in \mathbb{C}\backslash\mathbb{R} \tag{3.24}$$

and hence (3.11). Equation (3.12) is then obvious from (2.26) since $\widehat{H}_L$ is the operator of multiplication by λ in $\widehat{\mathcal{H}}_L$. □

If L_ℓ, $\ell = 1, 2$ are two bounded self-adjoint operators in $\mathcal{K}$ (with $\mathcal{H}, \mathcal{K}, H_0$, and K fixed, i.e., independent of $\ell = 1, 2$) one proves the following result relating $M_{L_1}(z)$ and $M_{L_2}(z)$.

Theorem 3.4. *Assume Hypothesis 3.1. Let $z \in \mathbb{C}\backslash\mathbb{R}$ and suppose H_{L_ℓ} and $M_{L_\ell}(z)$ are defined as in* (3.1) *and* (3.2) *with $\mathcal{H}, \mathcal{K}, H_0$ and K independent of $\ell = 1, 2$ and L_ℓ, $\ell = 1, 2$ bounded self-adjoint operators in $\mathcal{K}$. Then*

$$M_{L_2}(z) = M_{L_1}(z)(I_{\mathcal{K}} + (L_2 - L_1)M_{L_1}(z))^{-1} \tag{3.25a}$$

$$= (I_{\mathcal{K}} + M_{L_1}(z)(L_2 - L_1))^{-1}M_{L_1}(z). \tag{3.25b}$$

Proof. Using the resolvent equation for H_{L_2} and H_{L_1},

$$(H_{L_2} - z)^{-1} = (H_{L_1} - z)^{-1} - (H_{L_2} - z)^{-1}K(L_2 - L_1)K^*(H_{L_1} - z)^{-1} \tag{3.26a}$$

$$= (H_{L_1} - z)^{-1} - (H_{L_1} - z)^{-1}K(L_2 - L_1)K^*(H_{L_2} - z)^{-1}, \quad z \in \mathbb{C}\backslash\mathbb{R} \tag{3.26b}$$

and applying K^* on the left and K on the right of both sides of (3.26), results in

$$K^*(H_{L_1} - z)^{-1}K = K^*(H_{L_2} - z)^{-1}K(I + (L_2 - L_1)K^*(H_{L_1} - z)^{-1}K) \tag{3.27a}$$

$$= (I + K^*(H_{L_1} - z)^{-1}K(L_2 - L_1))K^*(H_{L_2} - z)^{-1}K \tag{3.27b}$$

and hence in (3.25). □

A comparison of (3.25) and (1.5), (1.6) then yields

$$A(L_1, L_2) = \begin{pmatrix} I_{\mathcal{K}} & L_2 - L_1 \\ 0 & I_{\mathcal{K}} \end{pmatrix} \in \mathcal{A}(\mathcal{K} \oplus \mathcal{K}) \tag{3.28}$$

for the corresponding matrix A in (1.5), (1.6).

We note that (3.25) also imply

$$(L_2 - L_1)M_{L_2}(z) - I_{\mathcal{K}} = -((L_2 - L_1)M_{L_1}(z) + I_{\mathcal{K}})^{-1}, \tag{3.29a}$$

$$M_{L_2}(z)(L_2 - L_1) - I_{\mathcal{K}} = -(M_{L_1}(z)(L_2 - L_1) + I_{\mathcal{K}})^{-1}. \tag{3.29b}$$

If $K\mathcal{K}$ is not a generating subspace for H_0 (i.e., (3.4) does not hold) then $\mathcal{H}$ decomposes into $\mathcal{H} = \mathcal{H}_{\mathcal{K}} \oplus \mathcal{H}_{\mathcal{K}}^{\perp}$, with

$$\mathcal{H}_{\mathcal{K}} = \overline{\text{lin.span}\{(H_0 - z)^{-1}Ke_n \in \mathcal{H} \,|\, n \in \mathcal{I},\, z \in \mathbb{C}\backslash\mathbb{R}\}} \tag{3.30a}$$

$$= \overline{\text{lin.span}\{E_0(\lambda)Ke_n \in \mathcal{H} \,|\, n \in \mathcal{I},\, \lambda \in \mathbb{R}\}} \tag{3.30b}$$

and $\mathcal{H}_{\mathcal{K}}$, $\mathcal{H}_{\mathcal{K}}^{\perp}$ both reducing subspaces for H_L ($\{e_n\}_{n\in\mathcal{I}}$ a complete orthonormal system in $\mathcal{K}$). Moreover, for all $L_\ell \in \mathcal{B}(\mathcal{K})$, $\ell = 1, 2$ self-adjoint,

$$H_{L_1} = H_{L_2} \text{ on } \operatorname{dom}(H_0) \cap \mathcal{H}_{\mathcal{K}}^{\perp} \tag{3.31}$$

and

$$H_0 = H_{0,\mathcal{K}} \oplus H_{0,\mathcal{K}}^{\perp}, \quad H_L = H_{L,\mathcal{K}} \oplus H_{0,\mathcal{K}}^{\perp}, \quad \operatorname{ran}(K) \subseteq \mathcal{H}_{\mathcal{K}}. \tag{3.32}$$

In particular,

$$M_L(z) = K^*(H_L - z)^{-1}K = K^*(H_{L,\mathcal{K}} - z)^{-1}K, \quad z \in \mathbb{C}\backslash\mathbb{R} \tag{3.33}$$

and the L-dependent spectral properties of H_L in $\mathcal{H}$ are effectively reduced to those of $H_{L,\mathcal{K}}$ in $\mathcal{H}_{\mathcal{K}}$.

In connection with our choice of KLK^* as a bounded self-adjoint perturbation of H_0, the following elementary observation might be of interest.

Lemma 3.5. *Let $V \in \mathcal{B}(\mathcal{H})$ be self-adjoint. Then V and $\mathcal{H}$ can be decomposed as*

$$V = K_0 L_0 K_0^* \oplus 0, \quad \mathcal{H} = \overline{\operatorname{ran}(V)} \oplus \ker(V), \tag{3.34}$$

where $K_0 : \mathcal{K} \to \mathcal{H}$, $L_0 = L_0^* \in \mathcal{B}(\mathcal{K})$, and $\mathcal{K} = \overline{\operatorname{ran}(V)}$.

Proof. Since $\overline{\operatorname{ran}(V)} = \ker(V)^{\perp}$, consider $V_0 = V|_{\overline{\operatorname{ran}(V)}} : \mathcal{K} \to \mathcal{K}$, $\mathcal{K} = \overline{\operatorname{ran}(V)}$. Then $V_0 = V_0^* \in \mathcal{B}(\mathcal{K})$ and V_0 admits the spectral representation $V_0 = \int_a^b dF_0(\lambda)\lambda$ for some $a, b \in \mathbb{R}$ and some family of self-adjoint spectral projections $\{F_0(\lambda)\}_{\lambda\in\mathbb{R}}$ of V_0 in $\mathcal{K}$. The decomposition (3.34) then follows upon introducing

$$K_0 = |V_0|^{1/2} = \int_a^b dF_0(\lambda)|\lambda|^{1/2}, \quad L_0 = \operatorname{sgn}(V_0) = \int_a^b dF_0(\lambda)\operatorname{sgn}(\lambda). \tag{3.35}$$

□

In (3.5)–(3.8) we showed that every collection $(H_0, K, L, \mathcal{H}, \mathcal{K})$ gives rise to an operator-valued Herglotz function $M_L(z) = \int_{\mathbb{R}} d\Omega_L(\lambda)(\lambda - z)^{-1}$ with certain properties recorded in (3.7) and (3.8). Conversely, introducing the following class $\mathcal{N}_1(\mathcal{K})$ of $\mathcal{B}(\mathcal{K})$-valued Herglotz functions (we use the symbol $\mathcal{N}_1(\mathcal{K})$ in honor of R. Nevanlinna)

(3.36)
$$\mathcal{N}_1(\mathcal{K}) = \{M \in \mathcal{B}(\mathcal{K}) \text{ Herglotz} \,|\, M(z) = \int_{\mathbb{R}} d\Omega(\lambda)(\lambda - z)^{-1};\ 0 \le \Omega(\mathbb{R}) \in \mathcal{B}(\mathcal{K})\},$$

we shall show in the remainder of this section that every element M of $\mathcal{N}_1(\mathcal{K})$ can be realized in terms of some collection $(H_0, K, \mathcal{H}, \mathcal{K})$ as in (3.6). (The operator Stieltjes integral in (3.36) converges in the norm of $\mathcal{B}(\mathcal{K})$ by Theorem I.4.2 of [17].) For this purpose we shall use a version of Naimark's dilation theorem [52], [53] as presented in Appendix I of [3] and Appendix I by Brodskii [17].

Theorem 3.6. ([17], App. I, [52].) *Suppose that $\Omega(\lambda)$, $\lambda \in \mathbb{R}$ is a strongly right-continuous nondecreasing function with values in $\mathcal{B}(\mathcal{K})$, $\mathcal{K}$ a complex separable Hilbert space, and assume* s-$\lim_{\lambda\downarrow-\infty} \Omega(\lambda) = 0$. *Then there exists a separable complex Hilbert space $\mathcal{H}$, a $K \in \mathcal{B}(\mathcal{K}, \mathcal{H})$, and an orthogonal family of strongly right-continuous spectral projections $\{E(\lambda)\}_{\lambda\in\mathbb{R}}$ in $\mathcal{H}$ such that* s-$\lim_{\lambda\downarrow-\infty} E(\lambda) = 0$, s-$\lim_{\lambda\uparrow\infty} E(\lambda) = I_{\mathcal{H}}$,

$$\Omega(\lambda) = K^* E(\lambda) K, \quad \lambda \in \mathbb{R}, \tag{3.37}$$

and

$$\overline{\{E(\lambda)K\xi \in \mathcal{H} \,|\, \xi \in \mathcal{K}, \lambda \in \mathbb{R}\}} = \mathcal{H}. \tag{3.38}$$

Moreover, if for some $\lambda_1, \lambda_2 \in \mathbb{R}$, $\Omega(\lambda_1) = \Omega(\lambda_2)$, then $E(\lambda_1) = E(\lambda_2)$.

The principal realization theorem for Herglotz operators of the type (3.36) then reads as follows

Theorem 3.7. *(i) Any $M \in \mathcal{N}_1(\mathcal{K})$ with associated measure Ω can be realized in the form*

$$M(z) = K^*(H - z)^{-1}K, \quad z \in \mathbb{C}\backslash\mathbb{R}, \tag{3.39}$$

where H represents a self-adjoint operator in some separable complex Hilbert space $\mathcal{H}$, $K \in \mathcal{B}(\mathcal{K}, \mathcal{H})$, and

$$\Omega(\mathbb{R}) = K^* K. \tag{3.40}$$

(ii) Suppose $M_\ell \in \mathcal{N}_1(\mathcal{K})$ with corresponding measures Ω_ℓ, $\ell = 1, 2$ and $M_1 \neq M_2$. Then M_1 and M_2 can be realized as

$$M_\ell(z) = K^*(H_{L_\ell} - z)^{-1}K, \quad z \in \mathbb{C}\backslash\mathbb{R}, \tag{3.41}$$

where H_{L_ℓ}, $\ell = 1, 2$ are self-adjoint perturbations of one and the same self-adjoint operator H_0 in some separable complex Hilbert space $\mathcal{H}$

$$H_{L_\ell} = H_0 + KL_\ell K^*, \quad \ell = 1, 2 \tag{3.42}$$

for some $L_\ell = L_\ell^* \in \mathcal{B}(\mathcal{K})$, $\ell = 1, 2$ *and some* $K \in \mathcal{B}(\mathcal{K}, \mathcal{H})$ *if and only if the following two conditions hold:*

$$\Omega_1(\mathbb{R}) = K^*K = \Omega_2(\mathbb{R}), \tag{3.43}$$

and for all $z \in \mathbb{C}\backslash\mathbb{R}$,

$$M_2(z) = M_1(z)(I_\mathcal{K} + (L_2 - L_1)M_1(z))^{-1}. \tag{3.44}$$

Proof. Applying Naimark's dilation theorem, Theorem 3.6, to $\Omega(\lambda)$, $\lambda \in \mathbb{R}$, (assuming s-$\lim_{\lambda\downarrow-\infty} \Omega(\lambda) = 0$ without loss of generality), yields $\Omega(\lambda) = K^*E(\lambda)K$, $\lambda \in \mathbb{R}$ and introducing the self-adjoint operator $H = \int_\mathbb{R} dE(\lambda)\lambda$ in $\mathcal{H}$ then proves (3.39). The normalization condition (3.40) then follows as discussed in (3.5)–(3.7). In exactly the same manner one proves the necessity of the normalization (3.43). The necessity of (3.44) was proven in Theorem 3.4. In order to prove sufficiency of (3.43) and (3.44) for (3.41) and (3.42) to hold, we argue as follows. Suppose s-$\lim_{\lambda\downarrow-\infty} \Omega_1(\lambda) = 0$ (otherwise, replace $\Omega_1(\lambda)$ by $\Omega_1(\lambda) -$ s-$\lim_{\nu\downarrow-\infty} \Omega_1(\nu)$) and represent $M_1(z)$ according to part (i) by

$$M_1(z) = K^*(H_1 - z)^{-1}K, \quad z \in \mathbb{C}\backslash\mathbb{R} \tag{3.45}$$

applying Naimark's dilation theorem and Theorem 3.6. Define

$$H_0 = H_1 - KL_1K^*, \quad \mathrm{dom}(H_0) = \mathrm{dom}(H_1) \tag{3.46}$$

for some $L_1 = L_1^* \in \mathcal{B}(\mathcal{K})$. Next, use $L_2 = L_2^* \in \mathcal{B}(\mathcal{K})$ in (3.44) to define

$$H_2 = H_0 + KL_2K^*, \quad \mathrm{dom}(H_2) = \mathrm{dom}(H_0) \tag{3.47}$$

and

$$M_{L_2}(z) = K^*(H_2 - z)^{-1}K. \tag{3.48}$$

By Theorem 3.4,

$$M_{L_2}(z) = M_{L_1}(z)(I_\mathcal{K} + (L_2 - L_1)M_1(z))^{-1} = M_2(z), \quad z \in \mathbb{C}\backslash\mathbb{R} \tag{3.49}$$

and the proof is complete. □

For a variety of results related to realization theorems of Herglotz operators we refer, for instance, to [10] and the literature cited therein. Fundamental results on nontangential boundary values of $M_L(z)$ as $z \to x \in \mathbb{R}$, under various conditions on K, can be found in [48]–[51]. Additional results on operators of the type $M_L(z)$ (including cases where K is a suitable unbounded operator) can be found, for instance, in [2], [46], [47] and the references therein.

4. On self-adjoint extensions of symmetric operators

In this section we consider self-adjoint extensions H of densely defined closed symmetric operators $\dot{H}$ with deficiency indices (k, k), $k \in \mathbb{N} \cup \{\infty\}$. We revisit Krein's formula relating self-adjoint extensions of $\dot{H}$, introduce the corresponding operator-valued Weyl m-functions and their linear fractional transformations, study a model for the pair $(\dot{H}, H)$, and consider Friedrichs H_F and Krein extensions H_K of $\dot{H}$ in the case where $\dot{H}$ is bounded from below.

In the special case $k = 1$, detailed investigation of this type were undertaken by Donoghue [25]. The case $k \in \mathbb{N}$ was recently discussed in depth in [30] (we also refer to [36] for another comprehensive treatment of this subject). Here we treat the general situation $k \in \mathbb{N} \cup \{\infty\}$ utilizing recent results in [28].

We start with a bit of notation and then recall some pertinent results of [28]. Let $\mathcal{H}$ be a separable complex Hilbert space and $\dot{H} : \mathrm{dom}(\dot{H}) \to \mathcal{H}$, $\overline{\mathrm{dom}(\dot{H})} = \mathcal{H}$ a densely defined closed symmetric linear operator with equal deficiency indices $\mathrm{def}(\dot{H}) = (k, k), k \in \mathbb{N} \cup \{\infty\}$. The deficiency subspaces $\mathcal{N}_\pm$ of $\dot{H}$ are given by

$$\mathcal{N}_\pm = \ker(\dot{H}^* \mp i), \quad \dim_{\mathbb{C}}(\mathcal{N}_\pm) = k \tag{4.1}$$

and for any self-adjoint extension H of $\dot{H}$ in $\mathcal{H}$, the corresponding Cayley transform C_H in $\mathcal{H}$ is defined by

$$C_H = (H + i)(H - i)^{-1}, \tag{4.2}$$

implying

$$C_H \mathcal{N}_- = \mathcal{N}_+. \tag{4.3}$$

Two self-adjoint extensions H_1 and H_2 of $\dot{H}$ are called *relatively prime* (w.r.t. $\dot{H}$) if $\mathrm{dom}(H_1) \cap \mathrm{dom}(H_2) = \mathrm{dom}(\dot{H})$. Associated with H_1 and H_2 we introduce $P_{1,2}(z) \in \mathcal{B}(\mathcal{H})$ by

$$\begin{aligned} P_{1,2}(z) &= (H_1 - z)(H_1 - i)^{-1}((H_2 - z)^{-1} - (H_1 - z)^{-1})(H_1 - z)(H_1 + i)^{-1}, \\ & \quad z \in \rho(H_1) \cap \rho(H_2). \end{aligned} \tag{4.4}$$

We refer to Lemma 2 of [28] and [58] for a detailed discussion of $P_{1,2}(z)$. Here we only mention the following properties of $P_{1,2}(z)$, $z \in \rho(H_1) \cap \rho(H_2)$,

$$P_{1,2}(z)\big|_{\mathcal{N}_+^\perp} = 0, \quad P_{1,2}(z)\mathcal{N}_+ \subseteq \mathcal{N}_+, \tag{4.5}$$

$$\overline{\mathrm{ran}(P_{1,2}(i))} = \mathcal{N}_+, \quad \mathrm{ran}(P_{1,2}(z)\big|_{\mathcal{N}_+}) \text{ is independent of } z \in \rho(A_1) \cap \rho(A_2), \tag{4.6}$$

$$P_{1,2}(i)\big|_{\mathcal{N}_+} = (i/2)(I - C_{H_2}C_{H_1}^{-1})\big|_{\mathcal{N}_+} = (i/2)(I_{\mathcal{N}_+} + e^{-2i\alpha_{1,2}}) \tag{4.7}$$

for some self-adjoint (possibly unbounded) operator $\alpha_{1,2}$ in $\mathcal{N}_+$.

Next, given a self-adjoint extension H of $\dot{H}$ and a closed linear subspace $\mathcal{N}$ of $\mathcal{N}_+$, $\mathcal{N} \subseteq \mathcal{N}_+$, the Weyl-Titchmarsh operator $M_{H,\mathcal{N}}(z) \in \mathcal{B}(\mathcal{N})$ associated with the pair $(H, \mathcal{N})$ is defined by

$$\begin{aligned} M_{H,\mathcal{N}}(z) &= P_{\mathcal{N}}(zH + I_{\mathcal{H}})(H - z)^{-1}P_{\mathcal{N}}\big|_{\mathcal{N}} \\ &= zI_{\mathcal{N}} + (1 + z^2)P_{\mathcal{N}}(H - z)^{-1}P_{\mathcal{N}}\big|_{\mathcal{N}}, \quad z \in \mathbb{C}\backslash\mathbb{R}, \end{aligned} \tag{4.8}$$

with $I_{\mathcal{N}}$ the identity operator in $\mathcal{N}$ and $P_{\mathcal{N}}$ the orthogonal projection in $\mathcal{H}$ onto $\mathcal{N}$.

One verifies (cf. Lemma 4 in [28]) for H_1 and H_2 relatively prime w.r.t. $\dot{H}$,

$$(P_{1,2}(z)\big|_{\mathcal{N}_+})^{-1} = (P_{1,2}(i)\big|_{\mathcal{N}_+})^{-1} - (z-i)P_{\mathcal{N}_+}(H_1+i)(H_1-z)^{-1}P_{\mathcal{N}_+} \tag{4.9a}$$

$$= \tan(\alpha_{1,2}) - M_{H_1,\mathcal{N}_+}(z), \quad z \in \rho(H_1), \tag{4.9b}$$

where

$$C_{H_2}C_{H_1}^{-1}\big|_{\mathcal{N}_+} = -e^{-2i\alpha_{1,2}}. \tag{4.10}$$

Following Saakjan [58] (in a version presented in Theorem 5 and Corollary 6 in [28]), Krein's formula then can be summarized as follows.

Theorem 4.1. ([28], [58].) *Let H_1 and H_2 be self-adjoint extensions of $\dot{H}$ and $z \in \rho(H_1) \cup \rho(H_2)$. Then*

$$(H_2-z)^{-1} = (H_1-z)^{-1} + (H_1-i)(H_1-z)^{-1}P_{1,2}(z)(H_1+i)(H_1-z)^{-1} \tag{4.11}$$

$$\begin{aligned} &= (H_1-z)^{-1} + (H_1-i)(H_1-z)^{-1}P_{\mathcal{N}_{1,2,+}} \times \\ &\quad \times (\tan(\alpha_{\mathcal{N}_{1,2,+}}) - M_{H_1,\mathcal{N}_{1,2,+}}(z))^{-1}P_{\mathcal{N}_{1,2,+}}(H_1+i)(H_1-z)^{-1}, \end{aligned} \tag{4.12}$$

where

$$\mathcal{N}_{1,2,+} = \ker((H_1\big|_{\mathcal{D}(H_1)\cap\mathcal{D}(H_2)})^* - i), \tag{4.13}$$

$$e^{-2i\alpha_{\mathcal{N}_{1,2,+}}} = -C_{H_2}C_{H_1}^{-1}\big|_{\mathcal{N}_{1,2,+}}, \tag{4.14}$$

and

$$P_{1,2}(i)\big|_{\mathcal{N}_{1,2,+}} = (i/2)(I - C_{H_2}C_{H_1}^{-1})\big|_{\mathcal{N}_{1,2,+}}. \tag{4.15}$$

Next we recall that $M_{H,\mathcal{N}}$ and hence $P_{1,2}(z)\big|_{\mathcal{N}_+}$ and $-(P_{1,2}(z)\big|_{\mathcal{N}_+})^{-1}$ (cf. (4.9)), if the latter exists, are operator-valued Herglotz functions.

Theorem 4.2. *Let H be a self-adjoint extension of $\dot{H}$ with orthogonal family of spectral projections $\{E_H(\lambda)\}_{\lambda\in\mathbb{R}}$, $\mathcal{N}$ a closed subspace of $\mathcal{N}_+$. Then the Weyl-Titchmarch operator $M_{H,\mathcal{N}}(z)$ is analytic for $z \in \mathbb{C}\backslash\mathbb{R}$ and*

$$\operatorname{Im}(z)\operatorname{Im}(M_{H,\mathcal{N}}(z)) \geq (\max(1,|z|^2) + |\operatorname{Re}(z)|)^{-1}, \quad z \in \mathbb{C}\backslash\mathbb{R}. \tag{4.16}$$

In particular, $M_{H,\mathcal{N}}(z)$ is a $\mathcal{B}(\mathcal{N})$-valued Herglotz function and admits the representation valid in the strong operator topology of $\mathcal{N}$,

$$M_{H,\mathcal{N}}(z) = \int_{\mathbb{R}} d\Omega_{H,\mathcal{N}}(\lambda)((\lambda-z)^{-1} - \lambda(1+\lambda^2)^{-1}), \quad z \in \mathbb{C}\backslash\mathbb{R}, \tag{4.17}$$

where

$$\Omega_{H,\mathcal{N}}(\lambda) = (1+\lambda^2)(P_{\mathcal{N}}E_H(\lambda)P_{\mathcal{N}}\big|_{\mathcal{N}}), \tag{4.18}$$

$$\int_{\mathbb{R}} d\Omega_{H,\mathcal{N}}(\lambda)(1+\lambda^2)^{-1} = I_{\mathcal{N}}, \tag{4.19}$$

$$\int_{\mathbb{R}} d(\xi, \Omega_{H,\mathcal{N}}(\lambda)\xi)_{\mathcal{H}} = \infty \text{ for all } \xi \in \mathcal{N}\backslash\{0\}. \tag{4.20}$$

Proof. (4.17) has been derived in Lemma 7 of [28], hence we confine ourselves to a few hints. An explicit computation yields

$$\begin{aligned}\operatorname{Im}(z)\operatorname{Im}(M_{H,\mathcal{N}}(z)) = P_{\mathcal{N}}(I_{\mathcal{H}}+H^2)^{1/2}((H-\operatorname{Re}(z))^2+\operatorname{Im}(z))^2)^{-1} \\ \times (I_{\mathcal{H}}+H^2)^{1/2}P_{\mathcal{N}}\big|_{\mathcal{N}}, \quad z\in\mathbb{C}\backslash\mathbb{R}.\end{aligned} \tag{4.21}$$

Together with

$$\frac{1+\lambda^2}{(\lambda-\operatorname{Re}(z))^2+(\operatorname{Im}(z))^2} \geq \frac{1}{\max(1,|z|^2)+|\operatorname{Re}(z)|} \tag{4.22}$$

and the Rayleigh-Ritz argument this yields (4.16). The representation (4.17) and the fact (4.18) follow from (4.8) and $(H-z)^{-1}\xi = \int_{\mathbb{R}} d(E_H(\lambda)\xi)(\lambda-z)^{-1}$, $\xi\in\mathcal{H}$. (4.19) then follows from

$$\begin{aligned}\int_{\mathbb{R}} d(\Omega_{H,\mathcal{N}}(\lambda)\xi)(1+\lambda^2)^{-1} = \int_{\mathbb{R}} d(P_{\mathcal{N}}E_H(\lambda)\xi) = P_{\mathcal{N}}\int_{\mathbb{R}} d(E_H(\lambda)\xi) \\ = P_{\mathcal{N}}\xi = \xi \text{ for all } \xi\in\mathcal{N}.\end{aligned} \tag{4.23}$$

Finally,

$$\int_{\mathbb{R}} d(\xi,\Omega_{H,\mathcal{N}}(\lambda)\xi)_{\mathcal{H}} = \int_{\mathbb{R}} d(\xi,E_H(\lambda)\xi)_{\mathcal{H}}(1+\lambda^2) = \infty \text{ for all } \xi\in\mathcal{N}\backslash\{0\} \tag{4.24}$$

since $\mathcal{N}\subseteq\mathcal{N}_+$ and $\mathcal{N}_+\cap\operatorname{dom}(H)=\{0\}$ by von Neumann's formula

$$\operatorname{dom}(H) = \operatorname{dom}(\dot{H})\dot{+}\mathcal{N}_+\dot{+}(-C_H)^{-1}\mathcal{N}_+. \tag{4.25}$$

□

We also recall without proof the principal result of [28], the linear fractional transformation relating the Weyl-Titchmarch operators associated with different self-adjoint extensions of $\dot{H}$.

Theorem 4.3. ([28].) *Let H_1 and H_2 be self-adjoint extensions of $\dot{H}$ and $z\in\rho(H_1)\cap\rho(H_2)$. Then*

$$\begin{aligned}M_{H_2,\mathcal{N}_+}(z) = (P_{1,2}(i)\big|_{\mathcal{N}_+} + (I_{\mathcal{N}_+}+iP_{1,2}(i)\big|_{\mathcal{N}_+})M_{H_1,\mathcal{N}_+}(z))\times \\ \times((I_{\mathcal{N}_+}+iP_{1,2}(i)\big|_{\mathcal{N}_+}) - P_{1,2}(i)\big|_{\mathcal{N}_+}M_{H_1,\mathcal{N}_+}(z))^{-1},\end{aligned} \tag{4.26}$$

where

$$P_{1,2}(i)\big|_{\mathcal{N}_+} = (i/2)(I_{\mathcal{H}} - C_{H_2}C_{H_1}^{-1})\big|_{\mathcal{N}_+}, \tag{4.27}$$

$$I_{\mathcal{N}_+}+iP_{1,2}(i)\big|_{\mathcal{N}_+} = (1/2)(I_{\mathcal{H}} + C_{H_2}C_{H_1}^{-1})\big|_{\mathcal{N}_+}. \tag{4.28}$$

Introducing

$$e^{-2i\alpha_{1,2}} = -C_{H_2}C_{H_1}^{-1}\big|_{\mathcal{N}_+}, \tag{4.29}$$

(4.26) can be rewritten as

$$\begin{aligned}M_{H_2,\mathcal{N}_+}(z) = e^{-i\alpha_{1,2}}(\cos(\alpha_{1,2})+\sin(\alpha_{1,2})M_{H_1,\mathcal{N}_+}(z))\times \\ \times(\sin(\alpha_{1,2})-\cos(\alpha_{1,2})M_{H_1,\mathcal{N}_+}(z))^{-1}e^{i\alpha_{1,2}}.\end{aligned} \tag{4.30}$$

A comparison of (4.30) and (1.5), (1.6) then yields

$$A(\alpha_{1,2}) = \begin{pmatrix} e^{-i\alpha_{1,2}}\sin(\alpha_{1,2}) & -e^{-i\alpha_{1,2}}\cos(\alpha_{1,2}) \\ e^{-i\alpha_{1,2}}\cos(\alpha_{1,2}) & e^{-i\alpha_{1,2}}\sin(\alpha_{1,2}) \end{pmatrix} \in \mathcal{A}(\mathcal{K}\oplus\mathcal{K}) \tag{4.31}$$

for the corresponding matrix A in (1.5), (1.6).

Weyl operators of the type $M_{H,\mathcal{N}}(z)$ have attracted considerable attention in the literature. The interested reader can find a variety of additional results, for instance, in [18], [21]–[24], [40], [41], [45], [46], [56].

Next we will prepare some material that eventually will lead to a model for the pair $(\dot{H}, H)$. Let $\mathcal{N}$ be a separable complex Hilbert space, $\{u_n\}_{n\in\mathcal{I}}$, $\mathcal{I}\subseteq\mathbb{N}$ a complete orthonormal system in $\mathcal{N}$, $\{\widetilde{\Omega}(\lambda)\}_{\lambda\in\mathbb{R}}$ a family of strongly right-continuous nondecreasing $\mathcal{B}(\mathcal{N})$-valued functions normalized by

$$\widetilde{\Omega}(\mathbb{R}) = I_{\mathcal{K}}, \tag{4.32}$$

with the property

$$\int_{\mathbb{R}} d(\xi,\widetilde{\Omega}(\lambda)\xi)_{\mathcal{N}}(1+\lambda^2) = \infty \text{ for all } \xi\in\mathcal{N}\backslash\{0\}. \tag{4.33}$$

Introducing the control measure $\tilde{\mu}(B) = \sum_{n\in\mathcal{I}} 2^{-n}(u_n,\widetilde{\Omega}(B)u_n)_{\mathcal{N}}$, $B\in\Sigma$, and $\underline{\Lambda}$ as in Theorem 2.5, we may define $L^p(\mathbb{R},\mathcal{N};wd\widetilde{\Omega})$, $p\geq 1$, $w\geq 0$ a weight function, as in Section 2. Of special importance in this section are weight functions of the type $w_r(\lambda) = (1+\lambda^2)^r$, $r\in\mathbb{R}$, $\lambda\in\mathbb{R}$. In particular, introducing

$$\Omega(B) = \int_B (1+\lambda^2)d\tilde{\mu}(\lambda)\frac{d\widetilde{\Omega}}{d\tilde{\mu}}(\lambda), \quad B\in\Sigma, \tag{4.34}$$

we abbreviate $\widehat{\mathcal{H}} = L^2(\mathbb{R},\mathcal{N};d\Omega)$ and define the self-adjoint operator $\widehat{H}$ in $\widehat{\mathcal{H}}$,

$$(\widehat{H}\hat{f})(\lambda) = \lambda\hat{f}(\lambda), \quad \hat{f}\in\operatorname{dom}(\widehat{H}) = L^2(\mathbb{R},\mathcal{N};(1+\lambda^2)d\Omega), \tag{4.35}$$

with corresponding family of strongly right-continuous orthogonal spectral projections

$$(E_{\widehat{H}}(\lambda)\hat{f})(\nu) = \theta(\lambda-\nu)\hat{f}(\nu) \text{ for } \Omega-\text{a.e. } \nu\in\mathbb{R}, \quad \hat{f}\in\widehat{\mathcal{H}}. \tag{4.36}$$

Associated with $\widehat{H}$ we consider the linear operator $\hat{\dot{H}}$ in $\widehat{\mathcal{H}}$ defined as the following restriction of $\widehat{H}$

$$\begin{aligned} &\operatorname{dom}(\hat{\dot{H}}) = \{\hat{f}\in\operatorname{dom}(\widehat{H}) \,|\, \textstyle\int_{\mathbb{R}}(1+\lambda^2)d\tilde{\mu}(\lambda)(\underline{\xi},\hat{f}(\lambda))_{\mathcal{N}_\lambda} = 0 \text{ for all } \underline{\xi}\in\underline{\Lambda}(\mathcal{N})\}, \\ &\hat{\dot{H}} = \widehat{H}\big|_{\operatorname{dom}(\hat{\dot{H}})}. \end{aligned} \tag{4.37}$$

(The integral in (4.37) is well defined, see the proof of Theorem 4.4 below.) Here we used the notation introduced in the proof of Theorem 2.5,

$$\underline{\xi} = \underline{\Lambda}\xi = \{\underline{\xi}(\lambda) = \xi\}_{\lambda\in\mathbb{R}}. \tag{4.38}$$

Moreover, introducing the scale of Hilbert spaces $\widehat{\mathcal{H}}_{2r} = L^2(\mathbb{R}, \mathcal{N}; (1+\lambda^2)^r d\Omega)$, $r \in \mathbb{R}$, $\widehat{\mathcal{H}}_0 = \widehat{\mathcal{H}}$, we consider the unitary operator R from $\widehat{\mathcal{H}}_2$ to $\widehat{\mathcal{H}}_{-2}$,

$$R : \widehat{\mathcal{H}}_2 \longrightarrow \widehat{\mathcal{H}}_{-2}, \quad \hat{f} \longrightarrow (1+\lambda^2)\hat{f}, \tag{4.39}$$

$$(\hat{f}, \hat{g})_{\widehat{\mathcal{H}}_2} = (\hat{f}, R\hat{g})_{\widehat{\mathcal{H}}} = (R\hat{f}, \hat{g})_{\widehat{\mathcal{H}}} = (R\hat{f}, R\hat{g})_{\widehat{\mathcal{H}}_{-2}}, \quad \hat{f}, \hat{g} \in \widehat{\mathcal{H}}_2, \tag{4.40}$$

$$(\hat{u}, \underline{v})_{\widehat{\mathcal{H}}_{-2}} = (\hat{u}, R^{-1}\underline{v})_{\widehat{\mathcal{H}}_2} = (R^{-1}\hat{u}, \underline{v})_{\widehat{\mathcal{H}}} = (R^{-1}\hat{u}, R^{-1}\underline{v})_{\widehat{\mathcal{H}}_2}, \quad \hat{u}, \underline{v} \in \widehat{\mathcal{H}}_{-2}. \tag{4.41}$$

In particular,

$$\underline{\underline{\Lambda}}(\mathcal{N}) \subset \widehat{\mathcal{H}}, \quad \underline{\Lambda}(\mathcal{N}) \subset \widehat{\mathcal{H}}_{-2}, \quad \underline{\xi} \in \underline{\Lambda}(\mathcal{N}) \backslash \{0\} \Rightarrow \underline{\xi} \notin \widehat{\mathcal{H}} \tag{4.42}$$

(cf. (2.51) and (4.32)–(4.34)).

Theorem 4.4. *The operator $\hat{\dot{H}}$ in* (4.37) *is densely defined symmetric and closed in $\widehat{\mathcal{H}}$. Its deficiency indices are given by*

$$\operatorname{def}(\hat{\dot{H}}) = (k, k), \quad k = \dim_{\mathbb{C}}(\mathcal{N}) \in \mathbb{N} \cup \{\infty\}, \tag{4.43}$$

and

$$\ker(\hat{\dot{H}}^* - z) = \overline{\mathit{lin.span}\{\{(\lambda - z)^{-1} e_n\}_{\lambda \in \mathbb{R}} \in \widehat{\mathcal{H}} \,|\, n \in \mathcal{I}\}}, \quad z \in \mathbb{C} \backslash \mathbb{R}. \tag{4.44}$$

Proof. Writing $||\hat{f}(\lambda)||_{\mathcal{N}_\lambda} = (1+\lambda^2)^{-1/2}(1+\lambda^2)^{1/2}||\hat{f}(\lambda)||_{\mathcal{N}_\lambda}$ one infers that $\hat{f} \in L^1(\mathbb{R}, \mathcal{N}; d\Omega)$ for $\hat{f} \in \widehat{\mathcal{H}}_2$. Thus the integral in (4.37) and hence $\operatorname{dom}(\hat{\dot{H}})$ is well defined. As a restriction of $\widehat{H}$, $\hat{\dot{H}}$ is clearly symmetric. By (4.37) and (4.39)–(4.41) one infers

$$\operatorname{dom}(\widehat{H}) = \operatorname{dom}(\hat{\dot{H}}) = \widehat{\mathcal{H}}_2 \ominus_{\widehat{\mathcal{H}}_2} R^{-1} \underline{\Lambda}(\mathcal{N}), \tag{4.45}$$

where, in obvious notation, $\ominus_{\widehat{\mathcal{H}}_2}$ indicates the orthogonal complement in $\widehat{\mathcal{H}}_2$. Thus $\hat{\dot{H}}$ has a closed graph.

Next, to prove that $\hat{\dot{H}}$ is densely defined in $\widehat{\mathcal{H}}$, suppose there is a $\hat{g} \in \widehat{\mathcal{H}}$ such that $\hat{g} \perp \operatorname{dom}(\hat{\dot{H}})$. Then

$$0 = (\hat{f}, \hat{g})_{\widehat{\mathcal{H}}} = (\hat{f}, R^{-1}\hat{g})_{\widehat{\mathcal{H}}_2} \text{ for all } \hat{f} \in \operatorname{dom}(\hat{\dot{H}}) \tag{4.46}$$

and hence $R^{-1}\hat{g} \in R^{-1}\underline{\Lambda}(\mathcal{N})$, that is, there is an $\xi \in \mathcal{N}$ such that $\hat{g} = \underline{\Lambda}\xi$ Ω-a.e. by (4.45). Since $\underline{\Lambda}\xi \in \underline{\Lambda}(\mathcal{N}) \backslash \{0\}$ implies $\underline{\Lambda}\xi \notin \widehat{\mathcal{H}}$ by (4.42), $\hat{g} \in \widehat{\mathcal{H}}$ if and only if $\underline{\Lambda}\xi = \hat{g} = 0$. Finally, since $\widehat{H}$ is self-adjoint, $\operatorname{ran}(\widehat{H} - z) = \widehat{\mathcal{H}}$ for all $z \in \mathbb{C} \backslash \mathbb{R}$, and $(\widehat{H} \pm i) : \widehat{\mathcal{H}}_2 \to \widehat{\mathcal{H}}$ is unitary,

$$((\widehat{H} \pm i)\hat{f}, (\widehat{H} \pm i)\hat{g})_{\widehat{\mathcal{H}}} = \int_{\mathbb{R}} (1+\lambda^2)^2 d\tilde{\mu}(\lambda) (\hat{f}(\lambda), \hat{g}(\lambda))_{\mathcal{N}_\lambda} = (\hat{f}, \hat{g})_{\widehat{\mathcal{H}}_2}, \quad \hat{f}, \hat{g} \in \widehat{\mathcal{H}}_2. \tag{4.47}$$

Thus (4.45) and (4.46) yield

$$\widehat{\mathcal{H}} = (\widehat{H} \pm i)\widehat{\mathcal{H}}_2 = (\widehat{H} \pm i)(\operatorname{dom}(\hat{\dot{H}}) \oplus_{\widehat{\mathcal{H}}_2} R^{-1}\underline{\Lambda}(\mathcal{N}))$$

$$= (\hat{\dot{H}} \pm i)\mathrm{dom}(\hat{\dot{H}}) \oplus_{\hat{\mathcal{H}}} \overline{\mathrm{lin.span}\{\{(\lambda \pm i)(1+\lambda^2)^{-1}u_n\}_{\lambda\in\mathbb{R}} \in \widehat{\mathcal{H}} \,|\, n \in \mathcal{I}\}}$$

$$= \mathrm{ran}(\hat{\dot{H}} \pm i) \oplus_{\hat{\mathcal{H}}} \overline{\mathrm{lin.span}\{\{(\lambda \mp i)^{-1}u_n\}_{\lambda\in\mathbb{R}} \in \widehat{\mathcal{H}} \,|\, n \in \mathcal{I}\}} \tag{4.48}$$

and hence

$$\ker(\hat{\dot{H}}^* \mp i) = \overline{\mathrm{lin.span}\{(\{\lambda \mp i)^{-1}u_n\}_{\lambda\in\mathbb{R}} \in \widehat{\mathcal{H}} \,|\, n \in \mathcal{I}\}}. \tag{4.49}$$

Since $(\lambda - z)^{-1}\xi = (\lambda - i)^{-1}\xi + (z-i)(\lambda - z)^{-1}(\lambda - i)^{-1}\xi$, with $\{(\lambda - z)^{-1}(\lambda - i)^{-1}\xi\}_{\lambda\in\mathbb{R}} \in \widehat{\mathcal{H}}_2 = \mathrm{dom}(\widehat{H})$ for all $\xi \in \mathcal{N}$, $z \in \mathbb{C}\backslash\mathbb{R}$, (4.49) yields (4.44). □

Lemma 4.5. *Let $\dot{H}$ be a densely defined linear closed symmetric operator in a separable complex Hilbert space $\mathcal{H}$ with deficiency indices (k,k), $k \in \mathbb{N}\cup\{\infty\}$. Then $\mathcal{H}$ decomposes into the direct orthogonal sum*

$$\mathcal{H} = \mathcal{H}_0 \oplus \mathcal{H}_0^\perp, \quad \ker(\dot{H}^* - i) \subset \mathcal{H}_0, \quad z \in \mathbb{C}\backslash\mathbb{R}, \tag{4.50}$$

where $\mathcal{H}_0$ and $\mathcal{H}_0^\perp$ are invariant subspaces for all self-adjoint extensions of $\dot{H}$, that is,

$$(H - z)^{-1}\mathcal{H}_0 \subseteq \mathcal{H}_0, \quad (H-z)^{-1}\mathcal{H}_0^\perp \subseteq \mathcal{H}_0^\perp, \quad z \in \mathbb{C}\backslash\mathbb{R}, \tag{4.51}$$

for all self-adjoint extensions H of $\dot{H}$ in $\mathcal{H}$. Moreover, all self-adjoint extensions $\dot{H}$ coincide on $\mathcal{H}_0^\perp$, that is, if $\{H_\alpha\}_{\alpha\in\mathcal{I}}$ ($\mathcal{I}$ an appropriate index set) denotes the set of all self-adjoint extensions of $\dot{H}$, then

$$H_\alpha = H_{0,\alpha} \oplus H_0^\perp, \quad \alpha \in \mathcal{I} \quad \text{in} \quad \mathcal{H} = \mathcal{H}_0 \oplus \mathcal{H}_0^\perp, \tag{4.52}$$

where

$$H_0^\perp \text{ is independent of } \alpha \in \mathcal{I}. \tag{4.53}$$

Proof. Let H be a fixed self-adjoint extension of $\dot{H}$, denote $\mathcal{N}_\pm = \ker(\dot{H}^* \mp i)$, and define

$$\mathcal{H}_H = \overline{\mathrm{lin.span}\{(H-z)^{-1}u_+ \in \mathcal{H} \,|\, u_+ \in \mathcal{N}_+,\, z \in \mathbb{C}\backslash\mathbb{R}\}}. \tag{4.54}$$

Since $(H-z_1)^{-1}(H-z_2)^{-1} = (z_1-z_2)^{-1}((H-z_1)^{-1}-(H-z_2)^{-1})$, $\mathcal{H}_H$ is invariant with respect to $(H-z)^{-1}$, $(H-z)^{-1}\mathcal{H}_H \subseteq \mathcal{H}_H$, and since $((H-z)^{-1})^* = (H-\bar{z})^{-1}$, also $\mathcal{H}_H^\perp$ is invariant under $(H-z)^{-1}$ for all $z \in \mathbb{C}\backslash\mathbb{R}$. Since $\text{w-lim}_{z\to i\infty}(-z)(H-z)^{-1}f = f$ for all $f \in \mathcal{H}$, one concludes

$$\mathcal{N}_+ \subset \mathcal{H}_H. \tag{4.55}$$

Next, let $v \in \mathcal{H}_H^\perp$. Then also

$$w = (H-z)^{-1}v \in \mathcal{H}_H^\perp, \quad z \in \mathbb{C}\backslash\mathbb{R} \tag{4.56}$$

and

$$(u_+, v)_\mathcal{H} = (u_+, w)_\mathcal{H} = 0, \quad u_+ \in \mathcal{N}_+. \tag{4.57}$$

Since $w \in \mathrm{dom}(H)$

$$w \notin \mathcal{N}_\pm \tag{4.58}$$

(otherwise $\dot{H}^*w = \pm iw$ yields $Hw = \pm iw$ which contradicts the self-adjointness of H). By von Neumann's formulas

$$\operatorname{dom}(\dot{H}^*) = \operatorname{dom}(\dot{H}) \oplus_{\mathcal{H}_+} \mathcal{N}_+ \oplus_{\mathcal{H}_+} \mathcal{N}_-, \tag{4.59}$$

where $\oplus_{\mathcal{H}_+}$ denotes the direct orthogonal sum in the Hilbert space $\mathcal{H}_+$ defined by

$$\mathcal{H}_+ = (\operatorname{dom}(\dot{H}^*), (\cdot,\cdot)_+), \quad (f,g)_+ = (\dot{H}^*f, \dot{H}^*g)_{\mathcal{H}} + (f,g)_{\mathcal{H}}, \quad f,g \in \operatorname{dom}(\dot{H}^*). \tag{4.60}$$

Using (4.55), $Hw = zw + v$ (cf. (4.56)), (4.57), and (4.60) one computes

$$\begin{aligned}(u_+, w)_+ &= (\dot{H}^*u_+, \dot{H}^*w)_{\mathcal{H}} + (u_+, w)_{\mathcal{H}} = -i(u_+, Hw)_{\mathcal{H}} + (u_+, w)_{\mathcal{H}} \\ &= (-iz + 1)(u_+, w)_{\mathcal{H}} - i(u_+, v)_{\mathcal{H}} = 0.\end{aligned} \tag{4.61}$$

(4.58), (4.59), and (4.61) then prove $w \in \operatorname{dom}(\dot{H})$ and hence

$$Hw = \dot{H}w = zw + v. \tag{4.62}$$

If $\widetilde{H}$ is any other self-adjoint extension of $\dot{H}$, then $w \in \operatorname{dom}(\dot{H})$ also yields

$$\widetilde{H}w = \dot{H}w = zw + v \tag{4.63}$$

and hence

$$w = (H - z)^{-1}v = (\widetilde{H} - z)^{-1}v, \quad v \in \mathcal{H}_H^\perp. \tag{4.64}$$

Thus the resolvents of all self-adjoint extensions of $\dot{H}$ coincide on $\mathcal{H}_H^\perp$. Moreover,

$$((\widetilde{H} - \bar{z})^{-1}u_+, v)_{\mathcal{H}} = (u_+, (\widetilde{H} - z)^{-1}v)_{\mathcal{H}} = (u_+, w)_{\mathcal{H}} = 0 \tag{4.65}$$

yields

$$(\widetilde{H} - z)^{-1}u_+ \perp \mathcal{H}_H^\perp, \quad z \in \mathbb{C}\backslash\mathbb{R} \tag{4.66}$$

and hence $\mathcal{H}_{\widetilde{H}} \subseteq \mathcal{H}_H$. By symmetry in H and $\widetilde{H}$, $\mathcal{H}_{\widetilde{H}} = \mathcal{H}_H = \mathcal{H}_0$ completing the proof. □

In the following we call a densely defined closed symmetric operator $\dot{H}$ with deficiency indices (k,k), $k \in \mathbb{N} \cup \{\infty\}$ *prime* if $\mathcal{H}_0^\perp = \{0\}$ in the decomposition (4.50).

Given these preliminaries we can now discuss a model for the pair $(\dot{H}, H)$.

Theorem 4.6. *Let $\dot{H}$ be a densely defined closed prime symmetric operator in a separable complex Hilbert space $\mathcal{H}$. Assume H to be a self-adjoint extension of $\dot{H}$ in $\mathcal{H}$ with $\{E_H(\lambda)\}_{\lambda\in\mathbb{R}}$ the associated family of strongly right-continuous orthogonal spectral projections of H and define the unitary operator $\widetilde{U} : \widehat{\mathcal{H}} = L^2(\mathbb{R}, \mathcal{N}_+; d\Omega_{H,\mathcal{N}_+}) \to \mathcal{H}$ as the operator $\widetilde{U}$ in Lemma 2.8, where*

$$\Omega_{H,\mathcal{N}_+}(\lambda) = (1 + \lambda^2)(P_{\mathcal{N}_+} E_H(\lambda) P_{\mathcal{N}_+}|_{\mathcal{N}_+}), \tag{4.67}$$

with $P_{\mathcal{N}_+}$ the orthogonal projection onto $\mathcal{N}_+ = \ker(\dot{H}^* - i)$. Then the pair $(\dot{H}, H)$ is unitarily equivalent to the pair $\widehat{\dot{H}}, \widehat{H})$,

$$\dot{H} = \widetilde{U}\widehat{\dot{H}}\widetilde{U}^{-1}, \quad H = \widetilde{U}\widehat{H}\widetilde{U}^{-1}, \tag{4.68}$$

where $\widehat{H}$ and $\widehat{\dot{H}}$ are defined in (4.32)–(4.37), and Theorem 4.4, and $\mathcal{N}$ is identified with $\mathcal{N}_+$, etc. Moreover,

$$\widetilde{U}\widehat{\mathcal{N}}_+ = \mathcal{N}_+, \tag{4.69}$$

where

$$\widehat{\mathcal{N}}_+ = \overline{\textit{lin.span}\{\underline{\underline{u}}_{+,n} \in \widehat{\mathcal{H}} \,|\, \underline{\underline{u}}_{+,n}(\lambda) = (\lambda - i)^{-1}u_{+,n},\, \lambda \in \mathbb{R},\, n \in \mathcal{I}\}}, \tag{4.70}$$

with $\{u_{+,n}\}_{n\in\mathcal{I}}$ a complete orthonormal system in $\mathcal{N}_+ = \ker(\dot{H}^* - i)$.

Proof. Consider $\underline{\underline{u}}_{+,n}(\lambda) = (\lambda - i)^{-1}u_{+,n}$, $n \in \mathcal{I}$, then

$$\widetilde{U}\underline{\underline{u}}_{+,n} = \int_{\mathbb{R}} dE_H(\lambda)u_{+,n} = u_{+,n}, \quad n \in \mathcal{I} \tag{4.71}$$

proves (4.69). Moreover,

$$((\widehat{H} - z)^{-1}\underline{\underline{u}}_{+,n})(\lambda) = (\lambda - z)^{-1}(\lambda - i)^{-1}u_{+,n}, \quad n \in \mathcal{I},\, z \in \mathbb{C}\backslash\mathbb{R} \tag{4.72}$$

yields

$$\widetilde{U}(\widehat{H} - z)^{-1}\underline{\underline{u}}_{+,n} = \int_{\mathbb{R}} dE_H(\lambda)(\lambda - z)^{-1}u_{+,n} = (H - z)^{-1}u_{+,n}, \quad n \in \mathcal{I}. \tag{4.73}$$

Since by hypothesis $\dot{H}$ is a prime symmetric operator, finite linear combinations of the right-hand side in (4.73) are dense in $\mathcal{H}$. Since $\widetilde{U}$ is unitary, also finite linear combinations of $(\widehat{H} - z)^{-1}\underline{\underline{u}}_{+,n}$ on the left-hand side of (4.73) are dense in $\widehat{\mathcal{H}}$. Using the first resolvent equation one computes from (4.73)

$$\begin{aligned} \widetilde{U}(\widehat{H} - z)^{-1}\widetilde{U}^{-1}\widetilde{U}(\widehat{H} - z')^{-1}\underline{\underline{u}}_{+,n} &= \widetilde{U}(\widehat{H} - z)^{-1}\widetilde{U}^{-1}(H - z')^{-1}u_{+,n} \\ &= (H - z)^{-1}(H - z')^{-1}u_{+,n}. \end{aligned} \tag{4.74}$$

Since finite linear combinations of the form $(H - z')^{-1}u_{+,n}$ are dense in $\mathcal{H}$ we get

$$\widetilde{U}(\widehat{H} - z)^{-1}\widetilde{U}^{-1} = (H - z)^{-1}, \quad z \in \mathbb{C}\backslash\mathbb{R}. \tag{4.75}$$

(4.69) and (4.75) then yield $\widetilde{U}\widehat{\dot{H}}\widetilde{U}^{-1} = \dot{H}$. □

If $\dot{H}$ is a densely defined closed non-prime symmetric operator in $\mathcal{H}$, then in addition to (4.50), (4.52), and (4.53) one obtains

$$\dot{H} = \dot{H}_0 \oplus H_0^{\perp}, \quad \mathcal{N}_+ = \mathcal{N}_{0,+} \oplus \{0\} \tag{4.76}$$

with respect to the decomposition $\mathcal{H} = \mathcal{H}_0 \oplus \mathcal{H}_0^\perp$. In particular, the part $H_0^\perp$ of $\dot{H}$ in $\mathcal{H}_0^\perp$ is self-adjoint. For any closed linear subspace $\mathcal{N}$ of $\mathcal{N}_+$, $\mathcal{N} \subseteq \mathcal{N}_+$, one then infers $\mathcal{N} = \mathcal{N}_0 \oplus \{0\}$, $P_\mathcal{N} = P_{\mathcal{N}_0} \oplus 0$ and hence

$$M_{H,\mathcal{N}}(z) = M_{H_0,\mathcal{N}_0}(z), \quad z \in \mathbb{C}\backslash\mathbb{R}. \tag{4.77}$$

This reduces the H-dependent spectral properties of the Weyl-Titchmarch operator effectively to that of H_0, where $H = H_0 \oplus H_0^\perp$ is a self-adjoint extension of $\dot{H}$ in $\mathcal{H}$.

Next we digress a bit to the special case where $\dot{H} \geq 0$ and characterize Friedrichs and Krein extensions, H_F and H_K, of $\dot{H}$ in $\mathcal{H}$. Assuming $\dot{H}$ to be densely defined in $\mathcal{H}$ we recall the definition of H_F and H_K (cf., e.g., [7]),

$$\begin{aligned} \operatorname{dom}(H_F^{1/2}) = \{f \in \mathcal{H} \,|\, &\text{there is a } \{f_n\}_{n\in\mathbb{N}} \subset \operatorname{dom}(\dot{H}) \text{ s.t. } \lim_{n\to\infty} \|f_n - f\|_\mathcal{H} = 0 \\ &\text{and } \lim_{m,n\to\infty} ((f_n - f_m), \dot{H}(f_n - f_m))_\mathcal{H} = 0\}, \end{aligned}$$

$$H_F = \dot{H}^*\big|_{\operatorname{dom}(\dot{H}^*)\cap\operatorname{dom}(H_F^{1/2})}, \tag{4.78}$$

$$\begin{aligned} \operatorname{dom}(H_K) = \{f \in \operatorname{dom}(\dot{H}^*) \,|\, &\text{there is a } \{f_n\}_{n\in\mathbb{N}} \subset \operatorname{dom}(\dot{H}) \text{ s.t.} \\ &\lim_{n\to\infty} \|\dot{H} f_n - \dot{H}^* f\|_\mathcal{H} = 0 \text{ and } \lim_{m,n\to\infty} ((f_n - f_m), \dot{H}(f_n - f_m))_\mathcal{H} = 0\}, \end{aligned}$$

$$H_K = \dot{H}^*\big|_{\operatorname{dom}(H_K)}. \tag{4.79}$$

Moreover, we recall that

$$\inf \operatorname{spec}(H_F) = \inf\{(g, \dot{H} g)_\mathcal{H} \in \mathbb{R} \,|\, g \in \operatorname{dom}(\dot{H}),\ \|g\|_\mathcal{H} = 1\} \geq 0, \tag{4.80}$$

$$\inf \operatorname{spec}(H_K) = 0, \tag{4.81}$$

and

$$0 \leq (H_F - \mu)^{-1} \leq (\widetilde{H} - \mu)^{-1} \leq (H_K - \mu)^{-1}, \quad \mu < 0 \tag{4.82}$$

for any nonnegative self-adjoint extension $\widetilde{H} \geq 0$ of $\dot{H}$.

Next we discuss a slight refinement of a result of Krein [39] (see also [8], [65], [66]). We will use an efficient summary of Krein's result due to Skau [61] (cf. also [43]), which appears most relevant in our context.

Theorem 4.7. *Let $\dot{H} \geq 0$ be a densely defined closed nonnegative operator in $\mathcal{H}$ with deficiency subspaces $\mathcal{N}_\pm = \ker(\dot{H}^* \mp i)$. Suppose H is a self-adjoint extension of $\dot{H}$ in $\mathcal{H}$ with corresponding family of orthogonal spectral projection $\{E_H(\lambda)\}_{\lambda\in\mathbb{R}}$ and define*

$$\Omega_{H,\mathcal{N}_+}(\lambda) = (1 + \lambda^2)(P_{\mathcal{N}_+} E_H(\lambda) P_{\mathcal{N}_+}\big|_{\mathcal{N}_+}). \tag{4.83}$$

Denote by H_F and H_K the Friedrichs and Krein extension of $\dot{H}$, respectively. Then (i) $H = H_F$ if and only if $\int_R^\infty d\|E_H(\lambda)u_+\|_\mathcal{H}^2 \lambda = \infty$, or equivalently, if and only if $\int_R^\infty d(u_+, \Omega_{H,\mathcal{N}_+}(\lambda)u_+)_{\mathcal{N}_+} \lambda^{-1} = \infty$ for all $R > 0$ and all $u_+ \in \mathcal{N}_+\backslash\{0\}$.

(ii) $H = H_K$ *if and only if* $\int_0^R d||E_H(\lambda)u_+||^2_{\mathcal{H}}\lambda^{-1} = \infty$, *or equivalently, if and only if* $\int_0^R d(u_+, \Omega_{H,\mathcal{N}_+}(\lambda)u_+)_{\mathcal{N}_+}\lambda^{-1} = \infty$ *for all* $R > 0$ *and all* $u_+ \in \mathcal{N}_+\backslash\{0\}$.
(iii) $H = H_F = H_K$ *if and only if* $\int_R^\infty d||E_H(\lambda)u_+||^2_{\mathcal{H}}\lambda = \int_0^R d||E_H(\lambda)u_+||^2_{\mathcal{H}}\lambda^{-1} = \infty$, *or equivalently, if and only if for all* $R > 0$ *and all* $u_+ \in \mathcal{N}_+ \in \mathcal{N}_+\backslash\{0\}$, $\int_R^\infty d(u_+, \Omega_{H,\mathcal{N}_+}(\lambda)u_+)_{\mathcal{N}_+}\lambda^{-1} = \int_R^\infty d(u_+, \Omega_{H,\mathcal{N}_+}(\lambda)u_+)_{\mathcal{N}_+}\lambda^{-1} = \infty$.

Proof. By Lemma 4.5 and (4.76) we may assume that $\dot{H}$ is a prime symmetric operator. Moreover, by Theorem 4.6 we may identify $(\dot{H}, H)$ in $\mathcal{H}$ with the model pair $(\widehat{\dot{H}}, \widehat{H})$ in $\widehat{\mathcal{H}} = L^2(\mathbb{R}, \mathcal{N}_+; d\Omega_{H,\mathcal{N}_+})$. Since by (4.70),

$$\widehat{\mathcal{N}}_+ = \overline{\text{lin.span}\{\underline{u}_{+,n} = \{(\lambda - i)^{-1}u_{+,n}\}_{\lambda\in\mathbb{R}} \in \widehat{\mathcal{H}} \,|\, n \in \mathcal{I}\}}, \tag{4.84}$$

statements (i)–(iii) are reduced to those in Krein [39], respectively Skau [61], who use $\ker(\dot{H}^* + 1)$ instead of $\mathcal{N}_+ = \ker(\dot{H}^* - i)$, by utilizing the elementary identity $(\lambda+1)^{-1} = (\lambda-i)^{-1} - (1+i)(\lambda+1)^{-1}(\lambda-i)^{-1}$ and the fact that $\{(\lambda+1)^{-1}(\lambda-i)^{-1}u_{+,n}\}_{\lambda\in\mathbb{R}} \in \widehat{\mathcal{H}} = L^2(\mathbb{R}, \mathcal{N}_+; d\Omega_{H,\mathcal{N}_+})$ for all $n \in \mathcal{I}$. □

Corollary 4.8. ([22], [23], [24], [41], [67].)
(i) $H = H_F$ *if and only if* $\lim_{\lambda\downarrow-\infty}(u_+, M_{H,\mathcal{N}_+}(\lambda)u_+)_{\mathcal{N}_+} = -\infty$ *for all* $u_+ \in \mathcal{N}_+\backslash\{0\}$.
(ii) $H = H_K$ *if and only if* $\lim_{\lambda\uparrow 0}(u_+, M_{H,\mathcal{N}_+}(\lambda)u_+)_{\mathcal{N}_+} = \infty$ *for all* $u_+ \in \mathcal{N}_+\backslash\{0\}$.
(iii) $H = H_F = H_K$ *if and only if* $\lim_{\lambda\downarrow-\infty}(u_+, M_{H,\mathcal{N}_+}(\lambda)u_+)_{\mathcal{N}_+} = -\infty$ *and* $\lim_{\lambda\uparrow 0}(u_+, M_{H,\mathcal{N}_+}(\lambda)u_+)_{\mathcal{N}_+} = \infty$ *for all* $u_+ \in \mathcal{N}_+\backslash\{0\}$.

Proof. Since

$$\begin{aligned} M_{H,\mathcal{N}_+}(z) &= zI_{\mathcal{N}_+} + (1+z^2)P_{\mathcal{N}_+}(H-z)^{-1}P_{\mathcal{N}_+}\big|_{\mathcal{N}_+} \\ &= \int_{\mathbb{R}} d\Omega_{H,\mathcal{N}_+}(\lambda)((\lambda - z)^{-1} - \lambda(1+\lambda^2)^{-1}), \quad z \in \mathbb{C}\backslash[0,\infty) \end{aligned} \tag{4.85}$$

by (4.83), it suffices to involve Theorem 4.7 (i)–(iii) and the monotone convergence theorem. □

As a simple illustration we mention the following

Example 4.9. Consider the following operator $\dot{H}$ in $L^2(\mathbb{R}^n; d^n x)$,

$$\dot{H} = \overline{-\Delta\big|_{C_0^\infty(\mathbb{R}^n\backslash\{0\})}} \geq 0, \quad n = 2, 3. \tag{4.86}$$

Then

$$H_F = H_K = -\Delta, \quad \text{dom}(-\Delta) = H^{2,2}(\mathbb{R}^2) \text{ if } n = 2 \tag{4.87}$$

is the unique nonnegative self-adjoint extension of $\dot{H}$ in $L^2(\mathbb{R}^2; d^2x)$ and

$$H_F = -\Delta, \quad \text{dom}(-\Delta) = H^{2,2}(\mathbb{R}^3) \text{ if } n = 3, \tag{4.88}$$

$$H_K = Uh_0^N U^{-1} \oplus \bigoplus_{\ell\in\mathbb{N}} Uh_\ell U^{-1} \text{ if } n = 3. \tag{4.89}$$

Here $H^{p,q}(\mathbb{R}^n)$, $p,q \in \mathbb{N}$ denote the usual Sobolev spaces,

(4.90)
$$h_0^N = -\frac{d^2}{dr^2}, \quad r>0,$$
$$\operatorname{dom}(h_0^N) = \{f \in L^2((0,\infty);dr) \,|\, f,f' \in AC([0,R]) \text{ for all } R>0;\ f'(0_+)=0; \\ f'' \in L^2((0,\infty);dr)\},$$

(4.91)
$$h_\ell = -\frac{d^2}{dr^2} + \frac{\ell(\ell+1)}{r^2}, \quad r>0,\ \ell\in\mathbb{N},$$
$$\operatorname{dom}(h_\ell) = \{f \in L^2((0,\infty);dr) \,|\, f,f' \in AC([0,R]) \text{ for all } R>0;\ f(0_+)=0; \\ -f''+\ell(\ell+1)r^{-2}f \in L^2((0,\infty);dr)\},$$

and U denotes the unitary operator,

$$U: L^2((0,\infty);dr) \to L^2((0,\infty);r^2dr), \quad f(r)\to r^{-1}f(r). \tag{4.92}$$

Equations (4.87)–(4.89) follow from Corollary 4.8 and the facts

$$(u_+, M_{H_F,\mathcal{N}_+}(z)u_+)_{L^2(\mathbb{R}^n;d^nx)} = \begin{cases} -(2/\pi)\ln(z)+2i, & n=2,\\ i(2z)^{1/2}+1, & n=3,\end{cases} \tag{4.93}$$

and

$$(u_+, M_{H_K,\mathcal{N}_+}(z)u_+)_{L^2(\mathbb{R}^3;d^3x)} = i(2/z)^{1/2}-1. \tag{4.94}$$

Here

(4.95)
$$\mathcal{N}_+ = \text{lin.span}\{u_+\}, \quad u_+(x) = G_0(i,x,0)/\|G_0(i,\cdot,0)\|_{L^2(\mathbb{R}^n;d^nx)}, \ x\in\mathbb{R}^n\backslash\{0\},$$

where

$$G_0(z,x,y) = \begin{cases} \frac{i}{4}H_0^{(1)}(z^{1/2}|x-y|), & x\neq y,\ n=2,\\ e^{iz^{1/2}|x-y|}/(4\pi|x-y|), & x\neq y,\ n=3\end{cases} \tag{4.96}$$

denotes the Green's function of $-\Delta$ on $H^{2,2}(\mathbb{R}^n)$, $n=2,3$ (i.e., the integral kernel of the resolvent $(-\Delta-z)^{-1}$) and $H_0^{(1)}(\zeta)$ abbreviates the Hankel function of the first kind and order zero (cf., [1], Sect. 9.1). Equation (4.93) then immediately follows from repeated use of the identity (the first resolvent equation),

$$\int_{\mathbb{R}^n} d^nx'\, G_0(z_1,x,x')G_0(z_2,x',0) = (z_1-z_2)^{-1}(G_0(z_1,x,0)-G_0(z_2,x,0)),$$

(4.97)
$$x\neq 0,\ z_1\neq z_2,\ n=2,3$$

and its limiting case as $x\to 0$. Finally, (4.94) follows from the following arguments. First one notices that $(-(d^2/dr^2)+\nu r^{-2})|_{C_0^\infty((0,\infty))}$ is essentially self-adjoint if and only if $\nu\geq 3/4$. Hence it suffices to consider the restriction of $\dot H$ to the centrally symmetric subspace of $L^2(\mathbb{R}^3; d^3x)$ corresponding to angular momentum $\ell = 0$. But then it is a well-known fact (cf. Lemma 5.3) that the Dirichlet Donoghue m-function $(u_+, M_{H_F,\mathcal{N}_+}(z)u_+)_{L^2(\mathbb{R}^n;d^nx)}$ corresponding to

$$\begin{aligned} &h_0^D = -\frac{d^2}{dr^2}, \quad r > 0, \\ &\operatorname{dom}(h_0^N) = \{f \in L^2((0,\infty); dr) \mid f, f' \in AC([0,R]) \text{ for all } R > 0;\ f(0_+) = 0; \\ &\qquad\qquad f'' \in L^2((0,\infty); dr)\}, \end{aligned} \tag{4.98}$$

and the Neumann Donoghue m-function $(u_+, M_{H_N,\mathcal{N}_+}(z)u_+)_{L^2(\mathbb{R}^n;d^nx)}$ corresponding to h_0^N in (4.90) are related to each other by (5.29), with $\alpha = \pi/2$, $\beta = \pi/4$, proving (4.94).

Further explicit examples of Krein extensions can be found in [6] and the references therein. All self-adjoint extensions of $\dot{H}$ are described in [5], Section I.1.1 and Ch.1.5. Generalized Friedrichs and Krein extensions in the case where $\dot{H}$ has deficiency indices $(1,1)$ and $\dot{H}$ is not necessarily assumed to be bounded from below, are studied in detail in [32]–[35]. Interesting inverse spectral problems associated with self-adjoint extensions of symmetric operators with gaps were studied in the series of papers [4], [13]–[16].

Finally we discuss some realization theorems for Herglotz operators of the form (4.85). For this purpose introduce the following set of Herglotz operators,

$$\begin{aligned} \mathcal{N}_0(\mathcal{N}) = \{&M \in \mathcal{B}(\mathcal{N}) \text{ Herglotz} \mid M(z) = \int_{\mathbb{R}} d\Omega(\lambda)((\lambda - z)^{-1} - \lambda(1+\lambda^2)^{-1}); \\ &\widetilde{\Omega}(\mathbb{R}) = I_{\mathcal{N}};\ \text{for all } \xi \in \mathcal{N}\backslash\{0\},\ \int_{\mathbb{R}} d(\xi, \Omega(\lambda)\xi)_{\mathcal{N}} = \infty\}, \end{aligned} \tag{4.99}$$

$$\begin{aligned} \mathcal{N}_{0,F}(\mathcal{N}) = \{&M \in \mathcal{N}_0(\mathcal{N}) \mid \operatorname{supp}(\Omega) \subseteq [0,\infty);\ \text{for all } \xi \in \mathcal{N}\backslash\{0\}, \\ &\int_R^\infty d(\xi, \Omega(\lambda)\xi)_{\mathcal{N}}\lambda^{-1} = \infty \text{ for some } R > 0\}, \end{aligned} \tag{4.100}$$

$$\begin{aligned} \mathcal{N}_{0,K}(\mathcal{N}) = \{&M \in \mathcal{N}_0(\mathcal{N}) \mid \operatorname{supp}(\Omega) \subseteq [0,\infty);\ \text{for all } \xi \in \mathcal{N}\backslash\{0\}, \\ &\int_0^R d(\xi, \Omega(\lambda)\xi)_{\mathcal{N}}\lambda^{-1} = \infty \text{ for some } R > 0\}, \end{aligned} \tag{4.101}$$

$$\begin{aligned} \mathcal{N}_{0,F,K}(\mathcal{N}) = \{&M \in \mathcal{N}_0(\mathcal{N}) \mid \operatorname{supp}(\Omega) \subseteq [0,\infty);\ \text{for all } \xi \in \mathcal{N}\backslash\{0\}, \\ &\int_R^\infty d(\xi, \Omega(\lambda)\xi)_{\mathcal{N}}\lambda^{-1} = \int_0^R d(\xi, \Omega(\lambda)\xi)_{\mathcal{N}}\lambda^{-1} = \infty \text{ for some } R > 0\} \\ = \mathcal{N}&_{0,F}(\mathcal{N}) \cap \mathcal{N}_{0,K}(\mathcal{N}), \end{aligned} \tag{4.102}$$

where $\mathcal{N}$ is a separable complex Hilbert space, $\mathrm{supp}(\Omega)$ denotes the topological support of Ω, and $\widetilde{\Omega}(\lambda) = (1+\lambda^2)^{-1}\Omega(\lambda)$, $\lambda \in \mathbb{R}$.

Theorem 4.10. *(i) Any $M \in \mathcal{N}_0(\mathcal{N})$ can be realized in the form*

$$M(z) = V^*(zI_{\mathcal{N}_+} + (1+z^2)P_{\mathcal{N}_+}(H-z)^{-1}P_{\mathcal{N}_+}|_{\mathcal{N}_+})V, \quad z \in \mathbb{C}\backslash\mathbb{R}, \tag{4.103}$$

where H denotes a self-adjoint extension of some densely defined closed symmetric operator $\dot{H}$ with deficiency subspaces $\mathcal{N}_\pm$ in some separable Hilbert space $\mathcal{H}$.
(ii) Any $M \in \mathcal{N}_{0,F(resp.K)}(\mathcal{N})$ can be realized in the form

$$M(z) = V^*(zI_{\mathcal{N}_+} + (1+z^2)P_{\mathcal{N}_+}(H_{F(resp.K)}-z)^{-1}P_{\mathcal{N}_+}|_{\mathcal{N}_+})V, \quad z \in \mathbb{C}\backslash\mathbb{R}, \tag{4.104}$$

where $H_{F(resp.K)} \geq 0$ denotes the Friedrichs (respectively, Krein) extension of some densely defined closed symmetric operator $\dot{H}$ with deficiency subspaces $\mathcal{N}_\pm$ in some separable complex Hilbert space $\mathcal{H}$.
(iii) Any $M \in \mathcal{N}_{0,F,K}(\mathcal{N})$ can be realized in the form

$$M(z) = V^*(zI_{\mathcal{N}_+} + (1+z^2)P_{\mathcal{N}_+}(H_{F,K}-z)^{-1}P_{\mathcal{N}_+}|_{\mathcal{N}_+})V, \quad z \in \mathbb{C}\backslash\mathbb{R}, \tag{4.105}$$

where $H_{F,K} \geq 0$ denotes the unique nonnegative self-adjoint extension of some densely defined closed symmetric operator $\dot{H}$ with deficiency subspaces $\mathcal{N}_\pm$ in some separable complex Hilbert space $\mathcal{H}$.

In all cases (i)–(iii), V denotes a unitary operator from $\mathcal{N}$ to $\mathcal{N}_+$.

Proof. (i) Define

$$V : \mathcal{N} \to \widehat{\mathcal{N}}_+, \quad \xi \longrightarrow (\cdot - i)^{-1}\xi \tag{4.106}$$

and use the notation developed for the model pair $(\widehat{\dot{H}}, \widehat{H})$ in (4.32)–(4.37), Theorem 4.4, and Theorem 4.6. Then

$$(V\xi, V\eta)_{\widehat{\mathcal{N}}_+} = \int_{\mathbb{R}} d(\xi, \Omega(\lambda)\eta)_{\mathcal{N}}(1+\lambda^2)^{-1} = (\xi,\eta)_{\mathcal{N}}, \quad \xi,\eta \in \mathcal{N} \tag{4.107}$$

shows that V is a linear isometry from $\mathcal{N}$ into $\widehat{\mathcal{H}}_+$,

$$V^*V = I_{\mathcal{N}}, \quad \mathrm{ran}(V^*) = \mathcal{N}. \tag{4.108}$$

By (4.84) (identifying $\mathcal{N}_+$ and $\mathcal{N}$),

$$V^{-1} : \widehat{\mathcal{N}}_+ \to \mathcal{N}, \quad (\cdot - i)^{-1}\xi \longrightarrow \xi \tag{4.109}$$

is also a linear isometry from $\widehat{\mathcal{N}}_+$ into $\mathcal{N}$, implying

$$VV^* = I_{\widehat{\mathcal{N}}_+}, \quad \mathrm{ran}(V) = \widehat{\mathcal{N}}_+. \tag{4.110}$$

Thus V is unitary and one computes

$$\begin{aligned}&(\xi, V^*(zI_{\widehat{\mathcal{N}}_+} + (1+z^2)P_{\widehat{\mathcal{N}}_+}(\widehat{H}-z)^{-1}P_{\widehat{\mathcal{N}}_+}|_{\widehat{\mathcal{N}}_+})V\eta)_{\mathcal{N}}\\ &= (V\xi, (zI_{\widehat{\mathcal{N}}_+} + (1+z^2)P_{\widehat{\mathcal{N}}_+}(\widehat{H}-z)^{-1}P_{\widehat{\mathcal{N}}_+}|_{\widehat{\mathcal{N}}_+})V\eta)_{\widehat{\mathcal{N}}_+}\end{aligned}$$

$$= ((\cdot - i)^{-1}\xi, (zI_{\hat{\mathcal{N}}_+} + (1+z^2)P_{\hat{\mathcal{N}}_+}(\widehat{H} - z)^{-1}P_{\hat{\mathcal{N}}_+}|_{\hat{\mathcal{N}}_+})(\cdot - i)^{-1}\eta)_{\hat{\mathcal{N}}_+}$$

$$= \int_{\mathbb{R}} d(\xi, \Omega(\lambda)\eta)_{\mathcal{N}} z(1+\lambda^2)^{-1} + \int_{\mathbb{R}} d(\xi, \Omega(\lambda)\eta)_{\mathcal{N}}(1+z^2)(1+\lambda^2)^{-1}(\lambda - z)^{-1}$$

$$= \int_{\mathbb{R}} d(\xi, \Omega(\lambda)\eta)_{\mathcal{N}}((\lambda - z)^{-1} - \lambda(1+\lambda^2)^{-1})$$

(4.111)

$$= (\xi, M(z)\eta)_{\mathcal{N}}, \quad \xi, \eta, \in \mathcal{N}, \ z \in \mathbb{C}\backslash\mathbb{R}.$$

(ii) and (iii) then follow in the same way using Theorem 4.7. □

For a whole scale of Nevanlinna classes in the case where $\dot{H}$ has deficiency indices $(1,1)$ we refer to [37].

Remark 4.11. In the special case where $\dim_{\mathbb{C}}(\mathcal{N}) \in \mathbb{N}$, treated in detail in [30], we also considered at length the case where H and H_F (respectively, H_K) were relatively prime operators with respect to $\dot{H}$. In this case the limiting behavior of $M(z)$ as $\lambda \downarrow -\infty$ (respectively, $\lambda \uparrow 0$) crucially entered the corresponding results in Theorems 7.5–7.7 of [30]. These limits are given in terms of $\operatorname{Re}((P_{1,2}(i)|_{\mathcal{N}_+})^{-1})$ (cf. (4.15)) identifying $H_1 = H$, $H_2 = H_F$ or H_K, etc. In the present infinite-dimensional case, $(P_{1,2}(i)|_{\mathcal{N}_+})^{-1}$ exists if H_1 and H_2 are relatively prime with respect to $\dot{H}$. However, $(P_{1,2}(i)|_{\mathcal{N}_+})^{-1}$ is not necessarily a bounded operator in $\mathcal{N}_+$. In fact,

$$\operatorname{Im}((P_{1,2}(i)|_{\mathcal{N}_+})^{-1}) = -I_{\mathcal{N}_+}, \tag{4.112}$$

$$\operatorname{Re}((P_{1,2}(i)|_{\mathcal{N}_+})^{-1}) \in \mathcal{B}(\mathcal{N}_+) \text{ if and only if } \operatorname{ran}(P_{1,2}(i)) = \mathcal{N}_+ \tag{4.113}$$

as shown in Lemma 2 of [28]. This complicates matters since now the limits of $M(\lambda)$ as $\lambda \downarrow -\infty$ (or $\lambda \uparrow 0$) may exist but possibly represent unbounded self-adjoint operators in $\mathcal{N}_+$ and thus convergence of $M(\lambda)$ as $\lambda \downarrow -\infty$ (or $\lambda \uparrow 0$) in these cases is understood in the strong resolvent sense. A detailed treatment of this topic goes beyond the scope of this paper and is thus postponed.

Theorem 4.12. *Suppose $M_\ell \in \mathcal{N}_0(\mathcal{N})$, $\ell = 1,2$ and $M_1 \neq M_2$. Then M_1 and M_2 can be realized as*

(4.114)

$$M_\ell(z) = V^*(zI_{\mathcal{N}_+} + (1+z^2)P_{\mathcal{N}_+}(H_\ell - z)^{-1}P_{\mathcal{N}_+}|_{\mathcal{N}_+})V, \quad \ell = 1,2, \ z \in \mathbb{C}\backslash\mathbb{R},$$

where H_ℓ, $\ell = 1,2$ are distinct self-adjoint extensions of one and the same densely defined closed symmetric operator $\dot{H}$ with deficiency subspaces $\mathcal{N}_\pm$ in some separable complex Hilbert space $\mathcal{H}$, and V denotes a unitary operator from $\mathcal{N}$ to $\mathcal{N}_+$, if and only if,

(4.115)

$$M_2(z) = e^{-i\alpha}(\cos(\alpha) + \sin(\alpha)M_1(z))(\sin(\alpha) - \cos(\alpha)M_1(z))^{-1}e^{i\alpha}, \quad z \in \mathbb{C}\backslash\mathbb{R}$$

for some self-adjoint operator α in $\mathcal{N}$.

Proof. Assuming (4.114), (4.115) is clear from (4.30). Conversely, assume (4.115). By Theorem 4.13 (i), we may realize $M_1(z)$ as

$$M_1(z) = V^*(zI_{\mathcal{N}_+} + (1+z^2)P_{\mathcal{N}_+}(H_1 - z)^{-1}P_{\mathcal{N}_+}|_{\mathcal{N}_+})V, \quad z \in \mathbb{C}\backslash\mathbb{R}. \tag{4.116}$$

If $\widetilde{H} \neq H_1$ is another self-adjoint extension of $\dot{H}$ we introduce

$$\widetilde{M}(z) = V^*(zI_{\mathcal{N}_+} + (1+z^2)P_{\mathcal{N}_+}(\widetilde{H} - z)^{-1}P_{\mathcal{N}_+}|_{\mathcal{N}_+})V, \quad z \in \mathbb{C}\backslash\mathbb{R}, \tag{4.117}$$

and infer from Theorem 4.3,

$$\widetilde{M}(z) = e^{-i\widetilde{\alpha}}(\cos(\widetilde{\alpha}) + \sin(\widetilde{\alpha})M_1(z))(\sin(\widetilde{\alpha}) - \cos(\widetilde{\alpha})M_1(z))^{-1}e^{i\widetilde{\alpha}}, \quad z \in \mathbb{C}\backslash\mathbb{R} \tag{4.118}$$

for some $\widetilde{\alpha} = \widetilde{\alpha}^*$ in $\mathcal{N}$.

Since $(H_1 - z)(H_1 \pm i)^{-1}$ are bounded and boundedly invertible, $P_{1,2}(z)$ in (4.4) uniquely characterizes all self-adjoint extensions $H_2 \neq H_1$ of $\dot{H}$. Moreover, by (4.5)–(4.7) and von Neumann's representation of self-adjoint extensions in terms of Cayley transforms, all self-adjoint extensions $H_2 \neq H_1$ of $\dot{H}$ are in a bijective correspondence to all self-adjoint (possibly unbounded) operators $\alpha_{1,2}$ ($\alpha_{1,2} \neq \pi/2$) in $\mathcal{N}_+$. Hence we may choose $\widetilde{H}$ such that $\widetilde{\alpha}$ equals α in (4.115) implying $\widetilde{M}(z) = M_2(z)$. □

We conclude with a result on analytic continuations of general Herglotz operators from $\mathbb{C}_+$ into a subset of $\mathbb{C}_-$ through an interval of the real line, which is independent of our emphasis of perturbation problems in Section 3 and self-adjoint extensions in the present Section 4. As is well known, the usual convention for $M|_{\mathbb{C}_-}$ by means of reflection as in (1.4), in general, does not represent the analytic continuation of $M|_{\mathbb{C}_+}$. The following result is an adaptation of a theorem of Greenstein [31] for scalar Herglotz functions to the present operator-valued context.

Lemma 4.13. *Let $\mathcal{K}$ be a separable complex Hilbert space and M be a Herglotz operator in $\mathcal{K}$ with representation (1.1)–(1.3). Suppose that the operator Stieltjes integral in (1.1) converges in the strong operator topology of $\mathcal{K}$ and let $(\lambda_1, \lambda_2) \subseteq \mathbb{R}$, $\lambda_1 < \lambda_2$. Then a necessary condition for M to have an analytic continuation from $\mathbb{C}_+$ into a subset of $\mathbb{C}_-$ through the interval (λ_1, λ_2) is that for all $\xi \in \mathcal{K}$, the associated scalar measures $\omega_\xi = (\xi, \Omega\,\xi)_\mathcal{K}$ are purely absolutely continuous on (λ_1, λ_2), $\omega_\xi|_{(\lambda_1,\lambda_2)} = \big(\omega_\xi|_{(\lambda_1,\lambda_2)}\big)_{ac}$, and the corresponding density $\omega'_\xi \geq 0$ of ω_ξ is real-analytic on (λ_1, λ_2). If $\mathcal{K}$ is finite-dimensional, this condition is also sufficient. If M has such an analytic continuation into some domain $\mathcal{D}_- \subseteq \mathbb{C}_-$, then it is given by*

$$M(z) = M(\overline{z})^* + 2\pi i\Omega'(z), \quad z \in \mathcal{D}_-, \tag{4.119}$$

where $\Omega'(z)$ denotes the complex-analytic extension of $\Omega'(\lambda)$ for $\lambda \in (\lambda_1, \lambda_2)$. In particular, M can be analytically continued through (λ_1, λ_2) by reflection, that is, $M(z) = M(\overline{z})^*$ for all $z \in \mathbb{C}_-$ if Ω has no support in (λ_1, λ_2).

Proof. Suppose M has an analytic continuation from $\mathbb{C}_+$ into a subset of $\mathbb{C}_-$ through the interval (λ_1, λ_2). Then for all $\xi \in \mathcal{K}$, Greenstein's result [31] applies to the scalar Herglotz function $m_\xi(z) = (\xi, M(z)\xi)_\mathcal{K}$, $\xi \in \mathcal{K}$ associated to the measure $\omega_\xi = (\xi, \Omega\,\xi)_\mathcal{K}$. Consequently, m_ξ has an analytic continuation from $\mathbb{C}_+$ into a subset of $\mathbb{C}_-$ through the interval (λ_1, λ_2) if and only if the associated scalar measure $\omega_\xi = (\xi, \Omega\,\xi)_\mathcal{K}$ is purely absolutely continuous on (λ_1, λ_2), $\omega_\xi\big|_{(\lambda_1,\lambda_2)} = \big(\omega_\xi\big|_{(\lambda_1,\lambda_2)}\big)_{ac}$, and the corresponding density $\omega'_\xi \geq 0$ of ω_ξ is real-analytic on (λ_1, λ_2). In this case the analytic continuation of m_ξ into some domain $\mathcal{D}_{-,\xi} \subseteq \mathbb{C}_-$ is given by

$$m_\xi(z) = m_\xi(\overline{z})^* + 2\pi i \omega'_\xi(z), \quad z \in \mathcal{D}_{-,\xi}, \tag{4.120}$$

where $\omega'_\xi(z)$ denotes the complex-analytic extension of $\omega'_\xi(\lambda)$ for $\lambda \in (\lambda_1, \lambda_2)$. This can be seen as follows: If m_x can be analytically continued through (λ_1, λ_2) into some region $\mathcal{D}_- \subseteq \mathbb{C}_-$, then $\widetilde{m}_\xi(z) := m_\xi(z) - \pi i \omega'_\xi(z)$ is real-analytic on (λ_1, λ_2) and hence can be continued through (λ_1, λ_2) by reflection. Similarly, $\omega'_\xi(z)$, being real-analytic, can be continued through (λ_1, λ_2) by reflection. Hence (4.120) follows from

$$m_\xi(z) - \pi i \omega'_\xi(z) = \widetilde{m}_\xi(z) = \overline{\widetilde{m}_\xi(\overline{z})} = \overline{m_\xi(\overline{z})} + \pi i \omega'_\xi(z), \quad z \in \mathcal{D}_-. \tag{4.121}$$

Applying a standard polarization argument, we obtain that the analytic continuation of $m_{\xi,\eta}(z) = (\xi, M(z)\eta)_\mathcal{K}$, $\xi, \eta \in \mathcal{K}$ into some domain $\mathcal{D}_{-,\xi,\eta} \subseteq \mathbb{C}_-$ is given by

$$m_{\xi,\eta}(z) = m_{\xi,\eta}(\overline{z})^* + 2\pi i \omega'_{\xi,\eta}(z), \quad z \in \mathcal{D}_{-,\xi,\eta}, \tag{4.122}$$

where $\omega'_{\xi,\eta}(z) = (\xi, \Omega'(z)\,\eta)_\mathcal{K}$ is related to $\omega'_{\xi\pm\eta}(z)$ and $\omega'_{\xi\pm i\eta}(z)$ by polarization. In particular, if $M(z)$ has such an analytic continuation through the interval (λ_1, λ_2) it is necessarily of the form stated in (4.119). If $\dim_\mathbb{C}(\mathcal{K}) < \infty$, then (4.120) and (4.121) yield the weak and hence $\mathcal{B}(\mathcal{K})$-analytic continuation of M through the interval (λ_1, λ_2). □

Formula (4.119) shows that any possible singularity behavior of $M\big|_{\mathbb{C}_-}$ is determined by that of $\Omega'\big|_{\mathbb{C}_-}$ since M, being Herglotz, has no singularities in $\mathbb{C}_+$. Moreover, analytic continuations through different intervals on $\mathbb{R}$ in general, will lead to different $\Omega'(z)$ and hence to branch cuts of $M\big|_{\mathbb{C}_-}$.

5. One-dimensional applications

In our final section we consider concrete applications of the formalism of Section 4 in the special case $\dim_\mathbb{C}(\mathcal{N}_+) = 1$. We study Schrödinger operators on a half-line, compare the corresponding Donoghue and Weyl-Titchmarsh m-functions, and prove some estimates on linear functionals associated with these Schrödinger operators. We conclude with two illustrations of Livšic's result [44] on quasi-hermitian extensions in the special case of densely defined closed prime symmetric operators with deficiency indices $(1,1)$ in connection with first-order differential expressions $-id/dx$.

First we specialize some of the abstract material in Section 4 to the case of a densely defined closed prime symmetric operator $\dot{H}$ in a separable complex Hilbert space $\mathcal{H}$ with deficiency indices $(1,1)$. This case has been studied in detail by Donoghue [25] (see also [30]) and we partly follow his analysis.

Choose $u_\pm \in \ker(\dot{H}^* \mp i)$ with $\|u_\pm\|_{\mathcal{H}} = 1$ and introduce the one-parameter family H_α, $\alpha \in [0,\pi)$ of self-adjoint extensions $\dot{H}$ in $\mathcal{H}$ by

$$H_\alpha(f + c(u_+ + e^{2i\alpha}u_-)) = \dot{H}f + c(iu_+ - ie^{2i\alpha}u_-),$$

(5.1)

$$\mathrm{dom}(H_\alpha) = \{(f + c(u_+ + e^{2i\alpha}u_-)) \in \mathrm{dom}(\dot{H}^*) \,|\, f \in \mathrm{dom}(\dot{H}),\, c \in \mathbb{C}\},\ \alpha \in [0,\pi).$$

Let $\{E_{H_\alpha}(\lambda)\}_{\lambda\in\mathbb{R}}$ be the family of orthogonal spectral projections of H_α and suppose that H_α has simple spectrum for one (and hence for all) $\alpha \in [0,\pi)$. (This is equivalent to the assumption that $\dot{H}$ is a prime symmetric operator and also equivalent to the fact that u_+ is a cyclic vector for H_α for all $\alpha \in [0,\pi)$.) Next we introduce the model representation $(\widehat{\dot{H}}_\alpha, \widehat{H}_\alpha)$ for $(\dot{H}, H_\alpha)$ discussed in (4.32)–(4.37), Theorem 4.4, and Theorem 4.6. However, since in the present context $\mathcal{N}_+$ is a one-dimensional subspace of $\mathcal{H}$,

$$\mathcal{N}_+ = \mathrm{lin.span}\{u_+\}, \tag{5.2}$$

the model Hilbert space $\widehat{\mathcal{H}}_\alpha = L^2(\mathbb{R}, \mathcal{N}_+; d\Omega_{H_\alpha,\mathcal{N}_+})$, $\alpha \in [0,\pi)$ with the operator (in fact, rank-one) valued measure $\Omega_{H_\alpha,\mathcal{N}_+}$,

$$\begin{aligned} &\Omega_{H_\alpha,\mathcal{N}_+}(\lambda) = \omega_\alpha(\lambda) P_{\mathcal{N}_+}|_{\mathcal{N}_+}, \quad P_{\mathcal{N}_+} = (u_+, \cdot)u_+, \\ &\omega_\alpha(\lambda) = (1+\lambda^2)\|E_{H_\alpha}(\lambda)u_+\|^2_{\mathcal{H}}, \quad \alpha \in [0,\pi), \end{aligned} \tag{5.3}$$

can be replaced by the model space $\widetilde{\mathcal{H}}_\alpha = L^2(\mathbb{R}; d\omega_\alpha)$ with scalar measure ω_α. In particular, $\omega_\alpha(\lambda)$ can be taken as the control measure in this special case and

$$\begin{aligned} &V : \widehat{\mathcal{H}}_\alpha = L^2(\mathbb{R}, \mathcal{N}_+; d\Omega_{H_\alpha,\mathcal{N}_+}) \to \widetilde{\mathcal{H}}_\alpha = L^2(\mathbb{R}; d\omega_\alpha) \\ &\hat{f} = \{\hat{f}(\lambda) = \tilde{f}(\lambda)u_+\}_{\lambda\in\mathbb{R}} \to V\hat{f} = \tilde{f} = \{\tilde{f}(\lambda)\}_{\lambda\in\mathbb{R}} \end{aligned} \tag{5.4}$$

represents the corresponding unitary operator from $\widehat{\mathcal{H}}_\alpha = L^2(\mathbb{R}, \mathcal{N}_+; d\Omega_{H_\alpha,\mathcal{N}_+})$ to $\widetilde{\mathcal{H}}_\alpha = L^2(\mathbb{R}; d\omega_\alpha)$. Hence we translate in the following some of the results of Theorems 4.4 and 4.6 from $\widehat{\mathcal{H}}_\alpha$ to $\widetilde{\mathcal{H}}_\alpha$. However, due to the trivial nature of the unitary operator V in (5.4), we will ignore this additional isomorphism and simply keep using our $\widehat{\ }$-notation of Section 4 instead of the new $\widetilde{\ }$-notation. Thus, we consider the model Hilbert space $\widehat{\mathcal{H}}_\alpha = L^2(\mathbb{R}; d\omega_\alpha)$, $\alpha \in [0,\pi)$, where

$$\omega_\alpha(\lambda) = (1+\lambda^2)\|E_{H_\alpha}(\lambda)u_+\|^2_{\mathcal{H}}, \quad \alpha \in [0,\pi), \tag{5.5}$$

$$\int_{\mathbb{R}} d\omega_\alpha(\lambda)(1+\lambda^2)^{-1} = 1, \quad \int_{\mathbb{R}} d\omega_\alpha(\lambda) = \infty, \quad \alpha \in [0,\pi) \tag{5.6}$$

and define in $\widehat{\mathcal{H}}_\alpha$ the self-adjoint operator $\widehat{H}_\alpha$,

$$(\widehat{H}_\alpha \hat{f})(\lambda) = \lambda \hat{f}(\lambda), \quad \hat{f} \in \operatorname{dom}(\widehat{H}_\alpha) = L^2(\mathbb{R}; (1+\lambda^2)d\omega_\alpha) \tag{5.7}$$

and its densely defined and closed restriction $\widehat{\dot{H}}_\alpha$,

$$\operatorname{dom}(\widehat{\dot{H}}_\alpha) = \{\hat{f} \in \operatorname{dom}(\widehat{H}_\alpha) \,|\, \textstyle\int_{\mathbb{R}} d\omega_\alpha(\lambda)\hat{f}(\lambda) = 0\}, \quad \widehat{\dot{H}}_\alpha = \widehat{H}_\alpha\big|_{\operatorname{dom}(\widehat{\dot{H}}_\alpha)}. \tag{5.8}$$

Then

$$\ker(\widehat{\dot{H}}^* - z) = \{c(\cdot - z)^{-1} \in \widehat{\mathcal{H}}_\alpha \,|\, c \in \mathbb{C}\} \tag{5.9}$$

and the pair $(\dot{H}, H_\alpha)$ in $\mathcal{H}$ is unitarily equivalent to the pair $(\widehat{\dot{H}}_\alpha, \widehat{H}_\alpha)$ in $\widehat{\mathcal{H}}_\alpha$ (cf. Theorem 4.6). This representation of $(\dot{H}, H_\alpha)$ in terms of $(\widehat{\dot{H}}_\alpha, \widehat{H}_\alpha)$ has the advantage of very simple definitions of $\widehat{H}_\alpha$ and $\widehat{\dot{H}}_\alpha$, however, one has to pay a price since different $H_\alpha, \widehat{\dot{H}}_\alpha$ act in different Hilbert spaces $\widehat{\mathcal{H}}_\alpha$. Hence it is desirable to determine the expression for all H_α, $\alpha \in [0,\pi)$ in connection with one fixed α say, $\alpha_0 \in [0,\pi)$, in the corresponding fixed Hilbert space $\widehat{\mathcal{H}}_{\alpha_0} = L^2(\mathbb{R}; d\omega_{\alpha_0})$ and we turn our attention to this task next.

Lemma 5.1. *Fix $\alpha_0 \in [0,\pi)$ and define*

$$U_{\alpha_0} : \widehat{\mathcal{H}}_{\alpha_0} \longrightarrow \mathcal{H}, \quad \hat{f} \to U_{\alpha_0}\hat{f} = \operatorname*{s-lim}_{N\to\infty} \int_{-N}^{N} d(E_{H_{\alpha_0}}(\lambda)u_+)(\lambda - i)\hat{f}(\lambda). \tag{5.10}$$

Then U_{α_0} is a unitary operator from $\widehat{\mathcal{H}}_{\alpha_0}$ to $\mathcal{H}$ and

$$\dot{H} = U_{\alpha_0}\widehat{\dot{H}}_{\alpha_0}U_{\alpha_0}^{-1}, \quad H_{\alpha_0} = U_{\alpha_0}\widehat{H}_{\alpha_0}U_{\alpha_0}^{-1}. \tag{5.11}$$

Moreover,

$$\hat{u}_+(\lambda) = (U_{\alpha_0}^{-1}u_+)(\lambda) = (\lambda - i)^{-1}, \tag{5.12}$$

$$\hat{u}_-(\lambda) = (U_{\alpha_0}^{-1}u_-)(\lambda) = -e^{-2i\alpha_0}(\lambda + i)^{-1}, \quad \lambda \in \mathbb{R}, \tag{5.13}$$

and hence

$$\begin{aligned} &(U_{\alpha_0}^{-1}(u_+ + e^{2i\alpha}u_-))(\lambda) = 2ie^{i(\alpha-\alpha_0)}(1+\lambda^2)^{-1}(-\lambda\sin(\alpha-\alpha_0) + \cos(\alpha-\alpha_0)), \\ &\qquad \alpha \in [0,\pi),\ \lambda \in \mathbb{R}. \end{aligned} \tag{5.14}$$

Proof. (5.10) and (5.11) have been discussed in Theorem 4.6, (5.12) is clear from (5.10). From

$$U_{\alpha_0}^{-1}H_{\alpha_0}(u_+ + e^{2i\alpha_0}u_-) = U_{\alpha_0}^{-1}\dot{H}^*(u_+ + e^{2i\alpha_0}u_-) = i\hat{u}_+ - ie^{2i\alpha_0}\hat{u}_-, \tag{5.15}$$

$$\widehat{H}_{\alpha_0}(\hat{u}_+ + e^{2i\alpha_0}\hat{u}_-) = \lambda(\hat{u}_+ + e^{2i\alpha_0}\hat{u}_-), \tag{5.16}$$

and (5.12) one infers

$$i(\lambda - i)^{-1} - ie^{2i\alpha_0}\hat{u}_-(\lambda) = \lambda(\lambda - i)^{-1} + e^{2i\alpha_0}\lambda\hat{u}_-(\lambda) \tag{5.17}$$

and hence (5.13). Equation (5.14) then immediately follows from (5.12) and (5.13). □

Equation (5.14) confirms the fact that any two different self-adjoint extensions of $\dot{H}$ are relatively prime

$$\operatorname{dom}(H_\alpha) \cap \operatorname{dom}(H_\beta) = \operatorname{dom}(\dot{H}), \quad \alpha, \beta \in [0, \pi), \quad \alpha \neq \beta \tag{5.18}$$

since $\int_{\mathbb{R}} d\omega_{\alpha_0}(\lambda) = \infty$ and hence

$$\int_{\mathbb{R}} d\omega_{\alpha_0}(\lambda)\lambda^2 |U_{\alpha_0}^{-1}(u_+ + e^{2i\alpha}u_-)(\lambda)|^2 = \infty \text{ for all } \alpha \neq \alpha_0. \tag{5.19}$$

This is of course an artifact of our special hypothesis $\operatorname{def}(\widehat{H}) = (1,1)$.

Next, consider the normalized element (cf. (5.14) for $\alpha = \alpha_0$)

$$\begin{aligned} &\hat{g}_\alpha \in \big(\ker(\widehat{\dot{H}}^* - i) \dot{+} \ker(\widehat{\dot{H}}^* + i)\big) \cap \operatorname{dom}(\widehat{H}_\alpha), \\ &\hat{g}_\alpha(\lambda) = \Big(\int_{\mathbb{R}} d\omega_\alpha(\nu)(1+\nu^2)^{-2}\Big)^{-1/2} (1+\lambda^2)^{-1}, \quad \|\hat{g}_\alpha\|_{\hat{\mathcal{H}}_\alpha} = 1. \end{aligned} \tag{5.20}$$

Then

$$\operatorname{dom}(\widehat{H}_\alpha) = \operatorname{lin.span}\{\hat{g}_\alpha\} \dot{+} \operatorname{dom}(\widehat{\dot{H}}_\alpha) \tag{5.21}$$

by von Neumann's theory of self-adjoint extensions of symmetric operators (cf., e.g., [3], Ch. VII, [26], Sect. II.4, [54], Sect. 14, [55], Sect. X.1, [69]) and we may consider the linear functional $\ell_{\hat{g}_\alpha}$ on $\operatorname{dom}(\widehat{H}_\alpha)$ defined by

$$\ell_{\hat{g}_\alpha} : \operatorname{dom}(\widehat{H}_\alpha) \to \mathbb{C}, \quad \ell_{\hat{g}_\alpha}(\hat{f}) = c, \tag{5.22}$$

where

$$\hat{f} \in \operatorname{dom}(\widehat{H}_\alpha), \quad \hat{f} = c\hat{g}_\alpha + \hat{h}, \quad \hat{h} \in \operatorname{dom}(\widehat{\dot{H}}_\alpha). \tag{5.23}$$

Lemma 5.2. *Let $\alpha \in [0, \pi)$. Then*

$$\sup_{\hat{f} \in \operatorname{dom}(\hat{H}_\alpha)} \left(\frac{|\ell_{\hat{g}_\alpha}(\hat{f})|^2}{\|\hat{f}\|^2_{\hat{\mathcal{H}}_\alpha} + \|\widehat{H}_\alpha \hat{f}\|^2_{\hat{\mathcal{H}}_\alpha}} \right) = \int_{\mathbb{R}} d\omega_\alpha(\lambda)(1+\lambda^2)^{-2}. \tag{5.24}$$

Proof. By (5.6) and (5.8) one computes

$$\int_{\mathbb{R}} d\omega_\alpha(\lambda)\hat{f}(\lambda) = c\int_{\mathbb{R}} d\omega_\alpha(\lambda)\hat{g}_\alpha(\lambda) = \ell_{\hat{g}_\alpha}(\hat{f})\Big(\int_{\mathbb{R}} d\omega_\alpha(\lambda)(1+\lambda^2)^{-2}\Big)^{-1/2} \tag{5.25}$$

and hence the Cauchy-Schwarz inequality applied to

$$\begin{aligned} \Big|\int_{\mathbb{R}} d\omega_\alpha(\lambda)\hat{f}(\lambda)\Big| &\leq \Big(\int_{\mathbb{R}} d\omega_\alpha(\lambda)(1+\lambda^2)|\hat{f}(\lambda)|^2\Big)^{1/2} \Big(\int_{\mathbb{R}} d\omega_\alpha(\lambda)(1+\lambda^2)^{-1}\Big)^{1/2} \\ &= (\|\hat{f}\|^2_{\hat{\mathcal{H}}_\alpha} + \|\widehat{H}_\alpha \hat{f}\|^2_{\hat{\mathcal{H}}_\alpha})^{1/2} \end{aligned} \tag{5.26}$$

yields

$$\frac{|\ell_{\hat{g}_\alpha}(\hat{f})|^2}{||\hat{f}||^2_{\hat{\mathcal{H}}_\alpha} + ||\widehat{H}_\alpha \hat{f}||^2_{\hat{\mathcal{H}}_\alpha}} \leq \int_{\mathbb{R}} d\omega_\alpha(\lambda)(1+\lambda^2)^{-2}. \tag{5.27}$$

Since inequality (5.27) saturates for $\hat{f}_0(\lambda) = (1+\lambda^2)^{-1}$, $\hat{f}_0 \in \mathrm{dom}(\widehat{H}_\alpha)$, (5.24) is proved. □

Introducing the Donoghue-type m-function

$$m^D_\alpha(z) = \int_{\mathbb{R}} d\omega_\alpha(\lambda)((\lambda - z)^{-1} - \lambda(1+\lambda^2)^{-1}), \quad \alpha \in [0,\pi), \ z \in \mathbb{C}_+, \tag{5.28}$$

the analog of (4.17), one can prove the following result.

Lemma 5.3. (Donoghue [25].)

$$m^D_\beta(z) = \frac{-\sin(\beta-\alpha) + \cos(\beta-\alpha) m^D_\alpha(z)}{\cos(\beta-\alpha) + \sin(\beta-\alpha) m^D_\alpha(z)}, \quad \alpha, \beta \in [0,\pi), \ z \in \mathbb{C}_+. \tag{5.29}$$

Next we turn to the Schrödinger operator on the half-line $[0,\infty)$. Let $q \in L^1([0,R])$ for all $R > 0$, q real-valued and introduce the fundamental system $\phi_\gamma(z,x)$, $\theta_\gamma(z,x)$, $z \in \mathbb{C}$ of solutions of

$$-\psi''(z,x) + (q(x) - z)\psi(z,x) = 0, \quad x > 0 \tag{5.30}$$

($'$ denotes d/dx) satisfying

(5.31)
$$\phi_\gamma(z,0_+) = -\theta'_\gamma(z,0_+) = -\sin(\gamma), \quad \phi'_\gamma(z,0_+) = \theta_\gamma(z,0_+) = \cos(\gamma), \quad \gamma \in [0,\pi).$$

Assuming that $-\frac{d^2}{dx^2} + q$ is in the limit point case at ∞, let $\psi_\gamma(z,x)$ be the unique solution of (5.30) satisfying

$$\psi_\gamma(z,\cdot) \in L^2([0,\infty);dx), \quad \sin(\gamma)\psi'_\gamma(z,0_+) + \cos(\gamma)\psi_\gamma(z,0_+) = 1, \\ \gamma \in [0,\pi), \ z \in \mathbb{C}_+. \tag{5.32}$$

Then $\psi_\gamma(z,x)$ is of the form (see, e.g., the discussion of Weyl's theory in Appendix A of [29])

$$\psi_\gamma(z,x) = \theta_\gamma(z,x) + m^W_\gamma(z)\phi_\gamma(z,x), \quad \gamma \in [0,\pi), \ z \in \mathbb{C}_+, \tag{5.33}$$

where $m^W_\gamma(z)$ denotes the Weyl-Titchmarsh m-function [64], Chs. II, III, [70] (as opposed to Donoghue's m-function $m^D_\alpha(z)$ in (5.28)) corresponding to the operator $\widetilde{H}_\gamma$ in $L^2([0,\infty);dx)$ defined by

(5.34)
$$(\widetilde{H}_\gamma f)(x) = -f''(x) + q(x)f(x), \quad x > 0,$$
$$f \in \mathrm{dom}(\widetilde{H}_\gamma) = \{g \in L^2([0,\infty);dx) \,|\, g, g' \in AC([0,R]) \text{ for all } R > 0; \\ \sin(\gamma)g'(0_+) + \cos(\gamma)g(0_+) = 0; \ -g'' + qg \in L^2([0,\infty);dx)\}, \quad \gamma \in [0,\pi).$$

The family $\widetilde{H}_\gamma$, $\gamma \in [0,\pi)$ represents all self-adjoint extensions of the densely defined closed prime symmetric operator $\dot{\widetilde{H}}$ in $L^2([0,\infty);dx)$ of deficiency indices $(1,1)$,

$$(5.35)\quad \begin{aligned} &(\dot{\widetilde{H}}f)(x) = -f''(x) + q(x)f(x), \quad x>0,\\ &f \in \mathrm{dom}(\dot{\widetilde{H}}_\gamma) = \{g \in L^2([0,\infty);dx)) \,|\, g,g' \in AC([0,R]) \text{ for all } R>0;\\ &\qquad g'(0_+) = g(0_+) = 0;\ -g''+qg \in L^2([0,\infty);dx)\}. \end{aligned}$$

(Here $AC([a,b])$ denotes the set of absolutely continuous functions on $[a,b]$.) Weyl's m-function is a Herglotz function with representation

$$(5.36)\quad m_\gamma^W(z) = \begin{cases} c_\gamma + \int_{\mathbb{R}} d\omega_\gamma^W(\lambda)((\lambda-z)^{-1} - \lambda(1+\lambda^2)^{-1}), & \gamma\in[0,\pi),\\ \cot(\gamma) + \int_{\mathbb{R}} d\omega_\gamma^W(\lambda)(\lambda-z)^{-1}, & \gamma \in (0,\pi), \end{cases}$$

for some $c_\gamma \in \mathbb{R}$, where

$$(5.37)\quad \int_{\mathbb{R}} d\omega_\gamma^W(\lambda)(1+|\lambda|)^{-1} \begin{cases} <\infty, & \gamma\in(0,\pi),\\ =\infty, & \gamma = 0. \end{cases}$$

Moreover, one can prove the following result.

Lemma 5.4. (See, e.g., Aronszajn [9], [27], Sect. 2.5.)

$$(5.38)\quad m_\delta^W(z) = \frac{-\sin(\delta-\gamma) + \cos(\delta-\gamma)m_\gamma^W(z)}{\cos(\delta-\gamma)+\sin(\delta-\gamma)m_\gamma^W(z)}, \quad \delta,\gamma\in[0,\pi),\ z\in\mathbb{C}_+.$$

Moreover,

$$(5.39)\quad m_\gamma^W(z) \underset{z\to i\infty}{=} \begin{cases} \cot(\gamma) + O(z^{-1/2}), & \gamma\in[0,\pi),\\ iz^{1/2} + o(1), & \gamma = 0. \end{cases}$$

In the following we denote by H_α in $L^2([0,\infty);dx)$ the Schrödinger operator on $[0,\infty)$ defined as in (5.1) but with $\dot{H}$ replaced by $\dot{\widetilde{H}}$ in (5.35). The connection between H_α and $\widetilde{H}_\gamma$ and $m_\alpha^D(z)$ and $m_\gamma^W(z)$ is then determined as follows.

Theorem 5.5. *Suppose* $\gamma(\alpha) \in [0,\pi)$ *satisfies*

$$(5.40)\quad \cot(\gamma(\alpha)) = -\mathrm{Re}(m_0^W(i)) - \mathrm{Im}(m_0^W(i))\tan(\alpha), \quad \alpha\in[0,\pi).$$

Then

$$(5.41)\quad H_\alpha = \widetilde{H}_{\gamma(\alpha)}, \quad \alpha\in[0,\pi).$$

and

$$(5.42)\quad m_\alpha^D(z) = (m_{\gamma(\alpha)}^W(z) - \mathrm{Re}(m_{\gamma(\alpha)}^W(i))/\mathrm{Im}(m_{\gamma(\alpha)}^W(i)), \quad \alpha\in[0,\pi),\ z\in\mathbb{C}_+.$$

Proof. Since $\psi_\gamma(z,x)$ are just constant multiples of $\psi_0(z,x)$, it suffices to focus on $\psi_0(z,x)$. In order to prove (5.41), subject to (5.40), we need

$$\eta_\alpha = ||\psi_0(i)||^{-1}_{L^2([0,\infty);dx)}\psi_0(i) + ||\psi_0(-i)||^{-1}_{L^2([0,\infty);dx)}e^{2i\alpha}\psi_0(-i) \in \mathrm{dom}(H_\alpha) \tag{5.43}$$

according to (5.1) and the fact (cf. (5.32))

$$u_\pm = ||\psi_0(\pm i)||^{-1}_{L^2([0,\infty);dx)}\psi_0(\pm i). \tag{5.44}$$

Since it is known (see, e.g., [20], Sect. 9.2, [27], Sect. 2.2) that

$$||\psi_\gamma(z)||^2_{L^2([0,\infty);dx)} = \mathrm{Im}(m^W_\gamma(z))/\mathrm{Im}(z), \quad z \in \mathbb{C}\backslash\mathbb{R}, \tag{5.45}$$

one obtains from (5.52) and (5.33)

$$-\cot(\gamma(\alpha)) = \eta'_\alpha(0_+)/\eta_\alpha(0_+) = (1+e^{2i\alpha})^{-1}(m^W_0(i) + e^{2i\alpha}m^W_0(-i)), \tag{5.46}$$

which yields (5.40) and at the same time proves (5.41). By (5.28) and (5.36),

$$m^D_\alpha(z) = A_\alpha m^W_{\gamma(\alpha)}(z) + B_\alpha, \quad \alpha \in [0,\pi), \ z \in \mathbb{C}_+ \tag{5.47}$$

for some $A_\alpha > 0$ and $B_\alpha \in \mathbb{R}$. The fact

$$m^D_\alpha(i) = i, \quad \alpha \in [0,\pi) \tag{5.48}$$

(use (4.8) or combine the normalization $\int_\mathbb{R} d\omega_\alpha(\lambda)(1+\lambda^2)^{-1} = 1$ with (5.28)) immediately yields (5.42). □

Corollary 5.6. *Assume in addition that $\dot{H} \geq 0$. Then the Friedrichs extension H_F of $\dot{H}$ corresponds to*

$$\alpha = \alpha_F = \pi/2 \quad \text{and} \quad \gamma = \gamma_F = 0 \tag{5.49}$$

and the Krein extension H_K of $\dot{H}$ corresponds to

$$\tan(\alpha) = \tan(\alpha_K) = m^D_{\pi/2}(0_-) \quad \text{and} \quad \cot(\gamma) = \cot(\gamma_K) = -m^W_0(0_-) \tag{5.50}$$

in (5.1) and (5.34). The right-hand sides in (5.50) are simultaneously infinite if and only if $H_F = H_K$.

Proof. Since $\lim_{\lambda\downarrow-\infty} m^W_0(\lambda) = -\infty$ by (5.39), (5.49) follows from Corollary 4.8 (i). Similarly, (5.50) follows from (5.38) (replacing $\delta \to \gamma$ and $\gamma \to 0$) and Corollary 4.8 (ii). □

Finally we return to the functional $\ell_{\hat{g}_\alpha}$ in (5.22) and establish its properties in connection with the Schrödinger operator $\widetilde{H}_\gamma$ on $[0,\infty)$.

Lemma 5.7. *Define $\hat{g}_\alpha$ by*

$$U^{-1}_\alpha \hat{g}_\alpha = ||\psi_0(i) + e^{2i\alpha}\psi_0(-i))||^{-1}_{L^2([0,\infty);dx)}(\psi_0(i) + e^{2i\alpha}\psi_0(-i)), \quad \alpha \in [0,\pi). \tag{5.51}$$

Then

$$\ell_{\hat{g}_\alpha}(\hat{f}) = \begin{cases} (2i\mathrm{Im}(m_0^W(i)))^{-1}||\psi_0(i)-\psi_0(-i)||_{L^2([0,\infty);dx)}(U_{\pi/2}^{-1}\hat{f})'(0_+), & \alpha=\frac{\pi}{2}, \\ (1+e^{2i\alpha})^{-1}||\psi_0(i)+e^{2i\alpha}\psi_0(-i)||_{L^2([0,\infty);dx)}(U_\alpha^{-1}\hat{f})(0_+), & \alpha\in[0,\pi)\backslash\{\frac{\pi}{2}\}, \end{cases} \tag{5.52}$$

$$\hat{f}\in\mathrm{dom}(\widehat{H}_\alpha).$$

Proof. By (5.43) and (5.45),

$$\psi_0(i)+e^{2i\alpha}\psi_0(-i)\in\mathrm{dom}(H_\alpha).$$

Hence

$$f = c||\psi_0(i)+e^{2i\alpha}\psi_0(-i)||^{-1}_{L^2([0,\infty);dx)}(\psi_0(i)+e^{2i\alpha}\psi_0(-i))+h, \quad f\in\mathrm{dom}(H_\alpha),\ h\in\mathrm{dom}(\dot{H}) \tag{5.53}$$

and

$$\ell_{\hat{g}_\alpha}(\hat{f}) = c, \quad \hat{f}\in\mathrm{dom}(\widehat{H}_\alpha). \tag{5.54}$$

Since by (5.34),

$$h'(0_+) = h(0_+) = 0, \tag{5.55}$$

one computes in the case $\alpha=\pi/2$

$$\begin{aligned} f'(0_+) &= c||\psi_0(i)-\psi_0(-i)||^{-1}_{L^2([0,\infty);dx)}(\psi_0'(i,0_+)-\psi_0'(-i,0_+)) \\ &= c||\psi_0(i)-\psi_0(-i)||^{-1}_{L^2([0,\infty);dx)}2i\mathrm{Im}(m_0^W(i)), \quad f\in\mathrm{dom}(H_{\pi/2}) \end{aligned} \tag{5.56}$$

using (5.31) and (5.33). Similarly, for $\alpha\in[0,\pi)\backslash\{\pi/2\}$ one computes

$$\begin{aligned} f(0_+) &= c||\psi_0(i)+e^{2i\alpha}\psi_0(-i)||^{-1}_{L^2([0,\infty);dx)}(\psi_0(i,0_+)+e^{2i\alpha}\psi_0(-i,0_+)) \\ &= c||\psi_0(i)+e^{2i\alpha}\psi_0(-i)||^{-1}_{L^2([0,\infty);dx)}(1+e^{2i\alpha}), \\ &\qquad f\in\mathrm{dom}(H_{\pi/2}),\ \alpha\in[0,\pi)\backslash\{\pi/2\}, \end{aligned} \tag{5.57}$$

since $\psi_0(z,0_+)=1$, $z\in\mathbb{C}\backslash\mathbb{R}$ by (5.31) and (5.33). Combining (5.54) and (5.56), (5.57) proves (5.52). □

Lemmas 5.2 and 5.3 then yield the principal result of this section:

Theorem 5.8. *Let $\alpha\in[0,\pi)$. Then*

$$\sup_{f\in\mathrm{dom}(H_{\pi/2})}\left(\frac{|f'(0_+)|^2}{||f||^2_{L^2([0,\infty);dx)}+||H_{\pi/2}f||^2_{L^2([0,\infty);dx)}}\right) = \mathrm{Im}(m_0^W(i)), \tag{5.58}$$

$$\sup_{f\in\mathrm{dom}(H_\alpha)}\left(\frac{|f(0_+)|^2}{||f||^2_{L^2([0,\infty);dx)}+||H_\alpha f||^2_{L^2([0,\infty);dx)}}\right) = \frac{\cos^2(\alpha)}{\mathrm{Im}(m_0^W(i))}. \tag{5.59}$$

Proof. Consider $\alpha = \pi/2$ first. Then Lemma 5.2 combined with (5.11), (5.44), and (5.52) yields

$$\sup_{f\in\mathrm{dom}(H_{\pi/2})}\left(\frac{|f'(0_+)|^2}{||f||^2_{L^2([0,\infty);dx)} + ||H_{\pi/2}f||^2_{L^2([0,\infty);dx)}}\right)$$
$$= \frac{4|\mathrm{Im}(m_0^W(i))|^2}{||\psi_0(i)||^2_{L^2([0,\infty);dx)}||u_+ - u_-||^2_{L^2([0,\infty);dx)}}\int_{\mathbb{R}} d\omega_{\pi/2}(\lambda)(1+\lambda^2)^{-2}. \tag{5.60}$$

Since

$$||u_+ - u_-||^2_{L^2([0,\infty);dx)} = ||\hat{u}_+ - \hat{u}_-||^2_{\hat{\mathcal{H}}_{\pi/2}} = 4\int_{\mathbb{R}} d\omega_{\pi/2}(1+\lambda^2)^{-2} \tag{5.61}$$

by (5.12) (taking $\alpha_0 = \pi/2$) and

$$||\psi_0(i)||^2_{L^2([0,\infty);dx)} = \mathrm{Im}(m_0^W(i)) \tag{5.62}$$

by (5.45), the right-hand side of (5.60) coincides with that in (5.58). Similarly, one computes from Lemma 5.2, (5.11), (5.44), and (5.52),

$$\sup_{f\in\mathrm{dom}(H_\alpha)}\left(\frac{|f(0_+)|^2}{||f||^2_{L^2([0,\infty);dx)} + ||H_\alpha f||^2_{L^2([0,\infty);dx)}}\right)$$
$$= \frac{4\cos^2(\alpha)}{||\psi_0(i)||^2_{L^2([0,\infty);dx)}||u_+ + e^{2i\alpha}u_-||^2_{L^2([0,\infty);dx)}}\int_{\mathbb{R}} d\omega_\alpha(\lambda)(1+\lambda^2)^{-2}. \tag{5.63}$$

Because of (5.62) and

$$||u_+ + e^{2i\alpha}u_-||^2_{L^2([0,\infty);dx)} = ||\hat{u}_+ + e^{2i\alpha}\hat{u}_-||^2_{\hat{\mathcal{H}}_\alpha} = 4\int_{\mathbb{R}} d\omega_\alpha(\lambda)(1+\lambda^2)^{-2}, \tag{5.64}$$

(5.63) coincides with (5.59). □

Remark 5.9. (i) In the special case $q(x) = 0$, $x \geq 0$ one has

$$m_0^W(z) = i(z)^{1/2} \tag{5.65}$$

(using the branch with $\mathrm{Im}((z)^{1/2}) \geq 0$, $z \in \mathbb{C}$) and hence (5.58) yields

$$|f'(0_+)| \leq 2^{-1/4}\left(\int_0^\infty dx(|f(x)|^2 + |f''(x)|^2)\right)^{1/2}, \quad f \in H_0^{2,2}((0,\infty)), \tag{5.66}$$

with $2^{-1/4}$ best possible and

$$\begin{aligned} H_0^{2,2}((0,\infty)) = \{f \in L^2([0,\infty);dx) \,|\, f, f' \in AC([0,R]) \text{ for all } R>0;\\ f(0_+) = 0;\ f, f'' \in L^2([0,\infty);dx)\} \end{aligned} \tag{5.67}$$

the familiar Sobolev space.

(ii) Multiplying the two results (5.58) and (5.59) reveals the curious fact,

(5.68)

$$\sup_{f\in\mathrm{dom}(H_{\pi/2})}\left(\frac{|f'(0_+)|^2}{||f||^2_{L^2([0,\infty);dx)} + ||H_{\pi/2}f||^2_{L^2([0,\infty);dx)}}\right)\times$$

$$\times \sup_{f\in\mathrm{dom}(H_\alpha)} \left(\frac{|f(0_+)|^2}{||f||^2_{L^2([0,\infty);dx)} + ||H_\alpha f||^2_{L^2([0,\infty);dx)}} \right) = \cos^2(\alpha), \quad \alpha \in [0,\pi).$$

Finally, we conclude with two illustrations of a well-known result of Livšic [44] on quasi-hermitian extensions in the special case of densely defined closed prime symmetric operators with deficiency indices $(1,1)$.

Following Livšic [44] one defines a closed operator H to be a *quasi-hermitian* extension of a densely defined closed prime symmetric operator $\dot{H}$ with deficiency indices $(1,1)$ if

$$\dot{H} \subsetneqq H \subsetneqq \dot{H}^* \tag{5.69}$$

and H is not self-adjoint.

A typical example of a quasi-hermitian extension is obtained as follows.

Let $\dot{T}$ denote the following first-order differential operator on the interval $[0,2a]$, $a>0$,

$$\begin{aligned}(\dot{T}f)(x) &= -if'(x), \quad \xi \in (0,2a),\\ f \in \mathrm{dom}(\dot{T}) &= \{g \in L^2([0,2a]) \,|\, g \in AC([0,2a]);\ g(0_+) = g(2a_-) = 0;\\ &\qquad g' \in L^2([0,2a])\}.\end{aligned} \tag{5.70}$$

Then for $\rho \in \mathbb{C}\cup\{\infty\}$, $|\rho| \neq 1$ the operator T_ρ

$$\begin{aligned}(T_\rho f)(x) &= -if'(x), \quad \xi \in (0,2a),\\ f \in \mathrm{dom}(T_\rho) &= \{g \in L^2([0,2a]) \,|\, g \in AC([0,2a]);\ g(0_+) = \rho g(2a_-);\\ &\qquad g' \in L^2([0,2a])\}\end{aligned} \tag{5.71}$$

is a quasi-hermitian extension of $\dot{T}$. (Here $\rho=\infty$ in (5.71), in obvious notation, denotes the boundary condition $g(2a_-)=0$.) Among all quasi-hermitian extensions of $\dot{T}$ there are two exceptional ones that have empty spectrum. In fact, the operator T_0 corresponding to the value $\rho=0$ in (5.71) as well as its adjoint, $T_0^* = T_\infty$, have empty spectra, that is,

$$\mathrm{spec}(T_0) = \mathrm{spec}(T_\infty) = \emptyset. \tag{5.72}$$

The following theorem proven by Livšic in 1946 provides an interesting characterization of this example.

Theorem 5.10. (Livšic [44].) *For a densely defined closed prime symmetric operator with deficiency indices* $(1,1)$ *to be unitarily equivalent to the differentiation operator* $\dot{T}$ *in* $L^2([0,2a])$ *for some* $a>0$ *it is necessary and sufficient that it admits a quasi-hermitian extension with empty spectrum.*

Using Livšic's result we are able to characterize the model representation for the pair $(\dot{H},H)$, where $\dot{H}$ is a densely defined prime closed symmetric operator with deficiency indices $(1,1)$ which admits a quasi-hermitian extension with empty spectrum, and H a self-adjoint extension of $\dot{H}$.

Theorem 5.11. *Let ω be a Borel measure on $\mathbb{R}$ such that*

$$\int_{\mathbb{R}} \frac{d\omega(\lambda)}{1+\lambda^2} = 1, \quad \int_{\mathbb{R}} d\omega(\lambda) = \infty, \tag{5.73}$$

H the self-adjoint operator of multiplication by λ in $L^2(\mathbb{R}; d\omega)$,

$$(Hf)(\lambda) = \lambda f(\lambda), \quad f \in \mathrm{dom}(H) = L^2(\mathbb{R}; (1+\lambda^2)d\omega). \tag{5.74}$$

Define $\dot{H}$ to be the densely defined closed prime symmetric restriction of H,

$$\dot{H} = H\big|_{\mathrm{dom}(\dot{H})}, \quad \mathrm{dom}(\dot{H}) = \{f \in \mathrm{dom}(H) \,|\, \int_{\mathbb{R}} d\omega(\lambda) f(\lambda) = 0\}, \tag{5.75}$$

with deficiency indices $(1,1)$. Then $\dot{H}$ admits a quasi-hermitian extension with empty spectrum if and only if for some $a > 0$ and some $\alpha \in [0, \pi)$ the following representation holds

$$\int_{\mathbb{R}} d\omega(\lambda)((z-\lambda)^{-1} - \lambda(1+\lambda^2)^{-1}) = \frac{\sin(\alpha) - \cos(\alpha)(\cot(az)/\coth(a))}{\cos(\alpha) + \sin(\alpha)(\cot(az)/\coth(a))}, \quad z \in \mathbb{C}\backslash\mathbb{R}. \tag{5.76}$$

In this case the measure ω is a pure point measure,

$$\omega = \frac{\coth(a)(1+\cot^2(\alpha))}{a(1+\cot^2(\alpha)\coth^2(a))} \sum_{n\in\mathbb{Z}} \mu_{\{(\beta+\pi n)/a\}}, \tag{5.77}$$

where $\mu_{\{x\}}$ denotes the Dirac measure supported at $\xi \in \mathbb{R}$ with mass one and $\beta = \beta(\alpha, a) \in [0, \pi)$ is the solution of the equation

$$\cot(\beta) + \cot(\alpha)\coth(a) = 1 \text{ if } \alpha \in (0,\pi) \text{ and } \beta = 0 \text{ if } \alpha = 0. \tag{5.78}$$

Moreover, the self-adjoint operator H given by (5.74) is unitarily equivalent to the differentiation operator T_ρ in (5.71) with

$$\rho = e^{2i\beta}. \tag{5.79}$$

Proof. That $\dot{H}$ is a densely defined closed prime symmetric operator with deficiency indices $(1,1)$ is proven in [25]. By Livšic's theorem, Theorem 5.10, the pair $(\dot{H}, H)$ is unitarily equivalent to the pair $(\dot{T}, T_\rho)$, where $\dot{T}$ is the operator (5.70) in $L^2([0,2a])$ for some $a > 0$ and T_ρ is some self-adjoint extension of $\dot{T}$ given by (5.71) for some ρ, $|\rho| = 1$. By (4.8) and (4.17) (cf. also (5.28)) we conclude

$$m^D_{T_\rho}(z) = \int_{\mathbb{R}} d\omega(\lambda)((z-\lambda)^{-1} - \lambda(1+\lambda^2)^{-1}) \tag{5.80}$$

$$= z + (1+z^2)(u_+, (T_\rho - z)^{-1}u_+)_{L^2(\mathbb{R};d\omega)}, \quad u_+ \in \ker(\dot{T}^* - i),\ \|u_+\|_{L^2(\mathbb{R};d\omega)} = 1, \tag{5.81}$$

where $m^D_{T_\rho}(z)$ denotes the Donoghue Weyl m-function of the operator T_ρ.

Let $\widetilde{T}$ be the self-adjoint extension of $\dot{T}$ corresponding to periodic boundary conditions,

(5.82)
$$\operatorname{dom}(\widetilde{T}) = \{g \in L^2([0,2a]) \,|\, g \in AC([0,2a]);\, g(0_+) = g(2a_-);\, g' \in L^2([0,2a])\}.$$

By Lemma 5.3 there exists an $\alpha \in [0,\pi)$ such that

$$m_{T_\rho}^D(z) = \frac{\sin(\alpha) + \cos(\alpha) m_{\widetilde{T}}^D(z)}{\cos(\alpha) - \sin(\alpha) m_{\widetilde{T}}^D(z)}, \tag{5.83}$$

where $m_{\widetilde{T}}^D(z)$ is the Donoghue Weyl m-function of the extension $\widetilde{T}$

$$m_{\widetilde{T}}^D(z) = z + (1+z^2)(u_+, (\widetilde{T}-z)^{-1}u_+)_{L^2([0,2a];dx)}, \quad z \in \mathbb{C}\backslash\mathbb{R}. \tag{5.84}$$

The assertion (5.76) then follows from the fact

$$m_{\widetilde{T}}^D(z) = -\frac{\cot(az)}{\coth(a)}. \tag{5.85}$$

Next we prove (5.85). First, we note that the resolvent of the operator $\widetilde{T}$ can be explicitly computed as

(5.86)
$$((\widetilde{T}-z)^{-1}f)(x) = ie^{izx}\bigg(\int_0^x e^{-izt}f(t)dt + \frac{e^{2iza}}{1-e^{2iza}}\int_0^{2a} e^{-izt}f(t)dt\bigg), \; z \in \mathbb{C}\backslash\mathbb{R}.$$

Next we calculate the quadratic form of the resolvent of $\widetilde{T}$ on the element $u_+(x) = 2^{1/2}(1-e^{-4a})^{-1/2}\exp(-x)$ generating $\ker(\dot{T}^* - i)$. By (5.86) we have

$$((\widetilde{T}-z)^{-1}u_+)(x) = \frac{2^{1/2}(1-e^{-4a})^{-1/2}}{i-z}\bigg(e^{-x} - e^{izx}\frac{1-e^{-2a}}{1-e^{2iza}}\bigg) \tag{5.87}$$

and therefore,

(5.88)
$$(u_+, (\widetilde{T}-z)^{-1}u_+)_{L^2([0,2a];dx)} = \frac{1}{i-z}\bigg(1 + \frac{2(1-e^{-2a})(1-e^{2iaz-2a})}{(iz-1)(1-e^{-4a})(1-e^{2iaz})}\bigg).$$

Equations (5.84) and (5.88) then prove (5.85).

In order to prove (5.77) we note that the right-hand side of (5.76) is a periodic Herglotz function with period π/a. Such Herglotz functions have simple poles at the points $\{(\beta+\pi n)/a\}_{n\in\mathbb{Z}}$ with residues

$$\begin{aligned} &\operatorname*{Res}_{z=(\beta+\pi n)/a}\bigg(\frac{\sin(\alpha) - \cos(\alpha)(\cot(az)/\coth(a))}{\cos(\alpha) + \sin(\alpha)(\cot(az)/\coth(a))}\bigg) \\ &\quad = -\frac{\coth(a)(1+\cot^2(\alpha))}{a(1+\cot^2(\alpha)\coth^2(a))}, \quad n \in \mathbb{Z}, \end{aligned} \tag{5.89}$$

proving (5.77).

The last assertion of the theorem follows from the fact that the support of the measure ω coincides with the spectrum of H and therefore with the one of the operator T_ρ which is unitarily equivalent to H. The spectrum of the self-adjoint operator T_ρ can explicitly be computed as

$$\operatorname{spec}(T_\rho) = \{\frac{1}{2a}\arg\rho + \frac{\pi}{a}n\}_{n\in\mathbb{Z}}. \tag{5.90}$$

Since the sets (5.90) and $\operatorname{supp}(\omega)$ coincide we conclude (5.77). □

Remark 5.12. We note that the weak limit as $a \to \infty$ of the measures $\omega = \omega(\alpha, a)$ (with α fixed) given by (5.77) coincides with $\pi^{-1}d\lambda$, where $d\lambda$ denotes the Lebesgue measure on $\mathbb{R}$.

The next result shows that this limiting case $d\omega = \pi^{-1}d\lambda$ is also rather exotic.

Theorem 5.13. *Let ω be a Borel measure on $\mathbb{R}$ such that*

$$\int_{\mathbb{R}} \frac{d\omega(\lambda)}{1+\lambda^2} = 1, \quad \int_{\mathbb{R}} d\omega(\lambda) = \infty, \tag{5.91}$$

H the self-adjoint operator of multiplication by λ in $L^2(\mathbb{R}; d\omega)$,

$$(Hf)(\lambda) = \lambda f(\lambda), \quad f \in \operatorname{dom}(H) = L^2(\mathbb{R}; (1+\lambda^2)d\omega). \tag{5.92}$$

Define $\dot H$ to be the densely defined closed prime symmetric restriction of H,

$$\dot H = H\big|_{\operatorname{dom}(\dot H)}, \quad \operatorname{dom}(\dot H) = \{f \in \operatorname{dom}(H) \,|\, \int_{\mathbb{R}} f(\lambda)d\omega(\lambda) = 0\}, \tag{5.93}$$

with deficiency indices $(1,1)$. Then $\dot H$ admits a quasi-hermitian extension with pure point spectrum the open upper (lower) half-plane and spectrum the closed upper (lower) half-plane if and only if the following representation holds,

$$\int_{\mathbb{R}} d\omega(\lambda)((z-\lambda)^{-1} - \lambda(1+\lambda^2)^{-1}) = \begin{cases} i, & \operatorname{Im}(z) > 0, \\ -i, & \operatorname{Im}(z) < 0. \end{cases} \tag{5.94}$$

In this case

$$d\omega = \pi^{-1}d\lambda. \tag{5.95}$$

Proof. The setup in (5.91)–(5.93) is identical to that in Theorem 5.11 and hence needs no further comments. The fact that $\dot H$ is unitarily equivalent to the differentiation operator $\dot T$ acting in $L^2(\mathbb{R}; dx)$,

$$\begin{aligned} &(\dot T f)(x) = -if'(x), \quad \xi \in \mathbb{R}, \\ &f \in \operatorname{dom}(\dot T) = \{g \in L^2(\mathbb{R}; dx) \,|\, g \in AC(\mathbb{R});\, g(0) = 0;\, g' \in L^2(\mathbb{R}; dx)\} \end{aligned} \tag{5.96}$$

goes back to Livšic (see, e.g., Appendix I.5 in [3]). In fact, the quasi-hermitian extension T of $\dot T$ defined by

(5.97)

$$\begin{aligned} &(Tf)(x) = -if'(x), \quad \xi \in \mathbb{R}\backslash\{0\}, \\ &f \in \operatorname{dom}(T) = \{g \in L^2(\mathbb{R}; dx) \,|\, g \in AC([-R,0]) \cup AC([0,R]) \text{ for all } R > 0; \\ &\qquad\qquad g(0_-) = 0;\, g' \in L^2(\mathbb{R}; dx)\}. \end{aligned}$$

(and its adjoint T^* with corresponding boundary condition $g(0_+) = 0$) has spectrum the closed upper (lower) half-plane with pure point spectrum the open upper (lower) half-plane, respectively. This is easily verified from an alternative expression for T given by

$$T = \dot{T}_- \oplus T_+ \text{ in } L^2(\mathbb{R}; dx) = L^2((-\infty, 0]; dx) \oplus L^2([0, \infty); dx), \tag{5.98}$$

where

$$\begin{aligned} &(\dot{T}_- f)(x) = -if'(x), \quad x < 0, \\ &f \in \mathrm{dom}(\dot{T}_-) = \{g \in L^2((-\infty, 0]; dx) \,|\, g \in AC([-R, 0]) \text{ for all } R > 0; \\ &\qquad g(0_-) = 0;\, g' \in L^2((-\infty, 0]; dx)\}, \end{aligned} \tag{5.99}$$

$$\begin{aligned} &(T_+ f)(x) = -if'(x), \quad x > 0, \\ &f \in \mathrm{dom}(T_+) = \{g \in L^2([0, \infty); dx) \,|\, g \in AC([0, R]) \text{ for all } R > 0; \\ &\qquad g' \in L^2([0, \infty); dx)\}. \end{aligned} \tag{5.100}$$

The explicit expressions for the resolvents of $\dot{T}_-$ and T_+ (see, e.g., [38], Example III.6.9) then show that both operators have spectrum the closed upper half-plane, that is,

$$\mathrm{spec}(\dot{T}_-) = \mathrm{spec}(T_+) = \overline{\mathbb{C}_+}. \tag{5.101}$$

Together with the aforementioned result of Livšic, this shows that the pair $(\dot{H}, H)$ is unitarily equivalent to the pair $(\dot{T}, T_\rho)$, where T_ρ, $|\rho| = 1$ is some self-adjoint extension of $\dot{T}$ in $L^2(\mathbb{R}; dx)$,

(5.102)

$$\begin{aligned} &(T_\rho f)(x) = -if'(x), \quad \xi \in \mathbb{R}\backslash\{0\},\ |\rho| = 1, \\ &f \in \mathrm{dom}(T_\rho) = \{g \in L^2(\mathbb{R}; dx) \,|\, g \in AC([-R, 0]) \cup AC([0, R]) \text{ for all } R > 0; \\ &\qquad g(0_-) = \rho g(0_+);\, g' \in L^2(\mathbb{R}; dx)\}. \end{aligned}$$

Since the pair $(\dot{T}, T_1)$ is unitarily equivalent to the model pair $(\dot{H}, H)$ in (5.92) and (5.93) (it suffices applying the Fourier transform), where $d\omega = \pi^{-1} d\lambda$, we can immediately compute the Donoghue Weyl m-function $m^D_{T_1}(z)$ of the self-adjoint extension T_1,

$$m^D_{T_1}(z) = \frac{1}{\pi} \int_{\mathbb{R}} d\lambda ((z - \lambda)^{-1} - \lambda(1 + \lambda^2)^{-1}) = \begin{cases} i, & \mathrm{Im}(z) > 0, \\ -i, & \mathrm{Im}(z) < 0. \end{cases} \tag{5.103}$$

Since

$$\pm i = \frac{\sin(\alpha) + \cos(\alpha)(\pm i)}{\cos(\alpha) - \sin(\alpha)(\pm i)} \text{ for all } \alpha \in [0, \pi), \tag{5.104}$$

Lemma 5.3 implies that the Donoghue Weyl m-function $m^D_{T_\rho}(z)$ of the extension T_ρ is independent of ρ, $|\rho| = 1$ and hence $m^D_{T_1}(z) = m^D_\rho(z)$. Therefore, the model representation for the pair $(\dot{T}, T_\rho)$ is given by (5.91)–(5.93) with $d\omega = \pi^{-1} d\lambda$, proving (5.95). Finally, (5.94) follows from (5.103). □

Acknowledgments. Eduard Tsekanovskii would like to thank Daniel Alpay and Victor Vinnikov for support and their kind invitation to a wonderful conference organized in honor of the 80th birthday of Moshe Livšic, his dear and inspiring teacher.

References

[1] M. Abramowitz and I. A. Stegun, *Handbook of Mathematical Functions*, Dover, New York, 1972.

[2] V. M. Adamjan and H. Langer, *Spectral Properties of a class of rational operator valued functions*, J. Operator Th. **33**, 259–277 (1995).

[3] N. I. Akhiezer and I. M. Glazman, *Theory of Linear Operators in Hilbert Space*, Dover, New York, 1993.

[4] S. Albeverio, J. Brasche, and H. Neidhardt, *On inverse spectral theory for self-adjoint extensions: mixed types of spectra*, preprint, 1996.

[5] S. Albeverio, F. Gesztesy, R. Høegh-Krohn, H. Holden, *Solvable Models in Quantum Mechanics*, Springer, Berlin, 1988.

[6] A. Alonso and B. Simon, *The Birman-Krein-Vishik theory of self-adjoint extensions of semibounded operators*, J. Operator Th. **4**, 251–270 (1980).

[7] T. Ando and K. Nishio, *Positive selfadjoint extensions of positive symmetric operators*, Tôhoku Math. J. **22**, 65–75 (1970).

[8] Yu. M. Arlinskii and E. R. Tsekanovskii, *Nonselfadjoint contractive extensions of a Hermitian contraction and theorems of Krein*, Russ. Math. Surv. **37:1**, 151–152 (1982).

[9] N. Aronszajn, *On a problem of Weyl in the theory of singular Sturm-Liouville equations*, Am. J. Math. **79**, 597-610 (1957).

[10] H. Bart, I. Gohberg, and M. A. Kaashoek, *Minimal Factorizations of Matrix and Operator Functions*, Operator Theory: Advances and Applications, Vol. 1, Birkhäuser, Basel, 1979.

[11] H. Baumgärtel and M. Wollenberg, *Mathematical Scattering Theory*, Birkhäuser, Basel, 1983.

[12] M. S. Birman and M. Z. Solomjak, *Spectral Theory of Self-Adjoint Operators in Hilbert Space*, Reidel, Dordrecht, 1987.

[13] J. F. Brasche and H. Neidhardt, *Some remarks on Krein's extension theory*, Math. Nachr. **165**, 159–181 (1994).

[14] J. Brasche and H. Neidhardt, *On the absolutely continuous spectrum of self-adjoint extensions*, J. Funct. Anal. **131**, 364–385 (1995).

[15] J. Brasche and H. Neidhardt, *On the singularly continuous spectrum of self-adjoint extensions*, Math. Z. **222**, 533–542 (1996).

[16] J. Brasche, H. Neidhardt, and J. Weidmann, *On the point spectrum of self-adjoint extensions*, Math. Z. **214**, 343–355 (1993).

[17] M. S. Brodskii, *Triangular and Jordan Representations of Linear Operators*, Amer. Math. Soc., Providence, RI, 1971.

[18] D. Buschmann, *Spektraltheorie verallgemeinerter Differentialausdrücke – Ein neuer Zugang*, Ph.D. Thesis, University of Frankfurt, Germany, 1997.

[19] R. W. Carey, *A unitary invariant for pairs of self-adjoint operators*, J. Reine Angew. Math. **283**, 294–312 (1976).

[20] E. A. Coddington and N. Levinson,*Theory of Ordinary Differential Equations*, Krieger, Malabar, 1985.

[21] V. A. Derkach and M. M. Malamud, *On the Weyl function and Hermitian operators with gaps*, Sov. Math. Dokl. **35**, 393–398 (1987).

[22] V. A. Derkach and M. M. Malamud, *Generalized resolvents and the boundary value problems for Hermitian operators with gaps*, J. Funct. Anal. **95**, 1–95 (1991).

[23] V. A. Derkach and M. M. Malamud, *The extension theory of Hermitean operators and the moment problem*, J. Math. Sci. **73**, 141–242 (1995).

[24] V. A. Derkach, M. M. Malamud, and E. R. Tsekanovskii, *Sectorial extensions of a positive operator, and the characteristic function*, Sov. Math. Dokl. **37**, 106-110 (1988).

[25] W. F. Donoghue, *On the perturbation of spectra*, Commun. Pure Appl. Math. **18**, 559-579 (1965).

[26] N. Dunford and J. T. Schwartz, *Linear Operators Part II: Spectral Theory*, Interscience, New York, 1988.

[27] M. S. P. Eastham and H. Kalf, *Schrödinger–Type Operators with Continuous Spectra*, Pitman, Boston, 1982.

[28] F. Gesztesy, K.A. Makarov, E. Tsekanovskii, *An Addendum to Krein's formula*, J. Math. Anal. Appl., to appear.

[29] F. Gesztesy and B. Simon, *Uniqueness theorems in inverse spectral theory for one-dimensinal Schrödinger operators*, Trans. Amer. Math. Soc. **348**, 349-373 (1996).

[30] F. Gesztesy and E. Tsekanovskii, *On matrix-valued Herglotz functions*, preprint, 1997.

[31] D. S. Greenstein, *On the analytic continuation of functions which map the upper half plane into itself*, J. Math. Anal. Appl. **1**, 355–362 (1960).

[32] S. Hassi and H. de Snoo, *One-dimensional graph perturbations of selfadjoint relations*, Ann. Acad. Sci. Fenn. A I Math. **22**, 123–164 (1997).

[33] S. Hassi and H. de Snoo, *On rank one perturbations of selfadjoint operators*, Integral Eqs. Operator Th. **29**, 288–300 (1997).

[34] S. Hassi, M. Kaltenbäck, and H. de Snoo, *Triplets of Hilbert spaces and Friedrichs extensions associated with the subclass N_1 of Nevanlinna functions*, J. Operator Th. **37**, 155–181 (1997).

[35] S. Hassi, M. Kaltenbäck, and H. de Snoo, *Generalized Krein-von Neumann extensions and associated operator models*, preprint, 1997.

[36] S. Hassi, M. Kaltenbäck, and H. de Snoo, *Generalized finite rank perturbations associated with Kac classes of matrix Nevanlinna functions*, in preparation.

[37] S. Hassi, H. S. V. de Snoo, and A. D. I. Willemsma, *Smooth rank one perturbations of selfadjoint operators*, Proc. Amer. Math. Soc., to appear.

[38] T. Kato, *Perturbation Theory for Linear Operators*, corr. 2nd ed., Springer, Berlin, 1980.

[39] M. G. Krein, *The theory of self-adjoint extensions of semibounded Hermitian transformations and its applications*, I. Mat. Sb. **20**, 431-495 (1947) (Russian).

[40] M. G. Krein and I. E. Ovcharenko, *Q-functions and sc-resolvents of nondensely defined hermitian contractions*, Sib. Math. J. **18**, 728–746 (1977).

[41] M. G. Krein and I. E. Ovčarenko, *Inverse problems for Q-functions and resolvent matrices of positive hermitian operators*, Sov. Math. Dokl. **19**, 1131–1134 (1978).

[42] M. G. Krein and Ju. L. Smul'jan, *On linear-fractional transformations with operator coefficients*, Amer. Math. Soc. Transl. (2) **103**, 125-152 (1974).

[43] H. Langer, B. Textorius, *On generalized resolvents and Q-functions of symmetric linear relations (subspaces) in Hilbert space*, Pacific J. Math. **72**, 135–165 (1977).

[44] M. S. Livšic, *On a class of linear operators in Hilbert space*, Amer. Math. Soc. Transl. (2) **13**, 61–83 (1960).

[45] M. M. Malamud, *Certain classes of extensions of a lacunary Hermitian operator*, Ukrain. Math. J. **44**, 190–204 (1992).

[46] M. M.Malamud, *On a formula of the generalized resolvents of a nondensely defined hermitian operator*, Ukrain. Math. J. **44**, 1522–1547 (1992).

[47] R. Mennicken and A. A. Shkalikov, *Spectral decomposition of symmetric operator matrices*, Math. Nachr. **179**, 259–273 (1996).

[48] S. N. Naboko, *Boundary values of analytic operator functions with a positive imaginary part*, J. Sov. Math. **44**, 786–795 (1989).

[49] S. N. Naboko, *Nontangential boundary values of operator-valued R-functions in a half-plane*, Leningrad Math. J. **1**, 1255–1278 (1990).

[50] S. N. Naboko, *Structure of the singularities of operator functions with a positive imaginary part*, Funct. Anal. Appl. **25**, 243–253 (1991).

[51] S. N. Naboko, *Zygmund's theorem and the boundary behavior of operator R-functions*, Funct. Anal. Appl. **30**, 211–213 (1996).

[52] M. A. Naimark, *Spectral functions of a symmetric operator*, Izv. Akad. Nauk SSSR **4**, 227–318 (1940). (Russian.)

[53] M. A. Naimark, *On a representation of additive operator set functions*, Dokl. Akad. Nauk SSSR **41**, 359–361 (1943). (Russian.)

[54] M. A. Naimark, *Linear Differential Operators II*, F. Ungar Publ., New York, 1968.

[55] M. Reed and B. Simon, *Methods of Modern Mathematical Physics, Vol. II: Fourier Analysis, Self-Adjointness*, Academic Press, New York, 1975.

[56] C. Remling, *Spectral analysis of higher order differential operators I: General properties of the M-function*, J. London Math. Soc., to appear.

[57] M. Rosenblum, J. Rovnyak, *Hardy Classes and Operator Theory*, Dover, New York, 1997.

[58] Sh. N. Saakjan, *On the theory of the resolvents of a symmetric operator with infinite deficiency indices*, Dokl. Akad. Nauk Arm. SSR **44**, 193–198 (1965). (Russian.)

[59] B. Simon, *Spectral analysis of rank one perturbations and applications*, CRM Proceedings and Lecture Notes **8**, 109-149 (1995).

[60] B. Simon and T. Wolff, *Singular continuous spectrum under rank one perturbations and localization for random Hamiltonians*, Commun. Pure Appl. Math. **39**, 75-90 (1986).

[61] C. F. Skau, *Positive self-adjoint extensions of operators affiliated with a von Neumann algebra*, Math. Scand. **44**, 171-195 (1979).

[62] Yu. L. Shmul'yan, *On operator R-functions*, Sib. Math. J. **12**, 315–322 (1971).

[63] B. Sz.-Nagy and C. Foias, *Harmonic Analysis of Operators on Hilbert Space*, North-Holland, Amsterdam, 1970.

[64] E. C. Titchmarsh, *Eigenfunction Expansions Associated with Second-Order Differential Equations*, Part I, 2nd ed., Oxford University Press, Oxford, 1962.

[65] E. R. Tsekanovskii, *Non-self-adjoint accretive extensions of positive operators and theorems of Friedrichs-Krein-Phillips*, Funct. Anal. Appl. **14**, 156–157 (1980).

[66] E. R. Tsekanovskii, *Friedrichs and Krein extensions of positive operators and holomorphic contraction semigroups*, Funct. Anal. Appl. **15**, 308–309 (1981).

[67] E. R. Tsekanovskii, *Accretive extensions and problems on the Stieltjes operator-valued functions realizations*, in *Operator Theory and Complex Analysis*, T. Ando and I. Gohberg (eds.), Operator Theory: Advances and Applications, Vol. 59, Birkhäuser, Basel, 1992, pp. 328–347.

[68] E. R. Tsekanovskii and Yu. L. Shmul'yan, *The theory of bi-extensions of operators on rigged Hilbert spaces. Unbounded operator colligations and characteristic functions*, Russ. Math. Surv. **32:5**, 73–131 (1977).

[69] J. von Neumann, *Allgemeine Eigenwerttheorie hermitescher Funktionaloperatoren*, Math. Ann. **102**, 49–131 (1929–30).

[70] H. Weyl, *Über gewöhnliche Differentialgleichungen mit Singularitäten und die zugehörigen Entwicklungen willkürlicher Funktionen*, Math. Ann. **68**, 220–269 (1910).

Department of Mathematics, University of Missouri, Columbia, MO 65211, USA
E-mail address: fritz@math.missouri.edu
URL: http://www.math.missouri.edu/people/faculty/fgesztesypt.html

Department of Mathematics, University of Missouri, Columbia, MO 65211, USA
E-mail address: nigel@math.missouri.edu

Department of Mathematics, University of Missouri, Columbia, MO 65211, USA
E-mail address: makarov@azure.math.missouri.edu

Department of Mathematics, University of Missouri, Columbia, MO 65211, USA
Current address: Department of Mathematics, P.O. Box 2044, Niagara University, NY 14109, USA
E-mail address: tsekanov@niagara.edu

Operator Theory:
Advances and Applications, Vol. 123

Analogues of a Theorem of Frostman on Linear Fractional Transformations of Inner Functions and the Typical Spectral Structure of Analytic Families of Weak Contractions

Yu. P. Ginzburg

H. Frostman [3] proved in 1935 the following remarkable theorem: if $x(\zeta)$ is an inner function defined in the open unit disk $\mathbb{D}$, then

$$f(\zeta, z) = \frac{x(\zeta) - z}{1 - z^* x(\zeta)} \tag{0.1}$$

is, as a function of ζ, a Blaschke product for all the values of the parameter $z \in \mathbb{D}$, with a possible exception of a set of points M_f of zero logarithmic capacity ($\text{cap}_0 M_f = 0$). In 1955 M. Heins [12] established the following: if $x(\zeta)$ is an arbitrary function holomorphic and strictly contractive in $\mathbb{D}$, then the function $f(\zeta, z)$ defined by (0.1) is the product of a Blaschke product and of an outer function for all $z \in \mathbb{D} \setminus M_f$ (M_f of zero logarithmic capacity).

In the first section of the present work we introduce a class Q_G (G a domain in $\mathbb{C}^n$), which contains functions of the form (0.1) (for $G = \mathbb{D}$), as well as arbitrary functions of (ζ, z) holomorphic and bounded in modulus by 1 in $\mathbb{D} \times G$. For a function $f \in Q_G$, we prove a result of the Frostman-Heins type under an additional assumption that there exists a point $z_0 \in G$ such that $f(\zeta, z_0)$ is an outer function multiplied by a Blaschke product.

As we will see in Section 2, the class Q_G contains determinants of operator analogues of the right-hand side of (0.1), allowing us to formulate in terms of the multiplicative representations of Potapov [19] generalizations of the theorems of Frostman and Heins to operator valued functions with a finer characterization of the "smallness" of the exceptional set (pluripolarity) than what was proved in the works [6], [7], [8], [9], [10], [25].

Also, as in these works, the assertions proved in Section 2 may be used to study the spectral structure characteristic of "most" of the elements of certain analytic families of contractive operators in a Hilbert space. Two results of this type are given in Section 3.

1. Some scalar analogues of Frostman's theorem

Let $\mathbb{D}$ denote the open unit disk and $\mathbb{T}$ the unit circle. Let G be a domain in $\mathbb{C}^n$. We denote by Q_G the class of all complex-valued functions $f(\zeta, z)$ $(\zeta \in \mathbb{D},\ z \in G)$ with the following properties:

1. The function $|f(\zeta, z)|$ is Borel-measurable and $|f(\zeta, z)| \leq 1$ in $\mathbb{D} \times G$.
2. For every fixed $z \in G$ the function $f(\zeta, z)$ is analytic with respect to ζ in $\mathbb{D}$.
3. There exists a function $\tilde{f}(\zeta, z)$ analytic with respect to z in G for every fixed $\zeta \in \mathbb{D}$ such that
   (a) $|f(\zeta, z)| = |\tilde{f}(\zeta, z)| \quad (\zeta \in \mathbb{D}, z \in G)$, and
   (b) for almost every $\xi \in \mathbb{T}$ and for all $z \in G$ there exists the nontangential limit $\lim_{\zeta\to\xi} \tilde{f}(\zeta, z) \stackrel{\text{def}}{=} \tilde{f}(\xi, z)$ (obviously, this function will be analytic with respect to z in G for fixed ξ).

The simplest example of an element in Q_G is given by the function (0.1); here, $G = \mathbb{D}$ and

$$\tilde{f}(\zeta, z) = \frac{x(\zeta) - z}{1 - x(\zeta)^* z}.$$

See Section 2 for an important generalization of this example.

We will also consider the class $Q_G^{(0)}$ consisting of all the functions analytic and bounded in modulus by 1 in $\mathbb{D} \times G$. It is easily checked that $Q_G^{(0)} \subset Q_G$. Indeed, let $f \in Q_G^{(0)}$ and set $\tilde{f} = f$. It is enough to show that condition (3b) holds. To do that, we take a polydisk $\mathcal{U} = \mathcal{D}_1 \times \cdots \times \mathcal{D}_n \subset G$ (where the $\mathcal{D}_j$ are open disks in $\mathbb{C}$), and for each $j = 1, \ldots, n$, a countable set S_j with closure $S_j \subset \mathcal{D}_j$. For every $z \in S = S_1 \times \cdots \times S_n$ there exists a set E_z of full measure on $\mathbb{T}$ such that for every $\xi \in E_z$ the nontangential limit $f(\xi, z)$ exists. Then the set $E = \cap_{z\in S} E_z$ is a set of full measure on $\mathbb{T}$. Using Vitali's theorem on compactness, it is easily proved by induction that $f(\xi, z)$ exists for every $\xi \in E$ and $z \in \mathcal{U}$. Representing G as an at most countable union of polydisks, one sees that $f \in Q_G$.

We denote by the symbol K the set of all functions $x(\zeta)$ holomorphic and bounded in modulus by 1 in $\mathbb{D}$, and by K_b, K_i and K_e the subsets of Blaschke products, of inner functions and of outer functions in K. The symbol K_{be} denotes the set of functions of the form $x(\zeta) = x_1(\zeta)x_2(\zeta)$ where $x_1 \in K_b$ and $x_2 \in K_e$.

We recall that a subset $M \subset \mathbb{C}^n$ is called pluripolar if there exists a pluri-subharmonic function[1] equal to $-\infty$ on M. We consider the empty set to be pluripolar. As is well known, for a pluripolar set we have (see [20]) Γ-cap $M = 0$ and $\mathrm{cap}_{2n-2} M = 0$[2], and therefore for every $\epsilon > 0$ the $(2n - 2 + \epsilon)$-Hausdorf measure of M is zero [14].

Theorem 1.1. *Let $f \in Q_G$ and assume that for some $z_0 \in G$ the function $f(\cdot, z_0) \in K_{be}$. Then the function $f(\cdot, z)$ belongs to K_{be} for all $z \in G$, with a possible exception of a pluripolar set M_f of the class $G_{\delta\sigma}$. If $f \in Q_G^{(0)}$, then M_f belongs to the class F_σ.*

---

[1] We do not consider the "constant" $-\infty$ to be a pluri-subharmonic function.

[2] We denote here by $\mathrm{cap}_\alpha M$ the capacity generated by the kernel $r^{-\alpha}$ if $\alpha > 0$ and $\ln r$ if $\alpha = 0$.

The proof of Theorem 1.1 will follow from a sequence of lemmas. For $f \in Q_G$, $f \not\equiv 0$, $z \in G$, we set

$$\begin{aligned} u_f(\rho, z) &= \int_0^{2\pi} \ln |f(\rho e^{i\theta}, z)| d\theta \qquad (0 < \rho < 1), \\ u_f(z) &= \lim_{\rho \to 1} u_f(\rho, z), \\ v_f(z) &= \int_0^{2\pi} \ln |f(e^{i\theta}, z)| d\theta, \end{aligned}$$

with $f(e^{i\theta}, z) = \lim_{\rho \to 1} f(\rho e^{i\theta}, z)$.

Lemma 1.2. *Let $f \in Q_G$ ($G \subset \mathbb{C}^n$), $f \not\equiv 0$. Then the functions $u_f(\rho, z)$, $0 < \rho < 1$, and $v_f(z)$ are plurisubharmonic in G.*

This result follows directly from [15, Proposition I.14] on the integration of plurisubharmonic functions with respect to a parameter and from the obvious equalities

$$u_f(\rho, z) = \int_0^{2\pi} \ln |\tilde{f}(\rho e^{i\theta}, z)| d\theta, \qquad v_f(z) = \int_0^{2\pi} \ln |\tilde{f}(e^{i\theta}, z)| d\theta.$$

Clearly, we can assume without loss of generality that $f(0, z_0) \neq 0$. Let N_f denote the set of points $z \in G$ for which $f(0, z) = 0$ (and so $\tilde{f}(0, z) = 0$). Obviously, N_f is a proper analytic subset of G and consequently it is pluripolar and belongs to the class F_σ, and $G \setminus N_f$ is a domain (see, for instance, [2, Proposition 2.2]). It follows that it is enough to prove the theorem for the class $\widetilde{Q}_G$ (resp. $\widetilde{Q}_G^{(0)}$) consisting of those $f \in Q_G$ (resp. $f \in Q_G^{(0)}$) for which the function $f(\zeta, z)$ does not vanish identically in $\zeta \in \mathbb{D}$ for any $z \in G$.

For $f \in \widetilde{Q}_G$,

$$-\infty < u_f(\rho', z) \leq u_f(\rho'', z) \leq u_f(z) \leq v_f(z) \leq 0 \tag{1.1}$$

for $0 < \rho' < \rho'' < 1$ and $z \in G$.

Lemma 1.3. *Let $(g_k(z))$ be a sequence of functions holomorphic in $\mathbb{D}$ converging uniformly in $\mathbb{D}$ to a function $g(z) \not\equiv 0$. Then, for every $r \in (0, 1)$*

$$\lim_{k \to \infty} \int_0^{2\pi} \ln |g_k(re^{it})| dt = \int_0^{2\pi} \ln |g(re^{it})| dt.$$

For a proof, it is enough to use Jensen's formula for the functions g_k and g. A somewhat more general result is proved in [9].

Lemma 1.4. *Let $f \in \widetilde{Q}_G$. Then M_f consists of exactly those $z \in G$ for which $u_f(z) < v_f(z)$. The set M_f belongs to the class $G_{\delta\sigma}$ and for $f \in \widetilde{Q}_G^{(0)}$ — to the class F_σ.*

The first claim follows from (1.1) and from the well-known fact that for $x \in K$, $x \neq 0$, $x \in K_{be}$ if and only if

$$\lim_{\rho \to 1} \int_0^{2\pi} \ln |x(\rho e^{i\theta})| d\theta = \int_0^{2\pi} \ln |x(e^{i\theta})| d\theta. \tag{1.2}$$

To prove the second assertion, one can use [20, Lemma 1.5.2] and the fact that, by Lemma 1.3, the functions $u_f(\rho, z)$ are continuous functions of $z \in G$ for $f \in \widetilde{Q}_G^{(0)}$.

The following lemma is a weak version of Theorem 1.1.

Lemma 1.5. *Let $f \in \widetilde{Q}_G$ and assume that $f(\cdot, z_0) \in K_{be}$ for some $z_0 \in G$. Then M_f is a set of $2n$-dimensional Lebesgue measure 0.*

It is enough to consider the case when G is a polydisk and then use the Heine-Borel lemma. By Fubini's theorem we can reduce the proof to the case $n = 1$. Without loss of generality we will thus assume that $G = \mathbb{D}$, $f(\cdot, 0) \in K_{be}$ and $f(0, 0) \neq 0$.

Let $r \in (0, 1)$ and let ρ_k be an increasing sequence in $(0, 1)$ converging to 1. By condition (3) in the definition of the class Q_G there is a set E_1 of measure 2π such that for every $\theta \in E_1$, the limit

$$\widetilde{f}(e^{i\theta}, re^{it}) = \lim_{k \to \infty} \widetilde{f}(\rho_k e^{i\theta}, re^{it})$$

exists uniformly for $t \in [0, 2\pi]$. From Lemma 1.3 for $\theta \in E_1$ we have

$$\lim_{k \to \infty} \int_0^{2\pi} \ln \widetilde{f}(\rho_k e^{i\theta}, re^{it})| dt = \int_0^{2\pi} \ln \widetilde{f}(e^{i\theta}, re^{it})| dt \tag{1.3}$$

Using the theorems of Beppo Levi and of Fubini, we obtain

$$\int_0^{2\pi} u_f(re^{it}) dt = \lim_{k \to \infty} \int_0^{2\pi} u_f(\rho_k, re^{it}) dt = \lim_{k \to \infty} \int_0^{2\pi} d\theta \int_0^{2\pi} \ln |\widetilde{f}(\rho_k e^{i\theta}, re^{it})| dt. \tag{1.4}$$

Since $x(\cdot) \stackrel{\text{def}}{=} f(\cdot, 0) \in K_{be}$, (1.2) holds and it follows from Vitali's theorem on the passage to a limit that the integrals

$$F_k(\phi) \stackrel{\text{def}}{=} \int_0^{\phi} |\ln |\widetilde{f}(\rho_k e^{i\theta}, 0)| d\theta = \int_0^{\phi} |\ln f(\rho_k e^{i\theta}, 0)| d\theta \qquad (k \in \mathbb{N})$$

are uniformly absolutely continuous. From this fact and from the inequalities

$$0 \geq \int_0^{2\pi} \ln |\widetilde{f}(\rho_k e^{i\theta}, re^{it})| dt \geq 2\pi \ln |\widetilde{f}(\rho_k e^{i\theta}, 0)|,$$

making use again of Vitali's theorem we obtain from (1.4) and (1.3)

$$\begin{aligned}\int_0^{2\pi} u_f(re^{it})dt &= \int_0^{2\pi} d\theta \left(\lim_{k\to\infty} \int_0^{2\pi} \ln|\tilde{f}(\rho_k e^{is}, re^{it})|dt \right) \\ &= \int_0^{2\pi} \int_0^{2\pi} |\ln \tilde{f}(e^{is}, re^{it})|dt \\ &= \int_0^{2\pi} dt \int_0^{2\pi} \ln|\tilde{f}(e^{is}, re^{it})|ds \\ &= \int_0^{2\pi} v_f(re^{it})dt.\end{aligned}$$

Consequently, it follows from (1.1) that $u_f(re^{it}) = v_f(re^{it})$ for every $r \in (0,1)$ for almost every $t \in [0, 2\pi]$. Using Fubini's theorem, Lemma 1.4, and taking into account what we said in beginning of the proof of Lemma 1.5, we obtain the conclusion of Lemma 1.5.

We recall (see [20]) that for an arbitrary function $w(z)$ $(z \in G \subset \mathbb{C}^n)$ with values in $[-\infty, \infty)$ there exists an upper semicontinuous least upper majorant $w^+(z)$, called the (upper) regularization of w.

Lemma 1.6. *If $f \in \widetilde{Q}_G$ and if the function $f(\cdot, z_0) \in K_{be}$ for some $z_0 \in G$, then $u^+(z) = v_f(z)$.*

Indeed, the upper semicontinuity of the function $v_f(z)$ (Lemma 1.2) and (1.1) imply the inequalities $u_f(z) \le u_f^+(z) \le v_f(z)$ $(z \in G)$. Therefore, by virtue of Lemma 1.5 we have $u_f^+(z) = v_f(z)$ for almost all $z \in G$. From the subharmonicity of the left and right parts of these equalities it follows (see [20, Chapter I, Section 1, No. 4]) that these equalities hold in fact for all $z \in G$. The subharmonicity of the function $u_f^+(z)$ follows in turn (see [20, Chapter I, Section 5, No. 1]) from the fact that $u_f(z)$ is the upper envelope of a uniformly bounded family of subharmonic functions in G:

$$u_f(z) = \sup_{0<\rho<1} u_f(\rho, z), \quad u_f(\rho, z) \le 0.$$

Now, to complete the proof of Theorem 1.1, it suffices to use Lemmas 1.2, 1.4 and 1.6, and the fact that the set of points at which the upper envelope of a family of plurisubharmonic functions locally bounded from above is strictly less than its regularization is pluripolar (Bedford and Taylor [1]; see also [21]).

Theorem 1.7. *Let $f \in Q_G$ $(G \subset \mathbb{C}^n)$ and assume that the function $f(\cdot, z_0)$ belongs to K_i for some $z_0 \in G$. Then, the function $f(\cdot, z)$ belongs to K_i for all $z \in G$. If $f(\cdot, z_0)$ belongs to K_b then $f(\cdot, z)$ belongs to K_b for all $z \in G \setminus M_f$, where M_f is a pluripolar set of the class G_δ.*

Proof. If $f(\cdot, z_0)$ belongs to K_i we have $v_f(z_0) = 0$. Since $v_f(z) \le 0$ and is subharmonic in G (Lemma 1.2), the maximum modulus principle implies that $v_f(z) \equiv 0$ on G, and the first assertion in the statement of the theorem follows. The second assertion follows then from Theorem 1.1 and from the fact that $K_{be} \cap K_i = K_b$. □

Remark 1.8. The second assertion in the theorem follows also easily from a result of Lelong (see [15, Proposition 1.39].

Remark 1.9. It is not difficult to see that for $f \in Q_G^{(0)}$ the assertions of the theorem become trivial since in this case f does not depend on z.

2. Operator analogues of Frostman's theorem

Let $\mathcal{G}$ be a separable Hilbert space and let $[\mathcal{G}]$ denote the Banach space of bounded operators on $\mathcal{G}$. We denote by $\mathcal{R} = \mathcal{R}[\mathcal{G}]$ and $\overline{\mathcal{R}} = \overline{\mathcal{R}}[\mathcal{G}]$ the open and closed unit ball of $[\mathcal{G}]$ respectively. $\gamma_1 = \gamma_1[\mathcal{G}]$ denotes the ideal of nuclear (trace class) operators[3] in $[\mathcal{G}]$ and $I + \gamma_1$ denotes the set of elements $T \in [\mathcal{G}]$ such that $T - I_{\mathcal{G}} \in \gamma_1$.

For $A \in \overline{\mathcal{R}}$ and $B \in \mathcal{R}$ we set

$$\Phi(A, B) = (I - BB^*)^{-\frac{1}{2}}(A - B)(I - B^*A)^{-1}(I - B^*B)^{\frac{1}{2}}. \tag{2.1}$$

As is well known (see [13]), $\Phi(A, B) \in \overline{\mathcal{R}}$ and the map

$$Z \mapsto U\Phi(A, B)V$$

(U, V being unitary operators) represents the general form of a one-to-one linear-fractional map from $\mathcal{R}$ onto itself.

We remark (this will be used in Section 3) that the right-hand side of (2.1) may be rewritten as

$$\Phi(A, B) = -B + (I - BB^*)^{\frac{1}{2}}A(I - B^*A)^{-1}(I - B^*B)^{-\frac{1}{2}}, \tag{2.2}$$

and this expression still makes sense for any $A, B \in \overline{\mathcal{R}}$ such that $(I - B^*A)$ is boundedly invertible.

Theorem 2.1. *Let $X(\zeta)$ ($\zeta \in \mathbb{D}$) and $F(z)$ ($z \in G \subset \mathbb{C}^n$) be holomorphic functions with values in $\overline{\mathcal{R}} \cap (I + \gamma_1)$ and $\mathcal{R} \cap \gamma_1$ respectively. If* $\det \Phi(X(\cdot), F(z_0))$ *belongs to K_{be} for some $z_0 \in G$ then* $\det \Phi(X(\cdot), F(z))$ *belongs to K_{be} for all $z \in G$, with a possible exception of a pluripolar set M of class $G_{\delta\epsilon}$. If moreover* $\det \Phi(X(\cdot), F(z_0)) \in K_b$*, then* $\det \Phi(X(\cdot, F(z) \in K_b$ *for all $z \in G \setminus M$.*

Proof. We set

$$f(\zeta, z) = \det \Phi(X(\zeta), F(z)) = \frac{\det(X(\zeta) - F(z))}{\det(I - F(z)^*X(\zeta))},$$
$$\tilde{f}(\zeta, z) = \frac{\det(X(\zeta) - F(z))}{\det(I - X(\zeta)^*F(z))}.$$

Obviously, $|\tilde{f}(\zeta, z)| = |f(\zeta, z)| \leq 1$ ($\zeta \in \mathbb{D}, z \in G$).

---

[3] In this section and the following one we use some facts from the theory of nuclear operators and infinite determinants; see, e.g., [11].

To prove the analyticity of the functions f and $\tilde{f}$ with respect to ζ and z respectively[4], we consider an increasing sequence E_m of finite dimensional projection converging strongly to the identity operator, with E_m of rank m. Let $F_m(z) = E_m F(z) E_m$ and $X_m(\zeta) = E_m X(\zeta) E_m + (I - E_m)$. Let furthermore

$$f_m(z) = \det \Phi(X_m(\zeta), F_m(z)) = \frac{\det(X_m(\zeta) - F_m(z))}{\det(I - F_m(z)^* X_m(\zeta))}, \tag{2.3}$$

and

$$\tilde{f}_m(\zeta, z) = \frac{\det(X_m(\zeta) - F_m(z))}{\det(I - X_m(\zeta)^* F_m(z))}. \tag{2.4}$$

Since the infinite determinants in (2.3) and (2.4) can be expressed in terms of determinants of order m, the functions f_m and $\tilde{f}_m$ are respectively analytic in $\mathbb{D}$ for every $z \in G$ and in G for every $\zeta \in \mathbb{D}$. Since $F_m(z)$ and $X_m(\zeta) - I$ converge in the trace norm to $F(z)$ and $X(\zeta) - I$ respectively (for $z \in G$ and $\zeta \in \mathbb{D}$), we have

$$\lim_{m\to\infty} f_m(\zeta, z) = f(\zeta, z), \qquad \lim_{m\to\infty} \tilde{f}_m(\zeta, z) = \tilde{f}(\zeta, z).$$

It follows from the inequalities $|f_m(\zeta, z)| \le 1$ and $|\tilde{f}_m(\zeta, z)| \le 1$ that f is analytic in $\mathbb{D}$ and $\tilde{f}$ s analytic in G. It also follows that $f(\zeta, z)$ is Borel measurable on $\mathbb{D} \times G$.

To complete the proof that f belongs to Q_G it remains to check that condition (3b) of the definition of the class Q_G is fulfilled. Considering instead of $X(\zeta)$ and $F(z)$ the functions $\Phi(X(\zeta), F(z_0))$ and $\Phi(F(z), F(z_0))$, we assume that $\det X(\cdot) \in K_{be}$, and thus $X(\zeta)^{-1}$ exists for all $\zeta \in \mathbb{D}$ except of the sequence N of the roots of the function $\det X$. We have $X(\zeta)^{-1} = (\det X(\zeta))^{-1} Y(\zeta)$, where $Y(\cdot)$ is holomorphic in $\mathbb{D}$ with values in $\overline{\mathcal{R}}$. For $\zeta \in \mathbb{D} \setminus N$ and $z \in G$ we have

$$\tilde{f}(\zeta, z) = \det X(\zeta) \cdot \frac{\det(I - X(\zeta)^{-1} F(z))}{\det(I - X(\zeta)^* F(z))};$$

since the functions $X(\zeta)^*$ and $X(\zeta)^{-1}$ have strong nontangential limits almost everywhere on $\mathbb{T}$, and the multiplication by a trace class operator transforms strong convergence into trace class convergence, we conclude that for almost all $\xi \in \mathbb{T}$ and all $z \in G$ the nontangential limit $\tilde{f}(\zeta, z)$ exists. Thus $f \in Q_G$ and it remains to use Theorems 1.1 and 1.7. □

We note that the functions $W(\zeta)$ analytic in $\mathbb{D}$ with values in $\overline{\mathcal{R}} \cap (I + \gamma_1)$ and such that $\det W(\cdot) \in K_{be}$, appearing in Theorem 2.1, are characterized by admitting a representation

$$W(\zeta) = W_b(\zeta) W_e(\zeta).$$

[4]We avoid referring to [11, Chapter IV, Section 1, Proposition 8] because of a gap in the proof which is easy to remove in the situation we consider by the argument that follows.

In this expression, $W_b(\zeta)$ is a Blaschke-Potapov product and $W_e(\zeta)$ is an outer operator function:

$$W_b(\zeta) = U \overleftarrow{\prod_{j=1}^{r}} \left(\frac{\zeta_j - \zeta}{1 - \zeta_j^* \zeta} \cdot \frac{|\zeta_j|}{\zeta_j} P_j + Q_j \right),$$

where U is a unitary operator belonging to $I + \gamma_1$, $0 \leq r \leq \infty$, P_j is an orthogonal projection of rank one, $Q_j = I - P_j$, and $\sum_j (1 - |\zeta_j|) < \infty$[5];

$$W_e(\zeta) = \overleftarrow{\int_0^{2\pi}} \exp\left(\frac{\zeta + e^{i\theta}}{\zeta - e^{i\theta}} dH(\theta) \right),$$

where the integral is the multiplicative integral of Stieltjes-Potapov [19], $H(\theta)$ is an hermitian increasing operator-valued function on $[0, 2\pi]$ such that $\operatorname{tr} H(\theta)$ is absolutely continuous.

In the case of a finite dimensional $\mathcal{G}$ and under certain conditions on F it is possible to avoid the cumbersome a priori requirement that for some $z_0 \in G$ the function $\det \Phi(X(\cdot), F(z_0))$ belongs to the class K_{be}. Simple examples show that this requirement is necessary in general, but we recall that such a requirement does not appear in the theorems of Frostman and Heins. As as example we prove a statement which can be viewed as a natural matrix generalization of these theorems.

First we note that for $\dim \mathcal{G} = m < \infty$ the set $[\mathcal{G}]$ can be identified with $\mathbb{C}^{m^2}$ by associating to an element $Z \in [\mathcal{G}]$ the collection $z = (z_{11}, \dots, z_{1m}, \dots, z_{mm})$ of the elements of the matrix representation of Z with respect to a given orthonormal basis $e_1, \dots e_m$ ($z_{jk} = \langle Ze_j, e_k \rangle$, $j, k = 1, \dots, m$). Thus, the concept of a pluripolar set makes sense in $[G]$, and it is obviously independent of the chosen orthonormal basis.

Theorem 2.2. *Let $\mathcal{G}$ be a finite dimensional Hilbert space and let $X(\zeta)$ be a holomorphic function on $\mathbb{D}$ with values in $\overline{\mathcal{R}}[\mathcal{G}]$. Then for all $Z \in \mathcal{R}[\mathcal{G}]$ with a possible exception of a pluripolar set $\mathcal{M}$ of the class F_σ, the function* $\det \Phi(X(\cdot), Z)$ *belongs to K_{be}. If* $\det X(\cdot)$ *belongs to K_i (that is, $X(\cdot)$ is an inner operator function), then* $\det \Phi(X(\cdot), Z)$ *is a Blaschke product (i.e., the function $\Phi(X(\cdot), Z)$ is a Blaschke-Potapov product) for all $Z \in \mathcal{R} \setminus \mathcal{M}$.*

Proof. We set $f(\zeta, z) = \det \frac{1}{4}(X(\zeta) - Z)$, $G = \{z : \|Z\| < 3\} \subset \mathbb{C}^{m^2}$, where $z \in \mathbb{C}^{m^2}$ and $Z \in \mathcal{G}$ correspond as above. Clearly, $f \in Q_G^{(0)}$. Let $z_0 \in \mathbb{C}^{m^2}$ be the vector corresponding to the operator $Z_0 = 2I_m$. Since $f(\zeta, z_0) = \det(\frac{1}{2}(\frac{1}{2}X(\zeta) - I))$, the function $1/f(\zeta, Z_0)$ is bounded in $\mathbb{D}$, i.e., $f(\cdot, Z_0) \in K_e$. Theorem 1.1 then implies that $f(\cdot, Z) \in K_{be}$ for all $z \in \mathcal{G}$ with a possible exception of a pluripolar set of class F_σ. Since the functions

$$h_z(\zeta)^{\pm 1} = \left(\det \frac{1}{4}(I - Z^* X(\zeta))^{\pm 1} \right) \qquad (Z \in \mathcal{R})$$

[5]To simplify the notation we assume that the operator $W(0)$ is invertible, so that $\zeta_j \neq 0$ for all j.

are bounded in $\mathbb{D}$, $h_z(\zeta)$ is an outer function. The first claim of the theorem is proved.

The proof of the second part of the theorem is now immediate from the fact that the transformation (2.1) maps outer functions into outer functions and from the equality $K_{be} \cap K_i = K_b$. □

3. Typical spectral properties of analytic families of contractions

3.1.

The considerations of this section coincide for the most part with the considerations carried out in a less general situation in [7], [8], [25].

Let T be a linear contraction in a separable Hilbert space $\mathcal{G}$ ($T \in \overline{\mathcal{R}}[\mathcal{G}]$). We set $D_T = \sqrt{I - T^*T}$ and $\mathcal{D}_T = \text{closure}\, D_T\mathcal{G}$.

Let $\mathcal{L}_G$ denote the set of functions analytic in $G \subset \mathbb{C}^m$ and with values in $\overline{\mathcal{R}}[\mathcal{G}]$. If $T(\cdot) \in \mathcal{L}_G$, then (see [24]):

(a) The spaces $\mathcal{D}_{T(z)}\mathcal{G}$ and $\mathcal{D}_{T(z)^*}\mathcal{G}$ do not depend on $z \in \mathcal{G}$;
(b) For every $z_0 \in G$ there exists a uniquely determined function $\Delta_{z_0}(z)$ holomorphic on G with values in $[\mathcal{G}]$ such that for all $z \in G$

$$\Delta_{z_0}(z)(\mathcal{G} \ominus \mathcal{D}) = 0, \qquad \Delta_{z_0}(z)^*(\mathcal{G} \ominus \mathcal{D}_*) = 0 \tag{3.1}$$

(where we have set $\mathcal{D} = \mathcal{D}_{T(z)}$ and $\mathcal{D}_* = \mathcal{D}_{T(z)^*}$), and

$$T(z) - T(z_0) = D_{T(z_0)^*}\Delta_{z_0}D_{T(z_0)}. \tag{3.2}$$

Clearly,

$$\Delta_{z_0}(z) = \text{closure}\left(D^{[-1]}_{T(z_0)^*}(T(z) - T(z_0))D^{[-1]}_{T(z_0)}\right),$$

where the symbol $H^{[-1]}$ denotes the pseudo inverse of the self-adjoint operator H.

In [23] it is proved that the operators $I - T(z_0)^*\Delta_{z_0}(z)$ and $I - \Delta_{z_0}(z)T(z_0)^*$ are boundedly invertible for $z \in G$ and that the operators

$$\Lambda_{z_0}(z) = \Delta_{z_0}(z)(I - T(z_0)^*\Delta_{z_0}(z))^{-1} = (I - \Delta_{z_0}(z)T(z_0)^*)^{-1}\Delta_{z_0}(z) \tag{3.3}$$

are strict contractions ($\Lambda_{z_0}(z) \in \mathcal{R}[\mathcal{G}]$) for all $z \in G$. From (3.1) we get

$$\Lambda_{z_0}(z)\,(\mathcal{G} \ominus \mathcal{D}) = 0, \qquad \Lambda_{z_0}(z)^*\,(\mathcal{G} \ominus \mathcal{D}_*) = 0 \qquad (z \in G). \tag{3.4}$$

From (3.2), (3.3) and (3.4) it follows that

$$T(z) = \Phi(\Lambda_{z_0}(z), -T(z)) \qquad (z \in G), \tag{3.5}$$

where Φ is the linear fractional transformation given by (2.2). The contraction $\Lambda_{z_0}(z)$ is uniquely determined by (3.4) and (3.5) and is called (for obvious reasons) the Lobachevsky-Poincaré (*LP*) increment of the function $T(z)$ at the point z_0.

In the sequel we will be interested in the case of nuclear (and in particular finite rank) *LP* increments. It follows from (3.3) that the operator $\Lambda_{z_0}(z)$ will be nuclear if and only if the weighted increment $\Delta_{z_0}(z)$ is nuclear (which implies the nuclearity of the "regular" increment $T(z)-T(z_0)$). It is also clear that

$$\dim \Lambda_{z_0}(z)\mathcal{G} = \dim \Delta_{z_0}(z)\mathcal{G} = \dim(T(z)-T(z_0))\mathcal{G} \leq \dim \mathcal{D}_*.$$

3.2.

In obtaining theorems on the typical spectral properties of analytic families from the results of Section 2, an important role is played by the characteristic operator function (see [18]) of a contraction T in $\mathcal{G}$:

$$\theta_T(\zeta) = \left(-T + \zeta D_{T^*}(I-\zeta T^*)^{-1} D_T\right)|_{\mathcal{D}_T} = \Phi(\zeta I, T)|_{\mathcal{D}_T}, \tag{3.6}$$

whose values are contractions from $\mathcal{D}_T$ into $\mathcal{D}_{T^*}$ (notation: $\theta_T(\zeta) \in \overline{\mathcal{R}}[\mathcal{D}_T, \mathcal{D}_{T^*}]$) for every $\zeta \in \mathbb{D}$. As usual, we will define the characteristic function only up to a multiplication on the left and on the right by unitary operators, i.e., we will call characteristic operator function any function of the form

$$X(\zeta) = V\theta_T(\zeta)U^{-1}, \tag{3.7}$$

where U and V are unitary maps from $\mathcal{D}_T$ and $\mathcal{D}_{T^*}$ onto some Hilbert spaces $\mathcal{G}$ and $\mathcal{G}_*$ respectively.

We recall that a contraction T is called weak if its spectrum $\sigma(T)$ does not fill $\mathbb{D}$ and if $D_T^2 \in \gamma^1$. As will be clear from the sequel, for our purposes we may assume without loss of generality that $0 \notin \sigma(T)$ (by replacing if necessary T by $(T-\alpha I)(I-\alpha^* T)^{-1}$ with $\alpha \in \mathbb{D}$, $\alpha \notin \sigma(T)$, $|\alpha| < 1$). We will call a characteristic function $X(\zeta)$ (3.7) of the contraction T standard if $\mathcal{G} = \mathcal{G}_*$ and if $X(\zeta) \in \overline{\mathcal{R}}[\mathcal{G}] \cap (I+\gamma^1)$ for all $\zeta \in \mathbb{D}$. It is easily seen that every weak contraction has a standard characteristic function. For example, the function (3.7) is standard, with

$$U = I_{\mathcal{D}_T}, \qquad V = -\left((T^*T)^{\frac{1}{2}} T^{-1}\right)|_{\mathcal{D}_T^*}.$$

We introduce the class $\mathcal{L}_G^{(1)}(z_0)$ ($z_0 \in G$) which consists of all the functions $T(\cdot) \in \mathcal{L}_G$ for which $T(z_0)$ is a weak contraction and all the LP-increments of $T(\cdot)$ at z_0 are nuclear.

Theorem 3.1. *Let $T(\cdot) \in \mathcal{L}_G^{(1)}(z_0)$ and let X_{z_0} be a standard characteristic function of the form (3.7) of the contraction $T = T(z_0)$. Let $\Lambda_{z_0}(z)$ be the LP-increment function of $T(\cdot)$ at the point z_0 and*

$$\mathcal{D} \stackrel{\text{def}}{=} \mathcal{D}_{T(z_0)} = \mathcal{D}_{T(z)} \qquad (z \in G), \qquad \widetilde{\Lambda}_{z_0}(z) \stackrel{\text{def}}{=} V(\Lambda_{z_0}(z)|_{\mathcal{D}})U^{-1}.$$

Then the function

$$X_z(\zeta) = \Phi(X_{z_0}(\zeta), \widetilde{\Lambda}_{z_0}(z)) \tag{3.8}$$

is a standard characteristic operator function of the contraction $T(z)$ for every $z \in G$. This contraction is weak for all points of G, with a possible exception of the points of a certain $(n-1)$-dimensional analytic manifold, i.e., of a pluripolar set.

The first assertion follows directly from (3.4), (3.5), (3.6), (3.7); the computations are quite cumbersome (for details, see [10]). For the second assertion, it is enough to take N to be the zero set of the analytic function $\det((X_{z_0}(0) - \widetilde{\Lambda}_{z_0}(z))(I - X_{z_0}(0)^*\widetilde{\Lambda}_{z_0}(z))^{-1})$ on G, and use the equivalence between the bounded invertibility of $\theta_{T(z)}(0)$ and of $T(z)$. The fact that $D^2_{T(z)}$ is nuclear is immediate from the nuclearity of the operators $D^2_{T(z_0)}$ and $\Lambda_{z_0}(z)$ $(z \in G)$.

3.3.

We will say that a weak contraction $T \in \overline{\mathcal{R}}[\mathcal{G}]$ has an absolutely continuous (resp. singular) spectrum if $T = T_0 \oplus T_1$ where T_0 is a unitary operator with absolutely continous (resp. singular) spectrum, and T_1 is a completely nonunitary contraction of class C_{11} (resp. of class C_0).[6] This terminology is justified, to a certain extent, as follows. We recall ([18], [5]) that for every weak contraction T there exists a uniquely determined function $\Omega_T(\beta)$, defined on the σ-algebra $\mathcal{L}$ of Borel subsets of $\mathbb{C}$ with values subspaces of the space $\mathcal{G}$, satisfying the following properties:

1. $\Omega_T(\emptyset) = 0$ and $\Omega_T(\mathbb{C}) = \mathcal{G}$.
2. $\Omega_T(\cup_j \beta_j) = \bigvee_j \Omega_T(\beta_j)$ and $\Omega_T(\cap_j \beta_j) = \cap_j \Omega_T(\beta_j)$, for any sequence of Borel subsets $\beta_j \subset \mathbb{C}$.
3. For every closed Borel subset $\beta \subset \mathbb{C}$, $\Omega_T(\beta)$ is the greatest T-invariant subspace with the spectrum of the induced operator contained in β.

It turns out that T possesses an absolutely continuous spectrum (resp. singular spectrum) if and only if it happens that $\Omega_T(\beta) = 0$ for all $\beta \in \mathcal{L}$ of linear Hausdorff measure zero (resp. there is a $\beta_0 \in \mathcal{L}$ of linear Hausdorff measure zero such that $\Omega_T(\beta_0) = \mathcal{G}$). See [5].

From the theorem on the C_0-C_{11} decomposition of completely nonunitary weak contractions and from well-known results on unitary operators, it follows that for an arbitrary weak contraction T there is a decomposition $\mathcal{G} = \text{closure}(\mathcal{G}^{(a)} \dot{+} \mathcal{G}^{(s)})$ where $T^{(a)} = T|_{\mathcal{G}^{(a)}}$ and $T^{(b)} = T|_{\mathcal{G}^{(b)}}$ are weak contractions with absolutely continuous and singular spectrum respectively, called the absolutely continuous and the singular parts of T.

An operator T is called complete if the family of finite dimensional invariant subspaces of T is complete in $\mathcal{G}$. Obviously, such an operator has a singular spectrum. We call a weak contraction T complete in a generalized sense if its singular part is complete, i.e., if $\mathcal{G} = \mathcal{G}_1 \dot{+} \mathcal{G}_2$, where $T|_{\mathcal{G}_1}$ has an absolutelely continuous spectrum and $T|_{\mathcal{G}_2}$ is a complete operator.

The main result of this section is:

Theorem 3.2. *Let $T(z) \in \mathcal{L}^{(1)}_{\mathcal{G}}(z_0)$ (with $z_0 \in G \subset \mathbb{C}^n$) and suppose that $T(z_0)$ is a complete (resp. complete in a generalized sense) contraction. Then the same property holds for $T(z)$ for all z in G, with a possible exception of a pluripolar set $M \subset G$.*

---

[6] We use the notations of [18].

The proof of this assertion follows immediately from Theorems 2.1, 3.1 and from the following statements.

Lemma 3.3. *Let T be a completely nonunitary contraction and let $X(\zeta)$ be a standard characteristic function of T. In order that (I) T has an absolutely continuous spectrum, (II) T has a singular spectrum, (III) T is a complete operator, (IV) T is a complete operator in the generalized sense, it is necessary and sufficient that (I) $\det X(\cdot) \in K_e$, (II) $\det X(\cdot) \in K_i$, (III) $\det X(\cdot) \in K_b$, (IV) $\det X(\cdot) \in K_{be}$.*

Lemma 3.4. *Let $T(\cdot)$ be in $\mathcal{L}_G$; then all the contractions $T(z)$ $(z \in G)$ have the same unitary part.*

Lemma 3.3 is essentially well known; see [18], [4]. In particular, the equivalence between the definition of the absolute continuity of the spectrum of a completely nonunitary weak contraction given here and the definition given in [22] follows from this lemma. Lemma 3.4 follows from the maximum modulus (norm) principle for analytic functions with values bounded operators in a Hilbert space (see [18, Proposition 2.1, p. 188]).

Let now V be a completely nonunitary isometry in $\mathcal{G}$, let $\mathcal{G}_1$ and $\mathcal{G}_2$ be its defect spaces, and assume that $\dim \mathcal{G}_1 = \dim \mathcal{G}_2 = m < \infty$. Let $\mathcal{R}$ denote the open unit ball of the set $[\mathcal{G}_1, \mathcal{G}_2]$ of bounded operators from $\mathcal{G}_1$ into $\mathcal{G}_2$. One can introduce in $[\mathcal{G}_1, \mathcal{G}_2]$ in a natural way the notion of a pluripolar set (see Section 2). For $B \in \mathcal{R}$ we denote by V_B the extension of V such that $V_B|_{\mathcal{G}_1} = B$.

The next theorem follows from Theorems 2.2, 3.1 and Lemma 3.3:

Theorem 3.5. *With the above assumptions, the operator V_B is complete in the generalized sense for all $V \in \mathcal{R}$, with a possible exception of a pluripolar set $\mathcal{M}$. If V_{B_0} has a singular spectrum for some $B_0 \in \mathcal{B}$, then the operator V_B is complete for all $B \in \mathcal{R} \setminus \mathcal{M}$.*

We note that instead of the singularity of the spectrum of V_{B_0} we may require the singularity of the spectrum of some unitary extension U_0 of the isometry V. To see this, we consider the orthogonal decomposition of the identity $E(\lambda)$ $(0 \leq \lambda \leq 2\pi)$ associated to U_0 and let P_1 be the orthogonal projection on $\mathcal{G}_1$, $F(\lambda) = P_1 E(\lambda)|_{\mathcal{G}_1}$, and

$$S(\zeta) = \int_0^{2\pi} \frac{e^{i\lambda} + \zeta}{e^{i\lambda} - \zeta} dF(\lambda).$$

Then (see [16]) the function $X(\zeta) = (I - S(\zeta))(I + S(\zeta))^{-1}$ is a standard characteristic function of V_B for $B = 0$ (also called the characteristic function of the isometry V).

We also note that, in the setting of Theorem 3.5, formula (3.8) reduces to the formula of Livšic-Potapov [17] connecting the characteristic function of the isometry V and the characteristic function of the contraction V_B with $B \in \mathcal{R}$.

In conclusion, it is the author's pleasure to thank S. Yu. Favorov for helpful discussions.

References

[1] Eric Bedford and B. A. Taylor. A new capacity for plurisubharmonic functions. *Acta Math.*, 149(1-2):1–40, 1982.

[2] E. M. Chirka. *Complex analytic sets.* Kluwer Academic Publishers Group, Dordrecht, 1989. Translated from the Russian by R. A. M. Hoksbergen.

[3] O. Frostman. Potentiel d'équilibre et capacité des ensembles avec quelques applications à la théorie des fonctions. *Medd. Lunds Univ.Math. Semin.*, 3:1–118, 1935.

[4] Yu. P. Ginzburg. The spectral subspaces of contractions with a slowly increasing resolvent. *Mat. Issled.*, 5(vyp. 4):45–62, 1970.

[5] Yu. P. Ginzburg. Spectral measures and the duality of spectral subspaces of contractions with a slowly growing resolvent. *Zap. Nauchn. Sem. Leningrad. Otdel. Mat. Inst. Steklov. (LOMI)*, 73:203–206, 235 (1978), 1977. Investigations on linear operators and the theory of functions, VIII.

[6] Yu. P. Ginzburg. Almost invariant properties of analytic operator families. *Funktsional. Anal. i Prilozhen.*, 16(1):68–69, 1982.

[7] Yu. P. Ginzburg and N. A. Talyush. Polynomial pencils of analytic matrix functions and families of contracting extensions of isometric operators. *Izv. Vyssh. Uchebn. Zaved. Mat.*, 4:19–27, 1982. English translation in *Sov. Math.*, 26(4):21–32, 21–32.

[8] Yu. P. Ginzburg and N. A. Talyush. On exceptional sets of analytic matrix functions, contractive and dissipative operators. *Izv. Vyssh. Uchebn. Zaved. Mat.*, 8:9–14, 1984. English translation in *Sov. Math.*, 28(8):10–16, 1984.

[9] Yu. P. Ginzburg and A. A. Tarasenko. Analytic families of fractional-linear transformations of analytic operator-valued functions. In *Functional analysis*, No. 20, pp. 51–59. Ul̆yanovsk. Gos. Ped. Inst., Ul̆yanovsk, 1983.

[10] Yu. P. Ginzburg and A. A. Tarasenko. On singular parts of analytic contractive operator functions,. *Zap. Nauchn. Sem. Leningrad. Otdel. Mat. Inst. Steklov. (LOMI)*, 157:30–44, 1987.

[11] I. C. Gohberg and M. G. Krein. *Introduction to the theory of linear nonselfadjoint operators.* American Mathematical Society, Providence, R.I., 1969. Translations of Mathematical Monographs, Vol. 18.

[12] Maurice Heins. On the Lindelöf principle. *Ann. of Math. (2)*, 61:440–473, 1955.

[13] M. G. Kreĭn and Ju. L. Šmul̆jan. Fractional linear transformations with operator coefficients. *Mat. Issled*, 2(vyp. 3):64–96, 1967.

[14] N. S. Landkof. *Foundations of modern potential theory.* Springer-Verlag, New York, 1972. Translated from the Russian by A. P. Doohovskoy, Die Grundlehren der Mathematischen Wissenschaften, Band 180.

[15] P. Lelong and L. Gruman. *Entire functions of several complex variables.* Springer-Verlag, Berlin, 1986. Die Grundlehren der Mathematischen Wissenschaften, Band 282.

[16] M. S. Livšic. Isometric operators with equal deficiency indices, quasi-unitary operators. *Mat. Sbornik N.S.*, 26(68):247–264, 1950.

[17] M. S. Livšic and V. P. Potapov. A theorem on the multiplication of characteristic matrix functions. *Doklady Akad. Nauk SSSR (N.S.)*, 72:625–628, 1950.

[18] B. Sz. Nagy and C. Foias. *Harmonic analysis of operators on Hilbert spaces.* Akademia Kiado, Budapest, 1966.

[19] V.P. Potapov. The multiplicative structure of J-contractive matrix–functions. *Trudy Moskow. Mat. Obs.*, 4:125–236, 1955. English translation in: *American Mathematical Society Translations (2)*, vol. 15, pp. 131–243 (1960).

[20] L. I. Ronkin. *Introduction to the theory of entire functions of several variables.* American Mathematical Society, Providence, R.I., 1974. Translated from the Russian by Israel Program for Scientific Translations, Translations of Mathematical Monographs, Vol. 44.

[21] A. Sadullaev. Plurisubharmonic functions. In *Current problems in mathematics. Fundamental directions, Vol. 8*, pp. 65–113, 274. Akad. Nauk SSSR Vsesoyuz. Inst. Nauchn. i Tekhn. Inform., Moscow, 1985.

[22] L. A. Sakhnovich. Dissipative operators with absolutely continuous spectrum. *Trudy Moskow. Mat. Obs.*, 19:211–270, 1968.

[23] Yu. L. Shmuĺyan. Generalized fractional-linear transformations of operator spheres. *Sibirsk. Mat. Zh.*, 21(5):114–131, 191, 1980.

[24] Yu. L. Šmuĺjan. Certain stability properties for analytic operator-valued functions. *Mat. Zametki*, 20(4):511–520, 1976.

[25] A. A. Tarasenko. On exceptional values of analytic contractions with derivatives in H_1. VINITI, Odessa, 1985.

Odessa State Academy of Food Technologies, ul. Kanatnaya, 112, Odessa, Ukraine 270039

Operator Theory:
Advances and Applications, Vol. 123

Right and Left Joint System Representation of a Rational Matrix Function in General Position
(System Representation Theory for Dummies)

V. E. Katsnelson

Dedicated to Moshe Livšic, Morenu and Rabenu

Abstract. For a rational $k \times k$ matrix function R of one variable in general position, the matrix functions $R(x) \cdot R^{-1}(y)$ and $R^{-1}(x) \cdot R(y)$ of two variables are considered. For these matrix functions of two variables, the representations which are analogous to the system representation (or realization) of a rational matrix function of one variable are constructed. This representation of the function $R(x) \cdot R^{-1}(y)$ (of the function $R^{-1}(x) \cdot R(y)$) is said to be the joint right (respectively the joint left) system representation of the matrix functions R, R^{-1}. In these representations there appear diagonal $n \times n$ matrices, $A_{\mathcal{P}} = \operatorname{diag}(\lambda_1, \dots, \lambda_n)$ (called the pole matrix for R) and $A_{\mathcal{N}} = \operatorname{diag}(\mu_1, \dots, \mu_n)$ (called the zero matrix for R), where $\lambda_l, \dots, \lambda_n$ are poles of R, $\mu_l, \dots, \mu_n$ are poles of R^{-1}; and $k \times n$ matrices $F_{\mathcal{P}}$ and $F_{\mathcal{N}}$ (called the left pole and zero semi-residual matrices) and $n \times k$ matrices $G_{\mathcal{P}}$ and $G_{\mathcal{N}}$ (called the right pole and zero semi-residual matrices) which can be introduced from the additive decompositions $R(z) = R(\infty) + F_{\mathcal{P}}(zI - A_{\mathcal{P}})^{-1}G_{\mathcal{P}}$, $R^{-1}(z) = R(\infty)^{-1} + F_{\mathcal{N}}(zI - A_{\mathcal{N}})^{-1}G_{\mathcal{N}}$. The right joint system representation has the form

$$R(x) \cdot R^{-1}(y) = I + (x-y)F_{\mathcal{P}}(xI - A_{\mathcal{P}})^{-1}(S^r)^{-1}(yI - A_{\mathcal{N}})^{-1}G_{\mathcal{N}},$$

the left one has the form

$$R^{-1}(x) \cdot R(y) = I + (x-y)F_{\mathcal{N}}(xI - A_{\mathcal{N}})^{-1}(S^l)^{-1}(yI - A_{\mathcal{P}})^{-1}G_{\mathcal{P}}.$$

The $n \times n$ matrices S^r and S^l (the so-called right and left zero-pole coupling matrices for R) are solutions of the appropriate Sylvester-Lyapunov equations. These matrices are mutually inverse: $S^r \cdot S^l = S^l \cdot S^r = I$.

These results are essentially not new: they could be easily derived from known results on realization of a rational matrix functions (for example, from results by L. Sakhnovich or J. Ball, I. Gohberg, L. Rodman), however the method is new, as well as the emphasis on "the left, the right and their relationships". The presentation is oriented to a "traditional" analyst. No previous knowledge in realization theory of matrix functions or its ideology is assumed. One of the purposes of this paper is to provide a realization

1991 *Mathematics Subject Classification.* 93B15.

theory background for investigations of the deformation theory of a Fuchsian differential system and of rational solutions of the Schlesinger system. As an application we also consider the spectral (Wiener-Hopf) factorization.

The concluding Section 5 contains some historical remarks highlighting the role of M.S. Livšic as the forefather of the system realization theory.

Notations

- $\mathbb{C}$ stands for the complex plane; $\overline{\mathbb{C}}$ is the extended complex plane: $\overline{\mathbb{C}} \stackrel{\text{def}}{=} \mathbb{C} \cup \infty$;
- $\mathfrak{M}_k$ stands for the set of all $k \times k$ matrices with complex entries;
- I stands for the unit matrix of the appropriate dimension;
- $\mathcal{R}(\mathfrak{M}_k)$ stands for the set of all rational $\mathfrak{M}_k$-valued functions R with $\det R(z) \not\equiv 0$.
- $\mathcal{P}(R)$ stands for the set of all poles of the function R, $\mathcal{N}(R)$ stands for the set of all poles of the function R^{-1}; $\mathcal{P}(R)$ is said to be the pole set of the function R, $\mathcal{N}(R)$ is said to be the zero set of the function R.

0. Preface

The problem which we set as a goal in this paper for scalar (i.e., complex valued) functions means *to restore a rational function from its poles and zeros.* The traditional solution of this problem uses products constructed from the poles and the zeros of the function. For rational functions *in general position*, this solution can be explained particularly clearly.

Namely, let r be a rational function in general position[1], with the pole set $\mathcal{P}(r)$ and the zero set $\mathcal{N}(r)$. These sets $\mathcal{P}(r)$ and $\mathcal{N}(r)$ do not intersect (i.e., $\mathcal{P}(r) \cap \mathcal{N}(r) = \emptyset$) and are of the same cardinality: $\#\mathcal{P}(r) = \#\mathcal{N}(r)$. The function r admits a representation

$$r(z) = c \left(\prod_{\mu_l \in \mathcal{N}(r)} (z - \mu_l) \right) \cdot \left(\prod_{\lambda_j \in \mathcal{P}(r)} (z - \lambda_j) \right)^{-1}, \tag{0.1}$$

where $c = r(\infty)$. This *multiplicative* representation recovers the function r from its pole and zero sets and from the value $r(\infty)$. Conversely, given two finite non-intersecting sets $\mathcal{P}$ and $\mathcal{N}$ ($\mathcal{P}, \mathcal{N} \subset \mathbb{C}$, $\mathcal{P} \cap \mathcal{N} = \emptyset$) of the same cardinality and a complex number $c \neq 0, \infty$, we define the function r by the formula (0.1). This function r is a rational function in general position, the given sets $\mathcal{P}$ and $\mathcal{N}$ are its pole and zero sets $\mathcal{P}(r)$ and $\mathcal{N}(r)$ and $c = r(\infty)$.

However, in view of the non-commutativity of the matrix multiplication, the multiplicative representation (0.1) seems to be unsuitable for generalization to matrix functions. We present now such a representation of a rational function (in general position) based on its poles and zeros which can be generalized to the matrix case. This is the so-called system representation of a rational function.

[1] This means that all the poles and all the zeros of the function r are simple and $r(\infty) \neq 0, \infty$.

So, let again r be a rational function in general position, with the pole and zero sets $\mathcal{P}(r)$ and $\mathcal{N}(r)$. We derive its system representation. We assume for simplicity that the function r is normalized by the condition $r(\infty) = 1$. We start from the additive decomposition of the matrix function r:

$$r(z) = 1 + \sum_{\lambda_q \in \mathcal{P}(r)} \frac{\xi_q}{z - \lambda_q}. \tag{0.2}$$

The condition

$$r(\mu_p) = 0 \qquad (\forall \mu_p \in \mathcal{N}(r)) \tag{0.3}$$

leads to the system of linear equations

$$\sum_{\lambda_q \in \mathcal{P}(r)} \frac{\xi_q}{\mu_p - \lambda_q} = -1 \qquad (\forall \mu_p \in \mathcal{N}(r)). \tag{0.4}$$

Thus to restore a rational function in general position from its poles and zeros, we have to solve the linear system (0.4) with respect to ξ_q and then to substitute these ξ_q into (0.2). Since $\#\mathcal{P}(r) = \#\mathcal{N}(r) \overset{\text{def}}{=} n$, the matrix S of the system (0.4) is square:

$$S = \|s_{p,q}\|_{1 \leq p,q \leq n}, \qquad s_{p,q} = \frac{1}{\mu_p - \lambda_q}. \tag{0.5}$$

The system (0.4) is uniquely solvable: its determinant (which is known as the *Cauchy determinant*) can be calculated explicitly (see, for example, [PS, Part VII, Section 1, No. 3]). From this explicit expression for the determinant it is evident that $\det S \neq 0$.

We can formulate this method of restoring the function r from $\mathcal{P}(r)$ and $\mathcal{N}(r)$ in the matrix form. Let $A_{\mathcal{P}}$ and $A_{\mathcal{N}}$ be the diagonal matrices constructed from $\mathcal{P}(r)$ and $\mathcal{N}(r)$:

$$A_{\mathcal{P}} = \operatorname{diag}(\lambda_1, \lambda_2, \dots \lambda_n), \qquad A_{\mathcal{N}} = \operatorname{diag}(\mu_1, \mu_2, \dots \mu_n). \tag{0.6}$$

Let e be the n-row (i.e., $1 \times n$ matrix):

$$e = [1, 1, \dots, 1]. \tag{0.7}$$

As usual, by e^* we denote the Hermitian conjugate to e: e^* is a n-column (i.e., $n \times 1$ matrix). The representation (0.2) can be written in the form[2]

$$r(z) = 1 + e\,(zI - A_{\mathcal{P}})^{-1}[\xi_1, \xi_2, \dots, \xi_n]^T. \tag{0.8}$$

The system (0.4) can be represented in a matrix form: $S[\xi_1, \xi_2, \dots, \xi_n]^T = -e^*$. Thus,

$$[\xi_1, \xi_2, \dots, \xi_n]^T = -S^{-1}e^*.$$

[2]As usual, for a matrix M, M^T denotes the transposed matrix.

Substituting this expression for $[\xi_1, \xi_2, \ldots, \xi_n]^T$ into (0.8), we arrive at the representation

$$r(z) = 1 - e\,(zI - A_{\mathcal{P}})^{-1} S^{-1} e^*. \tag{0.9}$$

In the same way we can obtain a representation for the function r^{-1}. Starting from the additive representation

$$r^{-1}(z) = 1 + \sum_{\mu_p \in \mathcal{N}(r)} \frac{\eta_p}{z - \mu_p}, \tag{0.10}$$

we arrive at the linear system (with respect to η_p)

$$\sum_{\mu_p \in \mathcal{N}(r)} \frac{\eta_p}{\mu_p - \lambda_q} = 1 \qquad (\forall \lambda_q \in \mathcal{P}(r)),$$

which can be written in a matrix form:

$$[\eta_1, \eta_2, \ldots, \eta_n] S = e,$$

or

$$[\eta_1, \eta_2, \ldots, \eta_n] = e\,S^{-1}.$$

Thus,

$$r^{-1}(z) = 1 + eS^{-1}(zI - A_{\mathcal{N}})^{-1} e^*. \tag{0.11}$$

Here the matrix S is the same as in (0.5)!

The matrices S, $A_{\mathcal{P}}$, $A_{\mathcal{N}}$ satisfy the equality

$$A_{\mathcal{N}} S - S A_{\mathcal{P}} = e^* e, \tag{0.12}$$

which can be directly obtained from (0.5), (0.6), (0.7). The equality (0.12) and its generalizations are of fundamental importance in the theory we are developping.

The equality of the form $AX - XB = C$ (where A, B, C are square matrices) is known as the Sylvester-Lyapunov equality (or as the Sylvester-Lyapunov equation, if it is considered as an equation with respect to X).

Multiplying the right-hand sides of the representations (0.9) and (0.11) term by term, we obtain the equality

$$r(x) r^{-1}(y) = 1 + (x - y)\,e\,(xI - A_{\mathcal{P}})^{-1} S^{-1} (yI - A_{\mathcal{N}})^{-1} e^* \tag{0.13}$$

after some calculations. These calculations are based on the Sylvester-Lyapunov equality (0.12). The representations (0.9) and (0.11) are exactly what we need. They are said to be the system representations of the function $\boldsymbol{r}$ and $\boldsymbol{r^{-1}}$ respectively. The representation (0.13) is said to be the joint system representation of the pair of the (mutually inverse) functions $\boldsymbol{r}$ and $\boldsymbol{r^{-1}}$.

We obtained the joint system realization (0.13) from the system representations (0.9) and (0.11). Conversely, the representations (0.9) and (0.11) are contained in (0.13): (0.9) is (0.13) for $y = \infty$, (0.11) is (0.13) for $x = \infty$.

Now we derive the joint system representation (0.13) in a different way. Let k be the function

$$k(x,y) = \frac{r(x)r^{-1}(y) - 1}{x - y} \tag{0.14}$$

of two variables. Fixing $y \notin (\mathcal{P}(r) \cup \mathcal{N}(r) \cup \infty)$, we consider k as a function of the variable x. This function is rational (with respect to x), with simple poles located at the points $\lambda_p \in \mathcal{P}(r)$, with the residues $\rho_{\lambda_p(y)}$:

$$\rho_{\lambda_p(y)} = \frac{1}{(r^{-1})\,'(\lambda_p)} \cdot \frac{r^{-1}(y)}{\lambda_p - y}. \tag{0.15}$$

It is clear that $k(\infty, y) = 0$ and that $k(x,y)$ is holomorphic for $x = y$. Thus $k(x,y)$ admits an expansion in simple fractions

$$k(x,y) = \sum_{1 \le p \le n} \frac{1}{x - \lambda_p} \cdot \rho_{\lambda_p(y)}. \tag{0.16}$$

Next, the residue $\rho_{\lambda_p(y)}$, considered as a function of y, is a rational function. Since $r^{-1}(\lambda_p) = 0$, the point λ_p is not a pole of the function ρ_{λ_p}, and the points $\mu_q \in \mathcal{N}(r^{-1})$ are the only poles of the function ρ_{λ_p}. All these poles are simple, with the residues $h_{p,q} = \operatorname{res} \rho_{\lambda_p}(y)\,_{|y=\mu_q}$:

$$h_{p,q} = \frac{1}{(r^{-1})\,'(\lambda_p)} \cdot \frac{1}{\lambda_p - \mu_q} \cdot \frac{1}{r\,'(\mu_q)}. \tag{0.17}$$

As $\rho_{\lambda_p}(\infty) = 0$, the function ρ_{λ_p} admits an expansion in simple fractions

$$\rho_{\lambda_p}(y) = \sum_{1 \le q \le n} h_{p,q} \, \frac{1}{y - \mu_q}. \tag{0.18}$$

Substituting (0.18) into (0.16), we arrive at the formula

$$k(x,y) = \sum_{1 \le p,q \le n} \frac{1}{x - \lambda_p} \cdot h_{p,q} \cdot \frac{1}{y - \mu_q}, \tag{0.19}$$

where $h_{p,q}$ are defined by (0.17). The last formula can be represented in a matrix form. Let us organize the numbers $h_{p,q}$ into the matrix

$$H = \|h_{p,q}\|_{1 \le p,\, q \le n}. \tag{0.20}$$

Thus

$$k(x,y) = e(xI - A_{\mathcal{P}})^{-1} H (yI - A_{\mathcal{N}})^{-1} e^*. \tag{0.21}$$

and

$$r(x)r^{-1}(y) = 1 + e(xI - A_{\mathcal{P}})^{-1} H (yI - A_{\mathcal{N}})^{-1} e^*. \tag{0.22}$$

The comparison of the formulas (0.13) and (0.22) suggests that $H = S^{-1}$. The equality

$$HS = I \tag{0.23}$$

may be verified starting from the formula

$$r^{-1}(x) = 1 + \sum_{1\le q\le n} \frac{1}{x-\mu_q}\cdot\frac{1}{r'(\mu_q)}\,. \tag{0.24}$$

The matrix identity (0.23) is equivalent to the system of n^2 scalar identities

$$\sum_{1\le q\le n} h_{p,q}\, s_{q,t} = 0 \qquad (1\le p,t\le n,\ p\ne t) \tag{0.25}$$

and

$$\sum_{1\le q\le n} h_{p,q}\, s_{q,p} = 1 \qquad (1\le p\le n). \tag{0.26}$$

From (0.24) (and, of course, from (0.5) and (0.17)) it follows that[3]

(0.27)

$$\begin{aligned}\sum_{1\le q\le n} h_{p,q}\, s_{q,t} &= \sum_{1\le q\le n} \frac{1}{\lambda_p-\mu_q}\cdot\frac{1}{\mu_q-\lambda_t}\cdot\frac{1}{(r^{-1})'(\lambda_p)}\cdot\frac{1}{r'(\mu_q)}\\ &= \frac{1}{(r^{-1})'(\lambda_p)(\lambda_p-\lambda_t)}\Big(r^{-1}(\lambda_p)-r^{-1}(\lambda_t)\Big)=0 \qquad \text{for } p\ne t,\end{aligned}$$

since $r^{-1}(\lambda_k)=0$ for all $\lambda_k\in\mathcal{P}(r)(=\mathcal{N}(r^{-1}))$. Thus (0.25) is verified. Analogously, the equality (0.26) may be verified using the formula

$$(r^{-1})'(x) = -\sum_{1\le q\le n} \frac{1}{(x-\mu_q)^2}\cdot\frac{1}{r'(\mu_q)}. \tag{0.28}$$

Thus the equality (0.23) holds. Finally, the formula (0.22) can be represented in the form (0.13), with S of the form (0.5).

Remark 0.1. By the way, we obtained the following rule for the inversion of the matrix S of the form (0.5), where $\lambda_1, \lambda_2, \dots, \lambda_n;\ \mu_1, \mu_2, \dots, \mu_n$, are pairwise different complex numbers (this matrix is known as the Cauchy matrix). Starting from these numbers, we have to construct the rational function r of the form (0.1) (with some $c:\ c\ne 0,\ \infty$). The entries $h_{p,q}$ of the inverse matrix $H=S^{-1}$ are of the form (0.17). The expression (0.17) for the inverse matrix S^{-1} may be represented in the matrix form:

$$S^{-1} = D_1 S D_2,\quad \text{where } D_1=\operatorname{diag}\left(\frac{1}{(r^{-1})'(\lambda_l)}\right),\ D_2=\operatorname{diag}\left(\frac{1}{r'(\mu_l)}\right). \tag{0.29}$$

In particular, we established the invertibility of the Cauchy matrix S. Of course, the expression (0.17) for the entries of the inverse matrix S^{-1} differs from the expression obtained in ([BGR1]) in form only. In ([BGR1]) (Lemma 6.3 there) this expression was obtained by means of the Kramer matrix inversion rule, using explicit expression for the Cauchy determinant. (The determinant of the matrix S as well as its cofactors are Cauchy determinants.)

---

[3] $\dfrac{1}{\lambda_p-\mu_q}\cdot\dfrac{1}{\mu_q-\lambda_t}=\dfrac{1}{\lambda_p-\lambda_t}\cdot\left(\dfrac{1}{\lambda_p-\mu_q}-\dfrac{1}{\lambda_t-\mu_q}\right).$

The square of the Cauchy determinant can be calculated from (0.29):

$$(\det S)^2 = \prod_{1\le p\le n} (r^{-1})'(\lambda_p) \cdot \prod_{1\le q\le n} r'(\mu_q). \tag{0.30}$$

Remark 0.2. This derivation of the joint system representation can be extended to some classes of meromorphic functions.

Obtaining "system" representations of "scalar" rational functions, we made no use of anything that can not be generalized to the matrix case. Generalizations of the representations (0.9) and (0.11) to rational matrix functions have been already obtained. (See [S3], [GKLR], [BGR1], [BGR2], [BGRa].)

The term "system representation" is related to system theory. In this theory one associates to each linear stationary time invariant dynamical system its transfer function, which is a matrix valued (or even an operator valued) function. If the state space of the system is finite dimensional, then this transfer function is rational. And if the system is a SISO system (Single Input, Single Output), then its transfer function is scalar (complex valued). This transfer function is expressed in terms of the input, the output and the state space operators of the system. Conversely, each rational matrix function may be represented as the transfer function of an appropriate linear time invariant system with finite dimensional state space. Such a representation[4] is said to be the system realization or the system representation of the given rational function.

The relationship between rational functions and linear systems can be exploited in both directions. We may apply results from matrix function theory to the study of linear systems. And we may use the system representation as a tool for the study of rational matrix functions or as a tool to specify them.

It should be mentioned that it was M. Livšic who discovered the characteristic matrix function of linear operators and operator colligations (or operators nodes). He was the first to relate invariant subspaces of a linear operator and factors of its characteristic matrix function, [L2], [L3], [LP]. In the system theory language, the characteristic function of an operator node is just the transfer function of an appropriate linear time invariant system. This was shown by M.S. Livšic in [L7], [L8], [L9], [BrL]. He also related the characteristic matrix function with the scattering theory ([L5], [L6], [L9], [BrL]).

The main goal of this paper is to present some basic results on the system representation of rational matrix function in a self-contained form. For the clarity of presentation, we restrict our consideration to the simple but important case of a rational matrix function in general position. The presentation is oriented to a "traditional" analyst. No previous knowledge in realization theory of matrix functions or its ideology is assumed. We leave detailed historical remarks for Section 5.

[4]It looks like the representation (0.9).

1. Rational matrix functions in general position

Definition 1.1. *Let $M(z)$ be a $k \times k$ matrix function holomorphic in a punctured neighborhood of some point a (i.e., the point a is an isolated singularity of the function M). The point a is said to be* a simple pole of the matrix function M *if*

$$M(z) = \frac{M_a}{z-a} + H(z), \tag{1.1}$$

where M_a is a constant matrix and the matrix function H is holomorphic at the point a. The matrix M_a is said to be the residue of the matrix function M at the point a.

Definition 1.2. *Let a point a be a simple pole of a $k \times k$ matrix function M and let the residue M_a of the function M at the point a be a matrix of rank one. As a matrix of rank one, the $k \times k$ matrix M_a can be factorized in the form*

$$M_a = f_a \cdot g_a \tag{1.2}$$

where f_a is a k-column-vector (i.e., a $k \times 1$ matrix), g_a is a k-row-vector (i.e., a $1 \times k$ matrix), $f_a \neq 0, g_a \neq 0$. The vectors f_a and g_a are said to be the left semi-residual vector *and* the right semi-residual vector at the point a, *respectively. The vectors f_a and g_a are defined uniquely, up to a constant factor: we can represent the number 1 in the form $1 = d \cdot d^{-1}$ and then redistribute the factors d and d^{-1}:*

$$f_a \to f_a \cdot d; \quad g_a \to d^{-1} \cdot g_a. \tag{1.3}$$

We emphasize that the notions of left and right semi-residual vectors are defined only for a simple pole with residue of rank one.

Definition 1.3. *A rational matrix function R ($R \in \mathcal{R}(\mathfrak{M}_k)$) is said to be* a rational matrix function in general position *if:*

1. *The pole set $\mathcal{P}(R)$ and the zero set $\mathcal{N}(R)$ do not intersect:*

$$\mathcal{P}(R) \cap \mathcal{N}(R) = 0; \tag{1.4}$$

2. *All the poles of the function R are simple, and the residues at these poles are matrices of rank one;*
3. *All the poles of the function R^{-1} are simple, and the residues at these poles are matrices of rank one;*
4. *Both functions R and R^{-1} are holomorphic at the point $z = \infty$.*

Let R be a rational matrix function in general position, with the pole set $\mathcal{P}(R)$ and the zero set $\mathcal{N}(R)$. For $\lambda \in \mathcal{P}(R)$, $\mu \in \mathcal{N}(R)$, let

$$R_\lambda = f_\lambda \cdot g_\lambda, \quad (f_\lambda \neq 0, g_\lambda \neq 0); \qquad R_\mu = f_\mu \cdot g_\mu \quad (f_\mu \neq 0, g_\mu \neq 0) \tag{1.5}$$

be factorizations of the residue R_λ of the function R at the point λ and the residue R_μ of the function R^{-1} at the point μ respectively, where $f_\lambda, g_\lambda; f_\mu, g_\mu$, are the appropriate semi-residual vectors. The additive expansions

$$R(z) = R(\infty) + \sum_{\lambda \in \mathcal{P}(R)} \frac{R_\lambda}{z-\lambda}, \tag{1.6}$$

$$R^{-1}(z) = R^{-1}(\infty) + \sum_{\mu \in \mathcal{N}(R)} \frac{R_\mu}{z-\mu} \tag{1.7}$$

can be rewritten in the form

$$R(z) = R(\infty) + \sum_{\lambda \in \mathcal{P}(R)} f_\lambda \cdot \frac{1}{z-\lambda} \cdot g_\lambda, \tag{1.8}$$

$$R^{-1}(z) = R^{-1}(\infty) + \sum_{\mu \in \mathcal{N}(R)} f_\mu \cdot \frac{1}{z-\mu} \cdot g_\mu. \tag{1.9}$$

Lemma 1.1. I. *Let a matrix function R be analytic in a punctured neighborhood of a point $\lambda \in \mathbb{C}$, let λ be a simple pole of the function R and a regular point of the function R^{-1}, and let R_λ be the residue of the function R at the point λ. Then the point λ is a simple pole for the "logarithmic derivative" $R' \cdot R^{-1}$, and for the residue P_λ of this "logarithmic derivative" at this point the conditions*

$$P_\lambda^2 = -P_\lambda, \quad \mathrm{rank} P_\lambda = \mathrm{rank} R_\lambda \tag{1.10}$$

hold; in particular,

$$\mathrm{trace} P_\lambda = -\mathrm{rank} R_\lambda. \tag{1.11}$$

II. *Let a matrix function R^{-1} be analytic in a punctured neighborhood of a point $\mu \in \mathbb{C}$, let μ be a simple pole of the function R^{-1} and a regular point of the function R, and let R_μ be the residue of the function R^{-1} at the point μ. Then the point μ is a simple pole for the "logarithmic derivative" $R' \cdot R^{-1}$, and for the residue P_μ of this "logarithmic derivative" at this point the conditions*

$$P_\mu^2 = P_\mu, \quad \mathrm{rank} P_\mu = \mathrm{rank} R_\mu \tag{1.12}$$

hold; in particular,

$$\mathrm{trace} P_\mu = \mathrm{rank} R_\mu. \tag{1.13}$$

Proof. We prove only Statement I of the Lemma. Statement II can be proved analogously. Let

$$R(z) = \frac{R_\lambda}{z-\lambda} + A_0 + A_1(z-\lambda) + A_2(z-\lambda)^2 + \cdots \tag{1.14}$$

and

$$R^{-1}(z) = B_0 + B_1(z-\lambda) + B_2(z-\lambda)^2 + \cdots \tag{1.15}$$

be the Laurent expansions of the functions R and R^{-1} respectively. Then

$$R'(z) = -\frac{R_\lambda}{(z-\lambda)^2} + A_1 + 2A_2(z-\lambda) + \cdots. \tag{1.16}$$

Multiplying the Laurent expansions term by term, we obtain from (1.15) and (1.16)

$$R'(z) \cdot R^{-1}(z) = -\frac{R_\lambda B_0}{(z-\lambda)^2} - \frac{R_\lambda B_1}{z-\lambda} + (-RB_2 + A_1 B_0) + O(z-\lambda). \tag{1.17}$$

Substituting the Laurent expansions (1.14), (1.15) into the identity $R^{-1}(z)R(z) = I$ and multiplying these expansions term by term, we obtain

$$B_0 R_\lambda = 0, \tag{1.18}$$

$$B_1 R_\lambda + B_0 A_0 = I. \tag{1.19}$$

Analogously, from the identity $R(z)R^{-1}(z) = I$ we derive

$$R_\lambda B_0 = 0, \tag{1.20}$$

$$R_\lambda B_1 + A_0 B_0 = I. \tag{1.21}$$

Let us examine the expansion (1.17). According to (1.20), the term $-\frac{R_\lambda B_0}{(z-\lambda)^2}$ vanishes. Thus the point λ is a simple pole for the function $R'(z)R^{-1}(z)$, with the residue P_λ,

$$P_\lambda = -R_\lambda B_1. \tag{1.22}$$

From (1.21) and (1.22) it follows that $I + P_\lambda = A_0 B_0$. Hence,

$$(I + P_\lambda)P_\lambda = (A_0 B_0) \cdot (-R_\lambda B_1) = -A_0(B_0 R_\lambda)B_1.$$

According to (1.18), $B_0 R_\lambda = 0$. Thus $(I + P_\lambda)P_\lambda = 0$, i.e., $P_\lambda^2 = -P_\lambda$. Finally,

$$P_\lambda R_\lambda = -(I - A_0 B_0)R_\lambda = -R_\lambda + A_0(B_0 R_\lambda).$$

Since $B_0 R_\lambda = 0$,

$$R_\lambda = -P_\lambda R_\lambda. \tag{1.23}$$

From (1.22) and (1.23) it follows that $\operatorname{rank} P_\lambda = \operatorname{rank} R_\lambda$. □

Remark 1.1. From (1.19) and (1.20) it follows that $R_\lambda B_1 R_\lambda = R_\lambda$. Since $B_1 = (R^{-1})'(\lambda)$ (see 1.15), it can be written as

$$R_\lambda \left(R^{-1}\right)'(\lambda)\, R_\lambda = R_\lambda \quad (\lambda \in \mathcal{P}). \tag{1.24}$$

Analogously, we derive

$$R_\mu\, R'(\mu)\, R_\mu = R_\mu \quad (\mu \in \mathcal{N}). \tag{1.25}$$

We shall use the relations (1.24) and (1.25) in the following section.

Lemma 1.2. *For a rational matrix function R in general position,*

$$\#\mathcal{P}(R) = \#\mathcal{N}(R) \quad \left(\overset{\text{def}}{=} n(R)\right). \tag{1.26}$$

Proof. To prove the statement of the Lemma, we have to consider, in some way or another, the "logarithmic derivative" $R'(z) \cdot R^{-1}(z)$ (or $R^{-1}(z) \cdot R'(z)$). For a rational matrix function R in general position, its logarithmic derivative $R' \cdot R^{-1}$ admits an expansion

$$R'(z) \cdot R^{-1}(z) = \sum_{\lambda \in \mathcal{P}(R)} \frac{P_\lambda}{z-\lambda} + \sum_{\mu \in \mathcal{P}(R)} \frac{P_\mu}{z-\mu}. \tag{1.27}$$

Indeed, the (rational) function $R'(z) \cdot R^{-1}(z)$ may have singularities only at the points of the set $\mathcal{P}(R) \cup \mathcal{N}(R)$. According to item 4 of Definition 1.3, the logarithmic derivative vanishes at the point ∞; moreover,

$$R'(z) \cdot R^{-1}(z) = O(|z|^{-2}) \quad (z \to \infty). \tag{1.28}$$

According to Lemma 1.1, all the singularities of the logarithmic derivative are simple poles with residues P_λ and P_μ of rank one ($\lambda \in \mathcal{P}(R)$, $\mu \in \mathcal{N}(R)$, respectively). Thus the expansion (1.27) holds, with $\operatorname{trace} P_\lambda = -1$, $\operatorname{trace} P_\mu = 1$. From (1.27) and (1.28) it follows that

$$\sum_{\lambda \in \mathcal{P}(R)} P_\lambda + \sum_{\mu \in \mathcal{N}(R)} P_\mu = 0. \tag{1.29}$$

Since $\operatorname{trace} P_\lambda = -1$ ($\lambda \in \mathcal{P}(R)$)), $\operatorname{trace} P_\mu = 1$ ($\mu \in \mathcal{N}(R)$)), according to Lemma 1.1, it follows from (1.29) that $\sum\limits_{\lambda \in \mathcal{P}(R)} (-1) + \sum\limits_{\mu \in \mathcal{N}(R)} 1 = 0$. This equality coincides with (1.26). Of course, this result could be also obtained using the operator version of Rouché theorem from [GS]. □

Definition 1.4. *For a rational $k \times k$ matrix function R in general position, let us order in some way its poles and "zeros": $\mathcal{P} = \{\lambda_1, \dots, \lambda_n\}, \mathcal{N} = \{\mu_1, \dots, \mu_n\}$. (We remind that $\#\mathcal{P}(R) = \#\mathcal{N}(R)$.) We introduce the $n \times n$ diagonal matrices*

$$A_\mathcal{P} = \operatorname{diag}(\lambda_1, \dots, \lambda_n), \quad A_\mathcal{N} = \operatorname{diag}(\mu_1, \dots, \mu_n), \tag{1.30}$$

$k \times n$ matrices $F_\mathcal{P}$, $F_\mathcal{N}$ and $n \times k$ matrices $G_\mathcal{P}$, $G_\mathcal{N}$:

$$F_\mathcal{P} = [f_{\lambda_1}, \dots, f_{\lambda_n}], \quad F_\mathcal{N} = [f_{\mu_1}, \dots, f_{\mu_n}], \tag{1.31}$$

$$G_\mathcal{P} = \begin{bmatrix} g_{\lambda_1} \\ \vdots \\ g_{\lambda_n} \end{bmatrix}, \qquad G_\mathcal{N} = \begin{bmatrix} g_{\mu_1} \\ \vdots \\ g_{\mu_n} \end{bmatrix}, \tag{1.32}$$

where f_{λ_j}, g_{λ_j} ($\lambda_j \in \mathcal{P}$) are the left and the right semi-residual vectors at the pole λ_j of R, f_{μ_j}, g_{μ_j} ($\mu_j \in \mathcal{N}$) are the left and the right semi-residual vectors at the pole μ_j of R^{-1}.

The matrices $A_\mathcal{P}$ and $A_\mathcal{N}$ are said to be the pole matrix *and* the zero matrix *respectively* for the matrix function R.

The matrices $F_\mathcal{P}$ and $G_\mathcal{P}$ are said to be the left *and* the right semi-residual matrices corresponding to the pole set $\mathcal{P}(R)$.

The matrices $F_{\mathcal{N}}$ *and* $G_{\mathcal{N}}$ *are said to be* the left *and* the right semi-residual matrices corresponding to the zero set $\mathcal{N}(R)$.

Remark 1.2. It should be mentioned that if we order in some way the poles and the zeros, then the pole and the zero matrices $A_{\mathcal{P}}$ and $A_{\mathcal{N}}$ are defined uniquely, and the semi-residual matrices $F_{\mathcal{P}}, G_{\mathcal{P}}, F_{\mathcal{N}}, G_{\mathcal{N}}$ are defined essentially uniquely, up to a multiplication by diagonal matrices with non-zero diagonal entries:

$$F_{\mathcal{P}} \to F_{\mathcal{P}} \cdot D_{\mathcal{P}}, \quad G_{\mathcal{P}} \to D_{\mathcal{P}}^{-1} \cdot G_{\mathcal{P}}, \tag{1.33}$$

$$F_{\mathcal{N}} \to F_{\mathcal{N}} \cdot D_{\mathcal{N}}, \quad G_{\mathcal{N}} \to D_{\mathcal{N}}^{-1} \cdot G_{\mathcal{N}}, \tag{1.34}$$

where

$$D_{\mathcal{P}} = \operatorname{diag}(d_{1,\mathcal{P}}, \dots, d_{n,\mathcal{P}}) \quad (d_{j,\mathcal{P}} \neq 0, \ j = 1, 2, \dots n), \tag{1.35}$$

$$D_{\mathcal{N}} = \operatorname{diag}(d_{1,\mathcal{N}}, \dots, d_{n,\mathcal{N}}) \quad (d_{j,\mathcal{N}} \neq 0, \ j = 1, 2, \dots n). \tag{1.36}$$

This freedom in the choice of the diagonal matrices $D_{\mathcal{P}}$, $D_{\mathcal{N}}$ can be used to simplify some formulas. Of course, for a given left semi-residual matrix $F_{\mathcal{P}}$, the right semi-residual matrix $G_{\mathcal{P}}$ is determined uniquely; for a given right semi-residual matrix $G_{\mathcal{P}}$, the left semi-residual matrix $F_{\mathcal{P}}$ is determined uniquely, etc.

It is clear that the additive expansions (1.8) and (1.9) may be rewritten in the matrix form

$$R(z) = R(\infty) + F_{\mathcal{P}} \cdot \big(zI - A_{\mathcal{P}}\big)^{-1} \cdot G_{\mathcal{P}}, \tag{1.37}$$

$$R^{-1}(z) = R^{-1}(\infty) + F_{\mathcal{N}} \cdot \big(zI - A_{\mathcal{N}}\big)^{-1} \cdot G_{\mathcal{N}}. \tag{1.38}$$

2. The joint representation of the kernels associated with a rational matrix function in general position

Definition 2.1. *Given a rational matrix function* R *of one variable, we associate with it two matrix function of two variables,* K_{R}^{r} *and* K_{R}^{l}*:*

$$K_{R}^{r}(x,y) = \frac{R(x)R^{-1}(y) - I}{x - y} \tag{2.1}$$

and

$$K_{R}^{l}(x,y) = \frac{R^{-1}(x)R(y) - I}{x - y} \tag{2.2}$$

The function K_{R}^{r} *is said to be* the right kernel associated with the function R.
The function K_{R}^{l} *is said to be* the left kernel associated with the function R.

Remark 2.1. If P is a polynomial, the expression $B(x,y) = \dfrac{P(x) - P(y)}{x - y}$ is said to be the Bezoutian of the polynomial P. The expressions (2.1) and (2.2) look like a Bezoutian.

Theorem 2.1. *Let R be a rational $k \times k$ matrix function in general position, let $A_{\mathcal{P}}$ and $A_{\mathcal{N}}$ be the pole and the zero matrices for R, let $F_{\mathcal{P}}$ and $G_{\mathcal{N}}$ be the left pole and the right zero semiresidual matrices respectively. Then:*

1. *The right kernel K_R^r can be represented in the form*

$$K_R^r(x,y) = F_{\mathcal{P}}(xI - A_{\mathcal{P}})^{-1} H^r (yI - A_{\mathcal{N}})^{-1} G_{\mathcal{N}}, \tag{2.3}$$

   where H^r is some $n \times n$ matrix.

2. *For given matrices* [5] *$F_{\mathcal{P}}$ and $G_{\mathcal{N}}$, the matrix H^r is defined uniquely:*

$$H^r = \|h^r_{p,q}\|_{1\le p,q\le n}, \quad h^r_{p,q} = \frac{g_{\lambda_p} \cdot f_{\mu_q}}{\lambda_p - \mu_q}. \tag{2.4}$$

3. *The matrix $H^r = \|h^r_{p,q}\|$ is invertible.*

The left version of this theorem holds as well.

Theorem 2.2. *Let R be a rational $k \times k$ matrix function in general position, let $A_{\mathcal{P}}$ and $A_{\mathcal{N}}$ be the pole and the zero matrices for R, let $F_{\mathcal{N}}$ and $G_{\mathcal{P}}$ be the left zero and the right pole semiresidual matrices respectively. Then:*

1. *The left kernel K_R^l can be represented in the form*

$$K_R^l(x,y) = F_{\mathcal{N}}(xI - A_{\mathcal{N}})^{-1} H^l (yI - A_{\mathcal{P}})^{-1} G_{\mathcal{P}}, \tag{2.5}$$

   where H^l is some $n \times n$ matrix.

2. *For given matrices* [5] *$F_{\mathcal{N}}$ and $G_{\mathcal{P}}$, the matrix H^l is defined uniquely:*

$$H^l = \|h^l_{p,q}\|_{1\le p,q\le n}, \quad h^l_{p,q} = \frac{g_{\mu_p} \cdot f_{\lambda_q}}{\mu_p - \lambda_q}. \tag{2.6}$$

3. *The matrix H^l is invertible.*

Definition 2.2. *The matrices H^r and H^l which appear in the representations (2.3) and (2.5) of the kernels K_R^r and K_R^l, are said to be* the right core matrix *and the* left core matrix *respectively.*

Corollary 2.1. *Let R be a rational $k \times k$ matrix function in general position, let $A_{\mathcal{P}}$ and $A_{\mathcal{N}}$ be the pole and the zero matrices for R, let $F_{\mathcal{P}}, F_{\mathcal{N}}, G_{\mathcal{P}} G_{\mathcal{N}}$ be the appropriate semi-residual matrices, and let H^r, H^l be the appropriate core matrices.*

Then the matrices $R(x) \cdot R^{-1}(y)$ and $R^{-1}(x) \cdot R(y)$ admit the representations

$$R(x)(R(y))^{-1} = I + (x-y)\, F_{\mathcal{P}}\,(xI - A_{\mathcal{P}})^{-1} H^r (yI - A_{\mathcal{N}})^{-1} G_{\mathcal{N}}, \tag{2.7}$$

$$(R(x))^{-1} R(y) = I + (x-y) F_{\mathcal{N}}\,(xI - A_{\mathcal{N}})^{-1} H^l (yI - A_{\mathcal{P}})^{-1} G_{\mathcal{P}}. \tag{2.8}$$

Under the normalizing condition

$$R(\infty) = I, \tag{2.9}$$

[5] We remind, that the matrices $F_{\mathcal{P}}$, $G_{\mathcal{P}}$, $F_{\mathcal{N}}$, $G_{\mathcal{N}}$ are defined only up to transformations (1.33), (1.34) with arbitrary diagonal invertible matrices $D_{\mathcal{P}}$, $D_{\mathcal{N}}$.

the matrix functions R, R^{-1} themselves admit the representations

$$R(z) = I - F_{\mathcal{P}}(zI - A_{\mathcal{P}})^{-1}H^r G_{\mathcal{N}}, \tag{2.10}$$
$$R^{-1}(z) = I + F_{\mathcal{P}}H^r(zI - A_{\mathcal{N}})^{-1}G_{\mathcal{N}}, \tag{2.11}$$
$$R(z) = I + F_{\mathcal{N}}H^l(zI - A_{\mathcal{P}})^{-1}G_{\mathcal{P}}, \tag{2.12}$$
$$R^{-1}(z) = I - F_{\mathcal{N}}(zI - A_{\mathcal{N}})^{-1}H^l G_{\mathcal{P}}. \tag{2.13}$$

Proof. The representations (2.7) and (2.8) are nothing more than the representations (2.3) and (2.5) rewritten in the terms of the functions $R(x) \cdot R^{-1}(y)$ and $R^{-1}(x) \cdot R(y)$.

Letting y tend to ∞ in (2.7), we obtain (2.10); letting x tend to ∞ in (2.7), we obtain (2.11); Letting y tend to ∞ in (2.8), we obtain (2.12); letting x tend to ∞ in (2.8), we obtain (2.13). □

Theorem 2.3. *Let R be a rational $k \times k$ matrix function in general position, let $A_{\mathcal{P}}$ and $A_{\mathcal{N}}$ be the pole and the zero matrices for R, let $F_{\mathcal{P}}, F_{\mathcal{N}}, G_{\mathcal{P}}\, G_{\mathcal{N}}$ be the appropriate semi-residual matrices, and let H^r, H^l be the appropriate core matrices.*

Then these matrices satisfy the Sylvester-Lyapunov equalities:

$$A_{\mathcal{P}}H^r - H^r A_{\mathcal{N}} = G_{\mathcal{P}}F_{\mathcal{N}}, \tag{2.14}$$
$$A_{\mathcal{N}}H^l - H^l A_{\mathcal{P}} = G_{\mathcal{N}}F_{\mathcal{P}}. \tag{2.15}$$

Proof. The matrices $G_{\mathcal{P}}F_{\mathcal{N}}$ and $G_{\mathcal{N}}F_{\mathcal{P}}$ are of the form

$$G_{\mathcal{P}} \cdot F_{\mathcal{N}} = \|g_{\lambda_p} \cdot f_{\mu_q}\|_{1\le p,q\le n}, \qquad G_{\mathcal{N}}F_{\mathcal{P}} = \|g_{\mu_p} \cdot f_{\lambda_q}\|_{1\le p,q\le n}. \tag{2.16}$$

The assertion of Theorem 2.3 follows from the explicit expressions (2.16), (2.4), (2.6), (1.30). □

Theorem 2.4. *Let R be a rational $k \times k$ matrix function in general position, which satisfies the normalizing condition (2.9); let $F_{\mathcal{P}}, F_{\mathcal{N}}, G_{\mathcal{P}}\, G_{\mathcal{N}}$ be the appropriate semi-residual matrices, and let H^r, H^l be the appropriate core matrices.*

Then these matrices satisfy the equalities

(2.17)
(a) $H^r G_{\mathcal{N}} = -G_{\mathcal{P}}$; (b) $H^l G_{\mathcal{P}} = -G_{\mathcal{N}}$; (c) $F_{\mathcal{P}}H^r = F_{\mathcal{N}}$; (d) $F_{\mathcal{N}}H^l = F_{\mathcal{P}}$.

We give two proofs of Theorem 2.4.

Proof I. We compare the formulas (1.37) and (2.10). In the additive representation (1.37), for given left semi-residual matrix $F_{\mathcal{P}}$, the right matrix $G_{\mathcal{P}}$ is determined uniquely. (See Remark 1.2.) Therefore, (2.17a) holds. Analogously, comparing the formulas (1.37) and (2.12), we obtain (2.17d). Comparing (1.38) and (2.11), (2.13), we obtain (2.17c) and (2.17b). □

Proof II. We prove only the equality (2.17a). The equality (2.17c) can be proved analogously. (2.17b) follows from (2.17a) and (2.31), etc. The matrix equality (2.17a) is the same as the system of n scalar equalities

$$\sum_{1\leq q\leq n} h^r_{p,q} g_{\mu_q} = -g_{\lambda_p}, \qquad p = 1, 2 \ldots, n. \tag{2.18}$$

Since $f_{\lambda_p} \neq 0$, the last equality is equivalent to the equality

$$f_{\lambda_p} \cdot \Big(\sum_{1\leq q\leq n} h^r_{p,q} \Big) \cdot g_{\mu_q} = -f_{\lambda_p} \cdot g_{\lambda_p}. \tag{2.19}$$

Substituting (2.4) into (2.19), we arrive at the equality

$$f_{\lambda_p} \cdot \Big(\sum_{1\leq q\leq n} \frac{g_{\lambda_p} f_{\mu_q}}{\lambda_p - \mu_q} \Big) \cdot g_{\mu_q} = -f_{\lambda_p} \cdot g_{\lambda_p}, \tag{2.20}$$

which is the same as the equality

$$f_{\lambda_p} g_{\lambda_p} \cdot \Big(\sum_{1\leq q\leq n} \frac{f_{\mu_q} g_{\mu_q}}{\lambda_p - \mu_q} \Big) = -f_{\lambda_p} g_{\lambda_p}. \tag{2.21}$$

Since $f_{\lambda_p} g_{\lambda_p} = R_{\lambda_p}$, $f_{\mu_q} g_{\mu_q} = R_{\mu_q}$, the last equality becomes

$$R_{\lambda_p} \cdot \Big(I + \sum_{1\leq q\leq n} \frac{R_{\mu_q}}{\lambda_p - \mu_q} \Big) = 0. \tag{2.22}$$

According to (1.7),

$$I + \sum_{1\leq q\leq n} \frac{R_{\mu_q}}{\lambda_p - \mu_q} = R^{-1}(\lambda_p).$$

Hence, (2.22) becomes $R_{\lambda_p} \cdot R^{-1}(\lambda_p) = 0$. According to (1.20), this equality is true. □

The proofs of Theorems 2.1 and 2.2 are analogous. We will prove only the first one of them.

Proof of Theorem 2.1. First of all, we shall obtain the representation (2.3). The main idea of the proof is to expand the function K^r_R of two variables into a double series of simple fractions and then to interpret this expansion as the matrix equality (2.3). Actually, we derive not a *double* expansion of a function of two variables, but an *iterated* one. Let us fix a point $y \in \overline{\mathbb{C}} \setminus \mathcal{N}(R)$. For this fixed value y, we consider the kernel $K^r_R(x, y)$ as a function of the variable x. This function is rational with respect to x, and $K^r_R(x, y) \to 0$ as $x \to \infty$. It may have singularities only at the points of the set $\mathcal{P}(R)$ and at the point y where the denominator $x - y$ vanishes. Actually this function is holomorphic at the point $x = y$ because the numerator vanishes at the point $x = y$ as well. At each of the points $\lambda_p \in \mathcal{P}(R)$ the function

$K^{\tau}_{R}(x,y)$ (considered as a function of x) is either holomorphic or has a simple pole with residue $K^{\tau}_{\lambda_p}(y)$ of rank one:

$$K^{\tau}_{\lambda_p}(y) = \frac{R_{\lambda_p} \cdot R^{-1}(y)}{\lambda_p - y} \qquad (1 \le p \le n). \tag{2.23}$$

Expanding the function $K^{\tau}_{R}(x,y)$ into a sum of simple fractions, we obtain

$$K^{\tau}_{R}(x,y) = \sum_{1\le p\le n} K^{\tau}_{\lambda_p}(y) \cdot \frac{1}{x-\lambda_p}. \tag{2.24}$$

Next, the residue $K^{\tau}_{\lambda_p}(y)$, considered as a function of y, is a rational function. It vanishes at the point ∞. This function is also holomorphic at the point λ_p because the numerator $R_{\lambda_p} \cdot R^{-1}(y)$ vanishes at the point λ_p: the equality $R_{\lambda_p} \cdot R^{-1}(\lambda_p) = 0$ is the same as the equality $R_{\lambda} B_0 = 0$ in (1.20). Thus the only possible singularities of the function $K^{\tau}_{\lambda_p}(y)$ are the points μ_q $(q = 1, \dots, n)$ of the set $\mathcal{N}(R)$. These singularities are simple poles, with the residues $K^{\tau}_{p,q}$:

$$K^{\tau}_{p,q} = \frac{R_{\lambda_p} \cdot R_{\mu_q}}{\lambda_p - \mu_q} \quad (1 \le p,q \le n). \tag{2.25}$$

Expanding the function $K^{\tau}_{\lambda_p}$ into a sum of simple fractions, we obtain

$$K^{\tau}_{\lambda_p}(y) = \sum_{1\le q\le n} K^{\tau}_{p,q} \cdot \frac{1}{y-\mu_q}. \tag{2.26}$$

Combining (2.24) and (2.26) (and transforming an iterated sum into a double sum), we obtain a double expansion

$$K^{\tau}_{R}(x,y) = \sum_{\substack{1\le p\le n\\ 1\le q\le n}} \frac{1}{x-\lambda_p} \cdot K^{\tau}_{p,q} \cdot \frac{1}{y-\mu_q}. \tag{2.27}$$

Substituting into (2.25) expressions (1.5), we get

$$K^{\tau}_{p,q} = f_{\lambda_p} \cdot h^{\tau}_{p,q} \cdot g_{\mu_q}, \tag{2.28}$$

where $h^{\tau}_{p,q}$ are defined by (2.4). Thus the expansion (2.27) becomes

$$K^{\tau}_{R}(x,y) = \sum_{\substack{1\le p\le n\\ 1\le q\le n}} \frac{f_{\lambda_p}}{x-\lambda_p} \cdot h^{\tau}_{p,q} \cdot \frac{g_{\mu_q}}{y-\mu_q}, \tag{2.29}$$

where f_{λ_p} and g_{μ_q} are the left semi-residual vector at the pole λ_p and the right semi-residual vector at the "zero" μ_q respectively.

The representation (2.3) is simply the representation (2.29) in the matrix form. Statement 1 of Theorem 2.1 is proved.

From (2.27) it follows that

$$K^{\tau}_{p,q} = \lim_{\substack{x\to\lambda_p\\ y\to\mu_p}} (x-\lambda_p)(y-\mu_q) \cdot K^{\tau}_{R}(x,y). \tag{2.30}$$

Thus the values $K_{p,q}$ are determined from the kernel $K_R^r(x,y)$ uniquely. From (2.28) it follows that (for given f_{λ_p}, g_{μ_q}) the values $h_{p,q}^r$ are determined uniquely. Statement 2 of Theorem 2.1 is proved.

Statement 3 of Theorem 2.1 follows immediately from Theorem 2.5 below, where we not only prove the invertibility of the matrices H_r and H_r, but also find their inverse matrices. $\square$

Theorem 2.5. *Let R be a rational matrix function in general position, and let the matrices H^r, H^l be defined from it according to (2.4), (2.6), where $\{\lambda_1, \dots \lambda_n\} = \mathcal{P}(R)$, $\{\mu_1, \dots \mu_n\} = \mathcal{N}(R)$, f_λ, g_λ, f_μ, g_μ are the appropriate semi-residual vectors. Then the following equality holds:*

$$H^r \cdot H^l = I, \qquad H^l \cdot H^r = I \tag{2.31}$$

Remark 2.2. If we already know from somewhere that the matrix H^l is invertible, then we can easily deduce that $\left(H^l\right)^{-1} = H^r$. Indeed, multiplying the equation (2.15) by the matrices $\left(H^l\right)^{-1}$ from both sides, from the right and from the left, and taking into account that

$$\left(H^l\right)^{-1} G_{\mathcal{N}} = -G_{\mathcal{P}}, \qquad F_{\mathcal{P}} \left(H^l\right)^{-1} = F_{\mathcal{N}}$$

(these are equalities (2.17b) and (2.17d)), we obtain the equality

$$A_{\mathcal{P}} \left(H^l\right)^{-1} - \left(H^l\right)^{-1} A_{\mathcal{N}} = G_{\mathcal{P}} F_{\mathcal{N}}. \tag{2.32}$$

Thus each of the matrices $\left(H^l\right)^{-1}$ and H^r is the solution of the same Sylvester-Lyapunov equation $A_{\mathcal{P}} X - X A_{\mathcal{N}} = G_{\mathcal{P}} F_{\mathcal{N}}$. The condition $\mathcal{P}(R) \cap \mathcal{N}(R) = \emptyset$ means that $\sigma_{A_{\mathcal{P}}} \cap \sigma_{A_{\mathcal{N}}} = \emptyset$. Under this condition, the solution X of the Sylvester-Lyapunov equation $A_{\mathcal{P}} X - X A_{\mathcal{N}} = G_{\mathcal{P}} F_{\mathcal{N}}$ is unique. Hence, $\left(H^l\right)^{-1} = H^r$.

We give two proofs of Theorem 2.5.

Proof I. We prove only the first equality in (2.31). Multiplying (2.14) by the matrix H^l from the right and (2.15) by the matrix H^r from the left, we arrive at the equalities

$$A_{\mathcal{P}} H^r H^l - H^r A_{\mathcal{N}} H^l = G_{\mathcal{P}} F_{\mathcal{N}} H^l$$

and

$$H^r A_{\mathcal{N}} H^l - H^r H^l A_{\mathcal{P}} = H^r G_{\mathcal{N}} F_{\mathcal{P}}.$$

Taking into account the equalities (2.17a) and (2.17d), we obtain

$$A_{\mathcal{P}} H^r H^l - H^r A_{\mathcal{N}} H^l = G_{\mathcal{P}} F_{\mathcal{P}}$$

and

$$H^r A_{\mathcal{N}} H^l - H^r H^l A_{\mathcal{P}} = -G_{\mathcal{P}} F_{\mathcal{P}}.$$

Adding two last equalities, we see that the matrices $H^r H^l$ and $A_{\mathcal{P}}$ commute:

$$\left(H^r H^l\right) A_{\mathcal{P}} = A_{\mathcal{P}}\, H^r H^l.$$

Hence,

$$(H^r H^l)\,\varphi(A_{\mathcal{P}}) = \varphi(A_{\mathcal{P}})\,(H^r H^l). \tag{2.33}$$

for every function φ which is holomorphic on the spectrum $\sigma_{A_{\mathcal{P}}}$ of the matrix $A_{\mathcal{P}}$. From (2.17a) and (2.17b) it follows that

$$(H^r H^l)\,G_{\mathcal{P}} = G_{\mathcal{P}}.$$

Multiplying this equality by $\varphi(A_{\mathcal{P}})$ from the left and taking into account the commutation relation (2.33), we obtain that

$$(H^r H^l - I)\,\varphi(A_{\mathcal{P}})\,G_{\mathcal{P}} = 0. \tag{2.34}$$

Let us fix an index $q \in [1, \ldots, n]$ and specify the function $\varphi:\ \varphi(\lambda_p) = \delta_{p,q}$, $p = 1, 2, \ldots, n$. For such a choice of φ, $\varphi(A_{\mathcal{P}}) = \operatorname{diag}[\,\delta_{1,q}, \delta_{2,q}, \ldots, \delta_{n,q}\,]$. Thus

$$\varphi(A_{\mathcal{P}})\,G_{\mathcal{P}} = \begin{bmatrix} \delta_{1,q}\, g_{\lambda_1} \\ \vdots \\ \delta_{n,q}\, g_{\lambda_n} \end{bmatrix}.$$

Therefore we obtain that $m_{p,q}\, g_{\lambda_q} = 0$ for $1 \le p, q \le n$, where $M \stackrel{\text{def}}{=} H^r H^l - I$, $M = \|m_{p,q}\|_{1 \le p,q \le n}$. Since $m_{p,q} \in \mathbb{C}$ and g_{λ_q} is a non-zero vector row, $m_{p,q} = 0$ for all $1 \le p, q \le n$, i.e., $M = 0$. Hence, $H^r H^l - I = 0$. □

Proof II. We prove only the first equality in (2.31). This matrix equality is equivalent to the system of n^2 scalar equations

$$\sum_{1 \le q \le n} h^r_{p,q} \cdot h^l_{q,p} = 1, \qquad p = 1, 2, \ldots, n. \tag{2.35}$$

and

$$\sum_{1 \le q \le n} h^r_{p,q} \cdot h^l_{q,s} = 0, \qquad p = 1, 2, \ldots, n;\ s = 1, 2, \ldots, n;\ p \ne s. \tag{2.36}$$

According to (2.4), (2.6) equality (2.35) means that

$$\sum_{1 \le q \le n} \frac{(g_{\lambda_p} f_{\mu_q}) \cdot (g_{\mu_q} f_{\lambda_p})}{(\lambda_p - \mu_q)^2} = -1. \tag{2.37}$$

Because $f_{\lambda_p} \ne 0$, $g_{\lambda_p} \ne 0$, the last equality is equivalent [6] to the equality

$$f_{\lambda_p} \cdot \sum_{1 \le q \le n} \frac{(g_{\lambda_p} f_{\mu_q}) \cdot (g_{\mu_q} f_{\lambda_p})}{(\lambda_p - \mu_q)^2} \cdot g_{\lambda_p} = -f_{\lambda_p} \cdot g_{\lambda_p}. \tag{2.38}$$

[6] If f is non-zero column-vectors, g is non-zero vector-row, then the equality $c_1 = c_2$, where c_1, c_2 are complex numbers, is equivalent to the equality $fc_1g = fc_2g$.

or, what is the same [7], to the equality

$$f_{\lambda_p} g_{\lambda_p} \cdot \sum_{1 \le q \le n} \frac{f_{\mu_q} g_{\mu_q}}{(\lambda_p - \mu_q)^2} \cdot f_{\lambda_p} g_{\lambda_p} = -f_{\lambda_p} g_{\lambda_p}. \tag{2.39}$$

Taking into account the factorization (1.5), we see that the equality (2.39) is equivalent to the equality

$$R_{\lambda_p} \cdot \sum_{1 \le q \le n} \frac{R_{\mu_q}}{(\lambda_p - \mu_q)^2} \cdot R_{\lambda_p} = -R_{\lambda_p}. \tag{2.40}$$

In view of (1.7),

$$\sum_{1 \le q \le n} \frac{R_{\mu_q}}{(\lambda_p - \mu_q)^2} = -(R^{-1})'(\lambda_p). \tag{2.41}$$

Thus the equality (2.40) becomes

$$R_{\lambda_p} \cdot (R^{-1})'(\lambda_p) \cdot R_{\lambda_p} = R_{\lambda_p}. \tag{2.42}$$

According to Remark 1.1, the equality (2.42) holds. (See (1.24).) Thus the equalities (2.35) are established.

The equalities (2.36) can be established in the same way. According to (2.4) and (2.6), the equality (2.36) means that

$$\sum_{1 \le q \le n} \frac{(g_{\lambda_p} f_{\mu_q}) \cdot (g_{\mu_q} f_{\lambda_s})}{(\lambda_p - \mu_q) \cdot (\lambda_s - \mu_q)} = 0, \qquad p \neq s. \tag{2.43}$$

Because $f_{\lambda_p} \neq 0$, $g_{\lambda_s} \neq 0$, the last equality is equivalent (see the footnote [6]) to the equality

$$f_{\lambda_p} \cdot \sum_{1 \le q \le n} \frac{(g_{\lambda_p} f_{\mu_q}) \cdot (g_{\mu_q} f_{\lambda_s})}{(\lambda_p - \mu_q) \cdot (\lambda_s - \mu_q)} \cdot g_{\lambda_s} = 0, \tag{2.44}$$

or (see the footnote [7]), what is the same, to the equality

$$f_{\lambda_p} g_{\lambda_p} \cdot \sum_{1 \le q \le n} \left(\frac{f_{\mu_q} g_{\mu_q}}{\lambda_p - \mu_q} - \frac{f_{\mu_q} g_{\mu_q}}{\lambda_s - \mu_q} \right) \cdot f_{\lambda_s} g_{\lambda_s} = 0, \qquad p \neq s, \tag{2.45}$$

Taking into account the factorization (1.5), we see that the equality (2.45) is equivalent to the equality

$$R_{\lambda_p} \cdot \sum_{1 \le q \le n} \left(\frac{R_{\mu_q}}{\lambda_p - \mu_q} - \frac{R_{\mu_q}}{\lambda_s - \mu_q} \right) \cdot R_{\lambda_s} = 0. \tag{2.46}$$

In view of (1.7),

$$\sum_{1 \le q \le n} \left(\frac{R_{\mu_q}}{\lambda_p - \mu_q} - \frac{R_{\mu_q}}{\lambda_s - \mu_q} \right) = R^{-1}(\lambda_p) - R^{-1}(\lambda_s), \qquad p \neq s. \tag{2.47}$$

---

[7] Here we use the associativity of the matrix multiplication.

Thus the equality (2.46) becomes

$$R_{\lambda_p} \cdot \Big(R^{-1}(\lambda_p) - R^{-1}(\lambda_s)\Big) \cdot R_{\lambda_s} = 0. \tag{2.48}$$

In view of (1.20) and (1.18),

$$R_{\lambda_p} \cdot R^{-1}(\lambda_p) = 0, \qquad R^{-1}(\lambda_s) \cdot R_{\lambda_s} = 0. \tag{2.49}$$

Thus (2.46) holds. The equalities (2.36) are established. □

The representations (2.7) and (2.8) are almost what we need. However, there is an essential disadvantage to these representations: each one of them contains explicitly all the four semi-residual matrices. For example, the representation (2.7) contains explicitly not only the semi-residual matrices $F_{\mathcal{P}}$ and $G_{\mathcal{N}}$ (this is evident), but also the matrices $F_{\mathcal{N}}$ and $G_{\mathcal{P}}$ (see the expression (2.4) for the the right core matrix H^r). But the four semi-residual matrices (together with zero and pole locations) are over-determined data: the matrix function R is completely determined by two of those semi-residual matrices only. For example, from the additive representation (1.37) it follows that (under the normalization $R(\infty) = I$) the zero and pole matrices $A_{\mathcal{P}}$, $A_{\mathcal{N}}$ together with the two semi-residual matrices $F_{\mathcal{P}}$, $G_{\mathcal{P}}$ determine completely the matrix function R, and hence the other two semi-residual matrices $F_{\mathcal{N}}$, $G_{\mathcal{N}}$. Because of this, the semi-residual matrices $F_{\mathcal{N}}$, $G_{\mathcal{P}}$ are (at least in principle) expressible in terms of the matrices $A_{\mathcal{P}}$, $A_{\mathcal{N}}$, $F_{\mathcal{P}}$, $G_{\mathcal{N}}$. Hence, we can hope to express the right core matrix H^r in terms of the matrices $A_{\mathcal{P}}$, $A_{\mathcal{N}}$, $F_{\mathcal{P}}$, $G_{\mathcal{N}}$. Indeed, this can be done easily and explicitly: on the one hand, the left core matrix H^l is expressible in terms of the entries of the matrices $A_{\mathcal{P}}$, $A_{\mathcal{N}}$, $F_{\mathcal{P}}$, $G_{\mathcal{N}}$ only (see (2.6); on the other hand, $H^r = (H^l)^{-1}$ (see (2.31)).

This suggests to us that it may be reasonable to use the inverse matrices S^r and S^l instead of the matrices[8] H^r and H^l:

$$S^r \stackrel{\text{def}}{=} \big(H^r\big)^{-1} \quad \text{and} \quad S^l \stackrel{\text{def}}{=} \big(H^l\big)^{-1}. \tag{2.50}$$

The equalities (2.17) and (2.14), (2.15) can be rewritten in term of the matrices S^r and S^l. Namely, equalities (2.17) become

(2.51)

$$\text{(a) } G_{\mathcal{N}} = -S^r G_{\mathcal{P}}; \quad \text{(b) } G_{\mathcal{P}} = -S^l G_{\mathcal{N}}; \quad \text{(c) } F_{\mathcal{P}} = F_{\mathcal{N}} S^r; \quad \text{(d) } F_{\mathcal{N}} = F_{\mathcal{P}} S^l.$$

Multiplying the equality (2.14) by the matrix $(H^r)^{-1}$ from the right and from the left and taking into account equalities (2.17a) and (2.17c), we transform (2.14) to the form

$$A_{\mathcal{N}} S^r - S^r A_{\mathcal{P}} = G_{\mathcal{N}} F_{\mathcal{P}}. \tag{2.52}$$

Analogously, from (2.15) and (2.17b), (2.17d) we derive the equality

$$A_{\mathcal{P}} S^l - S^l A_{\mathcal{N}} = G_{\mathcal{P}} F_{\mathcal{N}}. \tag{2.53}$$

[8] We remind, that, according to Theorem 2.3, the matrices H^r and H^l are mutually inverse, and hence, invertible.

Thus the matrices S^r and S^l are solutions of the Sylvester-Lyapunov equations

$$A_{\mathcal{N}}X - XA_{\mathcal{P}} = G_{\mathcal{N}}F_{\mathcal{P}}. \tag{2.54}$$

$$A_{\mathcal{P}}X - XA_{\mathcal{N}} = G_{\mathcal{P}}F_{\mathcal{N}}. \tag{2.55}$$

respectively.

Now we change our point of view and define the matrices S^r and S^l as solutions of Sylvester-Lyapunov equations (and not as the matrices inverse to the core matrices H^r and H^s; see (2.50)).

Definition 2.3. Let R be a rational matrix function in general position, let $A_{\mathcal{P}}$ and $A_{\mathcal{N}}$ be its pole and zero matrices, let $F_{\mathcal{P}}$, $G_{\mathcal{P}}$, $F_{\mathcal{N}}$, $G_{\mathcal{N}}$ be the appropriate semi-residual matrices.

1. *The matrices S^r and S^l which are the solutions of the Sylvester-Lyapunov equations (2.54) and (2.55), are said to be* the right zero-pole coupling matrix *and* the left zero-pole coupling matrix *respectively.*
2. *The relations* (2.51) *are said to be* the zero-pole coupling relations.

Remark 2.3. Since the spectra of the matrices $A_{\mathcal{P}}$ and $A_{\mathcal{N}}$ do not intersect, the Sylvester-Lyapunov equations (2.55) and (2.54) are uniquely solvable. (However, as the matrices $A_{\mathcal{P}}$ and $A_{\mathcal{N}}$ are diagonal, the solvability of these equations as well as the uniqueness is obvious). Moreover, it is possible to obtain the explicit expressions for the matrices S^r and S^l from (2.55) and (2.54):

$$\begin{aligned} &\text{(r) } S^r = \|s^r_{p,q}\|_{1\le p,q\le n}, \quad s^r_{p,q} = \frac{g_{\mu_p}\cdot f_{\lambda_q}}{\mu_p - \lambda_q}, \\ &\text{(l) } S^l = \|s^l_{p,q}\|_{1\le p,q\le n}, \quad s^l_{p,q} = \frac{g_{\lambda_p}\cdot f_{\mu_q}}{\lambda_p - \mu_q}. \end{aligned} \tag{2.56}$$

(Actually, we *derived* the Sylvester-Lyapunov equations from the explicit expressions for the matrices which we interpret now as solutions of these equations.)

According to (2.50), the relations (2.31) can be rewritten in the form

$$S^r \cdot S^l = S^l \cdot S^r = I. \tag{2.57}$$

We may also refer directly to the equalities (2.37) and (2.43): these equalities mean that the matrices (2.56r) and (2.56l) are mutually inverse.

Remark 2.4. The representations (2.3) and (2.5) may be rewritten in terms of the matrices S^r and S^l (instead of the matrices H^r and H^l):

$$R(x)(R(y))^{-1} = I + (x-y)\,F_{\mathcal{P}}\,(xI - A_{\mathcal{P}})^{-1}\cdot (S^r)^{-1}\cdot (yI - A_{\mathcal{N}})^{-1}G_{\mathcal{N}}, \tag{2.58}$$

$$(R(x))^{-1}R(y) = I + (x-y)F_{\mathcal{N}}\,(xI - A_{\mathcal{N}})^{-1}\cdot (S^l)^{-1}\cdot (yI - A_{\mathcal{P}})^{-1}G_{\mathcal{P}}. \tag{2.59}$$

Under the normalizing condition (2.9), the matrix functions R, R^{-1} themselves admits the representations

$$R(z) = I - F_{\mathcal{P}}\,(zI - A_{\mathcal{P}})^{-1}(S^r)^{-1}G_{\mathcal{N}}, \tag{2.60}$$

$$R^{-1}(z) = I + F_{\mathcal{P}}\,(S^r)^{-1}(zI - A_{\mathcal{N}})^{-1}G_{\mathcal{N}}, \tag{2.61}$$

$$R(z) = I + F_{\mathcal{N}}(S^l)^{-1}(zI - A_{\mathcal{P}})^{-1}G_{\mathcal{P}}, \tag{2.62}$$

$$R^{-1}(z) = I - F_{\mathcal{N}}(zI - A_{\mathcal{N}})^{-1}(S^l)^{-1}G_{\mathcal{P}}. \tag{2.63}$$

These formulas may be obtained of from (2.58), (2.59), letting x or y tend to ∞ there, or from (2.10)–(2.13), rewriting them in terms of the matrices S^r, S^l (instead of the matrices H^r, H^l).

Remark 2.5. Of course, we may obtain the zero-pole coupling relations (2.51) comparing the representations (1.37), (1.38) and (2.10)–(2.13).

Remark 2.6. The semi-residual vectors are defined not completely uniquely, but up to the transformations (1.33) and (1.34) only. If the semi-residual vectors are transformed according to (1.33) and (1.34), the right-hand sides of the Sylvester-Lyapunov equations (2.54) and (2.55) are transformed as:

$$G_{\mathcal{N}}F_{\mathcal{P}} \to (D_{\mathcal{N}})^{-1}\cdot G_{\mathcal{N}}F_{\mathcal{P}}\cdot D_{\mathcal{P}}; \qquad G_{\mathcal{P}}F_{\mathcal{N}} \to (D_{\mathcal{P}})^{-1}\cdot G_{\mathcal{P}}F_{\mathcal{N}}\cdot D_{\mathcal{N}}. \tag{2.64}$$

The solutions S^r and S^l of the Sylvester-Lyapunov equations (2.54) and (2.55) are transformed as:

$$S^r \to (D_{\mathcal{N}})^{-1}\cdot S^r\cdot D_{\mathcal{P}}; \qquad S^l \to (D_{\mathcal{P}})^{-1}\cdot S^l\cdot D_{\mathcal{N}}. \tag{2.65}$$

Of course, the expressions (2.58), (2.59) (for $R(x)(R(y))^{-1}$ and $(R(x))^{-1}R(y)$) are invariant with respect to the transformations (1.33), (1.34), (2.65) (of the semi-residual and the zero-pole coupling matrices).

The representations (2.58) and (2.59) are exactly what we need. Obtaining them is one of the main goal of this paper. Therefore we choose a special name for this representation:

Definition 2.4. *Let R be a rational matrix function in general position, let $A_{\mathcal{P}}, A_{\mathcal{N}}$ be its pole and zero matrices, let $F_{\mathcal{P}}, G_{\mathcal{P}}, F_{\mathcal{N}}, G_{\mathcal{N}}$ be the appropriate semi-residual matrices, and let S^r, S^l be the solutions of the Sylvester-Lyapunov equations (2.54) and (2.55) respectively. The formulas (2.58) and (2.59) are said to be* the right joint system representation of the pair R, R^{-1} *and* the left joint system representation of the pair R, R^{-1} *respectively.*

The formulas (2.60), (2.61), (2.62), (2.63) (which can be obtained from (2.58), (2.59) by passage to the limit) are said to be the right system representation of the function R, the right system representation of the function R^{-1}, the left system representation of the function R, and the left system representation of the function R^{-1}*, respectively.*

Remark 2.7. The terminology is motivated by the so-called *system theory* or, in more details, by *the theory of linear time invariant dynamical system.* In this theory, all the objects such as the zero and the pole matrices, the semi-residual matrices, the zero-pole coupling matrices are interpreted from the point of view of dynamical systems. This interpretation does not play any role in our considerations. We need the joint system representations as a tool to introduce a convenient coordinates in the set of all rational matrix functions (in general position).

Remark 2.8. In the realization theory one obtains formulas like (2.60)–(2.63) for matrix functions R and R^{-1} considered individually. In the representations (2.58), (2.59) the matrix function R, R^{-1} are considered jointly. This is the reason for using the terminology joint system representation.

Now we summarize the results of this section and formulate

Theorem 2.6. *Let R be a rational matrix function in general position, let $A_{\mathcal{P}}$ and $A_{\mathcal{N}}$ be its pole and zero matrices, let $F_{\mathcal{P}}$, $G_{\mathcal{P}}$ be its left and right pole semi-residual matrices, let $F_{\mathcal{N}}, G_{\mathcal{N}}$ be its left and right zero semi-residual matrices. Then:*

1. *The matrices S^r, which is a solution of the Sylvester-Lyapunov equation $A_{\mathcal{N}}X - XA_{\mathcal{P}} = G_{\mathcal{N}}F_{\mathcal{P}}$, and S^l, which is a solution of the Sylvester-Lyapunov equation $A_{\mathcal{P}}X - XA_{\mathcal{N}} = G_{\mathcal{P}}F_{\mathcal{N}}$:*

$$\text{(2.56)}\qquad \begin{aligned} &\text{(r)}\ \ S^r = \|s^r_{p,q}\|_{1\le p,q\le n},\ s^r_{p,q} = \frac{g_{\mu_p}\cdot f_{\lambda_q}}{\mu_p - \lambda_q},\\ &\text{(l)}\ \ S^l = \|s^l_{p,q}\|_{1\le p,q\le n},\ s^l_{p,q} = \frac{g_{\lambda_p}\cdot f_{\mu_q}}{\lambda_p - \mu_q}, \end{aligned}$$

are mutually inverse, i.e., the equalities

$$\text{(2.57)}\qquad S^r\cdot S^l = I, \qquad S^l\cdot S^r = I$$

hold. In particular, the matrices S^r and S^l are invertible.

2. *The matrix function $R(x)(R(y))^{-1}$ can be recovered from the data $A_{\mathcal{P}}$, $A_{\mathcal{N}}$, $F_{\mathcal{P}}$, $G_{\mathcal{N}}$ by the formula (2.58)*

$$\text{(2.58)}\quad (R(x)(R(y))^{-1} = I + (x-y)\,F_{\mathcal{P}}\,(xI - A_{\mathcal{P}})^{-1}\cdot (S^r)^{-1}\cdot (yI - A_{\mathcal{N}})^{-1}G_{\mathcal{N}},$$

i.e., the right joint system representation holds.

3. *The matrix function $(R(x))^{-1}R(y)$ can be recovered from the data $A_{\mathcal{P}}$, $A_{\mathcal{N}}$, $F_{\mathcal{N}}$, $G_{\mathcal{P}}$ by the formula (2.59),*

$$\text{(2.59)}\quad (R(x))^{-1}R(y) = I + (x-y)F_{\mathcal{N}}\,(xI - A_{\mathcal{N}})^{-1}\cdot (S^l)^{-1}\cdot (yI - A_{\mathcal{P}})^{-1}G_{\mathcal{P}}.$$

i.e., the left joint system representation holds.

4. *Under the normalizing condition (2.9), the matrix functions R, R^{-1} themselves admit the representations*

$$\text{(2.60)}\qquad R(z) = I - F_{\mathcal{P}}\,(zI - A_{\mathcal{P}})^{-1}(S^r)^{-1}G_{\mathcal{N}},$$

$$\text{(2.61)}\qquad R^{-1}(z) = I + F_{\mathcal{P}}\,(S^r)^{-1}(zI - A_{\mathcal{N}})^{-1}G_{\mathcal{N}},$$

(2.62) $$R(z) = I + F_{\mathcal{N}}(S^l)^{-1}(zI - A_{\mathcal{P}})^{-1}G_{\mathcal{P}},$$

(2.63) $$R^{-1}(z) = I - F_{\mathcal{N}}(zI - A_{\mathcal{N}})^{-1}(S^l)^{-1}G_{\mathcal{P}}.$$

5. *The zero-pole coupling relations hold:*

(2.51)
$$\text{(a) } G_{\mathcal{N}} = -S^r G_{\mathcal{P}}; \quad \text{(b) } G_{\mathcal{P}} = -S^l G_{\mathcal{N}}; \quad \text{(c) } F_{\mathcal{P}} = F_{\mathcal{N}} S^r; \quad \text{(d) } F_{\mathcal{N}} = F_{\mathcal{P}} S^l.$$

6. *If the representations (2.7) and (2.8) hold with some matrices H^r and H^s, then necessarily $H^r = (S^r)^{-1}$, $H^l = (S^l)^{-1}$.*

Remark 2.9. The representation (2.58) allows us to recover the matrix function $R(x)(R(y))^{-1}$ from the left pole and the right zero semi-residual matrices $F_{\mathcal{P}}$ and $G_{\mathcal{N}}$. whereas the representation (2.59) allows us to recover the matrix function $(R(x))^{-1}R(y)$ from the right pole and the left zero semi-residual matrices $F_{\mathcal{N}}$ and $G_{\mathcal{P}}$. However, sometimes one needs some "hybrid" formulas which allow to recover the matrix function $R(x)(R(y))^{-1}$ from the right pole- and the left zero- semi-residual matrices $F_{\mathcal{N}}$ and $G_{\mathcal{P}}$ and the matrix function $(R(x))^{-1}R(y)$ from the left pole- and the right zero- semi-residual matrices $F_{\mathcal{P}}$ and $G_{\mathcal{N}}$. Such formulas can be easily derived from the joint system realization formulas (2.58), (2.59) combined with the zero-pole coupling relations (2.51). These "hybrid" formulas are of the form:

(2.66)
$$R(x) \cdot (R(y))^{-1} = I - (x - y)\, F_{\mathcal{N}}\, (S^l)^{-1}\, (xI - A_{\mathcal{P}})^{-1}\, S^l\, (yI - A_{\mathcal{N}})^{-1}\, (S^l)^{-1} G_{P},$$

(2.67)
$$(R(x))^{-1} \cdot R(y) = I - (x - y)\, F_{\mathcal{P}}\, (S^r)^{-1}\, (xI - A_{\mathcal{N}})^{-1}\, S^r\, (yI - A_{\mathcal{P}})^{-1}\, (S^r)^{-1} G_{N}.$$

The matrix S^l can be calculated from the data: $F_{\mathcal{N}}$, G_{P}, $A_{\mathcal{P}}$, $A_{\mathcal{N}}$; the matrix S^r can be calculated from the data: $F_{\mathcal{P}}$, G_{N}, $A_{\mathcal{P}}$, $A_{\mathcal{N}}$.

3. From the chain identity to the Sylvester–Lyapunov equation and back

The considerations of this section are centered around the so-called *chain identity*. Let us give a number of definitions.

Let $T(\,.\,,\,.\,)$ be a $k \times k$ matrix function of two complex variables, with domain of definition [9] $\mathcal{D}_T$, $\mathcal{D}_T \subseteq \overline{\mathbb{C}} \times \overline{\mathbb{C}}$, $T : \mathcal{D}_T \to \mathfrak{M}_k$.

Definition 3.1. *A function $T(\,.\,,\,.\,)$ of two variables is said to satisfy* the chain identity *if*

(3.1) $$T(x, y) \cdot T(y, z) = T(x, z)$$

for every x, y, z for which $(x, y) \in \mathcal{D}_T$ and $(y, z) \in \mathcal{D}_T$. (In particular, (x, z) must belong to $\mathcal{D}_T$ if $(x, y) \in \mathcal{D}_T$ and $(y, z) \in \mathcal{D}_T$.)

[9] We recall that $\overline{\mathbb{C}} \overset{\text{def}}{=} \mathbb{C} \cup \infty$ is the extended complex plane.

Definition 3.2. *A function $T(\,.\,,\,.\,)$ of two variables is said to satisfy* the diagonal unity identity *if*

$$T(x,x) = I \tag{3.2}$$

for every point x for which (x,x) belongs to the domain of definition $\mathcal{D}_T$ of the function T.

A class of function T satisfying both the chain identity and the diagonal unity identity can be constructed in the following way.

Definition 3.3. *Let Φ and Φ^{-1} be $k \times k$ matrix functions of one variable with domains of definition $\mathcal{D}_\Phi$ and $\mathcal{D}_{\Phi^{-1}}$ respectively, $\mathcal{D}_\Phi \subseteq \overline{\mathbb{C}}$, $\mathcal{D}_{\Phi^{-1}} \subseteq \overline{\mathbb{C}}$. Let us define a matrix function T_Φ of two variables by the equality*

$$T_\Phi(x,y) \stackrel{\text{def}}{=} \Phi(x) \cdot \Phi^{-1}(y) \tag{3.3}$$

with a domain of definition

$$\mathcal{D}_T \stackrel{\text{def}}{=} \mathcal{D}_\Phi \times \mathcal{D}_{\Phi^{-1}}, \quad \text{i.e.,} \quad \Big((x,y) \in \mathcal{D}_T\Big) \Leftrightarrow \Big((x \in \mathcal{D}_\Phi)\,\&\,(y \in \mathcal{D}_{\Phi^{-1}})\Big). \tag{3.4}$$

The function T_Φ is said to be the chain function generated by the function Φ.

Lemma 3.1. *Let Φ and Φ^{-1} be $k \times k$ matrix functions of one variable with domains of definition $\mathcal{D}_\Phi$ and $\mathcal{D}_{\Phi^{-1}}$ respectively, $\mathcal{D}_\Phi \subseteq \overline{\mathbb{C}}$, $\mathcal{D}_{\Phi^{-1}} \subseteq \overline{\mathbb{C}}$. Let T_Φ be the chain function generated by the function Φ. If the matrix functions Φ and Φ^{-1} are mutually inverse, i.e., the identities $\Phi(x) \cdot \Phi^{-1}(x) = \Phi^{-1}(x) \cdot \Phi(x) = I$ hold for all $x \in \mathcal{D}_\Phi \cap \mathcal{D}_{\Phi^{-1}}$, then for the matrix function T both the chain identity and the diagonal unity identity hold.*

Remark 3.1. Of course, Lemma 3.1 has a content only under condition $\mathcal{D}_\Phi \cap \mathcal{D}_{\Phi^{-1}} \neq \emptyset$. If this condition fails then the values $T(x,x)$ and $T(x,y) \cdot T(y,z)$ are defined for the empty set of arguments.

Proof of Lemma 3.1. The diagonal unity identity expressed that the functions Φ and Φ^{-1} are mutually inverse. The chain identity for the function T_Φ is a consequence of two facts: 1) The function Φ and Φ^{-1} are mutually inverse; 2) The matrix multiplication is associative. □

It turns out that each function T of two variables satisfying both the chain identity and the diagonal unity identity is of the form T_Φ for some function Φ of one variable.

Theorem 3.1. *Let T be a $k \times k$ matrix function of two variables, with a domain of definition $\mathcal{D}_T \subseteq \overline{\mathbb{C}} \times \overline{\mathbb{C}}$ of the form $\mathcal{D}_T = \mathcal{D}_1 \times \mathcal{D}_2$, where $\mathcal{D}_1 \subseteq \overline{\mathbb{C}}$ and $\mathcal{D}_2 \subseteq \overline{\mathbb{C}}$, with $\mathcal{D}_1 \cap \mathcal{D}_2 \neq \emptyset$. If for the function T both the chain identity and the diagonal unity identity are satisfied, then the function T is of the form $T = T_\Phi$, (i.e., $T(x,y) = \Phi(x) \cdot \Phi^{-1}(y)$), where Φ and Φ^{-1} are mutually inverse $k \times k$ matrix functions of one variable, with $\mathcal{D}_\Phi = \mathcal{D}_1$ and $\mathcal{D}_{\Phi^{-1}} = \mathcal{D}_2$.*

Proof. Let us fix an arbitrary point a belonging to the set $\mathcal{D}_1 \cap \mathcal{D}_2$. (We will call this point a the distinguished point.) Let us define now

$$(3.5) \qquad \mathcal{D}_\Phi \overset{\text{def}}{=} \mathcal{D}_1, \; \Phi(x) \overset{\text{def}}{=} T(x,a); \quad \mathcal{D}_\Phi^{-1} \overset{\text{def}}{=} \mathcal{D}_2, \; \Phi^{-1}(y) \overset{\text{def}}{=} T(a,y).$$

The functions Φ and Φ^{-1} are mutually inverse: this follows from the chain and the diagonal unity identities. The equality $T(x,y) = \Phi(x) \cdot \Phi^{-1}(y)$ is the chain identity written down for the triple of the points x, a, y. In addition, we note that this function Φ satisfies *the normalizing condition* $\Phi(a) = I$. □

Definition 3.4. *Let R be a rational $k \times k$ matrix function of one variables, $\det R \not\equiv 0$, and let R^{-1} be the inverse (in the usual sense) matrix function; the domain of definition $\mathcal{D}_R$ is the set of analyticity of the function R; the domain of definition $\mathcal{D}_R^{-1}$ is the set of analyticity of the function R^{-1}. (In other words,[10] $\mathcal{D}_R = \overline{\mathbb{C}} \setminus \mathcal{P}(R)$, $\mathcal{D}_R^{-1} = \overline{\mathbb{C}} \setminus \mathcal{N}(R)$). We associate with the function R two functions of two variables, $T_R^r(x,y)$ and $T_R^l(x,y)$:*

$$(3.6) \qquad \text{(a) } T_R^r(x,y) \overset{\text{def}}{=} R(x) \cdot R^{-1}(y); \qquad \text{(b) } T_R^l(x,y) \overset{\text{def}}{=} R^{-1}(x) \cdot R(y).$$

The function T_R^r is said to be the right chain function generated by R. *The function T_R^l is said to be* the left chain function generated by R.

Remark 3.2. It is clear that that the function T_R^r is a function of the form T_Φ (in the sense of Definition 3.3) for $\Phi = R$, and the function T_R^l is the function T_Φ for $\Phi = R^{-1}$. Thus the right chain function generated by R is the left chain function generated by R^{-1}:

$$(3.7) \qquad T_R^r(x,y) = T_{R^{-1}}^l(x,y).$$

From Lemma 3.1 and from Remark 3.2 it follows

Corollary 3.1. *Let R be a matrix function, $\det R \not\equiv 0$. Then both matrix functions $T_R^r(x,y)$ and $T_R^l(x,y)$ (see (3.6)) satisfy the chain identity and the diagonal unity identity:*

$$(3.8) \qquad T_R^r(x,y) \cdot T_R^r(y,z) \equiv T_R^r(x,z); \qquad T_R^r(x,x) \equiv I.$$

$$(3.9) \qquad T_R^l(z,y) \cdot T_R^l(y,x) \equiv T_R^l(z,x); \qquad T_R^l(x,x) \equiv I.$$

From (2.1) and (2.2) it is clear that the right and the left chain functions T_R^r and $T_{R^{-1}}^l$ can be expressed in terms of the right and the left kernels:

$$(3.10) \qquad T_R^r(x,y) = I + (x-y) \cdot K_R^r(x,y);$$

$$(3.11) \qquad T_R^l(x,y) = I + (x-y) \cdot K_R^l(x,y).$$

[10] We recall that $\mathcal{P}(R)$ is the pole set of the function R, $\mathcal{N}(R)$ is the zero set of the function R, i.e., the pole set of the function R^{-1}.

Substituting in the chain identity (3.8) the expression (3.10) for $\overset{r}{T}_R(x,y)$, we obtain an identity for the right kernel K^r_R:

$$(3.12)\quad \Big(I + (x-y)K^r_R(x,y)\Big)\cdot\Big(I + (y-z)K^r_R(y,z)\Big) \equiv I + (x-z)K^r_R(x,z).$$

Removing the parentheses, we obtain the identity

$$(x-y)K^r_R(x,y) + (y-z)K^r_R(y,z) - (x-z)K^r_R(x,z) \equiv -(x-y)(y-z)K^r_R(x,y)K^r_R(y,z).$$

Dividing by $(x-y)(y-z)$, we arrive at the identity

$$(3.13)\quad \frac{K^r_R(x,y) - K^r_R(x,z)}{y-z} - \frac{K^r_R(x,z) - K^r_R(y,z)}{x-y} \equiv -K(x,y)K(y,z).$$

Assume now that the kernel K^r_R admits the representation (2.3), with some $k\times n$ matrix $F_{\mathcal{P}}$, $n\times k$ matrix $G_{\mathcal{N}}$, and $n\times n$ matrices $A_{\mathcal{P}}, A_{\mathcal{N}}, H^r$. Substituting the expressions (2.3) into (3.11), we arrive at the identity

$$\begin{aligned}(3.14)\quad & F_{\mathcal{P}}\cdot(xI-A_{\mathcal{P}})^{-1}\cdot H^r\cdot\frac{(yI-A_{\mathcal{N}})^{-1}-(zI-A_{\mathcal{N}})^{-1}}{y-z}\cdot G_{\mathcal{N}}\\ &- F_{\mathcal{P}}\cdot\frac{(xI-A_{\mathcal{P}})^{-1}-(yI-A_{\mathcal{P}})^{-1}}{x-y}\cdot H^r\cdot(zI-A_{\mathcal{N}})^{-1}\cdot G_{\mathcal{N}}\\ &\equiv -F_{\mathcal{P}}\cdot(xI-A_{\mathcal{P}})^{-1}\cdot H^r\cdot(yI-A_{\mathcal{N}})^{-1}\cdot G_{\mathcal{N}}F_{\mathcal{P}}\\ &\qquad\cdot(yI-A_{\mathcal{P}})^{-1}\cdot H^r\cdot(zI-A_{\mathcal{N}})^{-1}\cdot G_{\mathcal{N}}.\end{aligned}$$

Using the Hilbert identity for resolvents, we arrive at the identity

$$\begin{aligned}(3.15)\quad & F_{\mathcal{P}}\cdot(xI-A_{\mathcal{P}})^{-1}\cdot H^r\cdot(yI-A_{\mathcal{N}})^{-1}\cdot(zI-A_{\mathcal{N}})^{-1}\cdot G_{\mathcal{N}}\\ &- F_{\mathcal{P}}\cdot(xI-A_{\mathcal{P}})^{-1}\cdot(yI-A_{\mathcal{P}})^{-1}\cdot H^r\cdot(zI-A_{\mathcal{N}})^{-1}\cdot G_{\mathcal{N}}\\ &\equiv F_{\mathcal{P}}\cdot(xI-A_{\mathcal{P}})^{-1}\cdot H^r\cdot(yI-A_{\mathcal{N}})^{-1}\cdot G_{\mathcal{N}}F_{\mathcal{P}}\\ &\qquad\cdot(yI-A_{\mathcal{P}})^{-1}\cdot H^r\cdot(zI-A_{\mathcal{N}})^{-1}\cdot G_{\mathcal{N}}.\end{aligned}$$

Putting the common factors outside the parentheses, we obtain

$$(3.16)\quad F_{\mathcal{P}}\cdot(xI-A_{\mathcal{P}})^{-1}\cdot M\cdot(zI-A_{\mathcal{N}})^{-1}\cdot G_{\mathcal{N}}\equiv 0,$$

where

$$\begin{aligned}(3.17)\quad M = {} & H^r\cdot(yI-A_{\mathcal{N}})^{-1} - (yI-A_{\mathcal{P}})^{-1}\cdot H^r\\ &- H^r\cdot(yI-A_{\mathcal{N}})^{-1}\cdot G_{\mathcal{N}}F_{\mathcal{P}}\cdot(yI-A_{\mathcal{P}})^{-1}\cdot H^r\end{aligned}$$

Assume moreover that R is a rational matrix function in general position. Let $A_{\mathcal{P}}$ and $A_{\mathcal{N}}$ be its pole and zero matrices (i.e., these matrices are of the form (1.30), where all the numbers $\lambda_1, \dots, \lambda_n; \mu_1, \dots, \mu_n$ are pairwise different), and let $F_{\mathcal{P}}$ and $G_{\mathcal{N}}$ be the pole left semi-residual and the zero right semi-residual matrices (in particular, they have the form (1.31), (1.32) where f_{λ_p} and g_{μ_q} are non-zero k-column-vectors).

From (3.16) it follows that

$$F_{\mathcal{P}} \cdot \varphi(A_{\mathcal{P}}) \cdot M \cdot \psi(A_{\mathcal{N}}) \cdot G_{\mathcal{N}} \equiv 0, \tag{3.18}$$

where φ and ψ are arbitrary functions which are analytic on the spectra of the matrices $A_{\mathcal{P}}$ and $A_{\mathcal{N}}$ respectively.

Let us fix two indices, $p \in [1, \dots, n]$ and $q \in [1, \dots, n]$ and specify two functions φ and ψ:

$$\varphi(\lambda_j) = \delta_{j,p}, \qquad \psi(\mu_j) = \delta_{j,q} \qquad (j = 1, 2, \cdot, n).$$

Then

$$\varphi(A_{\mathcal{P}}) = \operatorname{diag}\left(\delta_{1,p}, \dots, \delta_{n,p}\right), \qquad \psi(A_{\mathcal{N}}) = \operatorname{diag}\left(\delta_{1,q}, \dots, \delta_{n,q}\right). \tag{3.19}$$

For such functions φ and ψ, (3.18) becomes:

$$f_{\lambda_p} \cdot m_{p,q} \cdot g_{\mu_q} = 0, \tag{3.20}$$

where $m_{p,q}$ is the (p, q) entry of the matrix M. Since f_{λ_p} and g_{μ_q} are non-zero column-vector and row-vector, $m_{p,q} = 0$. Thus $M = 0$, or,

$$\begin{aligned} H^r \cdot (yI - A_{\mathcal{N}})^{-1} - (yI - A_{\mathcal{P}})^{-1} \cdot H^r \\ - H^r \cdot (yI - A_{\mathcal{N}})^{-1} \cdot G_{\mathcal{N}} F_{\mathcal{P}} \cdot (yI - A_{\mathcal{P}})^{-1} \cdot H^r \equiv 0. \end{aligned} \tag{3.21}$$

The "left" version of the equation (3.21) has the form

$$\begin{aligned} H^l \cdot (yI - A_{\mathcal{P}})^{-1} - (yI - A_{\mathcal{N}})^{-1} \cdot H^l \\ - H^l \cdot (yI - A_{\mathcal{P}})^{-1} \cdot G_{\mathcal{P}} F_{\mathcal{N}} \cdot (yI - A_{\mathcal{N}})^{-1} \cdot H^l \equiv 0. \end{aligned} \tag{3.22}$$

To obtain (3.22), we have to use the chain identity (3.9), the expression (3.11) for the chain function T_R^l in terms of its associated kernel K_R^l, and the representation (2.5) of this kernel. However, we may just replace in (3.21):

$$H^r \to H^l; \qquad F_{\mathcal{P}} \to F_{\mathcal{N}}; \qquad G_{\mathcal{N}} \to G_{\mathcal{P}} : \qquad A_{\mathcal{N}} \to A_{\mathcal{P}}; \qquad A_{\mathcal{P}} \to A_{\mathcal{N}}.$$

Let us examine the Laurent expansion (with respect to y) of the function on the left-hand side of (3.21):

$$\begin{aligned} & H^r \cdot (yI - A_{\mathcal{N}})^{-1} - (yI - A_{\mathcal{P}})^{-1} \cdot H^r \\ & - H^r \cdot (yI - A_{\mathcal{N}})^{-1} \cdot G_{\mathcal{N}} F_{\mathcal{P}} \cdot (yI - A_{\mathcal{P}})^{-1} \cdot H^r \\ & = (H^r A_{\mathcal{N}} - A_{\mathcal{P}} H^r - H^r G_{\mathcal{N}} \cdot F_{\mathcal{P}} H^r) y^{-2} + O(y^{-3}) \qquad (y \to \infty) \end{aligned}$$

In view of (3.21), the Sylvester-Lyapunov equality holds:

$$H^r A_{\mathcal{N}} - A_{\mathcal{P}} H^r = H^r G_{\mathcal{N}} \cdot F_{\mathcal{P}} H^r. \tag{3.23}$$

Analogously, from (3.22) we can derive the equality

$$H^l A_{\mathcal{P}} - A_{\mathcal{N}} H^l = H^l G_{\mathcal{P}} \cdot F_{\mathcal{N}} H^l. \tag{3.24}$$

According to Theorem 2.1 (actually, according to Theorem 2.5), the *core matrix* H^r (see Definition 2.2) is invertible. (The first proof of the Theorem 2.5 is based on the Sylvester-Lyapunov equalities (2.14), (2.15), but the second one is independent of them.) Multiplying (3.21) by the matrix $\left(H^r\right)^{-1}$ from the right and by the matrix $\left(H^r\right)^{-1}$ from the left and denoting, as before (see (2.50)), $S^r \stackrel{\text{def}}{=} \left(H^r\right)^{-1}$, we arrive at the equality (2.52). From (3.22) we can derive the equality (2.53) in the same way as we already derived the equality (2.52) from (3.21).

Thus we obtained the equalities (2.52) and (2.53) in two different ways. The first one is based on the explicit expressions (2.4) and (2.6) for the core matrices H^r and H^s. This method uses essentially the specific character of a rational matrix function *in general position.* The second method works for much broader classes of rational matrix functions. Up to a certain point, the method works for an *arbitrary* rational matrix function. First of all, this method uses the chain identities (3.8) and (3.9). These identities are evidently true for arbitrary matrix functions R which are non-degenerate (i.e., $\det R \not\equiv 0$). Then we use the representations (2.3), (2.5) for the kernels associated with R. However, we use nowhere that the pole and the zero matrices $A_{\mathcal{P}}$ and $A_{\mathcal{N}}$ are diagonal, with disjoint simple spectra. Actually, we have obtained the equality (3.18) for *any* rational matrix function R such that the associated kernel K_R^r admits the representation (2.1) with *arbitrary* $A_{\mathcal{P}}, A_{\mathcal{N}}, F_{\mathcal{P}}, G_{\mathcal{N}}$ and H^r (or, what is the same, for *any* rational matrix function R, such that the function $R(x) \cdot (R(y))^{-1}$ admits the representation (2.7) with *arbitrary* $A_{\mathcal{P}}, A_{\mathcal{N}}, F_{\mathcal{P}}, G_{\mathcal{N}}$ and H^r). Then we have to conclude from (3.18) (under the assumption that this equality holds for arbitrary functions φ, ψ which are holomorphic on the spectra of $A_{\mathcal{P}}$ and $A_{\mathcal{N}}$ respectively), that (3.21) holds.

Definition 3.5. (Row Version.) *Let Γ be a $k \times n$ matrix (k rows, n columns), and let A be an $n \times n$ matrix. The pair (Γ, A) is said to be* obstrollable[11] *if the linear span of the set of n-row-vectors $\{v\,\Gamma(\lambda I - A)^{-1}\}$, where v runs over the space $\mathbb{C}^k$ of all k-row-vectors and λ runs over $\mathbb{C} \setminus \sigma_A$ (σ_A is the spectrum of A), coincides with the whole space $\mathbb{C}^n$ (of all n-row-vectors).*
Equivalent definitions:
I. *The pair (Γ, A) is said to be* obstrollable, *if the linear span of the set of vectors $\{v\,\Gamma A^m\}$, where v runs over the space $\mathbb{C}^k$ of all k-row-vectors and m runs over the set $\mathbb{N}$ of all natural numbers, coincides with the whole space $\mathbb{C}^n$ (of all n-row-vectors).*
II. *The pair (Γ, A) is said to be* obstrollable, *if the linear span of the set of vectors $\{v\,\Gamma\varphi(A)\}$, where v runs over the space $\mathbb{C}^k$ of all k-row-vectors and φ runs over the set of all functions holomorphic on σ_A, coincides with the whole space $\mathbb{C}^n$ (of all n-row-vectors).*

Definition 3.5. (Column Version.) *Let Γ be an $n \times k$ matrix (n rows, k columns), and let b be an $n \times n$ matrix. the pair (B, Γ) is said to be* obstrollable, *if the linear span of the set of n-column-vectors $\{(\mu I - B)^{-1}\Gamma v\}$, where v runs over the space $\mathbb{C}^k$ of all k-column-vectors and μ runs over $\mathbb{C} \setminus \sigma_B$ (σ_B is the spectrum of B), coincides with the whole space $\mathbb{C}^n$ (of all n-column-vectors).*
Equivalent definitions:

[11]The word **obstrollable** is a mixture of the words **obs**ervable and con**trollable.**

I. *The pair* (B, Γ) *is said to be* obstrollable, *if the linear span of the set of vectors* $\{\Gamma B^m v\}$, *where* v *runs over the space* $\mathbb{C}^k$ *of all* k *column-vectors and* m *runs over the set* $\mathbb{N}$ *of all natural numbers, coincides with the whole space* $\mathbb{C}^n$ *(of all* n*-column-vectors).*
II. *The pair* (B, Γ) *is said to be* obstrollable, *if the linear span of the set of vectors* $\{\varphi(B)\Gamma v\}$, *where* v *runs over the space* $\mathbb{C}^k$ *of all* k *column-vectors and* φ *runs over the set of all functions holomorphic on* σ_B, *coincides with the whole space* $\mathbb{C}^n$ *(of all* n*-column-vectors).*

Comments On Terminology. This terminology is motivated by system theory. (In more details, by the theory of linear time invariant dynamical systems.) In this theory structures like $F(\lambda I - A)^{-1}$ and $(\mu I - B)^{-1}G$ appear, where F and G are $k \times n$ and $n \times k$ matrices and A, B are $n \times n$ matrix, and usually n is much bigger then k $(n \gg k)$. If F is interpreted as the input operator of the system and G is interpreted as its output operator, then the notion of the controllability of the pair (F, A) and the notion of the observability of the pair (B, G) are introduced. If G is interpreted as the input operator of the system and F is interpreted as its output operator, then the notion of controllability of the pair (B, G) and the notion of observability of the pair (F, A) are introduced. Structures analogous to the structure $F(\lambda I - A)^{-1}$, $(\mu I - B)^{-1}G$ appear in the system representation problems as well. We need to formulate a property which is controllability if Γ is interpreted as the input operator, and is observability if Γ is interpreted as the output operator. However, we would not like to give a preference to one of the two: "in" or "out" interpretation of the matrix Γ. Because of this, we choose the "neutral" term *obstrollability*.

Lemma 3.2. *Let* F, G *be* $k \times n$ *and* $n \times k$ *matrices, and let* A, B, M *be* $n \times n$ *matrices. Assume that*

1. $F(\lambda I - A)^{-1}M(\mu I - B)^{-1}G \equiv 0 \qquad (\forall\lambda \in \mathbb{C} \setminus \sigma_A, \quad \forall\mu \in \mathbb{C} \setminus \sigma_B).$
2. *The pairs* (F, A) *and* (B, G) *are obstrollable.*

Then $M = 0$.

Proof. The proof follows immediately from the definition of obstrollability. □

Thus the foregoing reasonings (the reasoning of this section and the reasoning used for the first proof of Theorem 2.5) prove the following

Theorem 3.2. I. *Let* R *be a rational function such that*[12] *the chain matrix function* $R(x)(R(y))^{-1}$ *admits a representation of the form* (2.7) *with some* $k \times n$ *matrix* $F_{\mathcal{P}}$, $n \times k$ *matrix* $G_{\mathcal{N}}$, *and* $n \times n$ *matrices* $A_{\mathcal{P}}, A_{\mathcal{N}}$ *and* H^r. *If the pairs* $(F_{\mathcal{P}}, A_{\mathcal{P}})$ *and* $(A_{\mathcal{N}}, G_{\mathcal{N}})$ *are obstrollable, then the matrix* H^r *satisfies the equality (3.23).*
II. *Let* R *be a rational function such that*[12] *the chain matrix function* $(R(x))^{-1}R(y)$ *admits a representation of the form* (2.8) *with some* $k \times n$ *matrix* $F_{\mathcal{N}}$, $n \times k$ *matrix*

[12] We assume neither that the matrices $A_{\mathcal{P}}$, $A_{\mathcal{N}}$ are diagonal, nor that their spectra are simple or non-intersecting. We also do not assume a priori that the matrices H^r, H^l are invertible.

$G_{\mathcal{P}}$, and $n \times n$ matrices $A_{\mathcal{P}}, A_{\mathcal{N}}$ and H^l. If the pairs $(F_{\mathcal{N}},\, A_{\mathcal{N}})$ and $(A_{\mathcal{P}},\, G_{\mathcal{P}})$ are obstrollable, then the matrix H^l satisfies the equality (3.24).

III. *Let R be a rational function such that the chain matrix functions $R(x)(R(y))^{-1}$ and $(R(x))^{-1}R(y)$ admit representations of the form (2.7) and (2.8) respectively, with some $k \times n$ matrices $F_{\mathcal{P}}$, $F_{\mathcal{N}}$, some $n \times k$ matrices $G_{\mathcal{N}}$, $G_{\mathcal{P}}$, and some $n \times n$ matrices $A_{\mathcal{P}}, A_{\mathcal{N}}, H^r, H^l$. If all four pairs $(F_{\mathcal{P}},\, A_{\mathcal{P}})$, $(A_{\mathcal{N}},\, G_{\mathcal{N}})$, $(F_{\mathcal{N}},\, A_{\mathcal{N}})$ and $(A_{\mathcal{P}},\, G_{\mathcal{P}})$ are obstrollable, and if moreover the coupling relations (2.17) hold, then the matrices H^r, H^l are mutually inverse: $H^r \cdot H^l = I, H^l \cdot H^r = I$, and for their inverse matrices $S^r = (H^r)^{-1}$ and $S^l = (H^l)^{-1}$ the equalities (2.52), (2.53) holds, i.e., the matrices S^r and S^l are solutions of the Sylvester-Lyapunov equations (2.54) and (2.55).*

Theorem 3.3 *Let R be a rational function such that the chain function $R(x)(R(y))^{-1}$ admits a representation of the form (2.58) with [13] some $k \times n$ and $n \times k$ matrices $F_{\mathcal{P}}$, $G_{\mathcal{N}}$, $n \times n$ matrices $A_{\mathcal{P}}$ and $A_{\mathcal{N}}$, and some* invertible *$n \times n$ matrix S^r. If the pairs $(F_{\mathcal{P}},\, A_{\mathcal{P}})$ and $(A_{\mathcal{N}},\, G_{\mathcal{N}})$ are obstrollable, then the equality (2.52) holds, i.e., the matrix S^r is a solution of the Sylvester-Lyapunov equation (2.54).*

The "left" version of this theorem holds as well.

Theorem 3.4 *Let R be a rational function such that the chain function $(R(x))^{-1}R(y)$ admits a representation of the form (2.59) with[13] some $k \times n$ and $n \times k$ matrices $F_{\mathcal{N}}$, $G_{\mathcal{P}}$, $n \times n$ matrices $A_{\mathcal{P}}$ and $A_{\mathcal{N}}$, and some invertible $n \times n$ matrix S^l. If the pairs $(F_{\mathcal{N}},\, A_{\mathcal{N}})$ and $(A_{\mathcal{P}},\, G_{\mathcal{P}})$ are obstrollable, then the equality (2.53) holds, i.e., the matrix S^l is a solution of the Sylvester-Lyapunov equation (2.55).*

Lemma 3.3. *Let $F = (f_1, \dots, f_n)$ be a $k \times n$ matrix (i.e., $f_1, \dots, f_n$ are k-column-vectors), and $A = \operatorname{diag}(\alpha_1, \dots, \alpha_n)$ be a diagonal matrix with simple spectrum (i.e., all diagonal entries $\alpha_1, \dots \alpha_n$ are pairwise different). If no column $f_1, \dots, f_n$ of the matrix F is equal to zero, then the pair (F, A) is obstrollable.*

Proof. Actually, the proof of the statement was already done (see how we obtained the equality (3.20)). □

This lemma (together with Theorem 2.1) shows that Theorem 3.2 and Theorem 3.3 are applicable to rational matrix functions in general position.

Now we "inverse" our reasonings leading from a chain identity to a Sylvester-Lyapunov equation. Let F, G be $k \times n$ and $n \times k$ matrices, A, B be $n \times n$ matrices, with spectra σ_A and σ_B, and S be an *invertible* $n \times n$ matrix for which the equality

$$BS - SA = GF \tag{3.25}$$

holds. Evidently, this equality is equivalent to the identity (with respect to $y \in \mathbb{C}$):

$$S(yI - A) - (yI - B)S \equiv GF.$$

---

[13] We assume neither that the matrices $A_{\mathcal{P}}$, $A_{\mathcal{N}}$ are diagonal, nor that their spectra are simple or non-intersecting. Of course, the invertibility of the matrix S^r imposes implicitly some restrictions on the matrices $A_{\mathcal{P}}$, $A_{\mathcal{N}}$, $F_{\mathcal{P}}$, $G_{\mathcal{N}}$.

Multiplying the last identity by the matrix $S^{-1}(yI-B)^{-1}$ from the left and by the matrix $(yI-A)^{-1}S^{-1}$ from the right, we arrive at the identity

$$S^{-1}(yI-B)^{-1}-(yI-A)^{-1}S^{-1}-S^{-1}(yI-B)^{-1}G\cdot F(yI-A)^{-1}S^{-1}\equiv 0.$$

(This is nothing more than the equality $M=0$, where M is defined by (3.17).) Multiplying the last identity by the matrix $F(xI-A)^{-1}$ from the left and by the matrix $(zI-B)^{-1}G$ from the right, we arrive at the identity (with respect to $x\in\mathbb{C}$, $y\in\mathbb{C}$, $z\in\mathbb{C}$):

$$F(xI-A)^{-1}S^{-1}\cdot(yI-B)^{-1}(zI-B)^{-1}\cdot G-F(xI-A)^{-1}(yI-A)^{-1}S^{-1}(zI-B)^{-1}G \\ -F(xI-A)^{-1}S^{-1}(yI-B)^{-1}GF(yI-A)^{-1}S^{-1}(zI-B)^{-1}G\equiv 0.$$

Using Hilbert identities

$$(xI-A)^{-1}(yI-A)^{-1}\equiv-\frac{(xI-A)^{-1}-(yI-A)^{-1}}{x-y},$$

$$(yI-B)^{-1}(zI-B)^{-1}\equiv-\frac{(yI-B)^{-1}-(zI-B)^{-1}}{y-z},$$

we reduce the last identity to the form

$$\frac{K(x,z)-K(y,z)}{x-y}-\frac{K(x,y)-K(x,z)}{y-z}\equiv K(x,y)\cdot K(y,z),$$

or, what is the same, to the identity

(3.26)

$$(x-y)\,K(x,y)+(y-z)\,K(y,z)-(x-z)\,K(x,z)\equiv-(x-y)(y-z)\,K(x,z).$$

where now the function $K(\xi,\eta)$ of two variables is defined by the formula

$$K(x,y)\stackrel{\text{def}}{=}F(xI-A)^{-1}\cdot S^{-1}\cdot(yI-B)^{-1}G. \tag{3.27}$$

The identity (3.26) may be rewritten in the form a chain identity (3.1): $T(x,y)\cdot T(y,z)\equiv T(x,z)$, where the function $T(\cdot,\cdot)$ of two variables is defined as $T(x,y)\stackrel{\text{def}}{=}I+(x-y)K(x,y)$, or

$$T(x,y)\stackrel{\text{def}}{=}I+(x-y)F(xI-A)^{-1}S^{-1}(yI-B)^{-1}G,$$
$$\mathcal{D}_T\stackrel{\text{def}}{=}\left(\overline{\mathbb{C}}\setminus\sigma_A\right)\times\left(\overline{\mathbb{C}}\setminus\sigma_B\right).$$

Thus we proved the following

Theorem 3.5. *Let F, G be $k\times n$ and $n\times k$ matrices, let A, B be $n\times n$ matrices with spectra σ_A and σ_B, and let S be an invertible $n\times n$ matrix for which the equality $BS-SA=GF$ holds.*

Then the matrix function $T(\cdot,\cdot)$, which is defined by

$$T(x,y)\stackrel{\text{def}}{=}I+(x-y)F(xI-A)^{-1}S^{-1}(yI-B)^{-1}G, \tag{3.28}$$
$$\mathcal{D}_T\stackrel{\text{def}}{=}\left(\overline{\mathbb{C}}\setminus\sigma_A\right)\times\left(\overline{\mathbb{C}}\setminus\sigma_B\right),$$

satisfies the chain identity (3.1):

$$T(x,\, y) \cdot T(y,\, z) \equiv T(x,\, z)$$

and the diagonal unity identity (3.2):

$$T(x,x) \equiv I,$$

and hence[14], *is of the form*

$$T(x,y) = R(x)R^{-1}(y),$$

where $R(x) \stackrel{\text{def}}{=} T(x,\infty)$, $(R(y))^{-1} \stackrel{\text{def}}{=} T(\infty,y)$:

$$(3.29) \quad R(x) = I - F(xI - A)^{-1}S^{-1}G, \qquad R^{-1}(y) = I + FS^{-1}(yI - B)^{-1}G,$$

are mutually inverse (i.e., $R(x)R^{-1}(x) \equiv R^{-1}(x)R(x) \equiv I$*) rational matrix functions.*

Proof. The chain identity (3.1) for the function T, defined by (3.28), was proved immediately before. The diagonal unity identity (3.2) evidently follows from the expression (3.28). The equality $T(x,y) \equiv T(x,\infty) \cdot T(\infty,y))$ is the special case of the chain identity (3.1) (written for the triple of points $x,\infty,y)$). See Theorem 3.1.

Letting y tend to ∞, we obtain the expression (3.29) for the matrix function $R(x) \stackrel{\text{def}}{=} T(x,\infty)$. Letting x tend to ∞, we obtain the expression (3.29) for the matrix function $R^{-1} \stackrel{\text{def}}{=} T(\infty,y)$.

The fact that the functions R and R_{-1}, defined by (3.29)), are mutually inverse follows from the chain identity written for the triples x,∞,x and ∞,x,∞. The fact that the function R is rational is evident. □

However, Theorem 3.5 says nothing about the nature of the rational function R. Imposing restrictions on the data A, B, F, G, we can say more about the matrix function R.

Theorem 3.6. *Let F be a $k \times n$ matrix and G be an $n \times k$ matrix with non-zero columns and non-zero rows respectively, i.e.,*

$$F = [f_1\, f_2\, \dots\, f_n], \qquad G = \begin{bmatrix} g_1 \\ g_2 \\ \vdots \\ g_n \end{bmatrix},$$

where no column $f_1, f_2, \dots, f_n$ *and no row* $g_1, g_2, \dots, g_n$ *are zero, and let* A, B *be* $n \times n$ *be diagonal matrices with simple disjoint spectra, i.e.,*

$$A = \operatorname{diag}(\lambda_1, \dots, \lambda_n), \qquad B = \operatorname{diag}(\mu_1, \dots, \mu_n),$$

[14] According to Theorem 3.1, the set $(\overline{\mathbb{C}} \setminus \sigma_A) \cap (\overline{\mathbb{C}} \setminus \sigma_B) = \overline{\mathbb{C}} \setminus (\sigma_A \cup \sigma_B)$ is not only nonempty but also very rich. So, we have many possibilities for a choice of a distinguished point. (See the proof of the Theorem 3.1.) In particular, we can choose the point ∞ as a distinguished point.

where $\lambda_1, \dots, \lambda_n;\ \mu_1, \dots, \mu_n$, are pairwise different complex numbers. Assume that the $n \times n$ matrix S,

$$S = \|s_{p,q}\|_{1 \le p,q \le n}, \qquad s_{p,q} = \frac{g_p\, f_q}{\mu_p - \lambda_q} \tag{3.30}$$

(which can be obtained from the data F, G, A, B as the unique solution of the Sylvester-Lyapunov equation $BX - XA = GF$) is invertible. Then:

1. *The matrix function $T(\cdot,\cdot)$ of two variables, which is defined by*

$$T(x, y) \stackrel{\text{def}}{=} I + (x - y)F(xI - A)^{-1}S^{-1}(yI - B)^{-1}G, \tag{3.28}$$
$$\mathcal{D}_T \stackrel{\text{def}}{=} \left(\overline{\mathbb{C}} \setminus \sigma_A\right) \times \left(\overline{\mathbb{C}} \setminus \sigma_B\right),$$

   satisfies the chain identity (3.1):

$$T(x, y) \cdot T(y, z) \equiv T(x, z),$$

   and the diagonal unity identity (3.2):

$$T(x, x) \equiv I.$$

2. *The matrix function $T(x, y)$ is of the form*

$$T(x, y) = R(x)R^{-1}(y),$$

   where the matrix functions R, R^{-1} are defined by the formulas $R(x) \stackrel{\text{def}}{=} T(x, \infty)$, $R^{-1}(y) \stackrel{\text{def}}{=} T(\infty, y)$:

$$R(x) = I - F(xI - A)^{-1}S^{-1}G, \quad R^{-1}(y) = I + FS^{-1}(yI - B)^{-1}G. \tag{3.29}$$

   and are mutually inverse (i.e., $R(x)R^{-1}(x) \equiv R^{-1}(x)R(x) \equiv I$).
3. *The matrix functions R and R^{-1} are rational matrix functions in general position.*
4. *The pole set $\mathcal{P}(R)$ of the function R coincides with the set $\{\lambda_1, \dots \lambda_n\}$; the zero set $\mathcal{N}(R)$ of the function R coincides with the set $\{\mu_1, \dots \mu_n\}$, i.e.,*

$$A_{\mathcal{P}} = A, \quad A_{\mathcal{N}} = B, \tag{3.31}$$

   where $A_{\mathcal{P}}$ and $A_{\mathcal{N}}$ are the pole and the zero matrices of the matrix function R.
5. *The semi-residual matrices*[15] *$F_{\mathcal{P}}, G_{\mathcal{P}}, F_{\mathcal{N}}, G_{\mathcal{N}}$ of the matrix function R can be expressed in terms of the data F, G and of the matrix S (which in its turn can be expressed in terms of the data F, G, A, B):*

$$F_{\mathcal{P}} = F; \quad G_{\mathcal{N}} = G; \qquad F_{\mathcal{N}} = F \cdot S^{-1}; \quad G_{\mathcal{P}} = -S^{-1} \cdot G. \tag{3.32}$$

6. *The right zero-pole coupling matrix S^r and the left zero-pole coupling matrix S^l for the matrix function R can be expressed in terms of the matrix S:*

$$S^r = S; \qquad S^l = S^{-1}. \tag{3.33}$$

[15] More precisely, one of the representatives of the equivalence class of the set of semi-residual matrices of the matrix function R. See Remark 1.2.

Proof. • Items 1 and 2 of Theorem 3.6 have already been proved. (See Theorem 3.5.)

• Let us prove that the matrix functions R and R^{-1} are in general positions and investigate their singularities. The expression (3.29) for R may be written in the form

$$R(x) = I + \sum_{1 \le j \le n} \frac{R_{\lambda_j}}{x - \lambda_j},$$

with the matrix R_{λ_j} is of the form

$$R_{\lambda_j} = f_j \cdot v_j,$$

where v_j is the k-th row of the $n \times k$ matrix $V \stackrel{\text{def}}{=} -S^{-1}G$. From this expression it follows that R is holomorphic outside of the points $\{\lambda_1, \dots, \lambda_n\}$ and its inverse R^{-1} is holomorphic outside of the points $\{\mu_1, \dots, \mu_n\}$. Let us focus on the point λ_j. There are two possibilities: either $v_j = 0$, or $v_j \neq 0$. If $v_j = 0$ then $R_{\lambda_j} = 0$ and hence, the function R is holomorphic at the point λ_j. If $v_j \neq 0$ then the matrix R_{λ_j} is non-zero, and has rank one. (We recall that, according to the assumptions of Theorem 3.6, $f_j \neq 0$.) We show now that the equality $v_j = 0$ is impossible. This equality may be written in the form $E_j S^{-1} G = 0$, where $E_j = \text{diag}(\delta_{1j}, \delta_{2j}, \dots, \delta_{nj})$ (δ is the Kronecker symbol). Multiplying the identity[16] $BS - SA = GF$ by the matrix $E_j S^{-1}$ from the left and by the matrix S^{-1} from the right and taking into account that the matrices e_j and A commute (both are diagonal), we arrive at the equality $(E_j S^{-1})B - A(E_j S^{-1}) = 0$. Because the spectra of A and B are disjoint, we obtain that $E_j S^{-1} = 0$, and hence $E_j = 0$. The contradiction shows that the equality $v_j = 0$ is impossible. Thus each point λ_j, $j = 1, 2, \dots, n$, is a simple pole of the matrix function R, with residue matrices of rank one.

• Analogously, we can show that the matrix function R^{-1} is holomorphic outside of the points $\{\mu_1, \dots, \mu_n\}$ and that each point μ_j, $j = 1, 2, \dots, n$, is a simple pole of the matrix function R^{-1}, with residue matrices of rank one. Thus items 3 and 4 of the claim of Theorem 3.6 are proved.

• Item 5 of the claim follows from the representations (3.29). (Compare (3.29) with (1.37)–(1.38)).

• Now that we have established the relation (3.32) we may rewrite the equality $BS - SA = GF$ in the form $A_{\mathcal{N}} S - S A_{\mathcal{P}} = G_{\mathcal{N}} F_{\mathcal{P}}$. Comparing the last equality with equality (2.52), we conclude that $S^r = S$. From (2.57) it follows now that $S^l = S^{-1}$. □

The "hybrid" version of this theorem (see Remark 2.9 and formulas (2.66), (2.67)) can be formulated as well. This is the form which is convenient for applications in study of the Schlesinger system.

---

[16] Which, in fact, serves as the definition of the matrix S.

Theorem 3.7. *Let F be a $k \times n$ matrix and G be an $n \times k$ matrix with non-zero columns and non-zero rows respectively, i.e.,*

$$F = [f_1\, f_2\, \dots\, f_n], \qquad G = \begin{bmatrix} g_1 \\ g_2 \\ \vdots \\ g_n \end{bmatrix},$$

where no column $f_1, f_2, \dots, f_n$ and no row $g_1, g_2, \dots, g_n$ are zero, and let A, B be $n \times n$ be diagonal matrices with simple disjoint spectra, i.e.,

$$A = \operatorname{diag}(\lambda_1, \dots, \lambda_n), \qquad B = \operatorname{diag}(\mu_1, \dots, \mu_n),$$

where $\lambda_1, \dots, \lambda_n;\ \mu_1, \dots, \mu_n$, are pairwise different complex numbers. Assume that the $n \times n$ matrix S,

$$S = \|s_{p,q}\|_{1 \le p,q \le n}, \qquad s_{p,q} = \frac{g_p\, f_q}{\lambda_p - \mu_q} \tag{3.34}$$

(which can be obtained from the data F, G, A, B as the unique solution of the Sylvester-Lyapunov equation $AX - XB = GF$) is invertible. Then:

1. *The matrix function T of two variables, which is defined by the formula*

$$T(x,y) \stackrel{\text{def}}{=} I - (x-y)FS^{-1}(xI-A)^{-1}S(yI-B)^{-1}S^{-1}G, \tag{3.35}$$

satisfies the chain identity (3.1):

$$T(x,y)\cdot T(y,z) \equiv T(x,z),$$

and the diagonal unity identity (3.2):

$$T(x,x) \equiv I.$$

2. *The matrix function $T(x,y)$ is of the form*

$$T(x,y) = R(x)R^{-1}(y),$$

where the matrix functions R, R^{-1} are defined by the formulas $R(x) \stackrel{\text{def}}{=} T(x,\infty)$, $R^{-1}(y) \stackrel{\text{def}}{=} T(\infty,y)$:

$$R(x) = I + FS^{-1}(xI-A)^{-1}G, \quad R^{-1}(y) = I - F(yI-B)^{-1}S^{-1}G. \tag{3.29}$$

and are mutually inverse (i.e., $R(x)R^{-1}(x) \equiv R^{-1}(x)R(x) \equiv I$).

3. *The matrix functions R and R^{-1} are rational matrix functions in general position.*
4. *The pole set $\mathcal{P}(R)$ of the function R coincides with the set $\{\lambda_1, \dots \lambda_n\}$; the zero set $\mathcal{N}(R)$ of the function R coincides with the set $\{\mu_1, \dots \mu_n\}$, i.e.,*

$$A_{\mathcal{P}} = A, \quad A_{\mathcal{N}} = B, \tag{3.36}$$

where $A_{\mathcal{P}}$ and $A_{\mathcal{N}}$ are the pole and the zero matrices of the matrix function R.

5. *The semi-residual matrices*[17] $F_{\mathcal{P}}$, $G_{\mathcal{P}}$, $F_{\mathcal{N}}$, $G_{\mathcal{N}}$ *of the matrix function* R *can be expressed in terms of the data* F, G *and of the matrix* S *(which in its turn can be expressed in terms of the data* F, G, A, B*):*

$$F_{\mathcal{N}} = F; \quad G_{\mathcal{P}} = G; \qquad F_{\mathcal{P}} = F \cdot S^{-1}; \quad G_{\mathcal{N}} = -S^{-1} \cdot G; \tag{3.37}$$

6. *The right zero-pole coupling matrix* S^r *and the left zero-pole coupling matrix* S^l *for the matrix function* R *can be expressed in terms of the matrix* S*:*

$$S^l = S; \qquad S^r = S^{-1}. \tag{3.38}$$

Proof. Theorem 3.7 is nothing more then Theorem 3.6 in "other variables". Let A, B, F, G be the data of Theorem 3.7, and let S be the matrix (3.34) generated by this data. Let us introduce the matrices

$$\tilde{F} = FS^{-1}, \quad \tilde{G} = -S^{-1}G, \quad \tilde{S} = S^{-1}.$$

The equality $AS - SB = GF$ when rewritten in terms of $A, B, \tilde{F}, \tilde{G}, \tilde{S}$ becomes $B\tilde{S} - \tilde{S}A = \tilde{G}\tilde{F}$. From this last equality it is easy to see that no column of the matrix $\tilde{F}$ and no row of the matrix $\tilde{G}$ equals zero: the equality $E_j\tilde{G} = 0$ or $\tilde{F}E_j = 0$, where $E_j = \operatorname{diag}(\delta_{1j}, \delta_{2j}, \ldots, \delta_{nj})$ (δ is the Kronecker symbol), leads to the equality $E_j\tilde{S} = 0$ or $\tilde{S}E_j = 0$, contradicting the invertibility of $\tilde{S}$. Now Theorem 3.6, applied to the matrix function

$$T(x,y) = I + (x-y)\tilde{F}(xI - A)^{-1}(\tilde{S})^{-1}(yI - B)^{-1}\tilde{G},$$

gives the chain and the diagonal unity identities for this T and the factorization $T(x,y) = R(x)R^{-1}(y)$, as well as the expressions for the semi-residual matrices $F_{\mathcal{P}}, G_{\mathcal{P}}, F_{\mathcal{N}}, G_{\mathcal{N}}$ of the matrix function R:

$$F_{\mathcal{P}} = \tilde{F}, \quad G_{\mathcal{N}} = \tilde{G}, \quad F_{\mathcal{N}} = \tilde{F} \cdot \tilde{S}^{-1}, \quad G_{\mathcal{P}} = -\tilde{S}^{-1}\tilde{G}.$$

When rewritten in terms of F, G, S, these relations become (3.37). □

4. The system representation as a tool for the spectral (Wiener–Hopf) factorization of matrix functions

In this section we show that the system representation may be used as an efficient tool for the so called spectral factorization (or the Wiener-Hopf factorization) of a matrix function. The problem of the spectral factorization can be formulated in the following way.

Geometric Configuration. *In the extended complex plane* $\overline{\mathbb{C}}$, *a simple closed contour* Γ *is given. This contour separates* $\overline{\mathbb{C}}$ *inter two regions,* G_+ *and* G_-. *These regions* G_+ *and* G_- *are connected open sets. We assume that the point* $\{\infty\}$ *does not belong to the contour* Γ, *thus one of the components, say* G_-, *contains the point* $\{\infty\}$.

[17] See the footnote [15].

Definition 4.1. *Given a $k \times k$ matrix function Φ on the contour Γ, a factorization of the form*

$$\Phi(\zeta) = \Phi_+(\zeta) \cdot \Phi_-(\zeta) \quad (\zeta \in \Gamma), \tag{4.1}$$

where Φ_+ and Φ_- are $k \times k$ matrix functions, the matrix function Φ_+ and its inverse $(\Phi_+)^{-1}$ are holomorphic on $G_- \cup \Gamma$, and the matrix function Φ_- and its inverse $(\Phi_-)^{-1}$ are holomorphic[18] *on $G_+ \cup \Gamma$, is said to be* a spectral factorization (*or* a Wiener-Hopf factorization) of the matrix-function Φ with respect to Γ.

We impose the normalizing condition

$$\Phi_+(\infty) = I. \tag{4.2}$$

on the factor Φ_+. (The function Φ_+ is holomorphic and invertible at the point ∞, so the condition (4.2) makes sense.)

Under the normalizing condition (4.2), the spectral factorization (4.1) is unique.

Even in the scalar case $k = 1$ (i.e., Φ is a complex valued function) the factorization problem (4.1) is not always solvable: there is a topological obstacle for the solvability. For a smooth nonvanishing complex valued function Φ on Γ, the factorization problem (4.1) is solvable if and only if there exists a single valued continuous branch of the function $\ln \Phi(\zeta)$ on Γ. If this condition is fulfilled, the solution of the factorization problem may be expressed in terms of the data Φ by the formula

$$\Phi_\pm(\zeta) = \exp\left\{ \pm \frac{1}{2\pi i} \int\limits_\Gamma \frac{\ln \Phi(t)}{t - z}\, dt \right\} \qquad (z \in G_\pm). \tag{4.3}$$

The proof of the fact that the formula (4.3) gives the solution of the factorization problem (4.1) in the scalar case is based essentially on the Sokhotskiĭ-Plemelj formulas on the boundary behavior of the Cauchy integral. Actually, in the scalar case we solve the additive problem

$$\Psi_+(\zeta) + \Psi_-(\zeta) = \ln \Phi(\zeta) \qquad (\zeta \in \Gamma),$$

and then we exponentiate. In the matrix case we still can solve the appropriate additive problem using the Cauchy integral, but exponentiating does not lead to the desired result: in view of the noncommutativity of the matrix multiplication, $\exp\{A+B\} \neq \exp\{A\}\cdot\exp\{B\}$ for matrices A and B in general. In the matrix case, the situation with the factorization problem (4.1) is much more complicated than in the scalar case. There are not only topological obstacles to the solvability of this problem. The factorization problem (4.1) is equivalent to some system of singular integral equations on Γ, and to analyze this system is approximately as hard as to investigate the original factorization problem (4.1). The factorization problem (4.1) appeared first in the context of Hilbert's twenty-first problem: to construct a Fuchsian linear differential system with the prescribed monodromy group. See [Pl2], [Bo] and [Gah] for details and historical references. The factorization of the

[18] In particular, the functions Φ_+ and Φ_- are holomorphic on the common boundary Γ of the domains G_+ and G_-, so the relation (4.1) makes sense.

type (4.1) is used also for solving systems of singular integrals equations with the Cauchy kernel on the contour as well as for solving systems of integral equations with a kernel depending on the differences of the arguments on the half axis. See [Vek], [GoKr] and [ClGo] on this subject. It should be mentioned that the seminal papers [Bir1] and [Bir2] by G. Birkhoff had a profound impact on further investigations of the matrix factorization.

We confine ourself to the case the function Φ is a rational matrix function (or, more precisely, the restriction to Γ of a rational matrix function) such that the functions Φ and Φ^{-1} are holomorphic on the contour Γ. In this case the factorization (4.1) is global, i.e., the matrix functions Φ_+ and Φ_- are rational, and the equality

$$\Phi(z) = \Phi_+(z) \cdot \Phi_-(z) \qquad (\forall z \in \overline{\mathbb{C}}) \tag{4.4}$$

holds. Indeed, in this case the function $\Phi_+^{-1}\Phi$ is holomorphic within G_- except finitely many poles located on the set $\mathcal{P}(\Phi) \cap G_-$. In view of (4.1), this function continues analytically into the function Φ_- which is holomorphic on $G_+ \cup \Gamma$. Thus the matrix-function Φ_- has no singularities in $\overline{\mathbb{C}}$ other than finitely many poles and hence is rational. By the same reasoning, the matrix function Φ_+ is rational.

Thus in the case that the initial matrix function Φ is rational, the problem of the spectral factorization may be reformulated in the following manner:

Definition 4.1$'$. *Given a $k \times k$ rational matrix function Φ,* $\det\Phi \not\equiv 0$, *a factorization of the form* (4.4), *where Φ_+, Φ_- are rational matrix functions with zero and pole location*

$$\mathcal{P}(\Phi_+) \subset G_+, \quad \mathcal{N}(\Phi_+) \subset G_+, \quad \mathcal{P}(\Phi_-) \subset G_-, \quad \mathcal{N}(\Phi_-) \subset G_-, \tag{4.5}$$

is said to be a spectral factorization (*or* a Wiener-Hopf factorization) of the matrix-function Φ with respect to Γ.

We consider even a more special case: the function Φ is a rational function in general position. In this case the calculation of the factors Φ_+ and Φ_- can be performed completely by hand, in terms of poles and "zeros" of the matrix function Φ and its semiresidual vectors.

So, let Φ be a rational matrix function in general position, normalized by the condition

$$\Phi(\infty) = I. \tag{4.6}$$

Let $\mathcal{P}(\Phi)$ and $\mathcal{N}(\Phi)$ be its pole and zero sets, $A_{\mathcal{P}}(\Phi)$ and $A_{\mathcal{N}}(\Phi)$ be its pole and zero matrices, $F_{\mathcal{P}}(\Phi)$, $G_{\mathcal{P}}(\Phi)$, $F_{\mathcal{N}}(\Phi)$, $G_{\mathcal{N}}(\Phi)$ be the appropriate semiresidual matrices. According to the Theorem 2.6, the zero-pole coupling matrices $S^r(\Phi)$ and $S^l(\Phi)$ are invertible, and the matrix-function Φ admits representations of the form (2.60) and (2.62):

$$\Phi(z) = I - F_{\mathcal{P}}(\Phi)(zI - A_{\mathcal{P}(\Phi)})^{-1} S^r(\Phi)^{-1} G_{\mathcal{N}}(\Phi), \tag{4.7}$$

and

$$\Phi(z) = I + F_{\mathcal{N}}(\Phi)S^l(\Phi)^{-1}(zI - A_{\mathcal{P}(\Phi)})^{-1}G_{\mathcal{P}}(\Phi). \tag{4.8}$$

The Sylvester-Lyapunov equations for the matrices $S^r(\Phi)$ and $S^l(\Phi)$ (which actually are the definitions of these matrices) are of the form:

$$A_{\mathcal{N}}(\Phi)S^r(\Phi) - S^r(\Phi)A_{\mathcal{P}}(\Phi) = G_{\mathcal{N}}(\Phi)F_{\mathcal{P}}(\Phi), \tag{4.9}$$

and

$$A_{\mathcal{P}}(\Phi)S^l(\Phi) - S^l(\Phi)A_{\mathcal{N}}(\Phi) = G_{\mathcal{P}}(\Phi)F_{\mathcal{N}}(\Phi). \tag{4.10}$$

Moreover, the matrices $S^r(\Phi)$ and $S^l(\Phi)$ satisfy the equality

$$S^l(\Phi) \cdot S^r(\Phi) = I. \tag{4.11}$$

(This is (2.57) for the matrix function Φ.) The zero-pole coupling relation hold:

$$\begin{aligned} &\text{(a) } G_{\mathcal{N}}(\Phi) = -S^r(\Phi)G_{\mathcal{P}}(\Phi); \qquad &&\text{(b) } G_{\mathcal{P}}(\Phi) = -S^l(\Phi)G_{\mathcal{N}}(\Phi);\\ &\text{(c) } F_{\mathcal{P}}(\Phi) = F_{\mathcal{N}}(\Phi)S^r(\Phi); &&\text{(d) } F_{\mathcal{N}}(\Phi) = F_{\mathcal{P}}(\Phi)S^l(\Phi). \end{aligned} \tag{4.12}$$

(This is (2.51) for the matrix-function Φ.)

According to the assumptions,

$$\text{(a) } \mathcal{P}(\Phi) \cap \Gamma = \emptyset; \qquad \text{(b) } \mathcal{N}(\Phi) \cap \Gamma = \emptyset. \tag{4.13}$$

Assume that the factorization (4.4) holds, and that the normalizing condition (4.2) is satisfied.

Since the functions Φ_-, Φ_-^{-1} are holomorphic in G_+ and the functions Φ, Φ^{-1} have only simple poles in G_+, from the relations

$$\text{(a) } \Phi_+ = \Phi \cdot \Phi_-^{-1}, \qquad \text{(b) } \Phi_+^{-1} = \Phi_- \cdot \Phi^{-1}, \tag{4.14}$$

it follows that the functions Φ_+, Φ_+^{-1} have only simple poles in G_+. In G_- and on Γ the functions Φ_+ and Φ_+^{-1} don't have singularities at all. Thus

$$\mathcal{P}(\Phi_+) = \mathcal{P}(\Phi) \cap G_+, \qquad \mathcal{N}(\Phi_+) = \mathcal{N}(\Phi) \cap G_+. \tag{4.15}$$

Let $\lambda \in \mathcal{P}(\Phi_+)$. From (4.14a) it follows that the residues $R_\lambda(\Phi_+)$ and $R_\lambda(\Phi)$ of the matrix functions Φ_+ and Φ at the point λ are related by the equality

$$R_\lambda(\Phi_+) = R_\lambda(\Phi) \cdot (\Phi_-(\lambda))^{-1} \qquad (\,\forall \lambda \in \mathcal{P}(\Phi_+)\,). \tag{4.16}$$

Since Φ is a matrix function in general position, the rank of the residue matrix $R_\lambda(\Phi)$ is equal to one. According to the assumptions, the matrix $\Phi_-(\lambda)$ is invertible for $\lambda \in G_+$. From (4.16) it follows now that the rank of the residue matrix $R_\lambda(\Phi_+)$ is equal to one as well. Let now $\mu \in \mathcal{N}(\Phi_+)$. From (4.14b) it follows that the residues $R_\mu(\Phi_+)$ and $R_\mu(\Phi)$ of the matrix functions Φ_+^{-1} and Φ^{-1} at the point μ are related by the equality

$$R_\mu(\Phi_+) = \Phi_-(\mu) \cdot R_\mu(\Phi) \qquad (\,\forall \mu \in \mathcal{N}(\Phi_+)\,). \tag{4.17}$$

The rank of the matrix $R_\mu(\Phi)$ is equal to one (Φ is a matrix function in general position); the matrix $\Phi_-(\mu)$ is invertible (according to the assumptions, the matrix

$\Phi_-(\mu)$ is invertible for $\mu \in G_+$). From (4.17) it follows now that the rank of the residue matrix $R_\mu(\Phi_+)$ is equal to one as well. From (4.15) it follows that

$$\mathcal{P}(\Phi_+) \cap \mathcal{N}(\Phi_+) = \emptyset; \qquad \{\infty\} \notin \mathcal{P}(\Phi_+), \quad \{\infty\} \notin \mathcal{N}(\Phi_+).$$

Thus Φ_+ is a rational matrix function in general position.

In the same way we obtain that Φ_- is a rational matrix function in general position, and

$$\mathcal{P}(\Phi_-) = \mathcal{P}(\Phi) \cap G_-, \qquad \mathcal{N}(\Phi_-) = \mathcal{N}(\Phi) \cap G_-. \tag{4.18}$$

Moreover, the residues $R_\lambda(\Phi_-)$ and $R_\mu(\Phi_-)$ of the matrix functions Φ_- and Φ_-^{-1} at the poles $\lambda \in \mathcal{P}(\Phi_-)$ and $\mu \in \mathcal{N}(\Phi_-) \, (= \mathcal{P}(\Phi_-^{-1}))$ are related to the residues of the matrix functions Φ and Φ^{-1} at the same points by the equalities

$$R_\lambda(\Phi_-) = (\Phi_+(\lambda))^{-1} \cdot R_\lambda(\Phi) \qquad (\ \forall \lambda \in \mathcal{P}(\Phi_-)\ ) \tag{4.19}$$

and

$$R_\mu(\Phi_-) = R_\mu(\Phi) \cdot \Phi_+(\mu) \qquad (\ \forall \mu \in \mathcal{N}(\Phi_+). \tag{4.20}$$

From (4.16) and (4.17) it follows that the left semiresidual vectors of the matrix functions Φ_+ and Φ at the poles $\lambda \in \mathcal{P}(\Phi_+)$ coincide:

$$f_\lambda(\Phi_+) = f_\lambda(\Phi) \qquad (\forall \lambda \in \mathcal{P}(\Phi_+)). \tag{4.21}$$

and the right semiresidual vectors of the matrix functions Φ_+^{-1} and Φ^{-1} at the poles $\mu \in \mathcal{N}(\Phi_+)$ coincide:

$$g_\mu(\Phi_+^{-1}) = g_\mu(\Phi^{-1}) \qquad (\forall \mu \in \mathcal{N}(\Phi_+)). \tag{4.22}$$

In the same way we can obtain that the right semiresidual vectors of the matrix functions Φ_- and Φ coincide:

$$g_\lambda(\Phi_-) = g_\lambda(\Phi) \qquad (\forall \lambda \in \mathcal{P}(\Phi_-)). \tag{4.23}$$

and the left semiresidual vectors of the matrix functions Φ_-^{-1} and Φ^{-1} coincide:

$$f_\mu(\Phi_-^{-1}) = f_\mu(\Phi^{-1}) \qquad (\forall \mu \in \mathcal{N}(\Phi_-)). \tag{4.24}$$

The equalities (4.21)–(4.24) are crucial for solving the factorization problem we consider.

According to Lemma 1.2, the equality

$$\#\mathcal{P}(\Phi_+) = \#\mathcal{N}(\Phi_+)$$

holds for the rational matrix function in general position Φ_+. Taking into account the equality (4.15), we obtain the following equality

$$\#(\mathcal{P}(\Phi) \cap G_+) = \#(\mathcal{N}(\Phi) \cap G_+) \overset{\text{def}}{=} n_+. \tag{4.25}$$

Of course, the equality

$$\#(\mathcal{P}(\Phi) \cap G_-) = \#(\mathcal{N}(\Phi) \cap G_-) \overset{\text{def}}{=} n_-. \tag{4.26}$$

holds as well.

To simplify the notation, we denote

(4.27)
$$\mathcal{P}_+ \stackrel{\text{def}}{=} \mathcal{P}(\Phi)\cap G_+;\ \mathcal{P}_- \stackrel{\text{def}}{=} \mathcal{P}(\Phi)\cap G_-;\ \mathcal{N}_+ \stackrel{\text{def}}{=} \mathcal{N}(\Phi)\cap G_+;\ \mathcal{N}_- \stackrel{\text{def}}{=} \mathcal{N}(\Phi)\cap G_-.$$

To the decompositions

$$\begin{aligned}\mathcal{P}(\Phi) &= \mathcal{P}_+ \cup \mathcal{P}_-, \quad \mathcal{P}_+ \cap \mathcal{P}_- = \emptyset,\\ \mathcal{N}(\Phi) &= \mathcal{N}_+ \cup \mathcal{N}_-, \quad \mathcal{N}_+ \cap \mathcal{N}_- = \emptyset\end{aligned}$$

of the pole and the zero sets $\mathcal{P}(\Phi)$, $\mathcal{N}(\Phi)$ of the matrix function Φ there correspond natural block-decompositions of the matrices which appear in the system representations of the matrix functions Φ and Φ^{-1}: the pole and the zero matrices $A_\mathcal{P}(\Phi)$, $A_\mathcal{N}(\Phi)$, the semiresidual matrices $F_\mathcal{P}(\Phi), F_\mathcal{N}(\Phi), G_\mathcal{P}(\Phi), G_\mathcal{N}(\Phi)$, as well as the zero-pole coupling matrices $S^r(\Phi)$ and $S^l(\Phi)$.

Namely, the decompositions of the pole and the zero matrices

$$A_\mathcal{P}(\Phi) = \operatorname{diag}(\lambda_l)_{\lambda_l\in\mathcal{P}(\Phi)} \text{ and } A_\mathcal{N}(\Phi) = \operatorname{diag}(\mu_l)_{\mu_l\in\mathcal{N}(\Phi)} \quad (\text{of dimension } n\times n)$$

are of the form:

$$A_\mathcal{P}(\Phi) = \begin{bmatrix} A_\mathcal{P}(\Phi)_1 & 0\\ 0 & A_\mathcal{P}(\Phi)_2\end{bmatrix}, \tag{4.28}$$

$$A_\mathcal{N}(\Phi) = \begin{bmatrix} A_\mathcal{N}(\Phi)_1 & 0\\ 0 & A_\mathcal{N}(\Phi)_2\end{bmatrix}, \tag{4.29}$$

where

$$A_\mathcal{P}(\Phi)_1 = \operatorname{diag}(\lambda_l)_{\lambda_l\in\mathcal{P}_+}; \qquad A_\mathcal{P}(\Phi)_2 = \operatorname{diag}(\lambda_l)_{\lambda_l\in\mathcal{P}_-};$$

$$A_\mathcal{N}(\Phi)_1 = \operatorname{diag}(\mu_l)_{\mu_l\in\mathcal{N}_+}; \qquad A_\mathcal{N}(\Phi)_2 = \operatorname{diag}(\mu_l)_{\mu_l\in\mathcal{N}_-};$$

$A_\mathcal{P}(\Phi)_1$ and $A_\mathcal{N}(\Phi)_1$ are diagonal $n_+\times n_+$ matrices, $A_\mathcal{P}(\Phi)_2$ and $A_\mathcal{N}(\Phi)_2$ are diagonal $n_-\times n_-$ matrices (n_+ and n_- are defined in (4.25) and (4.26): $n_+ = \#\mathcal{P}_+ = \#\mathcal{N}_+;\ n_- = \#\mathcal{P}_- = \#\mathcal{N}_-$).

The block-decompositions of the semiresidual matrices $F_\mathcal{P}(\Phi)$, $F_\mathcal{N}(\Phi)$, $G_\mathcal{P}(\Phi)$, $G_\mathcal{N}(\Phi)$:

$$\begin{aligned} F_\mathcal{P}(\Phi) &= \operatorname{row}(f_{\lambda_l})_{\lambda_l\in\mathcal{P}(\Phi)}, \quad F_\mathcal{N}(\Phi) = \operatorname{row}(f_{\mu_l})_{\mu_l\in\mathcal{N}(\Phi)} \quad \text{are} \quad k\times n \quad \text{matrices},\\ G_\mathcal{P}(\Phi) &= \operatorname{col}(g_{\lambda_l})_{\lambda_l\in\mathcal{P}(\Phi)}, \quad G_\mathcal{N}(\Phi) = \operatorname{col}(g_{\mu_l})_{\mu_l\in\mathcal{N}(\Phi)} \quad \text{are} \quad n\times k \quad \text{matrices},\end{aligned}$$

corresponding to the matrix-function Φ, are of the form:

(4.30) (p) $F_\mathcal{P}(\Phi) = [F_\mathcal{P}(\Phi)_1\ \ F_\mathcal{P}(\Phi)_2]\,;$ (n) $F_\mathcal{N}(\Phi) = [F_\mathcal{N}(\Phi)_1\ \ F_\mathcal{N}(\Phi)_2]\,,$

and

$$(4.31)\qquad (\mathrm{p})\quad G_{\mathcal{P}}(\Phi)=\begin{bmatrix} G_{\mathcal{P}}(\Phi)_1 \\ \\ G_{\mathcal{P}}(\Phi)_2\end{bmatrix};\qquad (\mathrm{n})\quad G_{\mathcal{N}}(\Phi)=\begin{bmatrix} G_{\mathcal{N}}(\Phi)_1 \\ \\ G_{\mathcal{N}}(\Phi)_2\end{bmatrix},$$

where

$$\begin{array}{llll} F_{\mathcal{P}}(\Phi)_1=\mathrm{row}(f_{\lambda_l})_{\lambda_l\in\mathcal{P}_+}, & F_{\mathcal{N}}(\Phi)_1=\mathrm{row}(f_{\mu_l})_{\mu_l\in\mathcal{N}_+} & \text{are } k\times n_+ & \text{matrices;}\\ F_{\mathcal{P}}(\Phi)_2=\mathrm{row}(f_{\lambda_l})_{\lambda_l\in\mathcal{P}_-}, & F_{\mathcal{N}}(\Phi)_2=\mathrm{row}(f_{\mu_l})_{\mu_l\in\mathcal{N}_-} & \text{are } k\times n_- & \text{matrices;}\\ G_{\mathcal{P}}(\Phi)_1=\mathrm{col}(g_{\lambda_l})_{\lambda_l\in\mathcal{P}_+}, & G_{\mathcal{N}}(\Phi)_1=\mathrm{col}(g_{\mu_l})_{\mu_l\in\mathcal{N}_+} & \text{are } n_+\times k & \text{matrices;}\\ G_{\mathcal{P}}(\Phi)_2=\mathrm{col}(g_{\lambda_l})_{\lambda_l\in\mathcal{P}_-}, & G_{\mathcal{N}}(\Phi)_2=\mathrm{col}(g_{\mu_l})_{\mu_l\in\mathcal{N}_-} & \text{are } n_-\times k & \text{matrices.}\end{array}$$

From (4.15) and (4.18) it follows that

$$(4.32)\qquad (\mathrm{p})\ A_{\mathcal{P}}(\Phi_+)=A_{\mathcal{P}}(\Phi)_1,\qquad (\mathrm{n})\ A_{\mathcal{N}}(\Phi_+)=A_{\mathcal{N}}(\Phi)_1,$$

and

$$(4.33)\qquad (\mathrm{p})\ A_{\mathcal{P}}(\Phi_-)=A_{\mathcal{P}}(\Phi)_2,\qquad (\mathrm{n})\ A_{\mathcal{N}}(\Phi_-)=A_{\mathcal{N}}(\Phi)_2.$$

In view of (4.21) and (4.22),

$$(4.34)\qquad (\mathrm{p})\quad F_{\mathcal{P}}(\Phi_+)=F_{\mathcal{P}}(\Phi)_1,\qquad (\mathrm{n})\quad G_{\mathcal{N}}(\Phi_+)=G_{\mathcal{N}}(\Phi)_1.$$

In view of (4.23) and (4.24),

$$(4.35)\qquad (\mathrm{p})\quad F_{\mathcal{N}}(\Phi_-)=F_{\mathcal{N}}(\Phi)_2,\qquad (\mathrm{n})\quad G_{\mathcal{P}}(\Phi_-)=G_{\mathcal{P}}(\Phi)_2.$$

Thus we have expressed the pole matrix $A_{\mathcal{P}}(\Phi_+)$ and the semiresidual matrices $F_{\mathcal{P}}(\Phi_+)$, $G_{\mathcal{N}}(\Phi_+)$ for the left factor Φ_+ in terms of the blocks of an appropriate block-decomposition of the pole matrix $A_{\mathcal{P}}(\Phi)$ and and the semiresidual matrices $F_{\mathcal{P}}(\Phi)$, $G_{\mathcal{N}}(\Phi)$ for the matrix function Φ to be factorized. We also have expressed the zero matrix $A_{\mathcal{N}}(\Phi_-)$ and the semiresidual matrices $F_{\mathcal{N}}(\Phi_-)$, $G_{\mathcal{P}}(\Phi_-)$ for the right factor Φ_- in terms of the blocks of the matrices $A_{\mathcal{N}}(\Phi)$, $F_{\mathcal{N}}(\Phi)$, $G_{\mathcal{P}}(\Phi)$. In principle, these data are sufficient to recover the factors Φ_+ and Φ_- (from the appropriate blocks of the pole and the semiresidual matrices for the matrix function Φ to be factorized). To carry out the recovery, we have to solve the Sylvester-Lyapunov equations (2.52) and (2.53) for the matrix functions Φ_+ and Φ_- respectively to find from these equations the zero-pole coupling matrices $S^r(\Phi_+)$ and $S^l(\Phi_-)$:

$$(4.36)\qquad A_{\mathcal{N}}(\Phi_+)S^r(\Phi_+)-S^r(\Phi_+)A_{\mathcal{P}}(\Phi_+)=G_{\mathcal{N}}(\Phi_+)F_{\mathcal{P}}(\Phi_+),$$

$$(4.37)\qquad A_{\mathcal{P}}(\Phi_-)S^l(\Phi_-)-S^l(\Phi_-)A_{\mathcal{P}}(\Phi_-)=G_{\mathcal{P}}(\Phi_-)F_{\mathcal{N}}(\Phi_-).$$

According to Statement 1 of Theorem 2.6 (actually, according to Theorem 2.5: see (2.31)), their solutions $S^r(\Phi_+)$ and $S^r(\Phi_-)$ are invertible matrices. Then we construct the factors Φ_+ and Φ_- according to the formulas (2.60) and (2.62):

$$(4.38)\qquad \Phi_+(z)=I-F_{\mathcal{P}}(\Phi_+)\big(zI-A_{\mathcal{P}}(\Phi_+)\big)^{-1}(S^r(\Phi_+))^{-1}G_{\mathcal{N}}(\Phi_+)$$

and

$$\Phi_-(z) = I + F_{\mathcal{N}}(\Phi_-)(S^l(\Phi_-))^{-1}\big(zI - A_{\mathcal{P}}(\Phi_-)\big)^{-1}G_{\mathcal{P}}(\Phi_-). \tag{4.39}$$

We express now the matrices $S^r(\Phi_+)$ and $S^l(\Phi_-)$ in terms of blocks of the matrices $S^r(\Phi)$ and $S^l(\Phi)$. The block-decomposition of the zero-pole coupling matrices $S^r(\Phi)$ and $S^l(\Phi)$ are of the form:

$$S^r(\Phi) = \begin{bmatrix} S^r(\Phi)_{11} & S^r(\Phi)_{12} \\ S^r(\Phi)_{21} & S^r(\Phi)_{22} \end{bmatrix}, \tag{4.40}$$

$$S^l(\Phi) = \begin{bmatrix} S^l(\Phi)_{11} & S^l(\Phi)_{12} \\ S^l(\Phi)_{21} & S^l(\Phi)_{22} \end{bmatrix}, \tag{4.41}$$

where 11-block-entries $S^r(\Phi)_{11}$, $S^l(\Phi)_{11}$ are $n_+ \times n_+$ matrices, and 22-block-entries $S^r(\Phi)_{22}$, $S^l(\Phi)_{22}$ are $n_- \times n_-$ matrices (n_+, n_- are defined in (4.25) and (4.26)). The block-decompositions (4.40) and (4.41) are consistent with the block-decompositions (4.28)–(4.31) of the pole and the semiresidual matrices. The Sylvester-Lyapunov equation (4.9) for the matrix $S^r(\Phi)$, written in the block-matrix form

$$\begin{bmatrix} A_{\mathcal{N}}(\Phi)_1 & 0 \\ 0 & A_{\mathcal{N}}(\Phi)_2 \end{bmatrix}\begin{bmatrix} S^r(\Phi)_{11} & S^r(\Phi)_{12} \\ S^r(\Phi)_{21} & S^r(\Phi)_{22} \end{bmatrix} - \begin{bmatrix} S^r(\Phi)_{11} & S^r(\Phi)_{12} \\ S^r(\Phi)_{21} & S^r(\Phi)_{22} \end{bmatrix}\begin{bmatrix} A_{\mathcal{P}}(\Phi)_1 & 0 \\ 0 & A_{\mathcal{P}}(\Phi)_2 \end{bmatrix} = \begin{bmatrix} G_{\mathcal{N}}(\Phi)_1 \\ G_{\mathcal{N}}(\Phi)_2 \end{bmatrix}[F_{\mathcal{P}}(\Phi)_1 \;\; F_{\mathcal{P}}(\Phi)_2],$$

may be considered as a system of matrix equations for the block-entries of the matrix $S^r(\Phi)$. This system is decomposed into four equations for the block-entries of this matrix.

In particular, the equation for the entry $S^r(\Phi)_{11}$ is of the form

$$A_{\mathcal{N}}(\Phi)_1 S^r(\Phi)_{11} - S^r(\Phi)_{11}A_{\mathcal{P}}(\Phi)_1 = G_{\mathcal{N}}(\Phi)_1 F_{\mathcal{P}}(\Phi)_1. \tag{4.42}$$

The Sylvester-Lyapunov equation (4.10) for the matrix $S^l(\Phi)$, written in the block-matrix form, may be considered as a system of matrix equations for the block-entries of the matrix $S^l(\Phi)$. In particular, the equation for the entry $S^l(\Phi)_{22}$ is of the form

$$A_{\mathcal{P}}(\Phi)_2 S^l(\Phi)_{22} - S^l(\Phi)_{22}A_{\mathcal{N}}(\Phi)_2 = G_{\mathcal{P}}(\Phi)_2 F_{\mathcal{N}}(\Phi)_2. \tag{4.43}$$

We show that

$$S^r(\Phi_+) = S^r(\Phi)_{11} \tag{4.44}$$

and

$$S^l(\Phi_-) = S^l(\Phi)_{22}. \tag{4.45}$$

The easiest way to do this is to use the explicit formulas (2.56) for the matrices S^r, S^l in terms of the poles, the zeros and the semiresidual matrices of the matrix function. For the function Φ the formula (2.56r) becomes

$$S^r(\Phi) = \|s^r_{p,q}(\Phi)\|_{\mu_p \in \mathcal{N}(\Phi), \lambda_q \in \mathcal{P}(\Phi)}, \quad s^r_{p,q}(\Phi) = \frac{g_{\mu_p}(\Phi) \cdot f_{\lambda_q}(\Phi)}{\mu_p - \lambda_q}. \tag{4.46}$$

In particular,

$$S^r(\Phi)_{11} = \|s^r_{p,q}(\Phi)\|_{\mu_p \in \mathcal{N}(\Phi) \cap G_+, \, \lambda_q \in \mathcal{P}(\Phi) \cap G_+}, \quad s^r_{p,q}(\Phi) = \frac{g_{\mu_p}(\Phi) \cdot f_{\lambda_q}(\Phi)}{\mu_p - \lambda_q}. \tag{4.47}$$

For the function Φ_+ the formula (2.56r) becomes

$$S^r(\Phi_+) = \|s^r_{p,q}(\Phi_+)\|_{\mu_p \in \mathcal{N}(\Phi_+), \lambda_q \in \mathcal{P}(\Phi_+)}, \quad s^r_{p,q}(\Phi_+) = \frac{g_{\mu_p}(\Phi_+) \cdot f_{\lambda_q}(\Phi_+)}{\mu_p - \lambda_q}. \tag{4.48}$$

Comparing two last formulas and taking into account (4.15), (4.21) and (4.22), we conclude that (4.44) holds. In the same way, comparing the formulas

$$S^l(\Phi) = \|s^l_{p,q}(\Phi)\|_{\lambda_p \in \mathcal{P}(\Phi), \mu_q \in \mathcal{N}(\Phi)}, \quad s^l_{p,q}(\Phi) = \frac{g_{\lambda_p}(\Phi) \cdot f_{\mu_q}(\Phi)}{\lambda_p - \mu_q}, \tag{4.49}$$

and

$$S^l(\Phi_-) = \|s^l_{p,q}(\Phi_-)\|_{\lambda_p \in \mathcal{P}(\Phi_-), \mu_q \in \mathcal{N}(\Phi_-)}, \quad s^l_{p,q}(\Phi_-) = \frac{g_{\lambda_p}(\Phi_-) \cdot f_{\mu_q}(\Phi_-)}{\lambda_p - \mu_q}, \tag{4.50}$$

and taking into account (4.18), (4.23) and (4.24), we conclude that (4.45) holds.

The explicit expressions (4.48), (4.50) for the zero-pole coupling matrices are a consequence of the Sylvester-Lyapunov equalities (4.36), (4.37). It is also possible to derive the equations (4.44) and (4.45) directly from the equalities (4.36), (4.37), bypassing the explicit expressions (4.48), (4.50). (4.46), (4.49). The latter way is better, because it is applicable not only to matrix functions in general position.

Taking into account the equalities (4.32), and (4.34), we arrive from (4.36) at the equation

$$A_{\mathcal{N}}(\Phi)_1 S^r(\Phi_+) - S^r(\Phi_+) A_{\mathcal{P}}(\Phi)_1 = G_{\mathcal{N}}(\Phi)_1 F_{\mathcal{P}}(\Phi)_1. \tag{4.51}$$

Comparing (4.42) and (4.51) and taking into account the uniqueness of the solution of the Sylvester-Lyapunov equation, we conclude that the equality (4.44) holds.

In the same way, we can establish the equality (4.45). Taking into account the equalities (4.33), (4.35), we arrive from (4.37) at the equation

$$A_{\mathcal{P}}(\Phi)_2 S^l(\Phi_-) - S^l(\Phi_-) A_{\mathcal{N}}(\Phi)_2 = G_{\mathcal{P}}(\Phi)_2 F_{\mathcal{N}}(\Phi)_2. \tag{4.52}$$

Comparing (4.43) and (4.52) and taking into account the uniqueness of the solution of the Sylvester-Lyapunov equation, we conclude that the equality (4.45) holds.

According to Theorem 2.6 (applied to the rational matrix functions in general position Φ_+ and Φ_-), the matrices $S^r(\Phi_+)$ and $S^l(\Phi_-)$ are invertible. In view of (4.44) and (4.45), the block-entries $(S^r(\Phi))_{11}$ and $(S^l(\Phi))_{22}$ (of the matrices $S^r(\Phi)$ and $S^l(\Phi)$ respectively) are invertible.

Now we may rewrite the formulas (4.38), (4.39) for the factors Φ_+ and Φ_- in terms of the block-entries of the pole, the zero, the semiresidual and the zero-pole coupling matrices for the matrix function Φ to be factorized. Substituting the expressions (4.32p), (4.34) and (4.44) into (4.38), we obtain

$$\Phi_+(z) = I - F_{\mathcal{P}}(\Phi)_1 (zI - A_{\mathcal{P}}(\Phi)_1)^{-1} (S^r(\Phi)_{11})^{-1} G_{\mathcal{N}}(\Phi)_1. \tag{4.53}$$

Analogously, substituting the expressions (4.32n), (4.35) and (4.45) into (4.39), we obtain

$$\Phi_-(z) = I + F_{\mathcal{N}}(\Phi)_2 (S^l(\Phi)_{22})^{-1} (zI - A_{\mathcal{P}}(\Phi)_2)^{-1} G_{\mathcal{P}}(\Phi)_2. \tag{4.54}$$

The formula (4.53) expresses the factor Φ_+ in terms of the values $F_{\mathcal{P}}(\Phi)$, $G_{\mathcal{N}}(\Phi)$ and $S^r(\Phi)$, which appear in the representation (4.7), whereas the formula (4.54) expresses the factor Φ_- in terms of the values $F_{\mathcal{N}}(\Phi)$, $G_{\mathcal{P}}(\Phi)$ and $S^l(\Phi)$, which appear in the other representation (4.8). This disagreement is inconvenient for some calculations. Therefore we also give a formula which expresses the factor Φ_- in terms of the same values $F_{\mathcal{P}}(\Phi)$, $G_{\mathcal{N}}(\Phi)$ and $S^r(\Phi)$, which appear in the expression (4.53) for the factor Φ_+. To do it, we have to use the zero-pole coupling relations (4.12a), (4.12c). Together with (4.35p) and (4.34n), these relations mean:

$$F_{\mathcal{N}}(\Phi_-) = (F_{\mathcal{P}}(\Phi) S^r(\Phi)^{-1})_2; \quad G_{\mathcal{P}}(\Phi_-) = -((S^r(\Phi))^{-1} G_{\mathcal{N}}(\Phi))_2. \tag{4.55}$$

The equality (4.11) together with (4.45) means:

$$S^l(\Phi_-) = \left(S^r(\Phi)^{-1}\right)_{22}. \tag{4.56}$$

Substituting the expressions (4.55) and (4.56) into (4.39), we obtain

$$\Phi_-(z) = I - \left(F_{\mathcal{P}}(\Phi)\, S^r(\Phi)^{-1}\right)_2 \left(\left(S^r(\Phi)^{-1}\right)_{22}\right)^{-1} \cdot \left(zI - A_{\mathcal{P}}(\Phi)_2\right)^{-1} \left(S^r(\Phi)^{-1}\, G_{\mathcal{N}}(\Phi)\right)_2. \tag{4.57}$$

The last formula already expresses the factor Φ_- in terms of the values $F_{\mathcal{P}}(\Phi)$, $G_{\mathcal{N}}(\Phi)$ and $S^r(\Phi)$, i.e., in terms of the same values which appear in the representation (4.53) of the factor Φ_+.

However, for further considerations it will be useful to transform this formula, substituting into it the expression for the inverse matrix $S^r(\Phi)^{-1}$ in terms of the block-entries of the matrix $S^r(\Phi)$ itself. First of all we recall a formula for the inversion of a 2×2 block matrix with square diagonal block-entries. Let

$$M = \begin{bmatrix} m_{11} & m_{12} \\ m_{21} & m_{22} \end{bmatrix} \tag{4.58}$$

be a square $n \times n$ block-matrix matrix, with block-entries m_{11} and m_{22} square $n_1 \times n_1$ and $n_2 \times n_2$ matrices respectively ($n = n_1 + n_2$). We assume that the matrix m_{11} is invertible. To invert the matrix M, we first of all factor it:

$$\begin{bmatrix} m_{11} & m_{12} \\ m_{21} & m_{22} \end{bmatrix} = \begin{bmatrix} I & 0 \\ m_{21}\, m_{11}^{-1} & I \end{bmatrix} \cdot \begin{bmatrix} m_{11} & 0 \\ 0 & m_{22} - m_{21}\, m_{11}^{-1}\, m_{12} \end{bmatrix} \cdot \begin{bmatrix} I & m_{11}^{-1} m_{12} \\ 0 & I \end{bmatrix}. \tag{4.59}$$

The matrices $\begin{bmatrix} I & 0 \\ m_{21}\, m_{11}^{-1} & I \end{bmatrix}$ and $\begin{bmatrix} I & m_{11}^{-1} m_{12} \\ 0 & I \end{bmatrix}$ are invertible, and

$$\begin{bmatrix} I & 0 \\ m_{21}\, m_{11}^{-1} & I \end{bmatrix}^{-1} = \begin{bmatrix} I & 0 \\ -m_{21}\, m_{11}^{-1} & I \end{bmatrix};$$

$$\begin{bmatrix} I & m_{11}^{-1} m_{12} \\ 0 & I \end{bmatrix}^{-1} = \begin{bmatrix} I & -m_{11}^{-1} m_{12} \\ 0 & I \end{bmatrix}.$$

Therefore the matrix M is invertible if and only if the matrix

$$\begin{bmatrix} m_{11} & 0 \\ 0 & m_{22} - m_{21}\, m_{11}^{-1}\, m_{12} \end{bmatrix}$$

is invertible. The latter is invertible if and only if the matrix $(m_{22} - m_{21}\, m_{11}^{-1}\, m_{12})$ is invertible. If the matrix $(m_{22} - m_{21}\, m_{11}^{-1}\, m_{12})$ is invertible [19], then

$$\begin{bmatrix} m_{11} & m_{12} \\ m_{21} & m_{22} \end{bmatrix}^{-1} = \begin{bmatrix} I & -m_{11}^{-1} m_{12} \\ 0 & I \end{bmatrix} \cdot \begin{bmatrix} m_{11}^{-1} & 0 \\ 0 & (m_{22} - m_{21}\, m_{11}^{-1}\, m_{12})^{-1} \end{bmatrix} \cdot \begin{bmatrix} I & 0 \\ -m_{21}\, m_{11}^{-1} & I \end{bmatrix}.$$

[19] The invertibility of the matrix m_{11} was assumed from the very beginning.

Finally, for the matrix M with the decomposition (4.58),

$$(4.60)\quad M^{-1} = \begin{bmatrix} m_{11}^{-1} & 0 \\ 0 & 0 \end{bmatrix} + \begin{bmatrix} -m_{11}^{-1}m_{12} \\ I \end{bmatrix} \cdot \Big[m_{22} - m_{21}\, m_{11}^{-1}\, m_{12}\Big]^{-1} \cdot \begin{bmatrix} -m_{21}\, m_{11}^{-1} & I \end{bmatrix}.$$

In particular,

$$(4.61)\qquad (M^{-1})_{22} = (m_{22} - m_{21}\, m_{11}^{-1}\, m_{12})^{-1},$$

If F is a $k \times n$ matrix, with a block-decomposition $F = [F_1 \quad F_2]$, where F_1, F_2 are $k \times n_1$ and $k \times n_2$ matrices respectively, then the 2-entry of the matrix $F\, M^{-1}$ is of the form

$$(4.62)\qquad (F\, M^{-1})_2 = [F_1 \quad F_2] \cdot \begin{bmatrix} -m_{11}^{-1}m_{12} \\ I \end{bmatrix} \cdot \big(m_{22} - m_{21}\, m_{11}^{-1}\, m_{12}\big)^{-1}.$$

If G is an $n \times k$ matrix, with a block-decomposition $G = \begin{bmatrix} G_1 \\ G_2 \end{bmatrix}$, where G_1, G_2 are $n_1 \times k$ and $n_2 \times k$ matrices respectively, then the 2-entry of the matrix $M^{-1}\, G$ is of the form

$$(4.63)\qquad \big(M^{-1}G\big)_2 = \big(m_{22} - m_{21}\, m_{11}^{-1}\, m_{12}\big)^{-1} \cdot \big[-m_{21}m_{11}^{-1} \quad I\big] \cdot \begin{bmatrix} G_1 \\ G_2 \end{bmatrix}.$$

Let us take the matrix $S^r(\Phi)$ with the decomposition (4.40) as the matrix M $(n_1 = n_+, n_2 = n_-)$. Both the matrix $S^r(\Phi)$ and its block-entry the matrix $S^r(\Phi)_{11}$ are invertible. Therefore the entry $(S^r(\Phi)^{-1})_{22}$ of the inverse matrix $(S^r(\Phi))^{-1} = S^l(\Phi)$ is invertible as well, and, according to (4.61),

$$(4.64)\qquad ((S^r(\Phi)^{-1})_{22})^{-1} = S^r(\Phi)_{22} - S^r(\Phi)_{21}\, S^r(\Phi)_{11}^{-1}\, S^r(\Phi)_{12}.$$

In particular, the matrix $\boldsymbol{S^r(\Phi)_{22} - S^r(\Phi)_{21}\, S^r(\Phi)_{21}^{-1}\, S^r(\Phi)_{12}}$ is invertible. In view of (4.11), the equality (4.45) can be represented in the form

$$(4.65)\qquad S^l(\Phi_-)_{22} = \big(S^r(\Phi)_{22} - S^r(\Phi)_{21}\, S^r(\Phi)_{21}^{-1}\, S^r(\Phi)_{12}\big)^{-1}.$$

Taking the matrix $F_{\mathcal{P}}(\Phi)$ with the decomposition (4.30p) as the matrix F and the matrix $G_{\mathcal{N}}(\Phi)$ with the decomposition (4.31n) as the matrix G. we reduce the formula (4.57) to the form

$$(4.66)\quad \Phi_-(z) = I - [F_{\mathcal{N}}(\Phi)_1 \;\; F_{\mathcal{N}}(\Phi)_2] \cdot \begin{bmatrix} -S^r(\Phi)_{11}^{-1} S^r(\Phi)_{12} \\ I \end{bmatrix} \cdot \Big(zI - A_{\mathcal{P}}(\Phi)_2\Big)^{-1} \cdot \Big(S^r(\Phi)_{22} - S^r(\Phi)_{21} S^r(\Phi)_{11}^{-1} S^r(\Phi)_{12}\Big)^{-1} \cdot \big[-S^r(\Phi)_{21} S^r(\Phi)_{11}^{-1} \;\; I\big] \begin{bmatrix} G_{\mathcal{N}}(\Phi)_1 \\ G_{\mathcal{N}}(\Phi)_2 \end{bmatrix}.$$

Thus we proved the following

Theorem 4.1. *Let Φ be a rational $k \times k$ matrix function in general position, satisfying the normalizing condition (4.6): $\Phi(\infty) = I$, with the pole and the zero sets $\mathcal{P}(\Phi)$ and $\mathcal{N}(\Phi)$, the pole and the zero matrices $A_{\mathcal{P}}(\Phi)$ and $A_{\mathcal{N}}(\Phi)$, the semiresidual matrices $F_{\mathcal{P}}(\Phi)$, $F_{\mathcal{N}}(\Phi)$, $G_{\mathcal{P}}(\Phi)$, $G_{\mathcal{N}}(\Phi)$, and the zero-pole coupling matrices $S^r(\Phi)$ and $S^l(\Phi)$, which are decomposed into blocks as described above. Assume that the function Φ admits a spectral factorization with respect to Γ, i.e.,*

$$(4.4)\qquad \Phi(z) = \Phi_+(z) \cdot \Phi_-(z) \quad (z \in \overline{\mathbb{C}}),$$

where Φ_+, Φ_- are rational matrix functions, with the pole and the zero location:

$$(4.5)\qquad \mathcal{P}(\Phi_+) \subset G_+, \quad \mathcal{N}(\Phi_+) \subset G_+, \quad \mathcal{P}(\Phi_-) \subset G_-, \quad \mathcal{N}(\Phi_-) \subset G_-.$$

Let the normalizing condition (4.2): $\Phi_+(\infty) = I$ hold. Then:

1. *For the matrix function Φ, the number of poles and the number of "zeros" located in G_+ are equal: $\#(\mathcal{P}(\Phi)\cap G_+) = \#(\mathcal{N}(\Phi)\cap G_+)$; the number of poles and the number of "zeros" located in G_- are also equal: $\#(\mathcal{P}(\Phi) \cap G_-) = \#(\mathcal{N}(\Phi) \cap G_-)$.*
2. *The block-entries $S^r(\Phi)_{11}$ and $S^l(\Phi)_{22}$ (in the block decompositions (4.40), (4.41) described above of the matrices $S^r(\Phi)$ and $S^l(\Phi)$ respectively), as well as the matrix $S^r(\Phi)_{22} - S^r(\Phi)_{21} S^r(\Phi)_{11}^{-1} S^r(\Phi)_{12}$, are square invertible matrices.*
3. *The factors Φ_+ and Φ_- are rational matrix functions in general position, which are representable in the form (4.53) and (4.54) respectively.*
4. *The representation (4.53) is the right system representation of the factor Φ_+; the entries of this representation is expressible in terms of the entries of the right system representation for Φ: (4.15), (4.32), (4.34) and (4.44) hold. The representation (4.54) is the left system representation of the factor Φ_-; the entries of this representation is expressible in terms of the entries of the right system representation for Φ: (4.18), (4.33), (4.35) and (4.45) hold. The representation (4.54) may be rewritten in terms of the entries of the right system representation for Φ_+: (4.66) holds.*

The converse statement is true as well.

Theorem 4.2. *Let Φ be a rational $k \times k$ matrix function in general position, satisfying the normalizing condition (4.6): $\Phi(\infty) = I$, with the pole and the zero sets $\mathcal{P}(\Phi)$ and $\mathcal{N}(\Phi)$, the pole and the zero matrices $A_{\mathcal{P}}(\Phi)$ and $A_{\mathcal{N}}(\Phi)$, the semiresidual matrices $F_{\mathcal{P}}(\Phi)$, $F_{\mathcal{N}}(\Phi)$, $G_{\mathcal{P}}(\Phi)$, $G_{\mathcal{N}}(\Phi)$, and the zero-pole coupling matrices $S^r(\Phi)$ and $S^l(\Phi)$, which are decomposed into blocks as described above. Assume that the following conditions are satisfied:*

1. *None of the poles of Φ as well as none of its "zeros" belongs to Γ:*
$$\mathcal{P}(\Phi) \cap \Gamma = \emptyset, \qquad \mathcal{P}(\Phi^{-1}) \cap \Gamma = \emptyset.$$
The number of poles and the number of zeros of the matrix-function Φ, located in G_+, are equal:
$$\#(\mathcal{P}(\Phi) \cap G_+) = \#(\mathcal{N}(\Phi) \cap G_+),$$
or, what is the same, the number of poles and the number of zeros of the matrix-function Φ, located in G_-, are equal:
$$\#(\mathcal{P}(\Phi) \cap G_-) = \#(\mathcal{N}(\Phi) \cap G_-).$$
2. *The block-entry $S^r(\Phi)_{11}$ of the matrix $S^r(\Phi)$ (in the above described block decompositions (4.40)) is invertible matrix.*

Then:

1. *The matrix functions Φ_+ and Φ_-, which are* defined *by (4.53) and*[20] *(4.54), are rational matrix functions in general position. For these functions Φ_+ and Φ_-, the conditions (4.5) are satisfied.*
2. *The matrix function Φ admits a spectral factorization (4.4) with respect to Γ, with these matrix-functions Φ_+ and Φ_- as the factors.*

Proof. To investigate the properties of the matrix functions Φ_+, Φ_-, we will use Theorem 3.6. For this theorem be applicable to the functions Φ_+, Φ_-, we have to check whether its assumptions follow from the assumptions of Theorem 4.2. First of all, we have to be sure that none of the columns of the matrices $F_{\mathcal{P}}(\Phi)_1$, $F_{\mathcal{N}}(\Phi)_2$ is a zero-column and none of the rows of the matrices $G_{\mathcal{N}}(\Phi)_1$, $G_{\mathcal{P}}(\Phi)_2$ is a zero-row. This property holds indeed because it holds for the including matrices $F_{\mathcal{P}}(\Phi)$, $F_{\mathcal{N}}(\Phi)$, $G_{\mathcal{P}}(\Phi)$, $G_{\mathcal{N}}(\Phi)$ (as for semiresidual matrices for the rational matrix function Φ in general position). Then we have to check that the matrices $S^r(\Phi)_{11}$, which appears at the "core" of the representation (4.53), satisfy some Sylvester-Lyapunov equality of the form $B\, S^r(\Phi)_{11} - S^r(\Phi)_{11}\, A = G_{\mathcal{N}}(\Phi)_1 \cdot F_{\mathcal{P}}(\Phi)_1$, where $A = A_{\mathcal{P}}(\Phi)_1$ and B is a diagonal matrix. In the considered case, the identity (4.42) plays the role of such Sylvester-Lyapunov equality, with $B = A_{\mathcal{N}}(\Phi)_1$. As we are already established, the equality (4.42) is the consequence of the equality (4.9). The latter holds as the Sylvester-Lyapunov identity for the matrix function Φ. Thus the representation (4.53) of the function Φ_+ is its system representation. According to Theorem 3.6, the matrix-functions Φ_+ is a rational matrix-function in general position, which pole- and zero-sets are totalities of the diagonal entries

[20]If the block-entry $S^r(\Phi)_{11}$ is invertible, the block-entry $S^l(\Phi)_{22}$ is invertible as well.

of the matrices $A_{\mathcal{P}}(\Phi)_1$ and $A_{\mathcal{N}}(\Phi)_1$ respectively: $\mathcal{P}(\Phi_+) = \left\{\operatorname{diag} A_{\mathcal{P}}(\Phi)_1\right\}$, and $\mathcal{N}(\Phi_+) = \left\{\operatorname{diag} A_{\mathcal{N}}(\Phi)_1\right\}$. Hence (see the definition (4.28), (4.29) of the matrices $A(\mathcal{P})_1$, $A(\mathcal{N})_1$), the conditions $\mathcal{P}(\Phi_+) = \mathcal{P}(\Phi) \cap G_+$ $\mathcal{N}(\Phi_+) = \mathcal{N}(\Phi) \cap G_+$ hold.

The matrix $S^r(\Phi)$ is invertible as the Sylvester-Lyapunov matrix related to the matrix-function Φ (Theorem 2.5. See (4.11)). The corner block-entry $S^r(\Phi)_{11}$ is invertible by the assumptions of Theorem. Hence (see the identity (4.60) with $S^r(\Phi)$ as M), the matrix $S^r(\Phi)_{22} - S^r(\Phi)_{21}\, S^r(\Phi)_{21}^{-1}\, S^r(\Phi)_{12}$ is invertible as well, and (4.64) holds. In view of (4.11), (4.65) holds. In particular, the block-entry $S^l(\Phi)_{22}$ is invertible as well. Thus the matrix-function Φ_- is well defined by the formula (4.54). The Sylvester-Lyapunov equality (4.43) for the matrix $S^l(\Phi)_{22}$ follows from the equality (4.10). (The latter holds as the "left" Sylvester-Lyapunov equality for Φ). According to Theorem 3.6 (to be more precise, according to the "left" version of this Theorem), Φ_- is a rational matrix function in general position, which pole- and zero-sets are totalities of the diagonal entries of the matrices $A_{\mathcal{P}}(\Phi)_2$ and $A_{\mathcal{N}}(\Phi)_2$ respectively: $\mathcal{P}(\Phi_-) = \left\{\operatorname{diag} A_{\mathcal{P}}(\Phi)_2\right\}$, and $\mathcal{N}(\Phi_-) = \left\{\operatorname{diag} A_{\mathcal{N}}(\Phi)_2\right\}$. Hence, the conditions $\mathcal{P}(\Phi_-) = \mathcal{P}(\Phi) \cap G_-$ $\mathcal{N}(\Phi_-) = \mathcal{N}(\Phi) \cap G_-$ hold. All the more, the conditions (4.5) hold.

Thus if the equality (4.4) holds for Φ_+, Φ_- defined by (4.53), (4.65), it gives a spectral factorization of Φ.

It remains only to verify the equality (4.4). The representation (4.54) is convenient to investigate the properties of Φ_-. However, to verify the equality (4.4), it is more convenient to use the representation (4.66). To derive (4.66) from (4.54), we have to use the zero-pole coupling relations (4.12), the formula (4.60) for inversion of 2×2 block- matrix (applied to the matrix $S^r(\Phi)$ with the block-decomposition (4.40)) and, in particular, the equality (4.65).

To abbreviate the notation, we omit some notation entries, like indices etc. So, for example, we write S instead of $S^r(\Phi)$, S_{11} instead of $S^r(\Phi)_{11}$, $F = [F_1\ F_2]$ instead of (4.30p), $G = \begin{bmatrix} G_1 \\ G_2 \end{bmatrix}$ instead of (4.30n), $A = \begin{bmatrix} A_1 & 0 \\ 0 & A_2 \end{bmatrix}$ instead of (4.28), $B = \begin{bmatrix} B_1 & 0 \\ 0 & B_2 \end{bmatrix}$ instead of (4.29).

Using the 2×2-block-matrix inversion rule (4.60), we present the representation (4.7) for Φ in the form

$$\Phi(z) = I - F\,(zI - A)^{-1} \left(\begin{bmatrix} S_{11}^{-1} & 0 \\ 0 & 0 \end{bmatrix} - \begin{bmatrix} -S_{11}^{-1} S_{12} \\ I \end{bmatrix} \Delta^{-1} \begin{bmatrix} -S_{21} S_{11}^{-1} & I \end{bmatrix} \right) G, \tag{4.67}$$

where

$$\Delta = S_{22} - S_{21} S_{11}^{-1} S_{12}.$$

We present the representation (4.53) for the function Φ_+ in the form

$$\Phi_+(z) = I - F(zI - A)^{-1} \begin{bmatrix} S_{11}^{-1} & 0 \\ 0 & 0 \end{bmatrix} G. \tag{4.68}$$

The representation (4.54) for Φ_-, rewritten in the form (4.66), is:

$$\Phi_-(z) = I - F \begin{bmatrix} -S_{11}^{-1}S_{12} \\ I \end{bmatrix} (zI - A_2)^{-1}\Delta^{-1} \begin{bmatrix} -S_{21}S_{11}^{-1} & I \end{bmatrix} G. \tag{4.69}$$

Multiplying the expressions in the right-hand sides of (4.68) and (4.69) term by term, we obtain:

$$\begin{aligned} \Phi_+(z)\Phi_-(z) = I & \\ & - F(zI - A)^{-1} \begin{bmatrix} S_{11}^{-1} & 0 \\ 0 & 0 \end{bmatrix} G \\ & - F \begin{bmatrix} -S_{11}^{-1}S_{12} \\ I \end{bmatrix} (zI - A_2)^{-1}\Delta^{-1} \begin{bmatrix} -S_{21}S_{11}^{-1} & I \end{bmatrix} G \\ & + F(zI - A)^{-1} \begin{bmatrix} S_{11}^{-1} & 0 \\ 0 & 0 \end{bmatrix} G \\ & \cdot F \begin{bmatrix} -S_{11}^{-1}S_{12} \\ I \end{bmatrix} (zI - A_2)^{-1}\Delta^{-1} \begin{bmatrix} -S_{21}S_{11}^{-1} & I \end{bmatrix}. \end{aligned} \tag{4.70}$$

We denote the three summands following I on the right-hand side of (4.70) by (II), (III), and (IV), respectively. Substituting the expression for GF from the Sylvester-Lyapunov identity $GF = BS - SA$, or

$$GF = S(zI - A) - (zI - B)S,$$

into the expression (IV), we obtain:

$$\begin{aligned} \text{(IV)} = F(zI - A)^{-1} & \begin{bmatrix} S_{11}^{-1} & 0 \\ 0 & 0 \end{bmatrix} \cdot S(zI - A) \\ & \cdot \begin{bmatrix} -S_{11}^{-1}S_{12} \\ I \end{bmatrix} (zI - A_2)^{-1}\Delta^{-1} \begin{bmatrix} -S_{21}S_{11}^{-1} & I \end{bmatrix} \\ & - F(zI - A)^{-1} \begin{bmatrix} S_{11}^{-1} & 0 \\ 0 & 0 \end{bmatrix} \cdot (zI - B)S \\ & \cdot \begin{bmatrix} -S_{11}^{-1}S_{12} \\ I \end{bmatrix} (zI - A_2)^{-1}\Delta^{-1} \begin{bmatrix} -S_{21}S_{11}^{-1} & I \end{bmatrix}. \end{aligned} \tag{4.71}$$

It is clear that

$$\begin{bmatrix} S_{11}^{-1} & 0 \\ 0 & 0 \end{bmatrix} (zI - B)^{-1} = \begin{bmatrix} * & 0 \\ 0 & 0 \end{bmatrix}, \quad S \begin{bmatrix} -S_{11}^{-1} S_{12} \\ I \end{bmatrix} = \begin{bmatrix} 0 \\ * \end{bmatrix}.$$

Thus,

$$\begin{bmatrix} S_{11}^{-1} & 0 \\ 0 & 0 \end{bmatrix} (zI - B)^{-1} S \begin{bmatrix} -S_{11}^{-1} S_{12} \\ I \end{bmatrix} = \begin{bmatrix} 0 \\ 0 \end{bmatrix},$$

the second summand in the expression (4.71) for (IV) vanishes, and

$$\begin{aligned} \text{(IV)} = F(zI - A)^{-1} & \begin{bmatrix} S_{11}^{-1} & 0 \\ 0 & 0 \end{bmatrix} \cdot S(zI - A) \\ & \cdot \begin{bmatrix} -S_{11}^{-1} S_{12} \\ I \end{bmatrix} (zI - A_2)^{-1} \Delta^{-1} \begin{bmatrix} -S_{21} S_{11}^{-1} & I \end{bmatrix} G. \end{aligned} \tag{4.72}$$

As

$$\begin{bmatrix} S_{11}^{-1} & 0 \\ 0 & 0 \end{bmatrix} \cdot S = I - \begin{bmatrix} 0 & -S_{11}^{-1} S_{12} \\ 0 & I \end{bmatrix},$$

we get

$$\begin{aligned} \text{(IV)} = F & \begin{bmatrix} -S_{11}^{-1} S_{12} \\ I \end{bmatrix} (zI - A_2)^{-1} \Delta^{-1} \begin{bmatrix} -S_{21} S_{11}^{-1} & I \end{bmatrix} G \\ & - F(zI - A)^{-1} \begin{bmatrix} 0 & -S_{11}^{-1} S_{12} \\ 0 & I \end{bmatrix} \\ & \cdot (zI - A) \begin{bmatrix} -S_{11}^{-1} S_{12} \\ I \end{bmatrix} (zI - A_2)^{-1} \Delta^{-1} \begin{bmatrix} -S_{21} S_{11}^{-1} & I \end{bmatrix} G. \end{aligned}$$

It is clear that

$$(zI - A) \begin{bmatrix} -S_{11}^{-1} S_{12} \\ I \end{bmatrix} (zI - A_2)^{-1} = \begin{bmatrix} * \\ I \end{bmatrix},$$

hence

$$\begin{bmatrix} 0 & -S_{11}^{-1} S_{12} \\ 0 & I \end{bmatrix} (zI - A) \begin{bmatrix} -S_{11}^{-1} S_{12} \\ I \end{bmatrix} (zI - A_2)^{-1} = \begin{bmatrix} -S_{11}^{-1} S_{12} \\ I \end{bmatrix}.$$

Thus,

$$(4.73)\quad (\mathrm{IV}) = F\begin{bmatrix} -S_{11}^{-1}S_{12} \\ I \end{bmatrix}(zI-A_2)^{-1}\Delta^{-1}\begin{bmatrix} -S_{21}S_{11}^{-1} & I \end{bmatrix}G$$
$$- F(zI-A)^{-1}\begin{bmatrix} -S_{11}^{-1}S_{12} \\ I \end{bmatrix}\Delta^{-1}\begin{bmatrix} -S_{21}S_{11}^{-1} & I \end{bmatrix}G.$$

From (4.70) and (4.73) it follows that

$$\Phi_+(z)\,\Phi_-(z) = I - F(zI-A)^{-1}$$
$$\left(\begin{bmatrix} S_{11}^{-1} & 0 \\ 0 & 0 \end{bmatrix} + \begin{bmatrix} -S_{11}^{-1}S_{12} \\ I \end{bmatrix}\Delta^{-1}\begin{bmatrix} -S_{21}S_{11}^{-1} & I \end{bmatrix}\right)G.$$

Comparing the last expression to (4.67), we conclude that the factorization equality (4.4) holds. Theorem 4.2 is proved. □

Remark 4.1. Let Φ be a rational matrix function in general position, satisfying the normalizing condition (4.6): $\Phi(\infty) = I$. Assume that none of the poles of Φ and none of the "zeros" of Φ belongs to the contour Γ, and the condition (4.25) : $\#(\mathcal{P}(\Phi)\cap G_+) = \#(\mathcal{N}(\Phi)\cap G_+)$ is satisfied (or, what is the same, the condition (4.26): $\#(\mathcal{P}(\Phi)\cap G_-) = \#(\mathcal{N}(\Phi)\cap G_-)$ is satisfied). This means that the matrices $S^r(\Phi)_{11}$ and $S^l(\Phi)_{22}$ are square. According to Theorems 4.1 and 4.2, the matrix function Φ admits the spectral factorization with respect to Γ if and only if the matrix $S^r(\Phi)_{11}$ is invertible, or, what is the same, the matrix $S^l(\Phi)_{22}$ is invertible. What can we say if the condition of the invertibility of the matrix $S^r(\Phi)_{11}$ is violated? According to the matrix factorization theory, the matrix function Φ admits a factorization of the form

$$(4.74)\qquad \Phi(z) = \Phi_+(z)\,D(z)\,\Phi_-(z),$$

where the matrix functions Φ_+ and Φ_+^{-1} are holomorphic on the set $G_-\cup\Gamma$, the matrix functions Φ_- and Φ_-^{-1} are holomorphic on the set $G_+\cup\Gamma$ and $D(z)$ is the matrix function of the form

$$(4.75)\qquad D(z) = \operatorname{diag}\left((z-z_0)^{\kappa_1},\,(z-z_0)^{\kappa_2},\,\dots,\,(z-z_0)^{\kappa_k}\right),$$

$\kappa_1,\,\kappa_2,\,\dots,\,\kappa_k$ are integer numbers. The numbers $\kappa_1,\,\kappa_2,\,\dots,\,\kappa_k$ are said to be the partial indices of the matrix function Φ with respect to the contour Γ.

The point z_0 is an arbitrarily chosen fixed point from G_+. Of course, the factors Φ_+, Φ_- and D from the factorization (4.74) depend of the choice of the point z_0. However, the set $\{\kappa_1,\,\kappa_2,\,\dots,\,\kappa_k\}$ (and under the normalizing condition

$$(4.76)\qquad \kappa_1 \le \kappa_2 \le \dots \kappa_k$$

the matrix D) is determined uniquely by the function Φ. In particular, the partial indices do not depend on the choice of the distinguished point z_0. In contrast to this, even for a given distinguished point z_0 the matrices Φ_+ and Φ_- are determined non-uniquely. However, this non-uniqueneess can be easily described. (See Theorem 7.1 from [GoKr].)

The partial indices play a fundamental role; in the homogeneous Hilbert problem they were first introduced by N.I. Muskhelishvili and N.P. Vekua [MuVe]. (A factorization of another form: $\Phi(z) = \Phi_+(z)\,\Phi_-(z)\,D(z)$ was considered by G. Birkhoff much earlier, in 1913 (see [Bir1]). The matrices Φ_+, Φ_- and D have the same properties that in above described factorization (4.74) (in particular, the matrix D is of the form (4.75)), but the matrices themselves are different. In particular, Birkhoff's partial indices may be be different. The relation between these kinds of factorizations is studed in [FM].)

A natural question arises how to compute the factors Φ_+, Φ_-, D in terms of the pole, zero and semiresidual matrices for the given rational matrix function Φ in general position? Of course, there are general methods for performing such a factorization (see, for example, [ClGo]). However, we believe that for a rational matrix function in general position the factorization may be performed much more explicitly.

Remark 4.2. Let as assume now that a rational matrix function Φ in general position depends holomorphically on some parameter $\alpha \in \mathbb{C}$ and satisfies the normalizing condition (4.6): $\Phi(\infty, \alpha) \equiv I$. The factors Φ_+, Φ_-, D in the factorization (4.74):

$$\Phi(z, \alpha) = \Phi_+(z, \alpha)\, D(z, \alpha)\, \Phi_-(z, \alpha), \tag{4.74 $_\alpha$}$$

$$D(z, (\alpha)) = \operatorname{diag}\left((z - z_0)^{\kappa_1(\alpha)}, (z - z_0)^{\kappa_2(\alpha)}, \dots, (z - z_0)^{\kappa_k(\alpha)}\right) \tag{4.75 $_\alpha$}$$

depend now on α, and under the normalizing conditions (4.2): $\Phi_+(\infty, \alpha) \equiv I$ and (4.76): $\kappa_1(\alpha) \leq \kappa_2(\alpha) \leq \dots \kappa_k(\alpha)$, are determined uniquely.

Assume now that for all α from some neighborhood of some point $\alpha_0 \in \mathbb{C}$ the condition $\#(\mathcal{P}(\Phi(\alpha)) \cap G_+) = \#(\mathcal{N}(\Phi(\alpha)) \cap G_+)$ is satisfied (thus the block-entry $S^r(\Phi(\alpha))_{11}$ is a square matrix for all α which are close to α_0). Assume also that $\det(S^r(\Phi(\alpha))_{11}) \not\equiv 0$, but $\det(S^r(\Phi(\alpha_0))_{11}) = 0$. According to Theorem 4.2, applied to the matrix-function $\Phi(\alpha)$ with $\alpha \neq \alpha_0$ (and α which is close to α_0), the matrix function $\Phi(z, \alpha)$ $(\alpha \neq \alpha_0)$ admits a factorization of the form $\Phi(z, \alpha) = \Phi_+(z, \alpha)\, \Phi_-(z, \alpha)$, in other words, $D(z, \alpha) \equiv I$ for $\alpha \neq \alpha_0$. Moreover, the factors $\Phi_+(z, \alpha)$, $\Phi_-(z, \alpha)$ depend on α holomorphically for $\alpha \neq \alpha_0$ (this may be established from the explicit formulas (4.53), (4.54) for the factors). However, according to Theorem 4.1, the factorization of this form is impossible for $\alpha = \alpha_0$ (otherwise the Lyapunov-Sylvester matrix $S^r(\Phi(\alpha_0))$ would be invertible). Thus $D(\alpha_0)) \neq I$, i.e., not all $\kappa_j(\alpha_0), j = 1, 2, \dots, k$, vanish (although still $\sum_{1 \leq j \leq k} \kappa_j(\alpha_0) = 0$). So, the factorization (4.74) undergoes a bifurcation at the value of the parameter α such that $\det S^r(\Phi(\alpha)_{11})$ vanishes. A natural question arises how to describe this bifurcation in a clear way? We believe that for rational matrix function Φ in general position it may be done more or less explicitly.

5. Some historical remarks

There are many different ways to specify and represent analytic functions: for instance, Taylor series, decomposition in continuous fractions, representations by Cauchy integrals or by Fourier integrals, etc. (Of course, the distinction between different representation methods is often artificial and hard to make.) In the first half of 1970's this toolkit was enriched by an additional representation method: so called system realizations of analytic functions. The sources of the theory of system realizations belong to several different domains, in particular synthesis theory of linear electrical networks, the theory of linear control systems, and the theory of operator colligations (or nodes) and their characteristic functions. Investigations in these theories were carried out by representatives of different scientific disciplines. The investigations done by the mathematicians have their root in the pioneering work of M. S. Livšic, who is the forefather of the theory of system realizations.

In the middle of 1940's M. S. Livšic has introduced the notion of the characteristic function of a linear operator. This notion was first introduced for nonselfadjoint extensions of hermitian operators (or nonunitary extensions of isometric operators) with defect indices $(1,1)$ [L1] and more generally (n,n) [L2, L3] (for $n > 1$ the characteristic function is matrix valued), and later for general operators with finite nonhermitian (or nonunitary) rank [L4]. M. S. Livšic has discovered the following properties of the characteristic function [21].

1. The characteristic function determines the corresponding operator essentially uniquely up to unitary equivalence (first results of this kind are contained already in [L1]).
2. For each invariant subspace of the operator there is a decomposition of the characteristic function into a product of two factors: one of these factors is the characteristic function of the restriction of the original operator onto the given invariant subspace, and the other factor is the characteristic function of the compression of the original operator onto the corresponding coinvariant subspace.

Property 2 is the so called "multiplication theorem" for characteristic functions. Initially the multiplication theorem was established under various additional technical restrictions. The final formulation of the theorem has been obtained in the framework of the theory of operator colligations and their characteristic functions.

The theory of operator colligations that was created by M. S. Livšic (and that was further developped by his collaborator M. S. Brodskii, see the book [Br]) was a natural development of the theory of nonselfadjoint operators and their characteristic functions. This development was also intimately related with applications of the theory of commuting nonselfadjoint operators to physical problems, in particular to the problems of scattering and to elementary particles physics, and later

[21] we formulate these properties, on purpose, in a somewhat rough and therefore slightly imprecise form

to synthesis problems for electrical networks [L5, L6, LF]. The theory of open systems took a definitive shape in the works of M. S. Livšic in the early 1960's [L7, L8]; the contents of these works were incorporated in the monograph [L9].

A theory parallel to the theory of open systems of M. S. Livšic has been developped by several other authors under the name of the theory of linear stationary dinamical systems. (For an exposition of system theory see [Z], [Fu], [KFA]; we especially recommend the survey [Kaas].) The transfer matrix function of such a system is precisely the characteristic function of the corresponding operator colligation. As M. S. Livšic has shown in several important examples, for scattering systems the characteristic function coincides with the scattering matrix [L5, L6].

A different line of investigation leading to the theory of system realizations is connected with control theory and with the name of R. Kalman (see [KFA]). Here one also encounters the notion of the transfer matrix function. Let us emphasize in this connection one important circumstance. In physical problems that M. S. Livšic was motivated by there appeared always an "energy balance" condition implying J-contractiveness of the characteristic (transfer) function and its "symmetry" with respect to the unit circle (or the real axis). It also imposed considering the adjoint operator A^* together with the state operator A of the system. Energy balance condition does not play an important role in control theory and this leads naturally to considering a general pair of operators A, B on the state space instead of the pair A, A^*. (In fact, R. Kalman develops system theory over arbitrary fields rather than over the field of complex numbers.)

Many results of the analytic theory of electrical networks can be considered as realization results for matrix functions of various classes. (There are many expositions of the theory of electrical networks; we recommend especially the monograph of V. Belevitch [Bel] that seems as if written for a mathematician. See also the survey [EfPo].) The well-known Darlington's synthesis method for passive networks has been formulated for mathematicians by V. P. Potapov in 1966 as a realization problem for passive rational matrix functions [Pot3]; this problem was considered in details by Potapov's Ph. D. student E. Melamud [Me].

Already in [L4] M. S. Livšic used the theory of characteristic functions for the reduction of a nonselfadjoint operator to a triangular form, generalizing the theorem of I. Schur that an arbitrary matrix can be brought to an upper (or lower) triangular form by unitary equivalence. In this approach a multiplicative decomposition of the characteristic function corresponds to an "additive" decomposition of the operator itself over a linearly ordered chain of its invariant subspaces. This correspondence was used by M. S. Livšic in both directions. Using invariant subspaces of a finite-dimensional approximation of a given operator he decomposes the characteristic function of the approximating operator into factors, then passing to a limit he obtains a multiplicative decomposition of the characteristic function of the given operator, and finally using this multiplicative decomposition he constructs a triangular model of the given operator which is unitary equivalent to the operator itself.

A multiplicative decomposition of a meromorphic J-contractive matrix function on the unit disk (or on the upper half plane) has been obtained by V. P. Potapov in [Pot3] in a purely function theoretic way (the simpler case $J = I$ was handled much earlier in [Pot1]). However an important special case of V. P. Potapov's theorem has already been obtained by M. S. Livšic in [L4] using operator theoretic methods; V. P. Potapov used the investigations of M. S. Livšic as a guideline in his own function theoretic approach.

It is natural to ask whether one can reduce an operator to a diagonal form by a linear — no longer unitary — transformation. In the middle of 1950's M. S. Livšic has posed this problem to his then Ph. D. student L. A. Sakhnovich who has obtained numerous results in this direction. In particular the problem of the reduction of a nonselfadjoint operator to a diagonal form has lead L. A. Sakhnovich to a relation of the form

$$AX - XA^* = GJG^* \tag{5.1}$$

considered as an equation for X (formula (3) in [S1]). This relation appeared in a hidden form also in the theory of J-contractive matrix functions (see formula (22) and the following unnumbered formula in Chapter 2, Section 4 of [EfPo]) and in the theory of classical interpolation problems (formula (12) in Section 1 of [Kov]). A relation of the form (5.1) in [Kov] appears exactly in connection with what we called the chain identity; however the chain identity plays there a secondary role and is not highlighted. It should be mentionned that all the main ideas of the paper [Kov], published in 1983, have been suggested by V. P. Potapov some 10–12 years earlier; unfortunately, V. P. Potapov's contribution is not adequately reflected there. An identity of the form (5.1) has also been considered by L. de Branges in the framework of a certain generalized moment problem (in the language of Hilbert spaces of entire functions; see Theorem 27 of [Bran] and also formula (6.10) in [GolM]). A. A. Nudelman has used an identity of the form (5.1) at the basis of an abstract scheme that he developped for considering classical interpolation problems. It is interesting to note that an identity of the form (5.1) (and a related identity $X - AXA^* = GJG^*$) appears in an entirely different context as well, namely in connection with fast inversion algorithms for structured matrices (Toeplitz, Hankel, Vandermonde, etc.); see the survey [KS].

Everything needed for the theory of system realizations in its present form has thus been available by the middle of 1970's. An important step was taken by L. A. Sakhnovich in [S2] (a detailed exposition of these results is contained in [S3]): he studied the spectral factorization of a rational matrix function R with both R and R^{-1} given as transfer functions of the corresponding linear systems (operator colligations). The spectra of the state space operators A and B of these systems do not intersect. One considers a pair of Sylvester–Lyapunov equations $AT - TB = F_1G_1$, $SA - BS = F_2G_2$, where F_1, G_1, F_2, G_2 are the input and the output operators of the systems realizing R, R^{-1}. It is shown that if the corresponding blocks of the solutions T and S of these equations are invertible then the matrix function R admits a spectral factorization, and formulas for the spectral factors (having the form of the formulas (4.53–4.54) of the present paper) are obtained. N. M. Kostenko (a Ph. D. student of L. A. Sakhnovich) has shown in [Kos] that the invertibility of these blocks is necessary for the existence of a spectral factorization. Relations which are analogous to zero-pole coupling relations were considered already in [S1] and used in [S2].

Let us notice that the spectral factors constructed in [S2] are simply the chacteristic (transfer) matrix functions of subsystems arising by restricting the respective state space operators to the corresponding spectral subspaces. This is the form taken by the multiplication theorem of M. S. Livšic in the current situation (the matrices T and S define "metrics" which are now just bilinear functionals, neither positive definite nor even hermitian). Notice also that an important achievement of [S2] was to introduce a pair of Sylvester–Lyapunov equations determined by the two state space operators A and B instead of a single equation (5.1) determined by the operators A and A^*, for a general rational matrix function R possessing no symmetry properties with respect to the unit circle (or the real axis). As already remarked above this is related to considering general linear systems without any "energy balance" considerations.

Unfortunately the paper [S2] did not have the impact it deserved. The subsequent development of the theory of system realizations is connected with the name of I. Gohberg. I. Gohberg has also lead and inspired a coherent work of many mathematicians in the theory of system realizations and its applications and this theory experienced a fast growth from the late 1970's onward. Already in 1979 there appeared the monograph [BGK1] dealing with spectral factorizations of rational matrix functions given as transfer functions of linear systems (operator colligations). The paper [GKLR] considers the realization problem for matrix functions R and R^{-1} as transfer functions starting with "local data" (principal parts of Laurent series for R and R^{-1} at each pole); see also [BGR1]. In the beginning it was assumed that the pole sets for R and R^{-1} do not intersect; later the general case when these sets may intersect was considered as well. These and many other questions are considered in great details in the monograph [BGR2]; see also [KRR]. The collection of papers [CoMe] is dedicated to the spectral factorization for rational matrix functions based on the theory of system realizations; the papers [BGK2] and [BGK3] are especially related with our exposition in Section 4. There are results on system realization for rectangular (non-square) rational matrix functions given by local data [BGRa]. There are also realization results for matrix functions on a Riemann surface [BV], using deep new ideas of M. S. Livšic and his collaborators on characteristic functions for commuting tuples of nonselfadjoint operators [LKMV].

Factorization of matrix functions is a tool for many other problems, e.g., the theory of inverse problems for differential equations and prediction theory of stationary stochastic processes. If the corresponding matrix function is rational, this factorization (which is a technical tool for the original problem) may be carried out using system realizations which then become involved in the solution of the original problem as well. See, e.g., [AG1, AG2]. It is clear that the theory of system realizations can be successfully used also for the solution of the problems considered in [Yag]. It would be interesting to connect the questions considered in [Dei] with the theory of system realizations.

References

[AG1] Alpay, D. and I. Gohberg. Potentials associated to rational weights. In: *New results in operator theory and its applications*, pp. 23–40, Operator Theory: Advances and Applications, **OT 98**, Birkhäuser, Basel, 1997.

[AG2] Alpay, D. and I. Gohberg. Inverse problem for Sturm-Liouville operators with rational reflection-coefficient. Dedicated to the memory of Mark Grigorievich Krein (1907–1989). *Integral Equations Operator Theory*, **30**:3, (1998), pp. 317–325.

[BGRa] Ball, J., Gohberg, I. and M. Rakowski. *Reconstruction of a rational non-square matrix function from local data.* Integral Equations Operator Theory **20**:3 (1994), pp. 249–305.

[BGR1] Ball, J. A., I. Gohberg and L. Rodman. Realization and interpolation of rational matrix functions. In: *Topics in Interpolation Theory of Rational Matrix-valued Functions*, (I.Gohberg- ed.), pp. 1–72. Operator Theory: Advances and Applications, **OT 33**, Birkhäuser Verlag, Basel · Boston · Berlin, 1988.

[BGR2] Ball, J. A., I. Gohberg and L. Rodman. *Interpolation of Rational Matrix Functions.* Birkhäuser Verlag, Basel · Boston · Berlin, 1990.

[BV] Ball, J. A. and V. Vinnikov. Zero-pole interpolation for meromorphic matrix functions on an algebraic curve and transfer functions of 2D systems. *Acta Appl. Math.* **45**:3 (1996), pp. 239–316.

[BGK1] Bart, H., I. Gohberg and M.A. Kaashoek. *Minimal factorization of matrix and operator functions.* Operator Theory: Advances and Applications, **OT 1**, Birkhäuser Verlag, Basel, 1979.

[BGK2] Bart, H., I. Gohberg and M.A. Kaashoek. Explicite Wiener-Hopf factorization and realization. In: [CoMe], pp. 235–316.

[BGK3] Bart, H., I. Gohberg and M.A. Kaashoek. Multiplication by diagonals and reduction to canonical factorization. In: [CoMe], pp. 357–372.

[Bel] Belevitch, V. *Classical Network Theory.* Holden-Day, San Francisco,CA · Cambridge · Amsterdam, 1980.

[Bir1] Birkhoff, G. A theorem on matrices of analytic functions. *Mathematische Annalen.* **74**:1 (1913), pp.122–133. (Reprinted in [Bir3], pp. 240–251).

[Bir2] Birkhoff, G. The generalized Riemann problem for linear differential equations and the allied problems for linear difference and q-difference equations. *Proc. Amer. Acad. Arts and Sci.*, **49**, (1913), pp.521–568. (Reprinted in [Bir3], pp. 239–306).

[Bir3] Birkhoff, George David. *Collected Mathematical Papers*, Vol.1. Amer. Math. Soc., New York, 1950.

[Bo] Bolibrukh, A.A.. The Riemann-Hilbert problem. *Uspehi Mat. Nauk*, **45**:2 (1990), pp.3–47 (In Russian); English transl. in: *Russian Math. Surveys*, **45**:2 (1990), pp.1–58.

[Bran] Branges, Louis de *Hilbert Spaces of Entire Functions*, Prentice Hall, Englewood Cliffs, NJ, 1968.

[Br] Brodskiĭ, M.S. *Triangular and Jordan representations of linear operators.* Nauka, Moscow, 1969. (Russian). Engl. transl.: Translations of Mathematical Monographs, Vol. **32.** American Mathematical Society, Providence, R.I., 1971.

[BrL] Brodskiĭ, M.S. and M.S. Livšic, Spectral analysis of non-self-adjoint operators and intermediate systems. *Uspekhi Matem. Nauk*, **13**:1 (1958), pp. 3–85 (Russian). English Transl.: Amer. Math. Soc. Transl. (Ser.**2**), **13**, pp. 265–346.

[ClGo] Clancey, C. and I. Gohberg, *Factorization of Matrix Functions and Singular Integral Operators*, Operator Theory: Advances and Applications, **OT 3**, Birkhäuser Verlag, Basel · Boston · Stuttgart, 1981.

[CoMe] *Constructive Methods of Wiener-Hopf Factorization*, (Gohberg, I. and M.A. Kaashoek, editors.) Operator Theory: Advances and Applications, **OT 21**, Birkhäuser Verlag, Basel · Boston · Stuttgart, 1986.

[Dei] Deift, P. Integrable Operators. In: *Differential Operators and Spectral Theory (M.Sh. Birman's 70th Anniversary Collection)* – Buslaev V., Solomyak, M. and D. Yafaev, editors. Amer. Math. Soc. Transl. (**2**), vol. **189**, (1989), 69–84.

[EfPo] Efimov, A.V. and V.P. Potapov, J-expanding matrix functions and their role in the analytical theory of electrical circuits.(Russian) *Uspehi Matem. Nauk*, **28**:1 (1973), pp. 65–130. Engl. Transl. in: *Russian Math. Survays*, **28**:1 (1973), pp. 69–140.

[FM] Feldman, I. and A. Markus. On some properties of factorization indices. *Integral Equations and Operator Theory*, **30**:3 (1998), 326–337.

[Fu] Fuhrmann, P. *Linear systems and operators in Hilbert space.* McGraw-Hill, New York, 1981.

[Gah] Gahov, F.D. Riemann's boundary problem for a system of n pairs of functions. (Russian) *Uspehi Matem. Nauk (N.S.)* **7**:4, (1952), pp. 3–54.

[GKLR] Gohberg, I., M. A. Kaashoek, L. Lehrer and L. Rodman. Minimal divisors of rational matrix function with prescribed zero and pole structure. In: *Topics in Operator Theory Systems and Networks*, (Dym, H. and I. Gohberg -ed.), Operator Theory: Advances and Applications, **OT 12**, Birkhäuser Verlag, Basel · Boston · Stuttgart, 1984, pp. 241–275,

[GoKr] Gohberg, I. and M. Krein . Systems of integral equations on a half line with kernel depending on the difference of arguments. *Uspehi Matem. Nauk*, **13**:2, 1958, pp.3–72 (In Russian); English Transl.: Amer. Math. Soc. Transl. (ser. **2**), **14**, 1960, pp. 217–287.

[GS] Gohberg, I. and E.I. Segal, On operator generalizations of the logarithmic residue theorem and the theorem of Rouché, *Mat. Sbornik*, **84** (1971), pp. 607–629 (Russian); English transl.: *Math. USSR Sbornik* **13** (1971), pp. 603–625.

[GolM] Golinskiĭ, L.M. and I.V. Mikhailova. Hilbert spaces of entire functions as J-theory subject. *Preprint No. 28–80, Institute for Low Temperature Physics and Engineering*, Kharkov, 1980 (Russian). English transl. in: *Topics in Interpolation Theory* (Dym, H., B. Fritzsche, V. Katsnelson and B. Kirstein, editors), Operator Theory: Advances and Applications, **OT 95**, Birkhäuser, Basel, 1997, pp. 205–251.

[Kaas] Kaashoek, M.A. Minimal factorization, linear systems and integral operators. In: *Operator and Function Theory* (*Lancaster*, 1984), pp. 41–86. *NATO Adv. Sci. Inst. Ser. C: Math. Phys. Sci.*, **153**. Reidel, Dordrecht-Boston, Mass., 1985.

[KRR] Kaashoek, M.A., A.C.M. Ran and L. Rodman. Local minimal factorizations of rational matrix functiona in terms of null and pole data: formulas for factors. *Integral equations and Operator Theory*, **16**:1, (1993), pp. 98–130.

[KS] Kailath, T., and A.H. Sayed. Displacement structure: theory and applications. *SIAM Review*, **37**:3, 297–386.

[KFA] Kalman, R., P. Falb and M. Arbib. *Topics in Mathematical Systems Theory*, McGraw-Hill, New York, 1969. Russian Transl.: *Očerki po Matematičeskoĭ Teorii Sistem*, Mir, Moskwa, 1971.

[K] Katsnelson, V. Fuchsian differential systems related to rational matrix functions in general position and the joint system realization. In: *Proceedings of the Ashkelon Workshop on Complex Function Theory* (1996), (Israel Math. Conf. Proc., **11**), pp. 117–143.

[Kos] Kostenko, N.M. A necessaray and sufficient condition for the factorization of a rational operator-function. *Funkcional'nyĭ Analiz i Ego Prilozheniya*, **12**:4 (1978), pp. 87–88 (In Russian). English Transl in: *Functional Analysis and its Applications*, **12**:4 (1978), pp. 315–317.

[Kov] Kovalishina, I.V. Analytic theory of a class of interpolation problem. *Izvestiya Akad. Nauk SSSR Ser. Matem.*, **47**:3 (1983), pp. 455–497 (Russian). Engl. Translation: *Math. USSR Izvestiya*, **22**:3 (1984), pp. 419–463.

[L1] Livšic, M.S. On a class of linear operators in Hilbert space. *Matem. Sbornic*, **19**:2 (1946), pp. 239–262 (Russian). Engl. Transl. in: Amer. Math. Soc. Transl. (Ser.**2**), **13** (1960), pp. 61–83.

[L2] Livšic, M.S. On the theory of isometric operators with equal deficiency indices. *Doklady Akad. Nauk SSSR*, **58** (1947), pp. 13–15 (Russian).

[L3] Livšic, M.S. Isometric operators with equal deficiency indices, quasi-unitary operators *Matem. Sbornik*, **26** (1950), pp. 247–264 (Russian). English translation: Amer. Math. Soc. Transl. (Ser.**2**), **13** (1960), pp. 85–103.

[L4] Livšic, M.S. On spectral decompositian of linear nonselfadjoint operator. *Matem. Sbornik*, **34**, (1954), pp. 145–199 (Russian),. English transl. in: Amer. Math. Soc. Transl. (Ser.**2**), **5** (1957), pp. 67–114.

[L5] Livšic, M.S. The application of non-self-adjoint operators to scattering theory. *Journ. Experimental Theor. Phys. (USSR)*, **31**, pp. 121–131 (Russian). English Transl. in: *Soviet Physics JETP*, **4**:1 (1957), pp. 91–98.

[L6] Livšic, M.S. The method of non-self-adjoint operators in scattering theory. *Uspekhi Matem. Nauk*, **12**:1, (1957), pp. 212–218 (Russian). Engl. Transl. in: *Amer. Math. Soc. Transl.* (**2**), **16**, (1960), pp.427–434.

[L7] Livšic, M.S. On linear physical systems connected with the external world by coupling channels, *Izvestija Akad. Nauk SSSR Ser. Mat.*, **27** (1963), pp. 993–1030 (Russian).

[L8] Livšic, M.S. Open systems as linear automata. *Izvestiya Akad. Nauk SSSR Ser. Mat.*, **27**:6 (1963), pp. 1215–1228. (Russian).

[L9] Livšic, M.S.. *Operatory, Kolebanija, Volny (Otkrytye sistemy)*, Nauka, Moscow, 1966 (Russian). English transl.: *Operators, oscillations, waves (open systems)*. American Mathematical Society, Providence, RI, 1973.

[LF] Livšic, M.S. and M.Sh. Flekser. The expansion of a reactive four-terminal network into a chain of simplest four-terminal networks (Russian). *Dokl. Akad. Nauk SSSR*, **135** (1960), pp. 542–547. Engl. Transl.: *Soviet Phys. Dokl.* **4** (1961), 1150–1152.

[LKMV] Livšic, M.S, N. Kravitsky, A.S. Markus, and V. Vinnikov. *Theory of Commuting Nonselfadjoint Operators*, Kluwer, Dordrecht, 1995.

[LP] Livšic, M.S. and V.P. Potapov. A theorem on the multiplication of characteristic matrix functions. *Doklady Akad. Nauk SSSR*, **72** (1950) pp. 625–628.

[Me] Melamud, E. (= Melamed, E.) A certain generalization of Darlington's theorem. (Russian) *Izvestija Akad. Nauk Armjanskoĭ SSR Ser. Mat.* **7**:3 (1972), 183–195.

[MuVe] Mushelišvili, N.I. (=Muskhelishvili, N.I.) and N.P. Vekua. Riemann's boundary value problem for several unknovn functions and its application to system of singular integral equations. *Trudy Tbiliss. Mat. Inst.* **12** (1943), 1–46. (Russian, Georgian summary).

[Nu] Nudelman, A.A. A new problem of the type of the moment problem. *Dokl. Akad. Nauk SSSR*, **233**:5, (1977), pp 792–795 (Russian). Engl. Transl. in: *Soviet. Math. Dokl.*, **18**:2 (1977), pp. 507–510.

[Pl1] Plemelj, J. Riemannsche Funktionenscharen mit gegebener Monodromiegruppe, Monatsheft für Math. Phys., **19** (1908), 211–245.

[Pl2] Plemelj, J. *Problems in the Sense of Riemann and Klein.* J. Wiley & Sons Inc., New York·London·Sydney, 1964.

[PS] Pòlya, G. and G. Szegö. *Problems and Theorems in Analysis, Volume* II. Springer Verlag, Berlin·Heidelberg·New York, 1972.

[Pot1] Potapov, V.P. On holomorphic matrix functions bounded in the unit circle. (Russian) *Doklady Akad. Nauk SSSR* **72**:5 (1950). pp. 849–852.

[Pot2] Potapov, V.P. The multiplicative structure of J-contractive matrix-function. *Trudy Moskow. Matem. Obščestva*, **4** (1955), pp. 125–236 (Russian). Engl. transl. in: *Amer. Math. Soc. Transl.* (Ser.**2**), **15** (1960), pp. 131–243.

[Pot3] Potapov, V.P. Multiplicative representations of analytic matrix-valued functions. In: *Abstracts of Brief Scientific Communications. Internat. Congr. Math. Moscow,* 1966, *Section 4*, pp. 74–75.

[S1] Sahnovič, L.A.(=Sakhnovich, L.A.) Similarity of operators. *Sibirskiĭ Matematicheskiĭ Zhurnal*, **13**:4 (1972), pp. 868–883 (Russian). English transl. in: *Sibirian Mathematical Journal*, **13**:4 (1972), pp. 604–615.

[S2] Sahnovič, L.A.(=Sakhnovich, L.A.), On the factorization of an operator-valued transfer function. *Doklady Akad. Nauk SSSR*, **226**:4 (1976), pp. 781–784 (Russian). English Transl.: Soviet. Math. Dokl. **17** (1976), pp. 203–207.

[S3] Sahnovič, L.A.(=Sakhnovich, L.A.) Factorization problems and operator identities. *Uspekhi Matem. Nauk*, **41**:1 (1986), pp. 3–55 (Russian). English translation in: *Russian Math. Surveys* **41**:1 (1986), pp. 1–64.

[S4] Sahnovič, L.A.(=Sakhnovich, L.A.) *Integral Equations with Difference Kernela on Finite Intervals.* Operator Theory: Advances and Applications, **OT 84**, Birkhäuser Verlag, Basel · Boston · Stuttgart, 1996.

[Vek] Vekua, N.P.. *Systems of Singular Integral Equations*, GITTL, Moscow, 1950 (In Rissian); English transl.: P. Noordhoff (Ltd), Groningen, The Netherlands, 1967.

[Yag] Yaglom, A.M. (=Jaglom, A.M.) Extrapolation, interpolation and filtration of stationary random processes with rational spectral density. (Russian) *Trudy Moskowskogo Matematičeskogo Obščestva*, **4** (1955), pp. 333–374. Engl. transl. in: *Selected Translations in Mathematical Statistics and Probability*, **4**, Amer. Math. Soc, Providence, Rhode Island, 1963, pp. 345–387.

[Z] Zadeh, L. and A.Dezoer. *Linear System Theory*, McGraw-Hill, New York·San Francisco·Toronto· London, 1963.

Department of Mathematics, The Weizmann Institute of Science, Rehovot 76100, Israel
E-mail address: katze@wisdom.weizmann.ac.il

Operator Theory:
Advances and Applications, Vol. 123

One-parameter Semigroups of Fractional-linear Transformations

Victor Khatskevich, Simeon Reich, and David Shoikhet

1. Introduction

In 1964 F. Forelli [7] showed that to each strongly continuous one-parameter group of isometries on a Hardy space H^p, $p \neq 2$, $p \geq 1$, there corresponds a group of fractional-linear automorphisms of the unit disk in the complex plane. His study has been continued by many mathematicians (see, for example, [2], [3], [15], [8]). In particular, E. Berkson and H. Porta [3] developed a generation theory for one-parameter semigroups of holomorphic self-mappings of the unit disk and used their results to study the eigenvalue problem for composition operators in Hardy spaces.

Such an approach was also used in the case of the unit ball in $\mathbb{C}^k$ by C. C. Cowen and B. D. MacCluer [4].

In the infinite-dimensional case continuous semigroups of holomorphic mappings have been studied, for instance, in [16], [17], [18], [9] and [14]. In the present paper we concentrate our attention on the case of semigroups of fractional-linear transformations acting on the unit operator ball of the space $\mathcal{L}(\mathcal{H}_1, \mathcal{H}_2)$, where $\mathcal{H}_1, \mathcal{H}_2$ are two Hilbert spaces. Namely, we consider the fractional-linear transformations F of the following form:

$$F(K) := F_T(K) = (T_{21} + T_{22}K)(T_{11} + T_{12}K)^{-1},$$

where $T_{ij} \in \mathcal{L}(\mathcal{H}_j, \mathcal{H}_i)$, $i, j = 1, 2$, and

$$K \in \mathcal{L}(\mathcal{H}_1, \mathcal{H}_2), \quad \| K \| < 1, \; T = (T_{ij})_{i,j=1}^2 = \begin{pmatrix} T_{11} & T_{12} \\ T_{21} & T_{22} \end{pmatrix}.$$

It is well known (see, for example, [11], [1]) that to each such fractional-linear transformation there corresponds a so-called plus-operator in a Krein space. Using this idea E. Vesentini [16], [17], [18] established the connection between semigroups of fractional-linear transformations and semigroups of linear contractions with respect to an indefinite metric (see details below).

Following this approach we study in the present paper the generators of strongly continuous semigroups of fractional-linear transformations and their basic properties.

The second author was partially supported by the Fund for the Promotion of Research at the Technion and by the Technion VPR Fund.

2. Preliminaries

Let $\mathcal{H}_1, \mathcal{H}_2$ be two Hilbert spaces, $\mathcal{L}(\mathcal{H}_1, \mathcal{H}_2)$ the space of all linear bounded operators from $\mathcal{H}_1$ into $\mathcal{H}_2$ and $\mathcal{K}$ the closed unit ball of $\mathcal{L}(\mathcal{H}_1, \mathcal{H}_2)$ with respect to the norm topology:

$$\mathcal{K} = \{K \in \mathcal{L}(\mathcal{H}_1, \mathcal{H}_2) : \; \| K \| \le 1\}\,. \tag{1}$$

Let $\mathcal{K}^0$ be the interior of $\mathcal{K}$:

$$\mathcal{K}^0 = \{K \in \mathcal{K} : \; \| K \| < 1\}.$$

Let $\mathrm{Hol}(\mathcal{K}^0, \mathcal{K})$ be the set of all holomorphic mappings F on $\mathcal{K}^0$ such that

$$F(\mathcal{K}^0) \subset \mathcal{K}. \tag{2}$$

By $\mathrm{Hol}(\mathcal{K}^0)$ we denote the family of those mappings F in $\mathrm{Hol}(\mathcal{K}^0, \mathcal{K})$ which also satisfy

$$F(\mathcal{K}^0) \subset \mathcal{K}^0. \tag{3}$$

Our object of interest will be the subset $\mathcal{F}$ of $\mathrm{Hol}(\mathcal{K}^0, \mathcal{K})$ consisting of all fractional-linear transformations. In other words,

$$F \in \mathcal{F} \text{ if } F \in \mathrm{Hol}(\mathcal{K}^0, \mathcal{K})$$

and

$$F(K) = F_T(K) = (T_{21} + T_{22}K)(T_{11} + T_{12}K)^{-1},$$

where $T_{ij} \in \mathcal{L}(\mathcal{H}_j, \mathcal{H}_i)$, $i, j = 1, 2$, and $K \in \mathcal{K}^0$.

Consider the operator block-matrix

$$T = \begin{pmatrix} T_{11} & T_{12} \\ T_{21} & T_{22} \end{pmatrix}. \tag{4}$$

Evidently, T is a linear bounded operator on the Hilbert space

$$\mathcal{H} = \mathcal{H}_1 \oplus \mathcal{H}_2\,, \tag{5}$$

i.e., $T \in \mathcal{L}(\mathcal{H})$.

We now define in $\mathcal{H}$ an indefinite metric $[\cdot,\cdot]$ as follows: $[x, y] = (Jx, y)$, $x, y \in \mathcal{H}$, where $(\cdot,\cdot)$ is the usual scalar product in $\mathcal{H}$, and with respect to the decomposition (5) of $\mathcal{H}$,

$$J = \begin{pmatrix} -I_1 & 0 \\ 0 & I_2 \end{pmatrix},$$

where I_i is the identity operator on $\mathcal{H}_i$, $i = 1, 2$. A natural question at this point is the following one: what are the connections among the entries of the block-matrix T imposed by the requirement that $F_T \in \mathrm{Hol}(\mathcal{K}^0, \mathcal{K})$? To this end, let us recall the basic notions of the indefinite structure of $\mathcal{H}$ (see for example [1]). Set $\mathcal{H}_1 = \mathcal{H}_-$, $\mathcal{H}_2 = \mathcal{H}_+$ and let P_+, P_- be the two orthogonal projections on $\mathcal{H}_+, \mathcal{H}_-$ respectively, generated by the decomposition (5), i.e., for any $x \in \mathcal{H}$,

$$x = x_- + x_+ \equiv P_-x + P_+x.$$

Consider the following sets:

$$\mathcal{P}_- = \{x \in \mathcal{H} : [x, x] \le 0\},$$
$$\mathcal{P}_{--} = \{x \in \mathcal{P}_- : [x, x] < 0\},$$
$$\mathcal{P}_+ = \{x \in \mathcal{H} : [x, x] \ge 0\},$$

and

$$\mathcal{P}_{++} = \{x \in \mathcal{P}_+ : [x, x] > 0\}.$$

A subspace $L \subset \mathcal{H}$ is called non-positive (non-negative) if $L \subset \mathcal{P}_-$ (respectively, $L \subset \mathcal{P}_+$). A non-positive (non-negative) subspace L of $\mathcal{H}$ is called maximal non-positive (maximal non-negative) if it is not a proper subset of any non-positive (respectively, non-negative) subspace. The set of all maximal non-positive (maximal non-negative) subspaces will be denoted by $\mathcal{M}_-$ (respectively, $\mathcal{M}_+$).

We need some auxiliary results. Their proofs will be sketched for completeness.

Lemma 1 [1]. A subspace $L_- \in \mathcal{M}_-$ ($L_+ \in \mathcal{M}_+$) if and only if $L_- = (P_- + K)\mathcal{H}_-$ for some $K \in \mathcal{K}$ (respectively, $L_+ = (P_+ + K^*)\mathcal{H}_+$ for some $K \in \mathcal{K}$).

Proof. Let $L_- \in \mathcal{M}_-$. Take $x \in L_-$, $x = x_- + x_+$. Since $[x, x] \le 0$, it follows that $\| x_+ \| \le \| x_- \|$. We also have

$$\| x_- \|^2 \le \| x_- \|^2 + \| x_+ \|^2 = \| x \|^2 \le 2 \| x_- \|^2 .$$

Hence the mapping $P_-|L_- : L_- \to \mathcal{H}_-$ is one-to-one. Putting $K = P_+(P_-|L_-)^{-1}$, we obtain $Kx_- = x_+$, so $K \in \mathcal{K}$.

In the other direction, if $K \in \mathcal{K}$, then $L_- = (P_- + K)\mathcal{H}_- \subset \mathcal{P}_-$. Let us prove that $L_- \in \mathcal{M}_-$. Let $x^n = x^n_- + x^n_+ \in L_-$, $x^n \to x^0$ as $n \to \infty$. Since $x^n_- \to x^0_-$, $Kx^n_- \to Kx^0$. But $Kx^n_- = x^n_+$, and therefore $x^n_+ \to x^0_+$. Thus we have $x^0 \in L_-$ and so L_- is closed. Suppose there exists $\tilde{L}_-$ such that $\tilde{L}_- \subset \mathcal{P}_-$, $L_- \subset \tilde{L}_-$, $L_- \ne \tilde{L}_-$. Then there exists $\tilde{x} \in \tilde{L}_-$, $\tilde{x} \ne 0$, $\tilde{x} \notin L_-$. But $\tilde{x}_- = P_-\tilde{x} \in P_-L_- = \mathcal{H}_-$ and $\tilde{\tilde{x}} = \tilde{x}_- + K\tilde{x}_- \in L_-$, $\tilde{\tilde{x}}_+ = \tilde{x}_+ - K\tilde{x}_- \in \mathcal{H}_+$. Thus we have $\tilde{\tilde{L}}_- = L_- \dotplus \mathrm{Span}(\tilde{\tilde{x}}_+) \not\subset \mathcal{P}_-$, but $\tilde{\tilde{L}}_- \subset \tilde{L}_- \subset \mathcal{P}_-$. A contradiction. The second part of the lemma can be proved in the same way.

Now let us define the classes $\mathcal{M}^0_-$ and $\mathcal{M}^0_+$ of all maximal uniformly negative subspaces $L_- \in \mathcal{M}_-$, and of all maximal uniformly positive subspaces $L_+ \in \mathcal{M}_+$ as follows:

$$L_- \in \mathcal{M}^0_- \text{ if } L_- = (P_- + K)\mathcal{H}_- \text{ and } K \in \mathcal{K}^0;$$
$$L_+ \in \mathcal{M}^0_+ \text{ if } L_+ = (P_+ + K^*)\mathcal{H}_+ \text{ and } K \in \mathcal{K}^0.$$

Lemma 2 [1]. Let $x \in \mathcal{P}_-$ ($x \in \mathcal{P}_+$). Then there exists $L_- \in \mathcal{M}_-$ ($L_+ \in \mathcal{M}_+$) such that $x \in L_-$ (respectively, $x \in L_+$). If $x \in \mathcal{P}_{--}$ ($x \in \mathcal{P}_{++}$), then L_- (L_+) can be chosen such that $L_- \in \mathcal{M}^0_-$ ($L_+ \in \mathcal{M}^0_+$).

Proof. Take $x \in \mathcal{P}_-$, $x = x_- + x_+$, and let $\hat{K}$ be defined as a linear operator between Span (x_-) and Span (x_+) as follows: $\hat{K}x_- = x_+$. Evidently, $\| \hat{K} \| \le 1$ and $\| \hat{K} \| < 1$ when $x \in \mathcal{P}_{--}$. Taking an extension K of $\hat{K}$ from Span (x_-) to the whole of $\mathcal{H}_-$ with the same norm, we obtain $K \in \mathcal{K}$ and $K \in \mathcal{K}^0$ when $x \in \mathcal{P}_{--}$.

Put $L_- = (P_- + K)\mathcal{H}_-$. Then $L_- \in \mathcal{M}_-$ and $L_- \in \mathcal{M}_-^0$ when $x \in \mathcal{P}_{--}$. In the same way one can prove the rest of the lemma.

We recall that for a set $M \subset \mathcal{H}$ the subspace $M^\perp$ is defined as follows:

$$M^\perp = \{x \in \mathcal{H} :\ (x, y) = 0 \text{ for all } y \in M\}.$$

Lemma 3 [10]. *$L_- \in \mathcal{M}_-$ ($L_- \in \mathcal{M}_-^0$) if and only if $L_-^\perp \in \mathcal{M}_+$ ($L_-^\perp \in \mathcal{M}_+^0$).*

Proof. One can check in a straightforward manner that if $L_- = (P_- + K)\mathcal{H}_- \in \mathcal{M}_-$, then $L_-^\perp = (P_+ - K^*)\mathcal{H}_+$. Hence we immediately obtain the conclusion of the lemma.

3. Fractional-linear transformations

Now we are able to prove the following result.

Theorem 1. *For each $F_T \in \mathcal{F}$ there exist $\hat{K} \in \mathcal{K}$ and $\tilde{K} \in \mathcal{K}$ such that with respect to the decomposition (5), T is of the form:*

$$T = \begin{pmatrix} T_{11} & T_{11}\hat{K}^* \\ \tilde{K}T_{11} & T_{22} \end{pmatrix}, \tag{6}$$

where $T_{11} \in \mathcal{L}(\mathcal{H}_-)$ is invertible and $T_{22} \in \mathcal{L}(\mathcal{H}_+)$ satisfies the following condition

$$\| \tilde{K}T_{11}x_- + T_{22}x_+ \| \leq \| T_{11}x_- + T_{11}\hat{K}^*x_+ \| \tag{*}$$

for $x = x_- + x_+ \in \mathcal{P}_-$. Moreover, if either $\|\tilde{K}\| = 1$ or $\|\hat{K}\| = 1$, then $F_T(K) =$ const., $K \in \mathcal{K}^0$. Also, if $\|\hat{K}\| < 1$ and $\|\tilde{K}\| < 1$, then $F_T \in$ Hol$(\mathcal{K}^0)$.

Proof. Define $\tilde{K} := F_T(0) \in \mathcal{K}$. Since $F_T(0) = T_{21}T_{11}^{-1}$, $T_{21} = \tilde{K}T_{11}$ and

$$T = \begin{pmatrix} T_{11} & T_{12} \\ \tilde{K}T_{11} & T_{22} \end{pmatrix}.$$

Consider $T^* = \begin{pmatrix} T_{11}^* & T_{11}^*\tilde{K}^* \\ T_{12}^* & T_{22}^* \end{pmatrix}$. We claim that $T^*\mathcal{P}_{--} \subset \mathcal{P}_-$. To this end, take $x \in \mathcal{P}_{--}$, $x = x_- + x_+$, $\|x_+\| < \|x_-\|$. If $T^*x \in \mathcal{P}_{++}$, then $[T^*x,\ T^*x] > 0$, and according to Lemma 2 there exists $L_+ \in \mathcal{M}_+^0$ such that $T^*x \in L_+$. By Lemma 3, the subspace $L_+^\perp$ (the orthogonal complement of L_+ in $\mathcal{H}$) belongs to $\mathcal{M}_-^0$ and $L_+^\perp = \{y \in \mathcal{H} : y = y_- + K^0y_-,\ y_- \in \mathcal{H}_-\}$, where $K^0 \in \mathcal{K}^0$. We have

$$\{0\} = (T^*x, L_+^\perp) = (x, TL_+^\perp), \text{ and}$$

$$TL_+^\perp = (P_- + F_T(K^0))\mathcal{H}_-, \text{ where } F_T(K^0) \in \mathcal{K}.$$

Hence $TL_+^\perp \in \mathcal{M}_-$, $x \in (TL_+^\perp)^\perp$, and as a consequence we have (see Lemma 3) $x \in \mathcal{P}_+$. A contradiction. So $T^*\mathcal{P}_{--} \subset \mathcal{P}_-$.

Consider $T^*\mathcal{H}_-$. As it was proved above, $T^*\mathcal{H}_- \subset \mathcal{P}_-$. On the other hand,

$$T^*\mathcal{H}_- = \begin{pmatrix} T_{11}^* & T_{11}^*\tilde{K}^* \\ T_{12}^* & T_{22}^* \end{pmatrix}\begin{pmatrix} \mathcal{H}_- \\ 0 \end{pmatrix} = T_{11}^*\mathcal{H}_- + T_{12}^*\mathcal{H}_-.$$

As T_{11} is invertible, $T^*\mathcal{H}_- \in \mathcal{M}_-$, and for all $x_- \in \mathcal{H}_-$, $T_{11}^*x_- + T_{12}^*x_- = y_- + \hat{K}y_-$, where $y_- = T_{11}^*x_- \in \mathcal{H}_-$ and $\hat{K} \in \mathcal{K}$. This means that $T_{12}^* = \hat{K}T_{11}^*$, hence $T_{12} = T_{11}\hat{K}^*$. The last assertion $(*)$ of the theorem is equivalent to

$$T\mathcal{P}_- \subset \mathcal{P}_-.$$

Now, if either $\| \hat{K} \| = 1$ or $\| \tilde{K} \| = 1$, then it is evident that either

$$\inf_{[x,x]<o} \frac{[Tx,Tx]}{[x,x]} = 0 \quad \text{or} \quad \inf_{[x,x]<0} \frac{[T^*x,T^*x]}{[x,x]} = 0.$$

Hence (see [12] and also [1]) either $[Tx,Tx] \leq 0$ for all $x \in \mathcal{H}$, and $\operatorname{Im}T$ $(= T\mathcal{H}) = (P_- + \tilde{K})\mathcal{H}_-$ or $[T^*x,T^*x] \leq 0$ for all $x \in \mathcal{H}$, and $\operatorname{Im}T^*$ $(= T^*\mathcal{H}) = (P_- + \hat{K})\mathcal{H}_-$. Let us recall that $\operatorname{Ker}T^* = (\operatorname{Im}T)^\perp$ and $\operatorname{Ker}T = (\operatorname{Im}T^*)^\perp$. Hence either $\operatorname{Ker}T^* = (P_+ - \tilde{K}^*)\mathcal{H}_+ \in \mathcal{M}_+$ or $\operatorname{Ker}T = (P_+ - \hat{K}^*)\mathcal{H}_+ \in \mathcal{M}_+$, i.e., either $\begin{pmatrix} T_{11}^* & T_{11}^*\tilde{K}^* \\ \hat{K}T_{11}^* & T_{22}^* \end{pmatrix}\begin{pmatrix} -\tilde{K}^* \\ I_+ \end{pmatrix} = \begin{pmatrix} 0 \\ 0 \end{pmatrix}$ and $T_{22}^* = \hat{K}T_{11}^*\tilde{K}^*$, which means that $T_{22} = \tilde{K}T_{11}\hat{K}^*$, or $\begin{pmatrix} T_{11} & T_{11}\hat{K}^* \\ \tilde{K}T_{11} & T_{22} \end{pmatrix}\begin{pmatrix} -\hat{K}^* \\ I_+ \end{pmatrix} = \begin{pmatrix} 0 \\ 0 \end{pmatrix}$ and $T_{22} = \tilde{K}T_{11}\hat{K}^*$. So, in both these cases $F_T(K) = \tilde{K}$, $K \in \mathcal{K}^0$. Now suppose that both $\tilde{K}$ and $\hat{K}$ are in $\mathcal{K}^0$. In this case it can be proved that $\inf_{[x,x]<0} \frac{[Tx,Tx]}{[x,x]} = \mu(T) > 0$. Hence (see [12] and also [1]) $[Tx,Tx] \leq \mu(T)[x,x]$ for all $x \in \mathcal{H}$ and F_T satisfies condition (3).

Example 1. Consider $F_T(z) = \alpha\frac{z+1}{z+1}$, where $0 \leq |\alpha| \leq 1$, $z \in \mathbb{C}$. Then $T = \begin{pmatrix} 1 & 1 \\ \alpha & \alpha \end{pmatrix}$, F_T is defined and holomorphic on $\Delta^0 = \{z \in \mathbb{C} : |z| < 1\}$, but F_T does not satisfy (3) when $|\alpha| = 1$.

Now we would like to give a necessary condition for F_T to belong to $\mathcal{F}$.

Proposition 1. If $F_T \in \mathcal{F}$, then the linear operator B on $\mathcal{L}(\mathcal{H}_-, \mathcal{H}_+)$ defined by $BK = T_{22}K - \tilde{K}T_{11}\hat{K}^*KT_{11}^{-1}$ satisfies the following condition:

$$\| B \| \leq 1.$$

Proof. This assertion follows from the fact that $\|F_T'(0)\| \leq 1$ (see [5] and also [11]) and the equality $F_T'(0) = B$.

4. Generators of semigroups

From now on we will consider the set of all mappings F_T which do satisfy (3), i.e., T will be of the form (6) with $\tilde{K}, \hat{K} \in \mathcal{K}^0$.

The set of all such mappings will be denoted by Φ. It is easy to see that $F_T = F_{\alpha T}$ for an arbitrary $\alpha \neq 0$. So in the sequel we will assume that the block-matrix T defining $F_T \in \Phi$ is a J-contraction:

$$[Tx,Tx] \leq [x,x],\ x \in \mathcal{H}\ , \tag{7}$$

or equivalently

$$T^*JT \leq J\ .$$

Consider now a semigroup $S(t)$, $t \geq 0$, such that $S(t) \in \Phi$ for all $t \geq 0$. So we have $S(t) = F_{T(t)}$, $t \geq 0$, where $T(t)$ satisfies (7) for all $t \geq 0$. The semigroup $T(t)$ is assumed to be strongly continuous. We intend to describe the generator g of $S(t)$. In this direction consider the infinitesimal generator A of $T(t)$. This means that A is the strong limit of $\frac{T(t)-I}{t}$ when $t \to 0^+$. We would like to recall that A is a densely defined closed operator: $\overline{\mathcal{D}(A)} = \mathcal{H}$.

Lemma 4 [18]. The semigroup $T(t)$ satisfies condition (7) if and only if

$$\text{Re}[Ax, x] \leq 0 \text{ for all } x \in \mathcal{D}(A). \tag{8}$$

Proof. For $x \in \mathcal{D}(A)$ we have:

$[T(t)x, T(t)x] = [(T(t)-I)x, (T(t)-I)x] + [(T(t)-I)x, x] + [x, (T(t)-I)x] + [x, x]$.

From (7) we obtain

$$[(T(t)-I)x, (T(t)-I)x] + [(T(t)-I)x, x] + [x, (T(t)-I)x] \leq 0,$$

and hence $2\text{Re}[Ax, x] \leq 0$, or $\text{Re}[Ax, x] \leq 0$. In the other direction, if (8) holds, then

$$\frac{d}{dt}[T(t)x, T(t)x] = [AT(t)x, T(t)x] + [T(t)x, AT(t)x] = 2\text{Re}[AT(t)x, T(t)x] \leq 0,$$

where $t \geq 0$ and $x \in \mathcal{D}(A)$. Hence $[T(t)x, T(t)x] \leq [T(\tau)x, T(\tau)x]$ when $t \geq \tau \geq 0$. As a consequence, we have $[T(t), T(t)x] \leq [T(0)x, T(0)x] = [x, x]$, where $x \in \mathcal{D}(A)$ and $t \geq 0$. Hence $[T(t)x, T(t)x] \leq [x, x]$ for all $x \in \mathcal{H}$ and $t \geq 0$.

Now we intend to express the generator A as an operator block-matrix with respect to the decomposition (5) of $\mathcal{H}$. This is connected with the following question: Is it possible to decompose $\mathcal{D}(A)$ as follows: $\mathcal{D}(A) = \mathcal{D}(A)_+ \oplus \mathcal{D}(A)_-$ with respect to (5)? The answer in general is "no", as the following example shows.

Example 2. Let $\mathcal{H}_\pm$ be the closed span of $\{e^i_\pm : i \in \mathbb{N}\}$, $\mathcal{H} = \mathcal{H}_+ \oplus \mathcal{H}_-$. Let $\mathcal{L}^1_0$ and $\mathcal{L}^2_0$ be the closed spans of $\{e^i_+ + e^i_- : i \in \mathbb{N}\}$ and $\{e^i_+ - e^i_- : i \in \mathbb{N}\}$, respectively. We have

$$\mathcal{H} = \mathcal{L}^1_0 \dot{+} \mathcal{L}^2_0, \quad \mathcal{L}^1_0,\ \mathcal{L}^2_0 \subset \mathcal{P}_- \cap \mathcal{P}_+.$$

Take $M = \text{Span}\{e^i_+ + e^i_- : i \in \mathbb{N}\}$, and let $\mathcal{L} = M \dot{+} \mathcal{L}^2_0$. Evidently $\bar{\mathcal{L}} = \mathcal{H}$ and $\dim\mathcal{L} = c$. If we suppose that there exist $\mathcal{L}_\pm \subset \mathcal{P}_{\pm\pm}$ such that $\mathcal{L} = \mathcal{L}_-[\dot{+}]\mathcal{L}_+$ (the sum is J-orthogonal), then the mapping $Q : \mathcal{L}_\pm \to M$ is one-to-one, where the projection $Q : \mathcal{L} \to M$ is generated by the decomposition $\mathcal{L} = M \dot{+} \mathcal{L}^0_2$. So $\dim\mathcal{L}_\pm \leq \dim M = \aleph_0$, and $\dim\mathcal{L} = \aleph_0$. A contradiction.

Now suppose that $P_+\mathcal{D}(A) \subset \mathcal{D}(A)$ or $P_-\mathcal{D}(A) \subset \mathcal{D}(A)$. Then

$$\mathcal{D}(A) = P_+\mathcal{D}(A) \oplus P_-\mathcal{D}(A) \quad \text{and} \quad A = \begin{pmatrix} A_{11} & A_{12} \\ A_{21} & A_{22} \end{pmatrix},$$

where $A_{ij} = P_iAP_j$ is densely defined on $\mathcal{H}_j$, $i, j = 1, 2$.

By Lemma 4 we have

$$0 \geq \mathrm{Re}[Ax, x] = \mathrm{Re}([A_{11}x_-, x_-] + [A_{22}x_+, x_+] + [A_{12}x_+, x_-] + [A_{21}x_-, x_+]).$$

Taking $x_+ = 0$, we obtain $\mathrm{Re}[A_{11}x_-, x_-] \leq 0$, and taking $x_- = 0$, we obtain $\mathrm{Re}[A_{22}x_+, x_+] \leq 0$. This means that A_{ii}, $i = 1, 2$, is a J-dissipative operator, or equivalently, A_{22} is dissipative and A_{11} is accretive ($\mathrm{Re}(A_{11}x_-, x_-) \geq 0$).

If both A_{11} and A_{22} are skew-symmetric ($\mathrm{Re}[A_{11}x_-, x_-] = \mathrm{Re}[A_{22}x_+, x_+] = 0$), then for all $x_- \in \mathcal{D}(A_{11})$ and $x_+ \in \mathcal{D}(A_{22})$,

$$\mathrm{Re}([A_{12}x_+, x_-] + [A_{21}x_-, x_+]) \leq 0.$$

Hence $\mathrm{Re}([A_{12}x_+, x_-] + [A_{12}x_-, x_+]) = 0$ (one can take $-x_-$ instead of x_-). Consequently, $\mathrm{Re}[A_{12}x_+, x_-] = -\mathrm{Re}[x_+, A_{21}x_-]$ or $\mathrm{Re}(A_{12}x_+, x_-) = \mathrm{Re}(x_+, A_{21}x_-)$. Hence $(A_{12}x_+, x_-) = (x_+, A_{21}x_-)$ for all $x = x_- + x_+ \in \mathcal{D}(A)$. If $\dim\mathcal{H}_+ < \infty$, then A_{12} is a bounded linear operator and $(A_{12}x_+, x_-) = (x_+, A_{12}^* x_-)$ for all $x_- \in \mathcal{H}_-$ and $x_+ \in \mathcal{H}_+$. Thus $A_{21} = A_{12}^* | \mathcal{D}(A_{11})$. In the case when $\dim \mathcal{H}_- < \infty$, we obtain analogously $A_{12} = A_{21}^* | \mathcal{D}(A_{22})$.

All the above means that the following statement holds.

Theorem 2. *Let A be the infinitesimal generator of a strongly continuous J-contractive semigroup $T(t)$ and let either $P_+\mathcal{D}(A) \subset \mathcal{D}(A)$ or $P_-\mathcal{D}(A) \subset \mathcal{D}(A)$. Then $A = \begin{pmatrix} A_{11} & A_{12} \\ A_{21} & A_{22} \end{pmatrix}$ with respect to the decomposition (5), and $T(t)$ is a semigroup of J-isometries if and only if A_{ii} is skew-symmetric, $i = 1, 2$, and $T(t)$ is a group of J-unitary operators if and only if A_{ii} is skew-self-adjoint, $i = 1, 2$.*

Remark 1. In the case of a Pontryagin space $\mathcal{H} = \Pi_\kappa$ with $\min\{\dim \mathcal{H}_+, \dim\mathcal{H}_-\} = \kappa < \infty$, for any dense linear subspace $\mathcal{L} \subset \mathcal{H}$ we can take a new canonical decomposition $\mathcal{H} = \tilde{\mathcal{H}}_-[\dot{+}]\tilde{\mathcal{H}}_+$ of the type (5) such that $\tilde{P}_+\mathcal{L} = \tilde{\mathcal{H}}_+$ (in case $\dim\mathcal{H}_+ = \kappa < \infty$) or $\tilde{P}_-\mathcal{L} = \tilde{\mathcal{H}}_-$ (in case $\dim\mathcal{H}_- = \kappa < \infty$) (see, for example, [1]). This means that with respect to the new decomposition the matrix of the generator, say $\tilde{A}$, is of the type which was described in Theorem 2. Since $A = U^*\tilde{A}U$, where U is a J-unitary operator (the operator performing the passage to the new canonical decomposition of $\mathcal{H}$), A itself is of the same type as $\tilde{A}$.

The following result is devoted to the spectral properties of the generator A in a Pontryagin space Π_κ (see also [18]).

Theorem 3. The generator A *is a densely defined, closed maximal J-dissipative operator in Π_κ, such that $\sigma(A) \cap \mathbb{C}_-$ consists of no more than κ normal eigenvalues, where $\mathbb{C}_- = \{z \in \mathbb{C} : \mathrm{Re} z < 0\}$.*

Proof: The operator A is densely defined and closed as an infinitesimal generator of the semigroup $T(t)$. The inequality (8) means that A is J-dissipative. It is easy to prove that the J-conjugate operator A^c $(= JA^*J)$ is the infinitesimal generator of the J-conjugate semigroup $T^c(t)$. So A^c is J-dissipative too. According to [1], Proposition 2.7^0, page 116, A is maximal J-dissipative. Now our claim follows from [1] (Corollary 2.23, page 123, Lemma 2.8, page 116, and Theorem 2.9, page 117).

Let us return to the general case of a Krein space $\mathcal{H} = \mathcal{H}_- \oplus \mathcal{H}_+$.

Definition. A is called uniformly J-dissipative if

$$\operatorname{Re}[Ax, x] \leq -d \parallel x \parallel^2$$

for some $d > 0$ and for all $x \in \mathcal{D}(A)$.

Proposition 2. If A is a bounded uniformly J-dissipative operator, then for all $\lambda \in \rho(A)$, $0 \leq \lambda < \frac{2d}{\|A\|^2}$, the operator $R(\lambda, A) = (I - \lambda A)^{-1}$ is a bounded J-contractive operator.

Proof. Taking $\lambda \in \rho(A)$, $0 \leq \lambda < \frac{2d}{\|A\|^2}$, we obtain

$$[(I - \lambda A)x,\ (I - \lambda A)x] = [x, x] - 2\operatorname{Re}\lambda[Ax, x] + |\lambda|^2[Ax, Ax]$$

$= [x, x] - \lambda(2\operatorname{Re}[Ax, x] - \lambda[Ax, Ax])$. We also have $2\operatorname{Re}[Ax, x] - \lambda[Ax, Ax] \leq -2d \parallel x\|^2 + \lambda \|Ax\|^2 \leq (-2d + \lambda \parallel A \parallel^2) \parallel x \parallel^2 < 0$.
Hence $[(I - \lambda A)x,\ (I - \lambda A)x] \geq [x, x]$, and $[R(\lambda, A)x, R(\lambda, A)x] \leq [x, x]$.

Now we intend to describe the set of all generators for strongly continuous semigroups of fractional-linear transformations belonging to Φ, in the case of a Pontryagin space Π_κ.

Lemma 5 (See [1], Exercise 10, page 212). If A is a uniformly J-dissipative operator in a Pontryagin space Π_κ, with $\dim(\Pi_\kappa)_- = \kappa < \infty$, then there exists a pair of subspaces L_+^0, L_-^0 such that $L_\pm^0 \in \mathcal{M}_\pm^0$ and $AL_+^0 \subset L_+^0$, $AL_-^0 = L_-^0$.

Proof. By Theorem 3 there exists a complex number $z \neq \bar{z}$ such that $z, \bar{z} \in \rho(A)$. Consider the Cayley transform

$$(A - \bar{z}I)^{-1}(A - zI) = U.$$

It is easy to check that U is a bounded J-contraction. Take the block-matrix of U with respect to the decomposition (5):

$$U = \begin{pmatrix} U_{11} & U_{12} \\ U_{21} & U_{22} \end{pmatrix},$$

and consider the fractional-linear transformation $F_U(K) = (U_{21} + U_{22}K)(U_{11} + U_{12}K)^{-1} : \mathcal{K} \to \mathcal{K}$. Since A is uniformly dissipative, $F_U(\mathcal{K})$ is strictly inside $\mathcal{K}$, i.e.

$$\operatorname{dist}(F_U(\mathcal{K}), \partial\mathcal{K}) > 0.$$

Hence [6] F_U has a unique fixed point $K_0 \in \mathcal{K}_0$. Evidently the corresponding subspace $L_-^0 \in \mathcal{M}_-^0$ and $UL_-^0 = L_-^0$, $AL_-^0 = L_-^0$. It is easy to check that U^* has the same property: $F_{U^*}(\mathcal{K})$ is strictly inside $\mathcal{K}$. Consequently there exists $\hat{L}_- \in \mathcal{M}_-^0$ such that

$$U^*\hat{L}_- = \hat{L}_-.$$

Taking $L_+^0 = \hat{L}_-$, we obtain

$$UL_+^0 \subset L_+^0 \text{ and } AL_+^0 \subset L_+^0.$$

Theorem 5 (see [18]). *Let A be a maximal J-dissipative operator in Π_κ with a dense domain:*

$$\overline{\mathcal{D}(A)} = \Pi_\kappa, \ \operatorname{Re}[Ax, x] \leq 0 \ , \quad x \in \mathcal{D}(A).$$

Then A generates a strongly continuous J-contractive semigroup $T(t)$.

Proof. Together with A the J-conjugate operator A^c $(= JA^*J)$ is maximal J-dissipative ([1], Proposition 2.7^0, page 116). Consider the maximal uniformly J-dissipative operator $B = A^c - J$. We have

$$\begin{aligned}\operatorname{Re}[(A^c - J)x, x] &= \operatorname{Re}[A^c x, x] - [Jx, x] \\ &= \operatorname{Re}[A^c x, x] - \| x \|^2 \leq - \| x \|^2, \ x \in \mathcal{D}(A^c).\end{aligned}$$

Using Lemma 5 we see that there exists $L_- \in \mathcal{M}_-^0$ such that $B^c L_- = L_-$. Taking the canonical decomposition

$$\mathcal{H} = L_- \dot{+} L_-^{[\perp]},$$

we have

$$B = \begin{pmatrix} B_{11} & B_{12} \\ 0 & B_{22} \end{pmatrix}, \ B^c = \begin{pmatrix} B_{11}^* & 0 \\ B_{12}^* & B_{22}^* \end{pmatrix}.$$

Since A and A^{cc} are maximal J-dissipative and $A \subset A^{cc}$, we get $A = A^{cc}$. Keeping in mind the inclusion $A - J \subset B^c$, we conclude that $A - J$ is of the form

$$A - J = \begin{pmatrix} \tilde{A}_{11} & 0 \\ \tilde{A}_{21} & \tilde{A}_{22} \end{pmatrix}.$$

Here $\tilde{A}_{11}$, $\tilde{A}_{21}$ are of finite rank and $\tilde{A}_{22}$ is dissipative:

$$\operatorname{Re}(\tilde{A}_{22}x, x) \leq 0, \ x \in \mathcal{D}(\tilde{A}_{22}) \ (\subset \mathcal{D}(A)).$$

Hence $A - J = \begin{pmatrix} \tilde{A}_{11} & 0 \\ \tilde{A}_{21} & 0 \end{pmatrix} + \begin{pmatrix} 0 & 0 \\ 0 & \tilde{A}_{22} \end{pmatrix}$ is a bounded perturbation of the dissipative operator $\begin{pmatrix} 0 & 0 \\ 0 & \tilde{A}_{22} \end{pmatrix}$, which generates a contractive strongly continuous semigroup. Consequently, $A - J$ generates a strongly continuous semigroup, and so $A = (A - J) + J$ (J is bounded!) also generates a strongly continuous semigroup $T(t)$. As a consequence of Lemma 4 we obtain

$$[T(t)x, \ T(t)x] \leq [x, x], \ x \in \Pi_\kappa.$$

Remark 2. It is well known (see for example, [19]) that any bounded dissipative operator generates a uniformly continuous contractive semigroup. The following example shows that even in the case of bounded uniformly J-dissipative operators this is no longer true.

Example 3. Consider $A = -J$. Then with respect to the canonical decomposition (5), $\mathcal{H} = \mathcal{H}_- \oplus \mathcal{H}_+$,

$$A = \begin{pmatrix} I_- & 0 \\ 0 & -I_+ \end{pmatrix}.$$

From

$$\mathrm{Re}[Ax, x] = \mathrm{Re}[-Jx, x] = - \| x \|^2,$$

it follows that A is uniformly J-dissipative. But for each $\epsilon > 0$,

$$I - \epsilon A = \begin{pmatrix} (1-\epsilon)I_- & 0 \\ 0 & (1+\epsilon)I_+ \end{pmatrix}$$

and $\| (I - \epsilon A)^{\pm n} \|$ is not uniformly bounded. So the semigroup $T(t)$ generated by A is not even bounded.

Now we are able to prove the following result.

Theorem 6. *Let $\{S(t) : t \geq 0\} \subset \Phi$ be a strongly continuous semigroup, and let min$\{\dim\mathcal{H}_1, \dim\mathcal{H}_2\} < \infty$. Let $D = \{K \in \mathcal{K}^0 : Kx_- \in D(A)$ for all $x = x_- + x_+ \in D(A)\}$. Then the generator g of $S(t)$ exists, is densely defined on $\mathcal{K}^0$, and*

$$g(K) = A_{21} + A_{22}K - KA_{11} - KA_{12}K,$$

for all $K \in D$, where $A_{ij} : \mathcal{H}_j \to \mathcal{H}_i$, $i, j = 1, 2$. In the case when $S(t)$ is a semigroup of isometries with respect to the Carathéodory metric on $\mathcal{K}^0$, then the operators A_{11} and A_{22} are skew-symmetric and A_{12}, A_{21} are conjugate to each other.

Proof. It is possible to show that D is dense in $\mathcal{K}^0$. We have for any $x \in \mathcal{D}(A)$ and $K \in D$,

$$\begin{aligned}
&\lim_{t\to 0^+} \left(\frac{S(t)K - K}{t} \right) x \\
&= \lim_{t\to 0^+} \left(\frac{(T_{21}(t) + T_{22}(t)K)(T_{11}(t) + T_{12}(t)K)^{-1} - K}{t} \right) x \\
&= \lim_{t\to 0^+} \left(\frac{T_{21}(t) + T_{22}(t)K - KT_{11}(t) - KT_{12}(t)K}{t} (T_{11}(t) + T_{12}(t)K)^{-1} \right) x \\
&= \lim_{t\to 0^+} \left(\frac{T_{21}(t)}{t} + \frac{T_{22}(t)K - K}{t} + \frac{K - KT_{11}(t)}{t} - \frac{KT_{12}(t)K}{t} \right) x \\
&= A_{21}x_- + A_{22}Kx_- - KA_{11}x_- - KA_{12}Kx_-.
\end{aligned}$$

Thus $g(K) = A_{21} + A_{22}K - KA_{11} - KA_{12}K$, $K \in D$, and $\overline{\mathcal{D}(g)} = \mathcal{K}^0$.

Consider now the case when $S(t)$ is a semigroup of isometries with respect to the Carathéodory metric on $\mathcal{K}^0$. Then $T(t)$ is a semigroup of J-isometries with respect to the indefinite metric $[x, y]$ in $\mathcal{H}$. Using Theorem 2 and the discussion preceding it we conclude that A_{ii}, $i = 1, 2$, are skew-symmetric operators and A_{12}, A_{21} are conjugate to each other.

References

[1] Azizov, T. and Iohvidov I., Foundations of Linear Operator Theory in Spaces with Indefinite Metrics, Nauka, Moscow, 1986 (in Russian).

[2] Berkson, E., Kaufman, R. and Porta, H., Möbius transformation of the disk and one-parameter groups of isometries of H^p, Trans. Amer. Math. Soc. 199 (1974), 223–238.

[3] Berkson, E. and Porta, H., Semigroups of analytic functions and composition operators, Michigan Math. J. 25 (1978), 101–115.

[4] Cowen, C. C. and MacCluer, B. D., Composition Operators on Spaces of Analytic Functions, CRC Press, Boca Raton, FL, 1995.

[5] Dineen, S., The Schwarz Lemma, Clarendon Press, Oxford, 1989.

[6] Earle, C. J. and Hamilton, R. S., A fixed-point theorem for holomorphic mappings, Proc. Symp. Pure Math., Vol. 16, AMS, Providence, RI, 1970, 61–65.

[7] Forelli, F., The isometries of H^p, Canad. J. Math. 16 (1964), 721–728.

[8] Jafari, F., Tonev, T., Toneva, E. and Yale, K., Holomorphic flows, cocycles and coboundaries, Michigan Math. J. 44(1997), 239–253.

[9] Khatskevich, V., Reich, S. and Shoikhet, D., Fixed point theorems for holomorphic mappings and operator theory in indefinite metric spaces, Integral Equations Operator Theory 22 (1995), 305–316.

[10] Khatskevich, V., On the symmetry of properties of the plus-operator and its conjugate operator, Functional Analysis, Ulyanovsk, Vol. 14 (1980), 177–186 (in Russian).

[11] Khatskevich, V. and Shoikhet, D., Differentiable Operators and Nonlinear Equations, Birkhäuser, Basel, 1994.

[12] Krein, M. G. and Shmuljan, Ju. L., Plus-operators in a space with an indefinite metric, Amer. Math. Soc. Transl. (2) 85 (1969), 93–113.

[13] Krein, M. G. and Shmuljan, Ju. L., On linear-fractional transformations with operator coefficients, Amer. Math. Soc. Transl. (2) 103 (1974), 125–152.

[14] Reich, S. and Shoikhet, D., Generation theory for semigroups of holomorphic mappings in Banach spaces, Abstract Applied Analysis 1 (1996), 1–44.

[15] Siskakis, A., Composition semigroups and the Cesàro operator on H^p, J. London Math. Soc. 36(1987), 153–164.

[16] Vesentini, E., Semigroups of holomorphic isometries, Advances in Math. 65 (1987), 272–306.

[17] Vesentini, E., Semigroups of holomorphic isometries, Complex Potential Theory, Kluwer Academic Publishers, Dordrecht, 1994, 475–548.

[18] Vesentini, E., Semigroups of linear contractions for an indefinite metric, Mem. Mat. Accad. Lincei 2 (1994), 53–83.

[19] Yosida, K., Functional Analysis, Springer, Berlin, 1965.

Department of Applied Mathematics, International College of Technology, 20101 Karmiel, Israel

Department of Mathematics, The Technion – Israel Institute of Technology, 32000 Haifa, Israel

Department of Applied Mathematics, International College of Technology, 20101 Karmiel, Israel

Operator Theory:
Advances and Applications, Vol. 123

Scattering from an Impurity: Lax-Phillips Approach

P. Kurasov

Abstract. A difference equation describing scattering from an impurity in a crystal is investigated. The time evolution consistent with the stationary operator leads to a difference wave equation, or linear version of the Hirota equation. Lax-Phillips scattering theory is developed for this evolution equation using a dynamical approach. It is shown that the Lax-Phillips scattering matrix for this special choice of the evolution equation coincides with the standard stationary scattering matrix.

1. Introduction

Scattering from an impurity in a crystal can be described by the Schrödinger equation with a periodic potential $V(x)$ perturbed by a decreasing at infinity potential $W(x)$

$$(-\Delta + V(x) + W(x))\,\psi(x,t) = \frac{1}{i}\frac{\partial\psi}{\partial t}. \tag{1}$$

The scattering problem amounts to comparing the evolution given by (1) with the unperturbed evolution given by the same equation, but with $W(x) = 0$. The existence and completeness of the wave operators can be shown for suitable W [1, 5, 7, 20]. The scattering matrix is an unitary function on the absolutely continuous spectrum of the periodic problem. An important problem is to understand the analytical properties of the scattering matrix on the Riemann surface of the quasimomentum. This Riemann surface has complicated structure because the corresponding Schrödinger operator has continuous band spectrum. Harmonic analysis for such Riemann surfaces has been developed by B. Pavlov and S. Fedorov [4, 15, 17, 18]. Analytical properties of the scattering matrix for the problems without periodic background can be studied with the help of the Lax-Phillips scattering theory [11]. This theory can not be applied to the problem (1) directly. The main reason is the absence of incoming and outgoing subspaces for this evolution. Such subspaces can be observed if we substitute the first derivative operator in the *rhs* of the equation (1) by another differential operator, which has the same essential spectrum as the stationary operator. This idea was proposed first by V. Evstratov and the author in 1989, but has never been developed. We present here this approach applied to partial difference operators. Lax-Phillips scattering theory is modified for the case of the stationary operator with the band continuous spectrum.

The importance of the discrete equations is explained by their numerous applications in theoretical physics (see [12] for references). Close relations between the continuous limits of the nonlinear partial difference equations and nonlinear partial differential equations, such as KdV and sinh-Gordon equations, have been discovered [6]. It appears that solutions to the discrete partial differential equations can be constructed with the help of the inverse scattering method. These equations became a useful tool during investigation of the chaotic phenomena. Exact solutions to these equations can be used to study transitions to chaos. We believe that construction of the Lax-Phillips theory for the scattering problems on the lattice can be used during the investigation of the geometrical and analytical properties of the simplectic maps, defined by nonlinear partial difference equations [3, 14].

The discrete stationary operator defined by a periodic selfadjoint Jacobi matrix is considered in this paper. Such operator with the band continuous spectrum is a discrete analog of the unperturbed Schrödinger operator $-\frac{d^2}{dx^2} + V(x)$. The impurity is described by an additive perturbation. Spectral and resonance properties of these operators have been studied by B. Pavlov and S. Fedorov[15, 16, 18, 19]. Their approach is based on the harmonic analysis on the Riemann surface of finite genius. We develop here the dynamical approach, based on a special form of the time evolution consistent with the stationary operator. The unperturbed evolution is defined on the discrete space-time lattice by the difference analog of the wave equation. This equation can be considered as a linear analogue of the Hirota equation [3, 6]. It is shown, that the difference wave equation has d'Alembert solutions. Incoming-outgoing translational representation of the evolution operator is constructed explicitly.

The evolution, described by the perturbed discrete wave equation is defined in the space of initial data with indefinite energy form. The Lax-Phillips scattering theory for this evolution group is developed in the spirit of [10]. The case of indefinite energy form for the operators with the band spectrum has not been considered. It is shown, that Lax-Phillips scattering theory can be constructed for the evolution restricted on the subspace, where the energy form is positive definite. Incoming and outgoing translational representations of the evolution are constructed explicitly using the dynamical approach. The scattering matrix is calculated in the spectral representation.

We show in Section 5, that the Lax-Phillips and the standard stationary scattering matrices differ by an inessential factor.

2. Time evolution for periodic Jacobi matrices, difference wave equation

We consider in this paper the stationary operator defined by a periodic selfadjoint Jacobi matrix $\mathcal{A}$ in ℓ_2. Matrix coefficients satisfy the following conditions

$\mathcal{A}_{m,n} = 0$ if $|m-n| > M$ for certain M;
$\mathcal{A}_{m+q,n+q} = \mathcal{A}_{m,n}$ for some fixed q;
$\bar{\mathcal{A}}_{m,n} = \mathcal{A}_{n,m}$.

Every such infinite matrix $\mathcal{A}$ can be presented in the tridiagonal block form

$$\mathcal{A} = \begin{pmatrix} \cdots & \cdots & \cdots & \cdots & \cdots & \cdots & \cdots \\ \cdots & A & \Gamma & 0 & 0 & 0 & \cdots \\ \cdots & \Gamma^* & A & \Gamma & 0 & 0 & \cdots \\ \cdots & 0 & \Gamma^* & A & \Gamma & 0 & \cdots \\ \cdots & 0 & 0 & \Gamma^* & A & \Gamma & \cdots \\ \cdots & 0 & 0 & 0 & \Gamma^* & A & \cdots \\ \cdots & \cdots & \cdots & \cdots & \cdots & \cdots & \cdots \end{pmatrix}$$

with A, Γ $N \times N$ matrices. The dimension N of the matrices can be chosen equal to $N = Mq$. It is natural to consider this operator in the space $\ell_2(\mathbf{C}^N) = \ell_2 \otimes \mathbf{C}^N$. This operator can be written in the following form using the shift operator $(T_\pm \psi)(n) = \psi(n \mp 1)$

$$\mathcal{A} = A + T_-\Gamma + T_+\Gamma^*. \tag{2}$$

We restrict our consideration to the case of selfadjoint matrices $\Gamma = \Gamma^*$.

Additive perturbations of the operator $\mathcal{A}$ by finite dimensional matrices $\mathcal{B}$ will be investigated. Only one special one dimensional family of such matrices will be considered $\mathcal{B}_{n,m} = \delta_{n,0}\delta_{m,0}\alpha\Gamma$, $\alpha \in \mathbf{R}$. Thus the following stationary discrete operator will be studied

$$\mathcal{A}_\alpha = A + \alpha\Gamma\delta + (T_+ + T_-)\Gamma, \tag{3}$$

where δ denotes the indicator of the origin: $(\delta\psi)(n) = \delta_{n,0}\psi(0)$.

Consistent with the stationary operator (2) a time evolution can be defined on the discrete space-time lattice $\Lambda = \{(n,t), n \in \mathbf{Z}, t \in \mathbf{Z}\}$ by the following partial difference equation

$$\left(A + (T_- + T_+)\Gamma\right) f = \left(A + (\mathcal{T}_- + \mathcal{T}_+)\Gamma\right) f, \tag{4}$$

where $\mathcal{T}_\pm$ defines the shift operators in the time direction $(\mathcal{T}_\pm f)(n,t) = f(n, t \mp 1)$. The latter equation can be simplified to get the difference wave equation

$$(T_- + T_+)f = (\mathcal{T}_- + \mathcal{T}_+)f. \tag{5}$$

Different components of the N-dimensional vectors $f(n,t)$ on the space-time lattice are independent. That's why we are going to study this equation for one dimensional vectors $f(n,t)$ instead of working all the time in the tensor representation. This equation will be called *difference wave equation*. It is a linear version of the Hirota equation [3, 6]. Lax-Phillips scattering theory for this equation will be constructed in Section 3.

The time evolution for the perturbed problem will be introduced in a similar way using the same unperturbed operator to connect the wave function at different time moments

$$(A + \alpha\Gamma\delta + \Gamma(T_+ + T_-))\, f = (A + \Gamma(\mathcal{T}_+ + \mathcal{T}_-))\, f. \tag{6}$$

This equation can be simplified as follows

$$(\alpha\delta + (T_+ + T_-))f = (\mathcal{T}_+ + \mathcal{T}_-)f. \tag{7}$$

The latter equation will be called *the perturbed difference wave equation.* It is possible to simplify eq. (6) due to the special form of the chosen perturbation. Different components of the N-dimensional vectors $f(n,t)$ are independent again and we are going to study this equation in the case $N = 1$. The case of arbitrary N can be considered in a similar way. The Lax-Phillips approach to this equation will be developed in Section 4.

3. Lax-Phillips theory for the difference wave equation

We consider the Lax-Phillips scattering theory for the evolution governed by the discrete wave equation (5). This equation defines the unperturbed evolution group $\{U_0^T\}$ for the discrete values of the time parameter T. Initial data for the evolution at the moment t are two infinite vectors $F(t) = \begin{pmatrix} \{f(n,t)\} \\ \{f(n,t+1)\} \end{pmatrix}$. The evolution equation (5) connects values of the solution at the points (n,t) with the same evenness of $n+t$. The evolution can be considered independently on the even and odd sublattices $\Lambda = \Lambda^{even} \bigcup \Lambda^{odd}$, $\Lambda^{even} = \{(n,t) : (-1)^{n+t} = 1\}$, $\Lambda^{odd} = \{(n,t) : (-1)^{n+t} = -1\}$. Let us introduce the following indicator of the even lattice $\chi(n,m) = \frac{1}{2}((-1)^{n+m} + 1)$. The solution of the evolution equation can be written as

$$f(n,t+T) = \sum_{\alpha=0,1} (-1)^{\alpha+T+1} \sum_{|m-n|<T} \chi(n-m, T-\alpha) f(m, t+\alpha), \quad T \geq 2. \tag{8}$$

Only initial data at the points $(m,t), (m,t-1)$ with the same evenness as $(n,t+T)$ from the region $|m-n| < T$ define the value of the solution $f(n,t+T)$. This property of the solution can be called "hyperbolicity" [3]. The same equation describes the evolution in the space direction. The problem under consideration is similar to one studied in [2].

The energy norm for this evolution equation is given by the following expression

$$\| F \|^2_{\mathcal{E}_0}(t) = \sum_{n\in\mathbf{Z}} \left(|f(n,t+1) - f(n+1,t)|^2 + |f(n,t+1) - f(n-1,t)|^2 \right). \tag{9}$$

This quadratic form is nonnegative and vanishes on the functions which are equal to constants on the sublattices

$$K = \{F = \begin{pmatrix} f(n,0), \\ f(n,1) \end{pmatrix} : f(n,\alpha) = a^{even}\chi(n,\alpha) + a^{odd}\chi(n+1,\alpha), \alpha = 0,1\}.$$

We are going to consider the space $\mathcal{H}_0$ of all initial data with the finite energy factorized by the kernel K. The energy form defines positive functional which can be used as the norm in this space. The generator of the unperturbed evolution group

$$\mathbf{U}_0 = \begin{pmatrix} 0 & 1 \\ -1 & T_+ + T_- \end{pmatrix}. \tag{10}$$

is an unitary operator in the space $\mathcal{H}_0$ with the energy norm.

Lemma 1. *Every solution of the evolution equation (5) can be represented by the linear combination of d'Alembert waves*

$$f(n,t) = v^{\rightarrow}(t-n) + v^{\leftarrow}(t+n). \tag{11}$$

Proof. It is clear that every function $f(n,t)$ possessing the representation (11) satisfies equation (5). Substitution of the ansatz (11) into the initial conditions

$$\begin{cases} f(n,0) &= f^0(n) \\ f(n,1) &= f^1(n) \end{cases}$$

gives the following equation for the functions $v^{\rightarrow}$ and $v^{\leftarrow}$

$$v^{\rightarrow}(-n) + v^{\leftarrow}(n) = f^0(n); \qquad v^{\rightarrow}(1-n) + v^{\leftarrow}(1+n) = f^1(n). \tag{12}$$

The latter equation does not define the functions $v^{\rightarrow}$ and $v^{\leftarrow}$ uniquely and two constants can be chosen arbitrarily, since the kernel K is two dimensional. One can put for example

$$v^{\leftarrow}(0) = v^{\rightarrow}(0) = \frac{f^0(0)}{2}, \quad v^{\leftarrow}(1) = v^{\rightarrow}(1) = \frac{f^1(0)}{2} \tag{13}$$

Difference derivatives of the functional invariant solutions on the even and odd sublattices can be calculated using equation (12)

$$\begin{aligned} Dv^{\rightarrow}(n) &= v^{\rightarrow}(n+2) - v^{\rightarrow}(n) = f^1(-n-1) - f^0(-n); \\ Dv^{\leftarrow}(n) &= v^{\leftarrow}(n+2) - v^{\leftarrow}(n) = f^1(n+1) - f^0(n). \end{aligned} \tag{14}$$

Functions $v^{\rightarrow}$ and $v^{\leftarrow}$ can be reconstructed from their derivatives and values at the points $s = 0, 1$:

$$\begin{cases} v(2n) = v(0) + \sum_{l=1}^n (Dv)(2l-2); \\ v(-2n) = v(0) - \sum_{l=1}^n (Dv)(-2l); \\ v(2n+1) = v(1) + \sum_{l=1}^n (Dv)(2l-1); \\ v(-2n+1) = v(1) - \sum_{l=1}^n (Dv)(-2l+1); \end{cases} \qquad n > 0. \tag{15}$$

Lemma 1 is proven. □

The norm of the derivatives of the functional invariant solutions coincides with the energy norm of initial data:

$$\begin{aligned}(16)\quad \| Dv^{\rightarrow} \|^2_{\ell_2} + \| Dv^{\leftarrow} \|^2_{\ell_2} \\ &= \sum_{n\in\mathbf{Z}} \left(|f^1(-n-1) - f^0(-n)|^2 + |f^1(n+1) - f^0(n)|^2\right) \\ &= \sum_{n\in\mathbf{Z}} \left(|f^1(n) - f^0(n+1)|^2 + |f^1(n) - f^0(n-1)|^2\right) \\ &= \| (f^0, f^1) \|^2_{\mathcal{E}} .\end{aligned}$$

Theorem 1. *The mapping*

$$\Phi_0 : (f^0, f^1) \Rightarrow h(s) = \begin{pmatrix} (Dv^{\rightarrow})(-s) \\ (Dv^{\leftarrow})(-s) \end{pmatrix}$$

defines a unitary translational representation for the group $\{\mathbf{U}_0^t\}$, *which is both incoming and outgoing.*

Proof. Here $h(s)$ is an element of $\ell_2(\mathbf{Z}; \mathbf{C}^2)$. It is necessary to show that

$$\Phi_0 \ \mathbf{U}_0^T = \mathbf{T}_+^T \ \Phi_0$$

to prove that this representation is a translational one. We introduce the following notation

$$u_T(n,t) = u(n, t+T).$$

Then

$$u_T(n,t) = v_T^{\rightarrow}(t-n) + v_T^{\leftarrow}(t+n),$$

where functions $v_T^{\rightarrow}(s)$ and $v_T^{\leftarrow}(s)$ can be defined using the following formulas for even s

$$v_T^{\leftarrow}(s) = v^{\leftarrow}(s+T) + \frac{1}{2}\left(v^{\rightarrow}(T) - v^{\leftarrow}(T)\right);$$

$$v_T^{\rightarrow}(s) = v^{\rightarrow}(s+T) - \frac{1}{2}\left(v^{\rightarrow}(T) - v^{\leftarrow}(T)\right)$$

and for odd s

$$v_T^{\leftarrow}(s) = v^{\leftarrow}(s+T) + \frac{1}{2}\left(v^{\rightarrow}(1+T) - v^{\leftarrow}(1+T)\right);$$

$$v_T^{\rightarrow}(s) = v^{\rightarrow}(s+T) - \frac{1}{2}\left(v^{\rightarrow}(1+T) - v^{\leftarrow}(1+T)\right).$$

Functions $v_T^{\rightarrow}(s)$ and $v_T^{\leftarrow}(s)$ can be obtained from the functions $v^{\rightarrow}(s)$ and $v^{\leftarrow}(s)$ by shifting the argument and extracting some constant, which does not give any contribution into the derivative. It follows that $h_T(s) = h(s-T)$.

We shall prove now that subspace $\ell_2(1, \infty; \mathbf{C}^2)$ is outgoing. Consider first the even sublattice and suppose, that $h(0) = h(-2) = h(-4) = \cdots = 0$. Both d'Alembert waves are equal to a certain constant on the negative part of the even sublattice

$$v^{\rightarrow}(0) = v^{\leftarrow}(0) = v^{\rightarrow}(2) = v^{\leftarrow}(2) = v^{\rightarrow}(4) = \dots$$

and this common value can be denoted by a^{even}. It implies that $u(n,t) = 2a^{even}$ for $|n| \leq t$ and this solution is equivalent to the solution vanishing on the even sublattice in the region $|n| \leq t$. The odd sublattice can be considered in a similar way. The unitary character of the considered representation follows from the formula (16). □

Such defined incoming and outgoing subspaces lead to the trivial Lax-Phillips scattering matrix. It shows that the chosen time evolution is consistent with the stationary operator. We note that consistent time evolution is not defined in a unique way.

4. Lax-Phillips theory for the perturbed difference wave equation

We consider in this section the Lax-Phillips scattering theory for the evolution, described by the perturbed difference wave equation (7). The perturbed evolution group defined by this equation will be denoted by $\{\mathbf{U}^T\}$. Evolutions on the even and odd sublattices of the space-time lattice are not independent anymore.

We introduce the bilinear energy form

$$\begin{aligned}(17)\qquad \langle\!\langle F, G\rangle\!\rangle_{\mathcal{E}}(t) = \sum_{n\in\mathbf{Z}} \Big(&\langle f(n,t+1) - f(n+1,t), g(n,t+1) - g(n+1,t)\rangle \\ &+ \langle f(n,t+1) - f(n-1,t), g(n,t+1) - g(n-1,t)\rangle \Big) \\ &- \alpha\left(\langle f(0,t+1), g(0,t)\rangle + \langle f(0,t), g(0,t+1)\rangle\right).\end{aligned}$$

The corresponding quadratic form is preserved during the evolution, but it is not positive definite. We are going to consider the case of positive α only. Negative α can be studied in a similar way.

The generator of the perturbed evolution is equal to

$$(18)\qquad \mathbf{U} = \begin{pmatrix} 0 & 1 \\ -1 & T_+ + T_- + \alpha\{\delta_{n,0}\} \end{pmatrix}.$$

This operator has two eigenvalues with nonunit norm $z_\pm = e^{\mp b}$, where $e^b - e^{-b} = \alpha$. The corresponding eigenvectors are equal to

$$\Phi^\pm = \begin{pmatrix} \{e^{-b|n|}\} \\ e^{\mp b}\{e^{-b|n|}\} \end{pmatrix}.$$

These eigenvectors are related to the eigenvectors of the stationary operator (see Section 5).

Lemma 2. *Eigenvectors $\Phi^\pm$ satisfy the following conditions with respect to the energy form $\langle\langle *, * \rangle\rangle_{\mathcal{E}}$:*

$$\langle\langle \Phi^\pm, \Phi^\pm \rangle\rangle_{\mathcal{E}} = 0, \tag{19}$$

$$\langle\langle \Phi^\pm, \Phi^\mp \rangle\rangle_{\mathcal{E}} = -e^{2b} + e^{-2b} \neq 0. \tag{20}$$

Proof. The first two equalities (19) follow from the fact, that $\Phi^\pm$ are eigenvectors of the evolution operator corresponding to the nonunit eigenvalues.

Equality (20) follows from the following calculations

$$\langle\langle \Phi^+, \Phi^- \rangle\rangle_{\mathcal{E}} = \alpha(-2 + (e^{-b} - 1)(e^b - 1)) = -(e^{2b} - e^{-2b}) \neq 0.$$

The lemma is proven. □

We shall use the following definition

Definition 1. Subspace B of initial data is a finite dimensional subspace of all linear combinations of the vectors $\Phi^\pm$.

There is no element in this subspace which is orthogonal to the whole B, i.e., the energy form is nondegenerate on this subspace. Subspace B is invariant subspace for the evolution operator $\mathbf{U}^T$. The evolution operator restricted on this subspace is a multiplication by diagonal matrix. Let $F \in B$, then

$$F(t) = f^+(t)\Phi^+ + f^-(t)\Phi^-.$$

The operator in this basis is equal to

$$\begin{pmatrix} f^+(t+T) \\ f^-(t+T) \end{pmatrix} = \begin{pmatrix} e^{-bT} & 0 \\ 0 & e^{bT} \end{pmatrix} \begin{pmatrix} f^+(t) \\ f^-(t) \end{pmatrix}.$$

Lemma 3. *The subspace B consists of all elements from $\ell_2 \oplus \ell_2$ of the following form*

$$B = \{e^{-b|n|} \begin{pmatrix} g_0 \\ g_1 \end{pmatrix} : g_0, g_1 \in \mathbf{C}\}.$$

Proof. Every element from B possesses the desired representation. Consider an arbitrary element $F = e^{-b|n|} \begin{pmatrix} f_0 \\ f_1 \end{pmatrix}$. If it belongs to the subspace B , then the coordinates of this vector f^+, f^- should satisfy the following linear equation

$$\begin{cases} f^+ + f^- = f_0 \\ e^{-b} f^+ + e^b f^- = f_1 \end{cases}.$$

The determinant of the linear system is equal to $e^b - b^{-b} = |\alpha|$. Thus the coordinates $f^\pm$ can be calculated and F belongs to B. □

We are going to consider the evolution group restricted on the following subspace.

Definition 2. Subspace H' is the subspace of all initial data which are $\mathcal{E}$-orthogonal to B.

This subspace is invariant with respect to the evolution operator because the subspace B is invariant with respect to the reverse evolution. Every element from H has unique $\mathcal{E}$-projection on H'. Orthogonal projector on the space H' will be denoted by $\mathbf{P}$.

Lemma 4. *The invariant subspace H' consists of the elements from the space H satisfying the following conditions*

$$\sum_n f(n,0)e^{-b|n|} = 0, \qquad \sum_n f(n,1)e^{-b|n|} = 0. \tag{21}$$

Proof. Let $F \in H'$, then it is orthogonal to all elements G from B. This condition is equivalent to the orthogonality condition with the elements $G_0 = e^{-b|n|}(1,0)$ and $G_1 = e^{-b|n|}(0,1)$

$$2\sum_n f(n,0)e^{-b|n|} - (e^{-b}+e^{b})\sum_n f(n,1)e^{-b|n|} = 0, \tag{22}$$

$$-(e^{-b}+e^{b})\sum_n f(n,0)e^{-b|n|} + 2\sum_n f(n,1)e^{-b|n|} = 0. \tag{23}$$

We get a system of linear equations for the elements

$$\sum_n f(n,0)e^{-b|n|} \text{ and } \sum_n f(n,1)e^{-b|n|}.$$

The determinant of the linear system (22,23) is equal to $-\alpha^2 \neq 0$. It follows that the unique solution of this linear system is trivial, which leads to the relations (21). $\square$

Theorem 2. *The quadratic energy form is nonnegative on the subspace H' and it is equivalent to the unperturbed energy norm.*

Proof. Lemma 4 shows, that the values of the elements $F = (\{f(n,0), f(n,1)\})$ at the origin can be calculated from their difference derivatives. Conditions (21) can be written in the following form using notation $z = e^{-b}$

$$\begin{aligned} 0 &= f(0,0) + f(1,0)z + f(2,0)z^2 + \cdots + f(-1,0)z + f(-2,0)z^2 + \cdots \\ & f(0,0)(1+\frac{2z^2}{1-z^2}) + f(0,1)\frac{2z}{1-z^2} \\ & + \sum_{n\geq 1}(f(n,0) - f(n-1,1) + f(-n,0) - f(-n+1,1))\frac{z^n}{1-z^2} \\ & + \sum_{n\geq 1}(f(n,1) - f(n-1,0) + f(-n,1) - f(-n+1,0))\frac{z^{n+1}}{1-z^2}, \end{aligned} \tag{24}$$

$$0 = f(0,1)(1+\frac{2z^2}{1-z^2}) + f(0,0)\frac{2z}{1-z^2} \tag{25}$$

$$+\sum_{n\geq 1}(f(n,1)-f(n-1,0)+f(-n,1)-f(-n+1,0))\frac{z^n}{1-z^2}$$

$$+\sum_{n\geq 1}(f(n,0)-f(n-1,1)+f(-n,0)-f(-n+1,1))\frac{z^{n+1}}{1-z^2}.$$

These conditions are equivalent to the linear system

$$\begin{cases}(1+z^2)f(0,0)+2zf(0,1)=-zA-z^2B\\ 2zf(0,0)+(1+z^2)f(0,1)=-z^2A-zB\end{cases},$$

where the following notations have been used:

$$A=\sum_{n\geq 1}(f(n,1)-f(n-1,0)+f(-n,1)-f(-n+1,0))z^{n-1};$$

$$B=\sum_{n\geq 1}(f(n,0)-f(n-1,1)+f(-n,0)-f(-n+1,1))z^{n-1}.$$

This linear system for $f(0,0), f(1,0)$ can be easily solved

$$\begin{cases}f(0,0)=\dfrac{-zA+z^2B}{1-z^2}\\ f(0,1)=\dfrac{z^2A-zB}{1-z^2}\end{cases}.$$

Then the following is valid for the perturbed energy form

$$\begin{aligned}\langle\!\langle F,F\rangle\!\rangle_{\mathcal{E}} &= \langle\!\langle F,F\rangle\!\rangle_{\mathcal{E}_0}-\frac{1}{z(1-z^2)}\left(-2z^3(|A|^2+|B|^2)+z^2(z^2+1)(A\bar{B}+B\bar{A})\right)\\ &\geq \langle\!\langle F,F\rangle\!\rangle_{\mathcal{E}_0}-\frac{z(1-z)}{1+z}(|A|^2+|B|^2)\\ &\geq \left(1-\frac{2z(1-z)}{1+z}\right)\langle\!\langle F,F\rangle\!\rangle_{\mathcal{E}_0}\\ &= \frac{(4z-1)^2+7}{8(1+z)}\langle\!\langle F,F\rangle\!\rangle_{\mathcal{E}_0}\\ &\geq 0.\end{aligned}$$

We used here the following estimate: $|A|^2+|B|^2\leq 2\langle\!\langle F,F\rangle\!\rangle_{\mathcal{E}_0}$. The upper estimate for the quadratic form can be proven as follows

$$\begin{aligned}\langle\!\langle F,F\rangle\!\rangle_{\mathcal{E}} &\leq \langle\!\langle F,F\rangle\!\rangle_{\mathcal{E}_0}+\frac{z}{1-z^2}\left|-2z(|A|^2+|B|^2)+(z^2+1)(A\bar{B}+B\bar{A})\right|\\ &\leq \langle\!\langle F,F\rangle\!\rangle_{\mathcal{E}_0}+z\frac{(1+z)^2}{(1-z^2)}(|A|^2+|B|^2)\\ &\leq \frac{(4z+1)^2+7}{8(1-z)}\langle\!\langle F,F\rangle\!\rangle_{\mathcal{E}_0}.\end{aligned}$$

This finishes proof of the theorem. □

Consider the subspaces D_{inc} and $D_{out} \in H$ formed by the initial data which define solutions of the evolution equation, vanishing in the corresponding regions $|n| \leq -t,\ t < 0$ and $|n| \leq t,\ t > 0$. These subspaces do not belong to H'. Projections of these subspaces on the space H' will define the incoming and outgoing subspaces for the restricted evolution.

Lemma 5. *The subspace D_{inc} consists of all initial data satisfying the following conditions*

$$\begin{array}{ll} f(n,0) = f(n+1,1), n \leq -2; & f(n,0) = f(n-1,1), n \geq 2; \\ f(0,1) = f(-1,0) + f(1,0); & f(0,0) = 0. \end{array}$$

The subspace D_{out} consists of all initial data satisfying the conditions

$$\begin{array}{ll} f(n,0) = f(n-1,1), n \leq -1; & f(n,0) = f(n+1,1), n \geq 1; \\ f(0,0) = f(-1,1) + f(1,1); & f(0,1) = 0. \end{array}$$

Proof can be carried out by induction starting from the origin.

These two subspaces are invariant with respect to the evolution and reverse evolution operators correspondingly

$$\mathbf{U}D_{out} \subset D_{out}, \quad \mathbf{U}^{-1}D_{inc} \subset D_{inc}.$$

The energy norm is positive definite on each of these subspaces. It is possible to define special bases in the subspaces D_{inc} and D_{out}, such that the evolution operator will act as translation in the corresponding representations. Consider the orthogonal complement of the subspace $\mathbf{U}D_{out}$ in the subspace $D_{out} : N_{out} = D_{out} \ominus \mathbf{U}D_{out}$. Subspace N_{out} is a finite dimensional subspace. Vectors

$$E_{-1}^{out,r} = \begin{pmatrix} \Theta(m)\chi(0,m) \\ \Theta(m-1)\chi(1,m) \end{pmatrix}, \quad E_{-1}^{out,l} = \begin{pmatrix} \Theta(-m)\chi(0,m) \\ \Theta(-m-1)\chi(1,m) \end{pmatrix}$$

form a basis in the subspace N_{out}. Then the basis associated with the outgoing subspace can be chosen equal to the following one for all $n \geq -1$

$$E_n^{out,l} = \begin{pmatrix} e_n^{out,l}(m,0) \\ e_n^{out,l}(m,1) \end{pmatrix} = \begin{pmatrix} \Theta(-n-m-1)\chi(n,m+1) \\ \Theta(-n-m-2)\chi(n,m) \end{pmatrix}, n = -1,0,1,2,\dots;$$

$$E_n^{out,r} = \begin{pmatrix} e_n^{out,r}(m,0) \\ e_n^{out,r}(m,1) \end{pmatrix} = \begin{pmatrix} \Theta(-n+m-1)\chi(n,m+1) \\ \Theta(-n+m-2)\chi(n,m) \end{pmatrix}, n = -1,0,1,2,\dots.$$

This outgoing basis can be extended for all negative n using the translational property, which is valid for all $n \geq 0 :\ E_{n-1}^{out} = \mathbf{U}^{-1}E_n^{out}$.

Similar translational basis in the subspace D_{inc} is formed by the elements

$$E_n^{inc,l} = \begin{pmatrix} e_n^{inc,l}(m,0) \\ e_n^{inc,l}(m,1) \end{pmatrix} = \begin{pmatrix} \Theta(n-m-1)\chi(n,m+1) \\ \Theta(n-m)\chi(n,m) \end{pmatrix}, \ n = 0,-1,-2,\dots;$$

$$E_n^{inc,r} = \begin{pmatrix} e_n^{inc,r}(m,0) \\ e_n^{inc,r}(m,1) \end{pmatrix} = \begin{pmatrix} \Theta(n+m-1)\chi(n,m+1) \\ \Theta(n+m)\chi(n,m) \end{pmatrix}, \ n = 0,-1,-2,\dots.$$

One can use the translational property to define the incoming basis vectors for all $n \in \mathbf{Z}$.

Consider the projections of these subspaces and corresponding bases on H'. The following notations will be used in the future

$$D'_{out} = \mathbf{P}D_{out};\; D'_{inc} = \mathbf{P}D_{inc};$$

$$\mathcal{E}_n^{inc,l} = \mathbf{P}E_n^{inc,l};\; \mathcal{E}_n^{inc,r} = \mathbf{P}E_n^{inc,r};\; \mathcal{E}_n^{out,l} = \mathbf{P}E_n^{out,l};\; \mathcal{E}_n^{out,r} = \mathbf{P}E_n^{out,r}.$$

Lemma 6. *The subspaces D'_{inc} and D'_{out} span the space H'.*

Proof. To prove this Lemma, it is enough to show, that subspaces D_{inc}, D_{out} span the space H. Consider arbitrary element F from H and solution of the homogeneous equation with these initial data. Every such solution can be presented by the combination of the d'Alembert waves (Lemma 1) and certain vectors from the kernel $\mathbf{K}$

$$F = \begin{pmatrix} v^{\rightarrow}(-n) + v^{\leftarrow}(n) + a^{even}\chi(n,0) + a^{odd}\chi(n,1) \\ v^{\rightarrow}(1-n) + v^{\leftarrow}(1+n) + a^{even}\chi(n,1) + a^{odd}\chi(n,0) \end{pmatrix}.$$

Every element from the kernel $\mathbf{K}$ and d'Alembert waves are equal to the linear combination of the basis elements $E_{n-1}^{out}, E_{-n}^{inc}, n = 0,1,2,\cdots$:

$$\begin{aligned}(26)\quad F = {} & v^{\rightarrow}(0)E_{-1}^{out,r} + v^{\rightarrow}(-1)E_0^{out,r} + \sum_{n\geq 1}(v^{\rightarrow}(-n-1) - v^{\rightarrow}(-n+1))E_n^{out,r} \\ & + v^{\rightarrow}(1)E_0^{inc,l} + v^{\rightarrow}(2)E_{-1}^{inc,l} + \sum_{n\geq 2}(v^{\rightarrow}(n+1) - v^{\rightarrow}(n-1))E_{-n}^{inc,l} \\ & + v^{\leftarrow}(0)E_{-1}^{out,l} + v^{\leftarrow}(-1)E_0^{out,l} + \sum_{n\geq 1}(v^{\leftarrow}(-n-1) - v^{\leftarrow}(-n+1))E_n^{out,l} \\ & + v^{\leftarrow}(1)E_0^{inc,r} + v^{\leftarrow}(2)E_{-1}^{inc,r} + \sum_{n\geq 2}(v^{\leftarrow}(n+1) - v^{\leftarrow}(n-1))E_{-n}^{inc,r} \\ & + a^{odd}(E_0^{inc,l} + E_0^{out,r}) + a^{even}(E_{-1}^{out,r} + E_{-1}^{inc,l}).\end{aligned}$$

The last sum converges for every element with the finite energy norm. It follows, that the spaces D'_{out}, D'_{inc} span together the space H'. □

Theorem 3. *The subspaces D'_{inc} and D'_{out} are correspondingly incoming and outgoing subspaces for the perturbed evolution in H'.*

Proof. We are going to prove, that subspace D'_{out} is outgoing in the Lax-Phillips sense. We have to show, that

1. $\mathbf{U}D'_{out} \subset D'_{out}$,
2. $\cap \mathbf{U}^T D'_{out} = \{0\}$,
3. $\cup \mathbf{U}^T D'_{out}$ is dense in H'.

The first two properties can be proven using similar properties of the subspace D_{out}. The perturbed evolution restricted on the subspace D_{out} coincides with the unperturbed one. Hence the first two properties follow from the similar results for the unperturbed evolution.

Consider the third property now. The subspaces D'_{out} and D'_{inc} span the space H'. Hence it is enough to show that every vector from D'_{inc} is an element of $\cup \mathbf{U}^T D'_{out}$. We are going to show, that every basis element $\mathcal{E}_n^{inc,l}$ belongs to the closure of $\cup \mathbf{U}^T D'_{out}$ in the energy norm, i.e., that there exist constants a_n^{ll}, a_n^{lr}, such that the following representation converges in the energy norm

$$\mathcal{E}_n^{inc,l} = \sum_k \left(a_{n-k}^{ll} \mathcal{E}_k^{out,l} + a_{n-k}^{lr} \mathcal{E}_k^{out,r} \right). \tag{27}$$

Coefficients in the last representation depend on the difference $n-k$ because the evolution operator acts as translation in both incoming and outgoing bases. The same result for the vectors $\mathcal{E}_n^{inc,r}$ will follow from the symmetry of the problem with respect to the point zero. Suppose, that representation (27) is proven, then the following formula is valid also

$$\mathcal{E}_n^{inc,r} = \sum_k \left(a_{n-k}^{rr} \mathcal{E}_k^{out,r} + a_{n-k}^{rl} \mathcal{E}_k^{out,l} \right), \quad a_k^{ll} = a_k^{rr}, \; a_k^{rl} = a_k^{lr}. \tag{28}$$

We prove the existence of such representations for $n = 0$ first. The same property for arbitrary n will follow from the fact, that these bases define translational representations of the evolution operator.

Consider first the following property of the basis elements

$$E_0^{inc,l} - E_0^{inc,r} = E_0^{out,l} - E_0^{out,r}.$$

The same property is valid for their projections and we get the following condition on the constants a^{ll}, a^{rl}

$$a_k^{ll} - a_k^{rl} = \delta_{k,0}. \tag{29}$$

Consider now the elements from the subspace B. The following representations are valid for $\Phi^{\pm}$

$$\begin{aligned} \Phi^+ &= \Delta_0 - e^{-b}\Delta_1 + e^{-b}(E_0^{inc,r} + E_0^{inc,l}) \\ &\quad + e^{-2b}(E_{-1}^{inc,r} + E_{-1}^{inc,l}) + e^{-b} \sum_{m \geq 2} (E_{-m}^{inc,r} + E_{-m}^{inc,l}) e^{-mb}(1 - e^{2b}); \end{aligned} \tag{30}$$

$$\begin{aligned} \Phi^- &= -\Delta_0 + e^{b}\Delta_1 + (E_{-1}^{out,r} + E_{-1}^{out,l}) \\ &\quad + e^{-b}(E_0^{out,r} + E_0^{out,l}) + e^{-b} \sum_{m \geq 1} (E_m^{out,r} + E_m^{out,l}) e^{-mb}(1 - e^{2b}), \end{aligned} \tag{31}$$

where $\Delta_0 = \begin{pmatrix} \{\delta_{n,0}\} \\ 0 \end{pmatrix}; \Delta_1 = \begin{pmatrix} 0 \\ \{\delta_{n,0}\} \end{pmatrix}.$

Projections of $\Phi^\pm$ on H' are equal to zero and projections of the vectors Δ_0, Δ_1 can be calculated

$$\begin{aligned}(32)\quad \mathbf{P}\Delta_1 = \frac{-1}{\alpha}\Big\{& e^{-b}(\mathcal{E}_0^{inc,r} + \mathcal{E}_0^{inc,l}) + e^{-2b}(\mathcal{E}_{-1}^{inc,r} + \mathcal{E}_{-1}^{inc,l}) \\ &+ e^{-b}\sum_{m\geq 2}(\mathcal{E}_{-m}^{inc,r} + \mathcal{E}_{-m}^{inc,l})e^{-mb}(1-e^{2b}) + (\mathcal{E}_{-1}^{out,r} + \mathcal{E}_{-1}^{out,l}) \\ &+ e^{-b}(\mathcal{E}_0^{out,r} + \mathcal{E}_0^{out,l}) + e^{-b}\sum_{m\geq 1}(\mathcal{E}_m^{out,r} + \mathcal{E}_m^{out,l})e^{-mb}(1-e^{2b})\Big\};\end{aligned}$$

$$\begin{aligned}(33)\quad \mathbf{P}\Delta_0 = \frac{-1}{\alpha}\Big\{& (\mathcal{E}_0^{inc,r} + \mathcal{E}_0^{inc,l}) + e^{-b}(\mathcal{E}_{-1}^{inc,r} + \mathcal{E}_{-1}^{inc,l}) \\ &+ \sum_{m\geq 2}(\mathcal{E}_{-m}^{inc,r} + \mathcal{E}_{-m}^{inc,l})e^{-mb}(1-e^{2b}) + e^{-b}(\mathcal{E}_{-1}^{out,r} + \mathcal{E}_{-1}^{out,l}) \\ &+ e^{-2b}(\mathcal{E}_0^{out,r} + \mathcal{E}_0^{out,l}) + e^{-2b}\sum_{m\geq 1}(\mathcal{E}_m^{out,r} + \mathcal{E}_m^{out,l})e^{-mb}(1-e^{2b})\Big\}.\end{aligned}$$

We use the following identity for the basis elements

$$\begin{aligned}(34)\quad & E_0^{inc,l} + E_0^{inc,r} + E_{-1}^{inc,l} + E_{-1}^{inc,r} \\ &= E_{-1}^{out,l} + E_{-1}^{out,r} + E_0^{out,l} + E_0^{out,r} + 2(\Delta_1 - \Delta_0).\end{aligned}$$

The latter identity is valid also for the projections of basis elements. Substitution of the representation (27) into the last equality together with the property (29) leads to the following equation on the coefficients a_n^{rl}

(35)

$$\begin{aligned}&\sum_k(\delta_{k,0} + 2a_{-k}^{rl})(\mathcal{E}_k^{out,l} + \mathcal{E}_k^{out,r}) \\ &+ \sum_k(\delta_{k+1,0} + 2a_{-1-k}^{rl})(\mathcal{E}_k^{out,l} + \mathcal{E}_k^{out,r}) - (\mathcal{E}_{-1}^{out,l} + \mathcal{E}_{-1}^{out,r}) - (\mathcal{E}_0^{out,l} + \mathcal{E}_0^{out,r}) \\ &= \frac{2(1-e^b)}{\alpha}\Big\{\sum_k(\delta_{k,0} + 2a_{-k}^{rl})(\mathcal{E}_k^{out,l} + \mathcal{E}_k^{out,r}) \\ &+ e^{-b}\sum_k(\delta_{k+1,0} + 2a_{-1-k}^{rl})(\mathcal{E}_k^{out,l} + \mathcal{E}_k^{out,r}) \\ &+ \sum_{m\geq 2}(1-e^{2b})e^{-mb}\sum_k(\delta_{k+m,0} + 2a_{-m-k}^{rl})(\mathcal{E}_k^{out,l} + \mathcal{E}_k^{out,r}) \\ &- (\mathcal{E}_{-1}^{out,r} + \mathcal{E}_{-1}^{out,l}) - e^{-b}(\mathcal{E}_0^{out,r} + \mathcal{E}_0^{out,l}) \\ &- e^{-b}\sum_{m\geq 1}(\mathcal{E}_m^{out,r} + \mathcal{E}_m^{out,l})(1-e^{2b})e^{-mb}\Big\}.\end{aligned}$$

If the coefficients in front of all terms $(\mathcal{E}_n^{out,r} + \mathcal{E}_n^{out,l})$ at the *lhs* and at the *rhs* of the last equation are equal, then this equality holds. We get infinite linear system on $\{a_n^{rl}\}$

$$\sum_m Q_{nm}(\delta_{m,0} + 2a_m^{rl}) = q_n, \tag{36}$$

where

$$\begin{array}{lll} Q_{nm} = Q_{n-m}; & Q_m = 0,\ m \geq 1; & Q_0 = 1 - e^b; \\ Q_{-1} = 2e^{-b} - 1 - e^b; & Q_m = 2(1-e^{2b})e^{mb},\ m \leq -2; & q_m = Q_{-m-1}. \end{array}$$

This system can easily be solved using the Fourier representation

$$\hat{q}(\varphi) = \sum_n q_n e^{in\varphi}.$$

The infinite matrix Q in the Fourier representation is the multiplication operator and equation (36) is transformed to the following one

$$\frac{(1-e^b)(1+e^{-i\varphi})(1+e^{-b}e^{-i\varphi})}{(1-e^{-b}e^{-i\varphi})}(1 + 2\hat{a}^{rl}(\varphi)) = \frac{e^{-i\varphi}(1-e^b)(1+e^{i\varphi})(1+e^{-b}e^{i\varphi})}{(1-e^{-b}e^{i\varphi})}. \tag{37}$$

The coefficients $a_n^{ll}, a_n^{rl}, a_n^{lr}, a_n^{rr}$ can easily be calculated from their Fourier representations

$$\hat{a}^{rl}(\varphi) = \hat{a}^{lr}(\varphi) = \frac{2i\sin\varphi}{\alpha - 2i\sin\varphi} \hat{a}^{rr}(\varphi) = \hat{a}^{ll}(\varphi) = \frac{\alpha}{\alpha - 2i\sin\varphi}. \tag{38}$$

It is proven, that representation (27) holds and corresponding sums converge in the energy norm. It accomplishes the proof of the Theorem. □

Corrolary. The sets of elements $\{\mathcal{E}_n^{inc,l}, \mathcal{E}_n^{inc,r}\}_{n=-\infty}^{+\infty}$ and $\{\mathcal{E}_n^{out,l}, \mathcal{E}_n^{out,r}\}_{n=-\infty}^{+\infty}$ are orthogonal bases in the space H'.

Incomimg and outgoing translation representations of the evolution operator in the space H' can be defined.

Theorem 4. *The Lax-Phillips scattering matrix is meromorphic on the Riemann surface of the quasimomentum and it is equal to the following multiplication operator in the spectral representation*

$$\hat{\mathbf{S}} = \begin{pmatrix} \dfrac{-2i\sin\varphi}{\alpha + 2i\sin\varphi} & \dfrac{\alpha}{\alpha + 2i\sin\varphi} \\ \dfrac{\alpha}{\alpha + 2i\sin\varphi} & \dfrac{-2i\sin\varphi}{\alpha + 2i\sin\varphi} \end{pmatrix}. \tag{39}$$

Proof. The Lax-Phillips scattering matrix connects the incoming and outgoing translational representations for the evolution operator. Suppose $F \in H'$, then it can be presented by one of the following sums

$$F = \sum_n \left(f_n^{inc,l}\mathcal{E}_n^{inc,l} + f_n^{inc,r}\mathcal{E}_n^{inc,r}\right) = \sum_n \left(f_n^{out,r}\mathcal{E}_n^{out,r} + f_n^{out,l}\mathcal{E}_n^{out,l}\right).$$

The scattering matrix is the operator, which connects these two representations

$$\mathbf{S} : \begin{pmatrix} f^{inc,l} \\ f^{inc,r} \end{pmatrix} \Rightarrow \begin{pmatrix} f^{out,r} \\ f^{out,l} \end{pmatrix}.$$

This operator can be calculated using the decomposition of the basis vectors $\mathcal{E}_n^{inc,l}$ and $\mathcal{E}_n^{inc,r}$ in the outgoing basis $\mathcal{E}_n^{out,l}, \mathcal{E}_n^{out,l}$ (27,28). Proof of Theorem 3 shows, that it is more convenient to work in the spectral representation $\hat{f}(\varphi) = \sum_n f_n e^{in\varphi}$. The following formula connects the incoming and outgoing spectral representations

$$\begin{pmatrix} \hat{f}^{out,l}(\varphi) \\ \hat{f}^{out,r}(\varphi) \end{pmatrix} = \begin{pmatrix} \hat{a}^{lr}(-\varphi) & \hat{a}^{rr}(-\varphi) \\ \hat{a}^{ll}(-\varphi) & \hat{a}^{rl}(-\varphi) \end{pmatrix} \begin{pmatrix} \hat{f}^{inc,l}(\varphi) \\ \hat{f}^{inc,r}(\varphi) \end{pmatrix}.$$

It follows, that the scattering matrix in the spectral representation is given by the formula (39). Analytical properties of the scattering matrix follow immediately from this formula. □

5. Stationary scattering matrix

We consider in this section stationary approach to the scattering problem, described by the stationary operators $\mathcal{A}$ and $\mathcal{A}_\alpha$ (2,3). We show, that Lax-Phillips scattering matrix coincides with the stationary one in the simplest nontrivial case of $N = 2$ and following matrices A, Γ

$$A = \begin{pmatrix} \lambda & 0 \\ 0 & -\lambda \end{pmatrix}, \lambda \in \mathbf{R}_+; \;\; \Gamma = \begin{pmatrix} 0 & \gamma \\ \gamma & 0 \end{pmatrix}, \gamma \in \mathbf{R}.$$

Eigenfunctions of the operator $\mathcal{A}$ are of Bloch type

$$\psi(n+1) = e^{ip}\psi(n). \tag{40}$$

Dispersion equation, which connects the quasimomentum $p \in [0, 2\pi)$ and the energy E of the eigenfunction

$$E = \pm E(p), \;\; E(p) = \sqrt{\lambda^2 + 4\gamma^2 \cos^2 p}$$

can be obtained by substituting ansatz (40) into the eigenfunction equation

$$A\psi(n) + \Gamma(\psi(n+1) + \psi(n-1)) = E\psi(n).$$

The dispersion equation defines two bands of continuous spectrum situated symmetrically with respect to the origin $[-\sqrt{\lambda^2 + 4\gamma^2}, -\lambda] \bigcup [\lambda, \sqrt{\lambda^2 + 4\gamma^2}]$. Normalized eigenfunctions of the operator can be chosen with the real zero component

$$\begin{aligned} \psi_+(p,n) &= \frac{e^{inp}}{2\sqrt{\pi E(p)}} \begin{pmatrix} \sqrt{E(p)+\lambda} \\ \operatorname{sign}(\cos p)\sqrt{E(p)-\lambda} \end{pmatrix} \\ \psi_-(p,n) &= \frac{e^{inp}}{2\sqrt{\pi E(p)}} \begin{pmatrix} -\operatorname{sign}(\cos p)\sqrt{E(p)-\lambda} \\ \sqrt{E(p)+\lambda} \end{pmatrix}. \end{aligned} \tag{41}$$

These eigenfunctions satisfy the following normalization conditions

$$\langle\langle \psi_+(p), \psi_+(p') \rangle\rangle = \delta(p-p');$$
$$\langle\langle \psi_-(p), \psi_-(p') \rangle\rangle = \delta(p-p');$$
$$\langle\langle \psi_+(p), \psi_-(p') \rangle\rangle = 0.$$

The eigenfunction $\psi_+(p)$ corresponds to the positive band of the spectrum and $\psi_-(p)$ - to the negative one.

Eigenfunctions $\phi(p,n), 0 \le p \le \pi/2$ of the perturbed stationary operator $\mathcal{A}_\alpha$ are equal to the linear combinations of the unperturbed Bloch waves outside the origin

$$\phi_\pm(p,n) = \begin{cases} a_1\psi_\pm(p,n) + b_3\psi_\pm(-p,n) + a_2\psi_\pm(\pi-p,n) + b_4\psi_\pm(p-\pi,n), & n<0 \\ a_3\psi_\pm(-p,n) + b_1\psi_\pm(p,n) + a_4\psi_\pm(p-\pi,n) + b_2\psi_\pm(\pi-p,n), & n>0 \end{cases} \tag{42}$$

with certain coefficients $a_j, b_j \in \mathbf{C}, j = 1,2,3,4$. Incoming eigenfunctions $\phi_\pm^{inc,m}$, $m = 1,2,3,4$, are characterized by the following conditions on the coefficients $a_j^{inc,m} = \delta_{j,m}$, $j = 1,2,3,4$. Similar outgoing eigenfunctions $\phi_\pm^{out,m}$, $m = 1,2,3,4$, are defined by the conditions $b_j^{out,m} = \delta_{j,m}$, $j = 1,2,3,4$. Incoming and outgoing eigenfunctions define two different spectral decompositions for the absolutely continuous subspace of the operator $\mathcal{A}_\alpha$. The stationary scattering matrix is the operator, which connects these spectral representations.

We are going to calculate as example incoming eigenfunction of the first type. Corresponding coefficients in the representation (42) will be denoted by $a_j^{inc,1}$, $b_j^{inc,1}$. Special choice of the perturbation defines no interaction between the eigenfunctions with different absolute value of the quasimomenta and coefficients $b_2^{inc,1}$, $b_4^{inc,1}$ can be chosen equal to zero. The form of the ansatz is chosen in such a way, that the eigenfunction equation is obviously satisfied for all values of the parameters at every point except $n = 0, \pm 1$. Consideration of the equation at these points allows us to calculate unknown coefficients

$$\begin{aligned} b_1^{inc,1} = b_2^{inc,2} = b_3^{inc,3} = b_4^{inc,4} \equiv T(p) = \frac{2i\sin p}{\alpha + 2i\sin p} b_3^{inc,1} \\ = b_4^{inc,2} = b_1^{inc,3} = b_2^{inc,4} \equiv R(p) = \frac{-\alpha}{\alpha + 2i\sin p}. \end{aligned} \tag{43}$$

The transition $T(p)$ and reflection $R(p)$ coefficients posses the following property $T(-p+\pi) = T(p), R(-p+\pi) = R(p)$. These coefficients form the 4×4 stationary scattering matrix

$$S(E) = \begin{pmatrix} T(p) & R(p) & 0 & 0 \\ R(p) & T(p) & 0 & 0 \\ 0 & 0 & T(p) & R(p) \\ 0 & 0 & R(p) & T(p) \end{pmatrix} = S_0 \times \mathbf{I}, \tag{44}$$

$$S_0(E) = \begin{pmatrix} T(p) & R(p) \\ R(p) & T(p) \end{pmatrix}. \tag{45}$$

Here **I** denotes the unit 2×2 matrix.

The matrix $S_0(E)$ differs from the Lax-Phillips scattering matrix by the factor -1. It is due to the fact that incoming and outgoing bases have been chosen in such a way, that the Lax-Phillips scattering matrix for the unperturbed operator is equal to -1.

The stationary scattering matrix has singularities at the points $-2i \sin p = \alpha$ on the Riemann surface of the quasimomentum. These singularities correspond to the bound states. The stationary operator $\mathcal{A}_\alpha$ has two eigenvalues

$$E_\pm = \pm\sqrt{\lambda^2 + 4\gamma^2 + \alpha^2\gamma^2}, \tag{46}$$

which depend on the absolute value of the interaction parameter α only. These eigenvalues are situated outside the continuous spectrum. We have defined by b the real positive solution of the equation $e^b - e^{-b} = \alpha$. Then the eigenfunctions of the stationary operator have the form

$$\phi^\pm(n) = \phi^\pm(0)e^{-b|n|}.$$

The amplitudes $\phi^\pm(0)$ are equal to

$$\phi^\pm(0) = c_\pm \begin{pmatrix} 2\gamma \cosh b \\ E_\pm - \lambda \end{pmatrix}$$

with the normalizing coefficients $c_\pm$. Calculated eigenfunctions of the perturbed stationary operator are related to the eigenfunctions of the perturbed evolution group corresponding to the eigenvalues $z_\pm = e^{\mp b}$.

Acknowledgments. The author wants to thank Prof. B. Pavlov and S. Fedorov for pointing out the problem and fruitful discussions. Collaboration with V. Evstratov was the starting point for this investigation. Financial support from the Alexander von Humboldt foundation is gratefully acknowledged.

References

[1] F. Bentosela, Scattering from Impurities in a Crystal, *Commun. Math. Phys.* **46** (1976), 153–166.

[2] Yu. N. Demkov, P. B. Kurasov, V. N. Ostrovsky, Doubly periodical in time and energy exactly soluble system with two interacting systems of states, *J. Phys. A: Math. Gen.*, **28** (1996), 297–323.

[3] L. D. Faddeev, A. Yu. Volkov, Hirota Equation as an Example of an Integrable Symplectic Map, *Letters in Math. Phys.* **32** (1994), 125–135.

[4] S. Fedorov, Harmonic Analysis in a Multiply-connected Domain I, II, *Math. USSR Sbornik* **70** (1991), 263–296, 297–339.

[5] N. Firsova, Direct and inverse problem of scattering for one dimensional perturbed Hill operator, *Mathematics of the USSR-Sbornik*, **58** (1987), 351–388.

[6] R. Hirota, Nonlinear Partial Difference Equations III; Discrete Sine-Gordon Equation, *J. of Phys. Soc. Japan* **43** (1977), 2079–2086.

[7] Yu. E. Karpeshina, "Perturbation theory for the Schrödinger operator with a periodic potential", Springer, 1997 (Lecture Notes in Mathematics, N 1663).

[8] P. Kurasov, B. Pavlov, Nonphysical Sheet and Schrödinger Evolution, Preprint Mittag-Leffler Institute, Nr 20, 1990/91.

[9] P. Kurasov, N. Elander, B. Pavlov, Resonances and Irreversibility for the Schrödinger Evolution, *Int. J. Quantum Chemistry* **46** (1993), 401–414.

[10] P. D. Lax, R. S. Phillips, The Acoustic Equation with an Indefinite Energy Form and the Schrödinger Equation, *J. Func. Anal.* **1** (1967), 37–83.

[11] P. D. Lax, R. S. Phillips, "Scattering Theory", Revised edition, Academic Press, New York, 1989.

[12] D. C. Mattis, The Few.body Problem on a Lattice, *Rev. Mod. Phys.* **58** (1986), 361–379.

[13] R. G. Newton, Bloch-wave Scattering by Crystal Defects, *J. Math. Phys.* **32** (1991), 551–560.

[14] F. W. Nijhoff, H. W. Capel and V. G. Papageorgiou, Integrable Quantum Mappings, *Phys. Rev. A* **46** (1992), 2155–2158.

[15] B. Pavlov, Lax-Phillips Theory and Nonphysical Sheet for the Band Spectrum, Preprint Institut Mittag-Leffler, Report N36, 1990/91.

[16] B. Pavlov, S. Fedorov, The group of shifts and harmonic analysis on a Riemann surface of genus one, *Leningrad Math. J.* **1** (1990), 447–490.

[17] B. Pavlov, Nonphysical Sheet for the Perturbed Lattice Model, *in*: "Proc. of the Int. Workshop 'Mathematical Aspects of the Scattering Theory and Applications'", Saint-Petersburg, Russia, 1991, 13–124.

[18] B. Pavlov, Zero Index of Pair of Projectors and Expansions by Resonance States for Operators with Band Spectrum, Preprint Int. Erwin Schrödinger Institute for Math. Physics, to be published.

[19] B. Pavlov, Nonphysical sheet for perturbed Jacobian matrices, *St. Petersburg Math. J.* **6** (1994), 185–199.

[20] L. E. Thomas, Time Dependent Approach to Scattering from Impurities in a Crystal, *Comm. Math. Phys.* **33** (1973), 335–343.

Dept. of Mathematics, Stockholm University, 10691 Stockholm, Sweden;

Dept. of Mathematical and Computational Physics, St. Petersburg University, 198904 St.Petersburg, Russia;

Department of Mathematics, Luleå University, 97187 Luleå, Sweden;

Alexander von Humboldt fellow, Department of Mathematics, Ruhr University-Bochum, 44780 Bochum, Germany

Operator Theory:
Advances and Applications, Vol. 123

Spectral Theory of Wiener-Hopf Operators and Functional Model

J. P. MacCormick and B. S. Pavlov

Abstract. Using the Nagy-Foias functional model for contractions we reduce the spectral problem for Wiener-Hopf Operators with analytic and/or rational symbols to the spectral problem for finite matrices.

1. Introduction

The classical problem of asymptotic calculation of determinants of Wiener-Hopf operators T, defined on the finite interval $(0,a)$ as a restriction of an integral operator with difference kernel:

$$\begin{array}{llll} T \equiv T_a(g): & L_2(0,a) & \longrightarrow & L_2(0,a) \\ & u(x) & \longmapsto & u(x) + \int_0^a g(x-s)u(s)\,ds. \end{array} \tag{1}$$

was considered first for positive continuous symbols $\sigma = 1 + \int e^{iks}g(s)ds$ [1, 2, 4]. In [9, 11] the general approach for asymptotic calculation of the determinants was proposed. This approach permits investigation of the asymptotics of the Fredholm determinants of Wiener-Hopf operator with symbols possessing real zeroes. H. Widom investigated the case of piecewise continuous symbols and general case of matrix integral operators (see for instance [6, 7]). The corresponding abstract constructions which permit us to investigate similar operators in weighted L_2 spaces was developed by H. Dym, see for instance [10]. For complete review and more references see [13].

In the present paper we suggest a straightforward approach to spectral problems for Wiener-Hopf operators based on the Lax-Phillips version of analytic semigroup theory suggested first in [27] for scattering problems. This approach is equivalent to the construction of functional models of dissipative operators [5].

We consider Wiener-Hopf operators on the finite interval $(0,a)$ whose symbols σ are analytic in the upper half plane or rational functions. In the first case a straightforward procedure based on the Functional Model for calculating the determinant of the operator is suggested. In the second case we just reduce the spectral problem for Wiener-Hopf operator to the similar problem for a finite matrix. Our approach can be considered as an alternative to the classical approach to Wiener-Hopf operators with analytic and/or rational symbols [13] which differs from the prototype in two respects: our simple geometric mechanics remains practically unchanged (but perhaps a little less elegant) for matrix case and it permits considering in a similar way a wide class of generalized Wiener-Hopf operators connected to any divergent sequence of outgoing subspaces (in Lax-Phillips sense). These interesting subjects will be discussed soon elsewhere.

The authors are grateful to Dr. K. Makarov, who attracted their attention to the analytical problems relevant to the Wiener-Hopf operators and to Rowan Killip, who made inquisitive comments on an early draft of this paper.

The authors are grateful to Professor P. Fuhrmann who brought their attention to the paper [29], where some details of our general construction were anticipated, to Professor A. Laptev who pointed out to us the paper [30], and to Professor I. Gohberg for his friendly and inspiring discussion during the conference on Operator Theory in Groningen, July 1998.

We recognize the support from the Royal Society of New Zealand (Marsden Fund Grant 3368152) and partial support from the Russian Academy of Sciences (Grant RFFI 97-01-01149).

The material of the paper was presented as a talk at the conference in Beer-Sheva dedicated to Professor Moshe Livšic on his 80 birthday.

2. Determinants of Wiener-Hopf operators with analytic symbols – sketch of the plan and the highlights

Let T be the above Wiener-Hopf operator (1), and set $\rho = \sigma - 1 = \mathcal{F}^* g$, where $\mathcal{F}^*$ is the inverse Fourier transform. Our approach to the calculation of the determinant of T is based on the fact that T can be approximated by some other operators W_β whose eigenvalues are known exactly.

Recall that the inverse Fourier transform $\mathcal{F}^*$ maps $L_2[0,\infty]$ unitarily to the Hardy space H_2^+ of the upper half plane so that $\mathcal{F}^*(L_2[0,a]) = H_2^+ \ominus \theta_a H_2^+, \theta_a \equiv e^{ika}$ and the convolution operator becomes the multiplication operator. Denoting $H_2^+ \ominus \theta_a H_2^+$ by K_a, and writing P_a for orthogonal projection $H_2^+ \to K_a$, we see that T is unitarily equivalent to $W \equiv W_a(\sigma) = P_a \sigma P_a$. We reduce in this way our original problem to the calculation of

$$\det\,(P_a \sigma P_a). \tag{2}$$

The exponential θ_a is a singular function, so we can find, see for instance [3], a sequence of Blaschke products $\Pi_\beta(k)$ tending uniformly to it on the upper half plane, e.g.:

$$\Pi_\beta(k) = \frac{e^{ika} - e^{-\beta}}{1 - e^{ika}e^{-\beta}},$$

$$|e^{ika} - \Pi_\beta(k)| \leq \text{Const } e^{-\beta}, \ \beta > 1.$$

It is convenient to note here that the zeroes of $\Pi_\beta(k)$ occur at the points $k_l = 2\pi l/a + i\beta/a$, $l \in \mathbb{Z}$.

Let $K_\beta = H_2^+ \ominus \Pi_\beta H_2^+$, and $P_\beta =$ orthogonal projection $H_2^+ \to K_\beta$. We consider the operator

$$W_\beta = P_\beta \sigma P_\beta$$

as an approximation for W, because *in a sense to be made precise later*, $K_\beta \to K_a$ and $P_\beta \to P_a$ as $\beta \to \infty$. The idea is that instead of the initial determinant we can use

$$\lim_{\beta\to\infty} \det (P_\beta \sigma P_\beta), \tag{3}$$

provided that the W_β approximate W well enough. The whole point of this approach is that the operator $P_\beta \sigma P_\beta$ turns out to have a remarkably simple form, provided the function σ is *analytic in the upper half plane*: its eigenvectors form a complete set (and even a Riesz basis) in K_β, and the eigenvalues are just $\sigma(k_l)$. This fact immediately gives an explicit expression for (3).

The straightforward plan outlined here meets some hidden obstacles, such as the fact that the operators W and W_β are close in *operator norm* but not in *trace norm*. Therefore we need an intermediate operator, which is similar [1] to W_β (and therefore has the same determinant), but close to W in trace norm. This intermediate operator will be constructed as the image of W_β under similarity transform defined by the multiplication operators formed by an entire function f_β, $(f_\beta)^{-1}$ which are bounded and invertible as operators in $L_2(\mathbb{R})$. In summary, the plan is realized as the chain of Lemmas, Propositions and Theorems 1–8, Section 3, which sketch the way of using the functional model for calculating Szegö-Kac determinants for analytic symbol.

3. More general class of symbols

Note that using the standard tools of perturbation theory we may extend the approach described above to the case where the symbol has the form

$$\sigma(k)\pi^{-1}(k).$$

Here σ is a bounded analytic function in the upper half plane and π is a finite Blaschke product whose *zeroes* are in the upper half plane. The corresponding Wiener-Hopf operator on a finite interval $(0, a)$ is Fourier equivalent to the operator $P_a \sigma \pi^{-1}|_{K_a}$, hence it may be approximated in trace class by the operator $f_\beta P_\beta \sigma \pi^{-1} P_\beta f_\beta^{-1}$. But unfortunately this operator cannot be diagonalized using the semigroup approach as before. Nevertheless its determinant can be calculated by the method we sketch below.

Denoting the coinvariant subspace $H_+^2 \ominus \pi H_+^2$ by K_π and the corresponding orthogonal projection by P_π, we rewrite the operator $W_\beta = P_\beta \sigma \pi^{-1}|_{K_\beta}$ as

$$W_\beta = (P_\beta \sigma P_\beta)\,(P_\beta \pi^{-1} P_\beta)\,|_{K_\beta} + P_\beta \sigma \pi^{-1} P_\pi|_{K_\beta}.$$

[1] In the technical sense, i.e., equal to $A\,W_\beta\,A^{-1}$ for some A.

Using the orthogonal decomposition of the identity operator on $L_2(\mathbb{R})$:

$$I = P_{\pi^{-1}H_2^-} + \pi^{-1}P_\pi\pi + P_\beta + P_{\Pi_\beta H_2^+},$$

we have

$$\begin{aligned} W_\beta &= P_\beta\sigma I\pi^{-1}|_{K_\beta} \\ &= P_\beta\sigma\{\underbrace{P_{\pi^{-1}H_2^-}}_{\substack{\text{This is zero}\\ \text{on } \pi^{-1}K_\beta}} + \pi^{-1}P_\pi\pi + P_\beta + \underbrace{P_{\Pi_\beta H_2^+}}_{\substack{\text{This is zero}\\ \text{on } \Pi_{\beta H_+^2}}}\}\pi^{-1}|_{K_\beta} \\ &= P_\beta\sigma\{\pi^{-1}P_\pi\pi + P_\beta\}\pi^{-1}|_{K_\beta} \\ &= (P_\beta\sigma P_\beta)\left(P_\beta\pi^{-1}P_\beta\right)|_{K_\beta} + P_\beta\sigma\pi^{-1}P_\pi|_{K_\beta}, \end{aligned}$$

as claimed.

Now $\det(P_\beta\sigma P_\beta)$ can be calculated exactly as before, and $\det(P_\beta\ \pi^{-1}P_\beta)$ can be calculated using a parallel theory which exploits the fact that π^{-1} is analytic in the *lower* half plane. So the determinant of the first term in this expression for W_β can be calculated; we claim the second term contributes only a finite-dimensional perturbation. To see this, re-prove Proposition 1 with π in place of Π_β. The numbers k_l end up being the zeroes of π, and there are only finitely many of them! Hence K_π is finite-dimensional. Therefore, the standard theory of finite-dimensional perturbations can be used to obtain expressions for $\det W$ with this sort of symbol.

In Section 5 we develop a modified approach which permits to reduce the spectral problem for general Wiener-Hopf operator with rational symbol to the spectral problem for a finite matrix. Really, the rational symbol can be represented as a linear combination of reproducing kernels. Then each term can be analysed basing on explicit formula for the resolvents of generators of compressed shift semigroup and adjoint semigroup onto the coinvariant subspace $K_a = H^2 \ominus \theta_a H^2$:

$$P_a\frac{1}{k-p}u = \frac{u-u(p)}{k-p},\ \Im p > 0,\ u \in K_a$$

$$P_a\frac{1}{k-\bar{p}}u = \frac{u-\theta[\bar{\theta}u(\bar{p})]}{k-\bar{p}},\ \Im\bar{p} < 0,\ u \in K_a.$$

These formulas permit to express the Wiener-Hopf operator as a difference of multiplication operator and a finite-dimensional operator corresponding to the poles k_l of the symbol in upper and lower $\bar{k}_l$ halfplanes. The spectral analysis of the operator is reduced to the spectral analysis of the corresponding finite *paired Cauchy matrix*, see [19]

For some interesting operators with piecewise continuous symbols as Bessel kernel considered in ([8],[30]), the suggested procedure cannot be applied directly, but it is still possible to approximate the symbol with a rational function and develop the described procedure for the approximation. This combined approach is in fact equivalent to using a sort of functional model for operators with a rich complex spectrum and it may be applied to analytic symbols with branching points and cuts in upper halfplane thus reducing the initial problem back to singular integral equations with Cauchy kernels on relevant contours in complex plane.

4. Proof of the main theorem for analytic symbol

In all the following results, $a > 0$ and $\beta > 1$. We first formulate some result (Propositions 1, 2, 3) which follow directly from the version of the theory of the shift operator developed in ([23],[18])for investigation of exponential bases. Just for the sake of completeness of our text we supply them with short sketches of proofs.

Proposition 1. *Let Π_β be the family of Blaschke products*

$$\Pi_\beta(k) = \frac{e^{ika} - e^{-\beta}}{1 - e^{ika}e^{-\beta}},$$

approaching the singular function $e^{ika} \equiv \theta_a$ uniformly in upper half plane as $\beta \to \infty$. Consider the generators B_β of the contracting semigroup

$$Z_\beta(t) = P_\beta e^{ikt} P_\beta \equiv e^{iB_\beta t}, \ \ t > 0,$$

which arises as a compression of the shift group onto the coinvariant subspaces

$$K_\beta = H_+^2 \ominus \Pi_\beta H_+^2.$$

Then the generators B_β are simple dissipative operators, with eigenfunctions given by

$$\psi_l(k) = \frac{\Pi_\beta(k)}{k - k_l}, \ \ l \in \mathbb{Z}$$

and corresponding eigenvalues $k_l = 2\pi l/a + i\beta/a$.

Proof. Here we sketch a well-known proof of this result for completeness.

Denote by $\langle \cdot , \cdot \rangle$ the standard inner product on $L_2(\mathbb{R})$ (and inherited by H_2^+). First observe that $\frac{\Pi_\beta}{k-k_l} \in K_\beta$, since for any $g \in H_2^+$ we have

$$\begin{aligned} \left\langle \tfrac{\Pi_\beta}{k-k_l}, g\Pi_\beta \right\rangle &= \int_{-\infty}^{\infty} \tfrac{\Pi_\beta}{k-k_l} \overline{\Pi_\beta g}\, dk \\ &= \left\langle \tfrac{1}{k-k_l}, g \right\rangle && \text{as } \Pi_\beta \bar{\Pi}_\beta \equiv 1 \text{ on real axis} \\ &= 0 && \text{as } g \in H_2^+ \text{ and } \tfrac{1}{k-k_l} \in H_2^- \end{aligned}$$

So to prove ψ_l is an eigenvector with eigenvalue k_l, it remains to show that

$$P_{K_\beta}\left[(e^{ikt} - e^{ik_l t}) \frac{\Pi_\beta}{k - k_l} \right] = 0.$$

But this is equivalent to $(e^{ikt} - e^{ik_l t})\frac{\Pi_\beta}{k-k_l} \in \Pi_\beta H_2^+$, or $\frac{e^{ikt} - e^{ik_l t}}{k-k_l} \in H_2^+$, which is obviously true since $t > 0$. $\square$

A similar statement is valid for systems of eigenvectors of the adjoint operators B_β^*. Actually in this case the eigenvectors conveniently coincide with the H_2^+ reproducing kernels: $\varphi_l(k) = \frac{1}{k-\bar{k}_l}$. In other words, we have,

Proposition 2. *The eigenvectors of B_β^* are*

$$\varphi_l(k) = \frac{1}{k - \bar{k}_l}, \quad l \in \mathbb{Z}$$

with eigenvalues $\bar{k}_l = 2\pi l - i\beta$.

Proof. We need only check that $\{\varphi_l\}$ and $\{\psi_l\}$ are biorthogonal sets. To do this we need a simple result stating the reproducing kernel for H_2^+:

Let $c \in$ upper half plane, and $f \in H_2^+$. Then $\langle f, \frac{1}{k-\bar{c}}\rangle = 2\pi i f(c)$. So if $l \neq m$, $\langle \psi_l, \varphi_m\rangle = 2\pi i \frac{\Pi_\beta(k_m)}{k_m - k_l} = 0$. And if $l = m$, $\langle \psi_l, \varphi_l\rangle = 2\pi i \prod_{l\neq m}(k_m - k_l) \neq 0$ as an infinite Blaschke product converges to a non-zero complex number except at zeroes of its factors. □

We also have the following fact which will be crucial for our reasoning later on.

Proposition 3. *The sets $\{\varphi_l\}$ and $\{\psi_l\}$ both form Riesz bases*[2] *for the subspace K_β.*

Proof. Since we showed in Proposition 2 that $\{\varphi_l\}$ and $\{\psi_l\}$ are biorthogonal sets, it will be enough to show this for the $\{\varphi_l\}$ only.

First we show the $\{\varphi_l\}$ are complete in K_β. So suppose $f \in K_\beta$, and $\langle f, \frac{1}{k-\bar{k}_l}\rangle = 0$, each $l \in \mathbb{Z}$. Then by the result used in the proof of Proposition 2, $f(k_l) = 0$ for all $l \in \mathbb{Z}$. So $f \in \Pi_\beta H_2^+$. But $f \in K_\beta$, hence $f \equiv 0$.

Finally we show the complete set φ_l is in fact a Riesz basis. The well-known Carleson condition was discovered by ([15]) in the context of interpolation by analytic functions, and later rediscovered by different authors ([16],[17]) as a test for Riesz basis property for dissipative operators[3]:

$$\inf_m \prod_{l\neq m} \left| \frac{k_m - k_l}{k_m - \bar{k}_l} \right| > 0.$$

A quick calculation shows that this condition holds for our set $\{\varphi_l\}$. □

In the next theorem we describe an important automorphism of $L_2(\mathbb{R})$ which maps K_β to K_a.

---

[2]By a Riesz basis, we mean a basis obtained from an orthonormal basis by an invertible, bounded, linear transformation.

[3]Actually, the Carleson condition guarantees only that we have a so-called *unconditional basis.* A Riesz basis must also satisfy $\inf \|\varphi_l\| > 0$ and $\sup \|\varphi_l\| < \infty$, but these conditions are clearly fulfilled here.

Theorem 4. *Write $\theta(k) = e^{ika}$, and let $f_\beta(k)$ be the entire function of exponential type defined by*

$$f_\beta(k) = 1 - e^{-\beta} e^{ika}.$$

Then the multiplication operator $u \mapsto f_\beta u$ is a bounded and invertible operator on $L_2(\mathbb{R})$, transforming the orthogonal sum

$$L_2(\mathbb{R}) = H^2_- \oplus K_\beta \oplus \Pi_\beta H^2_+$$

into the direct sum

$$\overline{\Pi_\beta} \theta H^2_- + K_a + \Pi_\beta H^2_+,$$

where $K_a = H^2_+ \ominus \theta H^2_+$ is a coinvariant subspace of the shift group corresponding to the singular function θ. The entire functions

$$\Phi_l = f_\beta \, \varphi_l$$

form a Riesz basis in K_a for each $\beta > 1$.

Proof. Recall that $a > 0$ and $\beta > 1$ – we will use this fact repeatedly without mentioning it. Multiplication by f_β is clearly bounded and invertible since $1 - e^{-\beta} \leq |f_\beta| \leq 1 + e^{-\beta}$ on the real axis. We check each component of the claimed transformation separately.

1st component: To see $f_\beta H_2^- \subseteq \overline{\Pi_\beta} \theta H_2^-$, suppose $g \in H_2^-$. Then

$$f_\beta g = (1 - e^{-\beta} e^{ika}) g = \overline{\Pi_\beta} \theta (1 - e^{-\beta} e^{-ika}) g \in \overline{\Pi_\beta} \theta H_2^-.$$

To see $f_\beta^{-1} \overline{\Pi_\beta} \theta H_2^- \subseteq H_2^-$, suppose again that $g \in H_2^-$. Then

$$f_\beta^{-1} \overline{\Pi_\beta} \theta g = \frac{1}{1 - e^{-ika} e^{-\beta}} g \in H_2^-.$$

2nd component: To see $K_\beta \xrightarrow{f_\beta} K_a$ is a bijection, it will be enough (by Proposition 1) to show that $\{f_\beta \varphi_l\}_{l \in \mathbb{Z}}$ is a complete set for K_a. First we claim that $f_\beta \varphi_l \in K_a$:

$$f_\beta \varphi_l = \frac{1 - e^{-\beta} e^{ika}}{k - \overline{k}_l}$$

which is clearly in H_2^+, so suppose $g \in H_2^+$. Then

$$\begin{aligned} \langle f_\beta \varphi_l, \theta g \rangle &= \left\langle \frac{1 - e^{-\beta} e^{ika}}{k - \overline{k}_l}, e^{ika} g \right\rangle \\ &= \left\langle \frac{e^{-ika} - e^{-\beta}}{k - \overline{k}_l}, g \right\rangle \\ &= 0 \\ \text{as} \quad & \frac{e^{-ika} - e^{-\beta}}{k - \overline{k}_l} \in H_2^-. \end{aligned}$$

It is an entire function, with decreasing modulus as $\Im k \to -\infty$, whereas $g \in H_2^+$.

So we have shown $f_\beta \varphi_l \in K_a$; now we need to show $\{f_\beta \varphi_l\}$ is complete in K_a. Actually, it will be more convenient to show $\{\theta^{-1} f_\beta \varphi_l\}$ is complete in $\theta^{-1} K_a$, which amounts to the same thing since θ is unitary. Note[4] that $\theta^{-1} K_a = H_2^- \ominus e^{-ika} H_2^-$. So suppose $g \in H_2^-$, and $\langle g, \theta^{-1} f_\beta \varphi_l \rangle = 0$ for each $l \in \mathbb{Z}$. We need to show $g \in e^{-ika} H_2^-$. But

$$\begin{aligned} g \in e^{-ika} H_2^- \quad &\Leftrightarrow \quad e^{ika} g \in H_2^- \\ &\Leftrightarrow \quad \left\langle e^{ika} g, H_2^+ \right\rangle = 0 \\ &\Leftrightarrow \quad \left\langle e^{ika} g, K_\beta \right\rangle = 0 \text{ and } \left\langle e^{ika} g, \Pi_\beta H_2^+ \right\rangle = 0. \end{aligned}$$

We can check each of these last conditions separately. For the first, recall we assumed $\langle g, \theta^{-1} f_\beta \varphi_l \rangle = 0$, i.e., $\left\langle g, \frac{e^{-ika} - e^{-\beta}}{k - \bar{k}_l} \right\rangle = 0$. But $\left\langle g, \frac{e^{-\beta}}{k - \bar{k}_l} \right\rangle = 0$ since $g \in H_2^-$ and $\frac{1}{k - \bar{k}_l} \in H_2^+$, so we have $\left\langle g, \frac{e^{-ika}}{k - \bar{k}_l} \right\rangle = 0$ and hence $\langle g, e^{-ika} K_\beta \rangle = 0$ as $\left\{ \frac{1}{k - \bar{k}_l} \right\}_{l \in \mathbb{Z}}$ are complete in K_β.

We prove the second condition by contradiction. Suppose $\langle e^{ika} g, \Pi_\beta f \rangle \neq 0$ for some $f \in H_2^+$. Then $e^{ika} g = q + \Pi_\beta\, p$ for some $q \in H_2^-$ and some *non-zero* $p \in H_2^+$. It is easy to see that $q = e^{ika} r$ for some $r \in H_2^-$, so rearranging gives $\Pi_\beta\, p = e^{ika} s$, for some $s \in H_2^-$. Thus $s = e^{-ika} \Pi_\beta\, p = \frac{1 - e^{-ika} e^{-\beta}}{1 - e^{ika} e^{-\beta}} p$, or

$$\frac{s}{1 - e^{-ika} e^{-\beta}} = \frac{p}{1 - e^{ika} e^{-\beta}}.$$

But the left side here is in H_2^-, whereas the right side is in H_2^+. Thus $p \equiv s \equiv 0$, contradicting the fact that p was not identically zero.

3rd component: To see $f_\beta(\Pi_\beta H_2^+) \subseteq \Pi_\beta H_2^+$, suppose $g \in H_2^+$. Then $f_\beta \Pi_\beta g = (e^{ika} - e^{-\beta}) g \in H_2^+$. And for the reverse inclusion, $f_\beta^{-1} \Pi_\beta\, g = \Pi_\beta \frac{g}{1 - e^{-\beta} e^{ika}} \in \Pi_\beta H_2^+$.

Finally we remark that the $\{\Phi_l\}$ are certainly a Riesz basis of K_a as they were obtained from the Riesz basis $\{\varphi_l\}$ of K_β by a bounded invertible linear transformation. □

It is interesting to note here that in fact the functions Φ_l are Fourier images of the projections of exponentials $e^{-i\bar{k}_l x}$ in $L_2(\mathbb{R})$ onto $L_2(0, a)$.

As mentioned in the introduction, it turns out that the operators $W_\beta \equiv P_\beta \sigma P_\beta$ do not approximate the operator $W_a \equiv P_a \sigma P_a$ well enough for our purposes. Therefore in the next theorem we introduce the operators $\mathcal{W}_\beta$ which have the same determinants as the W_β. Then in Theorem 7 we show that $\mathcal{W}_\beta$ is a good enough approximation to W_a.

---

[4]To see this, observe that $K_a = L_2 \ominus e^{ika} H_2^+ \ominus H_2^-$, so $e^{-ika} K_a = L_2 - \ominus H_2^+ \ominus e^{-ika} H_2^- = H_2^- \ominus e^{-ika} H_2^-$.

Theorem 5. *The operator*

$$\mathcal{P}_a^\beta = f_\beta P_\beta f_\beta^{-1}$$

is a skew (i.e., nonorthogonal) projection onto K_a parallel to the sum of subspaces

$$\overline{\Pi}_\beta \theta H_-^2 + \Pi_\beta H_+^2.$$

For each essentially bounded function σ defined on the real axis, the operator $W_\beta \equiv P_\beta \sigma P_\beta$ is bounded and is similar to the operator $\mathcal{W}_\beta \equiv \mathcal{P}_a^\beta \sigma \mathcal{P}_a^\beta$ acting in the subspace K_a; in particular, $\mathcal{W}_\beta$ and W_β have the same determinants.

Proof. The property of f_β described in Theorem 4 means precisely that the linear operator $\mathcal{P}_a^\beta$ is zero on $\Pi_\beta \theta H_2^- + \Pi_\beta H_2^+$, and the identity on K_a; this is the definition of a skew projection so the first statement is proved.

This fact is not generally true for infinite-dimensional determinants, but the formula

$$\det(PQP^{-1}) = \det Q,$$

remains true if P maps some Riesz basis of dom(Q) to a Riesz basis of dom (PQP^{-1}). But we proved in Theorem 4 that f_β maps the Riesz basis $\{\varphi_l\}$ of K_β to the Riesz basis $\{\Phi_l\}$ of K_a. By the definition of $\mathcal{W}_\beta$ we have

$$\mathcal{W}_\beta|_{K_a} = f_\beta P_\beta \sigma P_\beta f_\beta^{-1}|_{K_a} = f_\beta W_a(\sigma) f_\beta^{-1}|K_a,$$

so the above remarks tell us that $\det \mathcal{W}_\beta = \det W_\beta$. □

Our next task is to estimate the norms of the various operators defined so far. Define

$$\epsilon_\beta = \sup_{k\in\mathbb{R}} |\Pi_\beta(k) - \theta(k)|.$$

It is easy to see these suprema are equal, and in fact that $\epsilon_\beta \leq \frac{2e^{-\beta}}{1-e^{-\beta}}$. In particular, we see that $\epsilon_\beta \to 0$ as $\beta \to \infty$.

We will need the following technical lemma.

Lemma 6. *(i) The operator norm of $\mathcal{P}_a^\beta$ is estimated by*

$$\|\mathcal{P}_a^\beta\|_{op} \leq \frac{1}{1-\sqrt{2}\epsilon_\beta.}$$

(ii) The following uniform estimate holds:

$$\|\mathcal{P}_a^\beta - P_a\|_{op} \leq \frac{\sqrt{2}\epsilon_\beta}{1-\sqrt{2}\epsilon_\beta}.$$

(iii) If $\rho \in L_\infty \cap L_2$, then $P_a \rho P_a$ is a Hilbert-Schmidt operator with Hilbert-Schmidt norm not greater than $\|\rho\|_2$.

(iv) If $\rho \in L_\infty \cap L_2$, then $\mathcal{P}_a^\beta \rho \mathcal{P}_a^\beta$ is a Hilbert-Schmidt operator with Hilbert-Schmidt norm not greater than

$$\left(\frac{1}{1-\sqrt{2}\epsilon_\beta}\right)^2 \|\rho\|_2.$$

Proof. (i) Given $f \in L_2$ with $\|f\| = 1$, write $f = \mathcal{P}_a^\beta f + \overline{\Pi}_\beta \theta f_- + \Pi_\beta f_+$, where $f_\pm \in H_2^\pm$. Let $g = \mathcal{P}_a^\beta f + f_- + \theta f_+$. ($g$ is the result of "straightening" our decomposition of f to make it an orthogonal decomposition.) Then

$$\begin{aligned}
\|f-g\| &= \left\| \left(\overline{\Pi}_\beta - 1\right) f_- + \left(\Pi_\beta - \theta\right) f_+ \right\| \\
&\leq \epsilon_\beta \left(\|f_-\| + \|f_+\|\right) \\
&\leq \sqrt{2}\epsilon_\beta \left(\|f_-\|^2 + \|f_+\|^2\right)^{1/2}.
\end{aligned}$$

So

$$\|g\| \leq 1 + \sqrt{2}\epsilon_\beta \left(\|f_-\|^2 + \|f_+\|^2\right)^{1/2} \quad \text{(recall } \|f\| = 1\text{)}.$$

Also, $\|g\|^2 = \|\mathcal{P}_a^\beta f\|^2 + \|f_-\|^2 + \|f_+\|^2$, so $\left(\|f_-\|^2 + \|f_+\|^2\right)^{1/2} \leq \|g\|$. Thus $\|g\| \leq 1 + \sqrt{2}\,\epsilon_\beta \|g\|$, i.e., $\|g\| \leq \frac{1}{1-\sqrt{2}\epsilon_\beta}$. Hence

$$\|\mathcal{P}_a^\beta f\| \leq \|g\| \leq \frac{1}{1-\sqrt{2}\epsilon_\beta},$$

as claimed.

(ii) Given $f \in L_2$ with $\|f\| = 1$, write $f = P_a f + f_- + \theta f_+$, where $f_\pm \in H_2^\pm$. Set $g = P_a f + \overline{\Pi}_\beta \theta f_- + \Pi_\beta f_+$. (We have "bent" the orthogonal components of f so that $\mathcal{P}_a^\beta g = P_a f$.) Then

$$\begin{aligned}
\|f-g\| &= \left\| \left(\overline{\Pi}_\beta \theta - 1\right) f_- + \left(e^{ika} - \Pi_\beta\right) f_+ \right\| \\
&\leq \epsilon_\beta (\|f_-\| + \|f_+\|) \\
&\leq \sqrt{2}\epsilon_\beta \left(\|f_-\|^2 + \|f_+\|^2\right)^{1/2}
\end{aligned}$$

Now we can obtain the desired estimate:

$$\begin{aligned}
\|(\mathcal{P}_a^\beta - P_a) f\| &= \left\|\mathcal{P}_a^\beta (f-g) + \mathcal{P}_a^\beta g - P_a f\right\| \\
&= \|\mathcal{P}_a^\beta (f-g)\| \qquad \text{(since } \mathcal{P}_a^\beta g = P_a f\text{)} \\
&\leq \|\mathcal{P}_a^\beta\|_{\text{op}} \|f-g\| \\
&\leq \frac{1}{1-\sqrt{2}\epsilon_\beta} \sqrt{2}\,\epsilon_\beta \left(\|f_-\|^2 + \|f_+\|^2\right)^{1/2} \\
&\leq \frac{\sqrt{2}\epsilon_\beta}{1-\sqrt{2}\epsilon_\beta} \|f\|.
\end{aligned}$$

(iii) Taking the Fourier transform of $P_a \rho P_a$, we have an integral operator whose kernel K is non-zero only on $[0,a] \times [0,a]$. Thus

$$\|k\|_{L_2 \times L_2} = \int_0^a \int_0^a |k(x-y)|^2 dx\, dy \leq a \int_{-\infty}^{\infty} |k|^2 dx < \infty,$$

as $\rho \in L_2$ so $k \in L_2$ too. A standard theorem on integral equations now tells us the operator is Hilbert-Schmidt[5].

(iv) Use a similar argument to (iii), together with (i). $\square$

The next theorem shows that the intermediate operator $\mathcal{W}_\beta$ is actually close to W in trace norm.

Theorem 7. *Let σ be a bounded analytic function in the upper half plane, set $\rho = \sigma - 1$ and suppose ρ can be expressed as the product of 3 functions, each in the intersection of the Hardy classes $H_\infty^+ \cap H_2^+$:*

$$\rho = \rho_1 \rho_2 \rho_3, \quad \rho_j \in H_\infty^+ \cap H_2^+ \tag{4}$$

Then

$$\left\| P_a \rho P_a - \mathcal{P}_a^\beta \rho \mathcal{P}_a^\beta \right\|_{Trace} \leq \frac{\sqrt{2}\epsilon_\beta}{\left(1 - \sqrt{2}\epsilon_\beta\right)^3} \, const,$$

where the constant depends only on the L_2 and L_∞ norms of the factors ρ_j. In particular, $\mathcal{P}_a^\beta \rho \mathcal{P}_a^\beta \to P_a \rho P_a$ in trace norm as $\beta \to \infty$.

Proof. We will need three basic facts about operators, see, for instance in ([20]):

Fact A. An operator is of trace class if and only if it can be expressed as the product of two Hilbert-Schmidt operators.

Fact B. The set of Hilbert-Schmidt operators is an ideal in the algebra of bounded operators. So if A is Hilbert-Schmidt, and B is bounded, then AB and BA are both Hilbert-Schmidt. Moreover, $\|BA\|_{\mathrm{HS}} \leq \|B\|_{\mathrm{op}} \|A\|_{\mathrm{HS}}$ and similarly for $\|AB\|_{\mathrm{HS}}$.

Fact C. Fact B holds if we replace Hilbert-Schmidt by trace class throughout.

To get the proof started, we also need the following result, which allows us to insert P_a and $\mathcal{P}_a^\beta$ in certain places without affecting anything. Let ρ_1, ρ_2 be functions in H_2^+. Then

$$P_a \rho_1 \rho_2 P_a = P_a \rho_1 P_a \rho_2 P_a, \qquad \text{and} \qquad \mathcal{P}_a^\beta \rho_1 \rho_2 \mathcal{P}_a^\beta = \mathcal{P}_a^\beta \rho_1 \mathcal{P}_a^\beta \rho_2 \mathcal{P}_a^\beta.$$

To see this, let $f \in L_2(\mathbb{R})$. Write $\rho_2 P_a f = P_a(\rho_2 P_a f) + \theta f_+$, where $f_+ \in H_2^+$. Then $P_a \rho_1 \rho_2 P_a f = P_a \rho_1 P_a(\rho_2 f) = P_a \rho_1 P_a \rho_2 f$. The second statement follows from a similar calculation.

[5]See, for example, Proposition 4.8 of ([12]) – actually the statement there is that the operator is compact, but exactly the same proof shows the operator is Hilbert-Schmidt.

Now we can proceed with the proof of the theorem. Applying the above result immediately, and then adding and subtracting some terms, we get

$$\begin{aligned}
&P_a \rho P_a - \mathcal{P}_a^\beta \rho \mathcal{P}_a^\beta \\
&\quad = P_a \rho_1 P_a \rho_2 P_a \rho_3 P_a - \mathcal{P}_a^\beta \rho_1 \mathcal{P}_a^\beta \rho_2 \mathcal{P}_a^\beta \rho_3 \mathcal{P}_a^\beta \\
&\quad = \left(P_a - \mathcal{P}_a^\beta\right) \rho_1 P_a \rho_2 P_a \rho_3 P_a \\
&\qquad + \mathcal{P}_a^\beta \rho_1 \left(P_a - \mathcal{P}_a^\beta\right) \rho_2 P_a \rho_3 P_a \\
&\qquad\quad + \mathcal{P}_a^\beta \rho_1 \mathcal{P}_a^\beta \rho_2 \left(P_a - \mathcal{P}_a^\beta\right) \rho_3 P_a \\
&\qquad\qquad + \mathcal{P}_a^\beta \rho_1 \mathcal{P}_a^\beta \rho_2 \mathcal{P}_a^\beta \rho_3 \left(P_a - \mathcal{P}_a^\beta\right)
\end{aligned}$$

We can estimate the trace norm of each term separately; we show only the second term here as an example.

$$\begin{aligned}
&\left\|\mathcal{P}_a^\beta \rho_1 \left(P_a - \mathcal{P}_a^\beta\right) \rho_2 P_a \rho_3 P_a\right\|_{\text{Trace}} \\
&\quad = \Big\| \underbrace{\left(\mathcal{P}_a^\beta \rho_1 \mathcal{P}_a^\beta\right)}_{\text{Hilbert-Schmidt}} \underbrace{\left(P_a - \mathcal{P}_a^\beta\right) \rho_2}_{\text{bounded}} \underbrace{\left(P_a \rho_3 P_a\right)}_{\text{Hilbert-Schmidt}} \Big\|_{\text{Trace}} \\
&\quad \le \left\|\mathcal{P}_a^\beta \rho_1 \mathcal{P}_a^\beta\right\|_{\text{HS}} \left\|\left(P_a - \mathcal{P}_a^\beta\right) \rho_2\right\|_{\text{op}} \left\|P_a \rho_3 P_a\right\|_{\text{HS}} \\
&\quad \le \left(\frac{1}{1-\sqrt{2}\epsilon_\beta}\right)^2 \|\rho_1\|_2 \left(\frac{\sqrt{2}\epsilon_\beta}{1-\sqrt{2}\epsilon_\beta}\right) \|\rho_2\|_\infty \|\rho_3\|_2
\end{aligned}$$

The first line here used Lemma 6 (iii) and (iv), together with a slight variant of the result above for inserting $\mathcal{P}_a^\beta$ and P_a. The second line uses Facts A, B and C, and the third line follows from Lemma 6 (ii)–(iv).

After estimating the remaining 3 terms in the same way and summarizing the results, we get precisely the formula announced in the theorem. □

At last we are in a position to prove our main theorem, obtaining a formula for $\det W$.

Theorem 8. *Suppose σ is an analytic function in the upper half plane, and $\sigma - 1$ satisfies the 3-factor condition (4) of the previous theorem. Then*

$$\det W_a(\sigma) = \lim_{\beta\to\infty} \prod_{l\in\mathbb{Z}} \sigma(k_l), \tag{5}$$

where $k_l = 2\pi l/a + i\beta/a$.

Proof. We apply our previous results to carry out the plan outlined in the introduction. Note that in the first line we will use the fact that det is continuous with respect to the trace norm. (This is proved in ([11]), for example.) We have

$$\begin{aligned}
\det W_a(\sigma) &= \lim_{\beta\to\infty} \det \mathcal{W}_\beta, && \text{by Theorem 7} \\
&= \lim_{\beta\to\infty} \det W_\beta, && \text{by Theorem 5} \\
&= \lim_{\beta\to\infty} \textstyle\prod(\text{evals of } W_\beta), && \text{by definition of det} \\
&= \lim_{\beta\to\infty} \textstyle\prod_{l\in\mathbb{Z}} \sigma(k_l), && \text{by Proposition 1.}
\end{aligned}$$
□

5. Wiener-Hopf Operators with rational symbols and functional model

The first part of our paper was aimed at revealing the connections between the Nagy-Foias Functional Model and the spectral properties of Wiener-Hopf operators with analytic symbols. In this part we develop a similar programme for rational symbols. In particular we show, that the spectral analysis of Wiener-Hopf operators with the real rational symbol with $2N$ poles can be reduced to the spectral problem for $2N \times 2N$ matrix.

The complete spectral theory of Wiener-Hopf operators with rational symbols was developed in ([13]). We suggest here an alternative approach, which can be easily generalized for matrix symbols and for operators acting in Hardy classes on multiconnected domains. Generally we do not need any conditions concerning the selfadjointness of the operators considered. Nevertheless, just to facilitate the formulation of our resuts and the comparison of them with classical ones we consider only Wiener-Hopf operators whose symbol σ are *real* rational functions

$$W \equiv W_a(\sigma) = P_a \sigma P_a,$$

where $P_a \equiv P_{K_a}, K_a = H_2^+ \ominus e^{ika} H_2^+$. The asymptotics of the corresponding determinants was calculated first in ([1],[4],[2]), and later for extended class of symbols by ([9], [11],[6]).

Note that the similar construction is generally used for calculating the resolvent of the model operator, see ([5]). In particular it was used in implicit form in the case of the simplest rational symbol $\sigma(k) = \frac{ak+b}{ck+d}$ and a special coinvariant subspace generated by the inner function containing only one Blashke factor and a singular factor generated by the trivial singular measure in ([29]) when calculating the sensitivity of some control problem. Our approach gives the general method of reduction of the Wiener-Hopf spectral problem with general rational symbol to the matrix problem on the basis of Theorem 9, when using characteristic functionals $u(k_l)$ $(\bar{\theta} u)(k_l)$ associated with poles of the symbol in upper and lower halfplanes.

Considering the real rational symbol σ we assume that it is represented in form of a finite sum of reproducing kernels (Cauchy kernels) with poles at prescribed complex points and the derivatives of the reproducing kernels:

$$\sigma(k) = 1 + \sum_{l=1,m=0}^{L,M_l} \frac{\alpha_l^m}{(k-k_l)^m} + \frac{\bar{\alpha}_l^m}{(k-\bar{k}_l)^m} \equiv 1 + \varphi(k), \ \Im k_l > 0.$$

It is easy to see that the corresponding Wiener-Hopf operator $P_a \varphi|_{K_a}$ is compact. We denote by k_n^λ,

$$n = -2\sum M_l, \dots, -2, -1, 1, 2, \dots, 2\sum M_l,$$

the roots of the auxilliary equation $\varphi(k) = \lambda$ (counting multiplicity) and assume that θ is any inner function, generating the "outgoing" invariant subspace θH_+^2.

Theorem 9. *The eigenvalues of the Wiener-Hopf operator in the coinvariant subspace $K_\theta = H^2 \ominus \theta H^2$, coincide with the zeroes of the determinant of a finite square matrix* [6] $\mathcal{A}$:

$$\begin{vmatrix} \cdots & \cdots & \cdots & \cdots \\ \cdots \; \alpha_l^m \begin{Bmatrix} s \\ m \end{Bmatrix} \dfrac{d^{m-s}}{dp^{m-s}} \dfrac{1}{k^{\lambda_n} - p}\Bigg|_{p=k^l} & & \theta(k_n^\lambda)\bar{\alpha}_l^m \begin{Bmatrix} t \\ m \end{Bmatrix} \dfrac{d^{m-t}}{d\bar{p}^{m-t}} \dfrac{1}{k_n^\lambda - \bar{p}}\Bigg|_{p=k_l^m} & \cdots \\ \cdots & \cdots & \cdots & \cdots \end{vmatrix} = 0.$$

Proof. It is based on the following statement, which is well known for $m = 1$ [5], and obviously true for $m > 1$.

Proposition 10. *The following representations are true for the resolvents of the generators of compressions shifts and adjoint shifts semigroups onto the coinvariant subspace $K_a = H^2 \ominus e^{ika} H^2$:*

$$P_a \frac{1}{k-p} u = \frac{u - u(p)}{k-p}, \; \Im p > 0, \; u \in K_a$$

$$P_a \frac{1}{k-\bar{p}} u = \frac{u - \theta[\bar{\theta} u(\bar{p})]}{k-\bar{p}}, \; \Im \bar{p} < 0, \; u \in K_a.$$

The similar statement are true for the derivatives order m, $m > 0$ of the resolvent in respect to the spectral parameter:

$$P_a \frac{1}{(k-p)^{m+1}} u = \frac{1}{m!} \frac{d^m}{dp^m} \frac{u - u(p)}{k-p}, \; \Im p > 0, \; u \in K_a$$

$$P_a \frac{1}{(k-\bar{p})^{m+1}} u = \frac{1}{m!} \frac{d^m}{d\bar{p}^m} \frac{u - \theta[\bar{\theta} u(\bar{p})]}{k-\bar{p}}, \; \Im \bar{p} < 0, \; u \in K_a.$$

The first two formulas give the analytic continuation of the resolvent of the shift generators across the absolutely-continuous spectrum R onto the "nonphysical sheet", which are just halfplanes $\Lambda_\pm$. In a way these formulas realize the connection between the elementary Wiener-Hopf operators and the functional model.

We begin the proof of the theorem noticing that each eigenfunction u of the Wiener-Hopf operator $P_a \varphi|_{K_a}$ fulfills the corresponding homogeneous equation:

$$[\varphi(k) - \lambda] u(k) = \sum_{l=1,m=0}^{N,M_l} \frac{1}{m!} \alpha_l^m \frac{d^m}{dp^m} \frac{u(p)}{k-p}\Bigg|_{p=k_l^m} + \frac{1}{m!} \theta(k) \bar{\alpha}_l^m \frac{d^m}{d\bar{p}^m} \frac{[\bar{\theta} u(\bar{p})]}{k-\bar{p}}\Bigg|_{p=k_l^m}.$$

Hence the zeroes of the left-hand side coincide *counting multiplicity* with the roots of the right side. This implies

$$\sum_{l=1,m=0}^{N,M_l} \frac{1}{m!} \alpha_l^m \frac{d^m}{dp^m} \frac{u(p)}{k-p}\Bigg|_{p=k_l^m} + \frac{1}{m!} \theta(k) \bar{\alpha}_l^m \frac{d^m}{d\bar{p}^m} \frac{[\bar{\theta} u(\bar{p})]}{k-\bar{p}}\Bigg|_{p=k_l^m} = 0$$

[6] The description of the matrix in more detail is given below.

for any $k = k_n^\lambda$. Considering $\left.\frac{d^s}{dp^s}u(p)\right|_{p=k_l} \equiv u_l^s$ and $\left.\frac{d^t}{d\bar{p}^t}\bar{\theta}u(p)\right|_{p=k_l} \equiv u_t^l$ as *new unknowns*,

$$\sum_{l=1,m=0}^{N,M_l} \frac{1}{m!}\alpha_l^m \frac{d^m}{dp^m}\left.\frac{u(p)}{k-p}\right|_{p=k_l^m} + \frac{1}{m!}\theta(k)\bar{\alpha}_l^m \frac{d^m}{d\bar{p}^m}\left.\frac{[\bar{\theta}u(\bar{p})]}{k-\bar{p}}\right|_{p=k_l^m} = \mathcal{A}\left(u_l^s, u_t^l\right),$$

we get the condition of existence of a nontrivial solution in the determinant form:

$$\det \mathcal{A} = \begin{vmatrix} \cdots & \cdots & \cdots & \cdots \\ \cdots\ \alpha_l^m & \left\{ \begin{matrix} s \\ m \end{matrix} \right\} \frac{d^{m-s}}{dp^{m-s}} \left.\frac{1}{k^{\lambda_n}-p}\right|_{p=k^l} & \theta(k_n^\lambda)\bar{\alpha}_l^m \left\{ \begin{matrix} t \\ m \end{matrix} \right\} \frac{d^{m-t}}{d\bar{p}^{m-t}} \left.\frac{1}{k_n^\lambda - \bar{p}}\right|_{p=k_l^m} & \cdots \\ \cdots & \cdots & \cdots & \cdots \end{vmatrix} = 0.$$

On the other hand, once the unknowns are defined as solutions of the corresponding homogeneous linear system, one can check that the functions constructed on the base of new unknowns as

$$u(k) = \frac{1}{[\varphi(k)-\lambda]} \sum_{l=1,m=0}^{N,M_l} \frac{1}{m!}\alpha_l^m \left\{ \begin{matrix} s \\ m \end{matrix} \right\} \left.\frac{u_l^s}{(k-p)^{m-s}}\right|_{p=k_l^m} + \frac{1}{m!}\theta(k)\bar{\alpha}_l^m \left\{ \begin{matrix} t \\ m \end{matrix} \right\} \left.\frac{u_m^t}{(k-\bar{p})^{m-t}}\right|_{p=k_l^m}$$

belong to K_θ and their derivatives take precisely the values u_l^r, u_r^l at the poles. Thus they coincide with the eigenfunctions of the original Wiener-Hopf operator with λ equal to the corresponding eigenvalue. Really, if the sum on the right side has a zero of the same multiplicity as the denominator $[\varphi(k)-\lambda]$ at the points k_n^λ, then the poles appearing from the factor $[\varphi(k)-\lambda]^{-1}$ are compensated by the numerator. On the other hand the poles of the numerator at the points k_i^m, $k_i^{\bar{m}}$ are compensated by the poles of the symbol φ. Thus the whole expression belongs to H_+^2. Vice versa after multiplication by $\bar{\theta}(k)$ we get $u \in H_-^2$, hence $u \in K_\theta$. □

Remark 1. The resolvent of the Wiener-Hopf operator with rational symbol can be calculated as well using the suggested finite-dimensional approach: the equation $[W_a - \lambda]u = f,\ f \in Ka$ is reduced for symbol with simple poles to

$$[\varphi(k)-\lambda]u(k) - \sum_{m=1}^{N} \alpha_m \left.\frac{u(p)}{k-p}\right|_{p=k_m} + \theta(k)\bar{\alpha}_m \left.\frac{[\bar{\theta}u(\bar{p})]}{k-\bar{p}}\right|_{p=k_m} = f,$$

which implies the corresponding matrix equation involving the values of the function f at the zeroes $k = k_n^\lambda$ of the shifted symbol $\varphi(k) - \lambda$:

$$\sum_{m=1}^{N} \alpha_m \left.\frac{u(p)}{k_n^\lambda - p}\right|_{p=k_m} + \theta(k_n^\lambda)\bar{\alpha}_m \left.\frac{[\bar{\theta}u(\bar{p})]}{k_n^\lambda - \bar{p}}\right|_{p=k_m} + f(k_n^\lambda) = 0.$$

The matrix of the last system is the so-called paired Cauchy matrix, see ([19]), where the properties of these matrices are discussed. If the determinant of it is different from zero the equation has unique solution $u(k_m)$, $\bar{\theta}\bar{u}(\bar{k_m})$. Then the expression for the resolvent of the operator W_a is given by the formula

$$u = \frac{1}{\varphi(k) - \lambda}\left[\sum_{m=1}^{N} \alpha_m \frac{u(p)}{k-p}\bigg|_{p=k_m} + \theta(k)\bar{\alpha}_m \frac{[\bar{\theta}u(\bar{p})]}{k-\bar{p}}\bigg|_{p=k_m} + f\right],$$

One can check in a similar manner to the preceeding theorem that in this case $u \in K_\theta$.

Remark 2. The asymptotic behaviour of the determinant of the matrix equations discussed above for small λ^δ is defined by the distribution of zeroes of the symbol $\varphi(k)$ and is described by a quasipolynomial in $\lambda^\delta, \exp^{\frac{d}{\lambda^\delta}}$, δ defined by the multiplicity of the zeroes of the symbol. The asymptotics of zeroes of these quasipolynomials for small λ may be derived by using some construction involving Newton polygones ([21], [22]).

Remark 3. The description of the set of all values $k^\lambda : \varphi(k^\lambda) = \lambda$ becomes essentially simpler if the symbol possesses some symmetry defined by a transformation group $G = \{g\} : \varphi(g(k)) = \varphi(k)$. In this case all values of k^λ may be restored from only one $k_g^\lambda = g(k_e^\lambda)$, see the example below.

6. An example

Consider the Wiener-Hopf operator with the symbol

$$\sigma(k) \equiv 1 - \mu\frac{1}{k^2+1}.$$

This operator was studied in details in [13]. Now we apply the proposed approach to calculate the spectral characteristics of it. We use the decomposition of the symbol

$$\sigma(k) = 1 + \frac{i\mu}{2}[\frac{1}{k-i} - \frac{1}{k+i}],$$

and the roots of $k_\pm = \pm i\sqrt{1 - \frac{\mu}{1-\lambda}}$. The eigenfunctions $u = u_\lambda$ of the operator $W_a = P_a\sigma|_{K_a}$ satisfy the equation:

$$[\sigma(k) - \lambda]u(k) = \frac{i\mu}{2}[\frac{u(k)-u(i)}{k-i} + \frac{\theta(k)[\bar{\theta}u(-i)]}{k+i}],$$

Which means that the eigenvalues λ serve as the zeroes of the corresponding determinant:

$$e^{ika}\frac{1}{(i+k)^2} - e^{-ika}\frac{1}{(i-k)^2} \equiv D(k),\ \lambda = 1 - \frac{\mu}{1+k^2},$$

or the following transcendental function of the spectral variable λ

$$e^{2\sqrt{1-\frac{\mu}{1-\lambda}}} = 1 - \frac{2}{\sqrt{1-\frac{\mu}{1-\lambda}} + \frac{1}{\sqrt{1-\frac{\mu}{1-\lambda}}} + 2}.$$

Reducing the last equation to the quasipolynomial form:

$$e^{2ika} = [\frac{k+i}{k-i}]^2.$$

one may calculate the asymptotics of eigenvalues λ or singular values of μ (corresponding to $\lambda = 0$) considering the relevant Newton polygons:

$$k_l = \frac{l\pi}{a} + \frac{1}{\pi l} + o\left(\frac{1}{l^2}\right);$$

$$\lambda_l = 1 - \frac{\mu}{k_\lambda^2 + 1},$$

$$\mu_l = k_\lambda^2 + 1 \approx (\frac{l\pi}{a})^2.$$

It is interesting to compare the determinant $D(k)$ of the finite matrix discussed above with the determinant of the corresponding Wiener-Hopf operator. Note that $k = 0$ is the zero of the determinant $D(k)$ of the auxilliary matrix, hence $\mu_0 = 1$ is the root of the corresponding determinant:

$$\mathrm{Det}[I_a - \mu P_a \frac{1}{1+k^2}|_{K_a}] = \prod_l (1 - \frac{\mu}{\mu_l}).$$

The last product converges uniformly on each compact domain in complex plane producing an entire function of the order $\frac{1}{2}$, which follows immediately from the asymptotics obtained above. One can represent the determinant as the function of the variable k, $\mu = k^2 + 1$:

$$\mathrm{Det}[I_a - (k^2+1) P_a \frac{1}{1+k^2}|_{K_a}] = \prod_{l \neq 0} \frac{k_l^2}{k_l^2 + 1} (-k^2) \prod_l (1 - \frac{k^2}{k_l^2}) \equiv d(k).$$

On the other hand we can *renormalize* the determinant $D(k)$ of the finite matrix multiplying it by $k(k^2+1)$ which will just double the multiplicity of the root $k = 0$. The resulting function of the variable k will be then

$$\tilde{D}(k) \equiv k[e^{ika}(i-k)^2 - e^{-ika}(i+k)^2].$$

It is an even entire function of exponential order 1 with the same zeroes as $d(k)$. Consequently it differs from $d(k)$ by the constant factor

$$-2i(a+1) \prod_{l \neq 0} \frac{1+k_l^2}{k_l^2}.$$

The even function $D(k)$ is an entire function of the order $\frac{1}{2}$ of the variable $k^2 = \mu - 1$. Substituting into $\tilde{D}(k)$ the spectral variable μ we obtain the explicit expression of the determinant of the corresponding Wiener-Hopf operator as an entire function of order $\frac{1}{2}$ of the variable μ.

The analysis of the last example shows that the renormalized determinant of the corresponding auxilliary matrix coincides with the determinant of the initial Wiener-Hopf operator represented as the function of the auxilliary variable k_λ, defined as one of roots of the equation $\sigma(k) = \lambda$. It it assumed that the remaining roots of it are expressed through k_λ via the Galois group of the corresponding polynomial. Sometimes this Galois group is trivial. For instance when considering the symbol

$$\rho(k) \equiv \frac{4k(k^2-1)}{(k^2+1)^2} \equiv \sin 4 \arctan k.$$

we see that using the main branch of the function $\arcsin \frac{\lambda}{4} \equiv \varphi(\lambda)$ we can enumerate the solutions $k_0, k_\infty, k_1, k_{-1}$ of the equation

$$\sigma(k) = \frac{4k(k^2-1)}{(k^2+1)^2} = \lambda = \frac{1}{\mu}$$

as the roots approaching $0, \infty, 1, -1$ respectively when $\lambda \to 0$

$$\begin{aligned} k_0(\lambda) &= \tan \frac{\varphi(\lambda)}{4}, \\ k_\infty(\lambda) &= -\frac{1}{\tan \frac{\varphi(\lambda)}{4}}, \\ k_1(\lambda) &= \frac{1 - \tan \frac{\varphi(\lambda)}{4}}{1 + \tan \frac{\varphi(\lambda)}{4}}, \\ k_{-1}(\lambda) &= \frac{-1}{k_1(\lambda)}. \end{aligned}$$

We see that the group is formed by two transforms $\zeta \to -\frac{1}{\zeta}$ and $\zeta \to \frac{1-\zeta}{1+\zeta}$, which permit us to represent the determinant of the finite matrix as a function of the variable $k_0(\lambda)$. The detailed analysis of the corresponding example with the use of "Mathematica" was done by J. M., see in [28].

References

[1] G. Szegö, *On certain Hermitian forms, associated with the Fourier series of a positive function.* Communications du séminaire mathématique de l'université de Lund, tome supplémentaire (1952) dédié à Marcel Riesz, pp. 228–238.

[2] M. Kac, *Toeplitz matrices, translation kernels and a related problem in probability theory.* Duke Math. J. **21** (1954), 501–509.

[3] K. Hoffman, *Banach spaces of analytic functions.* Prentice-Hall Ser. in Modern Analysis, Prentice Hall, Englewood Cliffs, N. J. (1962).

[4] N. I. Akhieser, *The continuous analogue of some theorems on Toeplitz Matrices.* Ukrain. Math. Zh. **16**:4 (1964) (in Russian); also in Amer. Math. Soc. Trans. (2) **50**, (1966), 295–316.

[5] B. Sz. Nagy, C. Foias. *Analyse Harmonique des Opérateurs de l'Espace de Hilbert.* Academie Kiado, Budapest (1967). English rev. translation North Holland, Amsterdam; American Elsevier, New York; Kiado, Budapest (1970).

[6] H. Widom, *Toeplitz determinants with singular generating function.* Amer. J. Math. **95** (1973), 333–383.

[7] H. Widom, *Szegö limit theorem: higher-dimensional matrix case* J. Func. Anal. **39** (1980), 182–198.

[8] C. Tracy, H. Widom, *Level spacing distribution and the Bessel kernel.* **161** (1994), 289–309.

[9] K. M. Day, *Toeplitz matrices generated by the Laurent series expansion of an arbitrary rational function.* Trans. Amer. Math. Soc. **206** (1975), 294–312.

[10] H. Dym, *On a Szegö formula for a class of generalized Toeplitz kernels.* Integral Equations Operator Theory **8** (1985), 428–431.

[11] A. Böttcher, *Wiener-Hopf Determinants with Rational Symbols.* Math. Nachrichten **144** (1989), 39–64.

[12] J. B. Conway, *A Course in Functional Analysis.* Springer (1985).

[13] I. Gohberg, M. A. Kaashoek, *Classes of linear operators* Vol.**1**, Operator Theory: Advances and Applications Vol. 49, Birkhäuser (1990).

[14] S. Albeverio, K. Makarov, S. Lakaev, *The Efimov effect and an extended Szegö-Kac limit theorem.* Lett. Math. Phys. **43** (1998), 73–85.

[15] L. Carleson, *Interpolation by bounded analytic functions and the corona problem.* Ann. Math. **76** (1962), 547–559.

[16] V. Katsnelson, *On conditions for the basis property of the system of root vectors of some classes of operators.* Funct. Anal. Prilozhenija. **1**, no. 2 (1967), 39–51.

[17] N. K. Nikolski, B. S. Pavlov, *Bases of eigenvectors of comletely nonunitary contractions and characteristic function.* Izvestia USSR Acad. Sc. **34** (1970), 90–133.

[18] S. V. Khruschev, N. K. Nikolski, B. S. Pavlov, *Unconditional bases of exponentials and reproducing kernels.* In: Complex analysis and spectral theory, Lecture notes in Mathematics **864**, Springer Verlag (1981), pp. 214–335.

[19] G. Heinig, L. A. Sakhnovich, I. F. Tidniuk, *Paired Cauchy Matrices.* Linear Algebra and its Applications. **249** (1996), 189–214.

[20] I. C. Gohberg, M. G. Krein, *Introduction to the Theory of Linear Nonselfadjoint Operators.* Trans. Math. Monographs **18**, Amer. Math. Soc., Providence, R.I. (1969).

[21] I. Horn, *Asymptotic distribution of roots of quasipolynomials.* Mathematische Annalen. Bd. **49** (1897) (in German).

[22] B. S. Pavlov, *Spectral Analysis of the differential operator with a spread boundary condition.* In: Problems of Mathematical Physics **6**, Leningrad University (1972), pp. 101–119 (in Russian).

[23] B. S. Pavlov, *Basicity conditions of an exponential system and Muckenhoupt's condition.* Soviet Math. Dokl. **20**:4 (1979), 655–659.

[24] M. Voichik, *Ideals and invariant subspaces of analytic functions.* Trans. Amer. Mat. Soc., Vol **111** (1964), pp. 193–512.

[25] B. Pavlov, S. Fedorov, *The group of shifts and harmonic analysis on a Riemann surface of genus one.* Leningrad Math. Journal **1** (1989), 132–168.

[26] B. Pavlov, *Zero index of pair of projections and expansions by resonance states for operators with band spectrum.* Wien, Preprint ESI **139** (1994).

[27] P. Lax, R. Phillips, *Scattering Theory.* Academic Press, New-York-London (1967).

[28] J. P. MacCormick, B. Pavlov, *A geometrical approach to calculation of determinants of Wiener-Hopf Operators.* In: Quantum Theory in the Rigged Hilbert Space – Irreversibility and Causality (Reviews and Contributions of the Semigroup Symposium at the XXI International Colloquium on Group Theoretical Methods in Physics), Ed. A. Bohm, H.-D. Doebner, P. Kielanowski, Springer Verlag (1998).

[29] C. Foias, A. Tannenbaum, G.Zames, *On the H^∞-optimal sensitivity problem for systems with Delays.* SIAM J. Control and Optimization **25**, No.1 (1987), 686–705.

[30] V. Buslaev, A. Budylin, *Quasiclassical asymptotics of the resolvent of an intergal convolution with a sine kernel on a finite interval.* St. Petersburg Math. Journal **7**:6 (1996), 925–942.

Department of Mathematics, University of Auckland, Private Bag 92019, Auckland, New Zealand

Operator Theory:
Advances and Applications, Vol. 123

Operator Dilations with Prescribed Commutators

Mihai Putinar

Paper dedicated to Moshe Livšic on the occasion of his eightieth birthday

Abstract. A dilation problem with prescribed commutator for pairs of self-adjoint matrices is related constructively to the truncated moment problems in the plane for positive measures as well as for bounded functions. A series of particular cases of this dilation problem reveals its deep connections with the structure of some classes of semi-normal operators.

Introduction

The recent work of Moshe Livšic and his collaborators was devoted to special classes of systems of commuting Hilbert space operators, cf. [L]. When represented in cartesian form and in a proper basis, these systems give rise to tuples of self-adjoint Jacobi matrices with commutator and rank constraints. We discuss below, in resonance with this subject, a particular problem issued from the study of pairs of commuting or almost commuting Jacobi matrices.

The last two decades progress in multivariate cubature formulas was recently translated and analyzed in terms of dilations of tuples of self-adjoint operators, [P]. The aim of the present note is to isolate from this work a basic dilation theory problem and to investigate it in the simplest case of two self-adjoint operators. By reversing the operator theory approach to cubature formulas we will find canonical parametrizations of all solutions of some truncated moment problems in two variables.

Let H_0, H_1 be two finite dimensional Hilbert spaces, let $A, B, C : H_0 \longrightarrow H_1$ be linear operators and let X_1, Y_1, D be self-adjoint operators acting on H_1. The problem to be investigated in the sequel is the following:

Find and describe all self-adjoint dilations (X, Y) of (X_1, Y_1), acting on the Hilbert space $H = H_1 \oplus H_2$, which satisfy:

$$2i[X,Y] = \begin{pmatrix} D & 0 \\ 0 & 0 \end{pmatrix}, \tag{1}$$

Paper partially supported by the National Science Foundation Grant DMS-9500954.

and

$$X\begin{pmatrix} A \\ 0 \end{pmatrix} + Y\begin{pmatrix} B \\ 0 \end{pmatrix} = \begin{pmatrix} C \\ 0 \end{pmatrix}. \tag{2}$$

By a *dilation* (X, Y) of the pair (X_1, Y_1) we mean a pair of bounded self-adjoint operators of the form:

$$X = \begin{pmatrix} X_1 & A_2{}^* \\ A_2 & X_2 \end{pmatrix}, \quad Y = \begin{pmatrix} Y_1 & B_2{}^* \\ B_2 & Y_2 \end{pmatrix}. \tag{3}$$

We call the pair (X, Y) a *minimal dilation* of (X_1, Y_1) whenever

$$H = \bigvee_{k,l=1}^{\infty} X^k Y^l H_1.$$

The order of X and Y in this definiton does not matter because the commutator $[X, Y]$ has its range contained in H_1. The first line in equation (2) is actually a relation among the given data: $X_1A + Y_1B = C$. Only the sceond line in (2) is imposing a condition on the matrix entries of X, Y: $A_2A + B_2B = 0$. Note that, although we start with finite dimensional Hilbert spaces H_0, H_1, we do not require the same for the Hilbert space H_2.

We will prove that the truncated moment problem for positive measures with compact support in the plane is equivalent to this dilation theory problem, in the sense that there exists a constructive bijection between all solutions of the moment problem and all minimal dilations of some associated pair (X_0, Y_0) with fixed data A, B, C. More surprising is the fact that the same dictionary, with appropriate data, can be established for the truncated moment problem of bounded measurable functions with compact support in $\mathbf{R}^2$ (the so-called L-problem of moments). A general explanation of this coincidence lies in the structure of the nearly commuting pairs of self-adjoint Jacobi matrices associated with the two problems. On the other hand, it is well known that in the case of one real variable the two moment problems are equivalent via an exponential of their generating functions, see [AK].

Commutator equations like (1) are classical in scattering theory. We will make explicit this connection in an example contained in Section 3. In particular, we prove there that equation (1) has always a solution, for arbitrary (finite dimensional) data X_1, Y_1, D, where $D > 0$.

By arranging the pair (X, Y) into a single operator $T = X + iY$ we are led to consider the above problem for normal, subnormal or hyponormal operators. In this way the solutions of our dilation problem will turn out to be natural linear parameters for quadrature domains in the plane, as explained in more detail in the paper [GP]. Relations (1) and (2) reflect the block structure of the (extremal) operators associated with quadrature domains. Connections with a series of very interesting works in progress can therefore be established, cf. [CF], [MY], [Xia], [Y].

Finally, and from a different perspective, we have to mention the work of Xu on multivariate cubature formulas, focused on the structure of commuting Jacobi matrices associated to a positive measure in the Euclidean space, [Xu]. The above dilation theory problem corresponds, at the abstract level of operator theory, to the generating recurrence relation satisfied by some Jacobi matrices.

1. The truncated moment problem

Let μ_0 be a positive measure with compact support in $\mathbf{C}$ and let $z = x+iy$ denote the complex coordinate. For a fixed degree d the power moments of the measure μ_0 are:

$$a_{mn}(\mu_0) = \int_{\mathbf{C}} z^m \overline{z}^n d\mu_0(z), \quad 0 \leq m \leq m+n \leq d. \tag{4}$$

We do not discuss here the intrinsinc characterization of the moment sequence, cf. [A], [CF]. We are rather interested in the set of all solutions of the corresponding truncated moment, namely:

$$\Sigma_d(\mu_0) = \{\mu \in \mathcal{M}_+; \quad a_{mn}(\mu) = a_{mn}(\mu_0), \quad 0 \leq m \leq m+n \leq d\}.$$

Above $\mathcal{M}_+$ stands for the set of positive measure with compact support in $\mathbf{C}$. The convexity properties of the set $\Sigma_d(\mu_0)$ are analyzed in detail in the monograph [KS].

Theorem 1. *Given the degree d and the measure μ_0 there are Hilbert spaces H_0, H_1 and linear operators A, B, C, X_0, Y_0, such that the minimal dilations (X,Y) satisfying (1), (2) are in bijective correspondence with the set $\Sigma_{2d+1}(\mu_0)$.*

The rest of thie section is devoted to the proof of Theorem 1 and to the constructive aspects of the relation between the data in its statement.

We begin with a few simple block-matrix computations. Let

$$M = \begin{pmatrix} X_0 & A_1{}^* & 0 \\ A_1 & X_1 & A_2{}^* \\ 0 & A_2 & X_2 \end{pmatrix}, \qquad N = \begin{pmatrix} Y_0 & B_1{}^* & 0 \\ B_1 & Y_1 & B_2{}^* \\ 0 & B_2 & Y_2 \end{pmatrix}, \tag{5}$$

be two bounded self-adjoint operators acting on the Hilbert space $K = H_0 \oplus H_1 \oplus H_2$. Then the commutator $[M,N]$ is a skew-symmetric operator of the form:

$$\begin{pmatrix} [X_0,Y_0] - B_1{}^*A_1 + A_1{}^*B_1 & * & * \\ A_1Y_0 - Y_1A_1 + X_1B_1 - B_1X_0 & A_1B_1{}^* - B_1A_1{}^* + [X_1,Y_1] + A_2{}^*B_2 - B_2{}^*A_2 & * \\ A_2B_1 - B_2A_1 & A_2Y_1 - Y_2A_2 - B_2X_1 + X_2B_2 & A_2B_2{}^* - B_2A_2{}^* + [X_2,Y_2] \end{pmatrix}. \tag{6}$$

Let P_j denote the orthogonal projection of K onto the subspace H_j, $j = 0,1,2$. With the notation introduced before, (3), we can state the following observation.

Lemma 1. *If the commutator $[M,N]$ has range in the subspace H_0, that is $[M,N] = P_0[M,N]P_0$, then there are operators $A, B, C : H_0 \longrightarrow H_1$ and $D : H_1 \longrightarrow H_1, D^* = D$, so that relations (1) and (2) are fulfilled.*

Proof. According to the above explicit computations we take:

$$A = B_1, \quad B = -A_1, \quad C = B_1X_0 - A_1Y_0, \quad D = 2i(B_1{A_1}^* - A_1{B_1}^*).$$

Corollary 1. *The set of all self-adjoint operators M, N above with PMP, PNP fixed and $[M,N] = P_0[M,N]P_0$, where $P = P_0 + P_1$, is given by the dilations (X,Y) of (X_1,Y_1) satisfying relations (1) and (2).*

In other terms we fix $X_0, X_1, Y_0, Y_1, A_1, B_1$ and consider all possible completions of the matrices M, N with prescribed commutator, which in addition is non-trivial only on the left-upper corner entry. This simple corollary will be the basic tool for the rest of the note.

Proof of Theorem 1. Fix the measure μ_0 and the degree d. Let $H_0 \subset {H_1}'$ be the subspaces of $L^2(\mathbf{C}, d\mu_0)$ generated by polynomials $p(z,\overline{z})$ of degree $\deg(p) \leq d-1$, respectively $\deg(p) \leq d$. Denote $H_1 = {H_1}' \ominus H_0$. Let M_0, N_0 denote the multiplication operators by x, respectively y, on $L^2(\mathbf{C}, d\mu_0)$. With respect to the direct sum decomposition $L^2(\mathbf{C}, d\mu_0) = H_0 \oplus H_1 \oplus {H_1}'^{\perp}$, the commuting self-adjoint operators M_0, N_0 have block-matrices of the form (5).

By prescribing the moments $a_{mn}(\mu_0), m+n \leq 2d+1$, is equivalent to give the ${H_1}'$-compressions of M_0, N_0. Moreover, any other solution $\mu \in \Sigma_{2d+1}(\mu_0)$ of the associated truncated moment problem corresponds to a pair of commuting self-adjoint operators (M,N), with cyclic vector $\mathbf{1} \in H_0$ and the same ${H_1}'$-compression. Thus M, N have similar block-matrices as M_0, N_0, with only the entries A_2, B_2, X_2, Y_2 possibly altered. See for details [P].

Since $[M,N] = [M_0,N_0] = 0$ and $L^2(\mathbf{C}, d\mu) = \bigvee_{k,l=0}^{\infty} M^kN^kH_0$, the pair (M,N) corresponds to a minimal normal dilation of (M_0,N_0). Thus Corollary 1 applies and Theorem 1 is proved.

Note that in the above proof we do not exclude the case $H_0 = {H_1}'$. In this situation $H_0 = L^2(\mathbf{C}, d\mu_0)$ and the set $\Sigma_{2d+1}(\mu_0)$ reduces to a single element. These are exactly the extremal solutions of the truncated moment problem, cf. [KS].

A parametrization of all elements $\mu \in \Sigma_{2d+1}(\mu_0)$ with the property that

$$\dim(L^2(\mathbf{C}, d\mu)) < \infty$$

is discussed in detail in the paper [P]. This subset of solutions to the truncated moment problem corresponds to all exact quadratures of degree $2d+1$ for the measure μ_0, see also [My], [Xu].

As for the relation between the moments $a_{mn}(\mu_0), m+n \leq 2d+1$, and the data in the statement of Theorem 1, we simply remark that the $L^2(\mathbf{C}, d\mu_0)$-scalar products:

$$\langle zp, q \rangle = \langle zp\overline{q}, \mathbf{1} \rangle, \qquad \max(\deg(p), \deg(q)) \leq d,$$

are determined by these moments. Thus, up to a choice of orthonormal bases in the finite dimensional Hilbert space $H_0 \oplus H_1$, all the matrix entries X_0, X_1, A_1, Y_0, Y_1, B_1 can be determined. According to Lemma 1, the matrices A, B, C, D are also known.

2. The truncated L-moment problem

In analogy with the previous section we can treat now the truncated L-problem in two variables. We normalize the constant $L = 1$ and consider measures of the form $d\mu_0 = g_0 dA$, where $g : \mathbf{C} \longrightarrow [0, 1]$ is a measurable function with compact support and dA stands for the area measure.

The inverse problem for Pincus' principal function was exploited in a series of papers in order to investigate the moment sequence of $g_0 dA$, see [GP] and the references there. From this framework we recall that the principal function g_0 is a complete unitary invariant for an irreducible hyponormal operator $T_0 : H \longrightarrow H$ with rank-one self-commutator: $[{T_0}^*, T_0] = \xi \otimes \xi$. By specifying the moments $a_{mn}(g_0 dA)$, $\ m+n \leq 2d+1$, is equivalent to give the compression of the operator T_0 to the subspace ${H_1}' = \bigvee_{k+l=0}^{d} {T_0}^k {T_0}^{*l} \xi$. (For a proof one uses a series of commutator identities based on [MP], Lemma 2.1.4.) The order of T_0 and ${T_0}^*$ does not count in generating the space ${H_1}'$ because their commutator is $\xi \otimes \xi$. Let us introduce also the spaces $H_0 = \bigvee_{k+l=0}^{d-1} {T_0}^k {T_0}^{*l} \xi$ and $H_1 = {H_1}' \ominus H_0$.

The self-adjoint operators $M_0 = \Re T_0, N_0 = \Im T_0$ satisfy:

$$2i[M_0, N_0] = \xi \otimes \xi, \qquad M_0 H_0 + N_0 H_0 \subset {H_1}'. \tag{7}$$

Therefore M_0, N_0 have matrices of the form (5), with respect to the decomposition $H = H_0 \oplus H_1 \oplus {H_1'}^{\perp}$.

Any other measurable function $g : \mathbf{C} \longrightarrow [0, 1]$ with the same moments

$$a_{mn}(g dA) = a_{mn}(g_0 dA), \quad m + n \leq 2d + 1,$$

corresponds to a hyponormal operator $T : H \longrightarrow H$ satisfying $[T^*, T] = \xi \otimes \xi$ and having the compression to ${H_1}'$ identical to that of the operator T_0. Thus the cartesian parts M, N of $T, T = M + iN$, have the same upper-left 2×2 entries as M_0, N_0. By repeating the proof of Theorem 1, we are thus led to the following result.

Theorem 2. *Let $d > 1$ be a fixed degree and let $g_0 : \mathbf{C} \longrightarrow [0, 1]$ be a measurable function with compact support. Then there are finite dimensional Hilbert spaces H_0, H_1, linear operators $A, B, C : H_0 \longrightarrow H_1$ and self-adjoint operators $X_1, Y_1, D : H_1 \longrightarrow H_1$, such that there exists a bijection between the minimal dilations (X, Y) of (X_1, Y_1) satisfying (1), (2) and the set of measurable functions with compact support:*

$$\{g : \mathbf{C} \longrightarrow [0, 1]; \quad g dA \in \Sigma_{2d+1}(g_0 dA)\}.$$

The connection between the moments $a_{mn}(g_0)$, $m+n \le 2d+1$, and the scalar products $\langle T_0 {T_0}^i {T_0^*}^j \xi, {T_0}^k {T_0^*}^l \xi \rangle; i+j, k+l \le d$ is more subtle but also algorithmic in its nature, see [MP], [GP]. We recall that the self-adjoint pair (X, Y) appearing in Theorem 2 has necessarily infinite rank.

3. Semi-normal dilations

The structure of pairs of self-adjoint operators with prescribed "small" commutator described in the previous sections goes beyond the applications to truncated moment problems. In this section we focus on relation (5), imposed on a few classes of semi-normal operators $T = X + iY$.

a. Normal dilations

The following result is a classical application of Naimark's dilation theorem, cf. [RN], Appendix 1.

Proposition 1. *Let (X_1, Y_1) be a pair of self-adjoint operators, acting on the finite dimensional Hilbert space H_1. Then there exists a commutative pair (X, Y) of self-adjoint operators which dilates (X_1, Y_1).*

Such a minimal dilation (X, Y) can always be realized in the canonical form:

$$Xf(x, y) = xf(x, y), \qquad Yf(x, y) = yf(x, y),$$

where f is a vector-valued function belonging to the direct integral space $H = \int_{\mathbf{C}}^{\oplus} H(x, y) d\sigma(x, y)$, $d\sigma$ is a positive scalar measure with compact support in $\mathbf{C}$ and $H(x, y)$ is a measurable family of subspaces of H_1; for details see for instance [MP], Chapter 6.

In this functional model, the additional condition (2) is easily implementable. Indeed, assume that the matrices A, B, C are given and the first line of condition (2) is satisfied. Then it suffices to remark that by changing the fibres $H(x, y)$ to:

$$K(x, y) = H(x, y) \ominus \{xA(v) + yB(v) - C(v); \quad v \in H_0\}$$

the new operators of multiplication by (x, y) on $\int_{\mathbf{C}}^{\oplus} K(x, y) d\sigma(x, y)$ are still dilating (X_1, Y_1) and they satisfy in addition condition (2).

b. Subnormal dilations

To give a dilation (X, Y) of the pair of self-adjoint operators (X_1, Y_1) is equivalent to dilate the operator $T_1 = X_1 + iY_1$ to $T = X + iY$. In addition, $[T^*, T] = 2i[X, Y]$.

Proposition 2. *Let T_1, D, be operators on the finite dimensional Hilbert space H_1 and assume that $D > 0$. Then there is at most one subnormal minimal dilation T of T_1 satisfying $[T^*, T] = D$.*

Proof. Let T be a subnormal operator as in the statement. Then T^* leaves invariant the space $Ran(D) = H_1$ and the pair $(T^*|H_1, D)$ is a complete system of unitary invariants for T, see [MP] Chapter 1. Hence T, if it exists, is unique.

The characterization of those pairs of operators (T_1, D) as in Proposition 2, for which a subnormal dilation T exists, is subtle and it makes the subject of several recent parallel investigations, [MY], [X], [Y]. We confine to remark that in this case the spectrum of T is the closure of a quadrature domain Ω, in the terminology of Aharonov and Shapiro, and that T can be realized as the multiplication with the complex variable on natural analytic function spaces carried by Ω.

c. Hyponormal dilations

Finally we are interested in dilations (X, Y) satisfying relation (1) with $D > 0$ on H_1. The operator $T = X + iY$ will in this case be hyponormal with finite-rank self-commutator equal to D.

Theorem 3. *Let $X_1, Y_1, D > 0$, be self-adjoint operators acting on the finite dimensional Hilbert space H_1. Then there is a pair of self-adjoint operators (X, Y) which dilates (X_1, Y_1) and satisfying condition (1).*

Proof. Assume that the pair (X, Y) is minimal. Then there exists a measurable function with compact support $B : \mathbf{C} \longrightarrow L(H_1)$, $0 \le B \le I$, called the *mosaic* of (X, Y), such that:

$$\int_{\mathbf{C}} B(x,y)dxdy = \pi D, \qquad \int_{\mathbf{C}} B(x,y)h(y)dxdy = \pi D^{1/2}h(Y_1)D^{1/2}, \tag{8}$$

for every bounded measurable function $h(y)$. For details see [MP], Chapter 9, and the references contained there to the original work of Carey and Pincus.

In addition, for every such function B there exists a pair of self-adjoint operators (X', Y) satisfying (8), where $D = 2i[X', Y]$ and Y_1 is the compression of Y to the range of D, cf [CP].

Equations (8) leave a lot of freedom in choosing the function B. In particular we can find a $B(x, y)$ satisfying (8) and such that the matrix entries $B_{jj}(x, y)$, $\Re B_{jk}(x, y)$, $\Im B_{jk}(x, y)$, $1 \le j, k \le \dim(H_1)$, $j \ne k$, are all linearly independent over $\mathbf{R}$. Indeed, there exists a positive ϵ such that there exists a solution B' of the equations (8) satisfying $\epsilon \le B'(x, y) \le (1-\epsilon)I$ for almost all $(x, y) \in S$ where $S = \mathrm{supp}(B')$. Then we define $B = B' + C$, where $C(x, y)$ is an appropriate self-adjoint matrix of norm less than ϵ and satisfying: $\int_S C(x, y)dxdy = \int_S yC(x, y)dxdy = 0$.

The operator X', which is subject to the constraint $2i[X', Y] = D$, can be changed into $X = X' + h(Y)$, with $h(y)$ an arbitrary bounded, real valued measurable function. But then the matrix elements of

$$\pi D^{1/2} X D^{1/2} = \pi D^{1/2} X' D^{1/2} + \int_S B(x,y)h(y)dxdy$$

can be chosen so that X has the prescribed compression X_1 on the subspace H_1. This finishes the proof of Theorem 3.

In parallel to Proposition 2, one can show that in the case $D \ge 0, rank(D) = 1$, the equation (1) has at most one minimal solution $T = X + iY$ satisfying $T^* H_1 \subset H_1$. In this situation the operator T is given by a two block-diagonal matrix whose entries depend only on the pair $\Lambda = (D, T^*|H_1)$. The spectrum of T is again the closure of a quadrature domain Ω. The paper [GP] is devoted to the dictionnary relating in this way quadrature domains Ω to their linear parameters Λ.

References

[A] Akhiezer, N. I., *The classical moment problem and some related questions in analysis*, Hafner, New York, 1965.

[AK] Akhiezer, N. I. and Krein, M. G., *Some questions in the theory of moments*, Transl. Amer. Math. Soc. vol. 2, Providence, R.I., 1962.

[CF] Curto, R. and Fialkow, L., *Solution of the truncated complex moment problem for flat data*, Mem. Amer. Math. Soc. vol. 568, Providence, R.I., 1996.

[CP] Carey, R. and Pincus, J. D., Construction of seminormal operators with prescribed mosaic, Indiana Univ. Math. J. 23 (1974), 1155–1165.

[GP] Gustafsson, B. and Putinar, M., Linear analysis of quadrature domains. II, preprint 1997.

[KS] Karlin, S. and Studden, W. J., *Tchebycheff systems: with applications in Analysis and Statistics*, Interscience Publ., New York, 1966.

[L] Livšic, M. et al., *Theory of commuting nonselfadjoint operators*, Kluwer, Dordrecht, 1995.

[MP] Martin, M. and Putinar, M., *Lectures on hyponormal operators*, Birkhäuser, Basel, 1989.

[MY] McCarthy, J. and Yang, L., Subnormal operators and quadrature domains, Adv. Math. 127 (1997), 52–72.

[My] Mysovskikh, I. P., *Interpolatory cubature formulas* (in Russian), Nauka, Moscow, 1981.

[P] Putinar, M., A dilation theory approach to cubature formulas. II, Math. Nachr., to appear.

[RN] Riesz, F. and Sz.-Nagy, B., *Functional analysis*, Dover, New York, 1990.

[X] Xia, D., Hyponormal operators with finite-rank self-commutator and quadrature domains, J. Math. Anal. Appl. 203 (1996), 540–559.

[Xu] Xu, Y., On orthogonal polynomials in several variables, Fields Inst. Comm. 14 (Special functions, q-series and related topics, Ismail, M. E. H. et al., eds.), American Mathematical Society, Providence, 1997, pp. 247–270.

[Y] Yakubovich, D., Subnormal operators of finite type. I: Xia's model and real algebraic curves in $\mathbf{C}^2$, Rev. Mat. Iberoamericana 14 (1998), 95–115.

Department of Mathematics, University of California, Santa Barbara, CA 93106, USA

E-mail address: mputinar@math.ucsb.edu

Operator Theory:
Advances and Applications, Vol. 123

Condition of Orthogonality of Spectral Matrix Function

L. A. Sakhnovich

This paper is dedicated to Professor M.S. Livšic on the occasion of his 80th birthday. I am proud and happy that I was a student of this outstanding mathematician.

Introduction

Let us consider the system of the differential equations

$$\begin{cases} \frac{dP_1}{dx} = izP_1 + A(x)P_2 \\ \frac{dP_2}{dx} = A^*(x)P_1 \end{cases} \qquad 0 \le x < \infty, \quad z \in \mathbb{C}. \tag{0.1}$$

Here $P_1(x,z), P_2(x,z)$ and $A(x)$ are $m \times m$ matrix functions. We shall further suppose that the matrix function $A(x)$ and the solution $P_1(x,z), P_2(x,z)$ of system (0.1) satisfy the conditions

$$P_1(0,z) = S_1, \qquad P_2(0,z) = S_2, \tag{0.2}$$

where S_1 and S_2 are constant matrices such that

$$S_1^*S_1 = S_2^*S_2, \qquad det\ S_k \neq 0 \ \ (k=1,2). \tag{0.3}$$

System (0.1) coincides with the well-known Krein system (see [1]) when

$$m = 1, \qquad S_1 = S_2 = 1. \tag{0.4}$$

Definition 0.1. A monotonically increasing $m \times m$ matrix function $\tau(u)(-\infty < u < \infty)$ is called a spectral matrix function of boundary problem (0.1)–(0.3) if the operator

$$F(u) = Vg = \int_0^\infty P_1^*(x,u)g(x)dx \tag{0.5}$$

isometrically maps the space $L_m^2(0,\infty)$ into the space $L_m^2(\tau)$, i.e., the Parseval equality

$$\int_{-\infty}^\infty F^*(u)[d\tau(u)]F(u) = \int_0^\infty g^*(x)g(x)dx \tag{0.6}$$

holds (see [1], [2]).

Definition 0.2. A spectral matrix function $\tau(u)$ is called an orthogonal spectral matrix function if the operator V unitarily maps the space $L^2_m(0,\infty)$ onto the space $L^2_m(\tau)$.

When relations (0.4) are fulfilled, the equality

$$\int_{-\infty}^{\infty} \frac{\ln \tau'(u)}{1+u^2}\, du = -\infty \tag{0.7}$$

is a necessary and sufficient condition of orthogonality of the spectral function $\tau(u)$ (M.G. Krein [1]).

In this paper we find a sufficient condition of orthogonality of $\tau(u)$ when $m \geq 1$. This condition is an analogue of relation (0.7).

Equality (0.7) is a necessary and sufficient condition of singularity of the corresponding stationary process with continuous time (M.G. Krein [3]). For the stationary process with discrete time an analogous condition was found earlier (A.N. Kolmogorov [4]).

In this article we prove that the condition of orthogonality of the spectral $m \times m$ matrix function $\tau(u)$ is at the same time a condition of singularity of the corresponding m-stationary process. When $m = 1$ both problems (orthogonality of $\tau(u)$ and the singularity of stationary process) are closely connected with the problem of the completeness of the functions system

$$\left(\frac{u-i}{u+i}\right)^k \frac{1}{u+i} \quad (k = 0, 1, 2, \dots)$$

in the space $L^2(\tau)$. In the case when $m \geq 1$ we shall consider the system of the vector functions

$$G_{kj}(u) = \left(\frac{u-i}{u+i}\right)^k \frac{1}{u+i}\, h_j (k = 0, 1, 2, \dots ;\ j = 1, 2, \dots , m), \tag{0.7}$$

where the $m \times 1$ vector columns $h_j (j = 1, 2, \dots , m)$ are linearly independent.

1. On completeness of exponential vector functions

1. Let $\tau(\theta)$ be monotonically increasing on the segment $[0,\ 2\pi]$ $m \times m$ matrix function such that

$$\int_0^{2\pi} d[\operatorname{Tr} \tau(\theta)] < \infty. \tag{1.1}$$

We denote by $\tilde{L}^2_m(\tau)$ the space of the $m \times 1$ vector columns $f(\theta)$ with the scalar product

$$(f, g) = \int_0^{2\pi} g^*(\theta)[d\tau(\theta)]f(\theta). \tag{1.2}$$

By U_h we denote the set of the vector functions

$$g_h(\theta, k) = e^{ik\theta} h; \quad k = 0, 1, 2, \dots , \tag{1.3}$$

where h is a constant $m \times 1$ vector column and $h \neq 0$. Let a vector function $\varphi(\theta)$ from $\tilde{L}^2_m(\tau)$ be orthogonal to all the vector functions from U_h. Then the following equalities

$$\frac{1}{2\pi} \int_0^{2\pi} \varphi^*(\theta)[d\tau(\theta)]e^{ik\theta}h = 0; \quad k = 0, 1, 2, \ldots \tag{1.4}$$

are valid.

The relations (1.4) are equivalent to the equality

$$\frac{1}{2\pi} \int_0^{2\pi} \frac{\varphi^*(\theta)[d\tau(\theta)]}{e^{i\theta} - z} h = 0, \ |z| > 1. \tag{1.5}$$

As in the case when $m = 1$ (see [5]) we introduce the notation

$$\lambda_h(z) = \frac{1}{2\pi} \int_0^{2\pi} \frac{e^{i\theta}}{e^{i\theta} - z} \, d\tau_h(\theta), \tag{1.6}$$

where

$$d\tau_h(\theta) = e^{-i\theta}\varphi^*(\theta)d\tau(\theta)h. \tag{1.7}$$

Since

$$\frac{1}{2\pi} \int_0^{2\pi} |d\tau_h(\theta)| = \frac{1}{2\pi} \int_0^{2\pi} |\varphi^*(\theta)d\tau(\theta)h| \leq \tag{1.8}$$

$$\leq \frac{1}{2\pi} \left\{ \int_0^{2\pi} \varphi^*(\theta)[d\tau(\theta)]\varphi(\theta) \right\}^{1/2} \left[\int_0^{2\pi} h^* d\tau(\theta)h \right]^{1/2},$$

$\tau_h(\theta)$ is a function of bounded variation. Hence the function $\lambda_h(z)$, defined by (1.6), is analytic in the circle $|z| < 1$. From relations (1.5) and (1.6) we deduce that

$$\lambda_h(z) = 0, \ \ |z| > 1. \tag{1.9}$$

Thus the right-hand side of (1.6) is the Cauchy-Stieltjes integral. It means that there exist almost everywhere on segment $[0, 2\pi]$ limit values

$$\lambda_h(e^{i\theta}) = \lim_{r \to 1-0} \lambda_h(re^{i\theta}) = \tau_h'(\theta) = e^{-i\theta}\varphi^*(\theta)\tau'(\theta)h \tag{1.10}$$

and

$$\int_0^{2\pi} |\lambda_h(e^{i\theta})| d\theta < \infty. \tag{1.11}$$

Here $\tau'(\theta)$ is the derivative of the absolutely continuous part of $\tau(\theta)$.

If

$$\lambda_h(z) \not\equiv 0 \tag{1.12}$$

the inequality

$$\int_0^{2\pi} \ln |\lambda_n(e^{i\theta})| d\theta > -\infty \tag{1.13}$$

holds (see [6], Chapter II, Section 5).

In view of (1.10) and (1.13) we have

$$\int_0^{2\pi} \ln|\varphi^*(\theta)\tau'(\theta)h|^2 d\theta > -\infty. \tag{1.14}$$

Using the inequality

$$|\varphi^*(\theta)\tau'(\theta)h|^2 \leq [\varphi^*(\theta)\tau'(\theta)\varphi(\theta)] \cdot [h^*\tau'(\theta)h]$$

we deduce from (1.14) that

$$\int_0^{2\pi} \ln[\varphi^*(\theta)\tau'(\theta)\varphi(\theta)]d\theta + \int_0^{2\pi} \ln[h^*\tau'(\theta)h]d\theta > -\infty. \tag{1.15}$$

It is easy to see that the inequalities

$$\int_0^{2\pi} \ln^+[h^*\tau'(\theta)h]d\theta \leq \int_0^{2\pi} h^*\tau'(\theta)hd\theta < \infty. \tag{1.16}$$

$$\int_0^{2\pi} \ln^+[\varphi^*(\theta)\tau'(\theta)\varphi(\theta)]d\theta \leq \int_0^{2\pi} \varphi^*(\theta)\tau'(\theta)\tau'\varphi(\theta)d\theta < \infty. \tag{1.17}$$

are true.

Comparing inequalities (1.15)–(1.17) we obtain

$$\int_0^{2\pi} \ln^-[h^*\tau'(\theta)h]d\theta + \int_0^{2\pi} \ln^-[\varphi^*(\theta)\tau'(\theta)\varphi(\theta)]d\theta > -\infty. \tag{1.18}$$

Due to (1.16) and (1.18) the following assertion is true.

Lemma 1.1. Let the vector function $\varphi(\theta)$ from $L^2_m(\tau)$ be orthogonal to all the vector functions from U_h. Then one of the following relations

$$\lambda_h(z) \equiv 0, \tag{1.19}$$

$$\int_0^{2\pi} \ln[h^*\tau'(\theta)h]d\theta > -\infty \tag{1.20}$$

is fulfilled.

(Here U_h and $\lambda_h(z)$ are defined by (1.3) and (1.6) respectively.)

Theorem 1.1. Assume that for all constant $m \times 1$ vectors h, $h \neq 0$, the relation

$$\int_0^{2\pi} \ln[h^*\tau'(\theta)h]d\theta = -\infty \tag{1.21}$$

holds. Then the system of vector functions

$$g_kj(\theta) = e^{ik\theta}h_j(k = 0, 1, 2, \dots ; \ j = 1, 2, \dots , m), \tag{1.22}$$

where $h_1, h_2, \dots , h_m$ are constant linearly independent $m \times 1$ vectors, is complete in the space $\tilde{L}^2_m(\tau)$.

Proof. We suppose that there exists in the space $\tilde{L}^2_m(\tau)$ a vector function $\varphi(\theta)$ which is orthogonal to all the sets U_h. Then according to Lemma 1.1 and condition (1.21) equality (1.19) is true for all h, i.e.,

$$\int_0^{2\pi} e^{ik\theta} d\tau_h(\theta) = 0; \quad k = 0, \pm 1, \pm 2, \dots . \tag{1.23}$$

It follows from (1.23) that $d\tau_h(\theta) = 0$. Hence the equality

$$\varphi^*(\theta) d\tau(\theta) = 0 \tag{1.24}$$

is valid. Due to (1.24) the vector function $\varphi(\theta)$ is equivalent to zero in the space $\tilde{L}^2_m$. The theorem is proved.

Remark 1.1. If the condition

$$\int_0^{2\pi} \ln \parallel \tau'(\theta) \parallel d\theta = -\infty \tag{1.25}$$

is fulfilled then condition (1.21) is also fulfilled and the system of vector functions (1.22) is complete in the space $\tilde{L}^2_m(\tau)$.

From Theorem 1.1 we obtain the following assertion.

Corollary 1.1. Let one of the conditions (1.21) or (1.25) be fulfilled. Then the m-stationary process with discrete time and with the spectral matrix function $\tau(\theta)$ is singular.

Remark 1.2. A.N. Kolmogorov [4] proved that in the case when $m = 1$ the equality

$$\int_0^{2\pi} \ln \tau'(\theta) d\theta = -\infty$$

is the necessary and sufficient condition of singularity of the corresponding stationary process with discrete time.

Let us consider the monotonically increasing $m \times m$ matrix function $\tau(u)$ on the axis $-\infty < u < \infty$. We assume that

$$\int_{-\infty}^{\infty} \frac{d\tau(u)}{1 + u^2} < \infty. \tag{1.26}$$

By $L^2_m(\tau)$ we denote the space of the $m \times 1$ vector functions $f(u)$ with the scalar product

$$(f, g) = \int_{-\infty}^{\infty} g^*(u)[d\tau(u)]f(u). \tag{1.27}$$

Setting

$$e^{i\theta} = \frac{i - u}{i + u}, \quad z = \frac{i - \zeta}{i + \zeta} \tag{1.28}$$

we deduce from Theorem 1.1 the following assertion.

Theorem 1.2. Assume that for all constant $m \times 1$ vectors h, $h \neq 0$, the relation

$$\int_{-\infty}^{\infty} \frac{\ln[h^* \tau'(u) h]}{1+u^2} = -\infty \tag{1.29}$$

holds. Then the system of vector functions

$$G_{kj}(u) = \left(\frac{u-i}{u+i}\right)^k \frac{1}{u+i} h_j (k = 0, 1, 2, \dots ; \ j = 1, 2, \dots , m), \tag{1.30}$$

where $h_1, h_2, \dots , h_m$ are constant linearly independent $m \times 1$ vectors, is complete in the space $L_m^2(\tau)$.

Remark 1.3. If the condition

$$\int_{-\infty}^{\infty} \frac{\ln \parallel \tau'(u) \parallel}{1+u^2} du = -\infty \tag{1.31}$$

is fulfilled then condition (1.29) is also fulfilled and the system of vector functions (1.30) is complete in the space $L_m^2(\tau)$.

Remark 1.4. M.G. Krein [1], [3] proved that in the case when $m = 1$ the equality

$$\int_{-\infty}^{\infty} \frac{\ln \tau(u)}{1+u^2} = -\infty$$

is the necessary and sufficient condition of the completeness of system (1.29) in the space $L^2(\tau)$.

Corollary 1.2. Let one of the conditions (1.29) or (1.31) be fulfilled. Then the m-stationary process with continuous time and with the spectral matrix function $\tau(u)$ is singular.

2. Transformation Operator

Let us consider the system of the differential equations

$$\frac{dW(x,z)}{dx} = i\frac{z}{2} jW(x,z) + R(x)W(x,z), W(0,z) = E_{2m} \tag{2.1}$$

where

$$j = \begin{bmatrix} E_m & 0 \\ 0 & -R_m \end{bmatrix}, \quad R(x) = \begin{bmatrix} 0 & A(x) \\ A^*(x) & 0 \end{bmatrix}. \tag{2.2}$$

(Here $A(x)$ is a $m \times m$ matrix). In the article [7] the following assertion is proved.

Theorem 2.1. Let for some M the inequality

$$\parallel A(x) \parallel \leq M, \qquad 0 \leq x \leq \ell \tag{2.3}$$

be valid. Then there exists such a bounded $2m \times 2m$ matrix function $N(x,t)$ that

$$W(x,z) = e^{i\frac{z}{2}jx} + \int_{-x}^{x} N(x,t) e^{i\frac{z}{2}t} dt, \ 0 \leq x \leq \ell. \tag{2.4}$$

We write $W(x,z)$ in the block form

$$W(x,z) = \begin{bmatrix} W_{11}(x,z) & W_{12}(x,z) \\ W_{21}(x,z) & W_{22}(x,z) \end{bmatrix}. \tag{2.5}$$

The solution of system (0.1), (0.2) can be expressed in the form

$$P_1(x,z) = [W_{11}(x,z)S_1 + W_{12}(x,z)S_2]e^{\frac{izx}{2}} \tag{2.6}$$

$$P_2(x,z) = [W_{21}(x,s)S_1 + W_{22}(x,z)S_2]e^{\frac{izx}{2}}. \tag{2.7}$$

From formulas (2.4), (2.6) we deduce the following assertion.

Corollary 2.1. Let condition (2.3) be fulfilled. Then there exists such a bounded $m \times m$ matrix function $N_1(x,t)$ that

$$P_1(x,z) = S_1[e^{izx} + \int_0^x N_1(x,t)e^{izt}dt]. \tag{2.8}$$

3. The condition of orthogonality

Let us denote by K the set of all continuous differentiable finitely supported $m \times 1$ vector functions.

Lemma 3.1. The spectral $m \times m$ matrix function $\tau(u)$ of system (0.1) is orthogonal if the set of the $m \times 1$ vector functions

$$F(\lambda) = \int_0^\infty e^{ix\lambda}f(x)dx, \ f(x) \in K \tag{3.1}$$

is dense in $L_m^2(\tau)$.

Proof. We integrate by parts (3.1) and obtain the estimate

$$\| F(\lambda) \| = O(\frac{1}{\lambda}), \ \lambda \to \infty. \tag{3.2}$$

It follows from (1.26) and (3.2) that

$$\int_{-\infty}^\infty F^*(\lambda)[d\tau(\lambda)]F(\lambda) < \infty, \tag{3.3}$$

i.e., $F(\lambda) \in L_m^2(\tau)$.

Using formula (2.8) we deduce the representation

$$e^{izx} = S_1^{-1}P_1(x,z) + \int_0^x N_2(x,t)P_1(t,z)dt, \tag{3.4}$$

where $N_2(x,t)$ is a bounded $m \times m$ matrix function. With the help of (3.4) we write $F(\lambda)$ in the form

$$F(\lambda) = \int_0^a e^{-ix\lambda}f(x)dx = \int_0^a P_1^*(x,\lambda)g(x)dx, \tag{3.5}$$

where

$$g(x) = S_1^{*-1} f(x) + \int_x^a N_2^*(t,x) f(t) dt.$$

From (3.5) and the condition of the lemma we deduce that the operator V defined by formula (0.5) maps the space $L_m^2(0,\infty)$ onto the space $L_m^2(\tau)$. The lemma is proved.

Theorem 3.1. The spectral $m \times m$ matrix function $\tau(u)$ of system (0.1) is orthogonal if one of conditions (1.29) or (1.31) is fulfilled.

Proof. We denote by $\mathcal{L}$ the set of vector functions (3.1). It is easy to see that the vector functions

$$F_{kj}(\lambda) = \left(\frac{\lambda+i}{\lambda-i}\right) \frac{1}{\lambda-i} h_j \ (k = 0,1,2,\dots\ ;\ j = 1,2,\dots,m)$$

(the vectors $h_1, h_2, \dots, h_m$ are linearly independent) admit the representation

$$F_{kj}(\lambda) = \int_0^\infty e^{-i\lambda x} f_{kj}(x) dx. \tag{3.6}$$

Due to (3.6) the vector functions $F_{kj}(\lambda)$ belong to the space $\bar{\mathcal{L}}_n \subset L_m^2(\tau)$. According to Theorem 1.2 the vector functions $F_{kj}(\lambda)$ form a complete system in the space $L_m^2(\tau)$. The assertion of the theorem follows from this fact.

References

[1] Krein M. G., Continuous Analogues of Propositions on Polynomials Orthogonal on the Unit Circle, Dokl. Akad. Nauk SSSR 105, 637–640, 1955.

[2] Sakhnovich L. A., Factorization Problems and Operator Identities, Russian Math. Surveys 41:1, 1–64, 1986.

[3] Krein M. G., On a Problem of Extrapolation of A.N. Kolmogorov, Dokl. Akad. Nauk SSSR 46, 1945, 306–309.

[4] Kolmogorov A. N., Stationary Sequences in Hilbert Space, Bull. Math. Univ. Moscow 2 (1941a), 1–40.

[5] Geronimus Ja. L., Polynomials Orthogonal on a Circle and on an Interval, New York, Pergamon Press, 1966.

[6] Privalov I. I., Boundary Properties of Analytic Functions, Gosizdat, 1950 (in Russian).

[7] Sakhnovich L. A., Spectral Analysis of Volterra Operators given in Vector Functions Space $L_m^2(0,\ell)$, Ukr. Math. Journal 16:2, 259–268, 1964.

Department of Higher Mathematics, Ukranian State Academy of Communications, 270021 Odessa, Ukraine

Operator Theory:
Advances and Applications, Vol. 123

Functional Models of Linear Operators

Abraham Strauss

Dedicated to Professor M. Livšic on his eightieth birthday

Abstract. Under some conditions a functional model of a linear operator acting in an abstract Hilbert space is determined. It is the multiplication operator by an independent complex variable in a Hilbert space of vector valued holomorphic functions.

1. Introduction

Our aim is to describe a construction of functional models of linear operators acting in abstract Hilbert spaces and satisfying some conditions. A functional model is understood here as a multiplication operator by an independent complex variable in a Hilbert space of vector valued functions holomorphic on some domain. Recall that M. G. Krein [K1], [K2] has proposed and investigated such functional models of completely nonselfadjoint symmetric densely defined operators which have finite equal defect numbers. Yet earlier A. I. Plessner [P] has given and studied a functional model of a completely nonselfadjoint maximal symmetric operator. Functional models of completely nonselfadjoint symmetric operators under certain conditions have been constructed and studied by L. de Branges [dB] and also by L. de Branges and J. Rovnyak [dBR]. Functional models of some linear operators with finite defect numbers have been given by D. Alpay and H. Dym [AD] and by D. Alpay [A]. We consider here, in particular, also the case of a symmetric operator. The functional model of this operator and its characteristic function are closely connected. As it is well known, the beginning and the development of the theory of characteristic functions of linear operators are due to M. S. Livsic [L1], [L2].

2. The space $\mathfrak{A}$

Let A be closed linear operator acting in a Hilbert space $\mathfrak{H}$. $\operatorname{Dom} A$ is not assumed to be dense in $\mathfrak{H}$. We suppose only that $\operatorname{Dom} A \neq \{0\}$. Denote by $\Lambda(A)$ the set of all $z \in \mathbb{C}$ such that the operator $A - zI$ has a bounded inverse with

---

1991 *Mathematics Subject Classification.* 47A10, 47A45, 47B25.

$\mathrm{Dom}(A - zI)^{-1} = \mathrm{Ran}(A - zI) \neq \mathfrak{H}$. Assume that $\Lambda(A) \neq \emptyset$. Thus $\Lambda(A)$ is an open set and for each $z \in \Lambda(A)$ $\mathrm{Ran}(A - zI)$ is a closed nontrivial subspace in $\mathfrak{H}$. The defect subspace $\mathfrak{N}_z(A)$ corresponding to $z \in \Lambda(A)$ is defined by

$$\mathfrak{N}_z(A) := \mathfrak{H} \ominus \mathrm{Ran}(A - zI) .$$

We suppose that for some nonempty open connected set $\Omega \subset \Lambda(A)$

$$\bigcap_{z \in \Omega} \mathrm{Ran}(A - zI) = \{0\} . \tag{1}$$

Let us fix such domain Ω.

Note that $\dim \mathfrak{N}_z(A)$ is constant on Ω. Designate this cardinal number by $\mathfrak{n}$. Fix some Hilbert space $\mathfrak{P}$ with $\dim \mathfrak{P} = \mathfrak{n}$.

Assume that on Ω a holomorphic operator valued function $z \mapsto T(z)$ is given such that for each $z \in \Omega$ $T(z) : \mathfrak{H} \to \mathfrak{P}$ is a bounded linear operator with

$$\mathrm{Ker}\, T(z) = \mathrm{Ran}(A - zI) \tag{2}$$

and

$$\mathrm{Ran}\, T(z) = \mathfrak{P} . \tag{3}$$

Then $T^*(z) : \mathfrak{P} \to \mathfrak{H}$,

$$\mathrm{Ker}\, T^*(z) = \{0\} ,$$

and

$$\mathrm{Ran}\, T^*(z) = \mathfrak{N}_z(A) .$$

Operator valued functions of this kind appear in the spectral theory of linear operators in Hilbert spaces [K1].

For each $h \in \mathfrak{H}$ denote by $\widehat{h}$ the $\mathfrak{P}$-valued function defined on Ω by

$$\widehat{h}(z) := T(z)h , \quad z \in \Omega .$$

Let $\mathfrak{A}$ be the linear space of all such functions. Define the operator $\Gamma : \mathfrak{H} \to \mathfrak{A}$ by

$$\Gamma h := \widehat{h}$$

for each $h \in \mathfrak{H}$. So, for each $z \in \Omega$ and $h \in \mathfrak{H}$

$$\Big(\Gamma h\Big)(z) = T(z)h . \tag{4}$$

Note that

$$\mathrm{Ker}\, \Gamma = \{0\} . \tag{5}$$

In fact, by (4)

$$\mathrm{Ker}\, \Gamma = \bigcap_{z \in \Omega} \mathrm{Ker}\, T(z) .$$

Thus (5) follows in view of (2) and (1).

Now taking into account (5) define the inner product in $\mathfrak{A}$ by

$$(\widehat{f}, \widehat{g})_{\mathfrak{A}} := (f, g)_{\mathfrak{H}}$$

for each $f, g \in \mathfrak{H}$. In this way $\mathfrak{A}$ becomes a Hilbert space and Γ is a unitary operator from $\mathfrak{H}$ onto $\mathfrak{A}$.

We shall say that $\mathfrak{A}$ is generated by the operator valued function $T(z)$, $z \in \Omega$.

Proposition 1. *The Hilbert space $\mathfrak{A}$ has a reproducing kernel*

$$K(z,\zeta) = T(z)T^*(\zeta) , \quad z, \zeta \in \Omega ,$$

so that for each $a \in \mathfrak{A}$, $\zeta \in \Omega$, $p \in \mathfrak{P}$

$$\Big(a(\zeta), p\Big)_{\mathfrak{P}} = \Big(a, \Gamma T^*(\zeta)p\Big)_{\mathfrak{A}} .$$

Proof. Let $a = \Gamma h$, where $h \in \mathfrak{H}$. Then

$$\Big(a(\zeta), p\Big)_{\mathfrak{P}} = \Big(T(\zeta)h, p\Big)_{\mathfrak{P}} = \Big(h, T^*(\zeta)p\Big)_{\mathfrak{H}} = \Big(\Gamma h, \Gamma T^*(\zeta)p\Big)_{\mathfrak{A}} = \Big(a, \Gamma T^*(\zeta)p\Big)_{\mathfrak{A}} .$$

□

3. The multiplication operator

Let M denote the multiplication operator by the independent complex variable z in $\mathfrak{A}$. $\operatorname{Dom} M$ is the set of all $a \in \mathfrak{A}$ for which the multiplication by z gives an element of $\mathfrak{A}$. For all $a \in \operatorname{Dom} M$ and $z \in \Omega$

$$\Big(Ma\Big)(z) := za(z) .$$

Proposition 2.

$$M = \Gamma A \Gamma^{-1} . \tag{6}$$

Proof. If $f \in \operatorname{Dom} A$, then for each $z \in \Omega$ by (2)

$$T(z)(A - zI)f = 0 ,$$

whence

$$T(z)Af = zT(z)f .$$

This means that $\Gamma f \in \operatorname{Dom} M$ and $\Gamma Af = M\Gamma f$.

Now it remains to prove that if $a \in \operatorname{Dom} M$, then $a \in \Gamma \operatorname{Dom} A$.

Let $a \in \operatorname{Dom} M$. For a fixed $\zeta \in \Omega$ set

$$b_\zeta = (M - \zeta I)a .$$

As $b_\zeta \in \mathfrak{A}$, we have

$$b_\zeta = \Gamma g_\zeta$$

for some $g_\zeta \in \mathfrak{H}$. Then for each $z \in \Omega$

$$T(z)g_\zeta = b_\zeta(z) = (z - \zeta)a(z) . \tag{7}$$

In particular, for $z = \zeta$

$$T(\zeta)g_\zeta = 0 .$$

Consequently, by (2)

$$g_\zeta = (A - \zeta I)f_\zeta$$

for some $f_\zeta \in \operatorname{Dom} A$. Then in virtue of the proved part of this proposition

$$T(z)g_\zeta = (z-\zeta)T(z)f_\zeta \tag{8}$$

for each $z \in \Omega$. From (7) and (8)

$$(z-\zeta)a(z) = (z-\zeta)T(z)f_\zeta \ .$$

Hence

$$a(z) = T(z)f_\zeta$$

for each $z \in \Omega \setminus \{\zeta\}$ and then also for $z = \zeta$ because the functions under consideration are holomorphic on Ω. Thus $a = \Gamma f_\zeta$, where $f_\zeta \in \operatorname{Dom} A$. □

Proposition 3. *For each $\zeta \in \Omega$*

$$\operatorname{Ran}(M - \zeta I) = \{a \in \mathfrak{A} : a(\zeta) = 0\} \ .$$

Proof. If $a \in \operatorname{Ran}(M-\zeta I)$, then obviously $a(\zeta) = 0$. Conversely, suppose that $a(\zeta) = 0$. As $a = \Gamma h$ for some $h \in \mathfrak{H}$, we see that $T(\zeta)h = 0$. Therefore by (2) $h \in \operatorname{Ran}(A - \zeta I)$ and hence $a \in \operatorname{Ran}(M - \zeta I)$. □

Note that if for some $\zeta \in \Omega$ $a \in \operatorname{Ran}(M - \zeta I)$, then

$$\Big((M-\zeta I)^{-1}a\Big)(z) = \frac{a(z)}{z-\zeta} \tag{9}$$

for each $z \in \Omega$; if $z = \zeta$, the expression on the right side of (9) is understood as $\left.\dfrac{da}{dz}\right|_{z=\zeta}$.

4. The R-property

Now we shall study the situation concerning the following possible additional property of $\mathfrak{A}$ which will be called the R-property: for each $\zeta \in \Omega$ and $a \in \mathfrak{A}$ the $\mathfrak{P}$-valued function a_ζ defined on Ω by

$$a_\zeta(z) := \frac{a(z) - a(\zeta)}{z - \zeta} \ , \quad z \in \Omega \ ,$$

also belongs to $\mathfrak{A}$ and the operator $R_\zeta : \mathfrak{A} \to \mathfrak{A}$ defined by

$$R_\zeta a = a_\zeta$$

is bounded.

In view of (9)

$$R_\zeta \supset (M - \zeta I)^{-1} \ . \tag{10}$$

Note that such spaces consisting of holomorphic vector valued functions have been investigated by L. de Branges [dB], L. de Branges and J. Rovnyak [dBR], and H. Dym [D].

It is easy to check that for each $\zeta_1, \zeta_2 \in \Omega$

$$R_{\zeta_1} - R_{\zeta_2} = (\zeta_1 - \zeta_2) R_{\zeta_1} R_{\zeta_2} \ . \tag{11}$$

Proposition 4. *The space $\mathfrak{A}$ generated by the operator valued function $z \mapsto T(z)$, $z \in \Omega$, has the R-property if and only if for each $\zeta \in \Omega$ there exists a bounded linear operator $R(\zeta) : \mathfrak{H} \to \mathfrak{H}$ such that*

$$R(\zeta) \supset (A - \zeta I)^{-1} \tag{12}$$

and for each $\zeta_1, \zeta_2 \in \Omega$

$$T(\zeta_1) - T(\zeta_2) = (\zeta_1 - \zeta_2) T(\zeta_1) R(\zeta_2) \ . \tag{13}$$

In this case such operator valued function $\zeta \mapsto R(\zeta)$, $\zeta \in \Omega$, is unique,

$$R(\zeta) = \Gamma^{-1} R_\zeta \Gamma \ , \tag{14}$$

and for each $\zeta_1, \zeta_2 \in \Omega$

$$R(\zeta_1) - R(\zeta_2) = (\zeta_1 - \zeta_2) R(\zeta_1) R(\zeta_2) \ . \tag{15}$$

Proof. Suppose that the space $\mathfrak{A}$ has the R-property. Define the operator valued function $\zeta \mapsto R(\zeta)$, $\zeta \in \Omega$, by (14). Then for each $\zeta \in \Omega$ $R(\zeta) : \mathfrak{H} \to \mathfrak{H}$ is a bounded linear operator. In view of (10) we obtain (12). As

$$\Gamma R(\zeta) = R_\zeta \Gamma$$

for each $\zeta \in \Omega$,

$$T(z) R(\zeta) h = \frac{1}{z - \zeta} \Big(T(z) h - T(\zeta) h \Big)$$

for each $z \in \Omega$ and $h \in \mathfrak{H}$. Hence (13) follows.

Conversely, suppose that there exists an operator valued function $\zeta \mapsto R(\zeta)$ on Ω such that for each $\zeta \in \Omega$ $R(\zeta) : \mathfrak{H} \to \mathfrak{H}$ is a bounded linear operator and the relations (12), (13) hold. By (13) for each $z, \zeta \in \Omega$

$$\frac{1}{z - \zeta} \Big(T(z) - T(\zeta) \Big) = T(z) R(\zeta) \ .$$

Therefore the space $\mathfrak{A}$ has the R-property and for each $\zeta \in \Omega$

$$R_\zeta \Gamma = \Gamma R(\zeta) \ .$$

Hence (14) follows and by (11) we obtain (15). □

Suppose that the space $\mathfrak{A}$ generated by the operator valued function $T(z)$, $z \in \Omega$, has the R-property. Taking into account Proposition 4, consider the corresponding operator valued function $R(\zeta)$, $\zeta \in \Omega$. By (15)

$$\operatorname{Ran} R(\zeta_1) = \operatorname{Ran} R(\zeta_2)$$

and

$$\operatorname{Ker} R(\zeta_1) = \operatorname{Ker} R(\zeta_2) \tag{16}$$

for each $\zeta_1, \zeta_2 \in \Omega$. Set

$$\mathfrak{N} := \operatorname{Ker} R(\zeta)\ , \tag{17}$$

$\zeta \in \Omega$, so that in view of (16) $\mathfrak{N}$ does not depend on ζ. Denote by $\mathfrak{A}_0$ the set consisting of all constant on Ω $\mathfrak{P}$-valued functions belonging to $\mathfrak{A}$, so that $\mathfrak{A}_0$ is a subspace of $\mathfrak{A}$.

Let us show that

$$\mathfrak{N} = \Gamma^{-1}\mathfrak{A}_0\ . \tag{18}$$

In fact, if $u \in \mathfrak{N}$, then by (13)

$$T(\zeta_1)u - T(\zeta_2)u = 0$$

for each $\zeta_1, \zeta_2 \in \Omega$. This implies that $\Gamma u \in \mathfrak{A}_0$ and $u \in \Gamma^{-1}\mathfrak{A}_0$. Conversely, assume that $\Gamma u \in \mathfrak{A}_0$. Then by (13)

$$T(z)R(\zeta)u = 0$$

for each $z, \zeta \in \Omega$. Hence by (2)

$$R(\zeta)u \in \bigcap_{z\in\Omega} \operatorname{Ran}(A - zI)\ .$$

Therefore by (1) $R(\zeta)u = 0$, whence $u \in \mathfrak{N}$.

We shall consider two cases concerning the space $\mathfrak{A}$ which has the R-property.

a) The space $\mathfrak{A}$ contains no $\mathfrak{P}$-valued constant different from zero.

b) The space $\mathfrak{A}$ contains all $\mathfrak{P}$-valued constants.

Of course, if $\dim \mathfrak{P} > 1$, then other cases also are possible, but we shall not discuss them here.

Proposition 5. *The space $\mathfrak{A}$ with the R-property contains no $\mathfrak{P}$-valued constant different from zero if and only if the operator A has an extension B such that $\rho(B) \supset \Omega$ and the resolvent of B coincides on Ω with the operator valued function $R(\zeta)$, $\zeta \in \Omega$, defined by (14), that is,*

$$(B - \zeta I)^{-1} = \Gamma^{-1}R_\zeta\Gamma$$

for each $\zeta \in \Omega$.

Proof. The space $\mathfrak{A}$ with the R-property contains no $\mathfrak{P}$-valued constant different from zero if and only if $\operatorname{Ker} R_\zeta = \{0\}$ for some $\zeta \in \Omega$ (and then for each $\zeta \in \Omega$). In this case the operator valued function $\zeta \mapsto R_\zeta$ on Ω coincides with the resolvent of an extension N of M defined by

$$N = R_\zeta^{-1} + \zeta I$$

which does not depend on $\zeta \in \Omega$. Thus the operator B defined by

$$B = \Gamma^{-1}N\Gamma$$

is the requisite extension of A. □

Proposition 6. *Given an extension $B \supset A$, such that $\rho(B) \supset \Omega$, and a bounded linear operator $T_0 : \mathfrak{H} \to \mathfrak{P}$ with*

$$\operatorname{Ker} T_0 = \operatorname{Ran}(A - \zeta_0 I) \tag{19}$$

for a fixed $\zeta_0 \in \Omega$ and

$$\operatorname{Ran} T_0 = \mathfrak{P} , \tag{20}$$

set

$$R(\zeta) = (B - \zeta I)^{-1} \tag{21}$$

and

$$T(\zeta) = T_0\Big(I + (\zeta - \zeta_0)R(\zeta)\Big) \tag{22}$$

for each $\zeta \in \Omega$. Then the operator valued function $\zeta \mapsto T(\zeta)$, $\zeta \in \Omega$, is holomorphic on Ω, satisfies the conditions (2), (3), the space $\mathfrak{A}$ generated by this function has the R-property, it contains no nonzero $\mathfrak{P}$-valued constant, and $R(\zeta)$ defined by (21) coincides with $\Gamma^{-1}R_\zeta\Gamma$ for each $\zeta \in \Omega$.

Proof. It is easy to check that in virtue of (19)–(22) the conditions (2), (3) are satisfied. Note that by (21) the equality (15) holds for each $\zeta_1, \zeta_2 \in \Omega$ as the resolvent identity. In view of (22) and (15) we have for each $\zeta_1, \zeta_2 \in \Omega$

$$\begin{aligned} T(\zeta_1) - T(\zeta_2) &= T_0\Big((\zeta_1 - \zeta_0)R(\zeta_1) - (\zeta_2 - \zeta_0)R(\zeta_2)\Big) \\ &= (\zeta_1 - \zeta_2)T_0\Big(R(\zeta_2) + (\zeta_1 - \zeta_0)R(\zeta_1)R(\zeta_2)\Big) = (\zeta_1 - \zeta_2)T(\zeta_1)R(\zeta_2). \end{aligned}$$

By Propositions 4 and 5 this completes the proof. □

Proposition 7. *Let the space $\mathfrak{A}$ be generated by some holomorphic operator valued function $z \mapsto T(z)$, $z \in \Omega$, satisfying the conditions (2), (3), and let $\mathfrak{A}$ have the R-property. $\mathfrak{A}$ contains all $\mathfrak{P}$-valued constants if and only if for some $\zeta_0 \in \Omega$ $\operatorname{Ran} R(\zeta_0) = \operatorname{Dom} A$ (and then for each $\zeta \in \Omega$ $\operatorname{Ran} R(\zeta) = \operatorname{Dom} A$), where $R(\zeta)$ is defined by (14). In this case for each $\zeta \in \Omega$ the space $\mathfrak{H}$ is a direct sum*

$$\mathfrak{H} = \operatorname{Ran}(A - \zeta I) \dotplus \mathfrak{N} , \tag{23}$$

where $\mathfrak{N}$ is defined by (17) and does not depend on $\zeta \in \Omega$,

$$R(\zeta) = (A - \zeta I)^{-1}P(\zeta) , \tag{24}$$

where $P(\zeta)$ is the projector of $\mathfrak{H}$ onto $\operatorname{Ran}(A - \zeta I)$ parallel to $\mathfrak{N}$, the subspace $\mathfrak{A}_0$ associated with $\mathfrak{N}$ by (18) contains all $\mathfrak{P}$-valued constants and therefore it may be identified with $\mathfrak{P}$, and

$$T(\zeta) = \Gamma_{\mathfrak{N}}Q(\zeta) , \quad \zeta \in \Omega , \tag{25}$$

where $Q(\zeta)$ is the projector of $\mathfrak{H}$ onto $\mathfrak{N}$ parallel to $\operatorname{Ran}(A - \zeta I)$, that is,

$$Q(\zeta) = I - P(\zeta) ,$$

and

$$\Gamma_{\mathfrak{N}} := \Gamma|\mathfrak{N} \tag{26}$$

is an isometric operator from $\mathfrak{N}$ *onto* $\mathfrak{P}$.

Proof. Suppose that $\mathfrak{A}$ contains all $\mathfrak{P}$-valued constants. Then for each $a \in \mathfrak{A}$ and each fixed $\zeta_0 \in \Omega$ $a - a(\zeta_0) \in \mathfrak{A}$. By Proposition 3 $a - a(\zeta_0) \in \operatorname{Ran}(M - \zeta_0 I)$, $R_{\zeta_0} a = \left(M - \zeta_0 I\right)^{-1}\left(a - a(\zeta_0)\right)$, and $\operatorname{Ran} R_{\zeta_0} = \operatorname{Dom} M$. Taking into account (6) and (14), we see that

$$\operatorname{Ran} R(\zeta_0) = \operatorname{Dom} A \ .$$

Conversely, suppose that for some $\zeta_0 \in \Omega$ $\operatorname{Ran} R(\zeta_0) = \operatorname{Dom} A$. Then

$$\operatorname{Ran} R(\zeta) = \operatorname{Dom} A$$

for each $\zeta \in \Omega$. Define the operator $P(\zeta)$ by

$$P(\zeta) := (A - \zeta I)R(\zeta) \ . \tag{27}$$

It is easy to check that

$$P^2(\zeta) = P(\zeta) \ .$$

As $\operatorname{Ran} P(\zeta) = \operatorname{Ran}(A - \zeta I)$ and $\operatorname{Ker} P(\zeta) = \mathfrak{N}$, we see that (23) holds and $P(\zeta)$ is the projector of $\mathfrak{H}$ onto $\operatorname{Ran}(A - \zeta I)$ parallel to $\mathfrak{N}$. Thus (24) follows from (27).

Now we shall show that $\mathfrak{A}_0$ contains all $\mathfrak{P}$-valued constants. Let $p \in \mathfrak{P}$. Choose $a \in \mathfrak{A}$ such that $a(\zeta_0) = p$. By (3) such an a exists. Set

$$h = \Gamma^{-1} a - (A - \zeta_0 I)R(\zeta_0)\Gamma^{-1} a \tag{28}$$

or

$$h = \left(I - P(\zeta_0)\right)\Gamma^{-1} a \ . \tag{29}$$

By (29) $h \in \mathfrak{N}$ and by (18) $\Gamma h \in \mathfrak{A}_0$. Therefore

$$T(\zeta)h = T(\zeta_0)h$$

for each $\zeta \in \Omega$. But by (28) and (2)

$$T(\zeta_0)h = T(\zeta_0)\Gamma^{-1} a = a(\zeta_0) = p \ .$$

Thus

$$T(\zeta)h = p$$

for each $\zeta \in \Omega$.

It remains to prove (25). By (2), (27), and (26) for each $h \in \mathfrak{H}$ and each $\zeta \in \Omega$ we have

$$T(\zeta)h = T(\zeta)\left(h - (A - \zeta I)R(\zeta)h\right) = T(\zeta)\left(I - P(\zeta)\right)h = T(\zeta)Q(\zeta)h = \Gamma_{\mathfrak{N}} Q(\zeta)h.$$

□

Proposition 8. *Let the operator $(A-\zeta_0 I)^{-1}$ for some $\zeta_0 \in \Omega$ have a bounded extension S such that* $\operatorname{Dom} S = \mathfrak{H}$, $\operatorname{Ran} S = \operatorname{Dom} A$ *and* $\frac{1}{\zeta-\zeta_0} \in \rho(S)$ *for each* $\zeta \in \Omega \setminus \{\zeta_0\}$. *Set* $\mathfrak{N}_S := \operatorname{Ker} S$. *Then for each* $\zeta \in \Omega$ $\mathfrak{H}$ *is a direct sum*

$$\mathfrak{H} = \operatorname{Ran}(A-\zeta I) \dotplus \mathfrak{N}_S \,. \tag{30}$$

Denote by $Q(\zeta)$ *the projector of* $\mathfrak{H}$ *onto* $\mathfrak{N}_S$ *parallel to* $\operatorname{Ran}(A-\zeta I)$. *Let* V *be an arbitrary isometric operator from* $\mathfrak{N}_S$ *onto* $\mathfrak{P}$. *Define the operator valued function* $\zeta \mapsto T(\zeta)$ *on* Ω *by*

$$T(\zeta) := VQ(\zeta) \,. \tag{31}$$

Then this function satisfies the conditions (2), (3), the space $\mathfrak{A}$ *generated by it has the R-property,* $\mathfrak{A}$ *contains all* $\mathfrak{P}$*-valued constants, and the operator valued function* $\zeta \mapsto R(\zeta)$ *defined by*

$$R(\zeta) := \Big(I - (\zeta-\zeta_0)S\Big)^{-1} S \,, \tag{32}$$

$\zeta \in \Omega$, *coincides with* $\Gamma^{-1} R_\zeta \Gamma$ *on* Ω.

Proof. Define the operator valued function $\zeta \mapsto R(\zeta)$, $\zeta \in \Omega$, by (32). Then

$$R(\zeta) \supset (A-\zeta I)^{-1} \,, \quad \operatorname{Ker} R(\zeta) = \mathfrak{N}_S \,, \quad \operatorname{Ran} R(\zeta) = \operatorname{Dom} A \,. \tag{33}$$

In view of (33) set

$$P(\zeta) := (A-\zeta I)R(\zeta)$$

for each $\zeta \in \Omega$. We have $P^2(\zeta) = P(\zeta)$, $\operatorname{Ker} P(\zeta) = \mathfrak{N}_S$, $\operatorname{Ran} P(\zeta) = \operatorname{Ran}(A-\zeta I)$. Therefore for each $\zeta \in \Omega$ the decomposition (30) is valid and $P(\zeta)$ is the projector of $\mathfrak{H}$ onto $\operatorname{Ran}(A-\zeta I)$ parallel to $\mathfrak{N}_S$. Set

$$Q(\zeta) := I - P(\zeta)$$

for each $\zeta \in \Omega$. Then $Q(\zeta)$ is the projector of $\mathfrak{H}$ onto $\mathfrak{N}_S$ parallel to $\operatorname{Ran}(A-\zeta I)$.

We shall verify that for each $\zeta_1, \zeta_2 \in \Omega$

$$Q(\zeta_1) - Q(\zeta_2) = (\zeta_1-\zeta_2)Q(\zeta_1)R(\zeta_2) \,. \tag{34}$$

Indeed,

$$(\zeta_1-\zeta_2)Q(\zeta_1)R(\zeta_2) = Q(\zeta_1)(A-\zeta_2 I)R(\zeta_2) - Q(\zeta_1)(A-\zeta_1 I)R(\zeta_2) = Q(\zeta_1)P(\zeta_2) \,. \tag{35}$$

Here we have taken into account that

$$Q(\zeta_1)(A-\zeta_1 I)R(\zeta_2) = 0$$

because $\operatorname{Ker} Q(\zeta_1) = \operatorname{Ran}(A-\zeta_1 I)$. Further,

$$Q(\zeta_1)P(\zeta_2) = Q(\zeta_1)\Big(I - Q(\zeta_2)\Big) = Q(\zeta_1) - Q(\zeta_2) \tag{36}$$

as $Q(\zeta_1)Q(\zeta_2) = Q(\zeta_2)$. Now (34) follows from (35) and (36).

Define $T(\zeta)$ by (31). In view of (34) we obtain (13).

To complete the proof it remains to use the Propositions 4 and 7. □

A subspace $\mathfrak{N} \subset \mathfrak{H}$ will be called a module subspace for A with respect to Ω if for each $\zeta \in \Omega$ the decomposition (23) is valid. In this case, according to Proposition 8, the operator valued function $\zeta \mapsto Q(\zeta)$ coinciding with $\zeta \mapsto T(\zeta)$ if $\mathfrak{P} = \mathfrak{N}$ and $V = I_{\mathfrak{N}}$ leads to a functional model of A in a space $\mathfrak{A}$ which contains all $\mathfrak{N}$-valued constants. Namely such approach to the theory of functional models of symmetric densely defined operators with finite equal defect numbers was originated by M. G. Krein [K1].

5. Entire operators

In this section we consider two cases when the linear operator A is unitarily equivalent to the multiplication operator by the independent complex variable in a Hilbert space which consists of entire vector valued functions and has the R-property. Such A is called an entire operator. Some functional models of entire symmetric operators have been investigated by M. G. Krein [K1], L. de Branges [dB], L. de Branges and J. Rovnyak [dBR]. For more details concerning the results of M. G. Krein we refer to the recent book of M. L. Gorbachuk and V. I. Gorbachuk [GG].

From Propositions 5, 6 and 7, 8 we obtain, respectively, the following two propositions.

Proposition 9. *Let $\Lambda(A) = \mathbb{C}$. The operator A is unitarily equivalent to the multiplication operator by the independent complex variable in some Hilbert space $\mathfrak{A}$ which consists of $\mathfrak{P}$-valued entire functions, has the R-property, and contains no nonzero $\mathfrak{P}$-valued constant, if and only if there exists an extension $B \supset A$ with $\rho(B) = \mathbb{C}$. In this case set $\mathfrak{P} = \mathfrak{N}_0(A)$, denote by P_0 the orthoprojector of $\mathfrak{H}$ onto $\mathfrak{N}_0(A)$, and define the entire operator valued function $z \mapsto T(z)$ by*

$$T(z) = P_0 B(B - zI)^{-1}\,, \quad z \in \mathbb{C}\,. \tag{37}$$

Then the space $\mathfrak{A}$ generated by this function has the requisite properties.

Example 1. Let $\mathfrak{P}$ be a Hilbert space of an arbitrary finite or infinite dimension. Consider the Hilbert space $\mathfrak{H} = \mathcal{L}_2([0,1], \mathfrak{P})$ which is a completion of the pre-Hilbert space of continuous $\mathfrak{P}$-valued functions on $[0,1]$ with the Lebesgue measure in the integral formula for the inner product. Let A be the closed linear operator in $\mathfrak{H}$ associated in a usual way with the derivation $\dfrac{d}{dx}$ and boundary conditions $f(0) = f(1) = 0$. Then $\Lambda(A) = \mathbb{C}$. The condition (1) is satisfied. The defect subspace $\mathfrak{N}_0(A)$ consists of $\mathfrak{P}$-valued constants and can be identified with $\mathfrak{P}$. Let B be the extension of A associated with the same derivation and the boundary condition $f(1) = 0$. Then $\rho(B) = \mathbb{C}$. B^* is an extension of $-A$ with the boundary condition $f(0) = 0$. We shall prove that in this case for the operator valued function

$z \mapsto T(z)$ defined by (37) for each z the following formula is valid:

$$T(z)h = \int_0^1 h(x)e^{-zx}\,dx \tag{38}$$

for each $h \in \mathfrak{H}$ with the Bochner integral on the right side.

Indeed, in view of (37) for each $h \in \mathfrak{H}$ and $p \in \mathfrak{P}$ we have

$$\Big(T(z)h, p\Big)_{\mathfrak{P}} = \Big(P_0 B(B - zI)^{-1}h, p\Big)_{\mathfrak{P}} = \Big(h, B^*(B^* - \bar{z}I)^{-1}p\Big)_{\mathfrak{H}} . \tag{39}$$

Observe that for each $z \in \mathbb{C}$

$$B^*(B^* - \bar{z}I)^{-1}\mathfrak{N}_0(A) = \mathfrak{N}_z(A) . \tag{40}$$

Set

$$u := B^*(B^* - \bar{z}I)^{-1}p . \tag{41}$$

Then

$$u = p + \bar{z}(B^* - \bar{z}I)^{-1}p . \tag{42}$$

In virtue of (40) and (41)

$$-\frac{du}{dx} = \bar{z}u . \tag{43}$$

By (42)

$$u(0) = p \tag{44}$$

because $(B^* - \bar{z}I)^{-1}p \in \operatorname{Dom} B^*$. From (43), (44) we conclude that

$$u(x) = e^{-\bar{z}x}p \tag{45}$$

for $x \in [0, 1]$. Thus in view of (39), (41), and (45)

$$\Big(T(z)h, p\Big)_{\mathfrak{P}} = \int_0^1 \Big(h(x), e^{-\bar{z}x}p\Big)_{\mathfrak{P}} dx = \Big(\int_0^1 h(x)e^{-zx}dx, p\Big)_{\mathfrak{P}}$$

and (38) follows.

Proposition 10. *Let $\Lambda(A) = \mathbb{C}$. The operator A is unitarily equivalent to the multiplication operator by the independent complex variable in some Hilbert space $\mathfrak{A}$ which consists of $\mathfrak{P}$-valued entire functions, has the R-property, and contains all constant $\mathfrak{P}$-valued functions, if and only if the operator A^{-1} has a quasinilpotent extension S with $\operatorname{Dom} S = \mathfrak{H}$ and $\operatorname{Ran} S = \operatorname{Dom} A$. In this case for each $z \in \mathbb{C}$ the direct sum decomposition*

$$\mathfrak{H} = \operatorname{Ran}(A - zI) \dotplus \operatorname{Ker} S$$

is valid. Denote by $Q(z)$ the projector of $\mathfrak{H}$ onto $\operatorname{Ker} S$ parallel to $\operatorname{Ran}(A - zI)$. The operator valued function $z \mapsto Q(z)$, $z \in \mathbb{C}$, is entire and the Hilbert space $\mathfrak{A}$ generated by this function has the requisite properties.

The Proposition 10 is near to our earlier result [S3].

Example 2. Let $\mathfrak{H}$ be an orthogonal sum,

$$\mathfrak{H} = \mathfrak{G}_1 \oplus \ldots \oplus \mathfrak{G}_n$$

with $\mathfrak{G}_k = E_k\mathfrak{G}$, $k = 1, \ldots, n$, where $\mathfrak{G}$ is a Hilbert space of an arbitrary finite or infinite dimension and E_k denotes the canonical isometric embedding of $\mathfrak{G}$ into $\mathfrak{H}$. Let L be a linear bounded invertible operator with $\operatorname{Dom} L = \operatorname{Ran} L = \mathfrak{H}$. Set $\mathfrak{M}_k = L\mathfrak{G}_k$, $J_k = LE_k$ for $k = 1, \ldots, n$. Then $\mathfrak{H}$ is a direct sum,

$$\mathfrak{H} = \mathfrak{M}_1 \dot{+} \ldots \dot{+} \mathfrak{M}_n$$

and $\mathfrak{M}_k = J_k\mathfrak{G}$.

Let A be a linear operator in $\mathfrak{H}$ with $\operatorname{Dom} A = \mathfrak{M}_1 \dot{+} \ldots \dot{+} \mathfrak{M}_{n-1}$ defined by

$$AJ_k = J_{k+1}\ , \qquad k = 1, \ldots, n-1\ .$$

Then $\Lambda(A) = \mathbb{C}$, A satisfies the condition (1), and for each $z \in \mathbb{C}$

$$\mathfrak{H} = \operatorname{Ran}(A - zI) \dot{+} \mathfrak{M}_1\ .$$

In this case $\mathfrak{M}_1 = \operatorname{Ker} S$, where S is a linear nilpotent extension of A^{-1} defined by

$$SJ_k = J_{k-1}$$

for $k = 2, \ldots, n$, and

$$SJ_1 = O|\mathfrak{G}\ .$$

For the projector $Q(z)$ of $\mathfrak{H}$ onto $\mathfrak{M}_1$ parallel to $\operatorname{Ran}(A - zI)$ for each $z \in \mathbb{C}$

$$Q(z)\Big(J_{k+1} - zJ_k\Big) = 0$$

for $k = 1, \ldots, n-1$, and

$$Q(z)J_1 = J_1\ .$$

Hence

$$Q(z)J_k = z^{k-1}J_1$$

for $k = 1, \ldots, n$.

The space $\mathfrak{A}$ consists of $\mathfrak{M}_1$-valued polynomials $p_0 + zp_1 + \ldots + z^{n-1}p_{n-1}$, where $p_k \in \mathfrak{M}_1$ for $k = 0, \ldots, n-1$. Observe that in this case not only (1) holds but also for each set of n different points $z_1, \ldots, z_n \in \mathbb{C}$

$$\bigcap_{k=1}^{n} \operatorname{Ran}(A - z_kI) = \{0\}\ .$$

Example 3. Let $\mathfrak{H}$ be the orthogonal sum,

$$\mathfrak{H} = \bigoplus_{0}^{\infty} \mathfrak{G}_k$$

with $\mathfrak{G}_k = E_k\mathfrak{F}$, $k = 0, 1, 2, \ldots$, where $\mathfrak{F}$ is a Hilbert space of an arbitrary finite or infinite dimension and E_k denotes the canonical isometric embedding of $\mathfrak{F}$ into $\mathfrak{H}$. Define the operator A in $\mathfrak{H}$ by

$$\operatorname{Dom} A = \left\{ \bigoplus_0^\infty E_k f_k \in \mathfrak{H} : f_k \in \mathfrak{F}, \sum_{k=0}^\infty (k+1)^2 ||f_k||^2 < \infty \right\},$$

$$A\Big(\bigoplus_0^\infty E_k f_k\Big) = \bigoplus_0^\infty (k+1) E_{k+1} f_k .$$

Then A is a closed linear operator, $\Lambda(A) = \mathbb{C}$, and $\mathfrak{N}_0(A) = \mathfrak{G}_0$. Define the linear extension S of A^{-1} by

$$S g_0 = 0$$

for each $g_0 \in \mathfrak{G}_0$. Then $S\mathfrak{H} = \operatorname{Dom} A$, $\operatorname{Ker} S = \mathfrak{G}_0$, and S is a quasinilpotent operator.

For each $z \in \mathfrak{H}$

$$\mathfrak{H} = \operatorname{Ran}(A - zI) \dot{+} \mathfrak{G}_0 .$$

Let $Q(z)$ be the projector of $\mathfrak{H}$ onto $\mathfrak{G}_0$ parallel to $\operatorname{Ran}(A - zI)$. Then for each $k = 0, 1, 2, \ldots$, $f \in \mathfrak{F}$ we have

$$Q(z)\Big((k+1)E_{k+1} f - zE_k f\Big) = 0 .$$

Hence

$$Q(z)E_{k+1} f = \frac{z}{k+1} Q(z) E_k f$$

and therefore

$$Q(z)E_{k+1} f = \frac{z^{k+1}}{(k+1)!} E_0 f .$$

So, for each $h = \bigoplus_0^\infty E_k f_k \in \mathfrak{H}$, where $f_k \in \mathfrak{F}$, and $\sum_{k=0}^\infty ||f_k||^2 < \infty$, we have

$$Q(z)h = \sum_{k=0}^\infty \frac{z^k}{k!} E_0 f_k . \tag{46}$$

In view of (46) it is easy to check that A satisfies the conditions (1). The space $\mathfrak{A}$ consists of all entire $\mathfrak{G}_0$-valued functions given by (46). Note that a generalization of the Example 3 analogous to Example 2 is not difficult.

6. Functional models of symmetric operators

Now let us suppose that A is a closed symmetric operator in $\mathfrak{H}$ and that in no nonzero subspace of $\mathfrak{H}$ the induced part of A is a maximal symmetric operator with the zero defect number in the upper halfplane $\mathbb{C}_+$. Note that the class of such operators is smaller then the class of symmetric completely nonselfadjoint operators.

Choose the defect subspace $\mathfrak{N}_i(A)$ as a module subspace for A. Then there exists a domain $\Omega \supset \mathbb{C}_+ \cup \Big(\mathbb{R} \cap \Lambda(A)\Big)$, such that for each $z \in \Omega$ we have the direct sum decomposition

$$\mathfrak{H} = \mathrm{Ran}(A - zI) \dotplus \mathfrak{N}_i(A) ,$$

see [S1], [S2]. Let $Q(z)$, $z \in \Omega$, be the projector of $\mathfrak{H}$ onto $\mathfrak{N}_i(A)$ parallel to $\mathrm{Ran}(A - zI)$. In this way we obtain a Hilbert space $\mathfrak{A}$ generated by the operator valued function $z \mapsto Q(z)$, $z \in \Omega$, and the corresponding functional model of A.

For each $z \in \Omega$ set

$$K(z) = Q(z)|\mathfrak{N}_{-i}(A) .$$

The operator valued function $z \mapsto C(z)$, $z \in \Omega \setminus \{-i\}$, defined by

$$C(z) = \frac{z - i}{z + i} K(z)$$

is the characteristic function of A in sense of the definition proposed in [S1] which may be considered as a generalization and modification of the first definition of this notion introduced by M. S. Livsic [L1]. Here we will only recall some result from [S4] which explains the role of the operator valued function $z \mapsto K(z)$, $z \in \Omega$, when we construct the space $\mathfrak{A}$ taking into account the Proposition 8 with $\mathfrak{N}_S = \mathfrak{N}_i(A)$ as a kernel of the extension $S \supset (A - iI)^{-1}$ defined by

$$S = (A - iI)^{-1} P_i ,$$

where P_i is the orthoprojector of $\mathfrak{H}$ onto $\mathrm{Ran}(A - iI)$. In the case under consideration we set $\mathfrak{P} = \mathfrak{N}_i(A)$ and $V = I_{\mathfrak{N}_i(A)}$, so that by (31)

$$T(\zeta) = Q(\zeta) , \quad \zeta \in \Omega .$$

According to [S4], the following equality is valid:

$$\begin{aligned} \Gamma\Big(\sum_{\mathrm{Im}\,\gamma \neq 0} \mathfrak{N}_\gamma(A)\Big) &= \Big\{u : u \in \mathfrak{N}_i(A)\Big\} + \Big\{K(\zeta)v : v \in \mathfrak{N}_{-i}(A)\Big\} + \\ &+ \sum_{\mathrm{Im}\,\alpha > 0} \left\{\frac{1}{\zeta - \alpha}\Big(K(\zeta) - K(\alpha)\Big)v : v \in \mathfrak{N}_{-i}(A)\right\} + \\ &+ \sum_{\mathrm{Im}\,\beta < 0} \left\{\frac{1}{\zeta - \beta}\Big(I - K(\zeta)K^*(\beta)\Big)u : u \in \mathfrak{N}_i(A)\right\} . \end{aligned} \tag{47}$$

Here the symbol $\sum$ is used to denote the linear sum or the linear span of the linear sets under consideration.

Suppose now that the defect number of A in the lower halfplane $\mathbb{C}_-$ is zero, so that A is a maximal nonselfadjoint symmetric operator. Then $\mathfrak{N}_{-i}(A) = \{0\}$, $\Omega = \mathbb{C}_+$, and by (37)

$$\Gamma\Big(\sum_{\alpha \in \mathbb{C}_+} \mathfrak{N}_\alpha\Big) = \Big\{u : u \in \mathfrak{N}_i(A)\Big\} + \sum_{\beta \in \mathbb{C}_-} \left\{\frac{1}{\zeta - \beta} u : u \in \mathfrak{N}_i(A)\right\} .$$

The functional model of such operator A that we obtain according to the general construction considered above coincides with the model proposed and investigated by A. I. Plessner [P].

Note that in [S5], [S6] we consider a functional model of an arbitrary symmetric completely nonselfadjoint operator A. In this general case together with $\mathfrak{N}_i(A)$ also $\mathfrak{N}_{-i}(A)$ is chosen as a module subspace. We construct a corresponding functional Hilbert space whose elements are pairs of $\mathfrak{N}_i(A)$-valued and $\mathfrak{N}_{-i}(A)$-valued functions holomorphic on some domains Ω and Ω_* containing the upper and lower halfplanes, respectively. There is also obtained a general integral formula for the inner product in this space.

References

[A] D. Alpay, A remark on the Cowen Douglas classes $B_n(\Omega)$, Arch. Math., 51 (1988), pp. 539–546.

[AD] D. Alpay, H. Dym, Hilbert spaces of analytic functions, inverse scattering and operator models. II. Integral Equations and Operator Theory, 8 (1985), pp. 145–180.

[dB] L. de Branges, Hilbert spaces of entire functions, Prentice Hall, Englewood Cliffs, N. J., 1968.

[dBR] L. de Branges, J. Rovnyak, Canonical models in quantum scattering theory, in Perturbation theory and its applications in quantum mechanics, Wiley, N. Y., 1966, pp. 295–391.

[D] H. Dym, Shifts, realizations and interpolation, redux, vol. 73 of Operator Theory: Advances and Applications, Birkhäuser Verlag, Basel, 1994, pp. 182–243.

[GG] M. L. Gorbachuk, V. I. Gorbachuk, M. G. Krein's Lectures on Entire Operators, Birkhäuser, Basel, 1997.

[K1] M. G. Krein, Fundamental aspects of the representation theory of Hermitian operators with deficiency index (m, m), Ukr. Mat. Z., 1 (1949), pp. 3–66 (Russian); English transl. in Amer. Math. Soc. Transl. (2) 97 (1970), pp. 75–143.

[K2] M. G. Krein, Analytic problems and results in the theory of linear operators in Hilbert space, Proc. Int. Congress Math. (Moscow 1966), Mir, Moscow, 1968, pp. 189–216 (Russian); English transl. in Amer. Math. Soc. Transl. (2) 90 (1970), pp. 181–210.

[L1] M. S. Livsic, On a class of linear operators in Hilbert space, Mat. Sbornik N. S. 19 (1946), pp. 239–262 (Russian); English transl. in Amer. Math. Soc. Transl. (2) 13 (1960), pp. 61–83.

[L2] M. S. Livsic, On the spectral decomposition of linear non-selfadjoint operators, Mat. Sbornik N. S. 34 (1954), pp. 145–199 (Russian); English transl. in Amer. Math. Soc. Transl. (2) 5 (1957), pp. 67–114.

[P] A. I. Plessner, Spectral analysis of maximal operators, Dokl. Akad. Nauk SSSR, 22 (1939), no. 5, pp. 225–228 (Russian).

[S1] A. V. Strauss, On the extensions and the characteristic function of a symmetric operator, Izv. Akad. Nauk SSSR, 32 (1968), pp. 186–207 (Russian); English transl. in Math. USSR — Izvestija, 2 (1968), pp. 181–204.

[S2] A. V. Strauss, On resolvents of extensions of a symmetric operator, Funct. Analysis, Ulyanovsk, 1977, 8, pp. 162–173 (Russian).

[S3] A. V. Strauss, Pseudo resolvents and representations of linear operators, Funct. Analysis, Ulyanovsk, 1989, 29, pp. 141–152 (Russian).

[S4] A. V. Strauss, Spectral representations of linear operators, Funct. Analysis, Ulyanovsk, 1993, 34, pp. 80–93 (Russian).

[S5] A. V. Strauss, Spectral representations and spectral functions of symmetric operators, vol. 87 of Operator Theory: Advances and Applications, Birkhäuser Verlag, Basel, 1996, pp. 399–412.

[S6] A. V. Strauss, Functional models and generalized spectral functions of symmetric operators, Ulyanovsk State Pedagogical University, Preprint, Ulyanovsk, 1997 (Russian).

Department of Physics and Mathematics, Ulyanovsk State Pedagogical University, Ulyanovsk, 432700, Russia

Operator Theory:
Advances and Applications, Vol. 123

Non-self-adjoint Periodic Dirac Operators

Vadim Tkachenko

1. Introduction

We consider the class $\mathcal{D}$ of Dirac operators

$$L = J\frac{d}{dx} + Q(x), \quad x \in \mathbb{R}, \tag{1.1}$$

with matrices

$$J = \left\| \begin{matrix} 0 & 1 \\ -1 & 0 \end{matrix} \right\|, \quad Q(x) = \left\| \begin{matrix} p(x) & q(x) \\ q(x) & -p(x) \end{matrix} \right\|, \tag{1.2}$$

where $p(x)$ and $q(x)$ are π-periodic complex functions satisfying the condition

$$\int_0^\pi (|p(x)|^2 + |q(x)|^2)\, dx < \infty. \tag{1.3}$$

Our aim is to describe spectral properties of such operators in the space $\mathcal{L}_2^2(\mathbb{R})$ of 2-coordinate complex vector functions $F(x) = \operatorname{col}\{f_1(x), f_2(x)\}$ with the norm

$$\|F\| = \left(\int_{-\infty}^{\infty} \left(|f_1(x)|^2 + |f_2(x)|^2\right) dx \right)^{1/2}.$$

The case of real functions $p(x)$ and $q(x)$ is one of the most well-known and well-studied. In this case

$$Q^*(x) = Q(x) \tag{1.4}$$

which implies that L is a selfadjoint operator (cf. [1], [2] for a general theory, and [3] for a solution of inverse problem). It was discovered by Zakharov and Shabat [4] that selfadjoint expressions (1.1) form the L-parts of Lax pairs for the defocusing nonlinear Schrödinger equation

$$iu_t(x,t) + u_{xx}(x,t) - |u(x,t)|^2 u(x,t) = 0, \quad x \in \mathbb{R}, \quad t > 0,$$
$$u(x,0) = u_0(x), \quad u_0(x) = p(x) + iq(x),$$

while the case of skew-symmetric potential matrices $Q(x)$, i.e.,

$$Q^*(x) = -Q(x) \tag{1.5}$$

Gvastela Fellow, partially supported by Israel Science Foundation and Israel Ministry of Science.

corresponds to the self-focusing nonlinear Schrödinger equation

$$iu_t(x,t) + u_{xx}(x,t) + |u(x,t)|^2 u(x,t) = 0, \quad x \in \mathbb{R}, \quad t > 0. \tag{1.6}$$

This explains and justifies the interest in the study of spectral properties of non-self-adjoint operators (1.1) both for decaying and periodic potential matrices (cf. [5]–[9]).

In what follows we denote by $\mathcal{D}_+$ the set of operators (1.1) with π-periodic potential matrices $Q(x)$ satisfying (1.1), (1.2) and (1.4), and by $\mathcal{D}_-$ the set of operators (1.1) with properties (1.1), (1.2) and (1.5).

Spectral theory of operator $L \in \mathcal{D}$ is based on the study of the Cauchy problem

$$\begin{cases} JY'(x,\lambda) + Q(x)Y(x,\lambda) = \lambda Y(x,\lambda), \\ Y(0,\lambda) = Y_0. \end{cases} \tag{1.7}$$

If $Q(x)$ is a continuous matrix function [1], or $\|Q(x)\| \in \mathcal{L}^1(0,\pi)$, see [2], then its solution $Y(x,\lambda)$ is representable via the transformation operator

$$Y(x,\lambda) = Y_0(x,\lambda) + \int_0^x K(x,s)Y_0(s,\lambda)\,ds, \quad 0 \le x \le \pi, \tag{1.8}$$

where $Y_0(x,\lambda)$ is the solution of (1.7) corresponding to $Q(x) \equiv 0$. The kernel $K(x,t)$ is the unique solution of the Gelfand-Levitan equation

$$F(x,t) + K(x,t) + \int_0^x K(x,s)F(s,t)\,ds = 0, \quad 0 \le t \le x \le \pi, \tag{1.9}$$

with a matrix function $F(x,t), 0 \le t \le x \le \pi$, constructed using spectral data of some operators in $\mathcal{L}^2_2(0,\pi)$ corresponding to (1.1) and boundary conditions at $x = 0$ and $x = \pi$. The integral operator in (1.9) turns out to be nuclear, and the potential matrix $Q(x)$ may be expressed via $K(x,t)$ by the formula

$$Q(x) = K(x,x)J - JK(x,x). \tag{1.10}$$

The most important for the spectral theory is the matrix solution

$$U(x,\lambda) = \left\| \begin{matrix} c_1(x,\lambda) & -s_2(x,\lambda) \\ s_1(x,\lambda) & c_2(x,\lambda) \end{matrix} \right\| \tag{1.11}$$

of (1.7) with $U(0,\lambda) = E$ where E is the unit matrix. Its value at $x = \pi$, i.e., $U(\pi,\lambda)$ is called the monodromy matrix of L. The Hill discriminant

$$\Delta(\lambda) = \frac{1}{2}\mathrm{Tr}\, U(\pi,\lambda)$$

completely determines the specrtum $\sigma(\lambda)$ of L in the space $\mathcal{L}^2_2(\mathbb{R})$. Namely,

$$\sigma(L) = \{\lambda : \Delta(\lambda) = t, \quad t \in [-1,\ 1]\}. \tag{1.12}$$

It is easy to see that (1.5) means that $p(x)$ and $q(x)$ in (1.2) are purely imaginary functions, i.e.,

$$Q(x) = i \left\| \begin{matrix} p(x) & q(x) \\ q(x) & -p(x) \end{matrix} \right\| \tag{1.13}$$

where $p(x)$ and $q(x)$ are real. It follows from (1.7) that, for such potential matrices $Q(x)$, the monodromy matrix is of the form

$$U(\pi, \lambda) = \left\| \begin{matrix} c(\lambda) & -s^*(\lambda) \\ s(\lambda) & c^*(\lambda) \end{matrix} \right\| \tag{1.14}$$

where $c(\lambda)$ and $s(\lambda)$ are entire functions of exponential type π,

$$\left(\sum_{k=0}^{\infty} c_k \lambda^k \right)^* = \sum_{k=0}^{\infty} \bar{c}_k \lambda^k,$$

and the identify holds

$$c(\lambda)c^*(\lambda) + s(\lambda)s^*(\lambda) \equiv 1. \tag{1.15}$$

The Hill discriminant (1.11) now has the form

$$\Delta(\lambda) = \frac{c(\lambda) + c^*(\lambda)}{2}, \tag{1.16}$$

and therefore $\Delta(\lambda)$ is real for $\lambda \in \mathbb{R}$. By virtue of (1.15), $|u(\lambda)| \leq |c(\lambda)| \leq 1$ for $\lambda \in \mathbb{R}$, and the entire real axis is contained in the spectrum of every operator $L \in \mathcal{D}_-$. Moreover, if $Q(x) \not\equiv 0$, and only in this case, this spectrum contains non-real points, in contrast to the selfadjoint case.

Our main goal is a spectral parametrization of operators (1.1) with potential matrices satisfying (1.2)–(1.3). Following the well-known scheme (cf. [1], [2]), we show in Section 2 that for $Q(x) \in \mathcal{D}$ there exists the transformation operator of the form (1.8), and derive $\mathcal{L}^2$-estimates for its kernel $K(x, s)$, $0 \leq s \leq x \leq \pi$. Equation (1.9) in the self-adjoint case is equivalent to the orthogonality of the spectral measure, which is not valid in the non-self-adjoint case. To derive (1.9) for $L \in \mathcal{D}$ we first prove in Section 3 expansion theorems for the corresponding boundary problems on $[0, \pi]$, and in Section 4, using the functions $c_1(\pi, \lambda)$ and $s_1(\pi, \lambda)$ from (1.11), we construct a matrix function $F(x, t)$, $0 \leq t \leq x \leq \pi$, such that $K(x, t)$ is the unique solution of (1.9). Generally speaking, in our case the integral operator with kernel $K(x, t)$ is not nuclear[1], and to recover the potential matrix $Q(x)$ from $K(x, t)$, we derive a formula different from (1.10) but, of course, reducible to (1.10) for continuous potential matrices $Q(x)$, as well as for $\|Q(x)\| \in \mathcal{L}^1(0, \pi)$. The functions $c_1(\pi, \lambda)$ and $s_2(\pi, \lambda)$ uniquely determine the entire monodromy matrices $U(\pi, \lambda)$ of operators $L \in \mathcal{D}$. In Section 5 a complete description of these functions is given. Our description gives one-to-one parametrization of potential matrices in spectral terms. In addition, we prove that such parametrization is Hölder-continuous with respect to an appropriate norm. In Section 6 we describe

[1] It "only" belongs to the Hilbert-Schmidt class.

all matrices (1.14) which are monodromy matrices of operators $L \in \mathcal{D}_-$. These results permit us to prove that the class of finite-band potential matrices is dense in the space of matrices $Q(x)$ satisfying (1.3). The corresponding density theorem will be published under a separate title.

In what follows, we introduce the Euclidean norm

$$\|F\| = \left(|f_1|^2 + |f_2|^2\right)^{1/2}$$

for vectors $F = \mathrm{col}\{f_1, f_2\} \in \mathbb{C}^2$, and set

$$< F, G >= f_1 g_1 + f_2 g_2.$$

If W is 2×2-matrix then we set $\|W\| = \sup_{\|F\| \le 1} \|WF\|$, and denote by $\mathcal{L}_2^2(0,x)$ and $\mathcal{L}_{2,2}^2(0,x)$, respectively, the spaces of 2-coordinate vector functions $F(t) = \mathrm{col}\{f_1(t), f_2(t)\}$ and 2×2-matrix functions $W(t)$ with finite norms

$$\|F\|_{\mathcal{L}_2^2(0,x)} = \left(\int_0^x \|F(t)\|^2 \, dt \right)^{1/2}, \quad \|W\|_{\mathcal{L}_{2,2}^2(0,x)} = \left(\int_0^x \|W(t)\|^2 \, dt \right)^{1/2}.$$

We denote by $\mathcal{PW}_\pi$ the Paley-Wiener class of all entire functions $f(\lambda)$ of exponential type π satisfying the condition

$$\int_{\mathbb{R}} |f(t)|^2 \, dt < \infty.$$

2. Transformation operator

In this section we shall construct a transformation operator for equation (1.7) with the potential matrix $Q(x)$ satisfying (1.2) and (1.3) and derive $\mathcal{L}^2$-estimates for its kernel.

Following [2], let us introduce a linear projection $P = P^2$ in $\mathbb{C}^2$ such that $JP + PJ = J$. Denote by $\Omega(x, \lambda; P)$ the solution of (1.7) satisfying the initial condition $\Omega(0, \lambda; P) = P$, and by $\Omega_0(x, \lambda; P)$ such solution corresponding to $Q(x) \equiv 0$.

Lemma 2.1. *If* (1.2) *and* (1.3) *are satisfied, then for each* $x \in [0, \pi]$ *the representation*

$$\Omega(x, \lambda; P) = \Omega_0(x, \lambda; P) + \int_0^x K(x, t; P)\Omega_0(0, \lambda; P)\, dt \tag{2.1}$$

holds and the estimate

$$\|K(x, \cdot; P)\|_{\mathcal{L}_{2,2}^2(0,x)} \le 4\|Q\|_{\mathcal{L}_{2,2}^2(0,x)} \exp \int_0^x \|Q(t)\|\, dt \tag{2.2}$$

is valid.

Proof. We shall use the scheme from [2], Ch.1, Section 2, making necessary changes to derive $\mathcal{L}^2$-estimates. First, let us introduce the matrix solution $U(x,\lambda)$ of (1.7) satisfying the initial condition $U(0,\lambda) = E$. The function

$$V(x,\lambda) = \exp(\lambda Jx)U(x,\lambda)$$

is a solution of integral equation

$$V(x,\lambda) = E + \int_0^x e^{2\lambda Jt} JQ(t)V(t,\lambda)\,dt, \quad 0 \le x \le \pi.$$

If

$$V(x,\lambda) = E + \int_0^x e^{2\lambda Jt} R(x,t)\,dt, \quad 0 \le x \le \pi, \tag{2.3}$$

then the kernel $R(x,t)$ satisfies the equation

$$R(x,t) = W(t) + \int_0^{x-t} JQ(t+s)R(t+s,s)\,ds, \quad 0 \le t \le x \le \pi, \tag{2.4}$$

where $W(t) = JQ(t)$. To solve (2.4), let us set

$$R_0(x,t) = W(t),$$

$$R_{n+1}(x,t) = \int_0^{x-t} JQ(t+s)R_n(t+s,s)\,ds = \int_t^x JQ(u)R_n(u,u-t)\,du, \tag{2.5}$$

and prove the estimate

$$\|R_n(x,\cdot)\|_{\mathcal{L}^2_{2,2}(0,x)} \le \|W\|_{\mathcal{L}^2_{2,2}(0,x)} \frac{1}{n!}\left(\int_0^\pi \|Q(s)\|\,ds\right)^n \tag{2.6}$$

for all $n \ge 0$. This estimate is trivial for $n = 0$, and to prove it for an arbitrary $n \ge 1$, we define the scalar functions

$$q(t) = \|W(t)\|, \quad r_n(x,t) = \|R_n(x,t)\|, \quad 0 \le t \le x \le \pi,$$

to arrive at the recurrent system of scalar inequalities

$$r_{n+1}(x,t) \le \int_t^x q(u)r_n(u,u-t)\,du. \tag{2.7}$$

For a fixed function $h(t) \in \mathcal{L}^2(0,x)$, $\|h\|_{\mathcal{L}^2(0,x)} = 1$, we have

$$\left|\int_0^x r_{n+1}(x,t)h(t)\,dt\right| \le \int_0^x |h(t)|\,dt \int_t^x q(u)r_n(u,u-t)\,du$$

$$= \int_0^x q(u)\,du \int_0^u |h(t)|r_n(u,u-t)\,dt \le \int_0^x q(u)\,du \left(\int_0^u r_n^2(u,t)\,dt\right)^{1/2}.$$

Assume now that the estimate

$$\left(\int_0^u r_n^2(u,s)\,ds\right)^{1/2} \le \frac{1}{n!}\|W\|_{\mathcal{L}^2_{2,2}(0,u)} \left(\int_0^u q(s)\,ds\right)^n \tag{2.8}$$

is proved for all $u \in [0,\pi]$ and some $n \ge 0$. Then

$$\left|\int_0^x r_{n+1}(x,t)h(t)\,dt\right| \le \frac{1}{n!}\|W\|_{\mathcal{L}^2_{2,2}(0,x)} \int_0^x q(u) \left(\int_0^u q(s)\,ds\right)^n du$$

$$= \frac{1}{(n+1)!}\|W\|_{\mathcal{L}^2_{2,2}(0,x)} \left(\int_0^x q(s)\,ds\right)^{n+1}.$$

Hence (2.8) is valid for all $n \in \mathbb{N}$, which proves (2.6). It follows now that the series

$$R(x,t) = \sum_{n=0}^{\infty} R_n(x,t) \tag{2.9}$$

converges in the space $\mathcal{L}^2(0,x)$ for every fixed $x \in [0,\pi]$, and according to (2.5), its sum is a solution of (2.4). The same arguments applied to the sequence (2.5) with $R_0(x,t) = 0$ prove that this solution is unique, while (2.6) implies the estimate

$$\|R(x,\cdot;P)\|_{\mathcal{L}^2_{2,2}(0,x)} \le \|W\|_{\mathcal{L}^2_{2,2}(0,x)} \exp \int_0^x \|q(t)\|\,dt. \tag{2.10}$$

Starting with the solution $R(x,t)$, we define $V(x,\lambda)$ by (2.3) and, using (2.4), check that the function $U(x,\lambda) = \exp(-\lambda Jx)V(x,\lambda)$ is a solution of (1.7), which is representable in the form

$$U(x,\lambda) = e^{-\lambda Jx} + \int_{-x}^{x} S(x,t)e^{-\lambda Jt}\,dt$$

where

$$S(x,t) = \frac{1}{4}R_+\left(x,\frac{x+t}{2}\right) + \frac{1}{4}R_-\left(x,\frac{x-t}{2}\right),$$

and $R_+(x,t) = R(x,t) + JR(x,t)J$, $R_-(x,t) = R(x,t) - JR(x,t)J$. Since $e^{-\lambda Jx} = \cos\lambda x - J\sin\lambda x$, we have $\Omega_0(x,\lambda;P) = P\cos\lambda x - JP\sin\lambda x$ and

$$\Omega(x,\lambda;P) = \Omega_0(x,\lambda;P) + \int_0^x (K_+(x,t)\cos\lambda t - K_-(x,t)J\sin\lambda t)\,dt,$$

where $K_+(x,t) = (S(x,t) + S(x,-t))\,P$, $K_-(x,t) = (S(x,t) - S(x,-t))\,(E-P)$. Since $P^2 = P$, $JP + PJ = J$, we have $P(E-P) = 0$, $PJP = 0$, and if we set $K(x,t;P) = K_+(x,t) + K_-(x,t)$, then we obtain

$$K(x,t;P)\Omega_0(t,\lambda;P) = (K_+(x,t) + K_-(x,t))\,(P\cos\lambda t - JP\sin\lambda t) =$$
$$= K_+(x,t)\cos\lambda t - K_-(x,t)J\sin\lambda t.$$

which proves (2.1). The estimate (2.2) follows from (2.10), which completes the proof of Lemma 2.1.

Let again $U(x,\lambda)$ be the solution of (1.7) satisfying the initial condition $U(0,\lambda) = E$ with the unit matrix E. It is evident that $U(x,\lambda) = \Omega(x,\lambda;P_1) + \Omega(x,\lambda;P_2)$, where P_1 and P_2 are projections on $e_1 = \{1,0\}$ and $e_2 = \{0,1\}$, respectively, such that $P_j e_i = \delta_{ij} e_j$. Lemma 2.1 yields the following proposition.

Theorem 2.1. *If $L \in \mathcal{D}$, then the monodromy matrix of L has the form*

$$U(\pi,\lambda) = \begin{Vmatrix} c_1(\lambda) & -s_2(\lambda) \\ s_1(\lambda) & c_2(\lambda) \end{Vmatrix} \tag{2.11}$$

where

$$c_k(\lambda) = \cos\pi\lambda + f_k(\lambda), \quad s_k(\lambda) = \sin\pi\lambda + g_k(\lambda), \tag{2.12}$$

all functions $f_k(\lambda)$, $g_k(\lambda)$ belong to the space $\mathcal{PW}_\pi$, and $\det U(\pi,\lambda) = 1$.

The following proposition shows that the kernel $K(x,t)$ depends continuously on Q in $\mathcal{L}^2_{2,2}(0,x)$ and, moreover, that this dependence satisfies the Hölder condition.

Lemma 2.2. *If $Q_1(x)$, $Q_2(x) \in \mathcal{L}^2_{2,2}(0,\pi)$, and $K_1(x,t;P)$, $K_2(x,t;P)$ are the corresponding kernels of transformation operators* (2.1)*, then*

$$\|K_1(x,\cdot;P) - K_2(x,\cdot;P)\|_{\mathcal{L}^2_{2,2}(0,x)} \le C\|Q_1 - Q_2\|_{\mathcal{L}^2_{2,2}(0,x)}\left(1 + \|q\|_{\mathcal{L}^2(0,x)}\right)\cdot\exp\int_0^x q(t)\,dt \tag{2.13}$$

where $q(t) = \max\{\|Q_1(t)\|, \|Q_2(t)\|\}$, $0 \le t \le \pi$, and C does not depend on x, $Q_1(x)$ and $Q_2(x)$.

Proof. If $R_1(x,t)$, $R_2(x,t)$ are functions from representations (2.3) corresponding to $Q_1(x)$, $Q_2(x)$, then (2.4) implies relation

$$R_1(x,t) - R_2(x,t) = W(x,t) + \int_0^{x-t} JQ_1(t+s)(R_1(t+s,s) - R_2(t+s,s))\,ds$$

with

$$W(x,t) = J(Q_1(t) - Q_2(t)) + \int_0^{x-t} J(Q_1(t+s) - Q_2(t+s))R_2(t+s,s)\, ds.$$

This is an equation of the type (2.4) with respect to $R(x,t) = R_1(x,t) - R_2(x,t)$. Since

$$\|W(x,\cdot)\|_{\mathcal{L}^2_{2,2}(0,x)} \leq \|Q_1 - Q_2\|_{\mathcal{L}^2_{2,2}(0,x)} \left(1 + \sqrt{\pi}\|q\|_{\mathcal{L}^2(0,x)}\right) \cdot \exp \int_0^x q(t)\, dt,$$

estimate (2.10) implies (2.13).

Lemma 2.3. *If $Q(t)$ is a smooth 2×2-matrix function on $[0,\pi]$ of the form (1.2) then the kernel $K(x,t;P)$ is a smooth solution of the problem*

$$JK_x(x,t;P) + K_t(x,t;P)J = -Q(x)K(x,t), \quad 0 \leq t \leq x \leq \pi, \tag{2.14}$$

$$Q(x) = K(x,x;P)J - JK(x,x;P), \quad 0 \leq x \leq \pi, \tag{2.15}$$

$$K(x,0;P)JP = 0, \quad 0 \leq x \leq \pi, \tag{2.16}$$

If, on the other hand, $Q(x)$ is a smooth 2×2-matrix function of the form (1.2), and $K(x,t;P)$ is a solution of problem (2.14)–(2.16), then the function $\Omega(x,\lambda;P)$ defined by (2.1) is a solution of (1.7) satisfying the initial condition $\Omega(0,\lambda;P) = P$.

Proof. If $M = \max_{0\leq t\leq\pi} \|Q(t)\|$, then the assumption

$$\max_{0\leq s\leq x} \|r_n(x,s)\| \leq \frac{M^n x^n}{n!}$$

yields, according to (2.7), the estimate

$$\max_{0\leq s\leq x} \|r_{n+1}(x,s)\| \leq \frac{M^{n+1}}{(n+1)!}x^{n+1}.$$

Hence the series (2.9) defines a continuous function $R(x,t)$. Moreover representation (2.4) shows that $R(x,t)$ is a smooth function and the partial derivatives $R_x(x,t)$ and $R_t(x,t)$ are continuous functions as well. Inserting (2.1) in (1.7), we obtain

$$(JK(x,x;P) - K(x,x;P)J + Q(x))\Omega_0(x,\lambda;P) + \int_0^x \{JK_x(x,t;P)$$
$$+ K_t(x,t;P)J + Q(x)K(x,t;P)\}\Omega_0(t,\lambda;P)\, dt + K(x,0;P)J\Omega_0(0,\lambda;P) = 0.$$

Since $\Omega_0(0,\lambda;P) \equiv P$, $K(x,0,P) = 2S(x,0)P$, and $PJP = 0$, the last term on the left-hand side of the latter equation vanishes proving (2.14). With λ tending to the point at infinity, we arrive at (2.14)–(2.15).

To prove the second statement of Lemma 2.3, we define $\Omega(x,\lambda;P)$ by (2.1), integrate by parts and, using (2.14)–(2.16), obtain

$$
\begin{aligned}
&J\Omega'(x,\lambda;P)+Q(x)\Omega(x,\lambda;P)-\lambda\Omega(x,\lambda;P)\\
&=JK(x,x;P)\Omega_0(x,\lambda;P)+Q(x)\Omega_0(x,\lambda;P)-\int_0^x K(x,s;P)J\Omega_0'(s,\lambda,P)\,ds\\
&+\int_0^x JK_x(x,t;P)\Omega_0(t,\lambda;P)\,dt+\int_0^x Q(x)K(x,t;P)\Omega_0(t,\lambda;P)\,dt\\
&=(JK(x,x;P)-K(x,x;P)J+Q(x))\Omega_0(x,\lambda;P)+K(x,0;P)JP\\
&=\int_0^x\{JK_x(x,t;P)+K_t(x,t;P)J+Q(x)K(x,t;P)\}\Omega_0(t,\lambda;P)\,dt=0
\end{aligned}
$$

which proves Lemma 2.3.

3. Expansion Theorem

In this section we consider two boundary problems related to the equation

$$JY'(x,\lambda)+Q(x)Y(x,\lambda)=\lambda Y(x,\lambda),\quad 0\le x\le\pi. \tag{3.1}$$

One of them is generated by (3.1) and boundary conditions

$$y_1(0)=y_1(\pi)=0, \tag{3.2}$$

and another by (3.1) and boundary conditions

$$y_2(0)=y_1(\pi)=0. \tag{3.3}$$

We will prove a theorem of expansion in series of their root vectors.

According to Lemma 2.1, the vector functions $Y(x,\lambda)=\Omega(x,\lambda;P_2)e_2$ and $Z(x,\lambda)=\Omega(x,\lambda;P_1)e_1$ are solutions of (3.1) representable in the form

$$
\begin{aligned}
Y(x,\lambda)&=Y_0(x,\lambda)+\int_0^x K(x,t;P_2)Y_0(t,\lambda)\,dt\\
Z(x,\lambda)&=Z_0(x,\lambda)+\int_0^x K(x,t;P_1)Z_0(t,\lambda)\,dt
\end{aligned} \tag{3.4}
$$

with

$$Y_0(x,\lambda)=\begin{Vmatrix}-\sin\lambda x\\ \cos\lambda x\end{Vmatrix},\qquad Z_0(x,\lambda)=\begin{Vmatrix}\cos\lambda x\\ \sin\lambda x\end{Vmatrix} \tag{3.5}$$

and

$$\|K(x,t;P_i)\|\le 4\|q\|_{\mathcal{L}^2(o,x)}\exp\int_0^x q(t)\,dt \tag{3.6}$$

where $q(t)=\|Q(t)\|$.

In particular it means that $Z(\pi,\lambda)$ and $Y(\pi,\lambda)$ are columns of the monodromy matrix (2.11) of equation (3.1).

Theorem 3.1. *Every vector function $H(x) \in \mathcal{L}_2^2(0,\pi)$ is representable in the form*

$$H(x) = \sum_k \operatorname*{res}_{s_2(\lambda_k)=0} \frac{c_1(\lambda)}{s_2(\lambda)} \int_0^\pi < H(t), Y(t,\lambda) > dt \; Y(x,\lambda) \tag{3.7}$$

corresponding to problem (3.1)+(3.2), *and in the form*

$$H(x) = -\sum_k \operatorname*{res}_{c_1(\mu_k)=0} \frac{s_2(\lambda)}{c_1(\lambda)} \int_0^\pi < H(t), Z(t,\lambda) > dt \; Z(x,\lambda) \tag{3.8}$$

corresponding to problem (3.1)+(3.3). *Here both series are converging to $H(x)$ with respect to the norm of $\mathcal{L}_2^2(0,\pi)$.*

Proof of both identities (3.7) and (3.8) is carried out in the same way, and we expose here a proof of (3.7) only.

First we note that the spectrum of problem (3.1)+(3.2) coincides with the zero set $\{\lambda_k\}$ of the function $s_2(\lambda)$, and if λ_k is such a zero of multiplicity $m_k \geq 1$, then the corresponding root subspace is spanned by the system of functions

$$Y(x,\lambda_k),\ \partial Y(x,\lambda_k)/\partial\lambda, \dots, \partial^{m_k-1} Y(x,\lambda_k)/\partial\lambda^{m_k-1}.$$

Since $s_2(\lambda) = \sin\lambda\pi + g_2(\lambda)$ with $g_2 \in \mathcal{PW}_\pi$, the asymptotic relation

$$\lambda_n = n + \delta_n, \quad |n| \to \infty, \quad \{\delta_n\} \in l^2, \tag{3.9}$$

is valid. The resolvent $R(\lambda)$ of problem (3.1)+(3.2) has the form

$$R(\lambda)H(x) = \frac{c_1(\lambda)}{s_2(\lambda)} \int_0^\pi < H(t), Y(t,\lambda) > dt \; Y(x,\lambda) + P(\lambda)H(x),$$

where $P(\lambda)$ is an entire operator function, and

$$< H(t), Y(t,\lambda) >= -h_1(t)s_2(t,\lambda) + h_2(t)c_2(t,\lambda).$$

For $H \in \mathcal{L}_2^2(0,\pi)$, let us set

$$H_*(x) = H(x) + \int_x^\pi K(t,x;P_2)H(t)\,dt.$$

It follows from Lemma 2.1 that $H_* \in \mathcal{L}_2^2(0,\pi)$ and by (3.9) we conclude that the sequence $h_n =< H(t), Y(t,\lambda_n) >=< H_*(t), Y_0(t,\lambda_n) >$ is square-integrable. Hence the series

$$F(x) = \sum_k \operatorname*{res}_{s_2(\lambda_k)=0} \frac{c_1(\lambda)}{s_2(\lambda)} \int_0^\pi < H(t), Y(t,\lambda) > dt \; Y_0(x,\lambda)$$

converges in the norm of space $\mathcal{L}_2^2(0,\pi)$, and by Lemma 2.1 the series in the right-hand side of (3.7) also converges. The sum of this series is

$$F_*(x) = F(x) + \int_0^x K(x,t;P_2)F(t)\,dt,$$

and since its terms have the form res $R(\lambda)H$, the series of the form (3.7) corresponding to $F_*(x)$ coincides with series (3.7). Therefore, the vector function $R(\lambda)V(x)$ with $V(x) = H(x) - F_*(x)$ is an entire function of λ with values in $\mathcal{L}_2^2(0,\pi)$, which means that

$$\begin{aligned} v(\lambda) &= \int_0^\pi < V(x), Y(x,\lambda) > dx\ s_2(\lambda)^{-1} \\ &= \int_0^\pi < V(x) + \int_x^\pi K(t,x;P_2)V(t)\,dt, Y_0(x,\lambda) > dx\ s_2(\lambda)^{-1} \end{aligned}$$

is an entire scalar function. Being a ration of entire functions of exponential type, $v(\lambda)$ is itself of exponential type. Moreover, as the representation $s_2(\lambda) = \sin\lambda\pi + g_2(\lambda)$ with $g_2 \in \mathcal{PW}_\pi$ shows, $v(\lambda)$ tends to zero on each ray $\{\lambda\ :\ \lambda = re^{i\theta},\ \theta \neq 0,\pi\}$. Hence, by the Phragmén-Lindelöf Theorem, $v(\lambda) = 0$,

$$V(x) + \int_x^\pi K(t,x;P_2)V(t)\,dt = 0,$$

and $V(x) = 0$. Therefore $F_*(x) = H(x)$, which completes the proof of Theorem 3.1 for the problem (3.1)+(3.2).

Corollary. *For every vector functions $H(x),\ G(x) \in \mathcal{L}_2^2(0,\pi)$ the Parseval identities*

$$\begin{aligned} &\int_0^\pi < H(x), G(x) > dx \\ &= \sum_k \operatorname*{res}_{s_2(\lambda_k)=0} \frac{c_1(\lambda)}{s_2(\lambda)} \int_0^\pi < H(x), Y(x,\lambda) > dx \int_0^\pi < G(x), Y(x,\lambda) > dx \\ &= -\sum_k \operatorname*{res}_{c_1(\lambda_k)=0} \frac{s_2(\lambda)}{c_1(\lambda)} \int_0^\pi < H(x), Z(x,\lambda) > dx \int_0^\pi < G(x), Z(x,\lambda) > dx \end{aligned}$$

are valid.

Assume now that $H(x)$ is a smooth function satisfying $h_1(0) = h_1(\pi) = 0$ and set $F(x) = JH'(x) + Q(x)H(x) - aH(x)$ where a does not belong to the set $\{\lambda_n\}$.

Then

$$\int_0^\pi < H(t), Y(t,\lambda_n) > dt = \frac{1}{\lambda_n - a}\int_0^\pi < H(t), (L-a)Y(t,\lambda_n) > dt$$

$$= \frac{1}{\lambda_n - a}\int_0^\pi < F(t), Y(t,\lambda_n) > dt.$$

By virtue of (3.9) the sequence

$$< F(t), Y(t,\lambda_n) >=< F(t) + \int_t^\pi K(u,t;P_2)F(u)\,du, Y_0(t,\lambda_n) >$$

is square-integrable, and therefore the following proposition is true.

Theorem 3.2. *If a vector function* $H(x) \in \mathcal{L}_2^2(0,\pi)$ *is such that* $JH'(x)+QH(x) \in \mathcal{L}_2^2(0,\pi)$, *and* $h_1(0) = h_1(\pi) = 0$ *(or* $h_2(0) = h_1(\pi) = 0$ *), then the series* (3.7) *(respectively,* (3.8)*) converges to* $H(x)$ *uniformly on* $[0,\ \pi]$.

The following proposition is a localization theorem for series (3.7) and (3.8).

Theorem 3.3. *If* $Q(x)$ *is a smooth matrix function and a vector function* $H(x) \in \mathcal{L}_2^2(0,\pi)$ *is differentiable in a neighborhood of a point* $a \in (0,\pi)$, *then series* (3.7) *and* (3.8) *converge to* $H(a)$.

Proof. We shall prove the statement for series (3.7). For large $|k|$, zeros of $s_2(\lambda)$ are simple and satisfy asymptotic relations (3.9). Now we have

(3.10)

$$\frac{c_1(\lambda_k)}{\dot{s}_2(\lambda_k)}\int_0^\pi < H(t), Y(t,\lambda_k) > dt\ Y(x,\lambda_k) - \frac{1}{\pi}\int_0^\pi < H(t), Y_0(t,k) > Y_0(x,k)\,dt$$

$$= \frac{1}{\dot{s}_2(\lambda_k)}(c_1(\lambda_k) - \frac{1}{\pi}\dot{s}_2(\lambda_k))\int_0^\pi < H(t), Y(t,\lambda_k) > dt\ Y(x,\lambda_k)$$

$$+ \frac{1}{\pi}\int_0^\pi < H(t), Y(t,\lambda_k) - Y_0(t,k) > dt\ Y(x,\lambda_k)$$

$$+ \frac{1}{\pi}\int_0^\pi < H(t), Y_0(t,k) > dt\ (Y(x,\lambda_k) - Y_0(x,k))$$

and

$$c_1(\lambda_k) - \frac{1}{\pi}\dot{s}_2(\lambda_k) = f_1(\lambda_k) - \frac{1}{\pi}\dot{g}_2(\lambda_k) \in l^2;$$

$$
\begin{aligned}
Y(t,\lambda_k) - Y_0(t,k) &= (Y(t,\lambda_k) - Y(t,k)) + (Y(t,k) - Y_0(t,k)) \\
&= Y_0(t,\lambda_k) - Y_0(t,k) + \int_0^t K(t,s;P_2)(Y_0(s,\lambda_k) \\
&\quad - Y_0(s,k))\, ds + \int_0^t K(t,s;P_2) Y_0(s,k)\, ds.
\end{aligned}
$$

Since $Q(x)$ is smooth, Lemma 2.3 implies that $K(x,s;P_2)$ is smooth as well. Integrating by parts, we obtain

$$
\begin{aligned}
&\int_0^\pi < H(t), \int_0^t K(t,s;P_2) Y_0(s,k)\, ds > dt \\
&\quad = k^{-1} \int_0^\pi < H(t), \int_0^t K(t,s;P_2) J Y_0'(s,k)\, ds > dt \\
&\quad = k^{-1} \int_0^\pi < H(t), K(t,t;P_2) J Y_0(t,k) \\
&\quad - K(t,0;P_2) J P_2 - \int_0^t K_s(t,s;P_2) J Y_0(s,k)\, ds > dt
\end{aligned}
$$

and since $K(t,0;P_2)JP_2 = 0$, the sequence

$$
\int_0^\pi < H(t), \int_0^t K(t,s;P_2) Y_0(s,k)\, ds > dt
$$

is absolutely converging. By virtue of asymptotic representation

$$
\begin{aligned}
Y_0(s,\lambda_k) - Y_0(s,k) &= \left\| \begin{matrix} -(\sin \lambda_k s - \sin ks) \\ \cos \lambda_k s - \cos ks \end{matrix} \right\| \\
&= -(\lambda_k - k)s \left\| \begin{matrix} \cos ks \\ \sin ks \end{matrix} \right\| + O((\lambda_k - k)^2)
\end{aligned}
$$

valid uniformly with respect to $s \in [0,\pi]$, we conclude that the left-hand side of (3.10) defines the series which is uniformly convergent in $[0,\pi]$. Hence the statement of Theorem 3.3 for series (3.7) is equivalent to the same statement for $Q(x) \equiv 0$, which is well known from the theory of Fourier series.

4. Representation of $K(x, y; P)$ via spectral data

Starting with entire functions $c_1(\lambda)$ and $s_2(\lambda)$ of exponential type π, satisfying (2.12), we define, for $t, x \in \mathbb{R}$, the matrix functions

$$F(x,t) = \sum_k \left\{ \operatorname*{res}_{s_2(\lambda_k)=0} \left(\frac{c_1(\lambda)}{s_2(\lambda)} Y_0(x,\lambda) Y_0^T(t,\lambda) \right) - \frac{1}{\pi} Y_0(x,k) Y_0^T(t,k) \right\}, \tag{4.1}$$

and

$$G(x,t) = -\sum_k \left\{ \operatorname*{res}_{c_1(\lambda_k)=0} \left(\frac{s_2(\lambda)}{c_1(\lambda)} Z_0(x,\lambda) Z_0^T(t,\lambda) \right) - \frac{1}{\pi} Z_0(x,k+\frac{1}{2}) Z_0^T(t,k+\frac{1}{2}) \right\}. \tag{4.2}$$

It is easy to see that

$$F(x,t) = \Phi(x-t) + \Psi(x+t) \tag{4.3}$$

where

$$\Phi(x) = \varphi(x)E - \psi(x)J, \quad \Psi(x) = \varphi(x)\sigma_1 - \psi(x)\sigma_2 \tag{4.4}$$

with the scalar functions

$$\varphi(x) = \frac{1}{2} \sum_k \left\{ \operatorname*{res}_{s_2(\lambda_k)=0} \left(\frac{c_1(\lambda)}{s_2(\lambda)} \cos \lambda x \right) - \frac{1}{\pi} \cos kx \right\}, \tag{4.5}$$

$$\psi(x) = \frac{1}{2} \sum_k \left\{ \operatorname*{res}_{s_2(\lambda_k)=0} \left(\frac{c_1(\lambda)}{s_2(\lambda)} \sin \lambda x \right) - \frac{1}{\pi} \sin kx \right\} \tag{4.6}$$

and matrices

$$\sigma_1 = \begin{Vmatrix} -1 & 0 \\ 0 & 1 \end{Vmatrix}, \quad \sigma_2 = \begin{Vmatrix} 0 & 1 \\ 1 & 0 \end{Vmatrix}.$$

For big $|k|$, zeros λ_k of $s_2(\lambda)$ are simple, and we have

$$\begin{aligned} &\frac{c_1(\lambda_k)}{\dot{s}_2(\lambda_k)} \cos \lambda_k x - \frac{1}{\pi} \cos kx \\ &\quad = \frac{c_1(\lambda_k) - \pi^{-1}\dot{s}_2(\lambda_k)}{\dot{s}_2(\lambda_k)} \cos \lambda_k x - \frac{1}{\pi}(\lambda_k - k) x \sin kx + O((\lambda_k - k)^2) \end{aligned} \tag{4.7}$$

and

$$\begin{aligned} &\frac{c_1(\lambda_k)}{\dot{s}_2(\lambda_k)} \sin \lambda_k x - \frac{1}{\pi} \sin kx \\ &\quad = \frac{c_1(\lambda_k) - \pi^{-1}\dot{s}_2(\lambda_k)}{\dot{s}_2(\lambda_k)} \sin \lambda_k x + \frac{1}{\pi}(\lambda_k - k) x \cos kx + O((\lambda_k - k)^2). \end{aligned} \tag{4.8}$$

With account taken of (2.12) and (3.9) it follows that $\varphi(x),\ \psi(x) \in \mathcal{L}^2(0, 2\pi)$, and hence

$$\|F(\cdot, x)\|_{\mathcal{L}^2_{2,2}(0,\pi)} + \|F(x, \cdot)\|_{\mathcal{L}^2_{2,2}(0,\pi)} \le C < \infty \tag{4.9}$$

with C not depending on x. The similar arguments prove the estimate

$$\|G(\cdot,x)\|_{\mathcal{L}^2_{2,2}(0,\pi)} + \|G(x,\cdot)\|_{\mathcal{L}^2_{2,2}(0,\pi)} \le C < \infty. \tag{4.10}$$

Theorem 4.1. *For each $x \in [0,\pi]$, the kernels $K(x,t;P_1)$ and $K(x,t;P_2)$ corresponding to an operator $L \in \mathcal{D}$ satisfy the Gelfand-Levitan equations*

$$F(x,t) + K(x,t;P_2) + \int_0^x K(x,s;P_2)F(s,t)\,ds = 0, \quad 0 \le t \le x \le \pi, \tag{4.11}$$

and

$$G(x,t) + K(x,t;P_1) + \int_0^x K(x,s;P_1)G(s,t)\,ds = 0, \quad 0 \le t \le x \le \pi, \tag{4.12}$$

where $F(x,t)$ and $G(x,t)$ are defined by (4.1) *and* (4.2) *with functions $c_1(\lambda) = c_1(\pi,\lambda)$ and $s_2(\lambda) = s_2(\pi,\lambda)$ from* (1.11).

Proof. According to (2.2), for each $x \in [0,\pi]$, we have $K(x,\cdot;P_2) \in \mathcal{L}^2_{2,2}(0,x)$, and representation (2.1) is valid. Hence

$$\begin{aligned}
&\int_0^x K(x,s;P_2)F(s,t)\,ds \\
&\quad = \sum_k \left\{ \operatorname*{res}_{s_2(\pi,\lambda_k)=0} \left(\frac{c_1(\pi,\lambda)}{s_2(\pi,\lambda)} Y(x,\lambda)Y_0^T(t,\lambda)\right) - \frac{1}{\pi} Y_0(x,k)Y_0^T(t,k) \right\} \\
&\quad - \sum_k \left\{ \operatorname*{res}_{s_2(\pi,\lambda_k)=0} \left(\frac{c_1(\pi,\lambda)}{s_2(\pi,\lambda)} Y_0(x,\lambda)Y_0^T(t,\lambda)\right) - \frac{1}{\pi} Y_0(x,k)Y_0^T(t,k) \right\} \\
&\quad - \sum_k \frac{1}{\pi} \int_0^x K(x,s;P_2)Y_0(s,k)Y_0^T(t,k)\,ds.
\end{aligned}$$

The third term on the right-hand side here is equal to $-K(x,t;P_2)$ by virtue of the convergence theorem for the Fourier series. According to definition (4.1), the second term coincides with $-F(x,t)$. Let us prove that

$$\Theta(x,t) \equiv \sum_k \left\{ \operatorname*{res}_{s_2(\pi,\lambda_k)=0} \frac{c_1(\pi,\lambda)}{s_2(\pi,\lambda)} Y(x,\lambda)Y_0^T(t,\lambda) - \frac{1}{\pi} Y_0(x,k)Y_0^T(t,k) \right\} = 0 \tag{4.13}$$

for $0 < t < x$. To this end, let $H(t)$ be an arbitrary smooth vector function on $[0,\pi]$, vanishing at $x = 0$ and in the interval $[x,\pi]$.

Then

$$\int_0^\pi \Theta(x,t)H(t)\,dt = \sum_k \operatorname*{res}_{s_2(\lambda_k)=0} \frac{c_1(\pi,\lambda)}{s_2(\pi,\lambda)} Y(x,\lambda) \int_0^x < H(t), Y_0(t,\lambda) > dt$$

$$- \frac{1}{\pi} \sum_k Y_0(x,k) \int_0^x < H(t), Y_0(t,k) > dt,$$

and since the last series converges to $H(x) = 0$, we have

$$\int_0^\pi \Theta(x,t)H(t)\,dt$$

$$= \sum_k \operatorname*{res}_{s_2(\pi,\lambda_k)=0} \frac{c_1(\pi,\lambda)}{s_2(\pi,\lambda)} Y(x,\lambda) \int_0^x < H(t) + \int_t^x L^T(u,t)H(u)\,du, Y(t,\lambda) > dt$$

where $L(s,t)$ is defined by the relation

$$Y_0(x,\lambda) = Y(x,\lambda) + \int_0^x L(x,u)Y(u,\lambda)\,du, \tag{4.14}$$

inverse to (2.1).

Let us assume for a moment that $Q(x)$ is a smooth matrix function. According to Lemma 2.3, $K(x,t;P_2)$ is smooth in x and t, and hence $L(x,u)$ is also smooth. Since

$$H(t) + \int_t^x L^T(u,t)H(u)\,du$$

is a smooth function vanishing at $t = x$, we conclude by Theorem 3.3 that

$$\int_0^\pi \Theta(x,t)H(t)\,dt = 0.$$

Therefore (4.13) is valid for smooth potential matrices $Q(x)$.

For an arbitrary matrix $Q(x) \in \mathcal{L}^2_{2,2}(0,\pi)$, let $Q_n(x)$ be a sequence of smooth matrix functions converging to $Q(x)$ in $\mathcal{L}^2_{2,2}(0,\pi)$. According to Lemma 2.2, the sequence of respective kernels $K_n(x,t;P_k)$ of transformation operators (2.1) is such that

$$\|K(x,\cdot;P_k) - K_n(x,\cdot;P_k)\|_{\mathcal{L}^2_{2,2}(0,x)} \le C\|Q - Q_n\|_{\mathcal{L}^2_{2,2}(0,x)}$$

with C not depending on n and x. It implies that if

$$U_n(\pi,\lambda) = \left\| \begin{array}{cc} c_{1,n}(\pi,\lambda) & -s_{2,n}(\pi,\lambda) \\ s_{1,n}(\pi,\lambda) & c_{2,n}(\pi,\lambda) \end{array} \right\| \tag{4.15}$$

is the sequence of monodromy matrices corresponding to potentials $Q_n(x)$, then

$$\int_{\mathbb{R}} \|U(\pi,\lambda) - U_n(\pi,\lambda)\|^2 \, d\lambda \le C\|Q - Q_n\|^2_{\mathcal{L}^2_{2,2}(0,x)}.$$

If $F_n(x,t)$ is the sequence of matrices of the form (4.1) corresponding to potentials $Q_n(x)$, then

$$\begin{aligned} F(x,t) - F_n(x,t) &= \sum_k \operatorname{res} \left(\frac{c_1(\pi,\lambda)}{s_2(\pi,\lambda)} - \frac{c_{1,n}(\pi,\lambda)}{s_{2,n}(\pi,\lambda)} \right) Y(x,\lambda) Y_0^T(t,\lambda) \\ &= \int\limits_{\mathbb{R}\pm ai} \left\{ \frac{c_1(\pi,\lambda)}{s_2(\pi,\lambda)} - \frac{c_{1,n}(\pi,\lambda)}{s_{2,n}(\pi,\lambda)} \right\} Y(x,\lambda) Y_0^T(t,\lambda) d\lambda \end{aligned}$$

with sufficiently large $a > 0$. The previous inequality shows that

$$\left\|\frac{c_1(\pi,\lambda)}{s_2(\pi,\lambda)} - \frac{c_{1,n}(\pi,\lambda)}{s_{2,n}(\pi,\lambda)}\right\|_{\mathcal{L}^2(\mathbb{R}\pm ai)} \le C\|Q - Q_n\|_{\mathcal{L}^2_{2,2}(0,x)}.$$

As a result, we obtain an estimate

$$\|F(x,\cdot) - F_n(x,\cdot)\|_{\mathcal{L}^2_{2,2}(0,x)} \le C\|Q - Q_n\|_{\mathcal{L}^2_{2,2}(0,x)}.$$

We have already proved the relation

$$F_n(x,t) + K_n(x,t;P_2) + \int\limits_0^x K_n(x,s;P_2) F_n(s,t) \, ds = 0, \quad 0 \le t \le x \le \pi.$$

Passing to the limit as $n \to \infty$, we arrive at (4.7). Relation (4.8) is proved in a similar way.

Theorem 4.2. *If $L \in \mathcal{D}$, then for each $x \in [0,\pi]$ the homogeneous equations*

$$K^T(t) + \int\limits_0^x K^T(s) F(s,t) \, ds = 0, \quad 0 \le t \le x, \tag{4.16}$$

and

$$K^T(t) + \int\limits_0^x K^T(s) G(s,t) \, ds = 0, \quad 0 \le t \le x, \tag{4.17}$$

have only the trivial solution in $\mathcal{L}^2_2(0,x)$.

Proof. According to (4.1), we have

$$K^T(t)+\int_0^x K^T(s)\sum_k\left\{\operatorname*{res}_{s_2(\pi,\lambda_k)=0}\left(\frac{c_1(\pi,\lambda)}{s_2(\pi,\lambda)}Y_0(s,\lambda)Y_0^T(t,\lambda)\right)-\frac{1}{\pi}Y_0(s,k)Y_0^T(t,k)\right\}$$

$$= K^T(t)-\frac{1}{\pi}\sum_k\int_0^x < K(s),Y_0(s,k) > Y_0^T(t,k)\,ds$$

$$+\int_0^x K^T(s)\sum_k\operatorname*{res}_{s_2(\pi,\lambda_k)=0}\left(\frac{c_1(\pi,\lambda)}{s_2(\pi,\lambda)}Y_0(s,\lambda)Y_0^T(t,\lambda)\right)\,ds$$

$$=\sum_k\int_0^x\operatorname*{res}_{s_2(\pi,\lambda_k)=0}\left(\frac{c_1(\pi,\lambda)}{s_2(\pi,\lambda)} < K(s,),Y_0(s,\lambda) > Y_0^T(t,\lambda)\right)\,ds=0.$$

Applying the transformation operator (2.1), we obtain

$$\sum_k\int_0^x\operatorname*{res}_{s_2(\pi,\lambda_k)=0}\left(\frac{c_1(\pi,\lambda)}{s_2(\pi,\lambda)} < K(s),Y_0(s,\lambda) > Y^T(t,\lambda)\right)=0,\quad 0\le t\le x,$$

and according to the expansion theorem

$$K(s)+\int_s^x L^T(u,s)K(u)\,du=0,\quad 0\le s\le x,$$

where $L(s,u)$ is the function from (4.14). The latter equation is of Volterra type, and hence $K(s)=0$. The same arguments are applicable to prove the trivial solvability of (4.17).

5. Inverse problem

Let $U(\pi,\lambda)$ be the monodromy matrix of some Dirac operator $L\in\mathcal{D}$. Its entries $c_1(\pi,\lambda)$ and $s_2(\pi,\lambda)$ are characteristic functions of boundary problems (3.1)+(3.3) and (3.1)+(3.2), respectively. Using these functions, it is easy to recover $c_2(\pi,\lambda)$ and $s_1(\pi,\lambda)$. Indeed, since $\det U(\pi,\lambda)\equiv 1$, for each zero λ_k of $s_1(\pi,\lambda)$ of multiplicity $m_k\ge 1$ we have equations

$$c_2(\pi,\lambda_k)c_1(\pi,\lambda_k)=1,$$
$$(c_2(\pi,\lambda)c_1(\pi,\lambda))^{(m)}\Big|_{\lambda=\lambda_k}=0,\quad m=1,\dots,m_k-1,$$

which completely determine $c_2(\pi,\lambda_k)$, $\dot c_2(\pi,\lambda_k),\dots,c_2^{(m_k-1)}(\pi,\lambda_k)$. Applying the Hermite interpolation formula, we can reconstruct the entire function $c_2(\pi,\lambda)=\cos\lambda\pi+g_2(\lambda)$ with $g_2(\lambda)\in\mathcal{PW}_\pi$ and the function

$$s_1(\pi,\lambda)=(1-c_1(\pi,\lambda)c_2(\pi,\lambda))s_2^{-1}(\pi,\lambda)=\sin\lambda\pi+f_1(\lambda),$$

$f_1(\lambda) \in \mathcal{PW}_\pi$. It means that the monodromy matrix $U(\pi,\lambda)$ of an operator $L \in \mathcal{D}$ is determined uniquely by its entries $c_1(\pi,\lambda)$ and $s_2(\pi,\lambda)$. In the present section we describe all functions which form the first line of monodromy matrices $U(\pi,\lambda)$ of operators $L \in \mathcal{D}$.

With each pair of entire functions $c_1(\lambda)$ and $s_2(\lambda)$ of exponential type representable in the form (2.12) we associate the matrix functions (4.1) and (4.2).

Theorem 5.1 *For $c_1(\lambda)$ and $-s_2(\lambda)$ to be the entries of the first line of monodromy matrix of some Dirac operator $L \in \mathcal{D}$, it is necessary and sufficient that they be entire functions of exponential type π with disjoint zero sets, and the following conditions be satisfied:*

i) the representations

$$c_1(\lambda) = \cos\lambda\pi + g_1(\lambda), \qquad s_2(\lambda) = \sin\lambda\pi + f_2(\lambda)$$

are valid with $g_1, f_2 \in \mathcal{PW}_\pi$;

ii) for each $x \in [0,\pi]$, the homogeneous equations (4.16) *and* (4.17) *in $\mathcal{L}^2_{2,2}(0,x)$ have the trivial solution only.*

The necessity of conditions of Theorem 5.1 follows from Theorems 2.1 and 4.2. Let us prove their sufficiency.

Let $F_m(s,t)$ be the m-th partial sum of series (4.1). Then by (3.9), (4.7) and (4.8) we have

$$\begin{aligned}(5.1)\quad \|F(\cdot,s) - F_m(\cdot,s)\|^2_{\mathcal{L}^2_{2,2}(0,\pi)} + \|F(s,\cdot) - F_m(s,\cdot)\|^2_{\mathcal{L}^2_{2,2}(0,\pi)} \le \\ \le C \sum_{|k|>m} (|\lambda_k - k|^2 + |\dot{f}_2(\lambda_k)|^2 + |g_1(\lambda_k|^2)\end{aligned}$$

with a constant C not depending on s and m.

For every fixed $x \in [0,\pi]$, let us consider the family of equations

$$K(t) + \mathbb{F}_m(x)K(t) + F_m(x,t) = 0, \quad 0 \le t \le \pi$$

with respect to $K(t) \in \mathcal{L}^2_{2,2}(0,\pi)$, where operators $\mathbb{F}_m(x)$ are of the form

$$(5.2) \qquad \mathbb{F}_m(x)K(t) = \begin{cases} \int_0^x K(s)F_m(s,t)\,ds, & 0 \le t \le x, \\ 0, & x \le t \le \pi. \end{cases}$$

If $0 \le x_1 \le x_2 \le \pi$, then

$$\|\mathbb{F}_m(x_1) - \mathbb{F}_m(x_2)\|^2 \le \int_0^\pi \int_{x_1}^{x_2} \|F_m(s,t)\|^2\,ds\,dt,$$

$$\|\mathbb{F}(x_1) - \mathbb{F}(x_2)\|^2 \le \int_0^\pi \int_{x_1}^{x_2} \|F(s,t))\|^2\,ds\,dt$$

and

$$\|\mathbb{F}(x) - \mathbb{F}_m(x)\|^2 \leq \int_0^\pi \int_0^x \|F(s,t) - F_m(s,t)\|^2 \, dt \, ds \tag{5.3}$$

where $\mathbb{F}(x)$ is the operator of the form (5.2) corresponding to $F(s,t)$. It means that the family $\mathbb{F}(x)$ depends continuously on $x \in [0,\pi]$, and the sequence $\{\mathbb{F}_m(x)\}$ converges to $\{\mathbb{F}(x)\}$ uniformly on $[0,\pi]$ in the norm of space of linear operators on $\mathcal{L}^2_{2,2}(0,\pi)$. According to assumptions of Theorem 5.1, for each $x \in [0,\pi]$, operator $\mathbb{E} + \mathbb{F}(x)$ is invertible. Therefore the family of inverse operators $\{(\mathbb{E} + \mathbb{F}(x))^{-1}\}$ is uniformly bounded, i.e., $\|(\mathbb{E} + \mathbb{F}(x))^{-1}\| \leq C$ with C not depending on x. We conclude, that for all big m each equation

$$K_m(x,t) + F_m(x,t) + \int_0^x K_m(x,s) F_m(s,t) \, ds = 0, \quad 0 \leq t \leq x \leq \pi, \tag{5.4}$$

is uniquely solvable with respect to $K_m(x,t)$ and its solution is a smooth function of x and t.

Let $K_m(x,t)$ be its solution for some $x \in [0,\pi]$. Matrices $F_m(x,t)$ are smooth in x and t, and since

$$J \frac{\partial}{\partial x} Y_0(x,\lambda) Y_0^T(t,\lambda) = \lambda Y_0(x,\lambda) Y_0^T(t,\lambda) = -\frac{\partial}{\partial t} Y_0(x,\lambda) Y_0^T(t,\lambda) J,$$

we have

$$J \frac{\partial F_m(x,t)}{\partial x} + \frac{\partial F_m(x,t)}{\partial t} J = 0.$$

Differentiating (5.4) with respect to x and t, we obtain

$$\begin{aligned} J \frac{\partial K_m(x,t)}{\partial x} \partial x + \frac{\partial K_m(x,t)}{\partial t} J + Q_m(x) K_m(x,t) \\ + \int_0^x \left(J \frac{\partial K_m(x,s)}{\partial x} + \frac{\partial K_m(x,s)}{\partial s} J + Q_m(x) K_m(x,s) \right) F_m(s,t) \, ds \\ + K_m(x,0) J F_m(0,t) = 0 \end{aligned} \tag{5.5}$$

where

$$Q_m(x) = K_m(x,x) J - J K_m(x,x). \tag{5.6}$$

If $t = 0$ in (5.4), then

$$K_m(x,0) + F_m(x,0) + \int_0^x K_m(x,s) F_m(s,0) \, ds = 0.$$

Since $Y_0(0,\lambda) = e_2$ and $e_2^T J e_2 = 0$, for all x in interval $[0,\pi]$ we have

$$F_m(x,0) J e_2 = \sum_{|k| \leq m} \left\{ \operatorname*{res}_{s_2(\lambda_k)=0} \left(\frac{c_1(\lambda)}{s_2(\lambda)} Y_0(x,\lambda) \right) - \frac{1}{\pi} Y_0(x,k) \right\} e_2^T J e_2 = 0,$$

and $F_m(x,0)JF_m(0,t) = 0$. Therefore

$$K_m(x,0)Je_2 = 0, \qquad K_m(x,0)JF_m(0,t) = 0, \tag{5.7}$$

and (5.5) is a homogeneous equation of the type (4.16). As we have already seen, for all sufficiently big m, it has the trivial solution only, and we find that $K_m(x,t)$ is a solution to the differential equation

$$J\frac{\partial K_m(x,t)}{\partial x} + \frac{\partial K_m(x,t)}{\partial t}J + Q_m(x)K_m(x,t) = 0, \quad 0 \le t \le x \le \pi,$$

satisfying boundary conditions (2.15) and (2.16). According to Lemma 2.3, we conclude that the vector function

$$Y_m(x,\lambda) = Y_0(x,\lambda) + \int_0^x K_m(x,t)Y_0(t,\lambda)\,dt \tag{5.8}$$

is a solution of (3.1) with $Q(x) = Q_m(x)$ defined by (5.6).

To prove that the sequence $\{Q_m(x)\}$ converges to a limit in the space $\mathcal{L}^2_{2,2}(0,\pi)$, we represent (5.6) in the form

$$\begin{aligned} Q_m(x) &= JF_m(x,x) - F_m(x,x)J \\ &+ \int_0^x (JK_m(x,s)F_m(s,x) - K_m(x,s)F_m(s,x)J)\,ds, \ 0 \le x \le \pi. \end{aligned} \tag{5.9}$$

It follows from (4.3) and (4.4) that

$$JF_m(x,x) - F_m(x,x)J = \varphi_m(2x)(J\sigma_1 - \sigma_1 J) - \psi_m(2x)(J\sigma_2 - \sigma_2 J) \tag{5.10}$$

where

$$\varphi_m(x) = \frac{1}{2}\sum_{|k|\le m}\left\{\operatorname*{res}_{s_2(\lambda_k)=0}\left(\frac{c_1(\lambda)}{s_2(\lambda)}\cos\lambda x\right) - \frac{1}{\pi}\cos kx\right\},$$

$$\psi_m(x) = \frac{1}{2}\sum_{|k|\le m}\left\{\operatorname*{res}_{s_2(\lambda_k)=0}\left(\frac{c_1(\lambda)}{s_2(\lambda)}\sin\lambda x\right) - \frac{1}{\pi}\sin kx\right\}.$$

According to (4.7) and (4.8) both sequences $\{\varphi_m(x)\}_1^\infty$ and $\{\psi_m(x)\}_1^\infty$ are converging in $\mathcal{L}^2(0,\pi)$ to the functions $\varphi(x)$ and $\psi(x)$, respectively. Using (5.1) we find that, for every $x \in [0,\pi]$, the sequence $K_m(x,\cdot)$ converges to the limit in the space $\mathcal{L}^2_{2,2}(0,x)$, which we denote by $K_F(x,\cdot)$. The limit transition in (5.9) shows that there exists the limit

$$\begin{aligned} Q(x) = \lim_{m\to\infty} Q_m(x) &= \varphi(2x)(J\sigma_1 - \sigma_1 J) - \psi(2x)(J\sigma_2 - \sigma_2 J) \\ &+ \int_0^x (JK_F(x,s)F(s,x) - K_F(x,s)F(s,x)J)\,ds \in \mathcal{L}^2_{2,2}(0,\pi), \end{aligned} \tag{5.11}$$

and it is evident that $JQ(x)+Q(x)J=0$. Therefore $Q(x)$ has the form (1.2), and condition (1.3) is fulfilled. It follows from (5.8), as $m\to\infty$, that

$$W(x,\lambda)=Y_0(x,\lambda)+\int_0^x K_F(x,t)Y_0(t,\lambda)\,dt$$

is the solution of (3.1) satisfying $W(0,\lambda)=e_2$. By virtue of the uniqueness theorem for (3.1) we conclude that $W(x,\lambda)\equiv Y(x,\lambda)$. Hence $K_F(x,t)=K(x,t;P_2)$ is the unique solution of (4.11). Let us prove that the first line of the monodromy matrix

$$\left\| \begin{array}{cc} c_1(\pi,\lambda) & -s_2(\pi,\lambda) \\ s_1(\pi,\lambda) & c_2(\pi,\lambda) \end{array} \right\|$$

corresponding to the potential $Q(x)$ is formed by the functions $c_1(\lambda)$ and $s_2(\lambda)$ which are generating the function $F(x,t)$ according to (4.1).

To this end, let us set $x=\pi$ in (4.11). Since $Y_0(\pi,k)=(-1)^k e_2$, the first line of $Y_0(\pi,k)Y_0^T(t,k)$ is zero, and hence, for an arbitrary vector function $H\in\mathcal{L}_2^2(0,\pi)$ we have

$$\sum_k \operatorname*{res}_{s_2(\lambda_k)=0}\frac{c_1(\lambda)}{s_2(\lambda)}s_2(\pi,\lambda)\int_0^\pi < Y_0(t,\lambda),H(t)>\,dt=0. \tag{5.12}$$

Let λ_k be a zero of $s_2(\lambda)$, and let m_k be its multiplicity. Every function

$$s_{k,m}(\lambda)=\frac{s_2(\lambda)}{(\lambda-\lambda_k)^m},\quad m=1,2,\dots,m_k,$$

belongs to the space $\mathcal{P}W_\pi$ and according to the Paley-Wiener theorem may be represented in the form

$$s_{k,m}(\lambda)=< Y_0(t,\lambda),H_{k,m}(t)>=\int_0^\pi (h_{k,m}^{(1)}(t)\sin\lambda t+h_{k,m}^{(2)}(t)\cos\lambda t)\,dt$$

where $H_{k,m}(t)=-h_{k,m}^{(1)}(t)e_1+h_{k,m}^{(2)}(t)e_2\in\mathcal{L}_2^2(0,\pi)$. If we substitute $H_{p,m}(t)$ in (5.12), we obtain

$$\sum_k \operatorname*{res}_{s_2(\lambda_k)=0}\left(\frac{c_1(\lambda)}{s_2(\lambda)}s_{p,m}(\lambda)s_2(\pi,\lambda)\right)=\sum_k \operatorname*{res}_{s_2(\lambda_k)=0}\left(\frac{c_1(\lambda)}{s_2(\lambda)}s_2(\pi,\lambda)\frac{s_2(\lambda)}{(\lambda-\lambda_p)^m}\right)$$

$$=\operatorname*{res}_{\lambda_p}\left(\frac{c_1(\lambda)s_2(\pi,\lambda)}{(\lambda-\lambda_p)^m}\right)=0,\quad m=1,\ 2,\dots,m_p.$$

Since $c_1(\lambda_p)\neq 0$, we conclude that $s_2^{(m)}(\pi,\lambda)=0,\quad m=0,\ 1,\dots,m_p-1$, and the function $s_2(\pi,\lambda)s_1^{-1}(\lambda)$ is entire. Both functions $s_2(\pi,\lambda)$ and $s_2(\lambda)$ satisfy the asymptotic representation (2.12), and the Phragmén-Lindelöf theorem implies that $s_2(\pi,\lambda)\equiv s_2(\lambda)$.

To prove that $c_1(\pi,\lambda) = c_1(\lambda)$, we shall use the identity

$$(5.13)\qquad \int_0^\pi < H(t), G(t) > dt = \sum_k \operatorname*{res}_{s_2(\lambda_k)=0} \left(\frac{c_1(\lambda)}{s_2(\lambda)} h(\lambda) g(\lambda) \right)$$

for arbitrary vector functions $H(t),\ G(t) \in \mathcal{L}_2^2(0,\pi)$ with

$$h(\lambda) = \int_0^\pi < H(t), Y(t,\lambda) > dt, \quad g(\lambda) = \int_0^\pi < G(t), Y(t,\lambda) > dt.$$

The proof of (5.13) is based on the trivial solvability of Eq. (4.16) only, and may be carried out in the same way as the proof of the same identity for the selfadjoint matrix $Q(x)$ given in [1]. On the other hand, according to the Corollary of Theorem 3.1, the Parseval identity

$$(5.14)\qquad \int_0^\pi < H(t), G(t) > dt = \sum_k \left(\operatorname*{res}_{s_2(\lambda_k)=0} \frac{c_1(\pi,\lambda)}{s_2(\lambda)} h(\lambda) g(\lambda) \right)$$

is valid. If $H(t) = Y(t,\mu),\ G(t) = Y(t,\nu)$, then

$$\int_0^\pi < Y(t,\mu), Y(t,\nu) > dt = -\frac{c_2(\pi,\mu)s_2(\pi,\nu) - s_2(\pi,\mu)c_2(\pi,\nu)}{\mu - \nu},$$

and it follows from (5.12) and (5.14)

$$\sum_k \operatorname*{res}_{s_2(\lambda_k)=0} \left(\frac{c_1(\lambda) - c_1(\pi,\lambda)}{s_2(\lambda)} \frac{c_2(\pi,\lambda)s_2(\pi,\nu)}{\lambda - \nu} \frac{\partial^p}{\partial \mu^p} \left(\frac{s_2(\pi,\lambda)c_2(\pi,\mu)}{\lambda - \mu} \right) \Bigg|_{\mu=\lambda_n} \right) = 0$$

for $p = 0,\ 1, \dots, m_n - 1$, and for all $\nu \notin \{\lambda_k\}$. All terms of the latter series with $k \neq n$ vanish, and hence

$$\operatorname*{res}_{\lambda_n} \frac{(c_1(\lambda) - c_1(\pi,\lambda))c_2(\pi,\lambda)s_2(\pi,\nu)}{(\lambda - \nu)(\lambda - \lambda_n)^{p+1}} = 0, \quad p = 0,\ 1, \dots, m_n - 1.$$

Since $s_2(\pi,\nu) \neq 0, c_2(\pi,\lambda_n) \neq 0$, we conclude that $c_1(\lambda) - c_1(\pi,\lambda) = O((\lambda - \lambda_n)^{m_n})$, and $\omega(\lambda) \equiv (c_1(\lambda) - c_1(\pi,\lambda))s_2^{-1}(\pi,\lambda)$ is an entire function of minimal exponential type. By virtue of (2.12) and condition $i)$ we have $c_1(\lambda) - c_1(\pi,\lambda) = o(\exp \pi|\mathrm{Im}\lambda|)$, as $|\lambda| \to \infty$, and $|s_2(\pi,\lambda)| \geq C \exp \pi|\mathrm{Im}\lambda|$ outside any set $\{\lambda : |\lambda| \geq R,\ |\mathrm{Im}\lambda| \geq 1\}$ with sufficient big $R > 0$. Therefore $\omega(\lambda) \equiv 0,\ c_1(\lambda) = c(\pi,\lambda)$, which proves that $i)$ and the trivial solvability of (4.16) are sufficient for the conclusion of Theorem 5.1.

The sufficiency of i) and trivial solvability of (4.17) is proved in the similar way. Here we note that if these conditions are fulfilled then

$$Q(x) = \tau(2x)(J\sigma_1 - \sigma_1 J) - \theta(2x)(J\sigma_2 - \sigma_2 J) + \int_0^x (JK_G(x,s)G(s,x) - K_G(x,s)G(s,x)J)\,ds \in \mathcal{L}^2_{2,2}(0,\pi), \tag{5.15}$$

where $G(x,s)$ is defined by (4.2),

$$\tau(x) = -\frac{1}{2}\sum_k \left\{ \operatorname*{res}_{c_1(\lambda_k)=0} \left(\frac{s_2(\lambda)}{c_1(\lambda)} \cos \lambda x \right) - \frac{1}{\pi} \cos kx \right\}, \tag{5.16}$$

$$\theta(x) = -\frac{1}{2}\sum_k \left\{ \operatorname*{res}_{c_1(\lambda_k)=0} \left(\frac{s_2(\lambda)}{c_1(\lambda)} \sin \lambda x \right) - \frac{1}{\pi} \sin kx \right\}, \tag{5.17}$$

and $K_G(x,t) = K(x,t;P_1)$ is the unique solution of (4.12).

Theorem 5.1 shows that the class $\mathcal{D}$ is one-to-one parametrized by the set of 2-coordinate entire vector functions $U(\lambda) = (c_1(\lambda), s_2(\lambda))$ with properties i)-ii). Let the set of such functions be endowed with the metric

$$||U_1 - U_2||_{\mathcal{PW}} = \Big(\int_{\mathbb{R}} (||U_1(\lambda) - U_2(\lambda)||^2) d\lambda \Big)^{1/2}.$$

The following proposition states that such a parametrization is Hölder-continuous.

Theorem 5.2. *If L_1 and L_2 are two operators from class $\mathcal{D}$ with potential matrices $Q_1(x)$ and $Q_1(x)$, and if $U_1(\lambda)$ and $U_2(\lambda)$ are the first lines of their monodromy matrices, respectively, then*

$$||U_1 - U_2||_{\mathcal{PW}} \le C||Q_1 - Q_2||_{\mathcal{L}^2_{2,2}(0,\pi)} \exp(C||q||)_{\mathcal{L}^2(0,\pi)} \tag{5.18}$$

where $q(t) = \max\{Q_1(t), Q_2(t)\}$ and C is independent of $L_i, Q_i, i = 1,2$.

If, on the other hand, $U(\lambda) = (c_1(\lambda), s_2(\lambda))$ is 2-coordinate entire vector function satisfying conditions i)–ii) of Theorem 5.1, then there exists sufficiently small number $\epsilon > 0$ such that every 2-coordinate entire vector function $V(\lambda)$ of exponential type π with $||U - V||_{\mathcal{PW}} \le \epsilon$ also satisfies the same conditions i)–ii), and if $Q_U(x)$ and $Q_V(x)$ are corresponding potential matrices, then

$$||Q_U - Q_V||_{\mathcal{L}^2_{2,2}(0,\pi)} \le C||U - V||_{\mathcal{PW}} \tag{5.19}$$

with a constant C independent of V.

Proof. Inequality (5.18) follows from representation (2.1) and estimate (2.13).

To prove (5.19) we use $U(\lambda)$ and $V(\lambda)$ to define the matrix functions $F_U(x,t)$ and $F_V(x,t)$ according to (4.1), and the scalar functions $\varphi_U(x)$, $\psi_U(x)$, $\varphi_V(x)$, $\psi_V(x)$ according to (4.5)–(4.6). Similar to (5.1) we prove that

$$\|F_U(\cdot,t) - F_V(\cdot,t)\|_{\mathcal{L}^2_{2,2}(0,\pi)} + \|F_U(x,\cdot) - F_V(x,\cdot)\|_{\mathcal{L}^2_{2,2}(0,\pi)} \le C\|U - V\|_{\mathcal{PW}}$$

and

$$\|\varphi_U - \varphi_V\|_{\mathcal{L}^2(0,2\pi)} + \|\psi_U - \psi_V\|_{\mathcal{L}^2(0,2\pi)} \le C\|U - V\|_{\mathcal{P}W} \tag{5.20}$$

with a constant C independent of U and V. According to Theorem 4.1, there exists the unique solution $K_U(x,t)$ of (4.11) with $F(x,t) = F_U(x,t)$. It follows now that, for all sufficiently small $\epsilon > 0$ and $\|U-V\|_{\mathcal{P}W} \le \epsilon$, Eq.(4.1) with $F(x,t) = F_V(x,t)$ has the unique solution $K_V(x,t)$ and, moreover,

$$\|K_U(\cdot,t)-K_V(\cdot,t)\|_{\mathcal{L}^2_{2,2}(0,\pi)}+\|K_U(x,\cdot)-K_V(x,\cdot)\|_{\mathcal{L}^2_{2,2}(0,\pi)} \le C\|U-V\|_{\mathcal{P}W} \tag{5.21}$$

with C independent of U and V. Representations (5.11) together with estimates (5.20) and (5.21) imply (5.19).

6. Periodic Dirac operators with skew-symmetric potential matrices

Theorem 5.1 permits us to give an explicit description of monodromy matrices of operators belonging to class $\mathcal{D}_-$.

Theorem 6.1. *For a 2×2-matrix $U(\lambda)$ to be the monodromy matrix of some operator $L \in \mathcal{D}_-$ it is necessary and sufficient that it has the form (1.14) where $c(\lambda)$ and $s(\lambda)$ are entire functions of exponential type π, satisfying the following conditions:*

i) the representations

$$c(\lambda) = \cos\lambda\pi + f(\lambda), \quad s(\lambda) = \sin\lambda\pi + g(\lambda) \tag{6.1}$$

are valid with $f,\ g \in \mathcal{P}W_\pi$;

ii) the identity holds

$$c(\lambda)c^*(\lambda) + s(\lambda)s^*(\lambda) \equiv 1;$$

iii) for each $x \in [0,\pi]$, the integral equations (4.16) and (4.17) with kernels (4.1) and (4.2) corresponding to $c_1(\lambda) = c(\lambda)$, $c_2(\lambda) = c^*(\lambda)$, $s_1(\lambda) = s(\lambda)$, $s_2(\lambda) = s^*(\lambda)$ have only the trivial solution in $\mathcal{L}^2_2(0,x)$.

Proof. Necessity. Representations (6.1) are obtained in Theorem 2.1. If $Q(x)$ has the form (1.13), then $(JQ(x))^* = -Q^*(x)J = Q(x)J = -JQ(x)$, and the shift operator $Y_0 \to Y_0(x,\lambda)$ is a unitary operator in $\mathbb{C}^2$ for every $x \in [0,\pi]$. Hence, for real λ, $U(\pi,\lambda)$ is a unitary matrix, which leads us to the system of equations

$$\begin{cases} c_1(\pi,\lambda)c_2(\pi,\lambda) + s_1(\pi,\lambda)s_2(\pi,\lambda) = 1 \\ -\overline{s_1(\pi,\lambda)}c_2(\pi,\lambda) + \overline{c_1(\pi,\lambda)}s_2(\pi,\lambda) = 0. \end{cases}$$

Solving this system with respect to $c_2(\pi,\lambda)$ and $s_2(\pi,\lambda)$ we find $c_2(\pi,\lambda) = \overline{c_1(\pi,\lambda)}$ and $s_2(\pi,\lambda) = \overline{s_1(\pi,\lambda)}$, $\lambda \in \mathbb{R}$. In other words, the identities $c_2(\pi,\lambda) = c_1^*(\pi,\lambda)$, $s_2(\pi,\lambda) = s_1^*(\pi,\lambda)$ are fulfilled for $\lambda \in \mathbb{R}$, and by the uniqueness theorem for analytic functions they are fulfilled for all $\lambda \in \mathbb{C}$, which proves *ii*). Property *iii*) is proved in Theorem 4.2.

Sufficiency. Given a matrix $U(\lambda)$ of the form (1.14) with $i) - iii)$ being fulfilled, we apply Theorem 5.1 with $c_1(\lambda) = c(\lambda)$ and $s_2(\lambda) = s^*(\lambda)$ and obtain a potential matrix $Q(x)$ of the form (1.2). The monodromy matrix of corresponding operator has the form

$$\begin{Vmatrix} c(\lambda) & -s^*(\lambda) \\ s_1(\lambda) & c_2(\lambda) \end{Vmatrix}$$

with $c(\lambda)c_2(\lambda)+s^*(\lambda)s_1(\lambda) \equiv 1$, and $s_1(\lambda)$ and $c_2(\lambda)$ satisfying (2.12). Comparison with conditions i) and ii) shows, as explained at the beginning of Section 5, that $s_1(\lambda) = s(\lambda)$, $c_2(\lambda) = c^*(\lambda)$. Let us prove that $Q^*(x) = -Q(x)$.

First of all, $JY_0(x, \lambda) = Z_0(x, \lambda)$, and by the complex conjugation of (4.1) we obtain

$$J\overline{F(x,t)}J = -\sum_k \left\{ \operatorname*{res}_{c^*(\mu_k)=0} \left(\frac{c^*(\lambda)}{s(\lambda)} Z_0(x,\lambda) Z_0^T(t,\lambda) \right) - \frac{1}{\pi} Z_0(x,k) Z_0^T(t,k) \right\}.$$

Since $ii)$ is fulfilled,

$$J\overline{F(x,t)}J$$

$$= \sum_k \left\{ \operatorname*{res}_{c(\lambda_k)=0} \left(\frac{s^*(\lambda)}{c(\lambda)} Z_0(x,\lambda) Z_0^T(t,\lambda) \right) - \frac{1}{\pi} Z_0(x, k+\frac{1}{2}) Z_0^T(t, k+\frac{1}{2}) \right\} + H(x,t)$$

where

$$H(x,t) = -\sum_k \left\{ \operatorname{res} \left(\frac{1}{s(\lambda)c(\lambda)} - \frac{1}{\sin\lambda\pi\cos\lambda\pi} \right) Z_0(x,\lambda) Z_0^T(t,\lambda) \right\}$$

$$= \lim_{k\to\infty} \int_{|\lambda|=k+\frac{1}{4}} \frac{s(\lambda)c(\lambda) - \sin\lambda\pi\cos\lambda\pi}{s(\lambda)c(\lambda)\sin\lambda\pi\cos\lambda\pi} Z_0(x,\lambda) Z_0^T(t,\lambda)\, d\lambda.$$

Representations (2.12) and (6.1) show that

$$|s(\lambda)c(\lambda) - \sin\pi\lambda\cos\pi\lambda| = o(1)e^{2\pi|\mathrm{Im}\lambda|},$$

$$|s(\pi,\lambda)c(\pi,\lambda)\sin\pi\lambda\cos\pi\lambda| \geq Ce^{4\pi|\mathrm{Im}\lambda|}$$

with C not depending on k. On the other hand, all entries of the matrix

$$Z_0(x,\lambda) Z_0^T(t,\lambda)$$

are bounded by $C\exp(x+t)|\mathrm{Im}\lambda|$. By virtue of the Jordan Lemma we conclude that $H(x,t) \equiv 0$, and hence $J\overline{F(x,t)}J = -G(x,t)$ where $G(x,t)$ is defined by (4.2) with $c_1(\lambda) = c(\lambda)$, $s_2(\lambda) = s^*(\lambda)$. If $B = iJ$, then $B^2 = E$ and

$$G(x,t) = B\overline{F(x,t)}B. \tag{6.2}$$

We multiply the complex conjugate of (4.11) by B from the left and right and find that (4.12) is uniquely solvable, and its unique solution has the form $K_G(x,t) =$

$\overline{BK_F(x,t)}B$ where $K_F(x,t)$ is the unique solution of (4.11). According to Theorem 5.1, $K_G(x,t)$ is the kernel of transformation operator of some operator $L \in \mathcal{D}$ with the same monodromy matrix $U(\pi,\lambda)$, and hence with the same potential matrix $Q(x)$. If the functions $\varphi(x)$, $\psi(x)$, $\tau(x)$, $\theta(x)$ are defined by (4.5), (4.6), (5.16), (5.17), respectively, with $c_1(\lambda) = c(\lambda)$ and $s_2(\lambda) = s^*(\lambda)$, then again using *ii)* we prove, similar to (6.2), that $\overline{\varphi(x)} = -\tau(x)$, $\overline{\psi(x)} = -\theta(x)$. With account taken of (5.15), we obtain

$$\begin{aligned}
\overline{Q(x)} &= \overline{\varphi(2x)}(J\sigma_1 - \sigma_1 J) - \overline{\psi(2x)}(J\sigma_2 - \sigma_2 J) \\
&\quad + \int_0^x (J\overline{K_F(x,s)}\,\overline{F(s,x)} - \overline{K_F(x,s)}\,\overline{F(s,x)}J)\,ds \\
&= -\tau(2x)(J\sigma_1 - \sigma_1 J) + \theta(2x)(J\sigma_2 - \sigma_2 J) \\
&\quad + \int_0^x (B\overline{K_F(x,s)})B)(B\overline{F(s,x)}B)J - J(B\overline{K_F(x,s)})B)(B\overline{F(s,x)}B))\,ds \\
&= -\tau(2x)(J\sigma_1 - \sigma_1 J) + \theta(2x)(J\sigma_2 - \sigma_2 J) \\
&\quad + \int_0^x (K_G(x,s)G(s,x)J - JK_G(x,s)G(s,x))\,ds = -Q(x).
\end{aligned}$$

The identity $\overline{Q(x)} = -Q(x)$ together with (1.2) is equivalent to representation (1.13) which completes the proof of Theorem 6.1.

References

[1] B. M. LEVITAN and I. S. SARGSYAN. Operators of Sturm-Liouville and Dirac, Nauka, Moscow, 1988.

[2] V. A. MARCHENKO. Sturm-Liouville operators and applications, Birkhäuser, Basel, 1986.

[3] T. V. MISYURA. Characteristics of spectra of periodic and anti-periodic problems generated by Dirac's operations, I. Theoriya funktsij, funkts. analiz i ikh prilozhen., 30, 1978, 90–101; II. ibid., 31, 1979, 102–109.

[4] V. E. ZAHAROV and V. B. SHABAT. Integration of nonlinear equations of mathematical physics by a method of inverse scattering problem, Functional Anal. and its Appl., 6:3, 1974, 43–53.

[5] Y. C. MA and M. J. Ablowitz. The periodic cubic Schrödinger equation, Studies in Applied Mathematics, 65, 1981, 113–158.

[6] R. K. DODD, J. C. EILBECK, J. D. GIBBON and H. C. MORRIS. Solitons and nonlinear wave equations, Academic Press Inc., London, 1982.

[7] L. D. FADDEEV and L. A. TAKHTADZHYAN. Hamiltonian methods in the theory of solitons, Springer Series in Soviet Mathematics, Springer-Verlag, Berlin-New York, 1987.

[8] D. W. McLAUGHLIN and C. M. SCHOEBER. Chaotic and homoclinic behavior for numerical discretizations of the nonlinear Schrödinger equation, Physica D, 57, 1992, 447–465.

[9] D. BÄTTIG, B. GRÉBERT, J. C. GUILLOT, T. KAPPELER. Foliation of phase space for the cubic non-linear Schrödinger equation, Compositio Mathematica, 85, 1993, 163–199.

Department of Mathematics and Computer Science, Ben-Gurion University of the Negev, Beer-Sheva 84105, Israel
E-mail address: tkachenk@math.bgu.ac.il

Operator Theory:
Advances and Applications, Vol. 123

A Note on Hyponormal Operators Associated with Quadrature Domains

Dmitry V. Yakubovich

Abstract. In a series of papers [15], [16], [10], Putinar and Gustafsson connect the properties of a quadrature domain $\mathcal{D}$ with the properties of the hyponormal operator T with rank one self-commutator, whose principal function coincides with the characteristic function of this domain. Xia's analytic model [19] of this hyponormal operator represents it as the operator of multiplication by the independent variable on a space of analytic functions on $\mathcal{D}$. We give a new formula for the norm in this model space. The relationship between quadrature domains and subnormal operators with finite rank self-commutator has been found and exploited in [20], [12], [23]. Our result shows that a hyponormal T as above can be obtained from a subnormal operator with finite rank self-commutator by a finite rank perturbation of the space and the norm in it.

1. Introduction and main result

Suppose $\mathcal{D}$ is a bounded domain in $\mathbb{C}$ and there are points z_k in $\mathcal{D}$ and constants c_{jk} such that an identity

$$\iint_{\mathcal{D}} f\,dx\,dy = \sum_{k=1}^{s} \sum_{j=1}^{r_k-1} c_{jk} f^{(j)}(z_k)$$

holds for all analytic functions f in $L^1(\mathcal{D})$. Then $\mathcal{D}$ is called a quadrature domain. Quadrature domains possess many interesting and intriguing properties, see [17], [18]. Their study is connected with such diverse areas as algebraic geometry, Hele-Shaw free boundary problems, subnormal operators. Let us outline the connection between quadrature domains and hyponormal operators, which was discovered by M. Putinar.

A (linear bounded) operator on a (complex) Hilbert space H is called hyponormal if the self-commutator $[T^*, T]$ is non-negative. Here $[A, B] = AB - BA$. We

This research was supported, in parts, by Russian RFFI Grant No. 98-01-00312, Russian Federal Aim Program "Integraciya" 326.53, Project 2.1, and the Grant by Russian Ministry of Education No. 97-0-1.4.

will always assume additionally that T has no nonzero reducing subspace on which it is normal.

As Carey and Pincus showed, to each T with trace $[T^*, T] < \infty$ one can associate a unique function $g_T \in L^1(dx\,dy)$ such that

$$\text{trace}\,[p(T,T^*),\, q(T,T^*)] = \frac{-1}{\pi} \iint_{\mathbb{C}} g_T(z) \det \begin{bmatrix} p'_z & p'_{\bar z} \\ q'_z & q'_{\bar z} \end{bmatrix} dx\,dy \tag{1.1}$$

for all polynomials $p(z, \bar z)$, $q(z, \bar z)$. It is called the principal function of T. It is nonnegative, and its support coincides with the spectrum of T. We refer to [11] for a background on hyponormal operators.

If we restrict ourselves to hyponormals T with rank one self-commutator, then the correspondence $T \leftrightarrow g_T$ becomes one-to-one. Namely, for each such T, g_T satisfies $0 \le g_T \le 1$, and each function $g_T \in L^\infty(dx\,dy)$ with $0 \le g_T \le 1$ and with compact support is principal function of a unique hyponormal operator with rank $[T^*, T] = 1$. The structure of hyponormal operators of this subclass is now relatively well understood, see [13], [3].

Putinar noticed that hyponormal operators T with rank $[T^*, T] = 1$ such that g_T is the characteristic function of a quadrature domain are of special interest. He proved the following result.

Theorem A [15]. *Let T be hyponormal and $[T^*, T] = \xi \otimes \xi$ for some $\xi \in H$. Then* span $\{T^{*k}\xi : k \ge 0\}$ *is finite-dimensional if and only if $g_T = \chi_{\mathcal{D}}$ for some quadrature domain $\mathcal{D}$.*

Following [15], let us call this operator T *the hyponormal operator associated with the quadrature domain $\mathcal{D}$*. In [16], [10], Gustafsson and Putinar show that the analysis of this operator helps in the study of the domain $\mathcal{D}$ itself.

Let $\mathcal{D}$ be a domain with piecewise smooth boundary and ρ a weight, that is a nonnegative function on $\partial\mathcal{D}$. Let ω be the harmonic measure for $\mathcal{D}$ (with respect to any point in $\mathcal{D}$). We always assume that $\log(\rho|dz|/d\omega(z))$ is in $L^1(d\omega)$, and that $\psi\rho|dz|$ is a finite measure for some polynomial $\psi(z)$. We will call such weights admissible. Define the Hardy class $H^2(\mathcal{D}, \rho\,|dz|)$ as the closure in $L^2(\rho|dz|)$ of those rational functions with poles off clos $\mathcal{D}$ that belong to $L^2(\rho|dz|)$. The functionals of evaluation in a point of $\mathcal{D}$ continue from the rational functions onto the whole $H^2(\mathcal{D}, \rho\,|dz|)$. Therefore one can associate with a function f in $H^2(\mathcal{D}, \rho\,|dz|)$ an analytic function $f^\sim$ in $\mathcal{D}$. The function f can be interpreted as the boundary values of $f^\sim$. From now on we will use the same notation for these two functions. We refer to [6] for the background on the Hardy classes of multiply connected domains. The criterion for a domain to be a quadrature domain in terms of the Hardy classes is as follows.

Theorem B (McCarthy, Yang [12]). *Let $\mathcal{D}$ be a bounded finitely connected domain in $\mathbb{C}$ and ρ an admissible weight on $\partial\mathcal{D}$. Then the operator*

$$M_z f(z) = z f(z)$$

on $H^2(\mathcal{D},\rho\,|dz|)$ *satisfies* $\operatorname{rank}[M_z^*,M_z]<\infty$ *if and only if* $\mathcal{D}$ *is a quadrature domain.*

By *a finite divisor* d in $\mathcal{D}$ we mean a finite subset $\{z_1,\dots,z_l\}$ in $\mathcal{D}$ with associated multiplicities $m_1,\dots,m_l\in\mathbb{N}$. For each such divisor d, we put $m(d)=\sum m_j$, and define $F_d: H^2(\mathcal{D},\rho\,|dz|)\to\mathbb{C}^{m(d)}$ by

$$F_d f=\big(f(z_1),\dots,f^{(m_1-1)}(z_1),\quad\dots\quad,f(z_l),\dots,f^{(m_l-1)}(z_l)\big).$$

An operator on a Hilbert space is called *subnormal* if it is a restriction of a normal operator to an invariant subspace. Each subnormal operator is hyponormal. It is obvious that for any bounded finitely connected domain $\mathcal{D}$ and any admissible ρ, M_z on $H^2(\mathcal{D},\rho\,|dz|)$ is subnormal.

For each quadrature domain $\mathcal{D}$, there is a unique polynomial $Q(z,w)$ of the form

$$Q(z,w)=\sum_{j=0}^{n}\sum_{k=0}^{n}a_{jk}z^jw^k, \tag{1.2}$$

with $a_{kj}=\bar a_{jk}$, $a_{nn}=1$, such that $\partial\mathcal{D}$, up to a finite number of points, coincides with the set $\{z\in\mathbb{C}:Q(z,\bar z)=0\}$. We put $P(z)=\sum_{j=0}^n a_{jn}z^j$. Define the weight ρ on $\partial\mathcal{D}$ as

$$\rho(z)=\frac{|P(z)|^2}{Q_2'(z,\bar z)}\,\frac{dz}{2\pi i|dz|} \tag{1.3}$$

(here $Q_2'=\frac{\partial Q}{\partial w}$). It will be shown that $\rho>0$ a.e. on $\partial\mathcal{D}$.

Theorems A and B show that to each quadrature domain one can associate a hyponormal operator with rank one self-commutator, on one hand, and a family of subnormal operators with finite rank self-commutator, on the other. Our main results show that these two classes of operators are very close to each other.

Theorem 1. *Let* $\mathcal{D}$ *be a quadrature domain and* T *the associated hyponormal operator with* $\operatorname{rank}[T^*,T]=1$*. Suppose that* ρ *is in* $L^1(|dz|)$*. Then there exist a finite divisor* d *in* $\mathcal{D}$ *and an* $m(d)\times m(d)$ *Hermitean symmetric matrix* A *such that* T *is unitarily equivalent to the operator*

$$M_zf(z)=zf(z)$$

on $H^2(\mathcal{D},\rho\,|dz|)$*, where* $H^2(\mathcal{D},\rho\,|dz|)$ *is given the new equivalent norm, generated by the scalar product*

$$\langle f,g\rangle_{\rho,d,A}=\int_{\partial\mathcal{D}}f(z)\bar g(z)\rho(z)\,|dz|-\langle AF_df,F_dg\rangle,\quad f,g\in H^2(\mathcal{D},\rho\,|dz|). \tag{1.4}$$

Now suppose that ρ is not in $L^1(|dz|)$. Let $\lambda_1^0,\dots,\lambda_\nu^0$ be all points of $\partial\mathcal{D}$ such that ρ is not locally in L^1 near these points. Then $\rho(z)\sim c_j|z-\lambda_j^0|^{-a_j}$, $z\in\partial\mathcal{D}$, $z\to\lambda_j^0$, where $a_j\ge1$ are rational and $c_j>0$ (see Proposition 3 below). Introduce integers

$$N_j=\min\{n:(z-\lambda_j^0)^n\in L^2(\rho|dz|)\}=[(a_j+1)/2]\ge1,$$

and put $N = \sum_1^\nu N_j$. Let Pol be the set of all (analytic) polynomials in $\mathbb{C}$ and Pol_N the set of polynomials in Pol of degree less than N. We set

$$\mathcal{H}(\rho) = \mathrm{Pol}_N + H^2(\mathcal{D}, \rho\,|dz|);$$

functions in $\mathcal{H}(\rho)$ are defined on $\mathcal{D}$. Note that each polynomial in $\mathrm{Pol} \cap H^2(\mathcal{D}, \rho\,|dz|)$ is a multiple of $\prod_j (z - \lambda_j^0)^{N_j}$. Hence $\mathrm{Pol}_N \cap H^2(\mathcal{D}, \rho\,|dz|) = 0$. We can equip $H^2(\mathcal{D}, \rho\,|dz|)$ with the Hilbert norm of direct sum. We remark that $\mathcal{H}(\rho)$ contains all polynomials.

By a Stoltz angle at a point λ of $\partial\mathcal{D}$ we mean an angle $\{z : 0 < |z - \lambda| < r, |\arg \zeta(z - \lambda)| < a\}$, where $r > 0$, $|\zeta| = 1$, $0 < a < \pi$, which is contained in $\mathcal{D}$ and whose sides are not tangential to $\partial\mathcal{D}$ at λ.

Proposition 1. *For each of the points λ_j^0 there exists a Stoltz angle at λ_j^0. If* St *is such angle, $0 \le k < N_j$ and $f = p + f_0 \in \mathcal{H}(\rho)$, $p \in \mathrm{Pol}_N$, $f_0 \in H^2(\mathcal{D}, \rho\,|dz|)$, then $\lim_{z \in \mathrm{St}, z \to \lambda_j^0} f^{(k)}(z) = p^{(k)}(\lambda_j^0)$.*

In view of this fact, for $f = p + f_0$ as above, we put $f^{(k)}(\lambda_j^0) = p^{(k)}(\lambda_j^0)$.

Theorem 2. *Suppose that ρ is not in $L^1(|dz|)$, and define $\mathcal{H}(\rho)$ as above. Then there exists an equivalent Hilbert norm $\|\cdot\|_{\mathcal{D}}$ on $\mathcal{H}(\rho)$ such that T is unitarily equivalent to the operator $M_z f(z) = zf(z)$ on $\mathcal{H}(\rho)$. There exist a finite divisor d in $\mathcal{D}$ and an $m(d) \times m(d)$ Hermitean symmetric matrix A such that the product, given in (1.4), coincides with the scalar product $\langle \cdot, \cdot \rangle_{\mathcal{D}}$ for $f, g \in H^2(\mathcal{D}, \rho\,|dz|) \subset \mathcal{H}(\rho)$.*

For each of the points λ_j^0, the functional $f \mapsto f(\lambda_j^0)$ is continuous on $\mathcal{H}(\rho)$ and non-zero. Hence $\bar{\lambda}_j^0$ is eigenvalue of the operator M_z^* on $\mathcal{H}(\rho)$, and therefore in this case for any admissible weight ψ, T cannot be similar to M_z on $H^2(\psi\,|dz|)$.

Let d_b be the divisor, defined by points $\{\lambda_1^0, \dots, \lambda_\nu^0\}$ and multiplicities $N_1, \dots, N_\nu$. The map $F_{d_b} : \mathcal{H}(\rho) \to \mathbb{C}^{m(d_b)}$ is well defined.

Remark. *In the situation of Theorem 2, there exists a divisor d_0 in $\mathcal{D}$ and a matrix B such that*

$$\langle f, q \rangle_{\mathcal{D}} = \langle F_{d_0} f, B F_{d_b} q \rangle, \quad f \in \mathcal{H}(\rho), q \in \mathrm{Pol}_N. \tag{1.5}$$

Formulas (1.4), (1.5) determine the scalar product on the whole $\mathcal{H}(\rho)$.

Our proof will involve dual analytic models of T by Xia (see [19, 13]). We refer to [21]–[23] for related and more general constructions for hyponormal, subnormal and Toeplitz operators.

2. Proofs

First we need some preparations. Let $Q(z, w)$ be an irreducible polynomial of the form (1.2), with $a_{kj} = \bar{a}_{jk}$ and $a_{nn} = 1$. Consider the algebraic curve

$$\Delta = \{(z, w) \in \mathbb{C}^2 : Q(z, w) = 0\}.$$

Let $\widehat{\Delta}$ be the blow-up of Δ, obtained by ungluing points of different sheets of Δ that coincide as points of $\mathbb{C}^2$ (see [7]). It is an abstract compact Riemann surface, and coordinate functions z, w are meromorphic functions on it. Since $Q(\bar{w}, \bar{z}) = \overline{Q(z, w)}$, the anti-analytic involution $\delta = (z, w) \mapsto \delta^* = (\bar{w}, \bar{z})$ maps Δ onto itself. It extends to $\widehat{\Delta}$. For any $\delta \in \widehat{\Delta}$ one can consider the corresponding point (z, w) in $\mathbb{C}^2$, where $z = z(\delta)$, $w = w(\delta)$, then we write $\delta \sim (z, w)$.

We call $\widehat{\Delta}$ *admissible* if the set of fixed points $\widehat{\Delta} \cap \{w = \bar{z}\}$ of the involution divides it into two connected components. One can call these components $\widehat{\Delta}_+$ and $\widehat{\Delta}_-$ in such a manner that $z(\widehat{\Delta}_+)$ is bounded and $z(\widehat{\Delta}_-)$ is not, see [23]. Then $\widehat{\Delta}$ is a disjoint union of $\widehat{\Delta}_+$, $\widehat{\Delta}_-$ and $\partial\widehat{\Delta}_+ = \partial\widehat{\Delta}_-$.

The connection between algebraic curves in $\mathbb{C}^2$ and quadrature domains now can be stated as follows.

Theorem C (Aharonov-Shapiro [1], Gustafsson [8]). *If Δ is admissible and $z(\cdot)|\Delta_+$ is injective, then the image $z(\Delta_+)$ is a quadrature domain. Each quadrature domain is formed in this way.*

For each $\zeta \in \mathbb{C}$, we put $w_1(\zeta), \ldots, w_n(\zeta)$ to be the roots of equation $Q(\zeta, w) = 0$. If $\zeta \in \mathcal{D}$, we agree that w_1 is the single-valued function $w_1(\zeta) = w\left(\left(z|\widehat{\Delta}_+\right)^{-1}(\zeta)\right)$. The function w_1 extends continuously to $\operatorname{clos}\mathcal{D}$; the extension satisfies $w_1(z) = \bar{z}$ on $\partial\mathcal{D}$. The other roots w_j, $j \geq 2$ are multivalued functions on $\mathcal{D}$. For $\zeta \notin \operatorname{clos}\mathcal{D}$, we fix the numbering arbitrarily.

Proposition 2. *If $z \in \mathbb{C} \setminus \operatorname{clos}\mathcal{D}$ and $1 \leq j \leq n$ or if $z \in \mathcal{D}$ and $j \geq 2$, then $\bar{w}_j(z) \in \mathcal{D}$.*

Proof. Let $\delta \in \widehat{\Delta}$, $\delta \sim (z, w_j(z))$. If $z \in \mathcal{D}$ and $j \geq 2$, then δ is in $\widehat{\Delta}_-$, by the definition of w_1. If $z \in \mathbb{C} \setminus \operatorname{clos}\mathcal{D}$ and j is arbitrary, then δ also is in $\widehat{\Delta}_-$, by Theorem C. In both cases, $\delta^* \sim (\bar{w}_j(z), \bar{z})$ is in $\widehat{\Delta}_+$, and so $\bar{w}_j(z) \in \mathcal{D}$. $\square$

Let $\Delta_+ \cap \Delta_- = \{\delta_1, \ldots, \delta_L\}$, $\delta_j = (\lambda_j, \mu_j)$ (here Δ_+ and Δ_- are subsets of $\mathbb{C}^2$ and not abstract Riemann surfaces). Then $\mu_j = w_1(\lambda_j) \in \mathcal{D}$, $1 \leq j \leq L$. The points $\delta_j^* = (\bar{\mu}_j, \bar{\lambda}_j)$ are also in $\Delta_+ \cap \Delta_-$. Therefore $\{\bar{\mu}_j\}$ is a rearrangement of $\{\lambda_j\}$.

We will prove that then Theorem 1 holds true for a certain divisor d, whose support is contained in the set $\{\lambda_1, \ldots, \lambda_L\}$.

Note that

$$\prod_{j=1}^{n}(w - w_j(z)) = \frac{Q(z, w)}{P(z)}. \tag{2.1}$$

Putting here $z = \infty$, we get

$$\overline{P(\bar{w})} = \prod_{j=1}^{n}(w - w_j(\infty)). \tag{2.2}$$

(see (1.2)). For each j, $(\infty, w_j(\infty)) \in \Delta_-$, therefore $(\bar{w}_j(\infty), \infty) \in \Delta_+$. So all points $\bar{w}_j(\infty)$ are in $\mathcal{D}$, and in particular, $P \neq 0$ on $\partial\mathcal{D}$.

By (2.1),

$$\frac{P(z)}{Q_2'(z,\bar{z})} = \prod_{j=2}^{n}(w_1(z) - w_j(z))^{-1}, \quad z \in \partial\mathcal{D}. \tag{2.3}$$

Hence we can rewrite (1.3) as follows:

$$\rho(z) = \frac{\bar{P}(z)}{\prod_{j=2}^{n}(w_1(z) - w_j(z))} \frac{dz}{2\pi i |dz|}. \tag{2.4}$$

We say that a point $\delta = (z_0, w_0) \in \Delta$ is regular if there are analytic functions ψ_1, ψ_2, defined in neighbourhoods of z_0, w_0, respectively, such that a neighbourhood of δ in $\mathbb{C}^2$ has both representations $w = \psi_1(z)$, $z = \psi_2(w)$. Each regular point has only one preimage on $\widehat{\Delta}$. We call a point $\lambda \in \partial\mathcal{D}$ regular point of $\partial\mathcal{D}$ if there is a real analytic function Ψ, defined in a neighbourhood of λ such that $\nabla\Psi(\lambda) \neq 0$ and locally near λ the domain $\mathcal{D}$ coincides with the set $\{z : \Psi(z) > 0\}$. Non-regular points of $\widehat{\Delta}$ or on $\partial\mathcal{D}$ are called singular.

For each $\delta_0 \in \widehat{\Delta}$ there is some natural number r such that some local chart η at δ_0 satisfies $\eta(\delta)^r = z(\delta) - z(\delta_0)$; then r is called the order of branching of δ_0.

Proposition 3. *1) If $\delta \in \widehat{\Delta}$ and $\delta \sim (\lambda, \bar{\lambda})$ for some $\lambda \in \partial\mathcal{D}$, then $\delta \in \partial\widehat{\Delta}_+$.*

2) $\partial\widehat{\Delta}_+$ is equal to the set of points $\delta \sim (z, w)$ of $\widehat{\Delta}$ such that $\bar{w} = z \in \partial\mathcal{D}$.

3) $\partial\mathcal{D} = z(\partial\widehat{\Delta}_+)$.

4) If $\lambda \in \partial\mathcal{D}$ and $(\lambda, \bar{\lambda})$ is a regular point of $\widehat{\Delta}$, then λ is a regular point of $\partial\mathcal{D}$.

5) If $\lambda \in \partial\mathcal{D}$ and $(\lambda, \bar{\lambda})$ is singular, then there are only two possibilities:

(a) There is only one point $\delta \in \widehat{\Delta}$ such that $\delta \sim (\lambda, \bar{\lambda})$. The order of branching of δ is two. $\mathcal{D}$ coincides with a curvilinear angle of size 2π locally near λ.

(b) There are exactly two non-branched points δ_1, δ_2 of $\widehat{\Delta}$ that are equivalent to $(\lambda, \bar{\lambda})$. There are disjoint domains $\mathcal{D}_1$, $\mathcal{D}_2$, which are regular at λ such that $\partial\mathcal{D}_1 \cap \partial\mathcal{D}_2 = \{\lambda\}$ and $\mathcal{D} \cap \mathcal{W} = (\mathcal{D}_1 \cup \mathcal{D}_2) \cap \mathcal{W}$ for a neighbourhood $\mathcal{W}$ of λ.

In each of these cases, there are a positive constant c and an integer a such that $|\rho(z)| \sim c|z - \lambda|^{a/2}$.

6) For each of the points λ_j^0 (see the Introduction), $(\lambda_j^0, \bar{\lambda}_j^0)$ is a singular point of $\widehat{\Delta}$.

Proof. Let $\delta \in \widehat{\Delta}$ and $\delta \sim (\lambda, \bar{\lambda}) \sim \delta^*$. If δ does not belong to $\partial\widehat{\Delta}_+$, then, say, $\delta \in \widehat{\Delta}_+$ and $\delta^* \in \widehat{\Delta}_-$. There are points $\delta_l \in \widehat{\Delta}$ such that $z(\delta_l) \notin \mathcal{D}$ and $\delta_l \to \delta$. It follows that $\delta_l \in \widehat{\Delta}_-$, $\delta \in \operatorname{clos}\widehat{\Delta}_-$, a contradiction. This proves 1). Since $\partial\widehat{\Delta}_+$ is contained in $\{\delta : w(\delta) = \bar{z}(\delta)\}$, 2) and 3) follow. The curve $\partial\widehat{\Delta}_+$ is real analytic without singularities when considered as a curve on $\widehat{\Delta}$. If $\lambda \in \partial\mathcal{D}$ and $\delta = (\lambda, \bar{\lambda})$ is regular, then δ is the only point of $\partial\widehat{\Delta}_+$ such that $z(\delta) = \lambda$. Hence $\partial\mathcal{D}$ is also regular at λ. Now let $(\lambda, \bar{\lambda})$ be singular, with $\lambda \in \partial\mathcal{D}$. Each of the points δ of the blow-up $\widehat{\Delta}$ that correspond to $(\lambda, \bar{\lambda})$ belongs to $\partial\widehat{\Delta}_+$, by 1). Since $z(\cdot)$ is univalent on $\widehat{\Delta}_+$, it follows that (a) and (b) are the only cases that can happen (see also [9]). In the case (a), $\prod_{j=2}^n(w_1(z)-w_j(z))$ has branching two in λ. In the case (b), this function is single-valued. Hence the asymptotics $|\rho(z)| \sim c|z-\lambda|^{a/2}$ follows.

At last, by (2.4), if $\lambda \in \partial\mathcal{D}$ and ρ is not bounded near λ, then necessarily two of w_j coincide at λ or λ is a projection of a branch point of $\widehat{\Delta}$. This gives 6). □

Proof of Proposition 1. Fix one of the points λ_j^0. It has to be proved that if $f \in \mathcal{H}(\rho)$ and $0 \le k < N_j$, then $\lim_{z\in \mathrm{St}\,, z\to\lambda_j^0} f^{(k)}(z) = 0$. Put $\psi(z) = h(z)^{1/2}\prod_l(z-\lambda_l^0)^{a_l/2}$, $z \in \mathcal{D}$, where h is a polynomial with roots in $\mathbb{C}\setminus\operatorname{clos}\mathcal{D}$, chosen in such a way that ψ is single-valued in $\mathcal{D}$. Then $f = \psi g$, where $g \in H^2(\mathcal{D}, |dz|)$. It easily follows from the Cauchy formula that $|g^{(s)}| = \bar{o}\big(|z-\lambda_j^0|^{-1/2-s}\big)$ in St as $z \to \lambda_j^0$. Since $|\psi^{(k-s)}| \le C_j|z-\lambda_j^0|^{a_j/2-k+s}$, it follows that $|f^{(k)}| = \bar{o}\big(|z-\lambda_j^0|^{-1/2+a_j/2-k}\big) = \bar{o}(1)$, $z \in \mathrm{St}\,, z \to \lambda_j^0$. □

Let $\mathcal{D}$ be any finitely connected domain with piecewise smooth boundary, and let T be the corresponding hyponormal operator. We cite the following result.

Theorem D (Pincus, D. Xia, J. Xia [19, 13]). *Define the kernel*

$$S(r,t) = \langle (r-T)^{-1}\xi, (t-T)^{-1}\xi\rangle, \quad r,t \in \mathbb{C}\setminus\operatorname{clos}\mathcal{D}. \tag{2.5}$$

Then

$$S(r,t) = \exp\,\frac{1}{\pi}\iint_{\mathcal{D}}\frac{dx\,dy}{(z-r)(\bar{z}-\bar{t})} - 1. \tag{2.6}$$

We will need the following

Lemma 1. *Let $\mathcal{D}$ be a quadrature domain and Δ the associated algebraic curve in $\mathbb{C}^2$. Then*

$$S(r,t) = \frac{P(r)\overline{P(t)}}{Q(r,\bar{t})} - 1 = \frac{P(r)}{\prod_{k=1}^n\big(r-\bar{w}_k(t)\big)} - 1, \quad r,t \in \mathbb{C}\setminus\operatorname{clos}\mathcal{D}. \tag{2.7}$$

Proof. For r, t in the unbounded component of $\mathbb{C}\setminus\operatorname{clos}\mathcal{D}$, (2.7) can be obtained by analytic continuation of formulas (7), (9) in [10]. Now suppose that, say, t is in a bounded component G_0 of $\mathbb{C}\setminus\operatorname{clos}\mathcal{D}$. Take any point α on the outer boundary of $\mathcal{D}$ and any $\beta \in \partial G_0$. Let γ be an arc in $\mathcal{D}$ that joins α and β. Put $\widetilde{\mathcal{D}} = \mathcal{D}\setminus\gamma$, then $\partial\widetilde{\mathcal{D}} = \partial\mathcal{D}\cup\gamma_+\cup\gamma_-$, where γ_+, γ_- are two sides of the cut γ, γ_+ starts in α and ends in β and γ_- starts in β and ends in α. If v is a continuous function on $\widetilde{\mathcal{D}}$, then we denote by v_+, v_- its boundary limit values on γ_+, γ_-, respectively. The function $\eta(z) = \log(\bar{z}-\bar{t})$ is single-valued on $\widetilde{\mathcal{D}}$, and $\log(\bar{z}-\bar{t})_+ - \log(\bar{z}-\bar{t})_- = -2\pi i$ on γ.

We put $\widetilde{\gamma} = \{\delta \in \widehat{\Delta}_- : \bar{w}(\delta) \in \gamma\}$ and $\widetilde{\Delta}_- = \{\delta \in \widehat{\Delta}_- : \bar{w}(\delta) \in \widetilde{\mathcal{D}}\} = \widehat{\Delta}_-\setminus\widetilde{\gamma}$; let us assume that $\widetilde{\gamma}$ does not contain points δ with $z(\delta) = r$. The function $\delta \mapsto \log\big(w(\delta)-\bar{t}\big) = \eta\big(\bar{z}(\delta^*)-\bar{t}\big)$ is single-valued on $\widetilde{\Delta}_-$, and $\log(w(\delta)-\bar{t})_+ - \log(w(\delta)-\bar{t})_- = -2\pi i$ on $\widetilde{\gamma}$. Note that $w(\cdot) = \bar{z}(\cdot)$ on $\partial\widehat{\Delta}_-$. Therefore by (2.6), the Cauchy-Green formula and the Cauchy residue theorem, applied to $\widetilde{\Delta}_-$,

$$\begin{aligned} S(r,t) &= \exp \frac{1}{2\pi i}\int_{\partial\widetilde{\mathcal{D}}} \frac{\log(\bar z - \bar t)}{z-r}\,dz \quad - \quad 1 \\ &= \exp -\int_\gamma \frac{dz}{z-r} \ \exp \frac{1}{2\pi i}\int_{\partial\widehat{\Delta}_+} \frac{\log(w(\delta)-\bar t)}{z(\delta)-r}\,dz(\delta) \quad - \quad 1 \\ &= \exp -\frac{1}{2\pi i}\int_{\partial\widetilde{\Delta}_-} \frac{\log(w(\delta)-\bar t)}{z(\delta)-r}\,dz(\delta) \quad - \quad 1 = \frac{\prod_{j=1}^n(\bar t - w_j(\infty))}{\prod_{j=1}^n(\bar t - w_j(r))} \quad - \quad 1\,. \end{aligned}$$

Since $S(r,t) = \overline{S(t,r)}$, (2.7) follows (see (2.1), (2.2)). □

Let $[T^*,T] = \xi\otimes\xi$. For every $\lambda\in\mathcal{D}$, there is a unique eigenvector k_λ of T^* such that

$$(T^* - \bar\lambda)k_\lambda = 0, \quad \langle k_\lambda, \xi\rangle \equiv 1.$$

Following Xia [19], consider the operator

$$U : H \to \operatorname{Hol}(\mathcal{D}), \quad \text{acting by} \quad (\mathrm{Ux})(\lambda) = \langle \mathrm{x}, \mathrm{k}_\lambda\rangle, \quad \lambda\in\mathcal{D}.$$

By a result of Clancey [2], U is one-to-one. It transforms T into the multiplication by the independent variable: $UT = M_zU$. Xia calls this model the analytic model of T.

Similarly, one can put

$$(Vx)(\lambda) = \langle x, (\lambda - T)^{-1}\xi\rangle, \quad \lambda\in\mathbb{C}\setminus\operatorname{clos}\mathcal{D}.$$

Functions Vx, $x\in H$ are antianalytic. Xia calls the function $Vx(\bar z)$ the dual analytic representation of vector x. He gives formulas for the action of T, T^* in both these models (for an arbitrary domain with analytic boundary, see also [13]).

Let $\mathcal{O}(\operatorname{clos}\mathcal{D})$ be the space of functions, analytic on $\operatorname{clos}\mathcal{D}$. If $f\in\mathcal{O}(\operatorname{clos}\mathcal{D})$ and, say, $f\in\operatorname{Hol}(\Omega)$, where Ω contains $\operatorname{clos}\mathcal{D}$ and has a good boundary, then we put

$$Wf = \frac{1}{2\pi i}\int_{\partial\Omega} f(\lambda)(\lambda-T)^{-1}\xi\,d\lambda = f(T)\xi. \tag{2.8}$$

Since $\langle(\lambda-T)^{-1}\xi, k_\mu\rangle = (\lambda-\mu)^{-1}$, it follows that $UWf = f$ for $f\in\mathcal{O}(\operatorname{clos}\mathcal{D})$.

We put $\theta(z) = \prod_j (z-\lambda_j^0)^{N_j+1}$ and

$$\mathcal{O}_0(\operatorname{clos}\mathcal{D}) = \{f\in\mathcal{O}(\operatorname{clos}\mathcal{D}) : \theta | f\}.$$

Let f, g be fixed functions in $\mathcal{O}(\operatorname{clos}\mathcal{D})$. Consider an antianalytic function

$$g^\dagger(t) = \frac{1}{2\pi i}\int_{\partial\mathcal{D}} g(r)S(r,t)\,dr, \quad t\in\mathbb{C}\setminus\operatorname{clos}\mathcal{D}$$

(note that $g^\dagger = VWg$). We define $g^\dagger|\mathcal{D}$ by

$$\bar g^\dagger(t) = \sum_{j=2}^n \frac{\bar g(\overline{w_j(t)})\overline{P(\overline{w_j(t)})}}{\prod_{1\le k\le n, k\ne j}(w_j(t)-w_k(t))}, \qquad t\in\mathcal{D}.$$

By Proposition 2, $\bar g^\dagger$ is well defined and meromorphic on $\mathcal{D}$.

Lemma 2. *The function* $\bar g^\dagger|\mathbb{C}\backslash \operatorname{clos}\mathcal{D}$ *extends analytically to* $\mathbb{C}\backslash\mathcal{D}_1$ *for some domain* $\mathcal{D}_1$ *such that* $\operatorname{clos}\mathcal{D}_1 \subset \mathcal{D}$ *and* $\lambda_j \in \mathcal{D}_1$. *The function* $\bar g^\dagger|\mathcal{D}$ *is meromorphic in* $\mathcal{D}$, *with the only possible poles in* $\lambda_1, \dots, \lambda_L$. *The function* $\theta\bar g^\dagger|\mathcal{D}$ *extends continuously to* $\partial\mathcal{D}$. *The coefficients of the main part of the Laurent expansion of* $\bar g^\dagger$ *at each* λ_j *linearly depend only on finitely many derivatives of* g *in* $\bar\mu_j$. *We have*

$$\bar g_e^\dagger(t) - \bar g_i^\dagger(t) = \frac{\bar g(t)|P(t)|^2}{Q_2'(t,\bar t)}, \qquad t \in \partial\mathcal{D}, \tag{2.9}$$

where $g_e^\dagger(t)$, $g_i^\dagger(t)$ *are boundary limit values of* $g^\dagger|\mathbb{C}\setminus\mathcal{D}$, $g^\dagger|\mathcal{D}$, *respectively.*

This lemma and formulas (2.5) and (2.8) imply that

$$\langle Wf, Wg\rangle = \frac{1}{2\pi i}\int_{\partial\mathcal{D}} f(z)\bar g_e^\dagger(z)\,dz. \tag{2.10}$$

Proof of Lemma 2. Put

$$h(t) = \frac{1}{2\pi i}\int_{\partial\mathcal{D}} g(r)\frac{P(r)}{\prod_{k=1}^n\big(r - \overline{w_k(t)}\big)}\,dr. \tag{2.11}$$

For $t \in \mathbb{C}\backslash\operatorname{clos}\mathcal{D}$, $g^\dagger(t) = h(t)$. Let $\mathcal{D}_2$ be a domain with smooth boundary, close to $\mathcal{D}$ and such that $\operatorname{clos}\mathcal{D} \subset \mathcal{D}_2$. Define h_1 by the same integral as in (2.11), with $\partial\mathcal{D}$ replaced by $\partial\mathcal{D}_2$. By Proposition 2 and the Cauchy residue theorem, $h_1 = h = g^\dagger$ on $\mathbb{C}\setminus\partial\mathcal{D}$. Note that $\bar h_1$ is analytic in $\mathbb{C}\setminus\partial\mathcal{D}_1$ for some domain $\mathcal{D}_1$ such that $\operatorname{clos}\mathcal{D}_1 \subset \mathcal{D}$.

For $t \in \mathbb{C}\setminus\operatorname{clos}\mathcal{D}$,

$$h(t) = \frac{g(\overline{w_1(t)})P(\overline{w_1(t)})}{\prod_{k=2}^n\big(\overline{w_1(t)} - \overline{w_k(t)}\big)} + \sum_{j=2}^n \frac{g(\overline{w_j(t)})P(\overline{w_j(t)})}{\prod_{1\le k\le n, k\ne j}\big(\overline{w_j(t)} - \overline{w_k(t)}\big)} \tag{2.12}$$

(see Proposition 2). The branch $w_1(t)$ is single-valued on $\mathcal{D}$ and satisfies $\overline{w_1(t)} = t$ on $\partial\mathcal{D}$. This and (2.3) give (2.9).

Put $\mathcal{A} = \{t \in \mathcal{D} : \overline{w_1(t)} \in \mathcal{D}\}$ and $\mathcal{B} = \{t \in \mathcal{D} : \overline{w_1(t)} \notin \mathcal{D}\}$, then $\mathcal{D} = \mathcal{A}\cup\mathcal{B}$. Let $\mathcal{P}$ be the set of poles of $\bar g^\dagger$ in $\mathcal{D}$.

For $t \in \mathcal{A}$, (2.12) also holds true. Since $\bar h$ is analytic in $\mathcal{A}$, the poles of the two summands of (2.12) in $\mathcal{A}$ coincide. Therefore $\mathcal{P}\cap\mathcal{A} \subset \{\lambda_1, \dots, \lambda_L\}$.

Let $t_0 \in \mathcal{B}$. Let $\mathcal{D}_{t_0}$ be a domain with good boundary such that $\operatorname{clos}\mathcal{D}_{t_0} \subset \mathcal{D}$ and $\bar w_j(t) \in \mathcal{D}_{t_0}$ for $j \ge 2$ and t in a neighbourhood $\mathcal{V}(t_0)$ of t_0. Then for $t \in \mathcal{V}(t_0)$,

$$g^\dagger(t) = \frac{1}{2\pi i}\int_{\partial\mathcal{D}_{t_0}} \frac{g(r)P(r)}{\prod_{j=1}^n\big(r - \overline{w_j(t)}\big)}\,dr.$$

It follows that $\bar g^\dagger$ is analytic at t_0. Hence $\mathcal{P}\cap\mathcal{B} = \emptyset$.

Next,

$$\bar g^\dagger(t) = \bar h(t) - \frac{\overline{g(\overline{w_1(t)})P(\overline{w_1(t)})}}{\prod_{j=2}^n (w_1(t) - w_j(t))}, \quad t \in \mathcal{A}, \tag{2.13}$$

and the analyticity of $\bar h$ on $\mathcal{A}$ implies the assertion about the main part of the Laurent expansion of $\bar g^\dagger$ in λ_j.

At last, if we replace $\bar h$ with $\bar h_1$ in (2.13), then (2.13) will hold in $\mathcal{D} \setminus \mathcal{D}_1$. This implies that $\theta \bar g^\dagger | \mathcal{D}$ extends continuously up to $\partial\mathcal{D}$. □

Lemma 3. *The weight ρ is positive a.e. on $\partial\mathcal{D}$. For suitable d and A, one has an identity*

$$\langle Wf, Wg\rangle = \langle f, g\rangle_{\rho,d,A}, \quad f, g \in \mathcal{O}_0(\operatorname{clos}\mathcal{D}). \tag{2.14}$$

Proof. By (2.10), Lemma 2 and (1.3),

$$\begin{aligned}
\langle Wf, Wg\rangle &= \frac{1}{2\pi i}\int_{\partial\mathcal{D}} f(z)\overline{g}_e^\dagger(z)\,dz \\
&= \frac{1}{2\pi i}\int_{\partial\mathcal{D}} f(z)\big(\overline{g}_e^\dagger(z) - \overline{g}_i^\dagger(z)\big)\,dz + \frac{1}{2\pi i}\int_{\partial\mathcal{D}} f(z)\overline{g}_i^\dagger(z)\,dz \\
&= \int_{\partial\mathcal{D}} f(z)\bar g(z)\rho(z)\,|dz| + \frac{1}{2\pi i}\int_{\partial\mathcal{D}} f(z)\overline{g}_i^\dagger(z)\,dz.
\end{aligned}$$

(all integrals converge). By Lemma 2, there are natural numbers m_j and constants α_{jks} such that the last integral rewrites as

$$\sum_{j=1}^{L}\sum_{k=0}^{m_j}\sum_{s=0}^{m_j} \alpha_{jks}\frac{d^k f}{dz^k}(\lambda_j)\frac{d^s g}{d\bar z^s}(\bar\mu_j).$$

Therefore one can define a divisor d, supported at points $\lambda_1, \dots, \lambda_L$, and a matrix A so that (2.14) holds.

Restrict W to functions $f \in \mathcal{O}_0(\operatorname{clos}\mathcal{D})$ such that $f^{(k)}(\lambda_j) = 0$, $0 \le k < m_j$. For these f's, $\|f\|^2_{\rho,d,A} = \int |f|^2\,\rho|dz|$, and it follows that ρ is positive a.e. Since the values of $f^{(k)}(\lambda_j)$, $0 \le k < m_j$ can be arbitrary for arbitrary f in $\mathcal{O}_0(\operatorname{clos}\mathcal{D})$, A is self-adjoint. □

Lemma 4. *Define the divisor d_b as in the introduction. There exists a divisor d_0, supported in zeros of P and an $m(d_b) \times m(d_0)$ matrix B such that*

$$\langle Wf, Wg\rangle_{\mathcal{D}} = \langle F_{d_0} f, BF_{d_b} g\rangle, \quad f \in \mathcal{O}(\operatorname{clos}\mathcal{D}), g \in \operatorname{Pol}_N. \tag{2.15}$$

Proof. Assuming $g \in \operatorname{Pol}_N$, let us define h on the whole $\mathbb{C}$ by (2.12). By symmetry, it is a single-valued function. It can only have poles of finite order at infinity and at points t where some $w_j(t) = \infty$ (replace $\partial\mathcal{D}$ in (2.11) by a large circle). Since $(t, \infty) \in \Delta$ is equivalent to $(\infty, \bar t) \in \Delta$, it follows that h only has poles at ∞ and at points $\overline{w_j(\infty)}$, $1 \le j \le n$. Points $\overline{w_j(\infty)}$ are exactly zeros of P (see (2.2)). So $\bar h$

is rational for every $g \in \mathrm{Pol}_N$. Take d_0 to be supported at the set of zeros of P, and put the multiplicity of d_0 at each of these points to be equal to the maximum of orders of poles of h's when g runs over Pol_N. By (2.10),

$$\langle Wf, Wg\rangle = \frac{1}{2\pi i}\int_{\mathcal{D}} f\bar{h}\,dz = \sum_{j=1}^{n} \mathrm{Res}_{\overline{w_j(\infty)}}(f\bar{h}).$$

One can restore g from $F_{d_b}g$ by the Lagrange interpolation formula. The coefficients of the Laurent expansion of $\bar{h}$ at points λ_j depend linearly on $F_{d_b}g$. It follows that the identity (2.15) holds for some matrix B. □

Proposition 4. *Let K be a Hilbert space and K_1 its subspace of finite codimension. Let $[\cdot,\cdot]$ be a Hermitean sesquilinear form on $K \times K$, which coincides with the original scalar product on K_1. If $[f, f] > 0$ for all f in K, then $[\cdot,\cdot]$ generates a Hilbert space norm on K, which is equivalent to the original one.*

Proof. There exists a subspace K_2 of K_1 of finite codimension such that $[f, f] = \langle Cf_1, f_1\rangle + \|f_2\|^2$ for all $f = f_1 + f_2$, $f_1 \in K_2^{\perp}$, $f_2 \in K_2$. The hypothesis implies that all eigenvalues of the self-adjoint matrix C are positive. □

Proof of Theorem 1. Since $\rho \in L^1(|dz|)$, $\mathcal{O}(\operatorname{clos}\mathcal{D})$ and $\mathcal{O}_0(\operatorname{clos}\mathcal{D})$ are contained in $H^2(\mathcal{D}, \rho\,|dz|)$ and are dense there. Hence W extends to whole $H^2(\mathcal{D}, \rho\,|dz|)$, and (2.14) still holds true for f, g in $H^2(\mathcal{D}, \rho\,|dz|)$. Since $\langle f, f\rangle_{\rho,d,A} = \|Wf\|^2$, Proposition 4 gives that $\|\cdot\|_{\rho,d,A}$ is a Hilbert space norm on $H^2(\mathcal{D}, \rho\,|dz|)$. Equip $H^2(\mathcal{D}, \rho\,|dz|)$ with this norm, then W is an isometry from $H^2(\mathcal{D}, \rho\,|dz|)$ into H. But $W\mathcal{O}(\operatorname{clos}\mathcal{D})$ is dense in H [3], hence $W : H^2(\mathcal{D}, \rho\,|dz|) \to H$ is unitary.

Since $U : H \to \mathrm{Hol}\,(\mathcal{D})$ is a closed operator and $UWf = f$ for $f \in \mathcal{O}(\operatorname{clos}\mathcal{D})$, we conclude that $U = W^{-1} : H \to H^2(\mathcal{D}, \rho\,|dz|)$ also is unitary. □

Proof of Theorem 2. Now $\mathcal{O}(\operatorname{clos}\mathcal{D}) \not\subset H^2(\mathcal{D}, \rho\,|dz|)$, but $\mathcal{O}_0(\operatorname{clos}\mathcal{D})$ still is a dense subset of $H^2(\mathcal{D}, \rho\,|dz|)$. As above, we extend W from $\mathcal{O}_0(\operatorname{clos}\mathcal{D})$ to $H^2(\mathcal{D}, \rho\,|dz|)$ by continuity. Formula (2.14) remains true for $f, g \in H^2(\mathcal{D}, \rho\,|dz|)$. We define W of $\mathcal{H}(\rho)$ by $W(q+g_0) = Wq + Wg_0$, $q \in \mathrm{Pol}_N$, $g_0 \in H^2(\mathcal{D}, \rho\,|dz|)$. Since $UWf = f$ for all $f \in \mathcal{H}(\rho)$, it follows that W is one-to-one on $\mathcal{H}(\rho)$. Define a pairing on $\mathcal{H}(\rho)$ by

$$\langle f_0 + p, g_0 + q\rangle_{\mathcal{D}} = \langle f_0, g_0\rangle_{\rho,d,A} + \langle BF_{d_b}p, F_{d_0}g_0\rangle + \langle F_{d_0}f, BF_{d_b}q\rangle,$$
$$f = f_0 + p,\ g = g_0 + q$$

for $f_0, g_0 \in H^2(\mathcal{D}, \rho\,|dz|)$, $p, q \in \mathrm{Pol}_N$. Fix any Euclidean norm on Pol_N and equip (temporarily) $\mathcal{H}(\rho)$ with the corresponding direct sum norm. By (2.14), (2.15),

$$\langle f, g\rangle_{\mathcal{D}} = \langle Wf, Wg\rangle_H$$

for $f, g \in \mathrm{Pol}_N + \mathcal{O}_0(\operatorname{clos}\mathcal{D})$. By continuity, this formula holds for all $f, g \in \mathcal{H}(\rho)$. Proposition 4, applied to $K = \mathcal{H}(\rho)$ and $K_1 = \{f \in H^2(\mathcal{D}, \rho\,|dz|) : F_d f = 0\}$, gives that $\langle\cdot,\cdot\rangle_{\mathcal{D}}$ is a Hilbert space pairing. Hence W is an isometry from the Hilbert space $\big(\mathcal{H}(\rho), \|\cdot\|_{\mathcal{D}}\big)$ into H. Since $\mathcal{O}(\operatorname{clos}\mathcal{D}) \subset \mathcal{H}(\rho)$ and $W\mathcal{O}(\operatorname{clos}\mathcal{D})$ is dense in

H, W is unitary. Since $U : H \to \mathrm{Hol}(\mathcal{D})$ is a closed operator and $UWf = f$ for $f \in \mathcal{O}(\operatorname{clos}\mathcal{D})$, we conclude that $U : H \to \mathcal{H}(\rho)$ also is unitary. □

Remark. As it was noted, $g^\dagger = VWg = VU^{-1}g$. Since $h = g^\dagger$ on $\mathbb{C} \setminus \operatorname{clos}\mathcal{D}$, (2.12) shows that $g^\dagger(t)$ depends only on values of g in points $z(\delta^*)$, where $\delta \in \Delta_-$, $z(\delta) = t$. The same kind of symmetry exists for models of subnormal operators with finite rank self-commutator, see [23].

References

[1] D. Aharonov, H. S. Shapiro, Domains on which analytic functions satisfy quadrature identities. J. d'Analyse Mathématique, 30, 1976, 39–73.

[2] K. F. Clancey, Completeness of eigenfunctions of seminormal operators. Acta Sci. Mathematics, 39, 1977, 31–37.

[3] K. F. Clancey, Hilbert space operators with one-dimensional self-commutators. J. Operator Theory, 13, No. 2, 1985, 265–289.

[4] J. B. Conway, Theory of Subnormal Operators. Amer. Math. Soc., Providence, 1991.

[5] P. L. Duren, Theory of H^p-spaces. Academic Press, New York, 1970.

[6] S. Fisher, Function Theory on Planar Domains. Wiley, 1983.

[7] W. Fulton, Algebraic curves. Addison-Wesley, Redwood City, etc., 1969.

[8] B. Gustafsson, Quadrature identities and the Schottky double. Acta Applicandae Math., 1, 1983, 209–240.

[9] B. Gustafsson, Singular and special points on quadrature domains from an algebraic geometric point of view, J. d'Analyse Math., 51 (1988), 91–117.

[10] B. Gustafsson, M. Putinar, Linear analysis of quadrature domains. II, to appear.

[11] M. Martin, M. Putinar, Lectures on Hyponormal Operators. Birkhäuser, Basel, Boston, Berlin, 1989 (Oper. Theory: Adv., Appl., vol. 39).

[12] J. E. McCarthy, L. Yang, Subnormal operators and quadrature domains. Advances in Math., 127, No. 1, 1997, 52-72.

[13] J.D. Pincus, D. Xia, J. Xia, The analytic model of a hyponormal operator with rank one self-commutator. Integral Eqs. Operator Theory, 7, No. 4, 1984, 516–535; 6, 1984, 893–894.

[14] I. I. Privalov, Boundary behavior of analytic functions. German translation in: Berlin, Deutscher Verlag der Wissenschaften, 1956; Moskow-Leningrad: GITTL, 1950.

[15] M. Putinar, Extremal solutions of the two-dimensional L-problem of moments. J. Funct. Analysis, 136, 1996, 331–364.

[16] M. Putinar, Linear analysis of quadrature domains. Ark. Math., 33, 1995, 357–376.

[17] M. Sakai, Quadrature domains. Lect. Notes in Math., Vol. 934, Springer-Verlag, Berlin-Heidelberg, 1982.

[18] H.S. Shapiro, The Schwartz function and its generalization to higher dimensions. Uni. of Arkansas Lect. Notes Math., Vol. 9, Wiley, N.Y. 1992.

[19] D. Xia, On the kernels associated with a class of hyponormal operators. Integral Eqs. Operator Theory, 6, No. 1, 1983, 444–452.

[20] D. Xia, On pure subnormal operators with finite rank self-commutators and related operator tuples. Integral Eqs. Operator Theory, 24, No. 1, 1996, 106–125.

[21] D. V. Yakubovich, Dual analytic models of seminormal operators. Integral Eqs. Operator Theory, 23, 1995, 353–371.

[22] D. V. Yakubovich, Dual piecewise analytic bundle shift models of linear operators. J. Funct. Analysis, 136, No. 2, 1996, 294–330.

[23] D. V. Yakubovich, Subnormal operators of finite type II. Structure theorems, Revista Matemática Iberoamericana, 14, No. 3, 1998.

Division of Mathematical Analysis, Dept. of Mathematics and Mechanics, St. Petersburg State University, Bibliotechnaya pl. 2, St. Peterhof, St. Petersburg, 198904, Russia

E-mail address: dm@yakub.niimm.spb.su, dmyakub@math.spbu.ru, diakubo@math.ec3m.es

Operator Theory:
Advances and Applications, Vol. 123

Two Remarks on Fuchsian Groups of Widom Type

Peter Yuditskii

Abstract. For groups of Widom type we prove that the direct integral of the spaces of character automorphic forms of weight 1 with the Haar measure on the group of characters is the space of automorphic forms of weight 1 with respect to commutator of the given group. It gives us an explanation why the whole set of ergodic Jacobi matrices with a fixed homogeneous (absolutely continuous) spectrum can be parameterized by the group of characters of the fundamental group of common resolvent set. Besides, it is shown that a Fuchsian group Γ is of Widom type if and only if the conditional entropy of the partition of the unit circle into singleton sets with respect to the partition into Γ-equivalent classes is finite.

In this note answers are given to questions that arose in our discussions with Victor Vinnikov concerning properties of Fuchsian groups of Widom type. These groups play an important role in spectral theory of almost periodic operators [8] and could be useful in spectral theory of pairs of non-selfadjoint commuting operators [6].

1. Groups of Widom type and conditional entropy

Let $\mathbb{D}$ denote the unit disk and $\mathbb{T}$ denote the unit circle

$$\mathbb{D} = \{z : \ |z| < 1\}, \quad \mathbb{T} = \{t : \ |t| = 1\}.$$

Let Γ be a discrete subgroup of $SU(1,1)$ consisting of elements of the form

$$\gamma = \begin{bmatrix} \gamma_{11} & \gamma_{12} \\ \gamma_{21} & \gamma_{22} \end{bmatrix}, \quad \gamma_{11} = \overline{\gamma_{22}}, \quad \gamma_{12} = \overline{\gamma_{21}}, \quad \det \gamma = 1.$$

For $\gamma \in \Gamma$, put $\gamma(z) = (\gamma_{11} z + \gamma_{12})/(\gamma_{21} z + \gamma_{22})$; as is well known, γ maps $\mathbb{D}$ and $\mathbb{T}$ onto themselves. For simplicity we assume that 0 is not an elliptic fixed point. Such a group Γ is called of convergence type [1, 9] if

$$\sum_{\gamma \in \Gamma} (1 - |\gamma(0)|^2) < \infty.$$

Then Green's function of Γ with respect to 0 is defined as the Blaschke product

$$b(z) = b(\Gamma, z) = z \prod_{\gamma \in \Gamma, \gamma \neq 1_2} \frac{z - \gamma(0)}{1 - z\overline{\gamma(0)}} \frac{|\gamma(0)|}{-\gamma(0)},$$

where 1_2 denotes the unit matrix.

There exist interesting relations between analytic properties of Green's function and dynamical properties of actions of the group Γ in the sense of measurable dynamics. First of all (see [1, Ch. VII]), the group Γ is of convergence type (that is Green's function exists) if and only if Γ acts dissipatively on $\mathbb{T}^2$ with respect to the Lebesgue measure m_2. Here, the action of the group Γ on $\mathbb{T}^2$ is defined by the equality $\gamma((t_1,t_2)) = (\gamma(t_1),\gamma(t_2))$, the action is said to be dissipative if there exists a measurable fundamental set $E_2 \subset \mathbb{T}^2$, that is

1) $E_2 \cap \gamma(E_2) = \emptyset$ for all $\gamma \neq 1_2$

2) $m_2(\cup_{\gamma\in\Gamma}\gamma(E_2)) = m_2(\mathbb{T}^2)$.

It was shown in [9] that Γ acts dissipatively on $\mathbb{T}$ (with respect to the Lebesgue measure m) if and only if b' has angular limit almost everywhere on $\mathbb{T}$.

The group is called of Widom type [9] if b' is a function of bounded characteristic, which is equivalent to the condition (we have to mention, that $|b'| \geq 1$ on $\mathbb{T}$)

$$\int_{\mathbb{T}} \log|b'|\, dm < \infty.$$

It seems interesting to give a dynamical interpretation of this property. In this note we propose a very simple way of such an interpretation. It should be mentioned that the much more difficult question on finiteness of entropy of the function b as a map $\mathbb{T} \to \mathbb{T}$ can be reduced to the finiteness of the same integral [2]. Such a reduction is a general property of inner functions and does not take into account specific features of the way of the construction of the function b.

Theorem 1. *Let Γ act dissipativly on $\mathbb{T}$, ϵ be the partition of $\mathbb{T}$ into singleton sets, and $\Gamma(\epsilon)$ be the partition of $\mathbb{T}$ into Γ-equivalence classes. Then Γ is of Widom type if and only if the conditional entropy $H(\epsilon|\Gamma(\epsilon))$ (with respect to the Lebesgue measure) is finite, moreover*

$$H(\epsilon|\Gamma(\epsilon)) = \int_{\mathbb{T}} \log|b'|\, dm.$$

We will remind the definition of the conditional entropy [7, Ch. 3] (see also [3, Ch. 3]) within the proof.

Proof. A partition is said to be measurable if its elements are level sets of some measurable function. Let us show that the partition $\Gamma(\epsilon)$ is measurable. Since Γ acts dissipatively, there exists a measurable fundamental set E. Define the function $\phi: \mathbb{T} \to \mathbb{T}$ to be identical on E and extend it automorphically to $\mathbb{T}$, $\phi(\gamma(t)) = \phi(t)$. We get a measurable function, whose level sets are elements of the partition $\Gamma(\epsilon)$.

For any measurable partition there exists a system of conditional measures uniquely determined by the following properties

1) any measure of the system is a probability measure supported on an element of the partition; in our case the support of a probability measure $\mu(\cdot|\Gamma(t))$ is the set $\Gamma(t) = \{\gamma(t)\}_{\gamma\in\Gamma}$;

2) if $f \in L^1$, then

$$\int_{\mathbb{T}} f\,dm = \int_{\mathbb{T}} \{\sum_{\gamma\in\Gamma} f(\gamma(t))\mu(\gamma(t)|\Gamma(t))\}\,dm. \tag{1}$$

Let us find these measures. Let $f \in L^1$. Since

$$\int_{\mathbb{T}} f\,dm = \int_{\bigcup_{\gamma\in\Gamma}\gamma(E)} f\,dm = \sum_{\gamma\in\Gamma}\int_{\gamma(E)} f\,dm,$$

we obtain that

$$\int_{\mathbb{T}} f\,dm = \sum_{\gamma\in\Gamma}\int_E f(\gamma(t))|\gamma'(t)|\,dm = \int_E \sum_{\gamma\in\Gamma} f(\gamma(t))|\gamma'(t)|\,dm.$$

Let us note (see [9]) that $\sum_{\gamma\in\Gamma}|\gamma'(t)| = |b'(t)|$. It is easy to see, that the function $\sum_{\gamma\in\Gamma} f(\gamma(t))\frac{|\gamma'(t)|}{|b'(t)|}$ is automorphic. So,

$$\int_{\mathbb{T}} f\,dm = \int_E \left\{\sum_{\gamma\in\Gamma} f(\gamma(t))\frac{|\gamma'(t)|}{|b'(t)|}\right\}|b'(t)|\,dm = \int_{\mathbb{T}} \left\{\sum_{\gamma\in\Gamma} f(\gamma(t))\frac{|\gamma'(t)|}{|b'(t)|}\right\}dm.$$

Hence,

$$\mu(\gamma(t)|\Gamma(t)) = \frac{|\gamma'(t)|}{|b'(t)|} = \frac{1}{|b'(\gamma(t))|}.$$

The entropy of a countable partition $\epsilon|\Gamma(t)$ (i.e. pointwise partition of $\Gamma(t)$) is defined to be

$$H(\epsilon|\Gamma(t)) = -\sum_{\gamma\in\Gamma}\log\mu(\gamma(t)|\Gamma(t))\mu(\gamma(t)|\Gamma(t)),$$

so we have

$$H(\epsilon|\Gamma(t)) = \sum_{\gamma\in\Gamma}\log|b'(\gamma(t))|\mu(\gamma(t)|\Gamma(t)).$$

The conditional entropy $H(\epsilon|\Gamma(\epsilon))$ is defined as the average of $H(\epsilon|\Gamma(t))$,

$$H(\epsilon|\Gamma(\epsilon)) = \int_{\mathbb{T}} H(\epsilon|\Gamma(t))\,dm.$$

Using (1) with $f(t) = \log|b'(t)|$, we then have the relations

$$H(\epsilon|\Gamma(\epsilon)) = \int_{\mathbb{T}}\left\{\sum_{\gamma\in\Gamma}\log|b'(\gamma(t))|\mu(\gamma(t)|\Gamma(t))\right\}dm = \int_{\mathbb{T}}\log|b'(t)|\,dm.$$

□

2. Direct integral of the spaces of character automorphic functions and forms

2.1. We use a standard terminology and notations of theory of functions of bounded characteristic in $\mathbb{D}$ [4]. In particular, H^2 denotes the standard Hardy space.

The second question concerns the following remarkable property of groups of Widom type [9, 10]. If Γ has no elliptic elements, then

$$\Gamma \text{ is of Widom type} \Longleftrightarrow H^2(\Gamma,\alpha) \neq \{\text{const}\} \text{ for all } \alpha \in \Gamma^*,$$

where Γ^* is the group of (unimodular) characters of Γ and for $\alpha \in \Gamma^*$ we put

$$H^2(\Gamma,\alpha) = \{f \in H^2 |\ f(\gamma(z)) = \alpha(\gamma) f(z)\}.$$

The question is what is the "union" of all $H^2(\Gamma,\alpha)$ on α. To formulate the result we need some extra definitions and notation.

Since b' is of bounded characteristic, one can define an outer function $\psi(z) = \psi(\Gamma, z)$ by the relation

$$|\psi(\Gamma,t)|^2 = |b'(\Gamma,t)|, \quad t \in \mathbb{T}.$$

The function ψ is defined up to a unimodular multiplier; we normalize it by the condition $\psi(0) > 0$. Since $|b'(\gamma(t))||\gamma'(t))| = |b'(t)|$, we have

$$\left|\frac{\psi(\gamma(t))}{\gamma_{21}t + \gamma_{22}}\right|^2 = |\psi(t)|^2.$$

Hence, the function ψ itself possesses the following property

$$\frac{\psi(\gamma(z))}{\gamma_{21}z + \gamma_{22}} = \alpha_\psi(\gamma)\psi(z), \quad \gamma \in \Gamma.$$

It is easy to see that α_ψ is a character of Γ.

Let Γ' be the commutator of the group Γ, i.e. the subgroup of Γ generated by all commutators $\gamma_1\gamma_2\gamma_1^{-1}\gamma_2^{-1}$ of pairs of elements of the group, $\gamma_1, \gamma_2 \in \Gamma$. The factor $\Gamma' \setminus \Gamma$ is an Abelian group. Due to duality theorem [5, Ch. Six], $(\Gamma^*)^* \simeq \Gamma' \setminus \Gamma$. Since

$$\sum_{\gamma_0 \in \Gamma'} |\gamma_0'(t)| < \sum_{\gamma \in \Gamma} |\gamma'(t)|,$$

Γ' is of Widom type if Γ is of Widom type. For brevity we introduce the notations $b_0(z) = b(\Gamma', z), \psi_0(z) = \psi(\Gamma', z)$, $E_0 \subset \mathbb{T}$ is a fundamental set of Γ', and $\alpha_{\psi_0} \in (\Gamma')^*$ is the character generated by ψ_0.

Let $d\mu(\alpha)$ be the Haar measure on Γ^* [5, Ch. Four]. Consider the direct integral of the spaces $H^2(\Gamma,\alpha)$,

$$\begin{aligned}&\int_{\Gamma^*} H^2(\Gamma,\alpha)\, d\mu(\alpha)\\ &\quad = \left\{F(t,\alpha) \in H^2(\Gamma,\alpha) \text{ for a.e. } \alpha \in \Gamma^* : \int_{\Gamma^*}\int_{\mathbb{T}} |F(t,\alpha)|^2\, dm\, d\mu(\alpha) < \infty\right\}.\end{aligned}$$

The following theorem shows that this space is unitarily isomorphic to the space $H^2(\Gamma', \alpha_{\psi_0}^{-1})$.

Theorem 2. *Let Γ be of Widom type. Then the transformations defined by*

$$(\mathcal{F}_1 F)(z) = \frac{\psi(z)}{\psi_0(z)} \int_{\Gamma^*} F(z,\alpha)\, d\mu(\alpha), \quad F \in \int_{\Gamma^*} H^2(\Gamma,\alpha)\, d\mu(\alpha)$$

and

$$(\mathcal{F}_2 f)(z,\alpha) = \sum_{\{\gamma\}\in\Gamma'\backslash\Gamma} \left(\frac{\psi_0}{\psi} f\right)(\gamma(z))\alpha^{-1}(\gamma), \quad f \in H^2(\Gamma', \alpha_{\psi_0}^{-1})$$

are mutually inverse and set a unitary isomorphism of the spaces $\int_{\Gamma^*} H^2(\Gamma,\alpha)\, d\mu(\alpha)$ and $H^2(\Gamma', \alpha_{\psi_0}^{-1})$.

Proof. Let $F(t,\alpha) \in \int_{\Gamma^*} H^2(\Gamma,\alpha)\, d\mu(\alpha)$. The system of the functions $e_{\{\gamma\}}(\alpha) = \alpha^{-1}(\gamma), \{\gamma\} \in \Gamma' \setminus \Gamma$, forms an orthogonal basis of the space $L^2_{d\mu(\alpha)}$ [5, Ch. Eight] and we can expand $F(t,\alpha)$ in the Fourier series,

$$F(t,\alpha) = \sum_{\{\gamma\}\in\Gamma'\backslash\Gamma} f_{\{\gamma\}}(t)\alpha^{-1}(\gamma),$$

where $f_{\{\gamma\}}(t) = \int_{\Gamma^*} F(t,\alpha)\alpha(\gamma)\, d\mu(\alpha)$. In addition,

$$||F||^2 = \int_{\mathbb{T}} \sum_{\{\gamma\}\in\Gamma'\backslash\Gamma} |f_{\{\gamma\}}(t)|^2\, dm. \tag{2}$$

Hence, any function $f_{\{\gamma\}}(t)$ belongs to L^2; moreover, since

$$\begin{aligned} \int_{\mathbb{T}} f_{\{\gamma\}}(t) t^n\, dm &= \int_{\mathbb{T}} \int_{\Gamma^*} F(t,\alpha)\alpha(\gamma)\, d\mu(\alpha)\; t^n\, dm \\ &= \int_{\mathbb{T}} \int_{\Gamma^*} F(t,\alpha) t^n\, dm\; \alpha(\gamma)\, d\mu(\alpha) \\ &= 0 \quad \text{for all } n > 0, \end{aligned}$$

$f_{\{\gamma\}}(t)$ belongs to H^2.

Observe the following automorphy property of $f_{\{\gamma\}}$,

$$\begin{aligned} f_{\{\gamma\}}(\gamma_0(t)) &= \int_{\Gamma^*} F(\gamma_0(t),\alpha)\alpha(\gamma)\, d\mu(\alpha) \\ &= \int_{\Gamma^*} F(t,\alpha)\alpha(\gamma_0)\alpha(\gamma)\, d\mu(\alpha) \\ &= f_{\{\gamma_0\gamma\}}(t). \end{aligned}$$

Hence, $f_{\{\gamma\}}$ is Γ' – automorphic and $f_{\{\gamma\}}(t) = f_{\{1_2\}}(\gamma(t))$.

Now we fix some system $\tilde{\Gamma}$ of representatives of $\Gamma' \setminus \Gamma$:

$$\bigcup_{\tilde{\gamma}\in\tilde{\Gamma}} \{\tilde{\gamma}\} = \Gamma' \setminus \Gamma, \quad \tilde{\gamma}_1\tilde{\gamma}_2^{-1} \notin \Gamma' \; (\tilde{\gamma}_1, \tilde{\gamma}_2 \in \tilde{\Gamma}, \tilde{\gamma}_1 \neq \tilde{\gamma}_2),$$

and, returning to (2),

$$\begin{aligned}||F||^2 &= \int_{\mathbb{T}} \sum_{\tilde\gamma\in\tilde\Gamma} |f_{\{\tilde\gamma\}}(t)|^2\, dm\\ &= \int_{\mathbb{T}} \sum_{\tilde\gamma\in\tilde\Gamma} |f_{\{1_2\}}(\tilde\gamma(t))|^2\, dm\\ &= \int_{\mathbb{T}} |f_{\{1_2\}}(t)|^2 \sum_{\tilde\gamma\in\tilde\Gamma} |\tilde\gamma'(t)|\, dm.\end{aligned}$$

The function $|f_{\{1_2\}}(t)|^2$ is Γ'-automorphic, so

$$\begin{aligned}||F||^2 &= \sum_{\gamma_0\in\Gamma'} \int_{E_0} |f_{\{1_2\}}(t)|^2 \sum_{\tilde\gamma\in\tilde\Gamma} |\tilde\gamma'(\gamma_0(t)||\gamma_0'(t)|\, dm\\ &= \int_{E_0} \frac{|f_{\{1_2\}}(t)|^2 \sum_{\gamma_0\in\Gamma',\tilde\gamma\in\tilde\Gamma} |(\tilde\gamma\gamma_0)'(t)|}{\sum_{\gamma_0\in\Gamma'} |\gamma_0'(t)|} \sum_{\gamma_0\in\Gamma'} |\gamma_0'(t)|\, dm.\end{aligned}$$

Since the quotient is Γ'-automorphic, we have

$$||F||^2 = \int_{\mathbb{T}} |f_{\{1_2\}}(t)|^2 \frac{\sum_{\gamma_0\in\Gamma',\tilde\gamma\in\tilde\Gamma} |(\tilde\gamma\gamma_0)'(t)|}{\sum_{\gamma_0\in\Gamma'} |\gamma_0'(t)|}\, dm.$$

Using the following relations

$$\sum_{\gamma_0\in\Gamma'} |\gamma_0'(t)| = |b_0'(t)| = |\psi_0(t)|^2$$

and

$$\sum_{\gamma_0\in\Gamma',\tilde\gamma\in\tilde\Gamma} |(\tilde\gamma\gamma_0)'(t)| = \sum_{\gamma\in\Gamma} |\gamma'(t)| = |b'(t)| = |\psi(t)|^2,$$

we conclude that

$$||F||^2 = \int_{\mathbb{T}} |f_{\{1_2\}}(t)\frac{\psi(t)}{\psi_0(t)}|^2\, dm.$$

So, $f_{\{1_2\}}(t)\frac{\psi(t)}{\psi_0(t)}$ belongs to L^2, and since $\frac{\psi}{\psi_0}(t)$ is an outer function, it belongs to H^2. Checking the automorphy property,

$$\begin{aligned}f_{\{1_2\}}(\gamma_0(t))\frac{\psi(\gamma_0(t))}{\psi_0(\gamma_0(t))} &= f_{\{\gamma_0\}}(t)\frac{\frac{\psi(\gamma_0(t))}{(\gamma_0)_{21}t+(\gamma_0)_{22}}}{\frac{\psi_0(\gamma_0(t))}{(\gamma_0)_{21}t+(\gamma_0)_{22}}}\\ &= f_{\{1_2\}}(t)\frac{\alpha_\psi(\gamma_0)\psi(t)}{\alpha_{\psi_0}(\gamma_0)\psi_0(t)} = \alpha_{\psi_0}^{-1}(\gamma_0)f_{\{1_2\}}(t)\frac{\psi(t)}{\psi_0(t)}, \quad \gamma_0\in\Gamma',\end{aligned}$$

we get $f_{\{1_2\}}(t)\frac{\psi(t)}{\psi_0(t)} \in H^2(\Gamma', \alpha_{\psi_0}^{-1})$.

We have thus proved that $\mathcal{F}_1$ maps $\int_{\Gamma^*} H^2(\Gamma,\alpha)\,d\mu(\alpha)$ into $H^2(\Gamma',\alpha_{\psi_0}^{-1})$ isometrically.

Now, let $f \in H^2(\Gamma',\alpha_{\psi_0}^{-1})$. Since, up to a set of measure 0,

$$\mathbb{T} = \cup_{\gamma\in\Gamma}\gamma(E) = \cup_{\gamma_0\in\Gamma'}\cup_{\tilde\gamma\in\tilde\Gamma}\gamma_0\tilde\gamma(E) = \cup_{\gamma_0\in\Gamma'}\gamma_0(\cup_{\tilde\gamma\in\tilde\Gamma}\tilde\gamma(E)),$$

the set $E_0 = \bigcup_{\tilde\gamma\in\tilde\Gamma}\tilde\gamma(E)$ is a fundamental set of Γ'. If it is so, we have

$$\begin{aligned}\int_{\mathbb{T}}|f(t)|^2\,dm &= \int_{E_0}|f(t)|^2|\psi_0(t)|^2\,dm\\ &= \int_{\bigcup_{\tilde\gamma\in\tilde\Gamma}\tilde\gamma(E)}|f(t)|^2|\psi_0(t)|^2\,dm\\ &= \int_E\{\sum_{\tilde\gamma\in\tilde\Gamma}|f\psi_0(\tilde\gamma(t))|^2|\tilde\gamma'(t)|\}\,dm\\ &= \int_E\{\sum_{\tilde\gamma\in\tilde\Gamma}|f\psi_0(\tilde\gamma(t))|^2\frac{|\tilde\gamma'(t)|}{|\psi(t)|^2}\}|\psi(t)|^2\,dm.\end{aligned}$$

Using the automorphy property of $\psi(t)$ we get the equality

$$\int_{\mathbb{T}}|f(t)|^2\,dm = \int_E\{\sum_{\tilde\gamma\in\tilde\Gamma}|f\frac{\psi_0}{\psi}(\tilde\gamma(t))|^2\}|\psi(t)|^2\,dm. \tag{3}$$

Let us check that the function

$$\sum_{\tilde\gamma\in\tilde\Gamma}|f\frac{\psi_0}{\psi}(\tilde\gamma(t))|^2$$

is Γ-automorphic.

For $\gamma\in\Gamma$ we put $\tilde\gamma\gamma = \gamma_{0,\gamma}\tilde\gamma_\gamma$, where $\gamma_{0,\gamma}\in\Gamma'$ and $\tilde\gamma_\gamma\in\tilde\Gamma$. Then

$$\sum_{\tilde\gamma\in\tilde\Gamma}|f\frac{\psi_0}{\psi}(\tilde\gamma(\gamma(t)))|^2 = \sum_{\tilde\gamma\in\tilde\Gamma}|f\frac{\psi_0}{\psi}(\gamma_{0,\gamma}(\tilde\gamma_\gamma(t)))|^2$$

The function $f\frac{\psi_0}{\psi}$ is Γ'-automorphic and the map $\tilde\gamma\mapsto\tilde\gamma_\gamma$ is a bijection, so

$$\sum_{\tilde\gamma\in\tilde\Gamma}|f\frac{\psi_0}{\psi}(\tilde\gamma(\gamma(t)))|^2 = \sum_{\tilde\gamma\in\tilde\Gamma}|f\frac{\psi_0}{\psi}(\tilde\gamma_\gamma(t))|^2 = \sum_{\tilde\gamma\in\tilde\Gamma}|f\frac{\psi_0}{\psi}(\tilde\gamma(t))|^2.$$

Therefore, instead of (3) we get

$$\int_{\mathbb{T}}|f(t)|^2\,dm = \int_{\mathbb{T}}\{\sum_{\tilde\gamma\in\tilde\Gamma}|f\frac{\psi_0}{\psi}(\tilde\gamma(t))|^2\}\,dm. \tag{4}$$

It shows us that for almost every t we can define a sequence $\{f_{\{\gamma\}}(t)\}_{\{\gamma\}\in\Gamma'\backslash\Gamma}$ as

$$f_{\{\gamma\}}(t) = f\frac{\psi_0}{\psi}(\gamma(t)) = f\frac{\psi_0}{\psi}(\gamma_0(\tilde\gamma(t))) = f\frac{\psi_0}{\psi}(\tilde\gamma(t))$$

which belongs to $l^2_{\Gamma'\backslash\Gamma} = \left\{\{a_{\{\gamma\}}\}_{\{\gamma\}\in\Gamma'\backslash\Gamma} : \sum_{\{\gamma\}\in\Gamma'\backslash\Gamma}|a_{\{\gamma\}}|^2 < \infty\right\}$.

Let $F(t,\alpha)$ be its Fourier transform:

$$F(t,\alpha) = \sum_{\{\gamma\}\in\Gamma'\backslash\Gamma} f_{\{\gamma\}}(t)\alpha^{-1}(\gamma) = \sum_{\tilde\gamma\in\tilde\Gamma} f\frac{\psi_0}{\psi}(\tilde\gamma(t))\alpha^{-1}(\tilde\gamma).$$

It is well defined for almost every $t\in\mathbb{T}$ and

$$\int_{\Gamma^*}|F(t,\alpha)|^2\,d\mu(\alpha) = \sum_{\tilde\gamma\in\tilde\Gamma}|f\frac{\psi_0}{\psi}(\tilde\gamma(t))|^2.$$

We have to prove that $F(t,\alpha)$ belongs to $H^2(\Gamma,\alpha)$ for almost every $\alpha\in\Gamma^*$. Let $\{\tilde\Gamma_n\}_n$ be some exhaustion of $\tilde\Gamma$ by finite subsets: $\tilde\Gamma_n\subset\tilde\Gamma_{n+1}, \bigcup_n\tilde\Gamma_n = \tilde\Gamma$. For any $\tilde\Gamma_n$ we have

$$\int_{\Gamma^*}|F(t,\alpha) - \sum_{\tilde\gamma\in\tilde\Gamma_n} f\frac{\psi_0}{\psi}(\tilde\gamma(t))\alpha^{-1}(\tilde\gamma)|^2\,d\mu(\alpha) = \sum_{\tilde\gamma\in\{\tilde\Gamma-\tilde\Gamma_n\}}|f\frac{\psi_0}{\psi}(\tilde\gamma(t))|^2$$

and therefore

$$\int_{\Gamma^*}\int_{\mathbb{T}}|F(t,\alpha) - \sum_{\tilde\gamma\in\tilde\Gamma_n} f\frac{\psi_0}{\psi}(\tilde\gamma(t))\alpha^{-1}(\tilde\gamma)|^2\,dm\,d\mu(\alpha) = \int_{\mathbb{T}}\sum_{\tilde\gamma\in\{\tilde\Gamma-\tilde\Gamma_n\}}|f\frac{\psi_0}{\psi}(\tilde\gamma(t))|^2\,dm.$$

Together with (4) this implies that the sequence

$$\int_{\mathbb{T}}|F(t,\alpha) - \sum_{\tilde\gamma\in\tilde\Gamma_n} f\frac{\psi_0}{\psi}(\tilde\gamma(t))\alpha^{-1}(\tilde\gamma)|^2\,dm$$

converges to 0 in $L^1_{d\mu(\alpha)}$ as $n\to\infty$. Hence, there exists a subsequence converging to 0 for a.e. $\alpha\in\Gamma^*$. Since any finite sum $\sum_{\tilde\gamma\in\tilde\Gamma_n} f\frac{\psi_0}{\psi}(\tilde\gamma(t))\alpha^{-1}(\tilde\gamma)$ belongs to H^2, the function $F(t,\alpha)$ belongs to H^2 for a.e. $\alpha\in\Gamma^*$. The automorphy property of F is evident.

As a result we have that $\mathcal{F}_2$ maps $H^2(\Gamma',\alpha^{-1}_{\psi_0})$ to $\int_{\Gamma^*}H^2(\Gamma,\alpha)\,d\mu(\alpha)$. Since $\mathcal{F}_1^*\mathcal{F}_1 = I$ and, evidently, $\mathcal{F}_1\mathcal{F}_2 = I$, the both operators are unitary. □

2.2. The preceding result will have more natural form when we make a passage from the spaces of character automorphic functions to the spaces of character automorphic forms of weight 1.

We remind that a function f is said to be of Smirnov class if it can be represented in the form $f = f_1/f_2$, where $f_1, f_2\in H^\infty$ and f_2 is an outer function (f is a function of bounded characteristic with an outer "denominator").

For a holomorphic function in $\mathbb{D}$ and $\gamma\in\Gamma$ we write

$$f|[\gamma] = \frac{f(\gamma(z))}{\gamma_{21}z + \gamma_{22}}.$$

Then it is easily verified that

$$f|[\gamma_1\gamma_2] = (f|[\gamma_1])|[\gamma_2].$$

Notice that two matrices γ and $-\gamma$ induce the same transformation on $\mathbb{D}$, however

$$f|[-\gamma] = -f|[\gamma].$$

Definition. *Let Γ be a group of Widom type and $E \subset \mathbb{T}$ be its fundamental set. For $\alpha \in \Gamma^*$ the space $A_1^2(\Gamma, \alpha)$ is formed by the holomorphic functions f on $\mathbb{D}$ that satisfy the following three conditions*

1) f is of Smirnov class

2) $f|[\gamma] = \alpha(\gamma)f$ for all $\gamma \in \Gamma$

3) the boundary value of f is square-integrable on E

$$\int_E |f|^2\, dm < \infty.$$

Lemma. $\psi(\Gamma, t)H^2(\Gamma, \alpha) = A_1^2(\Gamma, \alpha_\psi \alpha)$.

Proof. Let $f \in H^2(\Gamma, \alpha)$. Since

$$\int_{\mathbb{T}} |f|^2\, dm = \int_E |f\psi|^2\, dm,$$

we have $\psi f \in A_1^2(\Gamma, \alpha_\psi \alpha)$.

Conversely, let $f \in A_1^2(\Gamma, \alpha_\psi \alpha)$, then

$$\int_E |f|^2\, dm = \int_E |f/\psi|^2 |\psi|^2\, dm = \int_{\mathbb{T}} |f/\psi|^2\, dm.$$

Thus, $f/\psi \in L^2$, and since f/ψ is of Smirnov class, we get $f/\psi \in H^2$. □

Now we can reformulate Theorem 2 in the following form.

Theorem 2a. *Let Γ be a group of Widom type, Γ' be its commutator and $d\mu(\alpha)$ be the Haar measure on the group of characters Γ^*. Then*

$$A_1^2(\Gamma') = \int_{\Gamma^*} A_1^2(\Gamma, \alpha)\, d\mu(\alpha)$$

in the following sense:

If $F(t, \alpha) \in \int_{\Gamma^*} A_1^2(\Gamma, \alpha)\, d\mu(\alpha)$, i.e.

$$F(t, \alpha) \in A_1^2(\Gamma, \alpha) \text{ for a.e. } \alpha \in \Gamma^* \quad \text{and} \quad \int_{\Gamma^*} ||F(t, \alpha)||^2_{A_1^2(\Gamma,\alpha)}\, d\mu(\alpha) < \infty,$$

then

$$f(t) = \int_{\Gamma^*} F(t, \alpha)\, d\mu(\alpha) \in A_1^2(\Gamma'),$$

and

$$||f||_{A_1^2(\Gamma')} = ||F(t, \alpha)||_{\int_{\Gamma^*} A_1^2(\Gamma,\alpha)\, d\mu(\alpha)}.$$

If $f \in A^2_1(\Gamma')$, then

$$F(t,\alpha) = \sum_{\{\gamma\}\in\Gamma'\backslash\Gamma} f|[\gamma]\alpha^{-1}(\gamma) \in \int_{\Gamma^*} A^2_1(\Gamma,\alpha)\, d\mu(\alpha)$$

and

$$||F(t,\alpha)||_{\int_{\Gamma^*} A^2_1(\Gamma,\alpha)\, d\mu(\alpha)} = ||f||_{A^2_1(\Gamma')}.$$

In addition,

$$\int_{\Gamma^*} \{ \sum_{\{\gamma\}\in\Gamma'\backslash\Gamma} f|[\gamma]\alpha^{-1}(\gamma)\}\, d\mu(\alpha) = f.$$

To finish, we formulate a generalization of Theorem 2a.

Let Γ_0 be a normal subgroup of Γ such that $\Gamma' \subset \Gamma_0 \subset \Gamma$. In other words, $\Gamma_0 \setminus \Gamma$ is an Abelian group. Let $\alpha_0 \in \Gamma_0^*$ can be extended to a character on Γ, i.e.

$$\{\alpha \in \Gamma^* :\ \alpha|\Gamma_0 = \alpha_0\} \neq \emptyset.$$

Note that the set of characters

$$\{\beta \in \Gamma^* :\ \beta|\Gamma_0 = \iota\}$$

where $\iota(\gamma) = 1$ for all $\gamma \in \Gamma_0$, is canonically isomorphic to the set $(\Gamma_0 \setminus \Gamma)^*$.

Let us fix some element $\hat\alpha_0 \in \Gamma^* :\ \hat\alpha_0|\Gamma_0 = \alpha_0$. Since

$$\{\alpha \in \Gamma^* :\ \alpha|\Gamma_0 = \alpha_0\} = \{\hat\alpha_0\beta :\ \beta \in \Gamma^*,\ \beta|\Gamma_0 = \iota\},$$

we can define a measure $d\mu(\alpha)$ on $\{\alpha \in \Gamma^* :\ \alpha|\Gamma_0 = \alpha_0\}$ by the relation

$$d\mu(\alpha) = d\mu(\hat\alpha_0\beta) = d\nu(\beta),$$

where $d\nu(\beta)$ is the Haar measure on $(\Gamma_0 \setminus \Gamma)^*$ (the measure $d\mu(\alpha)$ does not depend on a choice of the element $\hat\alpha_0$).

Theorem 3. *Let Γ be a Fuchsian group of Widom type, Γ_0 be its normal subgroup containing the commutator Γ', and let $\alpha_0 \in \Gamma_0^*$ can be extended to a character on Γ. Then*

$$A^2_1(\Gamma_0,\alpha_0) = \int_{\{\alpha\in\Gamma^*:\ \alpha|\Gamma_0=\alpha_0\}} A^2_1(\Gamma,\alpha)\, d\mu(\alpha).$$

The proof is just a minor modification of the proof of Theorem 2.

Acknowledgement. This work was done during author's visit to the Institute of Mathematics, University Paris 7. The author is thankful to this institution for its generous hospitality, and to the participants of the seminar of A. Boutet de Monvel, especially V. Georgesku, L. Pastur, J.-B. Zuber for useful discussions.

References

[1] L. Ahlfors, *Möbius transformation in several dimensions*, Ordway Professorship Lectures in Math., University of Minnesota, 1980.

[2] M. Craizer, *Entropy of inner functions*, Israel J. Math. **74** (1991), 129–168.

[3] A. Eremenko and M. Lubich, *The dynamics of analytic transformation*, St. Petersburg Math. J. **1** (1990), 563–634.

[4] J. Garnett, *Bounded analytic functions*, Acadami press, 1981.

[5] E. Hewitt and K. Ross, *Abstract harmonic analysis I/II*, Springer-Verlag, 1963/1970.

[6] M. Livsic, N. Kravitsky, A. Markus, V. Vinnikov, *Theory of commuting nonselfadjoint operators*, Kluwer, 1995.

[7] Ya. Sinai (ed.), *Dynamical systems, II*, Springer-Verlag, 1988.

[8] M. Sodin and P. Yuditskii, *Almost periodic Jacobi matrices with homogeneous spectrum, infinite dimensional Jacobi inversion, and Hardy spaces of character-automorphic functions*, Journal of Geometric Analysis **7** (1997), 387–435.

[9] Ch. Pommerenke, *On the Green's function of Fuchsian groups*, Ann. Acad. Sci. Fenn. **2** (1976), 409–427.

[10] H. Widom, *The maximum principle for multiple valued analytic functions*, Acta Math. **126** (1971), 63–81.

Mathematical Division, Institute for Low Temperature Physics, Kharkov, Lenin pr. 47, 310164, Ukraine

Current address: Department of Mathematics, Michigan State University, East Lansing, MI 48824, USA

E-mail address: `yuditski@math.msu.edu`

Operator Theory:
Advances and Applications, Vol. 122

A Functional Model for the Lie Algebra $SL(2,\mathbb{R})$ of Linear Non-self-adjoint Operators

V. A. Zolotarev

Introduction

An important task of the spectral analysis is to construct the model representations for linear operators realizing them via the operator of multiplication by an independent variable in a particular function space. Unlike those spectral decompositions for self-adjoint (unitary) operators constructed by von Neumann, similar representations for non-self-adjoint (non-unitary) operators are rather arduous to obtain. In the fifties Livšic [20] undertook investigations in this direction and devised a theory of characteristic functions and a theory of triangular models of linear operators. Later, in mid-sixties, a theory of dilations for semigroups of contractions was created by Sz.-Nagy and Foias [1]. Simultaneously, Lax and Phillips [2] had shaped a geometric theory of scattering of acoustic waves at bounded obstacles. Further development in these three areas supplied a basis for creating a method of studying non-self-adjoint operators and building correspondent models.

A construction of a functional model, which is commonly accepted as an adequate substitute for a spectral decomposition of a dissipative densely defined operator A is accomplished as follows. It is well known that A is a generator of a strongly continuous contracting semigroup Z_t, governed by the Cauchy problem

$$(I)\qquad \begin{cases} \frac{d}{dt}h(t) = iAh(t) \\ h(0) = h \ (t \geq 0)\,, \end{cases}$$

where $h(t)$ is a vector-function in a Hilbert space H and $h(t) = Z_t h$. One constructs a unitary dilation for Z_t in the sense of Sz.-Nagy and Foias in the Hilbert space $\widetilde{H} \supset H : Z_t = P_H U_t\mid_H$ for $t \geq 0$ (P_H is the orthogonal projection onto H). It turns out that $\widetilde{H} = D_- \oplus H \oplus D_+$, where $D_+\,, D_-$ are the outgoing and incoming subspaces for the unitary group U_t in the sense of Lax and Phillips [2]. Then, following the Lax-Phillips scattering scheme [2] one constructs a translation model of the unitary group U_t the main parameter of which is a scattering S-operator. From the translation model one obtains a spectral decomposition of the dilation U_t by means of the Fourier transform. Finally, projecting the spectral decomposition of the dilation U_t onto H, one gets a functional model of the semigroup Z_t (and its generator A), which is an operator of multiplication by $\exp(it\lambda)$ (resp. by λ) in the certain particular space of functions $f(\lambda)$.

The scheme presented above for obtaining functional models came into being in the work of Adamyan and Arov, who gave a standard definition in terms of wave operators for the Lax-Phillips S-matrix and demonstrated that the kernel of S coincides with the Fourier transform of the Livšic characteristic function of the semigroup Z_t. The choice of particular form of notation for the spectral decomposition of U_t leads to the corresponding functional models of semigroups of contractions Z_t. Examples are: the Sz.-Nagy-Foias model; the Pavlov model; the de Branges-Rovnyak model, etc. [1], [4], [16].

The present study is a development of the previous work by the same author [11], [12] and is devoted to the construction of a functional model for a Lie algebra of linear non-self-adjoint operators $\{A_1, A_2, A_3\}$ defined by the commutation relations

$$[A_1, A_2] = -iA_3; \ [A_1, A_3] = 2iA_1; \ [A_2, A_3] = -2iA_2.$$

We will first present an abstract scheme.

Here is a method for constructing functional models for Lie algebras of linear operators. With each Lie algebra $\{A_k\}_1^n$ of linear operators acting in a Hilbert space H, $([A_k, A_s] = i\sum_p C_{k,s}^p A_p, C_{k,s}^p \in R,\ n \in N)$ one associates by analogy with (I) the multiparameter semigroup Z_x, governed by the Cauchy problem

$$(II) \qquad \begin{cases} (i\partial_k + A_k)h(x) = 0, \\ h(0) = h \ (1 \le k \le n), \end{cases}$$

where $x \in R^n$, $h(x) = Z_x h \in H$, and $\{\partial_k\}_1^n$ are first order differential operators with coefficients depending on $x \in R^n$. It is shown that the system (II) is consistent and admits a solution if and only if the operators $\{\partial_k\}_1^n$ form a Lie algebra of vector fields $([\partial_k, \partial_s] = \sum_p C_{k,s}^p \partial_p)$. Also, it is known that there exists a unique simply connected Lie group G corresponding to the Lie algebra $\{\partial_k\}_1^n$.

Following the scheme presented above, we first construct a dilation of multiparameter semigroup of contractions Z_x over a cone (the multidimensional analogue of R_+), and then we construct S, the Lax-Phillips scattering operator. The realization of the semigroup Z_x in an appropriate spectral representation of the unitary dilation leads to a functional model of the original Lie algebra. A typical feature of the construction of the dilation of the multidimensional semigroup of contractions is that as a first step one produces a dilation of a "distinguished" one-parameter semigroup of contractions. It is then extended along the remaining variables in the Lie group G with the help of compatibility conditions for a system of equations that is similar to the system (II) with the right side augmented. At the same time one chooses in G a convex salient cone K over which the corresponding dilation is constructed. Harmonic analysis performed in the framework of the multiparameter scattering scheme so constructed results in a functional model of the original Lie algebra $\{A_k\}_1^n$ of linear operators. The original Lie algebra $\{A_k\}_1^n$ turns out to be isomorphic to the restriction to a model subspace of a certain Lie algebra of self-adjoint operators, satisfying the same commutation relations. Typically, the model so obtained is realized in the class of meromorphic functions on a Riemann surface.

In spite of the generality of the suggested scheme, the process of constructing model representations for each particular Lie algebra is essentially unique, because harmonic analysis should be performed on the corresponding Lie group, determined by the given Lie algebra $\{A_k\}_1^n$. We point out that in each case one arrives at a spectral decomposition of a dilation of Z_x via a specific transformation on the corresponding Lie group. This is because the transformation sending the group translation operator on G to the operator of multiplication by an independent variable depends intrinsically on the properties of the particular group G.

For the Lie algebra $\{A_1, A_2, A_3\}$, $[A_1, A_2] = -iA_3$; $[A_1, A_3] = 2iA_1$; $[A_2, A_3] = -2iA_2$ under consideration in the present work, the vector fields $\{\partial_k\}$ are of the form $\partial_1 = e^{-2t_3}(\partial_{t_1} + t_2^2\partial_{t_2} + t_2\partial_{t_3}), \partial_2 = e^{2t_3}\partial_{t_2}$, $\partial_3 = \partial_{t_3}$, where $\partial_{t_k} = \partial/\partial_{t_3}$ $(k = 1, 2, 3)$ and G is the group SL(2,R). We note that the compatibility conditions for the system of equations $(7_\pm)$ which are basic for the dilation construction are overdetermined. The conditions (13), (14) presented below, which essentially mean that a certain differential form is closed, enable one to overcome this setback and to perform the construction of the dilation. The differential operators determining the extension to the cone K, $u_\pm(x_3) \to u_\pm(x)$ $(x = (x_1, x_2, x_3) \in R^3)$ are non-unitary and therefore the $u_\pm(x)$ are not in the space of the initial data $u_\pm(x_3)$. By selecting $u_\pm(x_3)$ from a narrower function space and taking the Fourier transform, we arrive at a functional model of the Lie algebra. The model space in this case is the Hardy class $H^2(\Pi)$ of functions holomorphic in the strip $\Pi = \{z \in C, \mid Im\ z \mid < 2\}$ (rather than the traditional L^2).

A functional model of the Lie algebra is realized in the class of meromorphic functions on a Riemann surface Q by operators of multiplication and weighted shift, which in the case of genus $g = 1$ of a curve Q lead to special functions on Q. We remark that in this case the curve Q is determined by a self-adjoint pencil of linear operators quadratically dependent on ξ.

1. Preliminaries

I. We consider a Lie algebra $\{A_1, A_2, A_3\}$ of linear operators in a Hilbert space H which are subject to the commutation relations

$$[A_1, A_2] = -iA_3;\ [A_1, A_3] = 2iA_1;\ [A_2, A_3] = -2iA_2 \tag{1}$$

We point out that the operators A_1, A_3 (or A_2, A_3) can not be bounded simultaneously. In fact, we imply from (1) that $[A_1^n, A_3] = 2inA_1^n$ and if A_1 is bounded then $\| A_3 \| \geq n$ $(\forall n \in N)$.

We write the following "dynamic system" generated by the Lie algebra (1)

$$\begin{cases} (i\partial_k + A_k)h(x) = 0 \\ 1 \leq k \leq 3; \end{cases}$$

where $h(x)$ is vector-function from H; $x \in R^3$; $\{\partial_k\}_1^3$ are first-order diff. This system of equations is consistent only if $\{\partial_k\}_1^3$ form the corresponding Lie algebra of vector fields [11, 12] :

$$(2) \qquad [\partial_1, \partial_2] = -\partial_3;\ [\partial_1, \partial_3] = 2\partial_1;\ [\partial_2, \partial_3] = -2\partial_2,$$

which, in particular, is realized as follows [13–16]:

$$(3) \qquad \partial_1 = e^{-2x_3}(\partial_{x_1} + x_2^2\partial_{x_2} + x_2\partial_{x_3});\ \partial_2 = e^{2x_3}\partial_{x_2};\ \partial_3 = \partial_{x_3},$$

where $\partial_{x_k} = \partial/\partial_{x_k}$, $(1 \le k \le 3)$; $x = (x_1, x_2, x_3) \in R^3$. It is known [13–16] that there exists a unique simply connected Lie group $G(= R^3)$ corresponding to the Lie algebra of vector fields $\{\partial_k\}_1^3$ (2),(3). The multiplication law in G is given by $\widetilde{g} = g \circ \widehat{g}$:

$$(4) \qquad \begin{cases} \widetilde{x}_1 = x_1 + \widehat{x}_1 e^{-2x_3}(1 + x_2\widehat{x}_1 e^{-2x_3}); \\ \widetilde{x}_2 = \frac{x_2}{1 + x_2\widehat{x}_1 e^{-2x_3}} + \widehat{x}_2 e^{2x_3}(1 + x_2\widehat{x}_1 e^{-2x_3}); \\ \widetilde{x}_3 = x_3 + \widehat{x}_3 + ln(1 + x_2\widehat{x}_1 e^{-2x_3}); \end{cases}$$

and $g = (x_1, x_2, x_3) \in G, \widehat{g} = (\widehat{x}_1, \widehat{x}_2, \widehat{x}_3) \in G, \widetilde{g} = (\widetilde{x}_1, \widetilde{x}_2, \widetilde{x}_3) \in G$.

It is easy to show that there is an isomorphism

$$(5) \qquad g = (x_1, x_2, x_3) \Longleftrightarrow B_g = \begin{bmatrix} e^{x_3} & x_2 e^{-x_3} \\ x_1 e^{x_3} & (1 + x_1 x_2)e^{-x_3} \end{bmatrix}$$

between the group G and the group of matrices $SL(2, R)$. It turns out that vector fields $\{\partial_k\}_1^3$ (3) are the right Lie derivatives [13–16]:

$$\partial_k f(x) = \frac{d}{dt} f\Big(x \circ \omega_k(t)\Big)\Big|_{t=0}, \quad (1 \le k \le 3)$$

along the one-parameter subgroups $\omega_k(t)$: $\omega_1(t) = (t, 0, 0)$, $\omega_2(t) = (0, t, 0)$, $\omega_3(t) = (0, 0, t)$. So one studies the evolution given by the Lie algebra (1) on the Lie group G (4).

II. The following system of differential equations forms the basis for the construction of the dilation of the Lie algebra (1) (see [1,4,11,12,25]):

$$(6) \qquad \begin{cases} (i\partial_k + A_k)h(g) = \varphi^*\sigma_k u_-(g) \\ (1 \le k \le 3)\ ;\ h(e) = h_0 \end{cases}$$

where $g = (x_1, x_2, x_3) \in G$; vector-functions $h(g)$ and $u_-(g)$ are from the Hilbert spaces H and E; the Hilbert space E and linear operators $\varphi : H \to E$, $\sigma_k : E \to E$ $(\sigma_k^* = \sigma_k)$ are such that $2Im\langle A_k h, h\rangle = \langle \sigma_k \varphi_h, \varphi_h\rangle$ for $\forall h \in D_{A_k}$ $[11 - 13, 24]$, and, finally, $e = (0, 0, 0)$ is the identity in G.

Remark 1. *The Hilbert space E and linear operators $\varphi, \{\sigma_k\}_1^3$ can be chosen as follows. If E is in an orthogonal complement with respect to the intersection of the kernels of the quadratic forms $2Im\langle A_k h, h\rangle$ then one can put $\varphi = P_E$, where P_E is an orthogonal projection on E, and σ_k, as usually, are determined by the restrictions onto E of the operators defined by the form $2Im\langle A_k h, h\rangle$.*

Theorem 1. *The system (6) is consistent if and only if the vector-function $u_-(g)$ is a solution of the system of equations*

$$(7_-) \qquad \begin{cases} [\sigma_k i\partial_s - \sigma_s i\partial_k + \gamma^-_{k,s}]u_-(g) = 0 \\ 1 \le k, s \le 3\,, \end{cases}$$

where $\gamma^-_{k,s}$ are linear operators in E such that $\gamma^-_{k,s}\varphi = \sigma_k\varphi A^*_s - \sigma_s\varphi A^*_k$, $\gamma^-_{k,s} = -\gamma^-_{s,k}$, $(1 \le k, s \le 3)$.

We define a map from the incoming waves $u_-(g)$ to the outgoing ones $u_+(g) \in E$ by the traditional formula [7,11–13,24]

$$(8) \qquad u_+(g) = u_-(g) - i\varphi h(g)$$

where $h(g)$ is a solution of the problem (6).

Theorem 2. *If $u_-(g)$ satisfies the system (7_-) and $h(g)$ is a solution of the Cauchy problem (6), then $u_+(g)$ (8) is a solution of the following system of equations*

$$(7_+) \qquad \begin{cases} [\sigma_k i\partial_s - \sigma_s i\partial_k + \gamma^+_{k,s}]u_+(g) = 0 \\ 1 \le k, s \le 3 \end{cases}$$

where $\gamma^+_{k,s} = \gamma^-_{k,s} + i(\sigma_k\varphi^*\varphi\sigma_s - \sigma_s\varphi^*\varphi\sigma_k)$; $1 \le k, s \le 3$.

The proof of Theorems 1 and 2 is an easy implication from the commutation relations for Lie algebras $\{A_k\}_1^3$ (1) and $\{\partial_k\}_1^3$ (2). We note that $\gamma^+_{k,s}\varphi = \sigma_k\varphi A_s - \sigma_s\varphi A_k$ and $\gamma^+_{k,s} = -\gamma^+_{s,k}$, $1 \le k, s \le 3$. The operators $\gamma^\pm_{k,s}$ have the following important properties:

$$(9) \qquad \gamma^\pm_{1,2} - (\gamma^\pm_{1,2})^* = -i\sigma_3\ ;\ \gamma^\pm_{3,1} - (\gamma^\pm_{3,1})^* = -2i\sigma_1\ ;\ \gamma^\pm_{3,2} - (\gamma^\pm_{3,2})^* = 2i\sigma_2$$

In addition we note that the operators $\{\gamma^\pm_{k,s}\}_1^3$ in the realization from Remark 1 are of the form

$$\gamma^-_{k,s} = \frac{1}{i}\left\{A_kA^*_s - A_sA^*_k + [A^*_s, A^*_k]\right\}\Big|_E\ ;\ \gamma^+_{k,s} = \frac{1}{i}\left\{A_sA^*_k - A_kA^*_s + [A_k, A_s]\right\}\Big|_E .$$

III. We single out the family of one-parameter subgroups G_p of the group G:

$$G_p = \left\{ g_p(t) = (x_1(t), x_2(t), x_3(t))\ ;\ x_1(t) = \frac{n}{\sqrt{nm+1}ch(t\sqrt{nm+1}) + 1}\ ;\right.$$

$$x_2(t) = \frac{m\ sh(2t\sqrt{nm+1}}{2\sqrt{nm+1}} + \frac{m\ sh^2(t\sqrt{nm+1}}{\sqrt{nm+1}}$$

$$(10) \qquad \left. x_3(t) = ln\ \left| ch(t\sqrt{nm+1}) + \frac{sh(t\sqrt{nm+1})}{\sqrt{nm+1}} \right| \right\}$$

which depends on the point $p = (n, m, 1) \in R^3$ and $t \in R$. It is easy to see that because of the multiplication law form (4) on G, G_p (10) is isomorphic to the additive group R: $g_p(t) \circ g_p(s) = g_p(t+s)$. By virtue of the relation (5) there is the correspondence between elements $g_p(t) \in G_p$ and matrices $B_p(t) \in SL(2,R)$:

$$g_p(t) \Leftrightarrow B_p(t) = \begin{bmatrix} ch(t\sqrt{nm+1}) + \frac{sh(t\sqrt{nm+1})}{\sqrt{nm+1}} & \frac{m\, sh(t\sqrt{nm+1}}{\sqrt{nm+1}} \\ \\ \frac{n\, sh(t\sqrt{nm+1})}{\sqrt{nm+1}} & ch(t\sqrt{nm+1}) - \frac{sh(t\sqrt{nm+1})}{\sqrt{nm+1}} \end{bmatrix}$$

It is easily seen that $g_p(t)$ as a curve in R^3 satisfies the relations

$$me^{2x_3}x_1 = nx_2 \ ; \ (mx_1+1)^2 + nme^{-2x_3} = nm+1$$

Thus, $g_p(t)$ lies on a cylindrical surface defined by the second relation. We note that if $n=0$, i.e. $p=(0,m,1)$ then the curve $g_p(t)$ is of the form $g_p(t) = (0, me^t sh\ t, t)$; if $m=0$, i.e. $p=(n,0,1)$, then $g_p(t) = (ne^{-t} sh\ t, 0, t)$. It is not hard to see that family G_p (10) fills up the entire group G in the sense that for each $x \in G$ there exists subgroup G_p passing through it, that is there exists $p=(n,m,1) \in R^3$ such that $g_p(t) = x$ for some $t \in R$.

It is known [16,17] that one-parameter subgroups G_p and these alone are the geodesics with respect to the Killing metric. The connection on G defined by the formula $\nabla_{L_x} = \frac{1}{2}[L_x, L_y]$, where L_x, L_y are the left-invariant vector fields corresponding to the elements X and Y of the Lie algebra of the group G, is symmetric and agrees with the Killing metric [11,12,16,17].

We find now the explicit form of the Killing metric on G using the isomorphism (5) between G and $SL(2,R)$. Since the Killing form on matrix Lie algebras is traditionally defined by the formula $\langle X,Y\rangle_k = tr\ XY$, [16,17],then for $b_g = \begin{bmatrix} y_3 & y_2 \\ y_1 & -y_3 \end{bmatrix} \in SL(2,R)$ from Lie algebra of the group $SL(2,R)$, $(y_k \in R)$ the Killing form is $\langle b_g, b_g\rangle_k = 2(y_3^2 + y_1y_2)$. We note that this form is indefinite and of the type (2,1).

Because of (5) the matrix $B_{g(t)}$ corresponds to the curve

$$g(t) = (x_1(t), x_2(t), x_3(t)) \in G.$$

The tangent vector to $g(t)$ in this matrix realization is of the form

$$\dot{b}_{g(t)} = B^{-1}_{g(t)} \frac{d}{dt} B_{g(t)} = \begin{bmatrix} \dot{x}_3 - x_2\dot{x}_1 & (\dot{x}_2 - x_2^2\dot{x}_1)e^{-2x_3} \\ \\ \dot{x}_1 e^{2x_3} & \dot{x}_1 x_2 - \dot{x}_3 \end{bmatrix} \tag{11}$$

where $\dot{x}_k = \dot{x}_k(t) = dx_k(t)/d(t)$, $(k=1,2,3)$. Hence, the Killing form for $\dot{b}_{g(t)}$ is $\langle \dot{b}_{g(t)}, \dot{b}_{g(t)}\rangle_k = 2(\dot{x}_3^2 + \dot{x}_1\dot{x}_2 - 2\dot{x}_3\dot{x}_1x_2)$. The Euler-Lagrange equations for the extremals of the functional

$$\int_P^Q \sqrt{\langle \dot{b}_{g(t)}, \dot{b}_{g(t)}\rangle_k} dt$$

lead to geodesic equations because of (5) (the parameter t is proportional to the natural one) which are of the form

$$\ddot{x}_k + \Gamma^k_{ij}\dot{x}_i\dot{x}_j = 0$$

where $\Gamma^1_{1,3} = 2,\ \Gamma^2_{2,3} = -2,\ \Gamma^2_{1,2} = -2x_2,\ \Gamma^2_{1,3} = 4x_2^2,\ \Gamma^3_{1,3} = -1,\ \Gamma^3_{1,3} = 2x_2$, and the remaining $\Gamma^k_{ij} = 0$. It is easily seen that the solution of this system of equations under the initial conditions $x_k(0) = 0,\ (k = 1,2,3),\ \dot{x}_1(0) = n,\ \dot{x}_2(0) = m, \dot{x}_3(0) = 1$ coincides with $g_p(t) = (x_1(t), x_2(t), x_3(t)) \in G_p$. The Christoffel symbols Γ^k_{ij} can be found as follows. Because of (11) $\dot{b}_{g(t)} = \sum \dot{x}_k a_k$, where $a_k \in SL(2,R)$. It turns out that Γ^k_{ij} coincide with the structure constants C^k_{ij} ($\Gamma^k_{ij} = C^k_{ij}$) of the Lie algebra $\{a_k\}$, $[a_i, a_j] = C^k_{ij}a_k$.

IV. We consider (in R^3) the path-independent line integral

$$h(x) = h(e) + \int_e^x \sum_k h_{\xi_k}(\xi)d\xi_k$$

from point $e = (0,0,0)$ to a point $x = (x_1, x_2, x_3)$, where $h_{\xi_k} = \partial h(\xi)/\sigma\xi_k$. This means that $h(x)$ as a solution of the Cauchy problem (6) when $u_-(x)$ satisfies (7_-) can be obtained along the corresponding subgroup G_p (10) joining e and x. It follows from the obvious relation

$$\frac{d}{dt}f(g_p(t)) = (n\partial_1 + m\partial_2 + \partial_3)f(g_p(t))$$

and from (6) that

$$i\frac{d}{dt}h(g_p(t)) + A_p h(g_p(t)) = \varphi^*\sigma_p u_-(g_p(t)),$$

where $A_p = nA_1 + mA_2 + A_3$ *and* $\sigma_p = n\sigma_1 + m\sigma_2 + \sigma_3$. It is now easy to derive for each $x = g_p(t)$ an explicit formula for the $h(x)$ along G_p:

$$h(x) = h(g_p(t)) = \exp(itA_p)h_0 - i\int_0^x \exp\Big[i(t-\tau)A_p\Big]\varphi^*\sigma_p u_-(g_p(\tau))d\tau.$$

2. Causality cones; dilation of the Lie algebra

V. In what follows we shall restrict our attention to the Lie algebras $\{A_k\}_1^3$ (1) such that

(12-1) A_3 is dissipative densely defined operator defect spaces of which coincide with each other and with $\widetilde{E}$ [4,6,12];

(12-2) if $f, g \perp E$ and $f \in D_{A_k}$, $g \in D_{A_k^*}$ then $Im\langle A_k f, f\rangle = 0$, $Im\langle A_k^* f, f\rangle = 0$, $(k = 1, 2)$;

(12-3) the defect operators of A_3 on E are boundedly invertible ;

(12-4) operator $A_t = \sum_1^3 A_k t_k$ is not self-adjoint for $\forall\, t \in R^3$.

It follows from 1) and 2) (12) that $\varphi : \widetilde{E} \to E$ can be chosen so as to make $\sigma_3 = I_E$ (see [11,12,24]), and we assume that such a choice has been made.

Remark 2. *From all the conditions 1)–4) (12) the essential ones are the dissipativity of A_3 1) and 4); the further constructions are impossible without them. The dilation theory of Sz.-Nagy and Foias [1,4,11–13] presupposes the contractiveness of the basic semigroup which means the dissipativity of the operator (for example A_3). If the condition 4) is not true then the intersection of the incoming and outgoing spaces D_+ and D_- from the Lax-Phillips scheme which follows can be nonempty.*

The systems of differential equations $(7_\pm)$ are overdetermined. We rewrite two of the equation $(7_\pm)$ as follows

$$\begin{cases} \partial_2 u_\pm = (\sigma_2\partial_3 + i\gamma_{3,2}^\pm)u_\pm; \\ \partial_1 u_\pm = (\sigma_1\partial_3 + i\gamma_{3,1}^\pm)u_\pm; \end{cases} \tag{13}$$

Taking into account the remaining relations in (2) one deduces from the condition $[\partial_1, \partial_2] = -\partial_3$ that

$$\begin{cases} 1) & [\sigma_1, \sigma_2] = 0; \\ 2) & [\sigma_1, \gamma_{3,2}] + [\gamma_{3,1}, \sigma_2] = -i(I + 4\sigma_1\sigma_2); \\ 3) & [\gamma_{3,1}, \gamma_{3,2}] = -2i(\sigma_1\gamma_{3,2} + \sigma_2\gamma_{3,1}); \end{cases} \tag{14}$$

where $\gamma_{k,s} = \gamma_{k,s}^\pm$, $(1 \le k, s \le 3)$. In view of (13) the equation

$$\left[\sigma_2 i\partial_1 - \sigma_1 i\partial_2 + \gamma_{2,1}^\pm\right] u_\pm = 0$$

implies that

$$\gamma_{2,1}^\pm = \sigma_2\gamma_{3,1}^\pm - \sigma_1\gamma_{3,2}^\pm. \tag{15}$$

We remark that the relation $\gamma_{2,1}^\pm - (\gamma_{2,1}^\pm)^* = i\sigma_3$ (9) follows from the remaining relations (9) and from 2) in (14).

It is therefore necessary for $\{\sigma_1, \sigma_2, \gamma_{3,1}^\pm, \gamma_{3,2}^\pm\}$ to satisfy the relations (9) and (14) and for operators $\gamma_{2,1}^\pm$ to be determined via formula (15).

VI. For the semigroup of contractions $Z_3(t) = \exp(itA_3)$, $(t \ge 0)$ in the Hilbert space H we will construct a minimal unitary dilation [1,4,11–13]. The Hilbert space H of the dilation [4,11–13] is formed by vector-functions $f(\xi) = (U_+(\xi), h, U_-(\xi))$ such that

$$\| f \|^2 = \int_{-\infty}^{0} \| u_+(\xi) \|^2 \, d\xi + \| h \|^2 + \int_{0}^{\infty} \| u_-(\xi) \|^2 \, d\xi < \infty \tag{16}$$

where $u_\pm(\xi) \in E,\ h \in H,\ Supp\ u_\pm(\xi) \subseteq R_\mp$. We embed $\xi \in R$ into the group G by identifying ξ with $(0,0,\xi) \in G$. We make use of the systems of differential equations $(7_\pm)$ to extend the incoming $u_-(x_3)$ and outgoing $u_+(x_3)$ waves [11–13] along the variables x_1 and x_2 in R^3. Namely, we consider the Cauchy problem

$$(17_\pm)\qquad \begin{cases} \partial_2 u_\pm(x) = (\sigma_2\partial_3 + i\gamma_{3,2}^{\pm})u_\pm(x); \\ \partial_1 u_\pm(x) = (\sigma_1\partial_3 + i\gamma_{3,1}^{\pm})u_\pm(x); \\ u_\pm(0,0,x_3) = (u_\pm(x_3);\ (\mp x_3 \geq 0); \end{cases}$$

where $x = (x_1, x_2, x_3) \in G = R^3$, assuming that (9) and (14) are valid. We consider first the scalar case $\dim E = 1$. The characteristic surfaces $\Phi(x_1, x_2, x_3) = 0$ of the system $(17_\pm)$ are the solutions of the differential equations

$$e^{2x_3}\Phi_{x_2} - \sigma_2\Phi_{x_3} = 0;\ \Phi_{x_1} + x_2^2\Phi_{x_2} + (x_2 - e^{2x_3}\sigma_1)\Phi_{x_3} = 0,$$

and $\Phi_{x_k} = \partial\Phi/\partial x_k$, $(1 \leq k \leq 3)$. Hence we easily find the characteristics for the system $(17_\pm)$ which are of course the same for $u_\pm(x)$:

$$(18)\qquad \begin{cases} 2\sigma_2 x_2 + e^{2x_3} = \xi; \\ x_1 = \eta \end{cases} \qquad \begin{cases} x_1 - \frac{1}{x_2} = \mu \\ \sigma_1 + \frac{3}{2}e^{-2x_3}x_2 = \nu x_2^3 \end{cases}$$

where the constants $\xi, \eta, \mu, \nu \in R$. Thus the functions $u_-(x_3)$ extend from the semiline R_+ along the characteristics (18) as the solutions of the Cauchy problem (17_-) to a domain K_- in R^3. Generally speaking, this set K_- does not form a cone because it is not true that for all $x \in K_-$ we have $\alpha\, x \in K_-$ for $\forall\alpha > 0$. Yet we will refer to K_- as a cone since evidently there exists a bijection from K_- onto cone.

In the case where $\dim E = r < \infty$ one has to choose a basis in E formed by the common eigenvectors of the self-adjoint operators σ_1 and σ_2 (which is possible by virtue of 1) (14)). One has to decompose each of the vector-functions $u_\pm(x_3)$ on this basis and to extend the Fourier coefficients (with respect to this basis) along the characteristics (18) as the solutions of the Cauchy problem $(17_\pm)$; in this case instead of the σ_k in (18) one must put the corresponding eigenvalues. We denote by K_- the intersection of all those domains for different components of the basis. The set K_- forms a salient cone. Indeed, assuming that K_- were not salient (that is, it contained a "straight" line), all the characteristics in (18) would coincide with each other (for instance, $2\sigma_2 x_2 + e^{2x_3} = \xi$); and therefore all the eigenvalues of the operator σ_2 would be the same, that is $\sigma_2 = \mu I_E$ $(\mu \in R)$ would be a scalar operator so that $\mu A_3 - A_2$ would be self-adjoint in contradiction to 4) (12).

In the case where $\dim E = \infty$, one has to consider a finite-dimensional approximation to the problems $(17_\pm)$. It is not hard to show, that the corresponding solutions to the finite-dimensional problems $(17_\pm)$ approximate the general solutions of the Cauchy problem $(17_\pm)$.

Theorem 3. *There exist salient centrally symmetric cones K_- and K_+ in R^3 which form the domains of existence and uniqueness for the Cauchy problems $(17_\pm)$ for* *the outgoing* $u_+(x)$ *and incoming* $u_-(x)$ *waves with the initial data lying on the semilines* R_- *and* R_+ *respectively.*

We denote by H the space of vector-functions $f = (u_+(x), h, u_-(x))$ such that $Supp\, u_\pm(x) \subset K_\pm$ and $u_\pm(x)$ are the solutions to the Cauchy problems $(17_\pm)$ with the initial data $u_\pm(0,0,x_3) \in L^2$. We define a norm in H, as before, via (16).

In the next subsection we give another description of the cones $K_\pm$ by means of the one-parameter subgroups G_p (10) which allow to consider $K_\pm$ as causality cones [26] of the Lie algebra (1).

VII. Theorem 4. *Let function* $u(x) = u_\pm(x)$ *be a solution of the problem* $(17_\pm)$ *and operators* $\{\sigma_1, \sigma_2, \gamma^{\pm}_{3,1}, \gamma^{\pm}_{3,2}\}$ *satisfy the relation (9),(14),(15). Then*

$$\oint_L \langle(\sigma_1 e^{2x_3} - \sigma_2 x_2^2 e^{-2x_3} - \sigma_3 x_2)u, u\rangle dx_1$$

$$+\langle\sigma_2 e^{2x_3} u, u\rangle dx_2 + \langle\sigma_3 u, u\rangle dx_3 \tag{19}$$

where L *is a piecewise-smooth contour in* R^3*, bordering a one-side surface.*

The proof consists of an application of Stokes's formula. Here one must bear in mind that $u(x)$ is a solution to the system $(17_\pm)$ and operators $\{\sigma_1, \sigma_2, \gamma^{\pm}_{3,1}, \gamma^{\pm}_{3,2}\}$ satisfy (9),(14),(15).

We fix a one-parameter subgroup G_p (10) and let $G_p^\pm = G_p \bigcap K_\mp$. The following statement is true.

Lemma 1. *For* $u_\pm(x)$ *as the solutions of* $(17_\pm)$ *the following formula is true*

$$\int_{R_\pm} \| u_\mp(0,0,\xi) \|^2 = \int_{G_p^\pm} \langle \sigma_p\, u_\mp, u_\mp\rangle dt, \tag{20}$$

where $\sigma_p = n\sigma_1 + m\sigma_2 + \sigma_3$, $p = (n, m, 1) \in R^3$.

Proof. We consider first the case where initial data $u_\pm(0,0,x_3)$ of the problem $(17_\pm)$ are compactly supported. We choose a closed contour

$$L_p^\pm(R) = G_0^\pm(R) \bigcup G_p^\pm(R) \bigcup C(R)$$

where $G_p^\pm(R) = G_p^\pm \bigcap \{x \in R;\ x_1^2 + x_2^2 + x_3^2 \le R^2\}$; $C(R)$ is the arc of the circle $x_1^2 + x_2^2 + x_3^2 = R^2$ joining $G_p^\pm(R)$ to the segment $G_0^\pm(R) = \{(0,0,x_3);\ | x_3 |\le R\}$, where $R \gg 1$.

From the form of the characteristics (18) and the formula (11) for the subgroups G_p (10) it is easily seen that the compactly supportedness of the initial data $u_\pm(0,0,x_1)$ implies that for large enough $R \gg 1$ the restrictions to G_p of the solutions $u_\pm(x)$ to the problems $(17_\pm)$ vanish for $t > R$. Therefore the line integral (19) along $C(R)$ vanishes. To prove (20) it remains to observe that the integral (19) when restricted to $G_0^\pm(R)$ and to $G_p^\pm(R)$ coincides with the corresponding expressions in (20). In the general case one has to resort to the density of the compactly supported functions in L.

It follows from this lemma that the quadratic form

$$\| f \|_p^r = \int\limits_{G_p^-} \langle \sigma_p u_+, u_+ \rangle dt + \| h \|^2 + \int\limits_{G_p^+} \langle \sigma_p u_-, u_- \rangle dt \tag{21}$$

coincides with the norm in the space H, $\| f \|_p = \| f \|$; $\forall f \in H$.

Theorem 5. *The cones $K_\pm$ which are the existence and uniqueness domains of the Cauchy problem $(17_\pm)$ coincide with the sets*

$$K_\pm = \bigcup_p \{G_p \,;\, \sigma_p > 0 \pm t \geq 0\} \tag{22}$$

where G_p are the distinguished one-parameter subgroups (10) and the operator $\sigma_p = n\sigma_1 + m\sigma_2 + \sigma_3$ is boundedly invertible, $p = (n, m, 1) \in R^3$.

It is not difficult to see that the proof of this theorem follows from (20). We consider first the case where $\dim E = r < \infty$. Since the system of equations $(17_\pm)$ is linear, the linear independence of the initial data $u_\pm(0,0,x_3)$ implies the linear independence of the solution to the Cauchy problem $(17_\pm)$ and therefore $u_\pm(x)$ fills in the entire space E; $E = \bigvee u_\pm(x)$ for each $x \in K_\mp$. To prove the invertibility of σ_p it is sufficient to show that $Ker\, \sigma_p = \{0\}$. Assuming the opposite we select $\vartheta \in E$ such that $\sigma_p \vartheta = 0$. Then we denote by $\vartheta(x)$ the function which is extended from G_p to $K_\pm$ $(x \in K_\pm)$ as a solution of the system of equations $(17_\pm)$. From (20) it follows that $\vartheta(0,0,x_3) = 0$ which means that $\vartheta(x) \equiv 0$ because of the uniqueness of the solution of the Cauchy problem $(17_\pm)$, and therefore $\vartheta = 0$.

In the case where $\dim E = \infty$ one must prove a) the invertibility σ_p in E and b) that the solutions $u_\pm(x)$ fill in the entire Hilbert space H. We note first that if n and m are small ($| n | \ll 1$, $| m | \ll 1$) then the invertibility as well as the positiveness of σ_p are evident. Assuming the opposite we select the sequence $\vartheta_n \in E$ such that $\| \vartheta_n \| = 1$ and $\sigma_p \vartheta_n \to 0$ if $n \to \infty$. After that we extend ϑ_n from the curve G_p into the domain $K_\pm$ as the solutions of the problem $(17_\pm)$. Then from (20) follows that $\vartheta_n(0,0,x_3) \to 0$ if $n \to \infty$. Because of the analiticity of $\vartheta_n(x)$ and the Cauchy-Kovalevskaya's theorem $[22-23]$ $\vartheta_n(x) \to 0$, $n \to \infty$ which is a contradiction to the condition that $\| \vartheta_n \| = 1$. The fact that $Ker\ \sigma_p = 0$ is proved in the same way as in the finite-dimensional case. We prove now that the linear span of $u_\pm(x)$ is dense in E. We assume that it is not true which means that there exists the vector $\vartheta \in E$ such that $\vartheta \perp u_\pm(x)$. We extend $\sigma_p^{-1}\vartheta$ from the subgroup G_p as the solution $(\sigma_p^{-1}\vartheta)(x)$, $x \in K_\pm$ of the problem $(17_\pm)$. Using the bilinear analog of the formula (20) we obtain that $(\sigma_p^{-1}\vartheta)(0,0,x_3) = 0$, and because of the uniqueness of the solution of the Cauchy problem $(17_\pm)$, $(\sigma_p^{-1}\vartheta)(x) = 0$, therefore $\sigma_p^{-1}\vartheta = 0$ and finally $\vartheta = 0$, which proves the theorem.

VIII. We now proceed to the construction of a dilation of the Lie algebra (1). The most convenient way of doing it is in terms of one-parameter subgroups. We choose a subgroup G_p (10) such that $G_p \subset K_+ \bigcup K_-$ and consider semigroup of contractions $Z_{g_p} = Z_p(t) = \exp(itA_p)$, where $g_p(t) \in G_p$, $t \geq 0$ and $A_p =$

$nA_1 + mA_2 + A_3$. We define the unitary dilation U_{g_p} of the semigroup Z_{g_p} in H as follows

$$u_{g_p} f = (u_+(x, g_p),\ h_{g_p},\ u_-(x, g_p)) \tag{22}$$

where $u_-(x, g_p) = P_{K_-} u_-(x \circ g_p)$; h_{g_p} is the solution of the Cauchy problem

$$\begin{cases} iy'(\xi) + A_p y(\xi) = \varphi^* \sigma_p u_-(g_p u_-(g_p(\xi)) \\ \qquad y(0) = h\ ,\ 0 \le \xi \le t \end{cases}$$

evaluted at $\xi = t$, $h_{g_p} = y(t)$; and finally $u_+(x, g_p) = u_+(x \circ g_p) + P_{K_+}\{u_-(x \circ g_p) - i\varphi y(x \circ g_p)\}$. Here $P_{K_\pm}$ are the operators of the restriction onto domains $K_\pm$. We denote by $P_{K_+}[u_-(x \circ g_p) - i\varphi y(x \circ g_p)]$ the function in K_+ which is obtained by the extension into K_+ of the function $P_{K_+}[u_-(g_p(\xi)) - i\varphi y(g_p(\xi))]$, $\xi \in [-t, 0]$ as a solution of the problem $(17_\pm)$ using Theorem 2. The unitarity of U_{g_p} (22) in the metric (21) is obvious.

Theorem 6. *Suppose that (12) holds for the Lie algebra (1) $\{A_k\}_1^3$. Then there exists a salient cone $K(= K_-)$ such that there is a unitary dilation U_{g_p} (22) of the semigroup of contractions Z_{g_p} along every one-parameter subgroup G_p such that $G_p \subset K$ (for $t \ge 0$); in addition a minimal unitary dilation [1,4,11–13] is uniquely defined up to an isomorphism.*

3. Scattering scheme of Lax-Phillips and translation model

IX. We write $\vartheta_\pm(x) \in N_\pm$ if $\vartheta_\pm(x)$ is a solution to the Cachy problem

$$\begin{cases} \partial_1 \vartheta_\pm(x) = (\sigma_1 \partial_3 + i\gamma_{3,1}^\pm)\vartheta_\pm(x); \\ \partial_2 \vartheta_\pm(x) = (\sigma_2 \partial_3 + i\gamma_{3,2}^\pm)\vartheta_\pm(x); \\ \vartheta_\pm(0,0,x_3) = \vartheta_\pm(x_3); (x_3 \in R); \end{cases} \tag{$23_\pm$}$$

where $\vartheta_\pm(x_3) \in L^2(E, dx_3)$ and $\{\sigma_k, \gamma_{3,k}^\pm\}_1^2$ satisfy the relations (14). In the Hilbert space

$$L^2(N_\pm, R) = \{\vartheta_\pm(x) \in N_\pm; \int_R \| \vartheta_\pm(x_3) \|^2\, dx_3 < \infty\}$$

we introduce the free group of translations

$$U_{g_p}^\pm \vartheta_\pm(x) = \vartheta_\pm(x \circ g_p) \tag{24}$$

along each admissible subgroup G_p (10), $G_p \in K_+ \cup K$.

It is obvious that $U_{g_p}^\pm$ (24) are unitary in the metric

$$\int_{G_p} < G_p u_\pm, u_\pm > dt < \infty, \tag{25}$$

which coincide with the norm in $L^2(N_\pm, R)$. The proof of this fact is essentially identical to the proof of Lemma 1.

The perturbed semigroup which is the dilation U_{g_p} (22), possesses outgoing D_+ and incoming D_- subspaces in the sense of Lax and Phillips [1–6,11–13], generated by functions $(u_+(x),0,0)$ and $(0,0,u_-(x))$ respectively. After the natural identification we can assume that $D_\pm \subset L^2(N_\pm, R)$. We define the wave operators [1–6, 11–13]

$$W_\pm(P) = \operatorname*{slim}_{t\to\pm\infty} U_{g_p}^{-1} P_{D_\pm} U_{g_p}^{\pm} \tag{26}$$

along each subgroup $G_p \subset K_+ \cup K_-$. It is not hard to prove the following theorem.

Theorem 7. *For each admissible subgroup $G_p \subset K_+ \cup K_-$ there exist wave operators $W_\pm(p)$ (26) that are isometries from $L^2(N_\pm, R)$ into $\stackrel{\simeq}{=}EE$ and have the following properties*

$$W_\pm(p)U_{g_p}^{\pm} = U_{g_p}W_\pm(p); \quad W_\pm(p)P_{D_\pm} = P_{D_\pm}. \tag{27}$$

As the wave operators $W_\pm(p)$(26) are known now we define the scattering operator [2–6, 11–13] in the usual fashion

$$S_p = W_+^*(p)W_-(p), \tag{28}$$

where the adjoint operator $W_+^*(p)$ is formed with respect to appropriate metrics (21) and (25).

Theorem 8. *The scattering operator S (28) is a non-expanding map between $L^2(N_-, R)$ and $L^2(N_+, R)$ where*

$$S_pU_{g_p}^- = U_{g_p}^+S_p; \quad S_pL^2(N_-, R) \subset L^2(N_+, R); \tag{29}$$

and also $\overline{S_pL^2(N_-, R)} = L^2(N_+, R)$, provided that $G_p \subset K_+ \cup K_-$.

The proofs of these theorem are similar to the arguments in [11].

X. A Lie algebra $\{A_k\}_1^3$ is called completely non-self-adjoint if there is no subspace reducing the Lie algebra $\{A_k\}_1^3$ (1) on which this Lie algebra $\{A_k\}_1^3$ induces the algebra of self-adjoint operators [11,12,24]. We consider the subspace

$$H_1 = S_{pan}\{Z_t\varphi^*E, Z_\tau^*\varphi^*E; t,\tau \geq 0\}$$

where

$$Z_t = \exp i(\sum t_kA_k); t = (t_1,t_2,t_3) \in R^3; (t \geq 0 \Leftrightarrow t_k \geq 0, \forall k).$$

It is obvious, that H_1 reduces each of the operators A_k and that every restriction $A_k|_{H_0}$ $(1 \leq k \geq 3)$ is self-adjoint; $H_0 = H \ominus H_1$. The Lie algebra of linear operators $\{A_k\}_1^3$ is called simple [11–13,24] if $H_0 = \{1\}$. We claim that

$$H_1 = S_{pan}\{Z_{t_3}\varphi^*E, Z_{\tau_3}^*\varphi^*E; t_3,\tau_3 \geq 0\} \tag{30}$$

where $t_3 = (0,0,t_3), \tau_3 = (0,0,\tau_3)$. It follows from the relation $A_k\varphi^* = A_3\varphi^*\sigma_k + \varphi^*(\gamma_{3,k}^-)^*(k = 1,2)$, that

$$A_k\varphi^*E = A_3\varphi^*\sigma_kE + \varphi^*(\gamma_{3,k}^-)^*E \subset H_1.$$

We show by induction that the relation $A_k A_3^n \varphi^* E \subset H_1$, $(k = 1, 2)$. Indeed, it follows from the commutation relations (1) that

$$A_1 A_3^{n+1} \varphi^* E = A_3 A_1 A_3^n \varphi^* E + 2i A_1 A_3^n \varphi^* E \subset H_1;$$

$$A_2 A_3^{n+1} \varphi^* E = A_3 A_2 A_3^n \varphi^* E - 2i A_2 A_3^n \varphi^* E \subset H_1;$$

Therefore $A_k Z_{t_3} \varphi^* E \subset H_1$, for $k = 1, 2$. In a similar way using the relations

$$A_k^* \varphi^* = A_3^* \varphi^* \sigma_k + \varphi^* (\gamma_{3,k}^+)^*, (k = 1, 2)$$

we prove that $A_k^* Z_{\tau_3}^* \varphi^* E \subset H_1$ for $k = 1, 2$. It follows from the obvious inclusions $A_k Z_{\tau_3}^* \varphi^* E = (A_k^* + i\varphi^* \sigma_k \varphi) Z_{\tau_3}^* \varphi^* E \subset H_1$ that H_1(30) reduces the operators $\{A_K\}_1^3$. Thus, the simple component H_1 is generated by the operator A_3.

Remark 3. *The above arguments are based on the fact that $\sigma_3 = I_E$. It is the reason to claim that the simple component H_1 of the Lie algebra $\{A_k\}_1^3$ is generated by the operator $A_p = nA_1 + mA_2 + A_3$ if $\sigma_p = n\sigma_1 + m\sigma_2 + \sigma_3$ is boundedly invertible* $(p = (n, m, 1) \in R^3)$.

Let $R_\pm(p) = \overline{W_\pm(p) L^2(N_\pm, R)}$. Then $\widetilde{H}_p = R_+(p) \vee R_-(p)$ in the case where the Lie algebra $\{A_k\}_1^3$ is simple; here $\widetilde{H}_p$ is the restriction of the vector-functions from $\widetilde{H}$ to the subgroup G_p. It is obvious that $\widetilde{H}_p = D_-(p) \oplus H \oplus D_+(p)$ where $D_\pm(p)$ are the restrictions of the components $u_\pm(x)$ $(17_\pm)$ to G_p. We define the map $\Psi_p : L^2(N_-, G_p) \oplus L^2(N_+, G_P) \to \widetilde{H}_p$ by the formula

$$\Psi_p \begin{bmatrix} g_- \\ g_+ \end{bmatrix} = \begin{bmatrix} W_-(p) & W_+(p) \\ 0 & 0 \end{bmatrix} \begin{bmatrix} g_- \\ g_+ \end{bmatrix}, \tag{31}$$

where $g_\pm \in L^2(N_\pm, G_p)$ are the restrictions of the solutions of the problems $(23_\pm)$ to G_p. Taking into account the fact that the $W_\pm(p)$ are isometries with respect to the metrics (21) and (25) we conclude that Ψ_p establishes an isomorphism between $\widetilde{H}_p$ and the Hilbert space

$$(32)\quad L^2 \begin{pmatrix} \sigma_p \, S_p^* \\ S_p \, \sigma_p \end{pmatrix}$$

$$= \left\{ \begin{bmatrix} g_- \\ g_+ \end{bmatrix} : g_\pm \in L^2(N_\pm, G_p), \int\limits_{G_p} < \begin{bmatrix} \sigma_p \, S_p^* \\ S_p \, \sigma_p \end{bmatrix} \begin{bmatrix} g_- \\ g_+ \end{bmatrix}, \begin{bmatrix} g_- \\ g_+ \end{bmatrix} > d\xi < \infty \right\}$$

Under this isomorphism the dilation U_{g_p} (22) by virtue of (27) turns to the translation model

$$\widehat{U}_{g_p} \begin{bmatrix} g_- \\ g_+ \end{bmatrix} = \begin{bmatrix} U_{g_p}^- \, g_- \\ U_{g_p}^+ \, g_+ \end{bmatrix}. \tag{33}$$

So the translation model of the semigroup of contractions $Z_{g_p} (g_p \in G_p^+)$ in the Hilbert space

$$\widehat{H}_p = L^2 \begin{pmatrix} \sigma_p \, S_p^* \\ S_p \, \sigma_p \end{pmatrix} \ominus [\widehat{D}_+(p) \oplus \widehat{D}_-(p)], \tag{34}$$

where

$$\widehat{D}_+(p) = \begin{bmatrix} 0 \\ L^2(N_+, G_p^-) \end{bmatrix}, \widehat{D}_-(p) = \begin{bmatrix} L^2(N_-, G_p^+) \\ 0 \end{bmatrix}$$

is given by the formula

$$Z_{g_p} = P_{H_p} \widehat{U}_{g_p} |_{\wedge H_p}, (g_p \in G_p^+) \tag{35}$$

Similarly, by defining the map

$$\Psi_p^- \begin{bmatrix} g_- \\ g_+ \end{bmatrix} = \begin{bmatrix} W_-(p) & W_+(p) - W_-(p)\sigma_p^{-1}S_p^* \\ 0 & 0 \end{bmatrix} \begin{bmatrix} g_- \\ g_+ \end{bmatrix} \tag{36}$$

from $L^2(N_-, G_p) \oplus L^2(N_+, G_p) to N_-(p) \oplus R_-^{\perp}(p) = \widetilde{H}_p$ we obtain an incoming translational representation [1, 4–6, 11–13] of the dilation $U_g(22)$ in the space

$$L^2 \begin{bmatrix} \sigma_p & 0 \\ 0 & \sigma_p - S_p\sigma_p^{-1}S_p^* \end{bmatrix},$$

given by the formula (33). In this case incoming and outgoing subspaces are of the form

$$\widehat{D}_-(p) = \begin{bmatrix} L^2(N_-, G_p^+) \\ 0 \end{bmatrix}, \widehat{D}_+(p) = \begin{bmatrix} \sigma_p^{-1}S_p^*L^2(N_+, G_p^-) \\ L^2(N_+, G_p^-) \end{bmatrix} \tag{37}$$

The translation model Z_{g_p} for $g_p \in G_p^+$ is again given by the formula (35) but now in the space

$$\widehat{H}_p = L^2 \begin{pmatrix} \sigma_p & 0 \\ 0 & \sigma_p - S_p\sigma_p^{-1}S_p^* \end{pmatrix} \ominus (\widehat{D}_+(p) \oplus \widehat{D}_-(p)), \tag{38}$$

where $\widehat{D}_\pm(p)$ from (37).

Theorem 9. *For each admissible subgroup $G_p(10)$ such that $G_p \subset (K_+ \cup K_-)$ the unitary dilation U_{g_p} (22) of the semigroup of contractions $Z_{g_p}(g_p \in G_p^+)$ has a symmetric and incoming translational representations $\widehat{U}_{g_p}$ (33) in the spaces $L^2 \begin{pmatrix} \sigma_p S_p^* \\ S_p\sigma_p \end{pmatrix}$ and $L^2 \begin{pmatrix} \sigma_p & 0 \\ 0 & \sigma_p - S_p\sigma_p^{-1}S_p^* \end{pmatrix}$ respectively. In addition the semigroup of contractions Z_{g_p} corresponds to the semigroup $\widehat{Z}_{g_p}$ (35) in H_p (34), (35).*

The translation model (32)–(38) generalizes the Pavlov model [4, 6, 11 − 13] and Sz.-Nagy-Foias model [1,4,11–13] for one-parameter semigroups to the case of the Lie algebra (1).

4. One-parameter functional model

XI. We fix an admissible subgroup G_p (10) from $K_+ \subset K_-$. The translationality along G_p enables us to proceed to new variables $y = (y_1, y_2, y_3), (x_k = x_k(y)); 1 \leq$

$k \leq 3$), where

$$(39)\quad \begin{cases} x_1 = y_1 + \frac{nsh(y_3\sqrt{nm+1})}{\sqrt{nm+1}ch(y_3\sqrt{nm+1})+(1+ny_2)sh(y_3\sqrt{nm+1})}; \\ x_2 = (y_2ch(y_3\sqrt{nm+1}) + \frac{m-y_2}{\sqrt{nm+1}}sh(y_3\sqrt{nm+1})) \\ (ch(y_3\sqrt{nm+1} + \frac{1+ny_2}{\sqrt{nm+1}}sh(y_3\sqrt{nm+1})); \\ x_3 = ln[ch(y_3\sqrt{nm+1}) + \frac{1+ny_2}{\sqrt{nm+1}}sh(y_3\sqrt{nm+1})]; \end{cases}$$

It is not hard to prove that the right translation induced by G_p acts additively only on variable $y_3 : g(y) \circ g_p(t) = g(y_1, y_2, y_3 + t), g_p(t) \in G_p$. Because of the isomorphism it is sufficient to note that

$$B_{g(y)} = \begin{bmatrix} 1 & y_2 \\ y_1 & 1+y_1y_2 \end{bmatrix} \cdot \begin{bmatrix} ch(y_3\sqrt{nm+1}) + \frac{sh(y_3\sqrt{nm+1})}{\sqrt{nm+1}} & \frac{msh(y_3\sqrt{nm+1})}{\sqrt{nm+1}} \\ \frac{nsh(y_3\sqrt{nm+1})}{\sqrt{nm+1}} & ch(y_3\sqrt{nm+1}) - \frac{sh(y_3\sqrt{nm+1})}{\sqrt{nm+1}} \end{bmatrix}$$

The Cauchy problems ($23_\pm$) in the new variables are of the form

$$(40_\pm)\quad \begin{cases} \sigma_p\widetilde{\partial}_1\vartheta_\pm(y) = (\sigma_1\partial_{y_3} + i\gamma_p, 1^\pm)\vartheta_\pm(y) \\ \sigma_p\widetilde{\partial}_2\vartheta_\pm(y) = (\sigma_2\partial_{y_3} + i\gamma_p, 2^\pm)\vartheta_\pm(y) \\ \vartheta_\pm(0,0,y_3) = \vartheta_\pm(y_3), (y_3 \in R), \end{cases}$$

where, as usual, $\sigma = n\sigma_1 + m\sigma_2 + \sigma_3$, $\gamma^\pm_{p,k} = n\gamma^\pm_{1,k} + m\gamma^\pm_{2,k} + \gamma^\pm_{3,k}$, $(k = 1,2)$. Here the differential operators $\widetilde{\partial}_1, \widetilde{\partial}_2$ are of the form

$$\begin{aligned} \widetilde{\partial}_1 = \frac{1}{2n\sqrt{nm+1}}\{&(\sqrt{nm+1}+1)^2 e^{-2y_3\sqrt{nm+1}}\partial_{y_1} \\ &+ [(e^{-2y_3\sqrt{nm+1}}(\sqrt{nm+1}+1)^2 y_2^2 - (\sqrt{nm+1}-1)^2 e^{2y_3\sqrt{nm+1}})]\partial_{y_2} \\ &+ ((\sqrt{nm+1}+1)^2 e^{-2y_3\sqrt{nm+1}} y_2 + 1)\partial_{y_3}\} \end{aligned}$$

$$\begin{aligned} \widetilde{\partial}_2 = \frac{n}{2\sqrt{nm+1}}\{&-e^{2y_3\sqrt{nm+1}}\partial_{y_1} + (e^{2y_3\sqrt{nm+1}} - y_2 e^{-2y_3\sqrt{nm+1}})\partial_{y_2} \\ &+ (1 - y_2 e^{-2y_3\sqrt{nm+1}})\partial_{y_3}\}. \end{aligned}$$

To obtain the functional model of the semigroup Z_{g_p} $(g_p \in G_p^+)$ it is necessary to take the Fourier transform with respect to the variable y_3

$$\widetilde{g}(y_1, y_2, \xi) = \frac{1}{\sqrt{2\pi}} \int_{-\infty}^{\infty} e^{-i\xi y_3} g(y_1, y_2, y_3) dy_3,$$

in the constructed above translation model (34), (35).

We restrict the functions $\vartheta_\pm(x) \in L^2(N_\pm, R)$ to the subgroup G_p. The properties (29) of the scattering operator S_p (28) imply [2,3,4,6,11,12] that

$$\vartheta_+(y_3) = (S_p\vartheta_-)(y_3) = \int_{\infty}^{\infty} S_p(y_3 - \xi)\vartheta_-(\xi)d\xi,$$

where $S_p(\xi)$ is a distribution from E to E with $S_p(\xi)=0$ for $\xi \in R_-$. The Fourier transformation sends S_p to the operator of multiplication by an operator-function

$$\widetilde{S}_p(\xi)\widetilde{\vartheta}_-(\xi)=\widetilde{\vartheta}_+(\xi).$$

By virtue of the Paley-Wiener theorem [4,6] the function $\widetilde{S_p}(\xi)$ can be holomorphically extended in C_- to a contracting (in the σ_p-metric) function $S_p(\lambda)$ for $\lambda \in C_-$. Furthermore, S_p is the characteristic function [4,6,11,12] of the operator $A_p=nA_1+mA_2+A_3$ and can be written as

$$S_p(\lambda)=I-i\varphi(A_p-\lambda I)^{-1}\varphi^*\sigma_p;\sigma_p=n\sigma_1+m\sigma_2+\sigma_3.$$

The space (32) now assumes the form

$$\text{(41)}\quad L^2\begin{bmatrix}\sigma_p & \widetilde{S}_p^* \\ \widetilde{S}_p & \sigma_p\end{bmatrix}$$

$$=\left\{\begin{bmatrix}\widetilde{g}_-(\xi)\\ \widetilde{g}_+(\xi)\end{bmatrix}:\int_R<\begin{bmatrix}\sigma_p & \widetilde{S}_p^*(\xi)\\ \widetilde{S}_p(\xi) & \sigma_p\end{bmatrix}\begin{bmatrix}\widetilde{g}_-\\ \widetilde{g}_+\end{bmatrix}(\xi);\begin{bmatrix}\widetilde{g}_-\\ \widetilde{g}_+\end{bmatrix}(\xi)>d\xi<\infty\right\}$$

while the subspaces $\widetilde{D}_\pm(p)$ are transformed into

$$\widetilde{D}_+(p)=\begin{bmatrix}0\\ H_+^2(E,\sigma_p d\xi)\end{bmatrix};\widetilde{D}_-(p)=\begin{bmatrix}H_-^2(E,\sigma_p d\xi)\\ 0\end{bmatrix}$$

where $H_\pm^2(E,\sigma_p d\xi)$ are the Hardy spaces of E-valued functions corresponding to the half-planes $C_\pm$ which belong to L^2 with respect to the measure $\sigma_p d\xi$ on the real axis. The dilation U_{g_p} now takes the form

$$\text{(42)}\qquad \widetilde{U}_{g_p}\begin{bmatrix}\widetilde{g}_-\\ \widetilde{g}_+\end{bmatrix}(\xi)=e^{it\xi}\begin{bmatrix}\widetilde{g}_-\\ \widetilde{g}_+\end{bmatrix}(\xi).$$

The space H in the symmetric spectral Pavlov representation (34) is isomorphic to

$$\text{(43)}\quad P_{S_p}=L^2\begin{bmatrix}\sigma_p\widetilde{S}_p^*\\ \widetilde{S}_p\sigma_p\end{bmatrix}\ominus[\widetilde{D}_+(p)\oplus\widetilde{D}_-(p)]$$

$$=\left\{\begin{bmatrix}\widetilde{g}_-\\ \widetilde{g}_+\end{bmatrix}(\xi)\in L^2\begin{bmatrix}\sigma_p\widetilde{S}_p^*\\ \widetilde{S}_p\sigma_p\end{bmatrix}:\begin{matrix}\widetilde{g}_-+\sigma_p^{-1}\widetilde{S}_p^*\widetilde{g}_+\in H_+^2(E,\sigma_p d\xi);\\ \sigma_p^{-1}\widetilde{S}_p\widetilde{g}_-+\widetilde{g}_+\in H_-^2(E,\sigma_p d\xi);\end{matrix}\right\}$$

and the action of Z_{g_p} (35) is transformed to the functional model

$$\text{(44)}\qquad \widetilde{Z}_{g_p}=P_{P_{S_p}}e^{it\xi}\,|_{P_{S_p}}$$

Theorem 10. *The semigroup of contractions $Z_{g_p}(g_p\in G_p^+)$ of a simple Lie algebra (1), satisfying (12), and its minimal dilation U_{g_p} (21) in H along each admissible subgroup $G_p\subset K_+\cup K_-$ are unitary equivalent to the functional model $\widetilde{Z}_{g_p}$ (44) in the space P_{S_p} (43) and $\widetilde{U}_{g_p}$ (42) in the space (41) respectively.*

The application of the Fourier transform to the incoming translation model (38) leads to the functional model of Sz.-Nagy-Foias which has the same form (42), (44) but in the another functional space.

XII. The characteristic function $S_p(\lambda)$ along with traditional properties [1, 4, 6, 11, 12, 24] of analyticity and contractibility in C_- has a number of important properties stemming from the commutation relations (1) of the Lie algebra.

Theorem 11. *The characteristic function* $S_p(\lambda)$ *satisfies the following relations:*

$$\begin{aligned} &i\{(2+nm)\partial_n + m^2\partial_m - m\lambda\partial_\lambda\}S_p(\lambda) = \sigma_p^{-1}(\sigma_1\lambda+\gamma_{p,1}^+)S_p(\lambda) \\ &\qquad -S_p(\lambda)\sigma_p^{-1}(\sigma_1\lambda+\gamma_{p,1}^-); \\ &i\{(2+nm)\partial_m + n^2\partial_n + m\lambda\partial_\lambda\}S_p(\lambda) = S_p(\lambda)\sigma_p^{-1}(\sigma_2\lambda+\gamma_{p,2}^-) \\ &\qquad -\sigma_{-1}(\sigma_2\lambda+\gamma_{p,2}^+)S_p(\lambda) \end{aligned} \tag{45}$$

where $p=(n,m,1)\in R^3$; $\partial_n=\partial/\partial_n$, $\partial_m=\partial/\partial_m$, $\partial_\lambda=\partial/\partial_\lambda$; *and the operator* σ_p *is invertible.*

Proof. Indeed, from (1) it follows that

$$[A_p,A_1] = imA_3 - 2iA_1 = imA_p - A_2im^2 - i(2+mn)A_1$$

therefore

$$[A_1,R_\lambda] = im\lambda R_\lambda^2 + imR_\lambda - im^2R_\lambda A_2R_\lambda - i(2+mn)R_\lambda A_1R_\lambda,$$

where $R_\lambda=(A_p-\lambda I)^{-1}$. Since

$$\partial_n S_p\lambda = i\varphi R_\lambda A_1R_\lambda\varphi^*\sigma_p - i\varphi R_\lambda\varphi^*\sigma_1$$

$$\partial_m S_p\lambda = i\varphi R_\lambda A_2R_\lambda\varphi^*\sigma_p - i\varphi R_\lambda\varphi^*\sigma_2$$

and taking into account the previous relation we obtain

$$\begin{aligned} &\{i(2+nm)\partial_n + im^2\partial_m\}S_p\lambda = i\varphi(R_\lambda A_1 - A_1R_\lambda \\ &+im\lambda R_\lambda^2 + imR_\lambda)\varphi^*\sigma_p + \varphi R_\lambda\varphi^*(\sigma_1(2+nm)+m^2\sigma_2). \end{aligned}$$

From the definitions of the operators $\gamma_{k,s}^-$ follows that $\gamma_{p,1}^-\varphi=\sigma_p\varphi A_1^*-\sigma_1\varphi A_p^*$ and using (9) we obtain that $R_\lambda A_1\varphi^*\sigma_p=\varphi^*\sigma_1+R_\lambda\varphi^*(\sigma_1\lambda+\gamma_{p,1}^-+2i\sigma_1-im\sigma_3)$. Similarly, from the definitions of operators $\gamma_{k,s}^+$ follows that $\varphi A_1R_\lambda=\sigma_p^{-1}(\sigma_1\lambda+\gamma_{p,1}^+)\varphi R_\lambda+\sigma_p^{-1}\sigma_1\varphi$.

Thus

$$\begin{aligned} &i\{(2+nm)\partial_n + m^2\partial_m\}S_p(\lambda) = i\varphi R_\lambda\varphi^*(\sigma_1\lambda+\gamma_{p,1}^-) \\ &-i\sigma_p^{-1}(\sigma_1\lambda+\gamma_{p,1}^+)\varphi R_\lambda\varphi^*\sigma_p = im\lambda\varphi R_\lambda^2\varphi^*\sigma_p + i\sigma_p^{-1}(\sigma_p\varphi\varphi^*\sigma_1 - \sigma_1\varphi\varphi^*\sigma_p) \end{aligned}$$

and to obtain the first equation in (45) it is necessary to use the connection between $\gamma_{k,s}^+$ *and* $\gamma_{k,s}^-$. To obtain the second equation it is necessary to take into account the second commutation relation

$$[A_p,A_2] = -inA_3 + 2iA_2 = -inA_p + in^2A_1 + i(2+nm)A_2$$

and to make the similar considerations.

Essentially, the equations (45) with the initial condition $S_p(\lambda)\mid_{n=m=0}= I - i\varphi(A_3-\lambda I)^{-1}\varphi^* = S_\lambda$ form the Cauchy problem for the function $S_p(\lambda)$. Thus, the operators $\{\sigma_k,\gamma^-_{3,k}\}$, satisfying (14), (15), (9) and the function $S(\lambda)$ form the total set of invariants of the Lie algebra $\{A_k\}_1^3$ (1).

5. A functional model of the Lie algebra

XIII. The functional model of the Lie algebra of Section 4 essentially depends on the one-parameter subgroup $G_p \subset G$. In order to construct a functional model of the Lie algebra $\{A_k\}_1^3$ (1) rather than that of linear combination $A_0 = nA_1 + mA_2 + A_3$ (for each $(p = (n,m,1) \in R^3)$ one has to take into consideration the dynamics of the way in which the function $S_p(\lambda)$ and the space (41) depend on p. We again consider the Cauchy problem $(23_\pm)$ and make the following important observation: generally speaking, $\vartheta_\pm(x)$ as a solution of the problem $(23_\pm)$ does not belong to the space $L^2(E,dx_3)$ for $x_1 \neq 0$, $x_2 \neq 0$ in spite of the fact that $\vartheta_\pm(x_3) \in L^2(E,dx_3)$. We rewrite the problem $(23_\pm)$ as follows

$(46_\pm)$

$$\begin{cases} \partial_{x_1}\vartheta_\pm(x) = \left\{[e^{2x_3}\sigma_1 - x_2^2e^{-2x_3}\sigma_2 - x_2]\partial_{x_3} + i[e^{2x_3}\gamma^\pm_{3,1} - x_2^2e^{-2x_3}\gamma^\pm_{3,2}]\right\}\vartheta_\pm(x) \\ \partial_{x_2}\vartheta_\pm(x) = [e^{-2x_3}\sigma_2\partial_{x_3} + ie^{-2x_3}\gamma^\pm_{3,2}]\vartheta_\pm(x) \\ \vartheta_\pm(0,0,x_3) = \vartheta_\pm(x_3), \end{cases}$$

where $\vartheta_\pm(x_3) \in L^2(E,dx_3)$. Taking into account (9) it is not difficult to see that differential operators on the right sides of the equations $(46_\pm)$ are self-adjoint in $L^2(E,dx_3)$. Nevertheless, in general, the operators of multiplication by e^{2x_3} and e^{-2x_3} appearing in the expressions $(42_\pm)$ lead out of the space $L^2(E,dx_3)$.

We introduce the Hilbert space

$$L^2_{(a,b)}(E,dx_3) = \left\{f(x_3) \in L^2(E,dx_3);\ \int_{-\infty}^{0} e^{2ax_3} \parallel f(x_3) \parallel^2 dx_3 + \int_0^\infty e^{2bx_3} \parallel f(x_3) \parallel^2 dx_3 < \infty\right\}$$

where $a,b \in R$; $a \le 0$, $b \ge 0$. The metric in $L^2_{(a,b)}(E,dx_3)$ is induced from $L^2(E,dx_3)$. It is easy to see that

$$L^2_{(\tilde a,\tilde b)}(E,dx_3) \subseteq L^2_{(a,b)}(E,dx_3) \subseteq L^2(E,dx_3)$$

if $[\tilde a,\tilde b] \supseteq [a,b]$; $a \le 0$, $b \ge 0$.

We apply the Fourier transform to the function space $L^2_{(a,b)}(E, dx_3)$ and consider the functions

$$(47) \qquad F_-(\xi) = \frac{1}{\sqrt{2\pi}} \int_0^\infty e^{-i\xi x_3} f(x_3) dx_3 \ ; \ F_+(\xi) = \frac{1}{\sqrt{2\pi}} \int_{-\infty}^0 e^{-i\xi x_3} f(x_3) dx_3.$$

Then in view of the properties of the function $f \in L_{(a,b)}(E, dx_3)$ we conclude that $F_-(\xi)$ can be holomorphically extended to the half-plane $Im\ \xi < b$ while $F_-(\xi)$ can be holomorphically extended to the half-plane $Im\ \xi > a$. These half-planes have a common part, the strip $\Pi_{(a,b)} = \{\xi \in C \ ; \ a < Im\ \xi < b\}$, therefore the function $F(\xi) = F_-(\xi) + F_+(\xi)$ is regular in the interior of this strip $\Pi_{(a,b)}$ and in addition

$$\sup_{\xi \in \Pi_{(a,b)}} \int_{-\infty}^{\infty} \| F(\xi) \|^2 \, d\alpha < \infty \quad (\alpha = Re\ \xi).$$

Consequently, the Fourier transform establishes an isomorphism between

$$L^2_{(a,b)}(E, dx_3)$$

and the Hardy space $H^2(E, \Pi_{(a,b)})$ of E-valued holomorphic in the strip $\Pi_{(a,b)}$ functions $F(\xi)$. It is obvious that

$$H^2(E, \Pi_{(a,b)}) = H^2_+(E, \Pi_{(a,b)}) \oplus H^2_-(E, \Pi_{(a,b)})$$

where $H^2_\mp(E, \Pi_{(a,b)})$ denote the Hardy spaces of functions from $H^2(E, \Pi_{(a,b)})$ that are extended analytically from $\Pi_{(a,b)}$ to the half-planes $\pm Im\ \xi > 0$.

We now return to the Cauchy problem $(46_\pm)$. The resolving operator $\vartheta_\pm(x_3) \to \vartheta_\pm(x, x_2, x_3)$ of this problem is a two-parameter semigroup on x_1 and x_2 the generators of which are the differential operators from the right sides of equations and commute with each other by virtue of (13). To confine the action of this semigroup to a single space we restrict ourselves to initial conditions $\vartheta_\pm(x_3)$ from $L^2_{(-2,2)}(E, dx_3)$ which imply in view of remarks made above that $\vartheta_\pm(x) \in L^2_{(-2,2)}(E, dx_3)$ for each fixed x_1, x_2. The Fourier transform establishes an isomorphism between $L^2_{(-2,2)}(E, dx_3)$ and the Hardy space $H^2(E, \Pi)$ in the strip $\Pi = \Pi_{(-2,2)}$. It is easy to see that Fourier transform (47) maps the functions $e^{\pm 2x_3} \frac{1}{i} f'(x_3)$ and $e^{\pm 2x_3} f(x_3) (f \in L^2_{(\pm)}(E, dx_3))$ to functions $(\xi \pm 2i)\widetilde{f}(\xi \pm 2i)$ and $\widetilde{f}(\xi \pm 2i)$ in $H^2(E, \Pi)$ respectively.

It is then not difficult to write down a functional model of the Lie algebra $\{A_k\}_1^3$ (1) in the spectral representation of the operator A_3 (subgroup G_0). We denote by P_Π the Hilbert space of vector-functions $\begin{bmatrix} \widetilde{g}_- \\ \widetilde{g}_+ \end{bmatrix}(\xi)$ which differs from P_{S_p} (43) for p=(0,0,1) in that $L^2(E)$ and $H^2_\pm(E)$ are replaced in a natural way by $H^2(E, \Pi)$ and $H^2_\pm(E, \Pi)$ respectively, $\widetilde{g}_\mp(\xi) \in H^2(E, \Pi)$. It is not hard to see that in this case in view of equations $(23_\pm)$ and $(46_\pm)$ and the form of differential operators $\{\partial_k\}_1^3$ (2) the operators A_1, A_2, A_3 (1) of Lie algebra correspond to operators $e^{2x_3}(-i\sigma_1\partial_{x_3}+\gamma_{3,1}^{\pm})$, $e^{-2x_3}(-i\sigma_2\partial_{x_3}+\gamma_{3,2}^{\pm})$ and $-i\sigma_{x_3}$ respectively. Thus the functional model of the Lie algebra assumes the following form

$$\widetilde{A}_1\begin{bmatrix}\widetilde{g}_-\\ \widetilde{g}_+\end{bmatrix}(\xi)=P_{P_\Pi}s\begin{bmatrix}\sigma_1\xi+\gamma_{3,1}^- & 0\\ 0 & \sigma_1\xi+\gamma_{3,1}^+\end{bmatrix}\begin{bmatrix}\widetilde{g}_-\\ \widetilde{g}_+\end{bmatrix}(\xi);$$

$$\widetilde{A}_2\begin{bmatrix}\widetilde{g}_-\\ \widetilde{g}_+\end{bmatrix}(\xi)=P_{P_\Pi}s^*\begin{bmatrix}\sigma_2\xi+\gamma_{3,2}^- & 0\\ 0 & \sigma_2\xi+\gamma_{3,2}^+\end{bmatrix}\begin{bmatrix}\widetilde{g}_-\\ \widetilde{g}_+\end{bmatrix}(\xi); \quad (48)$$

$$\widetilde{A}_3\begin{bmatrix}\widetilde{g}_-\\ \widetilde{g}_+\end{bmatrix}(\xi)=P_{P_\Pi}\xi=\begin{bmatrix}\widetilde{g}_-\\ \widetilde{g}_+\end{bmatrix}(\xi),$$

where s (s^*) is the shift operator $sf(\xi)=f(\xi+2i)$ ($s^*f(\xi)=f(\xi-2i)$).

Theorem 12. *The simple Lie algebra $\{A_k\}_1^3$ (1) satisfying (12) and under conditions (9) and (14) hold is unitarily equivalent to the functional model (48) in the Hilbert space P_Π.*

A similar theorem can be stated by using the incoming translation representation (38). In particular, the following theorem is true.

Theorem 13. *Let a Lie algebra of the linear operators $\{A_k\}_1^3$ satisfy the assumptions of Theorem 12 and let the characteristic function $S^*(\xi)$ of the operator A_3 be inner. Then the Lie algebra $\{A_1,A_2,A_3\}$ (1) is isomorphic to the functional model*

$$\begin{aligned}&\widetilde{A}_1\widetilde{g}_-(\xi)=P_k(\sigma_1(\xi+2i)+\gamma_{3,1}^-)\widetilde{g}_-(\xi+2i);\\ &\widetilde{A}_2\widetilde{g}_-(\xi)=P_k(\sigma_2(\xi-2i)+\gamma_{3,2}^-)\widetilde{g}_-(\xi-2i);\\ &\widetilde{A}_3\widetilde{g}_-(\xi)=P_k\xi\widetilde{g}_-(\xi)\end{aligned} \quad (49)$$

6. Functional model on a Riemann surface

XIV. We assume that $\dim E=r<\infty$ and consider in E following pencils of operators

(50) $L_1(\xi)=\sigma_1\xi+\gamma_{3,1},\ L_2(\xi)=\sigma_2\xi+\gamma_{3,2},$

$$N(\xi)=L_1^*(\xi)L_2(\xi)+\frac{i}{2}\xi I=\sigma_1\sigma_2\xi^2+(\sigma_1\gamma_{3,2}+\gamma_{3,1}^*\sigma_2+\frac{i}{2}I)\xi+\gamma_31^*\gamma_32,$$

where $\xi\in R$, $\gamma_3,k=\gamma_3,k^\pm$, $(k=1,2)$. We note that (9) implies that

$$L_1^*(\xi)=L_1(\xi+2i),\ L_2^*(\xi)=L_2(\xi-2i)$$

and in addition $\sigma_1\sigma_2\neq 0$, because of 2) in 14.

Lemma 1. The following relations hold for the operator pencils $L_1(\xi),L_2(\xi)$ and $N(\xi)$ (50):

$$\begin{aligned}&1.\ N(\xi)=N^*(\xi);\\ &2.\ N(\xi)L_1^*(\xi)=L_1^*(\xi)[N(\xi+2i)+i(\xi+i)I];\\ &3.\ N(\xi)L_2^*(\xi)=L_2^*(\xi)[N(\xi-2i)-i(\xi-i)I],\end{aligned}\qquad \text{where } \xi\in R. \quad (51)$$

Proof. To prove 1., it is necessary to show that

$$L_1(\xi+2i)L_2(\xi)+\frac{i}{2}\xi I=L_2(\xi-2i)L_1(\xi)-\frac{i\xi}{2}I$$

This equation after equating the corresponding coefficients at ξ gives 1)–3) (14). Taking into account this relation we obtain that

$$N(\xi)L_1(\xi+2i)=L_1(\xi+2i)[L_2(\xi)L_1(\xi+2i)+\frac{i}{2}I]$$

To prove 2. (51) it is sufficient to note that

$$L_2(\xi)L_1(\xi+2i)=N(\xi+2i)+(\frac{i}{2}\xi-1)I$$

The relation 3. (51) is proved in a similar way. We denote now by $L_k^{\pm}$ and $N_{\pm}(\xi)$ the operator pencils (50) where instead of $\gamma_{3,k}$ appear the operators $\gamma_{3,k}^{\pm}$, $(k=1,2)$. In the finite-dimensional space $E(\dim E=r<\infty)$ we choose the orthogonal basis of eigenvectors $\{h_k^{\pm}\}_1^r$ of the self-adjoint operator $N_{\pm}(\xi)$ for $\xi\in R$,

$$N_{\pm}(\xi)h_k^{\pm}(\xi)=\omega_k^{\pm}(\xi)h_k^{\pm}(\xi) \tag{52}$$

$(1\le k\le r)$. We normalize the vectors by the condition $h_{k,r}^{\pm}(\xi)=1$, where $h_{k,r}^{\pm}$ is the r-th component of the vector $h_k^{\pm}$. The vector $h_k^{\pm}(\xi)$ is non-trivial if $Q_{\pm}(\xi,\omega^{\pm})=0$, where

$$Q_{\pm}(\xi,\omega^{\pm})=\det[N_{\pm}(\xi)-\omega^{\pm}I]. \tag{53}$$

In what follows we assume the polynomials $Q_{\pm}(\xi,\omega^{\pm})$ (53) are nonsingular algebraic curves in C^2, that is their complex gradient grad $Q_{\pm}(\xi,\omega^{\pm})\neq 0$ for all $(\xi,\omega^{\pm})\in Q_{\pm}$.

$$Q_{\pm}=\{(\xi,\omega^{\pm})\in C^2;Q_{\pm}(\xi,\omega^{\pm})=0\} \tag{54}$$

This means that the polynomial $Q_{\pm}(\xi,\omega^{\pm})$ is irreducible that is its roots $\{\omega_k^{\pm}(\xi)\}_1^r$ are distinct with the natural exception of the branch points. Moreover, $\{\omega_k^{\pm}(\xi)\}_1^r$ are the branches of an r-valued algebraic function [17].

Note that $Q_{\pm}(\xi,\omega^{\pm})$ (53) determines an l-valued algebraic function $\xi(\omega^{\pm})$, where $l=2rank\sigma_1\sigma_2$ (for $\sigma_1\sigma_2\neq 0$) or $l=rank[(\gamma_3,1^{\pm})^*\sigma_2+\sigma_1\gamma_3,2^{\pm}+\frac{i}{2}I]$, naturally for $\sigma_1\sigma_2=0$. Therefore, the functions $\xi(P_{\pm})$ and $\omega^{\pm}(P_{\pm})$ on the surface $Q_{\pm}(54)$ have r and l poles respectively accounting for multiplicity.

The eigenvectors $h_k^{\pm}(\xi)(52)$ form one-dimensional bundles in E over $Q_{\pm}(54)$ (by virtue of the irreducibility of $Q_{\pm}(\xi,\omega^{\pm})(53)$).

$$h^{\pm}(P_{\pm})\in Ker[N_{\pm}(\xi)-\omega^{\pm}I] \tag{55}$$

which are rational functions of $P_{\pm}(\xi,\omega^{\pm})\in Q_{\pm}(54)$ and are normalized as indicated above.

We will compute the number (accounting for multiplicity) of poles $h^{\pm}(P_{\pm})$. To this end we consider $m_{\pm}(P_{\pm})=[\det M^{\pm}(P_{\pm})]^2$, where $M^{\pm}(P_{\pm})$ is the matrix whose columns are formed by the $h^{\pm}(P_{\pm})$ (52) which are the eigenvectors of $N_{\pm}(\xi)$ corresponding to the eigenvalues $\{\omega_k^{\pm}(\xi)\}_1^r$. Clearly, the poles of the meromorphic function $m_\pm(P_\pm)$ for $P_\pm \in Q_\pm$ coincide with the poles $h^\pm(P_\pm)$. The zeros of $m_\pm(P_\pm)$ are determined by those points ξ with merging eigenvalues $\{\omega_k^\pm(\xi)\}_1^r$ of the pencil $N_\pm(\xi)$. For finite ξ the coincidence is possible only at the branch points of $Q_\pm$, while for $\xi = \infty$ the number of 'merging' $\omega_k^\pm(\xi)$ equals $2(r-1)$ because rank $N_\pm(\xi) \to l = rank\ \sigma_1\sigma_2$ for $\xi \to \infty$ and $\sigma_1\sigma_2 \neq 0$ by virtue of 2) (14). Therefore $\nu + (2r - l_0) = 2N$, where N is a number of branch poles $h^\pm(P_\pm)$ and ν is a number of branch points $Q_\pm$ (accounting for multiplicity) because the number of zeros equals to the number of poles of meromorphic function $m_\pm(P_\pm)$ on $Q_\pm$. Using the Riemann-Hurwitz formula for a surface $Q_\pm$ of genus $g_\pm$ we obtain in the $2l$-sheeted case that $2g_\pm = \nu - 4l + 2$. Therefore, $2g_\pm + 4l = \nu + 2$, $2g_\pm + 4l = 2N - 4(r-l) + 2$ which means $N = g_\pm + 2r - 1$. Thus

$$N = g_\pm + 2r - 1$$

We identify on the Riemann surfaces $Q_\pm$ (54) the proper analogues of the half-planes $C_\pm$ of the real axis R and of the strip Π:

$$\begin{aligned} &C_\pm(Q_\pm) = \{P_\pm = (\xi,\omega^\pm) \in Q_\pm;\ Im\ \xi > 0\};\\ &R(Q_\pm) = \partial C_\pm(Q_\pm);\ \Pi(Q_\pm) = \{P_\pm = (\xi,\omega^\pm \in Q_\pm;\ \mid Im\ \xi \mid < 2\} \end{aligned} \tag{57}$$

For the reasons which will become clear below we shall refer to $R(Q_\pm)$ as cuts on the Riemann surface $Q_\pm$.

We expand each vector-function $f(\xi) \in H^2(E,\Pi)$ with respect to the orthogonal basis $h^\pm(P_\pm^K)$ (55), where $P_\pm^k = (\xi, \omega_k^\pm(\xi)) \in R(Q_\pm)$,

$$f(\xi) = \sum_k h^\pm(P_\pm^k) \parallel h^\pm(P_\pm^k \parallel_E^{-2} g^\pm(P_\pm^k),$$

where the scalar functions $g^\pm(P_\pm^k)$ are of the form $g_\pm(P_\pm^k) = \langle f(\xi), h^\pm(P_\pm^k)\rangle_E$ and are the branches of the same algebraic function $g^\pm(P_\pm)$, ($\|\cdot\|_E$, $\langle.,.\rangle_E$ in the sense of the Euclidean space E). We now define the Hilbert space $H^2(h^\pm,\Pi)$ formed by the vector-functions $f(P_\pm) = h^\pm(P_\pm) \parallel h^\pm(P_\pm) \parallel_E^{-2} g^\pm(P_\pm)$ such that $g^\pm(P_\pm)$ are holomorphic in $\Pi(Q_\pm)$ and also

$$\sup_{|t|<2} \int_{l_t} \mid g^\pm(P_\pm) \mid^2 \frac{dx}{\parallel h^\pm(P_\pm) \parallel^2} < \infty$$

where $x = Re\ \xi$, $l_t = \{P_\pm \in \Pi(Q_\pm);\ Im\ \xi(P_\pm) = t\}$; here the curves l_t contract to the slits $R(Q_\pm)$ as $t \to 0$ and functions $g^\pm(P_\pm)$ have the same singularities as $h^\pm(P_\pm)$ (with multiplicities accounted for).

XV. We will compute the action of the linear pencils $L_k^\pm(\xi)$ (50) on the eigenvectors $h_k^\pm(\xi)$ (52) of the pencil $N_\pm(\xi)$. It follows from Lemma 1 that

$$N_\pm(\xi)[L_1^\mp(\xi+2i)h_k^\pm(\xi+2i)] = (\omega_k^\pm(\xi+2i) + i\xi - 1)[L_1^\mp(\xi+2i)h_k^\pm(\xi+2i)].$$

Thus, $L_1^\mp(\xi+2i)h_k^\pm(\xi+2i)$ is the eigenvector of the pencil $N_\pm(\xi)$ and because of simplicity of the spectrum of $N_\pm(\xi)$ the vector $L_1^\mp(\xi+2i)h_k^\pm(\xi+2i)$ is proportional to some vector $h_s^{\pm}(\xi)$. Hence, we proved that

$$(58_1) \qquad L_1^{\pm}(\xi+2i)h_k^{\pm}(\xi+2i)=\mu_k^{\pm}(\xi)h_{\pi(k)}^{\pm}(\xi)$$

where π is an element of the permutation group $(1,2,\ \dots\ ,r)$, in particular $\pi(k)=S$ and $\mu_k^{\pm}(\xi)$ is some algebraic function. Similarly

$$(58_2) \qquad L_2^{\pm}(\xi-2i)h_k^{\pm}(\xi-2i)=\nu_k^{\pm}(\xi)h_{\sigma(k)}^{\pm}(\xi)$$

where $\nu_k^{\pm}(\xi)$ is algebraic function and σ_k is an element of the same permutation group $(1,2,\ \dots\ ,r)$. Using the form of $N_{\pm}(\xi)$, (52) and formulas $(58_{1,2})$ we obtain:

$$\begin{aligned}\omega_k^{\pm}h_k^{\pm}(\xi)&=(L_1^{\pm}(\xi+2i)L_2^{\pm}(\xi)+\frac{i}{2}\xi I)h_k^{\pm}(\xi)\\&=L_1^{\pm}(\xi+2i)\nu_k^{\pm}(\xi+2i)h_{\sigma(k)}^{\pm}(\xi)+\frac{i}{2}\xi h_k^{\pm}(\xi)\\&=\nu_k^{\pm}(\xi+2i)\mu_k^{\pm}(\xi)h_{\pi\sigma(k)}^{\pm}(\xi)+\frac{i}{2}\xi h_k^{\pm}(\xi)\end{aligned}$$

Hence $\sigma=\pi^{-1}$ and in addition,

$$\omega_k^{\pm}(\xi)\mu_k^{\pm}(\xi)\nu_k^{\pm}(\xi+2i)+\frac{i}{2}\xi==\mu_k^{\pm}(\xi-2i)\nu_k^{\pm}(\xi)-\frac{i}{2}\xi.$$

We also note that $\mu_k^{\pm}(\xi)$ and $\nu_k^{\pm}(\xi)$ are the branches of l_1- and l_2- valued algebraic function $\mu^{\pm}(\xi)$ and $\nu^{\pm}(\xi)$ where $l_k=rank\ \sigma_k$. Therefore

$$sL_1^{\pm}(\xi)f(\xi)=\sum_k\frac{h_k^{\pm}(\xi)}{\|h_k^{\pm}(\xi)\|_E^2}\cdot\frac{\|h_k^{\pm}(\xi)\|_E}{\|h_{\pi^{-1}(k)}^{\pm}(\xi+2i)\|_E^2}\mu_{\pi^{-1}(k)}^{\pm}(\xi)g_{\pi^{-1}(k)}^{\pm}(\xi+2i)$$

$$sL_2^{\pm}(\xi)f(\xi)=\sum_k\frac{h_k^{\pm}(\xi)}{\|h_k^{\pm}(\xi)\|_E^2}\cdot\frac{\|h_k^{\pm}(\xi)\|_E}{\|h_{\pi(k)}^{\pm}(\xi-2i)\|_E^2}\nu_{\pi(k)}^{\pm}(\xi)g_{\pi^{-1}(k)}^{\pm}(\xi-2i)$$

where $f(\xi)=\sum_k h_k^{\pm}(\xi)\|h_k^{\pm}(\xi)\|_E^{-2}\,g_k^{\pm}(\xi)$. Since the basis $h_k^{\pm}(\xi)$ is fixed, functions $f(\xi)$ are defined by their scalar components $g_k^{\pm}(\xi)$. Hence the operators $SL_1^{\pm}(\xi)$ and $SL_2^{\pm}(\xi)$ act on the scalar components as follows

$$\begin{aligned}sL_1^{\pm}(\xi(P_{\pm}))g^{\pm}(P_{\pm})&=\lambda^{\pm}(P_{\pm})\alpha Sg^{\pm}(P_{\pm})\\sL_2^{\pm}(\xi(P_{\pm}))g^{\pm}(P_{\pm})&=W^{\pm}(P_{\pm})\alpha^*S^*g^{\pm}(P_{\pm})\end{aligned}$$

where $P_{\pm}=(\xi,\omega^{\pm})\in Q_{\pm}$ (54); s and s^* are shifts

$$sg^{\pm}(P_{\pm})=g^{\pm}(\xi+2i,\omega^{\pm}(\xi+2i))\ ,\ s^*g^{\pm}(P_{\pm})=g^{\pm}(\xi-2i,\omega^{\pm}(\xi-2i))$$

α and α^* are automorphisms (commuting with s and s^*) induced by the permutations π^{-1} and π finally

$$\lambda^{\pm}(P_{\pm})=\|h^{\pm}(P_{\pm})\|_E^2\|\alpha h^{\pm}(\xi+2i,\omega^{\pm}(\xi+2i)\|_E^2\ \alpha\mu^{\pm}(P_{\pm});$$

$$W^{\pm}(P_{\pm})=\|h^{\pm}(P_{\pm})\|_E^2\|\alpha^*h^{\pm}(\xi-2i,\omega^{\pm}(\xi-2i)\|_E^2\ \alpha^*\nu^{\pm}(P_{\pm}).$$

Taking all above into account the functional model (48) can be now written down as follows

(59)
$$\begin{aligned}
\widetilde{A}_1 \begin{bmatrix} g^-(P_-) \\ g^+(P_+) \end{bmatrix} &= P_{P_\Pi} \begin{bmatrix} \lambda^-(P_-)\alpha S g^-(P_-) \\ \lambda^+(P_+)\alpha S g^+(P_+) \end{bmatrix};\\
\widetilde{A}_2 \begin{bmatrix} g^-(P_-) \\ g^+(P_+) \end{bmatrix} &= P_{P_\Pi} \begin{bmatrix} W^-(P_-)\alpha^* S^* g^-(P_-) \\ W^+(P_+)\alpha^* S^* g^+(P_+) \end{bmatrix};\\
\widetilde{A}_3 \begin{bmatrix} g^-(P_-) \\ g^+(P_+) \end{bmatrix} &= P_{P_\Pi} \begin{bmatrix} \xi(P_-)g^-(P_-) \\ \xi(P_+)g^+(P_+) \end{bmatrix};
\end{aligned}$$

where $P_\pm = (\xi, \omega^\pm(\xi)) \in Q_\pm$, (54). In order to state the final result we have to write down the action of the characteristic function $S(\xi)$ in terms of the components $g^\pm(P_\pm)$). We act by $S(\xi)$ on function

$$f(P_-) = h^-(P_-) \parallel h^-(P_-) \parallel_E^{-2} g^-(P_-) \in H^2(h^-, \Pi)$$

and obviously obtain that

$$s(\xi)f(P_-) = \sum_{\xi(P_-)=\xi(P_+)} g^-(P_-) \parallel h^-(P_-) \parallel_E^{-2} \parallel h^+(P_+) \parallel_E^{-2} \cdot$$
$$\cdot \langle S_p(\xi)h^-(P_-), h^+(P_+)\rangle_E h^+(P_+).$$

where $P_\pm = (\xi, \omega^\pm) \in Q_\pm$ (54). We introduce the kernel

$$\Theta(P_-, P_+) = \parallel h^-(P_-) \parallel_E^{-2} \langle S(\xi)h^-(P_-), h^+(P_+)\rangle_E.$$

Then the action of the characteristic function $S(\xi)$ written down in terms of the scalar functions $g^-(P_-) \to g^+(P_+)$ has the form

(60)
$$(\Theta g^-)(P_+) = \sum_{\xi(P_-)=\xi(P_+)} g^-(P_-)\Theta(P_-; P_+)$$

where $P_\pm = (\xi, \omega^\pm) \in Q_\pm$ (54) and $\xi \in R$. As a result we obtain the following theorem.

Theorem 14. *Let* $\dim E = r < \infty$ *and let the curves* $Q_\pm$ *(54) in* C^2 *be non-singular. Then there exist vector fields* $h^\pm(P_\pm)$ *(55) on* $Q_\pm$ *with non-special divisors of poles* $D_\pm$*, deg* $D_\pm N$ *(56) such that the functional model of the simple Lie algebra* $\{A_k\}_1^3$ *(1), satisfying (12), has the form (59) where* $\lambda^\pm(P_\pm)$, $\omega^\pm(P_\pm)$, $\xi^\pm(P_\pm)$ *are the meromorphic on* $Q_\pm$ *functions, operators* s *and* s^* *are shifts*

$$sg(\xi, \omega^\pm(\xi)) = g(\xi + 2i, \omega^\pm(\xi + 2i)); \; s^*g(\xi, \omega^\pm(\xi)) = g(\xi - 2i, \omega^\pm(\xi - 2i)),$$

and α *and* α^* *are automorphisms commuting with* s *and* s^**. The model space* $\widetilde{H}$ *is of the form*

$$\widetilde{H} = \left\{ \begin{bmatrix} g^-(P_-) \\ g^+(P_+) \end{bmatrix}; \begin{cases} g^-(P_-) \in H_+^2(h^-, \Pi) \ominus \Theta^* H_+^2(h^+, \Pi); \\ g^+(P_+) \in \Delta H^2(h^-, \Pi) \ominus \Delta H_+^2(h^+, \Pi); \end{cases} \right\}$$

where $\Delta^2 = 1 - \Theta\Theta^*$*, and* Θ *is given by the formula (60).*

By uniformizing [19] each curve $Q_\pm$ (53), we obtain functions $\xi(u_\pm)$ and $\omega^\pm(u_\pm)$, that are meromrphic in a disc K ($u_\pm \in K$), and K serves as the universal covering [19] for $Q_\pm$. The functions $\xi(u_\pm)$ and $\omega^\pm(u_\pm)$ are automorphic with respect to a certain group of fractional-linear transformations of K that is isomorphic to the fundamental group $F_\pm$ of the Riemann surface $Q_\pm$.

So we can assume that ξ and $\omega^\pm$ are defined in a fundamental domain $\Gamma_\pm$ (Poincaré polygon) with its sides "pasted together in a proper way" via the group $F_\pm$[19]. We denote by $\Gamma_\pm(Q_\pm), \Gamma_\circ(Q_\pm), \Gamma_\Pi(Q_\pm)$ the inverse images of $C_\pm(Q_\pm)$, $R(Q_\pm)$, $\Pi(Q_\pm)$ under the uniformization map $u_\pm \to (\xi(u_\pm), \omega^\pm(u_\pm))$. The natural reformulation of the definitions $H^2(h^\pm, \Pi)$ and $H^2_\pm(h^\pm, \Pi)$ leads to the following theorm which we state in the case where Θ^* is inner function (the corresponding analogue of Theorem 13).

Theorem 15. *Let Θ^* (60) be inner function and the assumptions of the theorem 14 hold. Then there exist vector fields $h^\pm(u_\pm)$ in the fundamental domains $\Gamma_\pm$ of the curves $Q_\pm(54)$ with non-special divisors of poles $D_\pm(degD_\pm = N(56))$ such that a functional model of Lie algebra $\{A_k\}_1^3$ (1) is of the form*

$$\begin{aligned} \widetilde{A}_1 f(u) &= P_{\widetilde{K}}\lambda(u) s\alpha f(u); \\ \widetilde{A}_2 f(u) &= P_{\widetilde{K}} w(u) s^*\alpha^* f(u); \\ \widetilde{A}_3 f(u) &= P_{\widetilde{K}}\xi(u) f(u), \end{aligned} \tag{61}$$

where the model space $\widetilde{K}$ has the form

$$K = H^2_+(h^-, \Gamma_\Pi) \ominus \Theta^* H^2_+(h^+, \Gamma_\Pi),$$

the functions $\lambda(u), w(u)$ are meromorphic in Γ, s and s^*- are "shifts" $sf(u) = f(u_i), s^* f(u) = f(u\_i)$ and in addition $\xi(u_i) = \xi(u) + 2i, \xi(u - i) = \xi(u) - 2i$ and, finally, α and α^* are the automorphisms of Γ commuting with s, s^*.

XVI. Here we consider a number of examples. We assume that the pencil $N(\xi)$ (50) is of the form

$$N(\xi) = \xi^2\sigma_1\sigma_2 + \xi(\sigma_1\gamma_{3,2} + \gamma^*_{3,1}\sigma_2 + \frac{i}{2}I) + \gamma^*_{3,1}\gamma_{3,2} \tag{62}$$

and is self-adjoint for $\xi \in R$ by virtue of (14). From these conditions in particular follows that

$$tr(\sigma_1\gamma_3, 2 + \sigma_2\gamma_{3,1}) = 0; tr\sigma_1\sigma_2 = -\frac{r}{4}; (r = \dim E). \tag{63}$$

Example 1. Let $\dim E = 1$. Then $\sigma_k \in R$, $\gamma^\pm_{3,k} = \gamma_{3,k} \in C, (k = 1, 2)$ are scalars and it follows from (9) that $\gamma_{3,1} = \alpha_1 - i\sigma_1, \gamma_{3,2} = \alpha_2 + i\sigma_2, (\alpha_1, \alpha_2 \in R)$. The relation (63) implies that $\sigma_1\sigma_2 = -\frac{1}{4}, \alpha_1\sigma_2 + \alpha_2\sigma_1 = 0$. The pencil $N(\xi)$(63) in this case has the form

$$N(\xi) = -\frac{\xi^2}{4} + \alpha_1\alpha_2 + \frac{1}{4}$$

which means that the curve $Q(54)$ is of the form

$$\xi^2 + 4\omega = 1 + 4\alpha_1\alpha_2$$

that is it forms the two-sheeted Riemann surface (of genus $g=0$), obtained by pasting together two ω-sheets C cross-wise along the slit $(-\infty, \alpha_1\alpha_2+\frac{1}{4})$. Since $Im\xi$ changes its sign on the slit alone, the $C_\pm(Q)$ are the corresponding sheets of Q, and $R(Q)$ are the proper banks of the slit. The curves

$$4y^2 - t^4 = (1+4\alpha_1\alpha_2 - 4x)t^2$$

form the analogue of the strip $\Pi(Q)$ in ω-plane C, $\omega = x+iy, t = Im\xi, \mid t \mid \le 2$, which contract to the slits $R(Q)$ as $t\to 0$. Thus, if the characteristic function $S(\xi)$ of the operator A_3, is inner then the functional model of $\{A_k\}_1^3$ (1) in the space

$$K = H_+^2(\Pi) \ominus S^*(\xi)H_+^2(\Pi)$$

has the form

$$\widetilde{A}_1 f(\xi) = P_k(\sigma_1(\xi+2i)+\gamma_{3,1})f(\xi+2i);$$
$$\widehat{A}_2 f(\xi) = P_k(\sigma_2(\xi-2i)+\gamma_{3,2})f(\xi-2i);$$
$$A_3 f(\xi) = P_k\xi f(\xi).$$

Example 2. Let $\dim E = 2$ and matricies $\sigma_1, \sigma_2, \gamma_{3,2}^{\pm} = \gamma_2 + i\sigma_2,\ \gamma_{3,1}^{\pm} = \gamma_1 - i\sigma_1$ are of the form

$$\sigma_1 = \alpha\begin{bmatrix} 1 & 0 \\ 0 & 1 \end{bmatrix}; \sigma_2 = -\frac{1}{4\alpha}\begin{bmatrix} 1 & 0 \\ 0 & 1 \end{bmatrix};$$

$$\gamma_2 = \begin{bmatrix} p & q \\ q & p \end{bmatrix}; \gamma_1 = 4\alpha\begin{bmatrix} \alpha p & \alpha q - b \\ \alpha q - b & \alpha p \end{bmatrix}$$

where $0<\alpha;\ \alpha, p, q, b \in R$. In this case the quadratic pencil $N(\xi)$ is of the form

$$N(\xi) = \begin{bmatrix} -\frac{\xi^2}{4}+n & \xi b + r \\ -\xi b + r & -\frac{\xi^2}{4}+n \end{bmatrix}$$

and

$$p^2 + \left(q - \frac{b}{2\alpha}\right)^2 = \frac{4n+4b^2-1}{16\alpha^2};\ p\left(q-\frac{b}{2\alpha}\right) = \frac{r}{8\alpha^2},\ 2\alpha\sqrt{4n+b^2-1} > r > o.$$

The algebraic curve Q is defined by the equation

$$\left(\frac{\xi^2}{4} - w - n\right)^2 = (b\xi + r)^2.$$

It means that the eigenvalues of the pencil $N(\xi)$ are

$$w_\pm = \frac{\xi^2}{4} - n \pm (b\xi + r).$$

The correspondent Riemann surface is of the genus $g=0$ and can be obtained from two w-sheets C patched together criss-cross along the cuts $[-n\pm r - b^2, \infty)$. An analogue of the strip $\Pi(Q)$ is given by the set formed by the family of curves

$$y^2 = t^4 + t^2x^2,$$

where $w = x+iy,\ |t| \le 2\,(Im\xi = t)$.

The eigenvectors of the pencil $N(\xi)$ are of the form

$$h(\xi, w) = \left[4\frac{b\xi + r}{\xi^2 + 8w - 4n}; \ 1\right]$$

where (ξ, w) is a point of the curve Q. It is not hard to see that the pencils $L_1(\xi)$ and $L_2(\xi)$ interchange the eigenvectors $h(\xi, w_\pm)$ after corresponding multiplying on scalars (they have rather cumbersome form to give it here).

References

[1] B.Sz.-Nagy and C. Foias, *Harmonic analysis of operators on Hilbert space*, North-Holland Publ. Co., Amsterdam, and Akademiai Kiado, Budapest, 1970.

[2] P.D. Lax and R.S. Phillips, *Scattering theory*, Academic Press, NY, 1967.

[3] P.D. Lax and R.S. Phillips, *Scattering theory for automorphic functions*, Annals of Math. Studies, no. 87, Princeton University Press, Princeton, NJ, 1976.

[4] B.S. Pavlov, Dilation theory and spectral analysis of non-self-adjoint differential operators, *Theory of operators in linear spaces* (Proc. Seventh Winter school, Drogobych, 1974), Tsentral. Ekonom.-Mat. Inst. Akad. Nauk SSSR, Moscow 1976, pp.3–69; English transl. in *Amer. Math. Soc. Transl.* (2) **115** (1980).

[5] B.S. Pavlov and L.D. Faddeev, Scattering theory and automorphic functions, *Zap. Nauchn. Sem. Leningrad. Otdel. Mat. Inst. Steklov* (LOMI) **27** (1972), 161–193; English transl. in *J. Soviet Math.* **5:4** (1975).

[6] B.S. Pavlov, Spectral analysis of a dissipative singular Schrödinger operator in terms of a functional model, *Itogi Nauki i Tekhniki: Ser. Sovr. Probl. Mat.: Fund. Napravleniya* vol. 65, VINITI, Moscow 1991, pp. 95–163 (in Russian).

[7] B.S. Pavlov and S.I. Fedorov, The group of shifts and harmonic analysis on a Riemann surface of genus one, *Algebra i Analiz* **1:2** (1989),132–169; Engish transl. in *Leningrad Math J.* **1** (1990).

[8] S.I. Fedorov, Harmonic analysis in a multiply connected domains. I, *Mat. Sb.* **181:6** (1990), 833–864 (in Russian).

[9] S.I. Fedorov, Harmonic analysis in a multiply connected domains, II. *Mat. Sb.* **181:7** (1990), 867–910 (in Russian).

[10] M.A. Semenov-Tyan-Shanskii, Harmonic analysis on Riemannian symmetric spaces of negative curvature, and scattering theory, *Izv. Akad. Nauk SSSR Ser. Mat.* **40** (1976), 562–592; English transl. in *Math. USSR-Izv.* **10** (1976).

[11] V.A. Zolotarev, Time cones and a functional model on a Riemann surface, *Mat. Sb.* **181** (1990), 965–995; English transl. in *Math. USSR-Sb.* **71** (1992).

[12] V.A. Zolotarev, The Lax-Phillips scattering scheme on groups and a functional model of a Lie algebra, *Mat. Sb.* **183**:5 (1992), 115–144; English transl. in *Russian Acad. Sci. Sb. Math.* **76** (1993).

[13] V.A. Zolotarev, A functional model for the Lie algebra ISO(1,1) of linear non-selfadjoint operators, *Mat. Sb.* **186:1** (1995), 79–106; English transl. in *Russian Acad. Sci. Sb. Math.* **186** (1995).

[14] D.P. Zhelobenko and A.I. Shtern, *Representations of Lie groups*, Nauka, Moscow 1983 (in Russian).

[15] N.Ya. Vilenkin and A.U. Klimuk, *Representations of Lie groups and special functions, Vol. I. Simplest Lie groups, special functions, and integral transforms. Itogi Nauki i Tekhniki: Ser. Sovr. Probl. Mat.: Fund. Napravleniya*, vol.59, VINITI Moscow 1990, pp. 145–264; English transl. in *Mathematics and its Applications (Soviet Series),* vol. 72, Kluwer, Dordrecht, 1991.

[16] N.Ya. Vilenkin, *Special functions and theory of group representations,* Nauka, Moscow 1965; English transl. Amer. Math. Soc., Providence, RI, 1968.

[17] B.Ya. Dubrovin, A.T. Fomenko, and S.P. Novikov, *Modern geometry. Methods and applications;* English transl. Graduate Texts in Math., vol. 93, Springer-Verlag, New York, 1984.

[18] A.T. Fomenko, *Symplectic geometry. Methods and applications,* Moscow, Izdat. Mosc. Univ., 1988 (in Russian).

[19] I. Springer, *Introduction to the theory of Riemann surfaces,* Moscow, 1960 (in Russian).

[20] B.A. Dubrovin, Theta-functions and non-linear equations, *Uspehi Mat. Nauk* **2(218):36** (1981), 11–80 (in Russian).

[21] N.K. Nikol'skii and S.V. Khrushchev, A functional model and some problems of the spectral theory of functions, *Trudy Mat. Inst. Steklov* **176** (1987), 97–210; English transl. in *Proc. Steklov Inst. Math.* **1988:3** (176).

[22] V.L. Ostrovskii and Yu.S. Samoilenko, Families of unbounded selfadjoint operators that are connected by non-Lie relations, *Funktsional. Anal. i Prilozhen.* **23:2** (1989), 67–68; English transl. in *Funct. Anal. Appl.* **23** (1989).

[23] Sh. Mizohata, *The theory of partial differential equations,* Cambridge University Press, 1973.

[24] K. Godunov, *Equations de la physique mathématique,* 2nd ed., Nauka, Moscow 1979; French transl. of lst ed., Editions Mir, Moscow 1973.

[25] M.S. Livsic and A.A. Yantsevich, *Operator colligations in Hilbert spaces,* Izdat. Khar'kov. Univ., Kharkov 1971; English transl. Wiley, New York, 1979.

[26] M.S. Livsic, Commuting nonselfadjoint operators and collective motions of systems, *Commuting non-self-adjoint operators in a Hilbert space,* Lecture Notes in Mathematics, no. 1272, Springer-Verlag, Berlin 1987, pp. 1–38.

[27] V.S. Vladimirov, *Equations of mathematical physics*, Nauka, Moscow, 1988 (in Russiasn).

[28] M.S. Livsic, N. Kravitsky, A. Markus, V. Vinnikov, *Theory of commuting nonselfadjoint operators*, Kluwer, Dordrecht, 1995.

[29] V. Vinnikov, Commuting nonselfadjoint operators and algebraic curves, Oper. Theory Adv. Appl. (1992), 348–371.

Department of Mathematics, Kharkov State University, Kharkov, Ukraine 310077